TEXTBOOK OF
Anatomy and Physiology

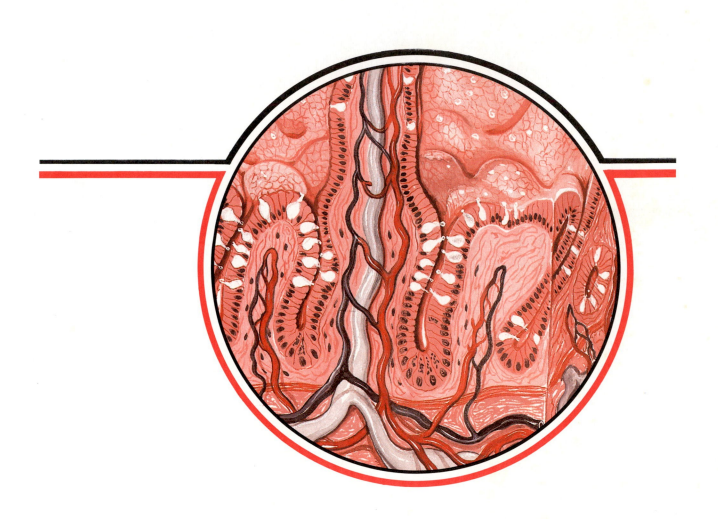

TEXTBOOK OF
Anatomy and Physiology

CATHERINE PARKER ANTHONY, R.N., B.A., M.S.

Formerly Assistant Professor of Nursing, Science Department, and Assistant Instructor of Anatomy and Physiology, Frances Payne Bolton School of Nursing, Case Western Reserve University, Cleveland, Ohio; formerly Instructor of Anatomy and Physiology, Lutheran Hospital and St. Luke's Hospital, Cleveland, Ohio

GARY A. THIBODEAU, Ph.D.

Associate Professor, South Dakota State University, Brookings, South Dakota

with Chapters **18** and **19** revised by
KATHLEEN SCHMIDT PREZBINDOWSKI, Ph.D.

Associate Professor, College of Mount St. Joseph, University of Cincinnati, Cincinnati, Ohio

and with **570** illustrations including **469** in color by
ERNEST W. BECK

TENTH EDITION

THE C. V. MOSBY COMPANY ST. LOUIS • TORONTO • LONDON 1979

TENTH EDITION

Copyright © 1979 by The C. V. Mosby Company

All rights reserved. No part of this book may be reproduced in any manner without written permission of the publisher.

Previous editions copyrighted 1944, 1946, 1950, 1955, 1959, 1963, 1967, 1971, 1975

Printed in the United States of America

The C. V. Mosby Company
11830 Westline Industrial Drive, St. Louis, Missouri 63141

Library of Congress Cataloging in Publication Data

Anthony, Catherine Parker, 1907-
    Textbook of anatomy and physiology.

    Includes bibliographies and index.
    1. Human physiology. 2. Anatomy, Human.
I. Thibodeau, Gary, 1938-   joint author.
II. Title.
QP34.5.A59   1979   612   78-11405
ISBN 0-8016-0255-6

GW/CA/K/B 9 8 7 6 5 4   03/B/307

Preface

A textbook should above all teach. This is its reason for being. To achieve this goal the tenth edition of *Textbook of Anatomy and Physiology* presents basic, up-to-date, accurate information about the body in a concise, clear, readable style. We have tried to create a textbook that students will want to pick up, one they will not put down with a sigh, a book they will enjoy because it stimulates them, and from which, therefore, they learn eagerly and easily. We have tried to make this a book that will exhilarate instructors also and that will help make their task easier.

The tenth edition of *Textbook of Anatomy and Physiology* is designed for college-level introductory courses in human anatomy and physiology. It should serve students in liberal arts, nursing, and other health-related programs equally well.

Begin your examination of this book by noting features of its organization. Outline surveys of the chapters that compose each unit appear immediately below the unit titles. Thus students are alerted, almost at a glance, to chapter contents. To help reinforce learning, summarizing outlines and thought-provoking questions complete each chapter. Summarizing tables and diagrams throughout the text serve a similar purpose.

But probably most important of all for understanding and learning the material in this book—and certainly for enjoying it—are its many illustrations. One of our most illustrious contemporary medical artists, Ernest W. Beck, has created them. To see whether you share our excitement over their unsurpassed clarity, accuracy, and beauty, please look at some of the bone and joint illustrations in Chapters 5 and 6 and at Figs. 7-7, 7-17, 8-9, 14-1, 16-6, 25-14, and 25-17. In addition to the numerous artist's drawings, this revision includes many electron micrographs.

A great many changes in content have been made in this edition. Outdated or inaccurate material has been deleted, and significant new material has been added. Examples: a new section on basic chemistry in the first chapter, new chapters on articulations and the immune system, and a new four-chapter unit on transportation. The latter includes new material about indirect methods of measuring blood volume, anemias, platelet functions, autonomic control of

v

the heart, the cardiac cycle, and vasomotor control mechanisms. Some other changes you will find in this book are new concepts about fever, an expanded discussion of renal function, new information about prostaglandins, pancreatic polypeptide, thymosin, recombinant DNA, embryology, antenatal diagnosis and treatment, and many more clinical applications.

The design of this book excels, we think, that of all previous editions. For example, look at the table starting on p. 120. Notice how the small red boxes attract your attention and cause your eye to automatically pick out major bone markings. As another example, glance at p. 676, and observe how quickly its design enables you to rapidly select supplementary readings for any chapter.

A host of people have produced this book. We wish to thank each and every one of them, and to thank, too, Dr. Kathleen Prezbindowski for revising the chapters on digestion and metabolism. With warmest appreciation, we acknowledge a deep indebtedness to the artist, Ernest W. Beck.

Catherine Parker Anthony
Gary A. Thibodeau

Contents

TEXTBOOK OF
Anatomy and Physiology

The body as a whole

chapter 1

Organization of the body

What a piece of work is man! how noble in reason! how infinite in faculty! in form and moving how express and admirable! in action how like an angel! in apprehension how like a god! the beauty of the world! the paragon of animals!

Hamlet, in *Hamlet, Prince of Denmark,*
Act II, Scene 2

Basic chemistry
Definitions
Atoms
Chemical reactions
Radioactivity

Selected facts from biochemistry
Definition of biochemistry
Biomolecules
Bioenergy
Biosynthesis

Generalizations about body structure

Terms used in describing body structure
Directional terms
Planes of body
Abdominal regions
Anatomical position

Generalizations about body function

Homeostasis

This is a book about that most wondrous of all structures—the human body. It presents information about both the body's structure and its functions. The science of the structure of organisms is called *anatomy* (from the Greek *ana*, up, and *temnein*, to cut). *Physiology* (from the Greek *physis*, nature, and *logos*, discourse) is the science of the functions of organisms. Many facets of anatomy and physiology are based on another science, that of chemistry. Therefore we shall begin this introductory chapter with some basic concepts of chemistry. If you are already familiar with these, you may wish to skim through this first section—or perhaps skip it entirely. The remaining pages of the chapter introduce some important ideas about the structure and functions of the body as a whole. They also contain definitions of terms used in describing the body's structures.

Basic chemistry
Definitions

matter—anything that occupies space, that has mass, whether it is large or small, living or nonliving, is matter.

elements—an element is a simple form of matter, a substance that cannot be broken down into two or more different substances. Twenty-four elements are present in the body. Of these, by far the most abundant are hydrogen, oxygen, carbon, and nitrogen.

atoms—atoms are the structural units that make up each element.

chemical symbols—chemical symbols are abbreviations for the names of elements. Symbols for 12 of the 24 elements in the body follow:

H hydrogen
O oxygen
C carbon
N nitrogen
P phosphorous
S sulfur
K potassium
(Latin, *kalium*)

Na sodium
(Latin, *natrium*)
Fe iron
(Latin, *ferrum*)
Ca calcium
Co cobalt
Mg magnesium

Atoms

The concept that matter is made up of atoms was proposed early in the 19th century by an English chemist, J. Dalton. He conceived atoms as solid, indivisible particles of matter. After a lapse of about 100 years, Lord Rutherford, another English scientist, disproved this idea and showed that still smaller particles compose atoms. Even the largest atoms are so small that a lineup of 100 million or more of them would measure barely an inch. A staggering fact, then, that even smaller particles can and do exist within such infinitesimal structures!

An atom has two main parts—a nucleus and one or more shells of space surrounding it. The nucleus consists of closely packed, positively charged particles called *protons* and uncharged or neutral particles named *neutrons*—therefore the nucleus as a whole carries a positive charge equal to the number of its protons. Moving around the nucleus in orbitals in each of the atom's shells of space are negatively charged particles called *electrons*. The number of electrically charged electrons in any one atom al-

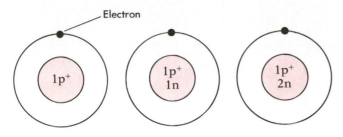

Electron

Fig. 1-1 Structure of hydrogen atoms. Left, ordinary hydrogen atom; nucleus consists of 1 proton, and its only shell contains 1 unpaired electron. Note differences in the other two atoms. Middle, rare type of hydrogen atom called deuterium. Right, another rare type of hydrogen atom called tritium.

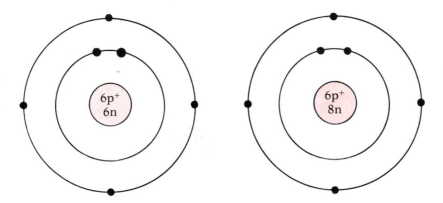

Fig. 1-2 Atomic structure of carbon isotopes. Left, ordinary carbon (atomic number, 6; atomic weight, 12; symbol C^{12}). Right, carbon isotope (atomic number, 6; atomic weight, 14; symbol C^{14}).

ways equals the number of positively charged protons. The opposite charges cancel or neutralize each other; all atoms, therefore, are electrically neutral structures.

The number of protons in an atom's nucleus identifies the kind of element it is and is known as an element's *atomic number*. Hydrogen, for example, has an atomic number of 1; this means that all hydrogen atoms—and only hydrogen atoms—have 1 proton in their nuclei. All carbon atoms, and only carbon atoms, contain 6 protons, so have an atomic number of 6. All oxygen atoms, and only oxygen atoms, have 8 protons and an atomic number of 8. In short, each element is identified by its own unique number of protons, that is, by its own unique atomic number. If two atoms contain a different number of

protons, they necessarily have different atomic numbers and are different elements.

Although all atoms of the same element have the same number of protons, they do not necessarily have the same number of neutrons in their nuclei. The nuclei of ordinary hydrogen atoms, for instance, contain no neutrons, but about 1 in 5,000 hydrogen atoms has 1 neutron in its nucleus, and a still smaller proportion of hydrogen atoms contains 2 neutrons. All hydrogen atoms have an atomic number of 1, so all contain 1 proton and 1 electron.

Protons, neutrons, and electrons are particles of extremely small *mass* or weight. (Mass equals weight under ordinary conditions.) A proton weighs almost exactly the same as a neutron, and, compared with an electron, they are heavy particles indeed—each weighs 1,836 times as much as an electron. For all practical purposes, therefore, protons plus neutrons constitute an atom's weight. Look now at the three atoms shown in Fig. 1-1. Note that each contains 1 proton, a fact that identifies each one as a hydrogen atom. How many neutrons does each of these hydrogen atoms contain? Apply the formula Atomic weight = $(p^+ + n)$ to each of these three types of hydrogen atoms. You will quickly discover that they have three different atomic weights. Why? Because each nucleus contains a different number of neutrons. Atoms such as these that contain the same number of protons but a different number of neutrons are called *isotopes*. Because isotopic atoms contain the same number of protons, they are necessarily atoms of the same element and possess identical chemical properties. But because the atoms of isotopes contain different numbers of neutrons, they also have different atomic weights. Fig. 1-2 shows how these facts apply to the structure of carbon isotopes.

Each shell of space around an atom's nucleus can accommodate a certain maximum number of electrons moving in orbitals in the shell. The innermost shell can hold the fewest electrons— 1 single electron in the hydrogen atom, but 1 pair of electrons in all other elements. Both the second and the outermost shells can hold a maximum of 4 pairs of electrons.

The number and arrangement of the electrons orbiting in the outermost shell of space around an atom's nucleus determines whether or not the atom is chemically active. When the outer shell contains 1 or more single, that is, unpaired, electrons, the atom is chemically active; it is able to take part in chemical reactions. But when the outer shell contains no single electrons but only pairs of electrons, the atom is chemically inactive or stable.

Chemical reactions

Chemical reactions involve unpaired electrons in the outer shells of atoms. A chemical reaction consists of either the transfer of unpaired electrons from the outer orbital of one atom to the outer orbital of another or the sharing of one atom's unpaired electrons with those of another atom. Fig. 1-3 illustrates an electron transfer type of chemical reaction. In examining this figure, pay attention first to the number and arrangement of electrons in the outer shells of the sodium and chlorine atoms. Note that sodium has only one unpaired electron in its outer shell in contrast to chlorine, which has 1 unpaired electron plus 3 paired electrons, or a total of 7 electrons. Sodium transfers or donates its 1 unpaired electron to chlorine. Chlorine accepts it and pairs it with its 1 unpaired electron, thereby filling its outer shell with the maximum of 4 electron pairs. The formation of this new electron pair in chlorine's outer orbit creates an attractive force that binds the sodium and chlorine atoms to each other. In short, it creates a *chemical bond* between them. The electron transfer constitutes a chemical reaction between the two atoms and produces one molecule of the compound sodium chloride (ordinary table salt). A chemical bond formed by the transfer of electrons is called an *electrovalent* or *ionic bond*. A *covalent bond* is formed by two atoms sharing a pair of electrons (Fig. 1-4). A *compound* is a substance made up of atoms of one or more elements joined to each other by chemical bonds to form units called molecules. *Molecules* are the structural units of compounds; *atoms* are the structural units of elements.

Valence is the number of unpaired electrons in an atom's outer shell, and therefore valence indicates the number of chemical bonds an atom can form with other atoms. For example, sodium and chlorine atoms each have 1 unpaired electron in their outer shells, so each has a valence of 1 and can form one chemical bond with another atom. Hydrogen also has 1 unpaired electron in its only shell, so it, too, has a valence of 1 and can form one chemical bond with another atom. Look again at Fig. 1-2. How many unpaired electrons are in the carbon atom? What

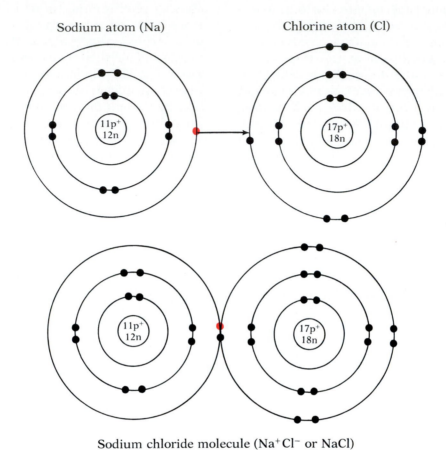

Sodium atom (Na)   Chlorine atom (Cl)

Sodium chloride molecule (Na⁺Cl⁻ or NaCl)

Fig. 1-3 Diagram of sodium and chlorine atoms and electron transfer to form an electrovalent or ionic bond between them and produce a molecule of sodium chloride.

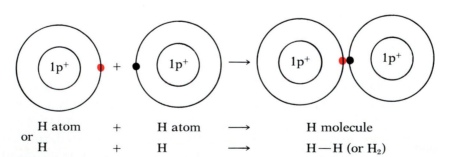

H atom	+	H atom	⟶	H molecule
H	+	H	⟶	H—H (or H_2)

Fig. 1-4 Formation of covalent bond, that is, the sharing of a pair of electrons, between 2 hydrogen atoms produces 1 hydrogen molecule.

is carbon's valence? How many chemical bonds can it form with other atoms? Answers: The carbon atom contains 4 unpaired electrons in its outer shell and so has a valence of 4 and can form chemical bonds with four other atoms. For example, one carbon atom can combine with four hydrogen atoms to form one molecule of methane (CH_4).

Four basic types of chemical reactions that you will learn to recognize as you study physiology are named synthesis, decomposition, exchange, and reversible reactions. *Synthesis* (from the Greek *syn*, together, and *thesis*, putting) reactions put together or combine two or more substances to form a more complex substance, and in so doing, synthesis reactions form new chemical bonds. Many such reactions occur in the body. Every one of its cells, for example, combines amino acid molecules to form complex protein compounds. The name *decomposition* suggests what these reactions do. They decompose or break down a substance into two or more simpler substances, and in so doing, decomposition reactions break chemical bonds. Decomposition and synthesis, in a word, are opposites. Synthesis builds up, decomposition breaks down. Synthesis forms chemical bonds, decomposition breaks chemical bonds. One of the many examples of decomposition reactions in the body is the digestion or breakdown of large fat molecules into two simpler substances, fatty acids and glycerol. The nature of *exchange reactions* is also suggested by their name. They break down or decompose two compounds and in exchange synthesize two new compounds. Certain exchange reactions take place in the blood. One example is the reaction between lactic acid and sodium bicarbonate. The decomposition of both substances is exchanged for the synthesis of sodium lactate and carbonic acid. Perhaps you can see these changes more easily in an equation.

$$H \cdot lactate + NaHCO_3 \rightarrow Na \cdot lactate + H \cdot HCO_3$$

The formula H $\cdot$ lactate represents lactic acid; $NaHCO_3$ is the formula for sodium bicarbonate;

Na $\cdot$ lactate represents sodium lactate; and H $\cdot$ HCO_3 represents carbonic acid. *Reversible reactions*, as the name suggests, proceed in both directions. A great many chemical reactions are reversible, and we shall cite a number of them in later chapters of the book.

Before we leave the subject of chemical reactions, we must state the following important principle about them: Every chemical change is coupled with an energy change. The formation of chemical bonds is coupled with energy use. The breaking of chemical bonds is coupled with energy release. Synthesis reactions, therefore, use energy, and decomposition reactions release energy. Example in living cells of synthesis with energy use: reactions that combine simple food compounds to form complex compounds. Example in living cells of decomposition with release of energy: reactions that break food compounds down into simpler compounds.

Radioactivity

Some atoms have the property not only of chemical activity but also of radioactivity. Whereas chemical activity consists of a change in an atom's outer shell—a transfer or a sharing of its single, unpaired electrons—radioactivity involves a change in an atom's nucleus. Specifically, *radioactivity* is the emission from the nucleus of particles or rays known as alpha, beta, and gamma rays. Alpha and beta rays are actually minute particles. *Alpha rays* consist of 2 protons and 2 neutrons ejected from the nucleus. *Beta rays* are electrons, not from an atom's shell, but from its nucleus, where they originate by a neutron breaking down into a proton and an electron. The proton remains behind in the nucleus, and the electron shoots out of it as a beta ray. *Gamma rays* are not particles but electromagnetic radiations emitted from an atom's nucleus and presumably produced by the shifting of particles in the nucleus after the ejection of alpha or beta particles.

How does radioactivity, the emission of alpha, beta, or gamma rays, change an atom? To answer this question, consider the following example. A synthetic radioactive isotope of iodine used in certain diagnostic tests and commonly called iodine-131, is represented by the symbol $_{53}I^{131}$. Interpreted, this symbol indicates iodine atoms that have an atomic number of 53 (53 protons in their nuclei) and an atomic weight of 131. By rewriting the formula Atomic weight = $(p^+ + n)$ as (Atomic weight $- p^+$) = n, we can determine that there are $(131 - 53)$ or 78 neutrons in the nucleus of iodine-131. Ordinary iodine atoms have an atomic weight of 127, with 53 protons in their nuclei but only 74 neutrons. Radioactive iodine-131 emits beta and gamma rays and thereby changes the nuclei of its atoms. One of its neutrons converts to 1 proton and 1 electron. The electron leaves the nucleus as a beta ray, but the proton stays in the nucleus. Result? The nucleus now contains one less neutron—it converted to a proton and electron—and one more proton than it had before its beta ray emission. Changing the number of protons in an atom's nucleus changes the atom into another element. A basic principle about atoms, you will recall, is that all atoms of the same element contain the same number of protons. A corollary stems from this principle: Atoms that contain different numbers of protons are necessarily different elements. Iodine-131 atoms that have emitted beta rays contain 54, not 53, protons, as do all iodine atoms. Atoms of the element xenon contain 54 protons. Radioactivity, therefore, changes iodine into xenon. An iodine atom that has participated in a chemical reaction is still an iodine atom; only the number of electrons in its outer shell has changed—the number of protons in its nucleus remains the same. We started this paragraph with a question and shall end it with an answer. Radioactivity, that is, the emission of alpha, beta, or gamma rays from an atom's nucleus, changes one element into another by changing the number of protons in the atomic nucleus. One more question and answer. Why do you, a student of anatomy and physiology, need to know anything about radioactivity? One rea-

son that seems valid, at least to us, is that radioactive isotopes have become widely used in studies of various aspects of physiology and in the diagnosis and treatment of diseases. The rays emitted by radioactive atoms serve as labels of the atoms, which can be detected by Geiger counters.

Selected facts from biochemistry
Definition of biochemistry

Biochemistry is the chemistry of living organisms. It is a young but extensive science—so extensive that some biochemistry textbooks contain more than a thousand pages. Out of this abundance, we have tried to select for inclusion in this chapter the facts and principles you will need most to know in order to understand the rest of the book. We have organized this information around three topics: biomolecules, bioenergy, and biosynthesis.

Biomolecules

A textbook by the distinguished biochemist Albert L. Lehninger opens with the sentence, "Living things are composed of lifeless molecules." Biomolecules are these lifeless molecules. By definition then, biomolecules are compounds that compose living organisms. Most abundant of them all is water. The thousands of other biomolecules fall into seven classes of compounds: acids, bases, salts, proteins, carbohydrates, lipids, and nucleic acids. To learn some basic facts about each of these kinds of biomolecules, study the following paragraphs.

Water serves a host of vital functions in the body, many of which are related to the fact that it dissolves so many substances. By dissolving oxygen and food substances, for instance, water enables these essential materials to enter and leave the blood capillaries and to enter cells. Also, by dissolving substances, water makes possible the thousands of chemical reactions that keep us alive. Dry substances are virtually nonreactive. Another important function of water, that of enabling the body to maintain a relatively constant temperature, stems from the fact that water both absorbs and gives up heat slowly.

Acids, bases, and salts belong to a large group of compounds called electrolytes. *Electrolytes*, by definition, are compounds whose molecules consist of positive ions (cations) and negative ions (anions) and that dissociate when dissolved in water into cations and anions. In short, electrolytes are substances that ionize in solution. Formation of an electrolyte takes place by electron transfer, as shown in Fig. 1-3. Here you see a sodium atom donating the single electron in its valence shell to the valence shell of a chlorine stem. Note again what this accomplishes. The transferred electron fills chlorine's outer shell by forming a fourth electron pair in it; this pair constitutes an electrovalent bond between the sodium and the chlorine. It also does something else. It converts both the sodium and chlorine atoms into ions or electrically charged particles. Since the sodium atom has donated one of its electrons, that is, one of its negative charges, it now contains one more positive charge than negative (11 p^+ and 10 e^-). It has become a *positive ion*, a *cation*, designated by the symbol Na^+. Chlorine, after accepting the electron, bears one more negative than positive charge. It has become a *negative ion*, an *anion*. The electrovalent bond between the two ions creates out of them

one molecule of the ionic compound sodium chloride. The formula Na^+Cl^- represents sodium chloride (ordinary table salt). Because chemically bound ions compose its molecules, it is an *ionic compound*. Because solutions of ionic compounds conduct an electric current, another name for them is electrolytes. Chapters 21 and 22 give more information about these essential biomolecules.

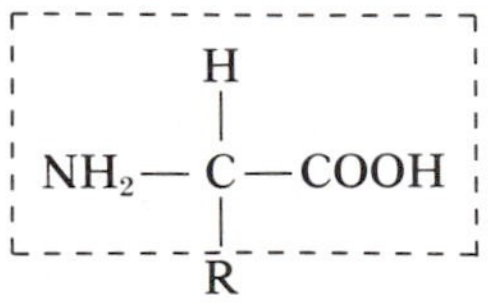

Acids, like all electrolytes, ionize when dissolved in water to form cations and anions. By definition, an *acid* is a compound that yields hydrogen ions (H^+) in solution. A *base* is a compound that yields hydroxyl or OH ions (OH^-) in solution. A *salt* is a compound that in water yields the positive ions of a base and the negative ions of an acid. A salt is formed when an acid reacts with a base. Here is an example:

$$HCl \quad + \quad NaOH \quad = \quad NaCl + Water$$

Hydrochloric acid   Sodium hydroxide, a base   Salt

Proteins (from the Greek *proteios*, of the first rank) are the most abundant of the carbon-containing or organic compounds in the body, and as their name implies, their functions are of first-rank importance. The units or building blocks that compose proteins are *amino acids*. Fig. 1-5 diagrams the basic structure of an amino acid. What four elements compose it? The same elements are present in all amino acids, plus sulfur in many of them and phosphorus or iron in some. Note that bonded to the first, or alpha, carbon are an amino group (NH_2), a carboxyl group (COOH), a hydrogen atom, and R, which stands for a side chain of one or more atoms. In 19 of the 20 different amino acids, R is a group of atoms—a different group in each one—but in the simplest amino acid (glycine), it is a hydrogen atom. The nature of the side chain identifies the amino acid.

Amino acids become linked together by *peptide bonds*, that is, chemical bonds that join the carboxyl group of one amino acid to the amino group of another by eliminating water from

$$NH_2 - \overset{\displaystyle \overset{H}{|}}{\underset{\displaystyle \underset{R}{|}}{C}} - COOH$$

Fig. 1-5 Basic structural formula for an amino acid. All amino acids contain the part of the molecule enclosed in the rectangle; *R* represents the part unique to each of the 20 different amino acids. Examples: In glycine, the simplest amino acid, *R* is a hydrogen atom; in alanine, it is the group CH_3.

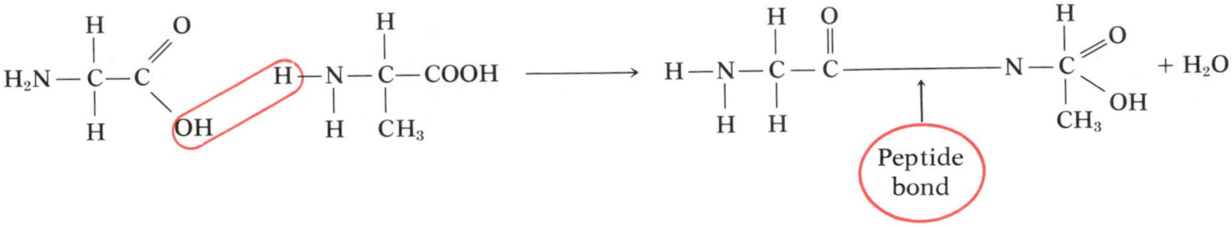

Fig. 1-6 Peptide bond formation. A peptide bond is one that joins the carboxyl group of one amino acid to the amino group of another. Note that the removal of water from these two groups produces the peptide bond between them and yields in addition to water a compound called a peptide. A peptide formed from two amino acids is a dipeptide. The reaction shown here forms the dipeptide glycylalanine from the two amino acids glycine and alanine.

these groups. Fig. 1-6 will help you form a mental picture of this. Two amino acids linked by a peptide bond form a *dipeptide;* many amino acids linked together constitute a *polypeptide.* A protein is made up of one or more polypeptides. The hormone commonly known as ACTH, for example, is a protein that consists of only one polypeptide chain made up of 39 amino acids. Insulin, perhaps the best known hormone, consists of two polypeptide chains, one of them 21 amino acids long and the other 30 amino acids long. Another of the body's many proteins is hemoglobin; it consists of four polypeptide chains with 141 amino acids in two of them and 146 amino acids in the other two.

Protein molecules come in various sizes and shapes. All of them, however, are large molecules (macromolecules) with molecular weights ranging from about 5,000 to a million or more. Each protein has a characteristic three-dimensional shape known as its *conformation;* the two major conformations are fibrous or threadlike and globular or spherical. Here are three important principles to remember about protein structure: The sequence of amino acids is different in each kind of protein molecule. The sequence of amino acids in a protein molecule determines its

conformation. The conformation of a protein molecule determines its function.

Proteins serve diverse and vital functions in the body. They catalyze chemical reactions, transport various substances, produce movements, regulate assorted functions, construct various parts, and last but certainly not least, they defend us against many potential killers. The largest group of proteins are those that catalyze reactions, namely, enzymes. Nearly 2,000 different enzymes are known, and each one, by virtue of its unique modified globular conformation, induces a different chemical reaction. Table 1-1 lists the names and functions of a few specific proteins.

Carbohydrates are the substances commonly called sugars and starches. The three types of carbohydrate compounds have long names—monosaccharides, disaccharides, and polysaccharides. In Fig. 1-7, you can see the structural formula of the body's main monosaccharide, that is, glucose. Fructose and galactose are other monosaccharides. All three of these sugars consist of the same number of atoms of the same elements, but the arrangement of them within their molecules differs slightly. Look now at Fig. 1-7 to note which three elements compose the glucose

Table 1-1 Names and functions of a few proteins

Name of protein	Function
Lactase	An enzyme in the intestinal digestive juice; catalyzes digestion of lactose (milk sugar) to glucose and galactose
Hemoglobin	Transports oxygen in blood
Myosin	Fibrous protein in muscle; acts with another protein, actin, to contract muscle and produce movement
Insulin	A hormone; regulates body's use of carbohydrates
Collagen	Part of structure of connective tissues
Antibodies	Defend against microorganisms

molecule. They also make up all other carbohydrate compounds. How many carbon atoms do you find in glucose? A monosaccharide with six carbon atoms in its molecule is called a *hexose*. The monosaccharides mentioned so far—glucose, fructose, and galactose—are all hexoses designated by the general formula $C_6H_{12}O_6$. Not all monosaccharides, however, are hexoses. Some are *pentoses*, so-named because they contain five carbon atoms. Ribose and deoxyribose are pentose monosaccharides of great importance in the body—more about them when we discuss nucleic acids. Like all monosaccharides, ribose and deoxyribose are sugars, strange sugars in that they are not sweet.

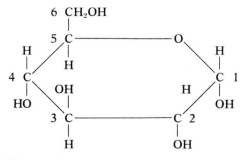

Fig. 1-7 Structural formula of glucose.

Disaccharides and polysaccharides are synthesized from monosaccharides. The following equation represents the synthesis of a disaccharide:

$$C_6H_{12}O_6 + C_6H_{12}O_6 \xrightarrow{\text{(synthesis)}} C_{12}H_{22}O_{11} + H_2O$$

Note that a disaccharide molecule consists of two monosaccharides minus a molecule of water. Sucrose (cane sugar), maltose, and lactose are disaccharides. Polysaccharides consist of many monosaccharides chemically joined by the removal of water. As you might guess, therefore, polysaccharides are large molecules. For instance, glycogen, the main polysaccharide in the body, has an estimated molecular weight of several million—truly a macromolecule. Starch and cellulose are the predominant plant polysaccharides.

What functions do carbohydrates serve in the body? Energy release and energy storage, to answer in the fewest words. Before we try to answer more fully, we shall relate a few facts about two other kinds of molecules, namely, lipids and nucleic acids.

Lipids, according to one definition, are water-insoluble organic biomolecules. They include a large assortment of compounds that have been classified in several different ways. For our purposes, classification of lipids under these headings—triglycerides, phosphoglycerides, steroids, and prostaglandins—seems sufficient.

Triglycerides (triacylglycerols, in international nomenclature; formerly called neutral fats) are the most abundant lipids, and they function as the body's most concentrated source of energy. They consist of three molecules of a fatty acid chemically joined to one molecule of glycerol (glycerin). Fig. 1-8 shows the formula for palmitic acid, a common fatty acid. Fig. 1-9 shows three molecules of palmitic acid combining with one molecule of glycerol to synthesize one molecule of the triglyceride named tripalmitin. Note the composition of a fatty acid as shown in Fig. 1-8. Reading the formula from the bottom up, a fatty acid consists of a methyl group (CH_3), a long chain of hydrocarbon groups, and a carboxyl group (the group that characterizes organic acids). Palmitic acid is a saturated fatty acid. By definition a *saturated fatty acid* is one in which

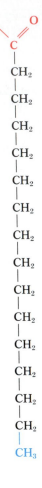

Fig. 1-8 Structural formula of palmitic acid, a common saturated fatty acid. Like all fatty acids, it consists of a carboxyl group (the group that characterizes all organic acids) and a taillike chain of hydrocarbons that terminates in a methyl group (CH_3).

Three other ways of writing the formula for palmitic acid are

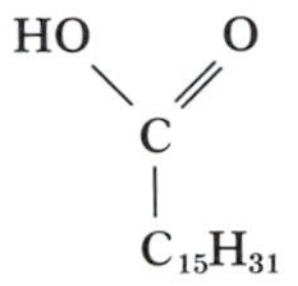

$$HO \diagdown \quad \diagup\!\!\diagup O$$
$$C$$
$$|$$
$$C_{15}H_{31} \quad or \quad CH_3(CH_2)^{14}COOH$$
$$or \quad C_{15}H_{31}COOH$$

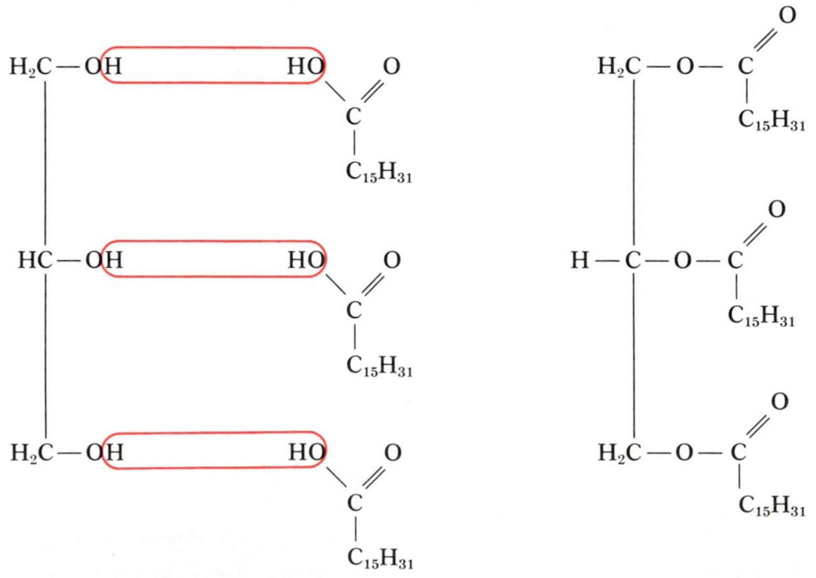

Fig. 1-9 Synthesis of the triglyceride tripalmitin from glycerol and palmitic acid.

1 Glycerol + 3 Palmitic acid ⟶ 1 Tripalmitin (a triglyceride)

all available bonds of its hydrocarbon chain are filled, that is, saturated, with hydrogen atoms; the chain contains, therefore, no double bonds. In contrast, an *unsaturated fatty acid* has one or more double bonds in its hydrocarbon chain because not all of the chain carbon atoms are saturated with hydrogen atoms. Everyone today knows the word "polyunsaturateds." Polyunsaturateds are triglycerides (fats) that contain fatty acids with two or more double bonds between chain carbons unsaturated with hydrogen. Newspapers, magazines, and advertisements recommend them for lowering one's blood cholesterol and preventing heart disease. Research studies support the claim that diets high in polyunsaturated and low in saturated fats lead to a lower content of cholesterol in the blood. But whether they help prevent heart disease is still an unanswered and controversial question.

Two polyunsaturated fatty acids that are classified as essential fatty acids are linoleic acid ($C_{17}H_{31}COOH$) and linolenic acid ($C_{17}H_{29}COOH$). These polyunsaturated fatty acids are essential for survival and must be present in one's diet because the body cannot synthesize them. The most common saturated fatty acids are palmitic acid ($C_{15}H_{31}COOH$) and stearic acid ($C_{17}H_{35}COOH$).

Phosphoglycerides are complex lipids composed of glycerol with two fatty acids and a phosphate group joined to it and an amino alcohol joined to the phosphate. Fig. 1-10 shows the structural formula for one of the most abundant phosphoglycerides, namely, choline phosphoglyceride (or phosphatidyl choline, or lecithin, now an out-of-date name). Choline is the amino alcohol present in choline phosphoglyceride, and palmitic and oleic acids are the fatty acids. Phosphoglycerides function as major structural components of cell membranes.

Steroids are an important and large group of compounds whose molecules have as their principle component the steroid nucleus (Fig. 1-11). Steroids and other simple lipids contain no fatty

acids in their molecules, whereas complex lipids such as triglycerides and phosphoglycerides do contain these. Steroid compounds in the body number a famous name among them—cholesterol—and also include both male and female sex hormones and the adrenocortical hormones.

Prostaglandins, according to one definition, are lipids composed of a 20-carbon fatty acid that contains a 5-carbon ring. Many different kinds of prostaglandins exist in the body—14 or more different prostaglandins in semen plus many other kinds in a number of tissues. Prostaglandins are reported to regulate hormone activity and metabolism in many tissues (discussed in Chapter 12).

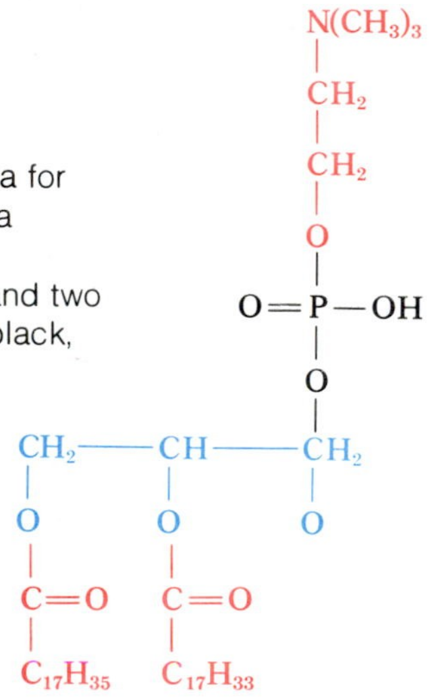

Fig. 1-10 Structural formula for choline phosphoglyceride, a compound present in cell membranes. Red, choline and two fatty acids; blue, glycerol; black, phosphate group.

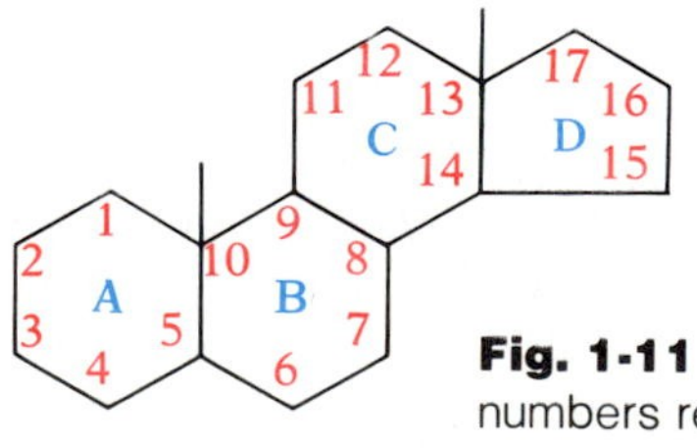

Fig. 1-11 Steroid nucleus; numbers represent carbon atoms.

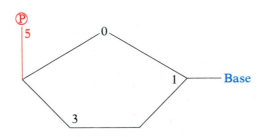

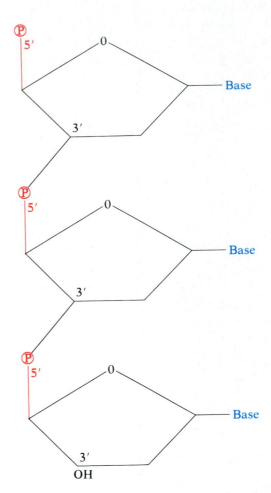

Ⓟ= Phosphate group

OH
|
H—O—P—O   (H₂PO₄)
‖
O

$$H-O-P-O \quad (H_2PO_4)$$

Base = Purine   or   Pyrimidine
       Guanine   Cytosine
       Adenine   Thymine

Fig. 1-12 Top, structural formula of the pentose sugar named deoxyribose. Below it is an abbreviated structural formula for a deoxyribonucleotide. Note that it consists of a sugar (deoxyribose) with a base bonded to its number 1' carbon atom and a phosphate group bonded to its 5' carbon atom. Deoxyribonucleotides are the structural units of DNA.

Fig. 1-13 Scheme to show linkage of nucleotides by phosphate groups.

Survival of man as a species—and also survival of every other species—depends largely on two kinds of *nucleic acid compounds*. Their abbreviated names, DNA and RNA, almost everyone has heard or seen, but their full names are much less familiar. They are deoxyribonucleic and ribonu-

cleic acids. In this chapter, we shall consider DNA only. Structural units called nucleotides compose it. A *nucleotide* consists of a phosphate group, a pentose sugar, and a nitrogenous base. Look at Fig. 1-12. Notice first the numbering of the carbon atoms in the pentose sugar, deoxyribose, shown at the top. Next, examine the abbreviated structural formula of the nucleotide shown below the deoxyribose formula; pay particular attention to the numbers of the carbon

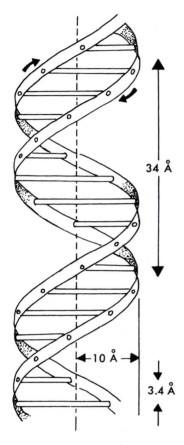

Fig. 1-14 Watson-Crick double-stranded helix configuration of DNA. (From Levine, L.: Biology of the gene, ed. 2, St. Louis, 1973, The C. V. Mosby Co.)

Adenine   Thymine

Guanine   Cytosine

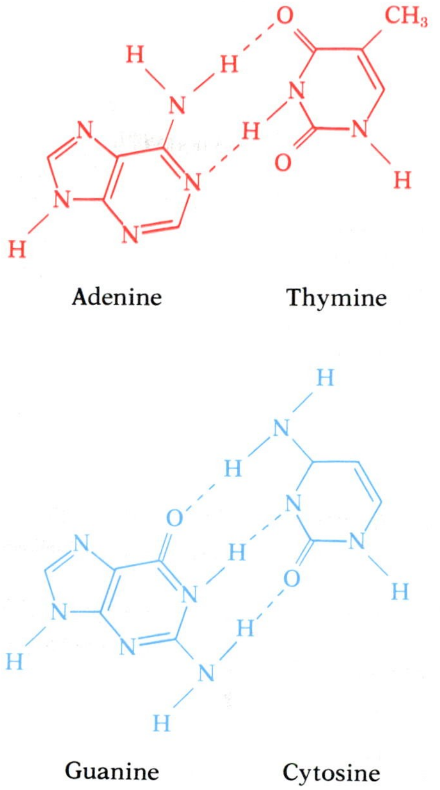

Fig. 1-15 Examples of base pairs formed between single DNA strands. (From Levine, L.: Biology of the gene, ed. 2, St. Louis, 1973, The C. V. Mosby Co.)

atoms where the phosphate group and base attach. Because the sugar in this nucleotide is deoxyribose, it is also called a *deoxyribonucleotide*. The nitrogenous base in a deoxyribonucleotide may be any of these four: cytosine, thymine, guanine, or adenine. Cytosine and thymine derive from a compound named pyrimidine and so are known as *pyrimidine bases*. Guanine and adenine derive from another compound, one named purine, so that they are called *purine*

bases. Nucleotides become linked together by the number 3' carbon atom of one sugar bonding to the phosphate group on the number 5' carbon of the sugar below it. You can see three nucleotides linked together in Fig. 1-13. As you look at this figure, try to visualize another nucleotide binding to the third one in the chain, and then another, and another, and another, to form a long chain of nucleotides—a *polynucleotide*, that is. One end of a polynucleotide chain terminates in the 5′ phosphate group and the other terminates in the 3′ hydroxyl group. Note that phosphate and sugar groups alternate to form the so-called backbone of a polynucleotide chain.

A DNA molecule consists of two polynucleotide chains. Each chain has a coiled shape similar to the shape of a wire in a spring. Such a shape is called a *helix*. Because both polynucleotide chains of DNA have this helix shape, the DNA molecule is described as a double helix (Fig. 1-14). Each helical chain in a DNA molecule has its phosphate-sugar backbone toward the outside and its bases pointing inward toward the bases of the other chain. More than that, each base in one chain is joined to a base in the other chain by means of a hydrogen bond to form what is known as a *base pair*. The two polynucleotide chains of a DNA molecule are thus held together by hydrogen bonds between the two members of each base pair. One important principle to remember is that only two kinds of base pairs are present in DNA—those shown in Fig. 1-15. What are they? Symbols used to represent them are A-T and G-C. Although a DNA molecule contains only these two kinds of base pairs, it contains millions of them—over 100 million pairs estimated in one human DNA molecule! Two other impressive facts are these: The millions of base pairs occur in the same sequence in all the millions of DNA molecules in one individual's body and in a different sequence in the DNA of all other individuals. In short, the base pair sequence in DNA is unique to each individual. This fact has momentous significance, but what it is we shall save to reveal later. For now, we shall merely state that DNA functions as the heredity molecule. It carries out a weighty responsibility, that of passing the traits of one generation on to the next. How it accomplishes this is a long story to be told in other portions of the book.

Bioenergy

By the term bioenergy, we mean energy that living organisms generate, store, and use to do the work of keeping themselves alive. The human body, like all other forms of animal life, generates energy from the foods it ingests. How it does this is one of biochemistry's major, best understood, and most extensive topics. We shall attempt only the briefest description of it here and discuss it more fully in Chapter 19. The name of the process by which living organisms generate and store energy is catabolism. *Catabolism* consists of chemical reactions—a very large number of them that take place in all cells in a highly coordinated fashion. They are essentially decomposition reactions; they break chemical bonds, and this not only breaks relatively complex compounds down into simpler ones but also releases energy from them. Catabolism breaks down the building blocks of food compounds—monosaccharides, fatty acids, glycerol, and amino acids—into simpler compounds, namely, carbon dioxide, water, and certain nitrogenous compounds. More than half of the released energy is immediately recaptured and put back into storage in one of the most important of all biomolecules—adenosine triphosphate (ATP). The rest of the energy released by catabolism is heat energy, the heat that keeps our bodies warm. The energy stored in ATP is used to do the body's work—the work of muscle contraction and movement, of active transport, and of biosynthesis.

Biosynthesis

Biosynthesis means a chemical reaction that puts simple molecules together to form complex biomolecules, notably, carbohydrates, lipids, proteins, nucleotides, and nucleic acids. Literally thousands of biosynthesis reactions take place continually in the body. They constitute the vast, highly integrated network of reactions that make up the process of *anabolism*. Together, catabolism and anabolism constitute the process of *metabolism*. Here are two short definitions of this term. Metabolism is all the chemical changes absorbed foods undergo in living cells. Metabolism is, most simply, the body's use of foods.

Generalizations about body structure

1 *The body is a single structure but it is made up of billions of smaller structures of four major kinds: cells, tissues, organs, and systems.* Smallest and most numerous of these are the cells. How many cells are there in the body? Far too many for our imaginations to visualize. One average-sized adult body, according to one estimate, consists of 100,000,000,000,000 cells. In case you cannot translate this number—1 with 14 zeroes after it—it is 100 trillions! or 100,000 billions! or 100 million millions!

Cells have long been recognized as the simplest units of living matter that can maintain life and reproduce themselves. In truth, however, they are far from simple. How extremely complex cells are you can discover by reading Chapter 2.

Tissues are somewhat more complex units than cells. By definition, a *tissue* is an organization of a great many similar cells with varying amounts and kinds of nonliving, intercellular substance between them.

Organs are more complex units than tissues. An *organ* is an organization of several different kinds of tissues so arranged that together they can perform a special function. For example, the stomach is an organization of muscle, connective, epithelial, and nervous tissues. Muscle and connective tissues form its wall, epithelial and connective tissues form its lining, and nervous tissue extends throughout both its wall and its lining.

Systems are the most complex of the component units of the body. A *system* is an organization of varying numbers and kinds of organs so arranged that together they can perform complex functions for the body. Nine major systems compose the human body: skeletal, muscular, nervous, endocrine, circulatory (cardiovascular and lymphatic), respiratory, digestive, urinary, and reproductive. (For names and descriptions of the organs of a particular system, see the chapter on that system.)

2 *A ventral and a dorsal cavity, each with subdivisions, house the numerous internal organs of the body.*

The *ventral cavity* consists of the *thoracic* or *chest cavity* and the *abdominopelvic cavity.* The thoracic cavity consists of a right and a left pleural cavity and a midportion called the mediastinum. Fibrous tissue forms a wall around the mediastinum, completely separating it from the right pleural sac, in which the right lung lies, and from the left pleural sac, which contains the left lung. Thus the only organs in the thoracic cavity that are not located in the mediastinum are the lungs. Organs located in the mediastinum are the following: the heart (enclosed in its pericardial sac), the trachea and right and left bronchi, the esophagus, the thymus, various blood vessels (for example, thoracic aorta, superior vena cava), the thoracic duct and other lymphatic vessels, various lymph nodes, and nerves (such as the phrenic and vagus). The abdominopelvic cavity consists of an upper portion, the *abdominal cavity*, and a lower portion, the *pelvic cavity.* The abdominal cavity contains the liver, gallbladder, stomach, pancreas, intestines, spleen, kidneys, and ureters. The bladder, certain reproductive organs (uterus, uterine tubes, and ovaries in the female; prostate gland, seminal vesicles, and part of the vas deferens in the male), and part of the large intestine (namely, the sigmoid colon and rectum) lie in the pelvic cavity.

The *dorsal cavity* consists of the cranial and spinal cavities. The *cranial cavity* lies in the skull and houses the brain. The *spinal cavity* lies in the spinal column and houses the spinal cord.

3 *Organization is an outstanding and, in fact, a vital characteristic of body structure.* Every one of its component units—every system, every organ, every tissue, and even every cell—consists of smaller parts arranged in an orderly and precise manner. Organization is surely one of the most important characteristics of life. Even the name *organism*, used to denote a living thing, implies organization.

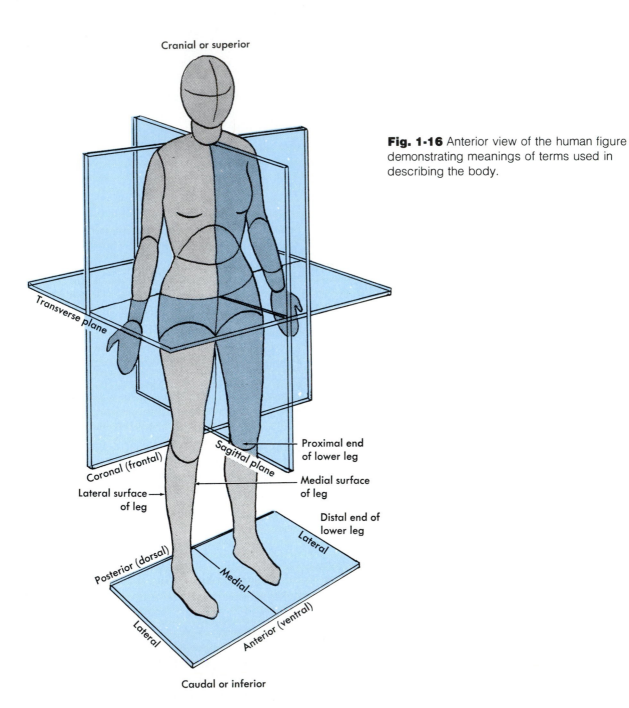

Fig. 1-16 Anterior view of the human figure demonstrating meanings of terms used in describing the body.

Fig. 1-17 The nine regions of the abdomen. The top horizontal line crosses the abdomen at the level of the ninth rib cartilages. The lower horizontal line crosses it at the level of the iliac crests. The vertical lines pass through the midpoints of the right and left inguinal (Poupart's) ligaments.

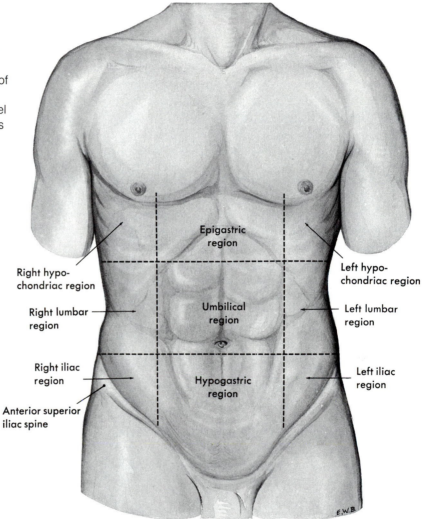

Epigastric region

Right hypo-chondriac region

Left hypo-chondriac region

Right lumbar region

Umbilical region

Left lumbar region

Right iliac region

Left iliac region

Hypogastric region

Anterior superior iliac spine

4 *Body structures change gradually over the years.* They develop and grow during the years before maturity (young adulthood). After maturity, they age and, in general, atrophy. We shall mention many specific examples of aging in the chapters that follow.

Terms used in describing body structure
Directional terms

superior or *cranial*—toward the head end of the body; upper (example, the hand is part of the superior extremity).

inferior or *caudal*—away from the head; lower (example, the foot is part of the inferior extremity).

anterior or *ventral*—front (example, the knee-cap is located on the anterior side of the leg).

posterior or *dorsal*—back (example, the shoulder blades are located on the posterior side of the body).

medial or *mesial*—toward the midline of the body (example, the great toe is located at the medial side of the foot).

lateral—away from the midline of the body (example, the little toe is located at the lateral side of the foot).

proximal—toward or nearest the trunk or the point of origin of a part (example, the elbow is located at the proximal end of the forearm).

distal—away from or farthest from the trunk or the point of origin of a part (example, the hand is located at the distal end of the forearm).

Planes of body

sagittal—a lengthwise plane running from front to back; divides the body or any of its parts into right and left sides (Fig. 1-16).

median—sagittal plane through midline; divides the body or any of its parts into right and left halves.

coronal or *frontal*—a lengthwise plane running from side to side; divides the body or any of its parts into anterior and posterior portions (Fig. 1-16).

transverse or *horizontal*—a crosswise plane; divides the body or any of its parts into upper and lower parts (Fig. 1-16).

Abdominal regions

For convenience in locating abdominal organs, anatomists divide the abdomen into nine imaginary regions. See Fig. 1-17 for their names and boundaries.

Anatomical position

The term anatomical position means an erect position of the body, with arms at sides, palms turned forward (supinated).

Generalizations about body function (Fig. 1-18)

1 Survival is the body's most important business—survival of itself and survival of the human species.

2 Survival depends on the body's maintaining or restoring homeostasis of its internal environment. (Homeostasis is discussed later.)

3 Homeostasis depends on the body's ceaselessly carrying on many activities. Its major activities or functions are responding to changes in its environment, exchanging materials between its environment and its cells, metabolizing foods, and integrating all of its diverse activities.

4 The body's functions are ultimately its cells' functions.

5 The body's ability to perform many of its functions changes gradually over the years. In general, the body performs its functions least well at both ends of life—in infancy and in old age. During childhood, body functions gradually become more and more efficient and effective. During late maturity and old age the opposite is true. They gradually become less and less effi-

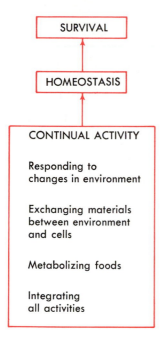

Fig. 1-18 Scheme to illustrate some generalizations about body function. Survival, the body's ultimate function, depends on homeostasis, which, in turn, depends on the functions listed in the lower box.

cient and effective. During young adulthood, they normally operate with maximum efficiency and effectiveness. The changes in functions that occur during the early years are called *developmental processes*. Those that occur during the late years are called *aging processes*. Developmental processes improve the functional capacities of the body. Aging processes, in contrast, diminish many functions and have no effect on others. Functional age changes will be noted in almost every chapter of this book.

Homeostasis

Homeostasis is a key word in physiology. It comes from two Greek words—*homoios*, meaning the same, and *stasis*, meaning standing.

"Standing or staying the same," then, is the literal meaning of the word. Walter B. Cannon, a noted physiologist, suggested homeostasis as the name for the steady states maintained by the body. He emphasized, however, that a steady state or homeostasis does not mean something set and immobile that stays exactly the same all the time. In his words, homeostasis "means a condition which may vary, but which is relatively constant."*

Human cells must have a relatively constant favorable environment if they are to survive in a healthy condition. This requirement applies to many aspects of the cellular environment—its chemical composition, its osmotic pressure, its hydrogen-ion concentration, its temperature, etc. Even a small shift away from homeostasis of any of these factors makes cells unable to function optimally. A larger deviation may cause death of cells and of the body.

Cellular environment, you must remember, is not synonymous with body environment. The body lives in the atmosphere surrounding it, the gaseous atmosphere of the external world. Body cells live in the atmosphere surrounding them, but this is the liquid atmosphere of an internal world.

The liquid environment around cells is called *extracellular fluid* for the obvious reason that it lies exterior to cells. Extracellular fluid occupies two main locations. It fills in the microscopic spaces between body cells, where it is called appropriately *intercellular fluid* (or *interstitial fluid*), and it flows through blood vessels, where it is called *blood plasma*.

More than a century ago, a great French physiologist, Claude Bernard, recognized that the extracellular fluid (or *milieu interne*, as he called it) constituted the internal environment, or, in other words, the cellular environment. He also noted that its physical and chemical composition must be maintained relatively uniform if

*From Cannon, W. B.: The wisdom of the body, rev. ed., New York, 1939, W. W. Norton & Co., Inc., p. 24.

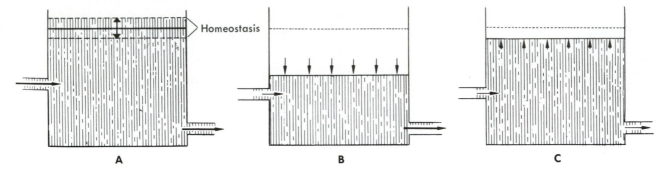

Fig. 1-19 A, Homeostasis as illustrated by constancy of a pool's water volume within a narrow range (indicated by double-headed arrow). When the volume of water entering the pool equals the volume leaving it, the pool volume remains constant at the level of the solid line, the midpoint of homeostasis for the pool's volume. **B,** The volume of water entering the pool has decreased. Because input is now less than output, the pool volume has decreased. **C,** The volume of water leaving the pool has decreased to compensate for the decreased volume entering it. Because output is now less than input, the pool volume has increased back to the original level; homeostasis of the pool volume is restored.

cells are to maintain their life and health. As stated previously, many years later the noted American physiologist Walter B. Cannon named the relative constancy of the internal environment *homeostasis.*

To maintain homeostasis, numerous complex processes called homeostatic mechanisms must operate. They involve the functioning of virtually all of the body's organs and systems. Hence homeostatic mechanisms constitute the major theme of physiology and of this book.

The following situation in the inanimate world may help you understand the meaning of homeostasis and homeostatic mechanisms. For example, imagine a swimming pool with water flowing into it at one end and an equal amount of water flowing out of it at the other end. Suppose that the depth of the water changes slightly from time to time, sometimes decreasing a little, sometimes increasing a little, but always staying within the same narrow range—say between 60 and 62 inches (Fig. 1-19, *A*). This narrow range represents homeostasis of the pool's water volume. Now suppose that the water inflow pipe be-

comes partially plugged by some object, causing less water to enter the pool than continues to leave it. The water's depth starts to decrease (Fig. 1-19, *B*). Suppose further that the pool contains a sensing device that is activated when the water's depth decreases to 61 inches and that the sensor, in turn, activates another device—one that partially closes a valve on the outflow pipe, so that the volume of water leaving the pool becomes less than the volume entering it. With less outflow than inflow, the water level increases back to its original 62 inches; our imaginary homeostatic mechanism has maintained homeostasis of the pool's water volume.

This example illustrates the following basic facts about homeostatic mechanisms:

1 A change in the body's internal environment acts as a stimulant to a sensor of one sort or another, and the stimulated sensor then initiates a complex response, a homeostatic mechanism.

2 A homeostatic mechanism consists of responses that reverse the initial change in the internal environment that "turned on" the homeostatic mechanism. Hence a homeostatic mechanism tends to maintain or to restore homeostasis.

3 The responses that make up a homeostatic mechanism are called appropriately *adaptive responses*. They enable the body to adapt to changes in its environment in ways that tend to maintain homeostasis and to promote healthy survival. Adaptive responses make it possible for the body to thrive as well as survive. Collectively, adaptive responses bring about adaptation. *Adaptation* consists of two kinds of changes: changes in the body induced by changes in the

environment and changes in the environment produced by changes in the body. Adaptation, in other words, is a great complex of interactions between the body and its environment. Together these interactions work toward the well-being of the body. In short, adaptation is the successful coping of an organism with its environment. Successful adaptation means healthy survival. Unsuccessful adaptation means disease or death.

■ ■ ■

Chapter 2 will describe cell structure and explain some of the intricacies of cell function.

Outline summary
Basic chemistry
A Definitions
1 Matter—anything that occupies space and has mass
2 Element—simple form of matter, cannot be broken down into two or more different substances
3 Atoms—structural units that make up an element
4 Chemical symbols—abbreviations for the names of elements

B Atoms
1 Consist of two main parts
   a Dense core called nucleus; consists of protons and neutrons
   b Shells of space around nucleus; serve as orbits in which electrons move
2 Protons—positively charged nuclear particles; each proton bears one positive charge; all atoms of same element contain same number of protons, and no two elements have same number of protons; in short, number of protons in atom's nucleus identifies kind of element it is; number of protons = (atomic weight minus number of neutrons)
3 Atomic number—number of protons in nucleus
4 Neutron—neutral or uncharged particle; number of neutrons = (atomic weight minus number of protons)
5 Atomic weight—number of protons plus number of neutrons
6 Isotopes—atoms of the same element that contain a different number of neutrons and that, therefore, have different atomic weights

7 Electrons—negatively charged nuclear particles; each electron bears one negative charge; each shell of space around the atom's nucleus can accommodate a certain maximum number of moving electrons; maximum for innermost shell is 2 electrons; for outermost shell is 4 pairs of electrons
8 One proton and 1 neutron weigh the same; each weighs 1,836 times as much as 1 electron

C Chemical reactions
1 Consist of either transfer of unpaired electrons from outer shell of one atom to outer shell of another or sharing of one atom's unpaired electrons with those of another atom (Fig. 1-3)
2 Electrovalent or ionic bond—formed by transfer of electrons
3 Covalent bond—formed by sharing of electron pairs of atoms
4 Compound—consists of one or more elements joined by chemical bonds to form molecules, the structural units of compounds
5 Valence—number of unpaired electrons in an atom's outer shell
6 Types of chemical reactions
   a Synthesis—combining of two or more substances to form more complex substance; formation of new chemical bonds
   b Decomposition—breaking down of substance into two or more simpler substances; breaking of chemical bonds
   c Exchange reactions—decomposition of two substances and, in exchange, synthesis of two new compounds from them
   d Reversible reactions—proceed in both direc-

tions, although not necessarily at the same rate in both directions

7 Chemical reactions and energy changes

   a Always coupled; formation of chemical bonds coupled with use of energy

   b Breaking of bonds coupled with release of free energy

D Radioactivity

  1 Radioactivity is emission from atom's nucleus of alpha, beta, or gamma rays

  2 Alpha rays are 2 protons plus 2 neutrons

  3 Beta rays are electrons originating from atom's nucleus by neutron breaking down into a proton and an electron; proton remains in nucleus, electron ejected from atom as beta ray

  4 Gamma rays are electromagnetic radiations

  5 Radioactivity changes number of protons in atom's nucleus, thereby changing it into a different element

Selected facts from biochemistry

A Biochemistry—chemistry of living organisms

B Biomolecules—compose living things; some are *inorganic compounds*, those which with few exceptions contain no carbon; most biomolecules are *organic* or *carbon-containing compounds* belonging to following classifications: protons, carbohydrates, lipids, and nucleic acids

  1 Electrolytes—compounds whose molecules consist of positive cations and negative anions held together by electrovalent chemical bond

  2 Proteins—compounds composed of amino acids linked together by peptide bonds; protein molecules are macromolecules; each protein has characteristic shape, that is, conformation; most are fibrous or globular; sequence of amino acids in protein molecule is different in each kind of protein and determines its conformation, which, in turn, determines its functions; largest group of proteins are enzymes; they function as catalysts for biochemical reactions (Table 1-1)

  3 Carbohydrates—compounds commonly called sugars and starches; three types of carbohydrates, that is, monosaccharides, disaccharides, and polysaccharides—monosaccharides with 6 carbon atoms are called hexoses (example, glucose), those with 5 carbon atoms called pentoses (example, ribose and deoxyribose); disaccharides are composed of two monosaccharides minus molecule of water (example, sucrose); polysaccharides are compounds composed of many

monosaccharides chemically joined by removal of water (example, glycogen); carbohydrate functions are energy release and energy storage

  4 Lipids—water-insoluble organic biomolecules of many types, chiefly, triglycerides, phosphoglycerides, steroids, and prostaglandins

   a Triglycerides—composed of three molecules of a fatty acid chemically joined to one molecule of glycerol (Figs. 1-8 and 1-9); most common saturated fatty acids are palmitic and stearic; polyunsaturated essential fatty acids are linoleic and linolenic

   b Phosphoglycerides—composed of glycerol with two fatty acids and a phosphate group joined to it and an amino alcohol (example, choline joined to the phosphate); choline phosphoglyceride (formerly called lecithin) is one of most abundant phosphoglycerides; phosphoglycerides function as major structural components of cell membranes.

   c Steroids—main component of steroid compounds is steroid nucleus (Fig. 1-11); include cholesterol, male and female sex hormones, and adrenocortical hormones

   d Prostaglandins—lipids composed of a 20-carbon fatty acid that contains 5-carbon ring; many different kinds of prostaglandins in body; reported function is to help regulate hormone activity and metabolism

  5 DNA (deoxyribonucleic acid)—composed of *deoxyribonucleotides*, that is, structural units composed of the pentose sugar (deoxyribose), phosphate group, and nitrogenous base—either cytosine or thymine or guanine or adenine; DNA molecule consists of two long chains of deoxyribonucleotides coiled into double-helix shape; alternating deoxyribose and phosphate units form backbone of the chains; base pairs, that is, either adenine-thymine or guanine-cytosine bonded together by hydrogen bonds, hold two chains of DNA molecule together; specific sequence of over 100 million base pairs constitute one human DNA molecule; sequence of base pairs is identical in all DNA of one individual and different in DNA of all other individuals; DNA functions as heredity molecule

C Bioenergy—energy that living organisms generate, store, and use to do work of muscle contraction, active transport, and biosynthesis; catabolism is process consisting of many highly coordinated chemical reactions that break down food compounds into carbon dioxide, water, and certain nitrogenous compounds, thereby releasing energy from them; immediately put more than half back into storage in ATP; rest of energy is released as heat energy

D Biosynthesis—chemical reaction that puts simple molecules together to form complex biomolecules, notably carbohydrates, lipids, proteins, nucleotides, and nucleic acids; anabolism is process consisting of thousands of highly integrated biosynthesis reactions; metabolism is catabolism plus anabolism: it is all chemical changes absorbed foods undergo in living cells; most simply, metabolism is body's use of foods

Generalizations about body structure

A The body is a large structural unit made up of countless smaller parts of four kinds

  1 Cells—smallest units that can maintain life and reproduce

  2 Tissues—organizations of many similar cells with nonliving intercellular substance between them

  3 Organs—organizations of several different kinds of tissues so arranged that they can perform a special function

  4 Systems—organizations of different kinds of organs so arranged that they can perform complex functions

B Ventral and dorsal cavities house the numerous internal organs (viscera) of the body

  1 Ventral cavity

    a Thoracic (or chest) cavity

      (1) Pleural cavities—right pleural cavity contains right lung; left pleural cavity contains left lung

      (2) Mediastinum—midportion of thoracic cavity, completely separated from pleural cavities by fibrous tissue wall; contains heart (enclosed in its pericardial sac), trachea, bronchi, esophagus, thymus, and various blood vessels, lymphatic vessels, and nerves

    b Abdominopelvic cavity

      (1) Abdominal portion—contains liver, gallbladder, stomach, pancreas, intestines, spleen, kidneys, and ureters

      (2) Pelvic portion—contains bladder, certain reproductive organs, and part of large intestine

  2 Dorsal cavity

    a Cranial cavity—contains brain

    b Spinal cavity—contains spinal cord

C Organization is a vital structural characteristic of the body and all its parts

D Body structures change gradually over years; in general, body structures develop and grow during the years before young adulthood; after the period of young adulthood, they age and, typically, atrophy

Terms used in describing body structure

A Directional terms

  1 Superior—upper

  2 Inferior—lower

  3 Anterior—front

  4 Posterior—back

  5 Medial—toward midline of body

  6 Lateral—away from midline of body

  7 Proximal—toward or nearest trunk or point of origin of part

  8 Distal—away from or farthest from trunk or point of origin of part

B Planes of body

  1 Sagittal—lengthwise, dividing body or any of its parts into right and left portions

  2 Median—sagittal plane through midline

  3 Frontal or coronal—lengthwise, dividing body or any of its parts into anterior and posterior portions

  4 Transverse or horizontal—crosswise, dividing body or any of its parts into upper and lower sections

C Abdominal regions—see Fig. 1-17

D Anatomical position—erect, arms at sides, palms forward

Generalizations about body function

A Survival is the body's most important business—survival of individual and of species

B Survival depends on the body's maintaining or restoring homeostasis

C Homeostasis depends on the body's continually carrying on many activities, notably, responding to changes in its environment, exchanging materials between its environment and its cells, metabolizing foods, and integrating all of its diverse activities

D Body functions are ultimately cell functions

E Body functions change gradually over the years; they are least efficient and effective during infancy and old age; changes in functions that occur during the years before young adulthood improve the body's ability to maintain homeostasis and to survive in a healthy condition and are called developmental processes; after young adulthood, changes in function lessen the body's ability to maintain homeostasis and healthy survival and are called aging processes

Homeostasis

A Homeostasis is relative constancy of extracellular fluid, notably with respect to its chemical composition, osmotic pressure, hydrogen ion concentration, and temperature

B Homeostatic mechanisms maintain or restore homeostasis

1 Changes in extracellular fluid activate homeostatic mechanisms by stimulating different types of sensors

2 Homeostatic mechanisms consist of responses that reverse the changes that activated them so they tend to maintain or restore homeostasis

3 Responses that make up homeostatic mechanisms are called adaptive responses because they enable the body to adapt to changes in its environment in ways that tend to maintain homeostasis and promote healthy survival

1 Define the following terms: element, compound, atom, molecule, proton, neutron, electron, isotopes.

2 Define and contrast atomic number and atomic weight.

3 Describe the change in an atom when it takes part in a chemical reaction.

4 Define and compare synthesis and decomposition reactions.

5 What relation, if any, exists between chemical reactions and energy changes?

6 Define radioactivity and compare with chemical activity, that is, chemical reactions.

7 What are electrolytes and how are they formed?

8 What is a cation? An ion?

9 What are the structural units or building blocks of proteins? Of carbohydrates? Of triglycerides? Of DNA?

10 What groups compose a nucleotide?

11 What sugar is present in a deoxyribonucleotide? Is it a hexose or a pentose?

12 Name the pyrimidine bases present in DNA. Name the purine bases in it.

13 Explain the chemical composition of DNA. Describe its shape and size.

14 What is a helix?

15 What base is thymine always paired with in the DNA molecule? What other two bases are always paired together in DNA?

16 In one word, what is the function of DNA?

17 What is catabolism? What function does it serve?

18 Compare catabolism, anabolism, and metabolism.

19 ATP is the abbreviation for what important bio-molecule? Why is it important?

20 Name and briefly define the four kinds of structural units that compose the body.

21 Name the two cavities on the dorsal surface of the body. What organs does each cavity contain?

22 Identify the divisions of the thoracic cavity and of the abdominopelvic cavity. Name the organs in each.

23 Define briefly each of the following terms: anatomical position, anatomy, anterior, distal, frontal plane, hypochondriac region, hypogastric region, iliac region, inferior, lateral, median plane, organ, physiology, posterior, proximal, sagittal plane, tissue, transverse plane.

24 Make a generalization about the changes in body structures over the years.

25 What is the body's one all-inclusive function?

26 Define the term homeostasis.

27 What is a homeostatic mechanism?

28 Make a generalization about changes in body functions over the years.

chapter 2

Cells

Cells, the smallest structures capable of maintaining life and reproducing, hold many mysteries. Small as they are, they have not yet yielded all their secrets to the probing tools of research, and a host of scientists are still at work exploring these deceivingly simple structures. The science that studies cells is a branch of biology called *cytology*. This chapter relates many facts and ideas from this science about the fascinating world of cells. It discusses cell structure first and then cell functions.

Cell structure

Protoplasm

A special kind of matter known as *protoplasm* composes all cells. A protoplasm is a complex organization of literally thousands of biomolecules formed out of only 24 elements. None of the elements is unique. All of them occur also in nonliving matter. Hydrogen, oxygen, carbon, and nitrogen contribute more than 99% of the atoms in protoplasm and, together with phosphorous and sulfur, make up its major compounds—water, proteins, carbohydrates, lipids, and nucleic acids. All of these compounds except water are unique to protoplasm, that is, found only in plant and animal cells. By far the most abundant compound in protoplasm (and in the inanimate world) is water. Salts are present in smaller amounts but are vitally important ingredients of cells. They exist mainly as ions. In cells, potassium ions are the chief cations, and phosphate ions are the main anions.

Ideas about cell structure have changed considerably since 1665, the year that Robert Hooke discovered cells. He visualized them as "hollow rooms." Later biologists thought of them as closed sacs containing a semiliquid material with little objects floating in it. They named the semiliquid material *cytoplasm* (meaning cell protoplasm) and called the little objects *organelles* (meaning little organs).

Today's biologists tend to think of a cell as a congested complex of membranes composed of molecules arranged in definite patterns. This concept has grown largely from what investigators have actually seen under the electron microscope. Its power to magnify objects several hundred thousand times has even made large molecules visible, for instance, those of DNA and proteins. More recently, the scanning electron microscope has provided three-dimensional pictures of cells.*

One of the remarkable techniques that has uncovered many secrets of both cell structure and function is radioautography. It combines the use of radioactive atoms with a photographic technique. And yet, however thoroughly we may manage to expose protoplasm's intricacy—millions of atoms in a single cell, organized into thousands of compounds, participating in numberless reactions—we shall probably never learn all of the cell's secrets.

Cell differences

Cell structure differs in certain respects in cells that perform different functions, notably in the size and shape of the cells and in the special structures that some of them possess. A human ovum (female sex cell), for instance, measures close to 1,000 micrometers* in diameter, but a human erythrocyte (red blood cell) has a diameter of only 7.5 micrometers. Cells differ even more markedly in shape than in size. As evidence, consider the cells illustrated in Fig. 2-1. The most evident special cell structures are different types of projections from a cell's surface. We shall describe these after we have discussed

---

*For example, see Everhart, T. E., and Hayes, T. L.: The scanning electron microscope, Sci. Am. **226:**65, 67, Jan., 1972. Almost surely you will find the three-dimensional pictures of human red blood cells and of the nerve cell of a sea hare shown on these pages dramatic and beautiful.

---

*By international agreement, it is now recommended that nanometers (nm) be used to indicate the size of molecules, and that the size of cells and cell structures be designated by micrometers (μm).

1 meter (m) = 39.37 inches
1 centimeter (cm) = 10^{-2} m (0.01 m)
1 millimeter (mm) = 10^{-3} m (0.001 m)
1 micrometer (μm) = 10^{-6} m (0.000001 m)
    or micron (μ)
1 nanometer (nm) = 10^{-9} m (0.000000001 m)
1 angstrom (Å) 10^{-10} m (0.0000000001 m)
1 inch = approximately
    2.5 cm
    25 mm
    25,000 μm
    25,000,000 nm
    250,000,000 Å

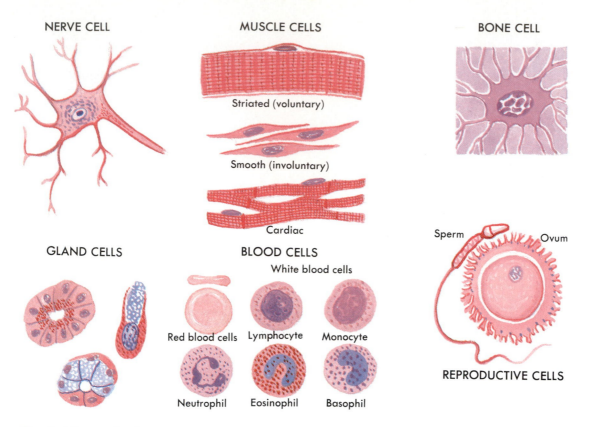

NERVE CELL

MUSCLE CELLS

Striated (voluntary)

Smooth (involuntary)

Cardiac

BONE CELL

GLAND CELLS

BLOOD CELLS

White blood cells

Red blood cells   Lymphocyte   Monocyte

Neutrophil   Eosinophil   Basophil

Sperm   Ovum

REPRODUCTIVE CELLS

Fig. 2-1 Types of cells.

the structures common to all cells—an outer, limiting membrane; cytoplasm; several kinds of organelles in the cytoplasm; and in almost all cells, a nucleus. An exception among human cells is the mature erythrocyte (red blood cell); it has no nucleus. Examine Fig. 2-2 carefully. Note the names and appearances of the structures shown in this cell and then learn more about them from the pages that follow.

Cell membranes

Membranes are a cell's most conspicuous structural feature. They form both the outer boundary of the cell itself and also the boundaries or walls of most of its internal parts. The membrane that constitutes the surface boundary or other limits of the cell is called the *plasma* *membrane,* or more descriptively, the *cytoplasmic membrane.* Internal cell structures that have membranous walls are the nucleus and the organelles named endoplasmic reticulum, Golgi complex, mitochondria, and lysosomes. Both the cytoplasmic membrane and the internal membranes have essentially the same structure. Both are composed largely of lipid and protein molecules. The arrangement of these molecules within the membrane has intrigued and challenged research scientists for many years. Today's most widely accepted concept postulates that two layers, that is, a bilayer, of phospholipid molecules form a fluid framework for cell membranes and that protein molecules also enter into their construction. Each phospholipid molecule has a "head" and a "tail." Its head contains phosphate, and its tail consists of two fatty acids. (See Fig. 1-10, p. 13.) A fascinating fact about

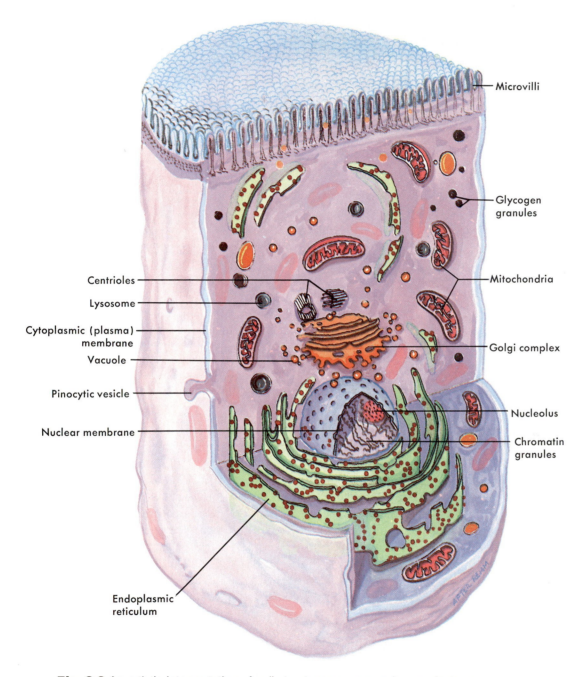

Microvilli

Glycogen granules

Centrioles

Lysosome

Cytoplasmic (plasma) membrane

Vacuole

Pinocytic vesicle

Nuclear membrane

Endoplasmic reticulum

Mitochondria

Golgi complex

Nucleolus

Chromatin granules

Fig. 2-2 An artist's interpretation of cell structure as seen under an electron microscope. Note the many mitochondria, popularly known as the "power plants of the cell." Note, too, the innumerable dots bordering the endoplasmic reticulum. These are ribosomes, the cell's "protein factories."

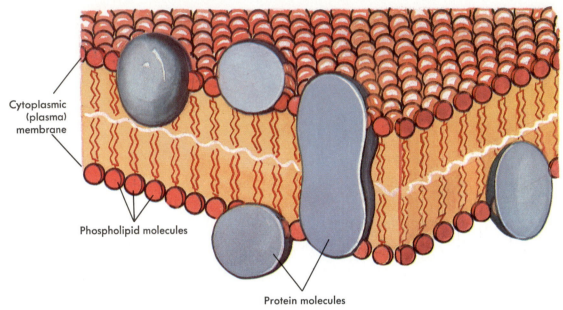

Cytoplasmic (plasma) membrane

Phospholipid molecules

Protein molecules

Fig. 2-3 Postulated structure of cytoplasmic and internal cell membranes. Note the following features of membrane's molecular structure: the bilayer of phospholipid molecules arranged with their "tails" pointing toward each other; the location of protein molecules in all possible positions with relation to the phospholipid bilayer—at the outer surface, at the inner surface, partially penetrating the bilayer, and extending all the way through the membrane.

phospholipid molecules is that their heads are hydrophilic and their tails are hydrophobic. (Hydrophilic: water-loving, hence, water-soluble. Hydrophobic: water-fearing, or water-insoluble.) This dual nature of phospholipid molecules causes them to position themselves in a curious way in membranes. Look at Fig. 2-3 to see what this position is. Observe that the water-fearing tails of phospholipid molecules point away both from the fluid outside the membrane and also away from the fluid inside it. How do they manage this? By lining up in two rows with their tails pointing at each other. This novel arrangement leaves the water-loving heads of the outer row of molecules facing into the fluid outside the membrane and the inner row of heads facing into the fluid inside the membrane.

Consider now the arrangement of protein molecules in cell membranes. Again, look at Fig.

2-3. Are the arrangements of protein molecules in the outer and inner surfaces of the membrane identical, that is, symmetrical? Are the same protein molecules present in both surfaces? The answer to both questions is "no." The arrangement of membrane proteins is not symmetrical, but asymmetrical. Different protein molecules are present in the outer and inner surfaces of membranes—a fact of functional significance, since different proteins perform different functions. Later in this chapter, we shall cite specific examples of this.

Cell membranes perform a variety of functions essential for the healthy survival of cells and of the body as a whole. The cytoplasmic membrane serves as the boundary between a cell's internal fluid and its external fluid environments. It is sturdy enough to maintain the cell's integrity, to preserve the arrangement of its internal structures necessary for their carrying on the activities that sustain the cell's life. This sounds impossible for a membrane less than a millionth of an inch thick, yet it is true. If a cell's cytoplasmic membrane becomes torn, the cell loses its integrity, its organized wholeness. Its inner structure becomes disorganized as the cell's contents leak out. Result? The cell dies. Another function of the cytoplasmic membrane is to determine what substances move through it to enter or leave the cell. It entirely excludes some substances. It allows some substances to move through it easily, accelerates the movement of others, and actively pumps through itself still other materials. Communication is another function of the cytoplasmic membrane. For example, the cytoplasmic membranes of some cells have proteins on their outer surfaces that serve as receptors for chemical messages. Certain hormones and certain chemicals called neurotransmitters bind to these surface proteins and thereby communicate their message to the cells. Some of the cytoplasmic membrane proteins on certain white blood cells (lymphocytes) function as antibodies. They combine with potentially harmful foreign proteins called antigens and thereby initiate a series of events that render the antigens harmless. Many of the surface proteins on cytoplasmic membranes serve as cell identification tags. Because they are unique for the cells of each individual, they can be used to distinguish one individual cells from another's.

Internal cell membranes serve as the boundaries of various internal structures and also perform other functions. Many of the proteins in them, for example, are enzymes that accelerate (catalyze) an enormous variety of chemical reactions. Enzymes make possible the chemical reactions that keep our cells and our bodies alive.

Cytoplasm and organelles

Cytoplasm is the part of a cell between its membrane and its nucleus. In other words, it is

all of a cell's protoplasm except its nucleus. Far from being the homogeneous substance once thought, it contains several different kinds of small structures. Not only that, in many cells, some of these number in the thousands. Collectively, they are known as *organelles*, that is, the "little organs" of cells. Each organelle consists of molecules arranged or organized in such a way that they can perform some function essential for maintaining the cell's life or for its reproduction. Membranes form the walls of most kinds of organelles. We shall discuss these membranous organelles—the endoplasmic reticulum, Golgi apparatus, mitochondria, and lysosomes—and then consider the nonmembranous ribosomes and the centrosome.*

Endoplasmic reticulum

Endoplasm means the cytoplasm located toward the center of a cell. Reticulum means network. Therefore the name endoplasmic reticulum (ER) means literally a network located deep inside the cytoplasm. And when first seen, it appeared to be just that. Later on, however, more highly magnified electron photomicrographs showed quite clearly that the endoplasmic reticulum has a much wider distribution than in the endoplasm alone. As you can observe in Fig. 2-2, it is scattered throughout the cytoplasm.

There are two types of endoplasmic reticulum: rough and smooth. Innumerable small granules—ribosomes, by name—dot the outer surface of the membranous wall of the rough type and give it its "rough" appearance. Ribosomes are themselves organelles. The rough endoplasmic reticulum seems to consist, as shown in Fig. 2-2, of flat, curving sacs arranged in parallel rows. Actually, it is a system of connected sacs and canals. The canals wind tortuously through the cytoplasm, extending all the way

* Recently, investigators using new techniques have identified complex, three-dimensional networks of very fine fibers (ranging from about 3 to 25 nm in diameter) extending throughout the cytoplasm. See Miller, J. A.: The bone and muscle of cells, Sci. News **112**:250-253, Oct. 15, 1977.

from the cytoplasmic membrane to the nucleus. The membrane forming the walls of the endoplasmic reticulum has essentially the same molecular structure as the cytoplasmic membrane.

The structural fact that the endoplasmic reticulum is an interconnected system of canals suggests that it might function as a miniature circulatory system for the cell. And in fact, proteins do move through the canals. The ribosomes attached to the rough endoplasmic reticulum synthesize proteins. These proteins enter the canals and move through them to the Golgi apparatus. Thus the rough endoplasmic reticulum functions in both protein synthesis and intracellular transportation.

No ribosomes border the membranous wall of the smooth endoplasmic reticulum—hence its smooth appearance and its name. Its functions are less well established and probably more varied than those of the rough type. In liver cells, for instance, it is thought that the smooth endoplasmic reticulum functions in lipid and cholesterol metabolism. On the other hand, in cells of the testes and adrenals, it apparently takes part in steroid hormone synthesis.

Golgi apparatus

The Golgi apparatus consists of tiny sacs stacked one on the other and located near the nucleus. Note in Fig. 2-4 that the sacs look more and more distended in successive layers of the pile—as if some material were filling them up to the bulging point. And this actually is the case. Several research teams have presented evidence that the Golgi sacs synthesize large carbohydrate molecules and then combine them with proteins (brought to them through the canals of the endoplasmic reticulum) to form compounds called glycoproteins. As the amount of glycoproteins in the sacs increases, their shape changes from flat to "fat." They turn into perfect little spheres or globules. Then one by one, they pinch away from the top of the stack. The Golgi apparatus, in other words, not only synthesizes

carbohydrate and combines it with protein, but it even packages the product! Neat little globules of glycoprotein migrate outward away from the Golgi apparatus to and through the cell membrane. Once outside the cell the globules break open, releasing their contents. The cell has secreted its product.

One might think that only in the secreting cells of glands would the Golgi apparatus function as just described. However, various other kinds of cells also make products for "export," products that move out of the cells that make them. In other words, various nonglandular cells secrete substances. Some examples: Liver cells secrete blood proteins, plasma cells secrete antibodies, and connective tissue cells of certain types secrete substances used in bone and cartilage formation. The Golgi apparatus in all of these cells is believed to synthesize carbohydrate, to combine it with protein, and to package the product in globules for secretion. The Golgi apparatus no longer seems mysterious or insignificant. Two scientists who have contributed greatly to our knowledge of this organelle make clear their high regard for it in the following words: "All in all, it looks as if the Golgi apparatus is a creative mechanism in the cell ranking in importance with the ribosome. Just as the ribosomes are responsible for the construction of proteins, so the Golgi apparatus seems to be the main agency for building a variety of large carbohydrates that serve many vital purposes."*

Mitochondria

Note the mitochondria shown in Fig. 2-1. Magnified thousands of times, as they are there, they look like small, partitioned sausages—if you can imagine sausages only 1.5 μm long and one half as wide. (In case you can visualize inches better than micrometers, 1.5 μm equals about 3/50,000 of an inch.) Yet, like all organelles, and tiny as

---

*From Neutra, M., and Leblond, C. P.: The Golgi apparatus, Sci. Am. **220**:100-107, Feb., 1969; copyright © 1969 by Scientific American, Inc.; all rights reserved.

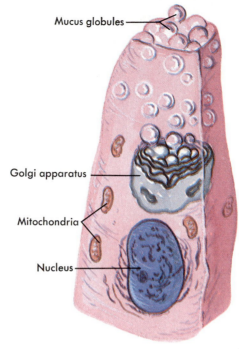

Mucus globules

Golgi apparatus

Mitochondria

Nucleus

Fig. 2-4

A, Artist's interpretation of a mucus-secretion cell, showing a typical Golgi apparatus, basal nucleus, and mitochondria.

Golgi vacuoles

Cisternae

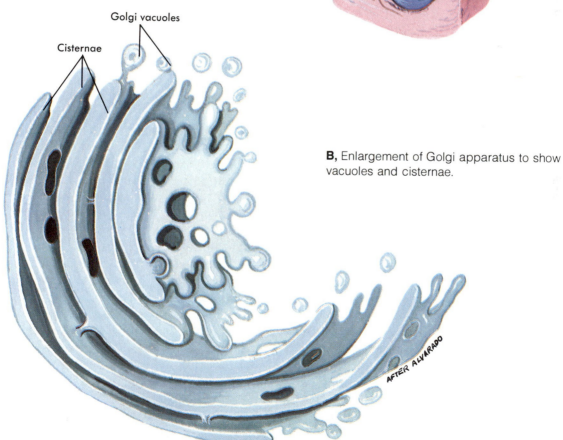

B, Enlargement of Golgi apparatus to show vacuoles and cisternae.

AFTER ALVARADO

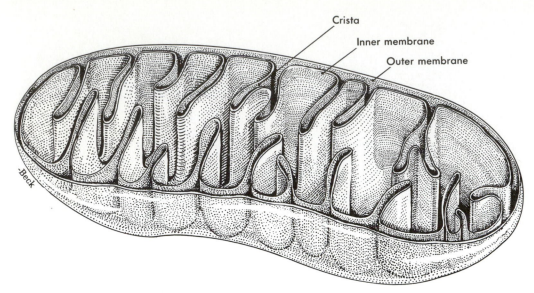

Crista
Inner membrane
Outer membrane

-Beck

Fig. 2-5 Mitochondrion. Cutaway diagram shows the two-layered structure of delicate membranous wall.

they are, mitochondria have a highly organized structure. The membranous wall of the mitochondrion consists not of one but of two delicate membranes. They form a sac within a sac. The mitochondrion's inner membrane folds into a number of extensions called *cristae*. Note in Fig. 2-5 how they jut into the interior of the mitochondrion like so many little partitions. With the very powerful magnification of an electron microscope, one can see small round knobs attached to the cristae by short stalks and projecting inward from them. A single mitochondrion may contain thousands of these tiny knobs. Each knob, in turn, contains enzymes essential for the making of one of the most important chemicals in the world. Without this compound, life cannot exist. Its long name, adenosine triphosphate, and its abbreviation, ATP, were mentioned in Chapter 1, and Chapter 19 presents more detailed information about this vital substance.

Both inner and outer membranes of the mitochondrion have essentially the same molecular structure as the cytoplasmic membrane. Many of the proteins in the membranes of the cristae are known to be enzymes. Also, all evidence so far indicates that they are arranged most precisely in the order of their functioning. This is another example, but surely an impressive one, of the principle that organization is a foundation stone and a vital characteristic of life.

Enzymes in the mitochondrial inner membrane catalyze oxidation reactions. These are the chemical reactions that provide cells with most of the energy that does all of the many kinds of work that keep them and the body alive. Thus do mitochondria earn their now familiar title, the "power plants" of cells (discussed in Chapter 19).

From the knowledge that mitochondria generate most of the power for cellular work, one might deduce that the number of mitochondria in a cell would be directly related to its amount of activity. This principle does seem to hold true. In general, the more work a cell does, the more mitochondria its cytoplasm contains. Liver cells, for example, do more work and have more mitochondria than sperm cells. Swanson* estimates that a single liver cell may contain as many as 1,000 mitochondria, whereas a single sperm cell has only about 25 mitochondria.

*Swanson, C. P.: The cell, ed. 3, Englewood Cliffs, N.J., 1969, Prentice-Hall, Inc.

Lysosomes

Lysosomes are membranous-walled organelles whose size and shape change with the stage of their activity. In their earliest, inactive stage, they look like mere granules. Later, as they become active, they take on the appearance of small vesicles or sacs (Fig. 2-2) and often contain tiny particles such as fragments of membranes or pigment granules. The interior of the lysosome contains various kinds of enzymes capable of breaking down all the main components of cells. These enzymes can, and under some circumstances actually do, destroy cells by digesting them. The graphic nickname "suicide bags," therefore, seems appropriate for lysosomes. Little wonder that these powerful and dangerous substances are usually kept sealed up in lysosomes. However, lysosomal enzymes more often protect than destroy cells. Large molecules and large particles (for example, bacteria) that find their way into cells enter lysosomes, and their enzymes dispose of them by digesting them. Therefore "digestive bags" and even "cellular garbage disposals" are other nicknames for lysosomes. White blood cells contain a great many lysosomes. They serve as scavenger cells for the body, engulfing bacteria and destroying them in their lysosomes.

Ribosomes

Every cell contains thousands of ribosomes. The small spherical organelles you can see attached to the endoplasmic reticulum and scattered through the cytoplasm in Fig. 2-2 are ribosomes. Because ribosomes are too small to be seen with a light microscope, no one knew they existed until the electron microscope revealed them. Now scientists even know a good deal about their molecular structure. For instance, they know that they consist of approximately two thirds ribonucleic acid (RNA) and one third protein. They also know that ribosomes consist of two subunits of different sizes.

The function of ribosomes is protein synthesis. Ribosomes are the molecular machines that make proteins. Or, to use a popular term, they are the cell's "protein factories." Ribosomes attached to the endoplasmic reticulum, as already mentioned, synthesize proteins for "export," whereas the ribosomes free in the cytoplasm make proteins for the cell's own domestic use. They make its structural proteins, in other words, and its enzymes. Working ribosomes, those that are actually making proteins, appear to function in groups called polyribosomes. Under the electron microscope, polyribosomes look like short strings of beads. In Chapter 19 we shall relate some current ideas about the complicated and still incompletely understood process of protein synthesis.

Centrosome

Centrosome, the name of this organelle, suggests its location near the center of the cell. Actually, this means that the centrosome is located near the nucleus, since the nucleus takes up the center space of most cells. With the light microscope and suitably prepared slides, one can see two dots (called *centrioles*) in the centrosome. The electron microscope, however, reveals them not as mere dots but as tiny cylinders (Fig. 2-2). The walls of the cylinders consist of nine bundles of fine tubules, with three tubules in each bundle. A curious fact about these two tubular-walled cylinders is their position at right angles to each other. One wonders why—a question as yet unanswered. Precisely how centrioles function is another unsolved riddle. They seem, however, to play some part in the formation of the spindle that appears during mitosis (cell division).

Nucleus

The nucleus, the largest cell structure, occupies the central portion of the cell. Both the shape of the nucleus and the number of nuclei present in a cell vary. One spherical nucleus per cell, however, is common. Electron micrographs show that two membranes perforated by openings or pores enclose the nucleoplasm (nuclear fluid). These nuclear membranes have essentially the same structure as the cytoplasmic membrane, but unlike the cytoplasmic membrane, the nuclear membranes disappear for a short time during the process of cell division. One or more small, dense bodies called *nucleoli* lie in the nucleoplasm, and in a nondividing cell, numerous chromatin granules are also visible in the nucleoplasm. Chromatin granules are composed partly of protein but chiefly of nucleic acid, the one named deoxyribonucleic acid, or DNA (see p. 14). Early in the process of cell division, chromatin granules disappear, and short rodlike structures take their place. These small structures bear a now-famous name—chromosomes. All human cells (except mature sex cells) contain 46 chromosomes, and each chromosome consists of one DNA molecule plus some protein molecules. Nucleoli, like chromosomes, also consist chiefly of a nucleic acid with some protein, but the nucleic acid in nucleoli is *not* DNA. It is ribonucleic acid, or RNA.

What functions does the nucleus perform? Briefly, it stores, transcribes, and transmits genetic information. Genetic information is stored in the DNA molecules present in the nucleus. It consists of the sequence of nucleotide base pairs in DNA molecules. The sequence of base pairs in DNA is transcribed into the sequence of base pairs in RNA molecules as they are being synthesized in the nucleus. This RNA, known as *messenger RNA*, leaves the nucleus through the pores in the nuclear membranes and attaches to ribosomes in the cytoplasm of the cell. The sequence of base pairs in the messenger RNA acts as a coded message that is translated in the ribosomes into the specific proteins they synthesize. Both the transcription of DNA into RNA and the translation of RNA into proteins are complex mechanisms, and we shall save our discussion of them for a later chapter (Chapter 23). Many of the proteins synthesized by the ribosomes are functional proteins, notably enzymes that make possible the numerous chemical reactions that maintain the cell's life and enable it to grow and reproduce. Other proteins are structural proteins that serve as integral parts of cell membranes and other cell structures. Thus the DNA molecules in a cell's nucleus, through the mechanisms of transcription and translation, determine both the cell's functions and its structure.

The nucleus does more than store genetic information and transcribe it. It also transmits genetic information. It does this in the following way. Before a cell divides to form two new cells, all of its DNA molecules replicate, that is, they make perfect copies of themselves. One set of the duplicated DNA molecules enters the nucleus of one of the new cells formed by cell division, and the other set enters the nucleus of the other new cell. In short, every cell inherits DNA molecules that are identical copies of the DNA in its parent cell. Therefore the inherited DNA, through the mechanisms of transcription and translation, orders the new cells to synthesize the same proteins as the parent cell made. And since proteins largely determine cell structure and function, the new cells have, in effect, inherited the structure and functions of the parent cell. In short, the nucleus has functioned to bring about cell heredity. Chapter 23 describes the mechanism of DNA replication.

Nucleoli play an essential role in the formation of ribosomes, the protein synthesizers of cells. You might guess, therefore, and correctly so, that the more protein a cell makes, the larger its nucleoli appear. Cells of the pancreas, to cite just one example, make large amounts of protein and have large nucleoli.

Special cell structures

Microvilli, cilia, and flagella are special cell structures, that is, they are present only in certain types of cells. *Microvilli*, for example, are special structures of epithelial cells that line the intestines. Microvilli consist of extensions of cytoplasm and cytoplasmic membrane. Like tiny fingers crowded close against each other, they project from the surface of the cell (Fig. 2-2). A single microvillus measures about 0.5 μm long and only 0.1 μm or less across. Since one cell has hundreds of these projections, the surface area of the cell is increased manyfold—a structural fact that enables the cell to perform its function of absorption at a faster rate.

Cilia, someone said, look like eyelashes attached to one surface of a cell. Among the cells possessing cilia are the epithelial cells forming the surface of the mucous membrane that lines some parts of the upper respiratory tract. The electron microscope has revealed the fine structure of a cilium—a fascinating example of the organization characteristic of living things. Each

cilium is essentially a projection of a cell's cytoplasm and cytoplasmic membrane; it consists of a very tiny cylinder (0.2 μm in diameter) made up of nine double microtubules arranged around two single microtubules in the center. One cell may have a hundred or more cilia. They move together in such a way that they propel a fluid in one direction over the surface of the cell. Ciliated cells in the respiratory mucous membrane, for example, propel mucus upward in the respiratory tract.

A *flagellum* is a single, hairlike projection from the surface of a cell. Each male sex cell (a spermatozoon) has a flagellum, commonly called a tail. A flagellum moves in such a way that it propels the spermatozoon forward in its fluid environment.

Table 2-1 Some major cell structures and their functions

Cell structures	Functions
Cytoplasmic membrane	Serves as the boundary of the cell, maintaining its integrity; protein molecules on outer surface of cytoplasmic membrane perform various functions, for example, they serve as markers that identify cells of each individual, receptor molecules for certain hormones and neurotransmitters provide means of communication between cells, receptor molecules for foreign proteins function to produce immunity
Endoplasmic reticulum (ER)	Serves as a cell's own circulatory system
Golgi apparatus	Synthesizes carbohydrate, combines it with protein, and packages the product as globules of glycoprotein
Mitochondria	Catabolism; ATP synthesis; a cell's "powerhouse"
Lysosomes	A cell's "digestive system"
Ribosomes	Synthesize proteins; a cell's "protein factory"
Nucleus	Dictates protein synthesis, thereby playing essential role in other cell activities, namely, active transport, metabolism, growth, and heredity
Nucleoli	Play essential role in the formation of ribosomes

Cell physiology—movement of substances through cell membranes

Every cell carries on a number of functions that maintain its own life—transportation and metabolism, for example. If a cell is to survive, it must continually move substances through its membranes and must metabolize foods (use them for energy and for building complex compounds). Also, from time to time, all but a few kinds of cells perform another function, that of reproducing themselves. In addition to self-serving activities, every cell in the body also performs some special function that serves the body as a whole. Muscle cells provide the function of movement, nerve cells contribute communication services, red blood cells transport oxygen, etc. In return the body performs vital functions for all of its cells. It brings food and oxygen to them and removes waste from them, to mention only two examples. In short, a relationship of mutual interdependence exists between the body as a whole and its various parts. Optimum health of the body depends on optimum health of each of its parts, down even to the smallest cell. Conversely, optimum health of each individual part depends on optimum health of the body as a whole. In equation form, this physiological principle of mutual interdependence might be expressed as follows:

cellular health ⇌ tissue health ⇌ organ health ⇌ system health ⇌ body health

Some parts of the body, of course, are more important for healthy survival than others. Obviously the heart is far more important than the appendix. And the nerve cells that control respiration are infinitely more important for survival than muscle cells that move the little finger.

Cell physiology deals with all the different kinds of functions cells perform. It is, as you might guess, an enormous subject. Therefore we have chosen to discuss only one aspect of it at this time—a cell's transportation activities. Later chapters will deal with the following cell functions: contractility, conductivity, metabolism, and reproduction. The rest of this chapter contains some known and some postulated answers to the question, "How do substances move through cell membranes?"

Types of processes

Heavy traffic moves continuously in both directions through cell membranes. Streaming in and out of all cells in endless procession go molecules of water, foods, gases, wastes, and many kinds of ions. Several processes carry on this mass transportation. They are classified under two general headings as physical (or passive) and physiological (or active) processes.

The main distinction between these two kinds of processes lies in the source of the energy that does the work of moving a substance through a membrane. If the energy comes from chemical reactions taking place in a living cell, the transport mechanism is classified as an *active* or *physiological process*. If the energy for moving a substance stems from some other source and not from a living cell's chemical reactions, the transport mechanism is classified as a *passive* or *physical process*. The two preceding sentences, you may have noticed, implied another distinction between active and passive transport mechanisms. Active mechanisms can move substances only through living cell membranes. Passive mechanisms, on the other hand, can move materials through either living or dead cell membranes and even through artificial membranes. Major passive or physical transport processes are diffusion, osmosis, and filtration. There are several kinds of active transport processes: those called "pumps," for example, and the mechanisms of phagocytosis and pinocytosis. First we shall consider these passive processes—diffusion (including dialysis and facilitated diffusion), osmosis, and filtration—and then we shall turn our attention to the active processes, namely, the so-called "pumps" and phagocytosis and pinocytosis.

Diffusion

Diffusion means scattering or spreading. It occurs because small particles such as molecules and ions are forever on the go. They move continuously, rapidly, and in all directions. Perhaps the easiest way for you to learn the essential facts and principles about diffusion is to discover them for yourself in the following example, illustrated in Fig. 2-6. Suppose a 10% sodium chloride (NaCl) solution is separated from a 20% NaCl solution by a membrane. Suppose further that the membrane is permeable to both NaCl and to water. This means that both these sub-

stances can and do pass through the membrane in both directions. NaCl particles and water molecules racing in all directions through each solution collide with each other and with the membrane. Some inevitably hit the membrane pores from the 20% side and some from the 10% side. Just as inevitably, some bound through the pores in both directions. For a while more NaCl particles enter the pores from the 20% side simply because they are more numerous there than on the 10% side. More of these particles, therefore, move through the membrane from the 20% solution into the 10% than diffuse through it in the opposite direction. Using different words for the same thought, *net diffusion* of NaCl takes places from the solution where its concentration is greater into the one where its concen-

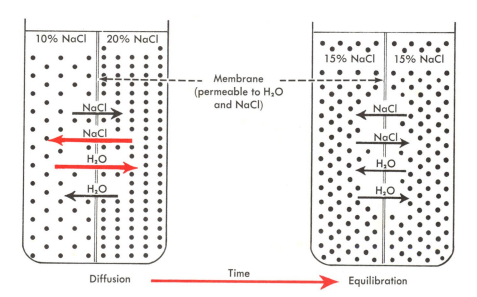

Fig. 2-6 Diffusion. Because the membrane separating the 10% NaCl from the 20% NaCl is freely permeable to both NaCl and H$_2$O, both substances diffuse rapidly through the membrane in both directions. But, as the red arrows indicate, more sodium and chloride ions move out of the 20% solution, where there are more of them, into the 10% solution, where there are fewer of them, than in the opposite direction. Simultaneously, more water molecules move from the 10% solution, where there are more of them, into the 20% solution, where there are fewer of them. Result: equilibration of the concentrations of the two solutions after an elapse of time. From then on, equal numbers of Na ions and Cl ions diffuse in both directions, as do equal numbers of H$_2$O molecules.

tration is lesser. Thus net diffusion of NaCl occurs "down" the *NaCl concentration gradient,* that is, from the higher concentration down to the lower concentration.

During the time that net diffusion of NaCl is taking place between the 20% and 10% solutions, net diffusion of water is also going on. The direction of net diffusion of any substance is always down that substance's concentration gradient. Applying this principle, net NaCl diffusion occurs down the NaCl concentration gradient, and net water diffusion occurs down the water concentration gradient. Since the greater concentration of water molecules lies in the more dilute 10% solution (where fewer water molecules have been displaced by NaCl molecules), more water molecules diffuse out of the 10% solution into the 20% solution than diffuse in the opposite direction. Thus the net diffusion of water removes water from the more dilute solution and adds it to the more concentrated one. How then does the net diffusion of water affect the concentrations of the two solutions? Does it tend to make them more equal or more different? How does the net diffusion of NaCl affect their concentrations? The answers are quite obvious. Both the net diffusion of water and the net diffusion of NaCl tend to equalize (equilibrate) the concentrations of the two solutions. Note that whereas net diffusion of NaCl and of water both go on at the same time, they go on in opposite directions.

Diffusion of NaCl and water continues even after equilibration has been achieved. But from that moment on, it is equal diffusion in both directions through the membrane and not net diffusion of either substance in either direction.

Dialysis

Dialysis is diffusion under certain conditions. It takes place when a solution that contains both crystalloids and colloids is separated from plain water by a membrane that is permeable to crystalloids but impermeable to colloids (Fig. 2-7). (Crystalloids or true solutes are solute particles with diameters less than 0.001 μm, for example,

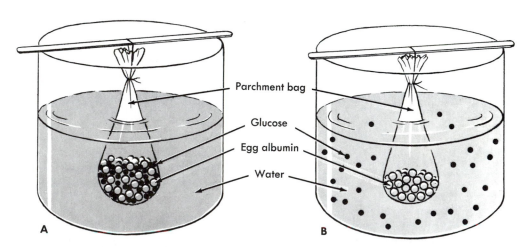

Fig. 2-7 Dialysis, the separation of crystalloids from colloids by means of a membrane permeable to crystalloids and impermeable to colloids. The parchment bag in **A,** which contains a solution of glucose and raw egg white, is permeable to glucose but not to albumin. Therefore glucose moves out of the bag into the surrounding water while albumin stays inside the bag. Time elapses between **A** and **B. B** shows the result of dialysis—separation of the crystalloid, glucose, from the colloid, albumin.

ions, glucose, or oxygen. Colloids are solute particles whose diameters range from about 0.001 to 0.1 μm. Enzymes and all other proteins are colloids.) When the membrane that separates a solution of crystalloids from water is permeable to crystalloids and impermeable to colloids, the crystalloids, as you would expect, diffuse through the membrane but the colloids do not. Net diffusion of crystalloids occurs down their concentration gradient, that is, out of the solution into the water. The colloids remain behind. Hence dialysis may be described as diffusion that separates crystalloids from colloids.

■ ■ ■

The following paragraphs summarize some facts and principles worth remembering about diffusion.

1 *Diffusion* is the movement of solute and solvent particles in all directions through a solution or in both directions through a membrane.

2 *Net diffusion* is the movement of more particles of a substance in one direction than in the opposite direction.

3 Net diffusion of any substance occurs down its own concentration gradient, which means from the higher to the lower concentration of that substance. Following are two applications of this principle: (a) net diffusion of solute particles occurs from the more concentrated to the less concentrated solution and (b) net diffusion of water molecules, in contrast, occurs from the more dilute to the less dilute solution.

4 Net diffusion of the solute in one direction through a membrane and of water in the opposite direction eventually makes the concentrations of the two solutions equal. In short, it *equilibrates* them. We can also state this principle in another way. Net diffusion of both solute and water causes the solute and water concentration gradients to gradually decrease. Eventually—at the point of equilibration—they disappear entirely. There is no difference in concentration (which is what concentration gradient means) between equilibrated solutions. Equilibrated solutions have the same, not different, concentrations.

5 Equal diffusion means that the number of solute and water particles diffusing in one direction equals the number diffusing in the opposite direction. Net diffusion means more particles diffusing in one direction than in the other.

6 Diffusion is a passive transport mechanism because cells are passive, not active and working in this process. Cellular chemical reactions do not supply the energy that moves diffusing particles. The continual random movements characteristic of all molecules and ions (molecular kinetic theory) furnish the energy for diffusion.

Think about these diffusion principles. Make sure that you understand them, for they have many applications in physiology. Our very lives, in fact, depend on diffusion. Evidence? Oxygen, the "breath of life," enters cells by diffusion through their membranes.

Facilitated diffusion

Facilitated diffusion resembles both ordinary diffusion and active transport. Like ordinary diffusion, facilitated diffusion is a passive process that moves a substance down its own concentration gradient. Like active transport, facilitated diffusion is a "carrier-mediated" process. By this we mean that in or near the outer surface of a cell membrane, a specific compound—the carrier—combines with the substance to be moved. The complex thus formed then rapidly diffuses through the membrane to its inner surface. There the carrier compound dissociates from the substance. It has fulfilled its function of facilitating (accelerating) the substance's diffusion through the membrane.

Osmosis

Osmosis is the diffusion of water through a selectively permeable membrane (Fig. 2-8). As its name suggests, a selectively permeable membrane is one that is not equally permeable to all solute particles present. It permits some solutes to diffuse through it freely but hinders or pre-

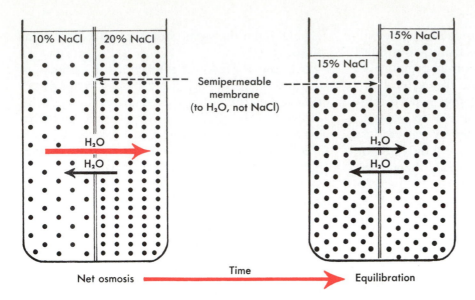

10% NaCl | 20% NaCl

Semipermeable membrane (to H₂O, not NaCl)

15% NaCl

15% NaCl

H_2O

H_2O

H_2O

H_2O

Net osmosis — Time → Equilibration

Fig. 2-8 Osmosis. Osmosis is the diffusion of water through a selectively permeable membrane. The membrane shown in this diagram is permeable to water but not to NaCl. Because there are relatively more water molecules in 10% NaCl than in 20% NaCl, more water molecules osmose from the more dilute into the more concentrated solution (as indicated by the red arrow in the left-hand diagram) than osmose in the opposite direction. The *net* direction of osmosis, in other words, is toward the more concentrated solution. Net osmosis produces the following changes in these solutions: their concentrations equilibrate, both the volume and the pressure of the originally more concentrated solution increase, and the volume and the pressure of the other solution decrease proportionately.

vents entirely the diffusion of others. Those solutes allowed to diffuse freely through the membrane obey the law of diffusion. Hence they eventually equilibrate across it. Their concentrations on both sides of the membrane become equal. But can particles not permitted to diffuse freely through a membrane also obey the law of diffusion? Can they, too, equilibrate across the membrane? The answer to both questions, as you can readily deduce, is "No." When a membrane hinders or prevents a substance from moving through it, the concentration of that substance necessarily remains higher on one side of the membrane than on the other. That solute cannot equilibrate across the membrane. In short, the selectively permeable membrane maintains a concentration gradient of the not freely diffusible solute.

Summarizing the preceding paragraph, a *selectively permeable membrane* may be defined either as one that does not permit free, unhampered diffusion of all the solutes present or as one that maintains at least one solute concentration gradient across itself. *Osmosis* is the diffusion of water through a selectively permeable membrane. Or, osmosis is the diffusion of water through a membrane that maintains at least one concentration gradient across itself.

Normal living cell membranes are selectively permeable membranes. Two results follow from this fact: These membranes maintain various solute concentration gradients, and water moves through them by osmosis. Here is an example that may help you understand this passive transport process. Imagine that you have a 20% NaCl solution separated from a 10% NaCl solution by a membrane. Assume that the membrane is impermeable to sodium and chloride ions but that it is freely permeable to water particles (Fig. 2-8). Obviously then, sodium and chloride ions will not diffuse through this membrane. It maintains an NaCl concentration gradient across itself, in other words. Therefore water will move through this membrane by the process of osmosis. It osmoses through the membrane in both directions but not in equal amounts.

To deduce the direction in which the greater volume of water will osmose (the direction, that is, of net osmosis), we must first decide which solution contains the greater concentration of water molecules. Pretty clearly the more dilute of two solutions contains the greater concentration of water molecules. In this case, then, water concentration is greater on the 10% side of the membrane than it is on the 20% side. Next we need to apply the principle already stated that "net diffusion of any substance occurs down the concentration gradient of that substance." Therefore net osmosis (net water diffusion through the selectively permeable membrane) occurs down the water concentration gradient. In our example, net osmosis takes place from the more dilute 10% salt solution into the more concentrated 20% solution. Thus the solution that was at first more concentrated gains water by net osmosis and becomes more dilute. And the solution that was at first more dilute loses water and becomes more concentrated. Net osmosis, in other words, tends to make the concentrations of the two solutions equal. In briefest form, the principle is this: net osmosis occurs down a water concentration gradient and tends to equilibrate solutions separated by a selectively permeable membrane.

Net osmosis into the originally more concentrated of our two salt solutions increases the volume of this solution. Further, the increases in its volume causes an increase in its pressure, called osmotic pressure—a logical term, since it is pressure caused by net osmosis. By definition, then, *osmotic pressure* is the pressure that de-

velops in a solution as a result of net osmosis into that solution. An important principle stems from this definition: Osmotic pressure develops in the solution that originally contains the higher concentration of the solute that does not diffuse freely through the membrane. For instance, in the example previously given, osmotic pressure would develop in the solution that originally had the 20% salt concentration.

Potential osmotic pressure is the maximum osmotic pressure that could develop in a solution if it were separated from distilled water by a selectively permeable membrane. (Actual osmotic pressure, on the other hand, is a pressure that already has developed, not just one that could develop.) What determines a solution's potential osmotic pressure? Answer: The number of solute particles in a unit volume of solution directly determines its potential osmotic pressure—the more solute particles per unit volume, the greater the potential osmotic pressure. The number of solute particles per unit volume (for example, per liter) of solution, in turn, is determined by the molar concentration of the solution and also, if the solute is an electrolyte, by the number of ions formed from each molecule of solutes. This concept can be expressed by a diagram (Fig. 2-9) or by a formula (see footnote to Fig. 2-9).

Since it is the number of solute particles per unit volume that directly determines a solution's potential osmotic pressure, one might at first jump to the conclusion that all solutions having the same percent concentration also have the same potential osmotic pressure. Obviously it is true that all solutions containing the same percent concentration of the same solute do also have the same potential osmotic pressures. All 5% glucose solutions, for example, have a potential osmotic pressure at body temperature of somewhat more than 5,300 mm Hg pressure. But all solutions with the same percent concentrations of different solutes do not have the same molar concentrations. Hence they do not have the same potential osmotic pressures. By applying the equation in the footnote, you will discover that 5% NaCl at body temperature has a potential osmotic pressure of approximately 33,000 mm Hg—quite different from 5% glucose's potential osmotic pressure of about 5,300 mm Hg.

Two solutions that have the same potential osmotic pressure are said to be *isosmotic* to each other. Because they have the same potential osmotic pressure, the same amount of water will osmose in both directions between them if they are separated by a selectively permeable membrane. In short, no net osmosis occurs in either direction between isosmotic solutions. Hence no actual osmotic pressure develops in either solution. Their pressures remain the same. And because they do, isosmotic solutions are also called isotonic (from the Greek *isos*, the same and *tonos*, tension or pressure).

By definition, *isotonic solutions* are those whose volumes and pressures will stay the same if the two solutions are separated by a membrane. For example, 0.85% NaCl is referred to as "isotonic saline," meaning that it is isotonic to the fluid inside human cells. Translated more fully, it means that if 0.85% NaCl is injected into human tissues or blood, no net osmosis occurs into or out of cells. Therefore no change in intracellular volume or pressure takes place. Cells in contact with isotonic solutions neither lose nor gain water. They become neither dehydrated nor hydrated, to use more technical terms. Physicians make use of this information frequently. For intravenous and intramuscular injections, they usually give isotonic solutions.

When two solutions have unequal potential osmotic pressures, their volumes and pressures will change if they are separated by a selectively permeable membrane. Net osmosis will occur into the solution that has the higher potential osmotic pressure. Net osmosis, as we have noted, always occurs down the water concentration gradient. Since the solution having the lower potential osmotic pressure has the lower concentration of solute particles, it necessarily has the

higher concentration of water molecules. Therefore net osmosis occurs out of this solution into the solution with the higher potential osmotic pressure. This increases the volume of the solution with the higher potential osmotic pressure. And the increase in its volume produces an increase in its pressure. This increase in pressure is osmotic pressure—but actual osmotic pressure and not just potential osmotic pressure.

When two solutions have unequal potential osmotic pressures, one of them is described as hypertonic and the other as hypotonic. Here are several definitions of the term *hypertonic solution*. A hypertonic solution is one that has a high-

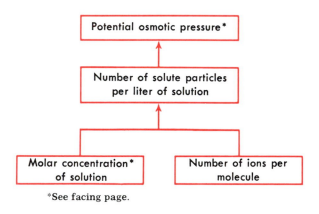

*See facing page.

Fig. 2-9 Factors that determine potential osmotic pressure. Interpreted, the diagram shows that the number of solute particles (ions and molecules) present in a liter of solution determines its potential osmotic pressure. Note, however, that the number of solute particles per liter is itself determined by two factors— the number of molecules in the solution (indicated by its molar concentration) and also, if the solute is an electrolyte, by the number of ions formed from each molecule.

er potential osmotic pressure than another. A hypertonic solution is one into which net osmosis occurs when a selectively permeable membrane separates it from another solution. A hypertonic solution is one whose volume and pressure increase when a selectively permeable membrane separates it from another solution. A hypertonic solution is one in which osmotic pressure develops when a selectively permeable membrane separates it from another solution (Fig. 2-10).

A *hypotonic solution* has characteristics opposite from those of a hypertonic solution. Example: The fluid inside human cells is hypertonic to distilled water and, conversely, distilled water is hypotonic to intracellular fluid. If, therefore, distilled water were injected into a vein, net osmosis would occur into blood cells. And eventually if intracellular volume and pressure increased beyond a certain limit, blood cell membranes would rupture and the cells would die. When this happens to red blood cells, their hemoglobin leaks out and they are said to be hemolyzed. (*Hemolysis* means the destruction of red blood cells with the escape of hemoglobin from them into the surrounding medium.)

Concentrate on remembering the following facts and principles about osmosis:

1 Osmosis is the diffusion of water through a selectively permeable membrane, a membrane

---

*Formulas:

$$\boxed{\begin{array}{c}\text{Potential osmotic} \\ \text{pressure of non-} \\ \text{electrolyte (in mm Hg)}\end{array}} = \boxed{\begin{array}{c}\text{Molar concentration} \\ \text{of solution}\end{array}} \times \quad 19,300\dagger$$

$$\boxed{\begin{array}{c}\text{Potential osmotic} \\ \text{pressure of elec-} \\ \text{trolyte (in mm Hg)}\end{array}} = \boxed{\begin{array}{c}\text{Molar concen-} \\ \text{tration of} \\ \text{solution}\end{array}} \times \boxed{\begin{array}{c}\text{Number ions} \\ \text{per mole-} \\ \text{cule}\end{array}} \times 19,300$$

†Experimentation has shown that a solution with a 1.0 molar concentration of any nonelectrolyte has a potential osmotic pressure of 19,300 mm Hg pressure (at body temperature, 37° C).

$$\boxed{\begin{array}{c}\text{Molar concen-} \\ \text{tration}\end{array}} = \boxed{\begin{array}{c}\text{Grams solute in 1 liter solution} \\ \text{divided by} \\ \text{Molecular weight of solute}\end{array}}$$

Example: Two solutions commonly used in hospitals are 0.85% NaCl and 5% glucose. What is the potential osmotic pressure of 0.85% NaCl at body temperature? (0.85% NaCl = 8.5 gm NaCl in 1 liter solution.)

Molecular weight of NaCl = 58 (NaCl yields 2 ions per molecule in solution)

Using the formula given for computing potential osmotic pressure of an electrolyte:

$$\boxed{\begin{array}{c}\text{Potential osmotic} \\ \text{pressure of} \\ \text{0.85\% NaCl}\end{array}} = \frac{8.5}{58} \times 2 \times 19,300 = \textit{5,658.6 mm Hg pressure}$$

Problem: What is potential osmotic pressure of 5% glucose at body temperature? Molecular weight glucose = 180. Glucose does not ionize. It is a nonelectrolyte.‡

‡A 5% glucose solution has potential osmotic pressure of 5,359.6 mm Hg pressure.

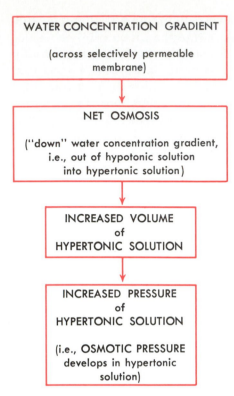

WATER CONCENTRATION GRADIENT

(across selectively permeable membrane)

↓

NET OSMOSIS

("down" water concentration gradient, i.e., out of hypotonic solution into hypertonic solution)

↓

INCREASED VOLUME of HYPERTONIC SOLUTION

↓

INCREASED PRESSURE of HYPERTONIC SOLUTION

(i.e., OSMOTIC PRESSURE develops in hypertonic solution)

Fig. 2-10 Effects produced by existence of a water concentration gradient across a selectively permeable membrane.

that maintains at least one concentration gradient across itself.

2 Net osmosis means the osmosis of more water in one direction through a membrane than in the opposite direction. The direction of net osmosis may be described in several ways. Net diffusion occurs out of the solution that contains the lower concentration of solute particles into the one that contains the higher solute concentration. Net diffusion occurs out of the solution that has the lower potential osmotic pressure into the solution with the higher potential osmotic pressure. Net diffusion occurs from a hypotonic solution into a hypertonic solution.

3 Net osmosis produces these results: it increases the volume of the hypertonic "more con-

centrated" solution, and this increased volume increases the pressure of the hypertonic solution.

Filtration

Filtration is the physical process by which water and solutes pass through a membrane when a hydrostatic pressure gradient exists across that membrane, that is, when the hydrostatic pressure on one side of the membrane is higher than on the other. (*Hydrostatic pressure* is the force or weight of a fluid pushing against some surface.) A principle about filtration that has great physiological importance is this: Filtration always occurs down a hydrostatic pressure gradient. This means that when two fluids have unequal hydrostatic pressures and are separated by a membrane, water and diffusible solutes (those to which the membrane is permeable) filter out of the solution that has the higher hydrostatic pressure into the solution that has the lower hydrostatic pressure.

How does filtration differ from diffusion and osmosis? Both water and solutes can filter or diffuse through a membrane, whereas only water, by definition, can osmose through a membrane. Filtration occurs in only one direction through a membrane—down a hydrostatic pressure gradient. Diffusion and osmosis go on in both directions through a membrane. Net diffusion and net osmosis, however, occur only in one direction. Net diffusion occurs down the concentration gradient of the substance diffusing. Net osmosis occurs down a water concentration gradient. Filtration is a major mechanism for moving substances through the membranous walls of the blood capillaries; water and true solutes filter out of blood into the interstitial fluid (in the microscopic spaces between cells).

Physiological pumps

The active transport of substances through cell membranes is a major and vital kind of cellular work. The energy that does this work comes from chemical reactions that take place within a living cell. Based on this criterion, we can define an active transport mechanism as a device that uses energy from cellular chemical reactions to move a substance through a cell membrane. One type of active transport mechanism is called a pump—a physiological or biological pump. By definition, a physiological pump is an active transport mechanism that moves molecules or ions through cell membranes in an uphill direction, meaning up their concentration gradients or against their natural tendency. Their natural tendency, as you know, is to diffuse down their concentration gradients. The law of diffusion requires that the net diffusion of any substance takes place from the area of its higher concentration to that of its lower concentration.

Every cell employs various physiological pumps. Of these, one of the most important for healthy human cell survival is the sodium pump. Fluid called interstitial fluid bathes our human cells. It contains a much higher concentration of sodium ions and a much lower concentration of potassium ions than does the fluid inside cells (intracellular fluid). In other words, both sodium and potassium concentration gradients exist across the cytoplasmic membrane. These are established and maintained by the sodium-potassium active transport mechanism. If diffusion alone moved these ions through the membrane, their concentration gradients could not be maintained. Why? Because diffusion always eventually results in equilibration of two solutions separated by a membrane permeable to both solute and solvent.

An enzyme named $Na^+K^+ATPase$ mediates the active transport of sodium and potassium ions through the cytoplasmic membrane. This enzyme molecule—a protein, as are all enzymes— is located in the cytoplasmic membrane and extends all the way through it. On its inner surface, it has a sodium-binding site, and on its outer surface, a potassium-binding site. The sodium-potassium active transport mechanism, according to one hypothesis, consists of the following events.

1 In the cytoplasm of a cell, in the presence of magnesium ions, sodium, ATP, and the enzyme ($Na^+K^+ATPase$) react to bind sodium to the enzyme and to transfer the terminal phosphate of ATP to it. This phosphorylation of the enzyme activates it. The following equation represents this reaction. (The symbol Ⓔ stands for the ATPase enzyme located in the cytoplasmic membrane with its sodium-binding site facing toward the inside of the cell.)

$$Na^+ + ATP + Ⓔ \xrightarrow{Mg^{++}} Na — Ⓔ \sim P + ADP$$

2 Phosphorylation of the enzyme ATPase activates it, that is, supplies it with energy to do the work of changing its shape so that its sodium-binding site faces the opposite direction, that is, toward the outside of the cell. The following equation represents this change. (The symbol 🄴 stands for the enzyme with its sodium-binding site facing outward.)

$$Na — Ⓔ \sim P \rightarrow Na — 🄴 \sim P$$

3 With the sodium-binding site on ATPase now facing outward, sodium unbinds from the enzyme and enters the extracellular fluid, as does phosphate. Immediately following the unbinding of sodium from the enzyme, potassium in the extracellular fluid binds to it. The following two equations represent these successive reactions.

$$Na — 🄴 \sim P \rightarrow Na^+ + P + 🄴$$
$$K^+ + 🄴 \rightarrow K — 🄴$$

4 The enzyme then changes back to its original shape, that is, with its sodium-binding site facing the cytoplasm inside the cell, and releases the potassium into it.

$$K — 🄴 \rightarrow K — Ⓔ \rightarrow K^+ + Ⓔ$$

The sodium-potassium transport mechanism just described illustrates certain features believed to characterize all active transport mechanisms and to distinguish them from passive mechanisms. An enzyme located in a cell membrane is postulated to serve as a carrier that combines with a substance and transports it through the membrane. (What is the name of the enzyme that transports sodium and potassium?) The enzyme receives energy for moving a substance through a membrane from a chemical reaction that occurs inside a living cell. (What chemical reaction energizes sodium-potassium transport? Step 1 describes it.) Energy that empowers passive transport mechanisms is not generated by chemical reactions in living cells. Instead, it is kinetic energy generated by ever-moving molecules and ions that supplies power for passive mechanisms. Another characteristic also distinguishes active from passive transport mechanisms. Active transport mechanisms always increase the concentration gradient of the transported substance across the membrane. In other words, they establish a higher concentration of the transported substance on one side of the membrane than on the other. (This last fact has given rise to the name *concentrative transport* as a synonym for active transport.) Passive transport mechanisms, in contrast, decrease the concentration gradient of the transported substance across the membrane—eventually to zero, since passive transport mechanisms equilibrate the concentrations of the transported substance on both sides of the membrane.

Phagocytosis and pinocytosis

Phagocytosis and pinocytosis, like the physiological pumps, are active transport mechanisms for moving substances through cell membranes —but only in one direction, namely, inward. About three quarters of a century ago, Elie Metchnikoff of the Pasteur Institute saw white blood cells engulf bacteria. It reminded him of eating. Therefore from the Greek words for eating, cell, and action, he coined the word phagocytosis. Some 30 years later, in 1931, W. H. Lewis of Johns Hopkins University saw something similar in time-lapse photographs of some tissue culture cells. These cells, however, were engulfing droplets of fluid instead of solid particles. They seemed to be drinking rather than eating, so he named the process pinocytosis (from the Greek word for drinking).

Both phagocytosis and pinocytosis consist of the same essential steps. A segment of a cell's plasma membrane forms a small pocket around a bit of solid or liquid material outside the cell, pinches off from the rest of the membrane, and migrates inward as a closed vacuole or vesicle. Later, it releases its contents into the cell's cytoplasm.

Summary

This chapter has presented information about the structure and some of the self-serving functions of cells. The next one will relate facts and principles about the structure and body-serving functions of tissues.

Outline summary
Cell structure

Protoplasm

A Definition—living matter; complex organization of thousands of biomolecules
B Composition
  1 Main elements—carbon, hydrogen, oxygen, nitrogen
  2 Main compounds—water, proteins, carbohydrates, lipids, and nucleic acids

Cell differences

A Size—for example, human ovum measures about 1,000 μm in diameter; erythrocyte diameter is about 7.5 μm
B Shape—see Fig. 2-1
C Special cell structures—for example, microvilli, cilia, flagella

Cell membranes

A Structure

1 Framework of cell membranes consists of a bilayer of phospholipid molecules with their hydrophobic tails pointing toward each other and away from both the outward and inner surfaces of the membrane; one row of hydrophilic heads of phospholipid molecules points toward outer surfaces of membrane and second row of heads points toward membrane's inner surface (Fig. 2-3)

2 Protein molecules, arranged asymmetrically as shown in Fig. 2-3, also form part of cell membranes

B Function

1 Serves as cell's boundary; maintains its integrity

2 Transports some substances through itself into or out of cell's cytoplasm; prevents other substances from moving into or out of cytoplasm

3 Communication—some proteins on outer surface of cytoplasmic membrane serve as receptors for chemical messages; certain hormones and neurotransmitters bind to these proteins and thereby communicate their message to cells

4 Immunity—some membrane proteins on lymphocytes combine with potentially harmful foreign proteins, thereby leading to events that render them harmless

5 Identification—many membrane proteins serve as markers that distinguish cells of one individual from those of another

6 Catalysts—many membrane proteins are enzymes that catalyze cell's chemical reactions

Cytoplasm and organelles

A Definition—cytoplasm is protoplasm located between cell membrane and nucleus

B Thousands of organelles ("little organs") present

1 Membranous organelles—endoplasmic reticulum, Golgi apparatus, mitochondria, and lysosomes

2 Nonmembranous organelles—ribosomes and centrosome

C Endoplasmic reticulum

1 Structure—complicated network of canals and sacs extending through cytoplasm and opening at surface of cell; many ribosomes attached to membranes of rough endoplasmic reticulum but not to smooth

2 Functions—ribosomes attached to rough endoplasmic reticulum synthesize proteins; canals of reticulum serve cell as its inner circulatory system, for example, proteins move through canals on way to Golgi apparatus

D Golgi apparatus

1 Structure—membranous vesicles near nucleus

2 Function—synthesizes large carbohydrate molecules, combines them with proteins, and secretes product (glycoproteins)

E Mitochondria

1 Structure—microscopic sacs; walls composed of inner and outer membranes separated by fluid; thousands of particles made up of enzyme molecules attached to both membranes

2 Function—"power plants" of cells; mitochondrial enzymes catalyze series of oxidation reactions that provide about 95% of cell's energy supply

F Lysosomes

1 Structure—microscopic membranous sacs

2 Function—cell's own digestive system; enzymes in lysosomes digest particles or large molecules that enter them; under some conditions, digest and thereby destroy cells

G Ribosomes

1 Structure—microscopic spheres composed of RNA and protein; large numbers of them are attached to endoplasmic reticulum

2 Function—"protein factories"; ribosomes attached to endoplasmic reticulum synthesize proteins to be secreted by cell, and those lying free in cytoplasm make proteins for cell's own use, that is, its structural proteins and enzymes; groups of ribosomes, called polysomes, synthesize proteins

H Centrosome

1 Structure

a Centrosome is spherical body near center of cell, that is, near nucleus

b Centrioles, located in centrosome, are tiny cylinders, walls of which are composed of nine groups of tubules, three tiny tubules in each group

2 Function—centrioles thought to play some part in formation of mitotic spindle

Nucleus

A Definition—spherical body in center of cell; enclosed by pore-containing membrane

B Structure—consists of

1 Nuclear membrane (two-layered, with essential-

ly same molecular structure as cytoplasmic membrane)

2 Numerous chromatin granules—consist chiefly of DNA and protein molecules; visible in nondividing cells, replaced early in process of cell division by chromosomes; 46 chromosomes in all human cells except mature sex cells, which contain 23 chromosomes

3 Nucleoli—consist chiefly of RNA and protein molecules; nucleoli not enclosed by membrane

C Functions

1 Stores genetic information in DNA molecules present in nucleus; information consists of the specific sequence of nucleotide base pairs in DNA

2 Transcribes genetic information—sequence of base pairs in DNA is transcribed into sequence of base pairs in messenger RNA as it is being synthesized in the nucleus; messenger RNA leaves nucleus through pores in nuclear membranes and attaches to ribosomes in cytoplasm, where it functions as a coded message that is translated in the ribosomes into specific proteins; complex mechanisms of DNA transcription into RNA and RNA translation into proteins discussed in Chapter 23; since some proteins synthesized at direction of RNA are functional proteins (notably, enzymes), nucleus by means of DNA and RNA directs cell functions, including growth and reproduction; also directs structure of cell, since many proteins synthesized are integral parts of cell structures

3 Transmits genetic information—DNA molecules replicate before cell divides; each new cell receives identical set of DNA molecules, thereby indirectly inheriting structure and functions of parent cell

Special cell structures

A Microvilli—projections of cytoplasm and cytoplasmic membrane; increase surface area of cells whose function is absorption

B Cilia—hairlike projections of cytoplasm and cytoplasmic membrane; each cilium is a tiny cylinder made up of nine double microtubules arranged around two single microtubules; one cell may have a hundred or more cilia; they propel fluid in one direction over surface of cell, for example, upward in respiratory tract

C Flagellum—single hairlike projection from cell's surface, for example, flagellum of spermatozoon propels it forward in its fluid environment

Cell physiology—movement of substances through cell membranes
Types of processes

A Physical—processes in which random, never-ceasing movements of solute particles supply energy for moving substances; names of physical processes: diffusion (including dialysis and facilitated diffusion), osmosis, and filtration

B Physiological or active—energy that moves substances comes from chemical reactions in living cell; names of physiological processes: pumps and phagocytosis and pinocytosis

C Diffusion

1 Movement of solute and solvent particles in all directions through solution or in both directions through membrane

a Net diffusion of solute particles—down solute concentration gradient, that is, from more to less concentrated solution

b Net diffusion of water—down water concentration gradient, that is, from less to more concentrated solution

2 Diffusion tends to produce equilibration of solutions on opposite sides of membrane

3 Dialysis—separation of crystalloids from colloids by diffusion of crystalloids through membrane permeable to them but impermeable to colloids

4 Facilitated diffusion—diffusion in which carrier substance combines with particle being moved

D Osmosis

1 In living systems, movement of water in both directions through membrane that maintains at least one solute concentration gradient across it

2 Net osmosis—more water osmoses in one direction through membrane than in opposite; net osmosis occurs down water concentration gradient (which is up solute concentration gradient and up potential osmotic pressure gradient); net osmosis tends to equilibrate two solutions separated by selectively permeable membrane

3 Osmotic pressure—pressure that develops in solution as result of net osmosis into it

4 Isotonic solution—one that has same potential osmotic pressure as solution to which it is isotonic; no net osmosis between isotonic solutions

5 Hypertonic solution—has greater potential osmotic pressure and higher solute concentration but lower water concentration than solution to which it is hypertonic; net osmosis into hypertonic solution from hypotonic solution

6 Hypotonic solution—has lower potential osmotic pressure and lower solute concentration but higher water concentration than solution to which it is hypotonic; net osmosis out of hypotonic solution into hypertonic solution

E Filtration

1 Definition—movement of solvent (water in body) and solutes through membrane in one direction only, that is, down hydrostatic pressure gradient

2 Comparison with diffusion

a Diffusion, like filtration, moves both water and solutes through membrane

b Diffusion occurs in both directions through membrane, filtration in only one

c Net diffusion of solute occurs down solute concentration gradient; net diffusion of water occurs down water concentration gradient; filtration occurs down hydrostatic pressure gradient

3 Comparison with osmosis

a Water is only substance that osmoses through membrane; occurs in both directions

b Net osmosis occurs down water concentration gradient; filtration occurs down hydrostatic pressure gradient

F Physiological pumps—active transport mechanisms that move ions or molecules through cell membranes against their concentration gradient, that is, in direction opposite from net diffusion or net osmosis; energy supplied by cellular chemical reactions

G Phagocytosis and pinocytosis

1 Phagocytosis—physiological process that moves solid particles into cell; segment of cell's plasma membrane forms pocket around particle outside cell, then pinches off from rest of membrane and migrates inward

2 Pinocytosis—physiological process that moves fluid into cell; process similar to phagocytosis

Review questions

1 Define briefly or give a synonym for each of the following terms: cytology, protoplasm, cytoplasm, cytoplasmic membrane.

2 What is the most abundant compound in protoplasm?

3 What four kinds of compounds occur naturally only in protoplasm?

4 Describe the currently most accepted concept of cell membrane structure.

5 The size of cells and cell structures is measured in micrometers. What symbol represents this measurement? What part of a meter is a micrometer? One thousandth? One millionth? One billionth?

6 Nanometers are used to measure the size of molecules. What part of a meter is 1 nm? How many nanometers is 1 μm?

7 The cytoplasmic membrane is said to be about 75 Å thick. How many nanometers thick is it?

8 One inch equals approximately how many angstroms? Micrometers? Nanometers?

9 One millimeter equals approximately what part of an inch?

10 Describe several functions of the cytoplasmic membrane.

11 Contrast active and passive transport mechanisms.

12 Describe briefly functions of a cell nucleus.

13 Identify the following organelles with a brief statement about the structure and function of each: endoplasmic reticulum, Golgi apparatus, mitochondria, lysosomes, ribosomes, centrioles.

14 Explain briefly what each of the following terms means: diffusion, net diffusion, facilitated diffusion, dialysis, osmosis, filtration, phagocytosis, pinocytosis.

15 State the principle about the direction in which net osmosis occurs.

16 Differentiate between actual osmotic pressure and potential osmotic pressure.

17 What factor directly determines the potential osmotic pressure of a solution?

18 Explain the terms isotonic, hypotonic, and hypertonic.

19 State the principle about which solution develops an osmotic pressure, given appropriate conditions.

chapter 3

Tissues

Tissues are organizations of cells that specialize in one or more functions that serve the body as a whole. In addition, tissue cells also carry on the various self-serving activities necessary for their own survival. There are four primary kinds of tissues: epithelial, connective (including blood and blood-forming tissues), muscle, and nervous. *Epithelial tissue* specializes in moving substances into and out of the blood (absorption and secretion) and in serving as a protective barrier. *Connective tissue's* special functions are to support the body and its parts, to connect and hold them together, to transport substances through the body, and to protect it from foreign invaders. *Muscle tissue* specializes in movement; it moves the body and its parts. *Nervous tissue* specializes in communication between the various parts of the body and in integration of their activities. Not only do cell size, shape, and arrangement vary in different kinds of tissues, but so, too, does the amount and kind of intercellular substance present. Some tissues contain almost no intercellular material. Others consist predominantly of it. Some intercellular substance contains many fibers, some is unformed gel, and some is fluid, the interstitial fluid that bathes most living human cells.

Epithelial tissue
Locations, functions, and types

Epithelial tissue, or epithelium, is widely distributed in the body. Defined by its locations, *epithelium* is the tissue that covers the body and some of its parts; that lines its serous cavities (pleural, pericardial, and peritoneal), its blood and lymphatic vessels, and the respiratory, digestive, and genitourinary tracts; and that forms its glands. Defined by functions, epithelium is the tissue that specializes in absorption, secretion, and provision of protective barriers. Two-word names identify several of the types of epithelial tissue. The first word is "simple" if only one layer of epithelial cells makes up the tissue and is "stratified" if two or more layers of cells compose it. The second word of the name of an epithelial tissue indicates the shape of its surface cells: squamous, if they appear flat and scale-like; cuboidal, if they are cubelike in shape; and columnar, if they are taller than they are long or wide. Here, then, are the names of six types of epithelial tissue: simple squamous epithelium, simple cuboidal epithelium, simple columnar epithelium, stratified squamous epithelium, stratified cuboidal epithelium, and stratified columnar epithelium. In addition, there are some other types of epithelial tissue, notably, pseudostratified and transitional. Pseudostratified epithelium consists of only one layer of cells. It appears, however, to consist of more than one layer because its cells are so crowded together that some of them do not reach to the surface of the tissue. Transitional epithelium (or uroepithelium) consists of several layers, but the shape of its surface cells is not strictly squamous, cuboidal, or columnar.

Generalizations about epithelial tissue

1 Probably the feature of epithelial tissue that you would notice first if you viewed it under the microscope is the close arrangement of its cells. They form continuous sheets of material with little or no intercellular material coming between the cells. At intervals between adjacent cells, their cytoplasmic membranes are modified in ways that hold the cells together. One such modification is called a *desmosome* (*desmo*, bond or fastening, and *soma*, bond). A desmosome consists of two little bodies or plaques—small thickened sections of the adjacent cytoplasmic membranes—separated by a space about 20 μm wide. Fibrils have been reported to extend from the plaques into the cytoplasm of each cell. Another type of epithelial cell junction appears as a fusion of the two cytoplasmic membranes with no space between them—a "tight junction," this is called.

2 Another prominent feature of epithelial tissues is that they contain no blood vessels. Oxygen and foods have to reach epithelial cells by diffusing up to them from capillaries located in the connective tissue beneath epithelial tissue. Usually, epithelial tissue rests on a thin layer of nonliving material, formed partly by epithelial and partly by connective tissue cells and called the *basement membrane*.

3 Another common characteristic of all types of epithelial cells is that they undergo cell division (mitosis)—a fact of practical importance, since it means that old or destroyed epithelial cells can be replaced by new ones. Surface cells of the mucous lining of the intestine, subjected as they are to considerable wear and tear, are estimated to reproduce every few days. In contrast, some epithelial cells in the mucous lining of the respiratory tract reportedly reproduce every few weeks. Of the several types of epithelial tissue, three are described in the following paragraphs.

Simple squamous epithelium

Simple squamous epithelium consists of only one layer of flat, scalelike cells. Consequently, substances can readily diffuse or filter through this type of tissue. The microscopic air sacs of the lungs, for example, are composed of this kind of tissue, as are the linings of blood and lymphatic vessels and the surfaces of the pleura, pericardium, and peritoneum. (Blood and lymphatic vessel linings are called *endothelium*, and the surfaces of the pleura, pericardium, and peritoneum are called *mesothelium*. Some histologists classify these as connective tissue.)

Stratified squamous epithelium

Stratified squamous epithelium, as shown in Fig. 3-1, lines the mouth and esophagus. Its several layers of cells serve a protective function. The surface of the skin is composed of a special kind of stratified squamous epithelium (p. 71).

Simple columnar epithelium

Simple columnar epithelium lines the stomach and intestine and parts of the respiratory tract. A single layer of cells composes this tissue, but two types of cells—goblet and columnar—may be present (Fig. 3-2). Goblet cells are the mucus-secreting specialists of the body. Columnar cells are its absorption specialists. Fig. 3-3 shows pseudostratified ciliated columnar epithelium, the type lining most of the upper respiratory tract.

Fig. 3-1

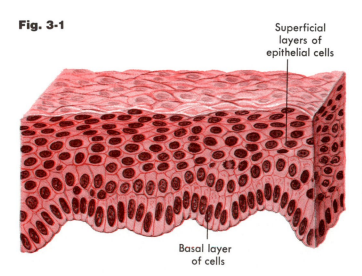

Superficial layers of epithelial cells

Basal layer of cells

A, Stratified squamous epithelium such as lines the mouth.

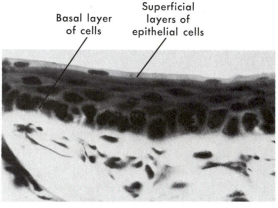

Basal layer of cells

Superficial layers of epithelial cells

B, Stratified squamous epithelium.

Courtesy Dr. C. R. McMullen, Department of Biology South Dakota State University.

Fig. 3-2

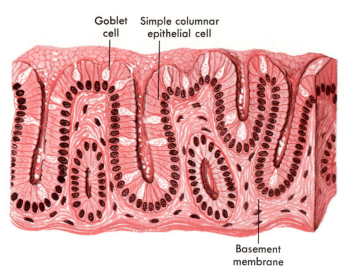

Goblet cell   Simple columnar epithelial cell

Basement membrane

A, Simple columnar epithelium with goblet cells such as lines intestines.

Simple columnar epithelial cell   Basement membrane

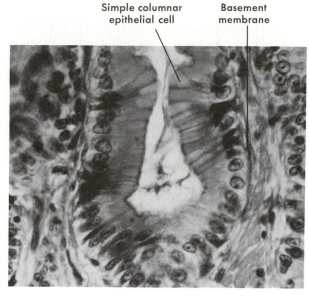

B, Simple columnar epithelium.

Courtesy Dr. C. R. McMullen, Department of Biology, South Dakota State University.

Fig. 3-3

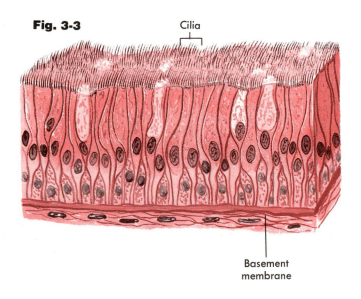

Cilia

Basement membrane

A, Pseudostratified ciliated columnar epithelium with goblet cells. Stratified tissue consists of two or more layers of cells. Pseudostratified tissue appears, in sections cut at certain angles, to meet this requirement but actually does not.

Basement membrane   Cilia

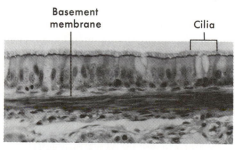

B, Pseudostratified (ciliated) epithelium.

Courtesy Dr. C. R. McMullen, Department of Biology, South Dakota State University.

Muscle tissue

The main specialty of muscle tissue is contraction. Because not all muscle tissue is alike—in location, microscopic appearance, and nervous control—these criteria are used to classify its types. Thus using location as the criterion, there are three kinds of muscle tissue:

1 *Skeletal muscle*—attached to bones
2 *Visceral muscle*—in the walls of hollow internal structures such as blood vessels, intestines, uterus, and many others
3 *Cardiac muscle*—composes the wall of the heart

With microscopic appearance as the basis of classification, there are only two types of muscle tissue: striated (named for cross striations seen in these cells) and nonstriated or smooth (no cross striations in cells).

On the basis of nervous control, there are also two kinds of muscle tissue: voluntary and involuntary.

Voluntary muscle receives nerve fibers from the somatic nervous system. Therefore its contraction can be voluntarily controlled. Involuntary muscle, on the other hand, receives nerve fibers from the autonomic nervous system so that voluntary control of its contraction is not ordinarily possible. In recent years, however, many individuals have used biofeedback devices to learn some voluntary control over smooth muscle contraction. Skeletal muscle is voluntary muscle. Visceral and cardiac muscles are involuntary muscles. Visceral and cardiac muscles are also automatic, meaning that even without nervous stimulation they continue to contract. Skeletal muscle, in contrast, cannot contract

Fig. 3-4

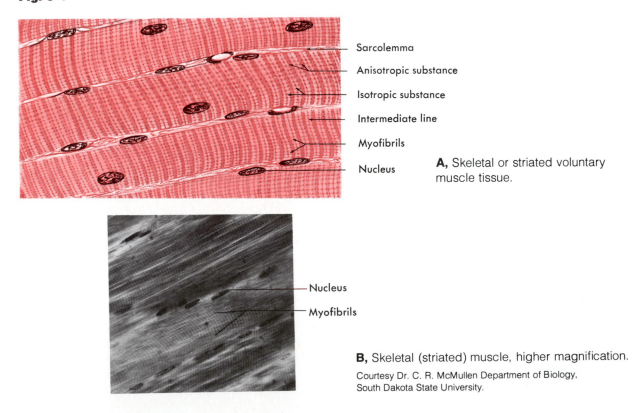

Sarcolemma
Anisotropic substance
Isotropic substance
Intermediate line
Myofibrils
Nucleus

A, Skeletal or striated voluntary muscle tissue.

Nucleus
Myofibrils

B, Skeletal (striated) muscle, higher magnification.
Courtesy Dr. C. R. McMullen Department of Biology, South Dakota State University.

automatically. Anything that cuts off its nerve impulses paralyzes it, that is, puts it immediately out of working order. Poliomyelitis acts this way, for example. It damages nerve cells that conduct impulses to skeletal muscles so that they no longer conduct, and, deprived of stimulation, the muscles are paralyzed.

Finally, combining these classifications, we have the following:

1 *Skeletal* or striated voluntary muscle
2 *Cardiac* or striated involuntary muscle
3 *Visceral* or nonstriated (smooth) involuntary muscle

Look at Fig. 3-4, and you can observe the following structural characteristics of skeletal muscle cells: They have many cross striations; each cell has many nuclei; the cells are long and narrow in shape, for example, as much as 40,000

μm (more than 1½ inches) long but only 10 to 100 μm in diameter. Because such measurements give muscle cells a threadlike appearance, they are probably more often called muscle fibers than cells. Chapter 7 describes the fine structure of skeletal muscle tissue.

Smooth muscle cells are also long narrow fibers but not nearly so long as striated fibers. One can see the full length of a smooth muscle fiber in a microscopic field but only part of a striated fiber. (According to one estimate, the longest smooth muscle fibers measure about 500 μm and the longest striated fibers about 40,000 μm. Can you translate these measurements into millimeters and inches? If not, and if you are curious, see footnote on p. 29.) As Fig. 3-5 shows, smooth muscle fibers have only one nucleus per fiber and are nonstriated or smooth in appearance.

Fig. 3-5

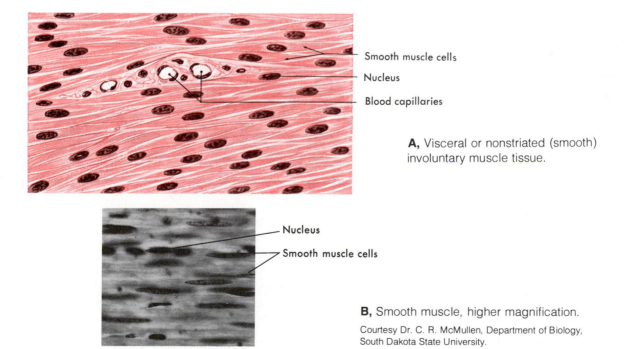

Smooth muscle cells

Nucleus

Blood capillaries

A, Visceral or nonstriated (smooth) involuntary muscle tissue.

Nucleus

Smooth muscle cells

B, Smooth muscle, higher magnification.

Courtesy Dr. C. R. McMullen, Department of Biology, South Dakota State University.

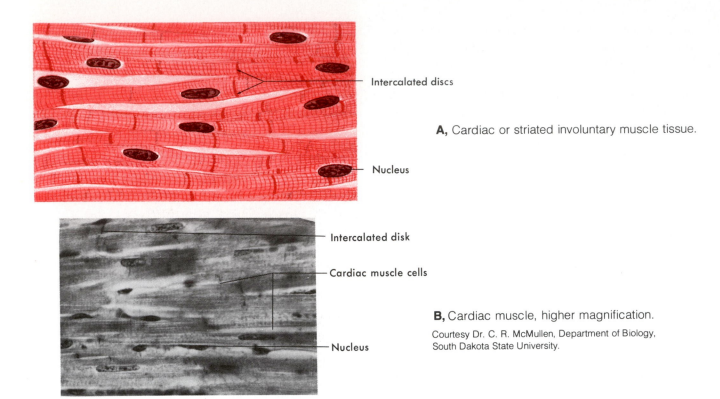

A, Cardiac or striated involuntary muscle tissue.

Intercalated discs

Nucleus

Intercalated disk

Cardiac muscle cells

B, Cardiac muscle, higher magnification.
Courtesy Dr. C. R. McMullen, Department of Biology, South Dakota State University.

Nucleus

Fig. 3-6

Under the light microscope, cardiac muscle fibers (Fig. 3-6) have cross striations and unique dark bands (intercalated disks). They also seem to be incomplete cells that branch into each other to form a big continuous mass of protoplasm. The electron microscope, however, has revealed that the intercalated disks are actually places where the cytoplasmic membranes of two cardiac fibers abut (Fig. 3-6, *B*). Cardiac fibers do branch and anastomose, but a complete cytoplasmic membrane encloses each cardiac fiber—around its end (at intercalated disks) as well as its sides.

Connective tissue

Connective tissue is the most widespread and abundant tissue in the body. It exists in more varied forms than the other three basic tissues. Delicate tissue paper webs, strong, tough cords, rigid bones, and a fluid, namely, blood—all are forms of connective tissue.

One scheme of classification lists the following main types of connective tissue:
1 Reticular
2 Loose, ordinary (areolar)
3 Adipose
4 Dense fibrous
5 Bone
6 Cartilage
7 Hemopoietic
8 Blood

Connective tissue connects, supports, transports, and defends. It connects tissues to each other, for example. It also connects muscles to muscles, muscles to bones, and bones to bones. It forms a supporting framework for the body

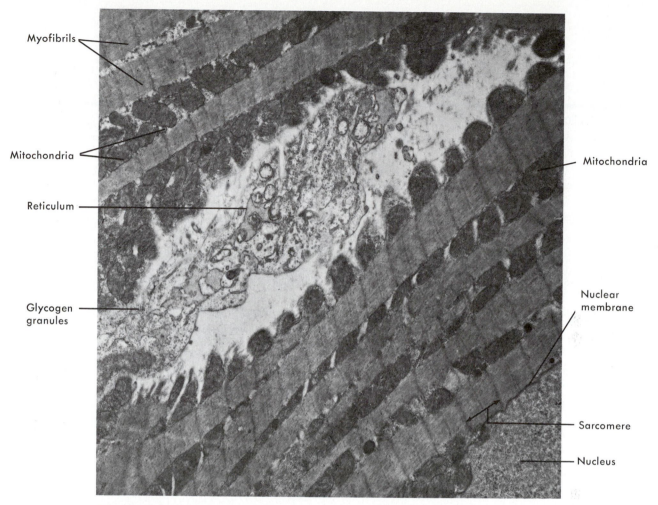

Myofibrils

Mitochondria

Reticulum

Glycogen granules

Mitochondria

Nuclear membrane

Sarcomere

Nucleus

Fig. 3-7 Electron photomicrograph of rabbit heart muscle (×12,500). Note the large number of regularly arranged mitochondria between the contractile components of the cell.

Courtesy Department of Pathology, Case Western Reserve University.

as a whole and for its organs individually. One kind of connective tissue—blood—transports a large array of substances between parts of the body. And finally, several kinds of connective tissue cells defend us against microbes and other invaders.

Connective tissue consists predominantly of intercellular material and relatively few cells. Hence the qualities of the intercellular material

largely determine the qualities of each type of connective tissue. Blood, for example, is a fluid connective tissue because its intercellular material is a fluid. Some connective tissues have the consistency of a soft gel, some are firm but flexible, some are hard and rigid, some are tough, others are delicate—and in each case it is their intercellular substance that makes them so.

Table 3-1 Tissues

Tissue	Location	Function
Epithelial		
Simple squamous	Alveoli of lungs	Absorption by diffusion of respiratory gases between alveolar air and blood
	Lining of blood and lymphatic vessels (called endothelium; classed as connective tissue by some histologists)	Absorption by diffusion; filtration; osmosis
	Surface layer of pleura, pericardium, peritoneum (called mesothelium; classed as connective tissue by some histologists)	Absorption by diffusion; osmosis; also, secretion
Stratified squamous	Surface of lining of mouth and esophagus	Protection
	Surface of skin (epidermis)	Protection
Simple columnar	Surface layer of lining of stomach, intestines, and part of respiratory tract	Protection; secretion; absorption; moving of mucus (by ciliated columnar)
Muscle		
Skeletal (striated voluntary)	Muscles that attach to bones	Movement of bones
	Extrinsic eyeball muscles	Eye movements
	Upper third of esophagus	First part of swallowing
Visceral (nonstriated involuntary or smooth)	In walls of tubular viscera of digestive, respiratory, and genitourinary tracts	Movement of substances along respective tracts
	In walls of blood vessels and large lymphatic vessels	Change diameter of blood vessels, thereby aiding in regulation of blood pressure
	In ducts of glands	Movement of substances along ducts
	Intrinsic eye muscles (iris and ciliary body)	Change diameter of pupils and shape of lens
	Arrector muscles of hairs	Erection of hairs (gooseflesh)
Cardiac (striated involuntary)	Wall of heart	Contraction of heart
Connective (most widely distributed of all tissues)		
Reticular tissue	Spleen, lymph nodes, bone marrow	Defense against microbes and harmful substances by filtration from blood and lymph by reticular network; phagocytosis by reticular cells; synthesis of reticular fibers by reticular cells
Loose, ordinary (areolar)	Between other tissues and organs	Connection
	Superficial fascia	Connection

Tissue	Location	Function
Connective—cont'd		
Adipose (fat)	Under skin	Protection
	Padding at various points	Insulation
		Support
		Reserve food
Dense fibrous	Tendons	Flexible but strong connection
	Ligaments	
	Aponeuroses	
	Deep fascia	
	Dermis	
	Scars	
	Capsule of kidney, etc.	
Bone	Skeleton	Support
		Protection
Cartilage		
Hyaline	Part of nasal septum	Firm but flexible support
	Covering articular surfaces of bones	
	Larynx	
	Rings in trachea and bronchi	
Fibrous	Disks between vertebrae	
	Symphysis pubis	
Elastic	External ear	
	Eustachian tube	
Hemopoietic		
Myeloid (bone marrow)	Marrow spaces of bones	Formation of red blood cells, granular leukocytes, platelets; also reticuloendothelial cells and some other connective tissue cells
Lymphatic	Lymph nodes	Formation of lymphocytes and monocytes; also plasma cells and some other connective tissue cells
	Spleen	
	Tonsils and adenoids	
	Thymus gland	
Blood	In blood vessels	Transportation
		Protection
Nervous	Brain	Irritability; conduction
	Spinal cord	
	Nerves	

Intercellular substance may contain one or more of the following kinds of fibers: collagenous (or white), reticular, or elastic. Fibroblasts and some other cells form these fibers. Collagenous fibers are tough and strong, reticular fibers are delicate, and elastic fibers are extensible and elastic. Collagenous or white fibers often occur in bundles—an arrangement that provides great tensile strength. Reticular fibers, in contrast, occur in networks and, although delicate, support small structures such as capillaries and nerve fibers. Collagen in its hydrated form you know as gelatin. Of all the hundreds of different protein compounds in the body, collagen is the most abundant. Biologists now estimate that it constitutes somewhat over one fourth of all the protein in the body. And interestingly, one of the most basic factors in the aging process, according to some researchers,* is the change in the molecular structure of collagen that occurs gradually with the passage of the years.

Like intercellular fibers, intercellular ground substance also varies in properties. The ground substance (matrix) in loose, ordinary connective tissue, for example, is a soft, viscous gel, whereas the matrix of bone is a very hard substance—as "hard as bone," in fact. A compound called hyaluronic acid gives the viscous quality to intercellular gels, but it can be converted to a watery consistency by the enzyme hyaluronidase. Physicians have made use of this latter fact for some time now. They frequently give a commercial preparation of hyaluronidase intramuscularly or subcutaneously with drugs or fluids. By decreasing the viscosity of intercellular material, the enzyme hastens diffusion and absorption the thereby lessens tissue tension and pain.

Reticular tissue

A three-dimensional web, or a reticular network, identifies reticular-type tissue. Slender, branching reticular fibers with reticular cells, that is, fixed or nonmotile macrophages, over-

*Kohn, R. R.: Principles of mammalian aging, Englewood Cliffs, N.J., 1971, Prentice-Hall, Inc., pp. 22-24.

lying them compose the reticular meshwork. Branches of the cytoplasm of reticular cells follow the branching reticular fibers.

Reticular tissue forms the framework of the spleen, lymph nodes, and bone marrow. It functions as part of the body's complex mechanism for defending itself against microbes and injurious substances. The reticular meshwork filters injurious substances out of the blood and lymph, and fixed macrophages and certain other kinds of cells phagocytose (engulf and destroy) them. Another and probably the major function of reticular cells is to make reticular fibers.

Loose, ordinary connective tissue (areolar)

First, a few words of explanation about the name "loose, ordinary connective tissue." It is loose because it is stretchable and ordinary because it is one of the most widely distributed of all tissues. It is common and ordinary, not special like some kinds of connective tissue (bone and cartilage, for example) that help form comparatively few structures. Areolar was the early name for the loose, ordinary connective tissue that connects many adjacent structures of the body. It acts like a glue spread between them but an elastic glue that permits movement. The word areolar means "like a small space" and refers to the bubbles that appear as areolar tissue is pulled apart during dissection.

Although intercellular substance is prominent in loose, ordinary connective tissue, cells also are numerous and varied (Fig. 3-8). Collagenous and elastic fibers are interwoven loosely and embedded in a soft viscous ground substance. Of the half dozen or so kinds of cells present, *fibroblasts* are the most common and *macrophages* are second. Fibroblasts synthesize intercellular substances of both types, that is, both fibers and gels. Macrophages (also known by several other names, for example, histiocytes and resting wandering cells) carry on phagocytosis and so are classified as phagocytes. Phagocytosis is part of the body's vital complex of defense mechanisms. Other kinds of cells found in loose, ordinary connective tissue are mast cells, some white blood

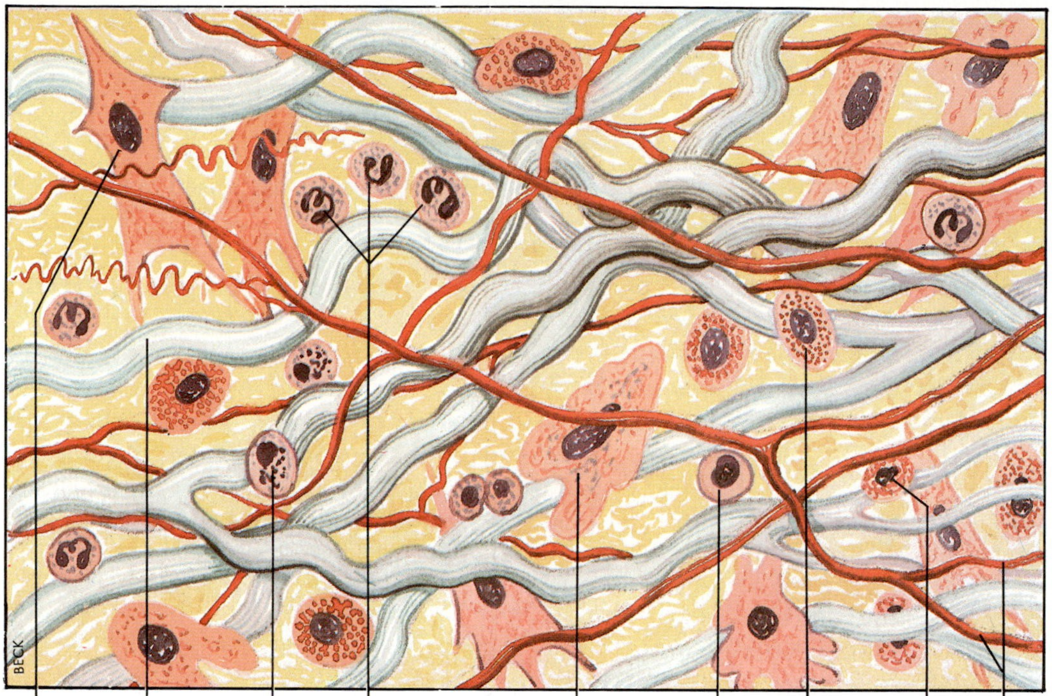

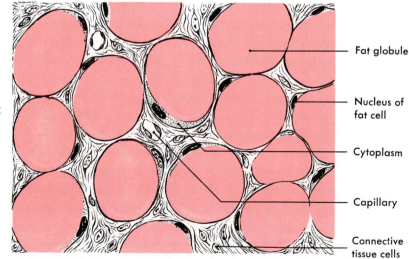

Fibrocyte (fibroblast) | Collagenous fibers | Plasma cell | Polymorphonuclear leukocytes | Macrophage | Monocyte | Eosinophil | Mast cell | Elastic fibers

Fig. 3-8 Areolar connective tissue. The large white fibers are collagenous fibers. Each of the red strands consists of a bundle of elastic fibers. Several fibroblasts are shown between the fibers. Also shown are macrophages, a plasma cell, a mast cell, and three types of white blood cells: polymorphonuclear leukocytes, eosinophils, and a monocyte.

Fig. 3-9 Adipose tissue. Fat droplet occupies nearly the entire area of the cell. Cytoplasm and nucleus are forced to the periphery of the cell.

Fat globule

Nucleus of fat cell

Cytoplasm

Capillary

Connective tissue cells

cells (leukocytes), an occasional fat cell, and some plasma cells.

Adipose tissue

Adipose tissue differs from loose, ordinary connective tissue mainly in that it contains predominantly fat cells and many fewer fibroblasts, macrophages, and mast cells (Fig. 3-9). Adipose tissue forms protective pads around the kidneys and various other structures. It also serves two other functions—it constitutes a storage depot for excess food and it acts as an insulating material to conserve body heat.

Dense fibrous tissue

Dense fibrous tissue consists mainly of bundles of fibers arranged in parallel rows in a fluid matrix. It contains relatively few fibroblast cells. It composes tendons and ligaments. Bundles of collagenous fibers endow tendons with great tensile strength and nonstretchability—desirable characteristics for these structures that anchor our muscles to bones. In ligaments, on the other hand, bundles of elastic fibers predominate. Hence ligaments exhibit some degree of elasticity.

Bone and cartilage

Bone and cartilage are discussed in Chapter 5.

Hemopoietic tissue

Hemopoietic tissue is discussed in Chapters 5 and 13.

Blood

Blood consists of billions of cells afloat in a fluid intercellular substance called plasma. We shall discuss both the cells and the intercellular substance of this vital connective tissue in considerable detail in Chapter 13.

Nervous tissue

Nervous tissue is discussed in Chapter 8.

Outline summary

Epithelial tissue

A Locations—covers and lines various parts of body
B Functions—specializes in absorption, secretion, and providing protective barriers
C Types
  1 Classified according to number of layers of cells—simple epithelium consists of one layer of cells; stratified epithelium consists of more than one layer of cells
  2 Classified according to cell shape—squamous epithelial cells are flat and scalelike; cuboidal cells are cubelike; columnar cells are taller than they are wide or long
D Generalizations
  1 Cells of epithelial tissue are arranged close together with little or no intercellular material between them; modifications of cytoplasmic membranes of adjacent cells hold them together
  2 No blood vessels in epithelial tissue; oxygen and food diffuse to epithelial cells from capillaries in connective tissue beneath epithelium; basement membrane (thin layer of nonliving material) attaches epithelial tissue to underlying connective tissue
  3 Epithelial cells undergo cell division from time to time
E Simple squamous epithelium
  1 Single layers of flat cells
  2 Function—diffusion and filtration
F Stratified squamous epithelium
  1 Several layers of cells
  2 Function—protection
G Simple columnar epithelium
  1 Single layer of columnar and goblet-shaped cells and, in some places, ciliated cells
  2 Functions—absorption, secretion, and moving mucus

Muscle tissue

A General functions—contraction and thereby movement

B Types
  1 Skeletal; also called striated voluntary
  2 Visceral; also called nonstriated or smooth involuntary
  3 Cardiac; also called striated involuntary

Connective tissue

A General functions—connection and support
B Main types
  1 Reticular
  2 Loose, ordinary connective (areolar)
  3 Adipose
  4 Dense fibrous
  5 Bone
  6 Cartilage
  7 Hemopoietic
  8 Blood
C General characteristics—intercellular material predominates in most connective tissues and determines their physical characteristics; consists of fluid, gel, or solid matrix, with or without fibers (collagenous, reticular, and elastic)
D Reticular tissue
E Loose, ordinary connective tissue (areolar)
  1 One of the most widely distributed of all tissues, intercellular substance is prominent and consists of collagenous and elastic fibers loosely interwoven and embedded in soft viscous ground substance; several kinds of cells present, notably, fibroblasts and macrophages, also mast cells, plasma cells, fat cells, and some white blood cells
  2 Function—connection
F Adipose tissue
  1 Similar to loose, ordinary connective tissue but contains mainly fat cells
  2 Functions—protection, insulation, support, and reserve food
G Dense fibrous tissue
  1 Fibrous intercellular substance (collagenous and elastic fibers); few fibroblast cells
  2 Function—furnishes flexible but strong connection

Review questions

  1 Name the four basic types of tissue.
  2 What are the main functions of each basic type of tissue?
  3 What are the names of the subtypes of connective tissue?
  4 Describe intercellular substance.
  5 Explain the scientific basis for giving hyaluronidase with fluids or certain injected drugs.
  6 Name several kinds of connective tissue cells.
  7 The intercellular gel and fibers in areolar tissue are produced by what kind of cells?
  8 Name three subtypes of muscle tissue. Give more than one name for each.
  9 What special function do reticuloendothelial cells perform?
 10 What are the main locations of reticuloendothelial cells?
 11 Make a list of terms you have encountered for the first time in this chapter. Define each in your own words.
 12 Recent evidence suggests that one of the most basic factors in the aging process is a molecular change in what intercellular protein?

chapter 4

Membranes and glands

Membranes

Definition and types

Membranes constitute a special class of organs in that they are merely thin sheets of tissues covering or lining various parts of the body. Of the numerous membranes in the body, four kinds are particularly important: mucous, serous, synovial, and cutaneous (skin). Other miscellaneous membranes (periosteum, fascia, dura mater, sclera, etc.) will be described from time to time.

Mucous membrane

Mucous membrane lines cavities or passageways of the body that open to the exterior, such as the lining of the mouth and entire digestive tract, the respiratory passages, and the genito-urinary tract. It consists, as do serous, synovial, and cutaneous membranes, of a surface layer of epithelial tissue over a deeper layer of connective tissue. Mucous membrane performs the functions of protection, secretion, and absorption—protection, for example, against bacterial invasion; secretion of mucus; and absorption of water, salts, and other solutes.

Serous and synovial membranes

Serous and synovial membranes line the closed cavities of the body, that is, those that do not open to the exterior. Serous membrane that lines the thoracic cavity is called *pleura,* that which lines the abdominal cavity is called *peritoneum,* and that which lines the sac in which the heart lies is called *pericardium.*

Not only does serous membrane line the thoracic and abdominal cavities and the pericardial sac, but it also covers the organs lying in these spaces. The term *visceral layer* is applied to the part of the membrane that covers the organs, whereas that which lines the cavity is called *parietal layer* (Fig. 4-1). Between the two layers, there is a potential space kept moist by a small amount of serous fluid. When an organ moves against the body wall (as the lungs do in respiration, or when the heart beats in its serous sac), friction between the moving parts is prevented by the presence of the smooth moist serous sheets lining the wall surface of the cavity and covering the organ surfaces. The mechanical principle that moving parts must have lubricated surfaces is thereby carried out in the body.

Synovial membrane lines joint cavities, tendon sheaths, and bursae. Its smooth moist surfaces protect against friction.

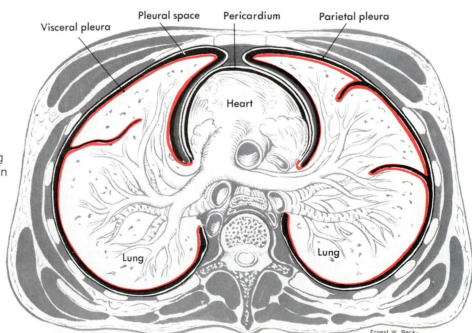

Fig. 4-1 Transverse section through the chest showing the parietal pleura in white (lining walls of chest cavity), the visceral pleura in red (covering lungs), and the pleural space in black (between parietal and visceral pleuras).

Visceral pleura   Pleural space   Pericardium   Parietal pleura

Heart

Lung   Lung

Ernest W. Beck

Cutaneous membrane

Vital, diverse, complex, extensive—these adjectives describe in part the body's largest and one of its most important organs, the skin. In terms of surface area the skin is as large as the body itself—probably 1.6 to 1.9 square meters (m²) in most adults. Skin functions are crucial to survival. They are also diverse, including such different functions as protection, excretion, and sensation. The skin also plays a part in maintaining fluid and electrolyte balance and normal body temperature. It protects us against entry of unconquerable hordes of microorganisms. It minimizes mechanical injury of underlying structures. It bars entry of excess sunlight and of most chemicals. Even water does not penetrate it under most circumstances.

Millions of microscopic nerve endings are distributed throughout the skin. These serve as antennas or receivers for the body, keeping it informed of changes in its environment—information essential for health and at times even vital for survival.

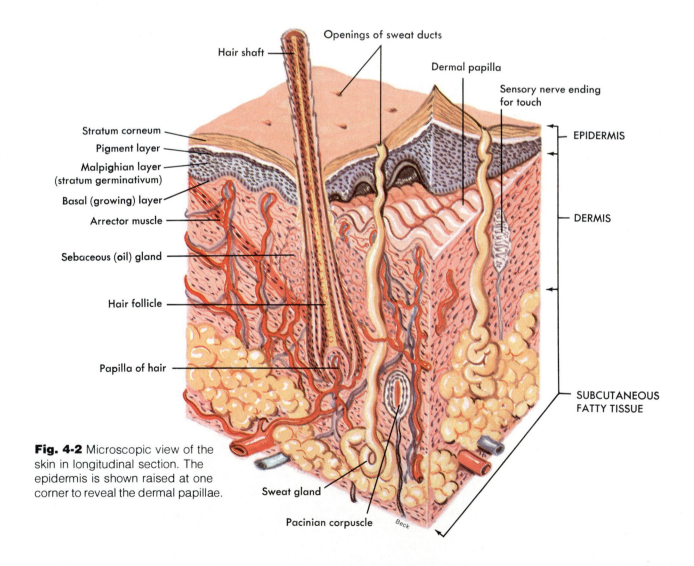

Fig. 4-2 Microscopic view of the skin in longitudinal section. The epidermis is shown raised at one corner to reveal the dermal papillae.

Hair shaft
Openings of sweat ducts
Dermal papilla
Sensory nerve ending for touch
Stratum corneum
Pigment layer
Malpighian layer (stratum germinativum)
Basal (growing) layer
Arrector muscle
Sebaceous (oil) gland
Hair follicle
Papilla of hair
Sweat gland
Pacinian corpuscle
EPIDERMIS
DERMIS
SUBCUTANEOUS FATTY TISSUE

Epidermis and dermis

Two main layers compose the skin: an outer and thinner layer, the *epidermis,* and an inner and thicker layer, the *dermis* (Fig. 4-2). The epidermis consists of stratified squamous epithelial tissue and the dermis of fibrous connective tissue. Underlying the dermis is subcutaneous tissue or *superficial fascia* made of areolar and, in many areas, adipose tissue. The epidermis in all parts of the body except the palms of the hands and soles of the feet has four layers. In the skin of the palms and soles, there are five layers of epidermis. From the outside in, they are as follows:

1 *Stratum corneum (horny layer)*—dead cells converted to a water-repellant protein called keratin that continually flakes off (desquamates)
2 *Stratum lucidum*—so named because of the presence of a translucent compound (eleidin) from which keratin forms; present only in thick skin of palms and soles
3 *Stratum granulosum*—so named because of granules visible in the cytoplasm of cells (cells die in this layer)

4 *Stratum spinosum (prickle cell layer)*—several layers of irregularly shaped cells
5 *Stratum germinativum (basal layer)*—columnar-shaped cells, the only cells in the epidermis that undergo mitosis; new cells are produced in this deepest stratum at the rate old keratinized cells are lost from the stratum corneum; new cells continually push surfaceward from the stratum germinativum into each successive layer, only to die, become keratinized, and eventually flake off, as did their predecessors. Incidentally, this fact illustrates nicely the physiological principle that while life continues the body's work is never done; even at rest it is producing new cells to replace millions of old ones.

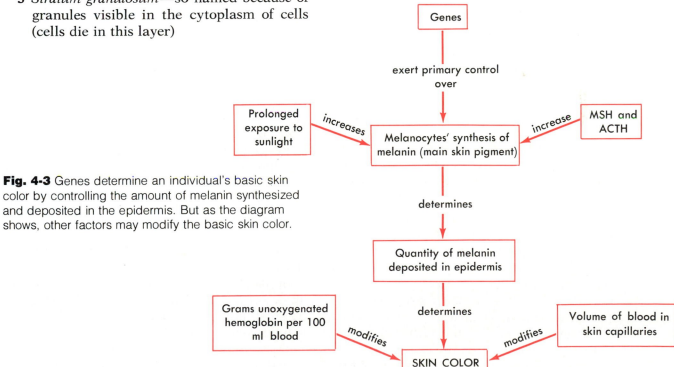

Fig. 4-3 Genes determine an individual's basic skin color by controlling the amount of melanin synthesized and deposited in the epidermis. But as the diagram shows, other factors may modify the basic skin color.

An interesting characteristic of the dermis or deep layer of the skin is its parallel ridges, suggestive on a miniature scale of the ridges of a contour plowed field. Epidermal ridges, the ones made famous by the art of fingerprinting, exist because the epidermis conforms to the underlying dermal ridges.

As everyone knows, human skin comes in a wide assortment of colors. Explanation for this fact lies in a constellation of factors that act and interact in complex ways (Fig. 4-3). The basic determinant, however, of skin color is the quantity of *melanin*—the main skin pigment—deposited in the epidermis. Heredity, first of all, determines this. Geneticists now tell us that four to six pairs of genes exert the primary control over the amount of melanin formed by special cells called *melanocytes*. Thus heredity determines how dark or light one's basic skin color will be. But other factors can modify the genetic effect. Sunlight is the obvious example. Prolonged exposure of the skin to sunlight causes melanocytes to increase melanin production and darken skin color. So, too, does an excess of adrenocorticotropic hormones (ACTH) or an excess of melanocyte-stimulating hormone (MSH), two of the hormones secreted by the anterior pituitary gland.

An individual's basic skin color changes, as we have just observed, whenever its melanin content changes appreciably. But skin color can also change without any change in melanin. In this case the change is usually temporary and most often stems from a change in the volume of blood flowing through skin capillaries. A marked increase in skin blood volume may cause the skin to take on a pink or red hue, and an appreciable decrease in skin blood volume may turn it frighteningly pale. In general, the sparser the pigments in the epidermis, the more transparent the skin is and, therefore, the more vivid the change in skin color will be with a change in skin blood volume. Conversely, the richer the pigmentation, the more opaque the skin is and the less skin color will change with a change in skin blood volume.

In some abnormal conditions, skin color changes because of an excess amount of unoxygenated hemoglobin in the skin capillary blood. If skin contains relatively little melanin, it will develop a bluish cast known as *cyanosis* when 100 ml of blood contains about 5 gm of unoxygenated hemoglobin. In general, the darker the skin pigmentation, the more unoxygenated hemoglobin must be present before cyanosis becomes visible.

Accessory organs of skin

The accessory organs of the skin consist of hair, nails, and microscopic glands.

Hair. Hair is distributed over the entire body except the palms and soles. The structure of a hair has several points of similarity to that of the epidermis. Just as the epidermis is formed by the cells of its deepest layer reproducing and forcing the daughter cells, which become horny in character, upward, so a hair is formed by a group of cells at its base multiplying and pushing upward and in so doing becoming keratinized. The part of the hair that is visible is the *shaft*, whereas that which is embedded in the dermis is the *root*. The root, together with its coverings (an outer connective tissue sheath and an inner epithelial coating that is a continuation of the stratum germinativum), forms the hair *follicle*. At the bottom of the follicle is a loop of capillaries enclosed in a connective tissue covering called the hair *papilla*. The clusters of epithelial cells lying over the papilla are the ones that reproduce and eventually form the hair shaft. As long as these cells remain alive, hair will regenerate even though it be cut or plucked or otherwise removed.

Each hair is kept soft and pliable by two or more *sebaceous glands*, which secrete varying amounts of oily *sebum* into the follicle near the surface of the skin. Attached to the follicle, too, are small bundles of involuntary muscle known as the *arrector pili muscles*. These muscles are of interest because when they contract, the hair "stands on end," as it does in extreme fright or cold, for example. This mechanism is responsible also for gooseflesh. As the hair is pulled into an upright position, it raises the skin around it into the familiar little goose pimples.

Hair color results from different amounts of melanin pigments in the outer layer (cortex) of the hair. White hair contains little or no melanin.

Some hair, notably that around the eyes and in the nose and ears, performs a protective function in that it keeps out some dust and insects. For the hair on the bulk of the skin, however, no function seems apparent.

Nails. The nails are epidermal cells that have been converted to keratin. They grow from epithelial cells lying under the white crescent (lunula) at the proximal end of each nail.

Skin glands. The skin glands include three kinds of microscopic glands, namely, sebaceous, sweat, and ceruminous.

Sebaceous glands secrete oil for the hair. Wherever hairs grow from the skin, there are sebaceous glands, at least two for each hair. The oil, or *sebum*, secreted by these tiny glands has value not only because it keeps the hair supple but also because it keeps the skin soft and pliant. It serves as nature's own cold cream. Moreover, it prevents excessive water evaporation from the skin and water absorption through the skin. And because fat is a poor conductor of heat, the sebum secreted into the skin lessens the amount of heat lost from this large surface.

Sweat glands, although very small structures, are very important and very numerous—especially on the palms, soles, forehead, and axillae (armpits). Histologists estimate, for example, that a single square inch of skin on the palms of the hands contains about 3,000 sweat glands.

Sweat secretion helps maintain homeostasis of fluid and electrolytes and of body temperature. For example, if too much heat is being produced, as in strenuous exercise, or if the environmental temperature is high, these glands secrete more sweat, which, in evaporating, cools the body surface. Inasmuch as sweat contains some nitrogenous wastes, the sweat glands also function as excretory organs.

Ceruminous glands are thought to be modified sweat glands. They are located in the external ear canal. Instead of watery sweat, they secrete a waxy, pigmented substance, the *cerumen*.

Terms used in connection with skin

hypodermic—beneath or under the skin
subcutaneous—same as hypodermic
intracutaneous—within the layers of the skin
diaphoresis—profuse perspiration
pores—minute openings of the sweat gland ducts on the surface of the skin; do not "open" or "close," since no muscle tissue enters into their formation; however, any agent that has an astringent action on the skin shrinks so, in effect, closes the pores
furuncle—a boil, an infection of a hair follicle

Glands

Glands consist of epithelial cells specialized for synthesizing compounds that they secrete into either ducts or blood. Relatively simple compounds enter a gland's cells from the blood, and chemical reactions within the cells build them up into the more complex compounds that make up the gland's secretion. A single cell may constitute a gland. Goblet cells in the mucous lining of the intestines and part of the respiratory tract are *unicellular glands*. Most glands, however, are multicellular, that is, composed of many cells. *Multicellular glands* are of two main types—endocrine (or ductless) and exocrine (or duct) glands. *Endocrine glands* secrete into blood capillaries, and *exocrine glands* secrete into ducts that open on the surface of the epithelium. Exocrine glands are of two types: *simple glands*, if they have only one duct, and *compound glands*, if they have more than one duct. Simple and compound glands are further classified, according to the shape of the secretory portion of the gland, as either tubular (tube-shaped) or alveolar (sac-shaped).

Fig. 4-4 Tubular exocrine glands. **A, B,** and **C** represent simple or single duct tubular glands. **D,** a compound gland, has more than one duct.

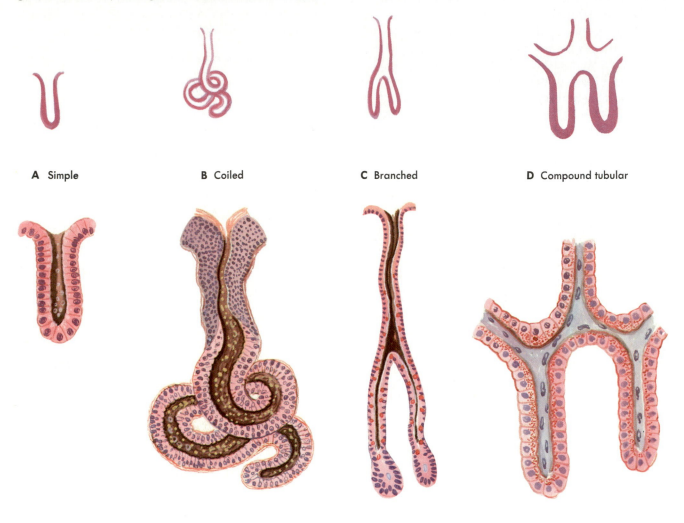

A Simple   B Coiled   C Branched   D Compound tubular

Hence there are the following eight types of multicellular glands (Figs. 4-4 to 4-6):

1 *Simple tubular glands*—one duct; tube-shaped secretory portion (example, intestinal glands)

2 *Simple coiled tubular glands*—one duct; coiled, tube-shaped secretory portion (example, sweat glands)

3 *Simple branched tubular glands*—one duct branches into two tube-shaped secretory portions (example, gastric glands)

4 *Simple alveolar glands*—one duct; sac-shaped secretory portion (example, sebaceous glands)

5 *Simple branched alveolar glands*—one duct; branching sac-shaped secretory portion (example, sebaceous glands)

6 *Compound tubular glands*—more than one

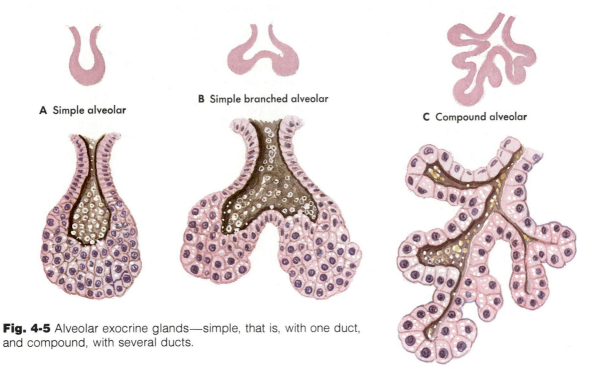

A Simple alveolar

B Simple branched alveolar

C Compound alveolar

Fig. 4-5 Alveolar exocrine glands—simple, that is, with one duct, and compound, with several ducts.

Fig. 4-6 Compound tubuloalveolar gland—more than one duct with both tube-shaped and sac-shaped secretory portions.

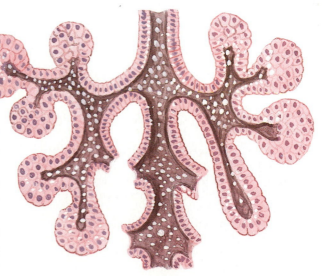

duct; tube-shaped secretory portion (example, male sex glands [testes])

7 *Compound alveolar glands*—more than one duct; sac-shaped secretory portion (example, mammary glands)

8 *Compound tubuloalveolar glands*—more than one duct; both tube-shaped and sac-shaped secretory portions (example, salivary glands)

Outline summary

Membranes

A Definition—thin sheet of tissues that either covers or lines part of body or divides organ

B Types—mucous, serous, synovial, cutaneous, and miscellaneous

Mucous membrane

A Location—lines cavities and passages that open to exterior

B Structure—surface layer of epithelial tissue over connective tissue

C Functions—protection, secretion, and absorption

Serous membrane

A Location—lines cavities that do not open to exterior

B Functions—protection and secretion

Synovial membrane

A Location—lines joint cavities, tendon sheaths, and bursae

B Functions—protection and secretion

Cutaneous membrane

A Functions
1 Protection against various factors, for example, microorganisms, sunlight, and chemicals
2 Excretion of sweat
3 Sensations
4 Fluid and electrolyte balance—skin helps maintain this by secreting varying amounts of sweat
5 Normal body temperature—skin helps maintain this by varying amounts of blood flow through it and also by varying amounts of sweat secretion

B Structure—two main layers: epidermis, outer, thinner layer of stratified squamous epithelium, and dermis, inner, thicker layer of connective tissue
1 Epidermis
  a Outer layer of stratified squamous epithelial cells
  b Surface cells dead, keratinized, and practically waterproof
  c Only deepest layer of cells undergoes mitosis to replace surface cells that continually desquamate
  d Melanin pigments mainly in deeper layer
2 Dermis
  a Dense fibrous connective tissue layer underlying epidermis
  b Dermis of palms and soles has numerous parallel ridges
  c Subcutaneous tissue (also called superficial fascia) underlying dermis, composed of areolar tissue or areolar and adipose tissue

C Accessory organs of skin
1 Hair
  a Distribution—over entire body except palms and soles
  b Shaft—visible part of hair
  c Root—part of hair embedded in dermis
  d Follicle—root with coverings
  e Papilla—loop of capillaries enclosed in connective tissue covering
  f Germinal matrix—cluster of epithelial cells lying over papilla; these cells undergo mitosis to form hair; must be intact in order for hair to regenerate
  g Sebaceous glands and arrector pili muscles—attach to follicle; contraction of latter produces gooseflesh
  h Color—results from different amounts of melanin pigments in cortex of hair
2 Nails
  a Epidermal cells converted to hard keratin
  b Grow from epithelial cells under lunula ("moons")
3 Skin glands
  a Sebaceous—secrete oil (sebum) that keeps hair and skin soft; helps prevent excess evaporation and absorption of water and excess heat loss
  b Sweat—numerous throughout skin, especially on palms, soles, forehead, and axillae; important in heat regulation
  c Ceruminous—thought to be modified sweat glands; located in external ear canal; secrete earwax or cerumen

Glands

A Composed of epithelial cells specialized for synthesizing compounds that they secrete either into ducts or blood capillaries

B Types
1 Exocrine glands—secrete into ducts
  a Simple glands—have single ducts
  b Compound glands—have more than one duct
  c Tubular glands—tubular-shaped secretory units
  d Alveolar glands—flask-shaped secretory units
2 Endocrine glands—secrete into blood

Review questions

1 What membranes line closed cavities? Cavities that open to the exterior?
2 What general functions do membranes serve?
3 What functions does the skin perform?
4 Describe the epidermis.
5 Describe the dermis.
6 What is keratin?
7 What is melanin and where is it found?
8 What is superficial fascia?
9 Discuss the factors that determine basic skin color and variations in it.
10 Name and describe the skin glands.
11 Define the following terms: alveolar gland, exocrine gland, endocrine gland, compound gland.
12 What organelle(s) would you expect to be prominent in secreting cells?

Support and movement

chapter 5

The skeletal system—bones and cartilage

Support and movement of the body are primary functions of the skeletal and muscular systems. Movements not only contribute immeasurably to the enjoyment of life, but they also are essential for maintaining life. They help the body maintain homeostasis by enabling it to respond to changes in both its external and internal environments. The skeletal system consists of bones and articulations, that is, joints between bones. This chapter discusses bones; Chapter 6 considers articulations.

Functions

Bones perform five functions for the body. Three of these you probably already know, but two of them may come as a surprise.

1 *Support*. Bones serve as the supporting framework of the body much as steel girders function as the supporting framework of our modern buildings.

2 *Protection*. Hard, bony "boxes" protect delicate structures enclosed by them. For example, the skull protects the brain, and the rib cage protects the lungs and heart.

3 *Movement*. Bones with their joints constitute levers. Muscles are anchored firmly to bones. As muscles contract and shorten, force is applied to the bony levers and movement results.

4 *Reservoir*. Bones serve as the major reservoir into which calcium is deposited or from which it is withdrawn as needed to maintain the homeostasis of blood calcium, a vitally important condition.

5 *Hemopoiesis.* Hemopoiesis means the process of forming blood cells. Tissues that carry on this process are specialized connective tissues of two kinds, namely, myeloid tissue and lymphatic tissue. Red bone marrow is another name for myeloid tissue. In the adult, it is found only in a few bones—in the sternum and ribs, in the bodies of the vertebrae, in the diploë of the cranial bones, and in the proximal epiphyses of the femur and humerus. Red marrow occurs in many more bones in newborn infants and children. Because it forms blood cells, the red bone marrow is one of the most important tissues of the body. Its very location, hidden in the bones, like valuables in a safety deposit box, suggests its vital importance.

Types of bones

The names of the four types of bones suggest their shapes, as follows:

1 *Long bones*—bones of upper and lower arm (humerus, ulna), of thigh and leg (femur, tibia and fibula), and of fingers and toes (phalanges)
2 *Short bones*—wrist and ankle bones (carpals and tarsals)
3 *Flat bones*—certain bones of the cranium (examples, frontal and parietal), ribs, and shoulder bones (scapulae)
4 *Irregular bones*—bones of spinal column (vertebrae, sacrum, coccyx) and certain bones of the skull (examples, sphenoid, ethmoid, mandible)

Macroscopic structure
Long bones

A long bone consists of the following structures visible to the naked eye: diaphysis, epiphyses, articular cartilage, periosteum, medullary (marrow) cavity, and endosteum. Identify each of these parts in Fig. 5-1 as you read about it.

1 *Diaphysis*—main shaftlike portion. Its hollow, cylindrical shape and the thick compact bone that composes it.adapt the diaphysis well to its function of providing strong support without cumbersome weight.

2 *Epiphyses*—extremities of long bones. Note their somewhat bulbous shape; it provides generous space for muscle attachments near joints and gives joints greater stability. Note, in Fig. 5-2, the appearance of the bone in the epiphysis. Like a sponge, it is permeated by innumerable small spaces—hence its name, spongy or cancellous bone. Because of the porous nature of this bone and because the epiphyses have only a thin outer layer of dense, compact bone, they are lightweight structures for their size. Marrow fills the spaces of cancellous bone—yellow marrow in most adult epiphyses but red marrow in the proximal epiphyses of the humerus and femur.

3 *Articular cartilage*—thin layer of hyaline cartilage that covers articular or joint surfaces of epiphyses. Resiliency of this material cushions jars and blows.

4 *Periosteum*—dense white fibrous membrane that covers bone except at joint surfaces, where articular cartilage forms the covering. Many of the periosteum's fibers penetrate the underlying bone, welding these two structures to each other. (The penetrating fibers are called Sharpey's fibers.) Muscle tendon fibers interlace with periosteal fibers, thereby anchoring muscles firmly to bone. The periosteum contains many small blood vessels, and numerous *osteoblasts* (bone-forming cells) form the inner layer of the periosteum of growing bones. Because of its blood vessels, which send branches into the bone, the periosteum is essential for the nutrition of bone cells and therefore for their survival. Because of its osteoblasts, the periosteum is essential for both growth and repair of bones.

5 *Medullary (or marrow) cavity*—tubelike hollow in diaphysis of long bone. In the adult, it contains yellow or fatty marrow.

6 *Endosteum*—membrane that lines medullary cavity of long bones.

Short, flat, and irregular bones

Short bones, flat bones, and irregular bones all have an inner portion of cancellous bone covered over on the outside with compact bone. Red marrow fills the spaces in the cancellous bone inside a few irregular and flat bones, for example, in the vertebrae and sternum. To help in the diagnosis of leukemia and certain other diseases, a physician may decide to perform a needle puncture of one of these bones. He inserts a needle and punctures through the skin and compact bone into the red marrow. (Stewart Alsop, a famous journalist, described this procedure as feeling like the bite of "a bloody great wasp,"* despite the use of a local anesthetic.) A

*Alsop, S.: Stay of execution, Philadelphia, 1973, J. B. Lippincott Co., p. 18.

small amount of red marrow is aspirated and examined under the microscope for normal or abnormal blood cells.

Microscopic structure
Bone

Bone, like other tissues, consists of living cells and nonliving intercellular substance. The intercellular material, or *matrix* of bone, predominates. It is much more abundant than the bone cells, and it contains many fibers of collagen (the body's most abundant protein). In these respects, bone's matrix resembles that of other connective tissues. But it also differs strikingly from them. For one thing, it is hard and rigid, not soft and flexible. It is calcified. In other words, calcium salts impregnate the intercellular material of bone—they make it "as hard as bone." Another unique feature is the way in which bone's intercellular material is arranged with relation to its cells. You can observe this extraordinary arrangement in compact bone in Fig. 5-3. Concentric, cylinder-shaped layers (usually fewer than six of them) of calcified intercellular material enclose a central longitudinal canal that contains a blood vessel. Bone cells, or *osteocytes*, lie imprisoned between the hard layers in small spaces called *lacunae;* the layers are called *lamellae*. The central canal enclosed by the concentric lamellae is an *haversian canal*. Great numbers of microscopic-sized canals—*canaliculi*—radiate in all directions from the lacunae. They provide routes for tissue fluid to move the short distance (not more than 0.1 mm) from an haversian canal to the lacunae and bone cells of an haversian system. The unit composed of an haversian canal, its surrounding lamellae, osteocytes, lacunae, and canaliculi is called an *haversian system*. Haversian systems are the structural units of compact bone, and thousands of them make up the diaphysis of a single long bone. Because of their cylindrical shape, haversian systems do not fit together

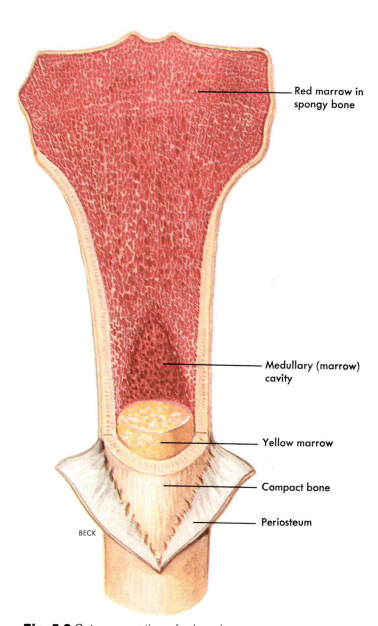

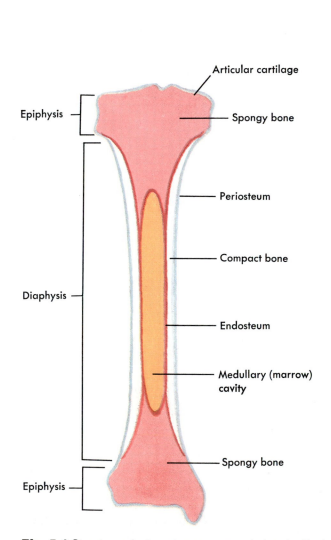

Fig. 5-1 Structure of a long bone as seen in longitudinal section.

Fig. 5-2 Cutaway section of a long bone.

Articular cartilage

Epiphysis

Spongy bone

Periosteum

Diaphysis

Compact bone

Endosteum

Medullary (marrow) cavity

Epiphysis

Spongy bone

Red marrow in spongy bone

Medullary (marrow) cavity

Yellow marrow

Compact bone

Periosteum

BECK

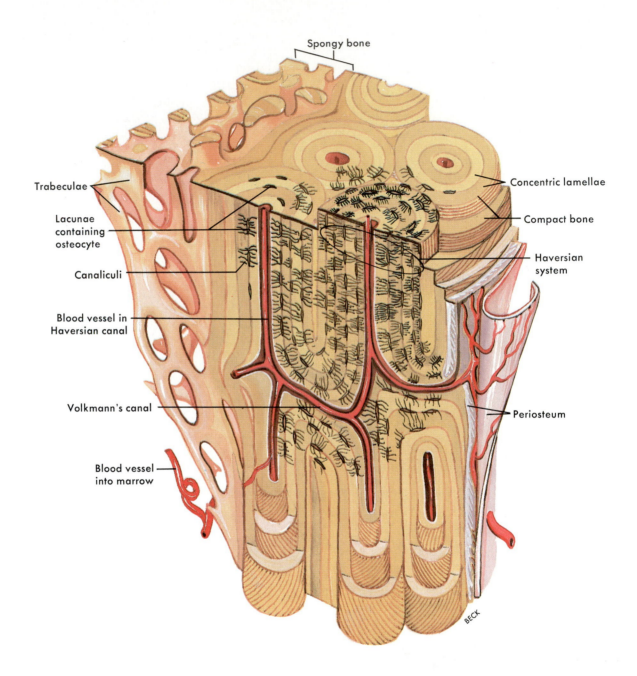

Spongy bone

Trabeculae

Lacunae
containing
osteocyte

Canaliculi

Blood vessel in
Haversian canal

Volkmann's canal

Blood vessel
into marrow

Concentric lamellae

Compact bone

Haversian
system

Periosteum

BECK

Fig. 5-3

A, Drawing of microscopic structure of bone. Haversian systems, several of which are shown here, compose compact bone. Note the structures that make up one haversian system: concentric lamellae, lacunae, canaliculi, and an haversian canal. Shown bordering the compact bone on the left is spongy bone, a name descriptive of the many open spaces that characterize it.

perfectly. The potential spaces between them are filled in by *interstitial* and *circumferential lamellae.*

Cancellous bone and compact bone differ in microscopic structure. In compact bone, interstitial lamellae fill in the spaces between adjacent haversian systems. In cancellous bone, many open spaces are present between thin processes of bone called *trabeculae.* Somewhat as beams of wood are joined in a scaffold, trabeculae are joined in cancellous bone. The arrangement of trabeculae along the lines of strain on individual bones gives bones added structural strength.

Bones are not the lifeless structures they seem to be. We tend to think of them as lifeless, perhaps because what we see when we look at a bone is its nonliving intercellular substance. But within this hard, lifeless material lie many living bone cells that must continually receive food and oxygen and excrete their wastes. So blood supply to bone is important and abundant. For example, numerous blood vessels from the periosteum penetrate bone by way of Volkmann's canals to connect with blood vessels of an haversian canal (Fig. 5-3). Also, one or more arteries supply the bone marrow in the internal medullary cavity of long bones.

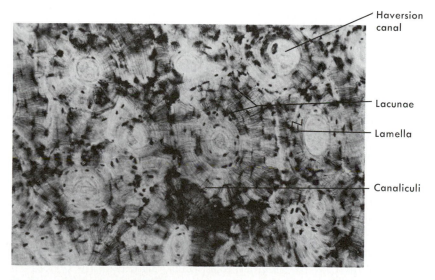

Haversion canal

Lacunae

Lamella

Canaliculi

B, Compact bone micrograph.

Courtesy Dr. C. R. McMullen, Department of Biology, South Dakota State University

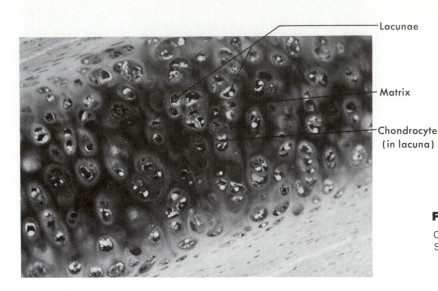

Lacunae

Matrix

Chondrocyte
(in lacuna)

Fig. 5-4 Micrograph of hyaline cartilage.

Courtesy Dr. C. R. McMullen, Department of Biology,
South Dakota State University

Cartilage

Cartilage both resembles and differs from bone. Like bone, cartilage consists more of intercellular substance than of cells. Innumerable collagenous fibers reinforce the matrix of both tissues. But in cartilage the fibers are embedded in a firm gel instead of in a calcified cement substance as they are in bone. Hence cartilage has the flexibility of a firm plastic material rather than the rigidity of bone. Another difference is this—no canal system and no blood vessels penetrate cartilage matrix. Cartilage is avascular, and bone is abundantly vascular. Cartilage cells, like bone cells, lie in lacunae. However, because no canals and blood vessels interlace cartilage matrix, nutrients and oxygen can reach the scattered, isolated chondrocytes (cartilage cells) only by diffusion. They diffuse through the matrix gel from capillaries in the fibrous covering of cartilage (perichondrium) or from synovial fluid in the case of articular cartilage.

Three types of cartilage are hyaline, fibrous, and elastic. They differ structurally mainly as to matrix fibers. Collagenous fibers are present in all three types but are most numerous in fibro-

cartilage. Hence it has the greatest tensile strength. Elastic cartilage matrix contains elastic fibers as well as collagenous fibers and so has elasticity as well as firmness. Hyaline is the most common type of cartilage (Fig. 5-4). It resembles milk glass in appearance. In fact, its name is derived from the Greek word meaning glassy. A thin layer of hyaline cartilage covers articular surfaces of bones, where it helps to cushion jolts. Fibrocartilage disks between the vertebrae also serve this purpose. For other locations of cartilage, see Table 3-1, p. 63.

Formation and growth of bone

The embryo skeleton, when first formed, consists of "bones" that are not really bones at all but hyaline cartilage or fibrous membrane structures shaped like bones. Gradually the process of *ossification* (or *osteogenesis*) replaces these prebone structures with bone. Because ossification is an extraordinarily complex process, we shall describe only its basic aspects, not its details. In either cartilage or fibrous membranes, ossification begins with the appearance of centers of ossification. These consist of clusters of special bone-forming cells called *osteoblasts*. The Golgi apparatus in an osteoblast specializes in synthesizing and secreting carbohydrate compounds of the type named mucopolysaccharides, and its endoplasmic reticulum makes and secretes collagen, a protein. In time, relatively large amounts of the mucopolysaccharide substance (or cement substance, as it is more commonly called) accumulate around each individual osteoblast, and numerous bundles of collagenous fibers become embedded in the cement substance. Together, the cement substance and collagenous fibers constitute the organic intercellular substance of bone called the *bone matrix*. If you like nicknames, you might call bone matrix the reinforced concrete of the body. Bundles of collagenous fibers in bone increase its strength much as iron rods in reinforced concrete strengthen it.

About as fast as the organic bone matrix forms, inorganic compounds—complex calcium salts, not yet positively identified—begin depositing in it. This calcification of its matrix is what makes bone one of our hardest tissues.

In summary then, ossification consists of two complex processes—synthesis of bone's organic matrix by osteoblasts, immediately followed by calcification of this matrix.

Endochondral ossification starts in the diaphysis and in both epiphyses of a long bone. It then progresses from the diaphysis toward each epiphysis and from each epiphysis toward the diaphysis. Until bone growth in length is complete, a layer of cartilage known as *epiphyseal cartilage* remains between each epiphysis and the diaphysis. Proliferation of epiphyseal cartilage cells brings about a thickening of the layer of epiphyseal cartilage from time to time. Ossification of this additional cartilage then follows, that is, osteoblasts synthesize organic bone matrix and the matrix undergoes calcification. As a result, the bone becomes longer. When epiphyseal cartilage cells stop multiplying and the cartilage has become completely ossified, bone growth has ended. X-ray pictures can reveal any epiphyseal cartilage still present. When bones have grown their full length, the epiphyseal cartilage disappears. Bone has replaced it and is then continuous between epiphyses and diaphysis.

Bones grow in diameter by the combined action of two special kinds of cells: osteoblasts and osteoclasts. Osteoclasts enlarge the diameter of the medullary cavity by eating away the bone of its walls. At the same time, osteoblasts from the periosteum build new bone around the outside of the bone. By this dual process a bone with a larger diameter and larger medullary cavity has been produced from a smaller bone with a smaller medullary cavity.

The formation of bone tissue continues long after bones have stopped growing in size. Throughout life, bone formation (osteogenesis) and bone destruction (resorption) go on concurrently. These opposing processes balance each other during adulthood's early to middle years, so in this period of life bones neither grow nor shrink in size. During childhood and adolescence, osteogenesis goes on at a faster rate than bone resorption. Bone gain outstrips bone loss, and therefore bones grow larger. But between the ages of 35 and 40 years, the tables turn, and from that time on bone loss exceeds bone gain. Bone gain occurs slowly at the outer or periosteal surfaces of bones. Bone loss, on the other hand, occurs at the inner or endosteal surfaces and goes on at a somewhat faster pace. More bone is lost on the inside than gained on the outside, and inevitably bones become remodeled as the years go by. For instance, the thickness of the compact bone in the diaphyses of long bones decreases, and the diameter of their medullary cavities increases. Gradually they become more like hollow shells and are less able to resist compression and bending. Vertebrae, to some degree, collapse, and height decreases. Femurs frequently break, and this ominous event often leads indirectly to death.

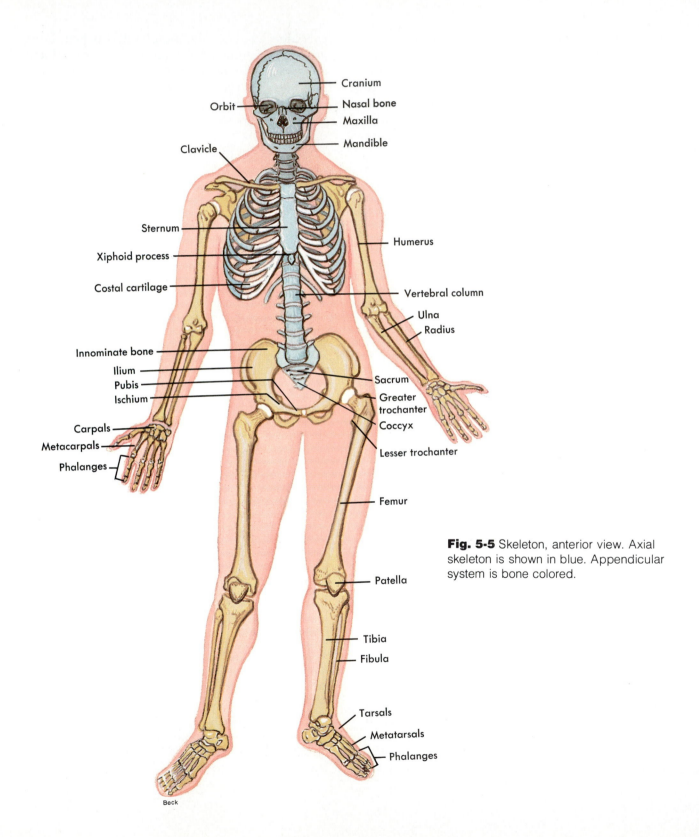

Cranium

Orbit

Nasal bone

Maxilla

Mandible

Clavicle

Sternum

Humerus

Xiphoid process

Costal cartilage

Vertebral column

Ulna

Radius

Innominate bone

Ilium

Pubis

Sacrum

Ischium

Greater trochanter

Carpals

Coccyx

Metacarpals

Lesser trochanter

Phalanges

Femur

Patella

Tibia

Fibula

Tarsals

Metatarsals

Phalanges

Beck

Fig. 5-5 Skeleton, anterior view. Axial skeleton is shown in blue. Appendicular system is bone colored.

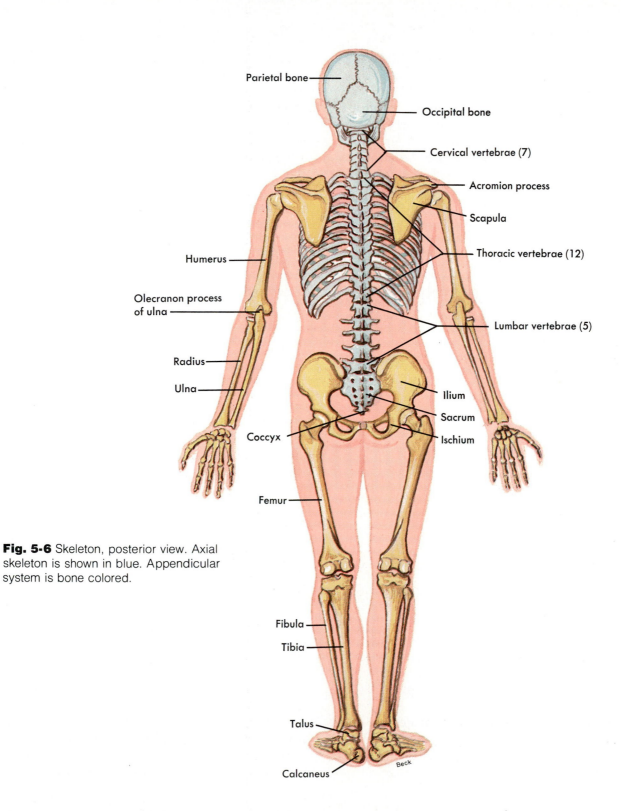

Parietal bone

Occipital bone

Cervical vertebrae (7)

Acromion process

Scapula

Thoracic vertebrae (12)

Humerus

Olecranon process of ulna

Lumbar vertebrae (5)

Radius

Ulna

Ilium

Sacrum

Coccyx

Ischium

Femur

Fibula

Tibia

Talus

Beck

Calcaneus

Fig. 5-6 Skeleton, posterior view. Axial skeleton is shown in blue. Appendicular system is bone colored.

Bone markings

Various points on bones are labeled according to the nature of their structure. This method of identifying definite parts of different bones proves helpful when locating other structures such as muscles, blood vessels, and nerves. Definitions of some common bone markings follow. After you have studied these, see Table 5-2, p. 119, for identification of markings on individual bones.

Depressions and openings

fossa—a hollow or depression (example, mandibular fossa of the temporal bone that serves as the socket for the lower jawbone)
sinus—a cavity or spongelike space in a bone (example, the frontal sinus)
foramen—a hole (example, foramen magnum of the occipital bone)
meatus—a tube-shaped opening (example, the external auditory meatus)

Projections or processes

The following projections or processes enter into the formation of joints:

condyle—a rounded projection (example, condyles of the femur)
head—a rounded projection beyond a narrow neckline portion (example, head of the femur)

Muscles attach to the following projections:

trochanter—a very large projection (example, greater trochanter of the femur)
crest—a ridge (example, the iliac crest); a less prominent ridge is called a line (example, ileopectineal line)
spinous process or *spine*—a sharp projection (example, anterior superior spine of the ilium)
tuberosity—a large, rounded projection (example, ischial tuberosity)
tubercle—a small, rounded projection (example, rib tubercles)

Divisions of skeleton

The human skeleton consists of two main divisions, namely, the axial skeleton and the appendicular skeleton. Eighty bones make up the *axial skeleton.* This includes 74 bones that form the upright axis of the body and 6 tiny middle ear bones. The *appendicular skeleton* consists of 126 bones—more than half again as many as in the axial skeleton. Bones of the appendicular skeleton form the appendages to the axial skeleton, that is, the shoulder girdles, arms, wrists, and hands and the hip girdles, legs, ankles, and feet. One of the first things you should do in studying bones is to learn their names. Table 5-1, p. 117, lists both the names and numbers of the bones and identifies them.

Description of major bones in skeleton
Axial skeleton

Skull. Twenty-eight irregularly shaped bones form the skull (Figs. 5-7 to 5-17). It consists of two major divisions: the cranium or brain case and the face. Eight bones form the cranium, namely, frontal, 2 parietal, 2 temporal, occipital, sphenoid, and ethmoid. The 14 bones that form the face are these: 2 maxillary, 2 zygomatic (malar), 2 nasal, mandible, 2 lacrimal, 2 palatine, 2 inferior nasal conchae (turbinates), and vomer. Note that all the face bones are paired except the mandible and vomer. The frontal and ethmoid bones of the skull help shape the face but are not numbered among the face bones.

The *frontal bone* forms the forehead and the anterior part of the top of the cranium. It contains mucosa-lined air-filled spaces, the *frontal sinuses*, and it forms the upper part of the orbits. It unites with the two parietal bones posteriorly in an immovable joint, the *coronal suture.* Several of the more prominent frontal bone markings are described in Table 5-2, p. 120.

The two *parietal bones* give shape to the bulging topside of the cranium. They form immovable joints with several bones: the *lambdoidal suture* with the occipital bone, the *squamous suture* with the temporal bone and part of the sphenoid, and the *coronal suture*, mentioned before, with the frontal bone.

The lower sides of the cranium and part of its floor are fashioned from two *temporal bones*. They house the middle and inner ear structures and contain the *mastoid sinuses*, notable because of the occurrence of mastoiditis, an inflammation of the mucous lining of these spaces. A description of several other temporal bone markings is included in Table 5-2, p. 120.

The *occipital bone* makes the framework of the lower, posterior part of the skull. It forms immovable joints with three other cranial bones—

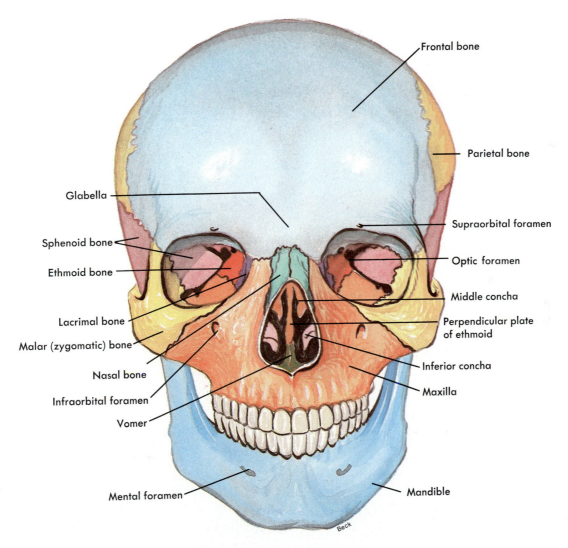

Fig. 5-7 Skull viewed from the front.

the parietal, temporal, and sphenoid—and a movable joint with the first cervical vertebra. Consult Table 5-2, p. 121, for a description of some of its markings.

The *sphenoid bone* resembles a bat with its wings outstretched and legs extended downward posteriorly. Note in Fig. 5-9 the location of the sphenoid bone in the central portion of the cranial floor. Here it serves as the keystone in the architecture of the cranium, anchoring the frontal, parietal, occipital, and ethmoid bones. The sphenoid bone also forms part of the lateral wall and floor of each orbit (Fig. 5-7). The sphenoid bone contains fairly large mucosa-lined air-filled spaces, the *sphenoid sinuses.* Several prominent sphenoid markings are described in Table 5-2, p. 120 (also see Figs. 5-9 to 5-11).

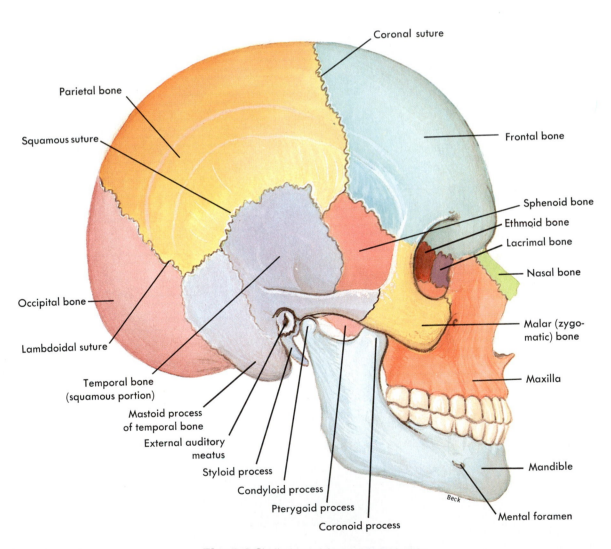

Fig. 5-8 Skull viewed from the right side.

The *ethmoid*, a complicated, irregular bone, lies anterior to the sphenoid but posterior to the nasal bones. It helps fashion the anterior part of the cranial floor (Fig. 5-9), the medial walls of the orbits (Figs. 5-7 and 5-12), the upper parts of the nasal septum (Fig. 5-7) and of the sidewalls of the nasal cavity (Figs. 5-11 and 5-17), and the part of the nasal roof perforated by small foramina through which olfactory nerve branches reach the brain. The lateral masses of the ethmoid bone are honeycombed with sinus spaces (Fig. 5-15). For more ethmoid bone markings see Table 5-2, p. 121.

The two *maxillae* together serve as the keystone in the architecture of the face just as the sphenoid bone acts as the keystone of the cranium. Nasal, malar, lacrimal, and palatine bones and the inferior nasal conchae all articu-

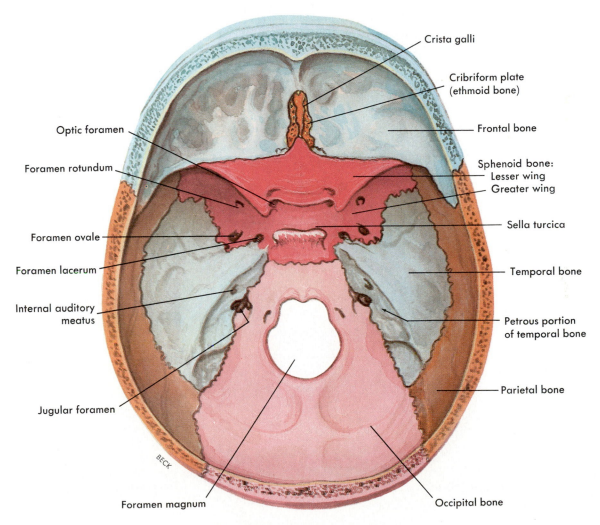

Optic foramen

Foramen rotundum

Foramen ovale

Foramen lacerum

Internal auditory meatus

Jugular foramen

BECK

Foramen magnum

Crista galli

Cribriform plate (ethmoid bone)

Frontal bone

Sphenoid bone:
Lesser wing
Greater wing

Sella turcica

Temporal bone

Petrous portion of temporal bone

Parietal bone

Occipital bone

Fig. 5-9 Floor of the cranial cavity.

late with a maxilla that articulates with the other maxilla in the midline. Of all the face bones, only the mandible does not articulate with the maxillae. The maxillae form part of the floor of the orbits, part of the roof of the mouth, and part of the floor and sidewalls of the nose. Each maxilla contains a mucosa-lined space, the maxillary sinus or *antrum of Highmore.* This sinus is the largest of the paranasal sinuses, that is, sinuses connected by channels to the nasal

cavity. For other markings of the maxillae, see Table 5-2, p. 121.

Unlike the upper jaw, which is formed by a pair of bones, the maxillae, the lower jaw, because of fusion of its halves during infancy, consists of a single bone, the *mandible.* It is the largest, strongest bone of the face. It articulates with the temporal bone in the only movable joint of the skull. Its major markings are identified in Table 5-2, p. 121.

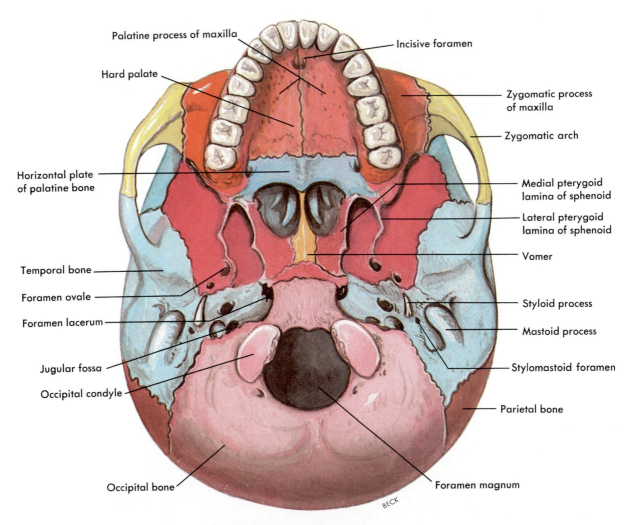

Fig. 5-10 Skull viewed from below.

The cheek is shaped by the underlying *zygomatic* or malar bone. This bone also forms the outer margin of the orbit and, with the zygomatic process of the temporal bone, makes the zygomatic arch. It articulates with four other face bones: the maxillae and the temporal, frontal, and sphenoid bones.

Shape is given to the nose by the two *nasal bones*, which form the upper part of the bridge of the nose, and by *cartilage*, which forms the lower part. Although small in size, the nasal bones enter into several articulations: with the perpendicular plate of the ethmoid bone, the cartilaginous part of the nasal septum, the frontal bone, the maxillae, and each other.

An almost paper-thin bone, shaped and sized about like a fingernail, lies just posterior and lateral to each nasal bone. It helps form the sidewall of the nasal cavity and the medial wall of the orbit. Because it contains a groove for the

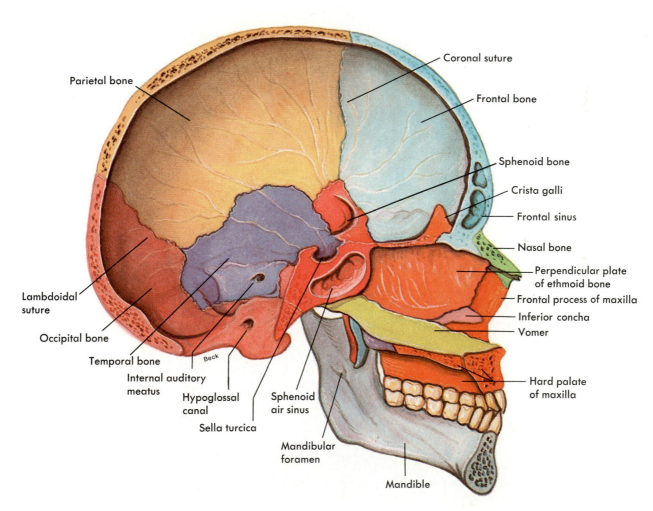

Fig. 5-11 Left half of skull viewed from within.

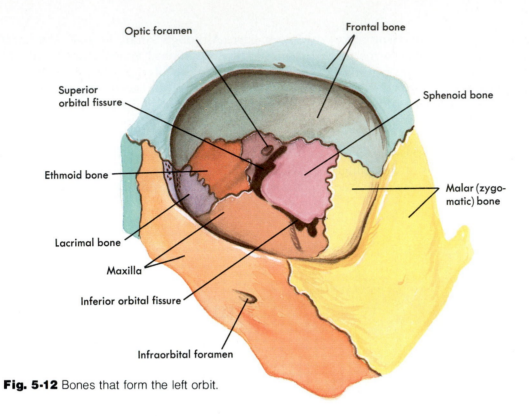

Fig. 5-12 Bones that form the left orbit.

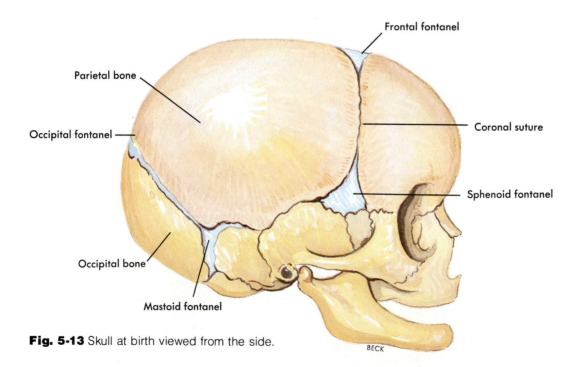

Fig. 5-13 Skull at birth viewed from the side.

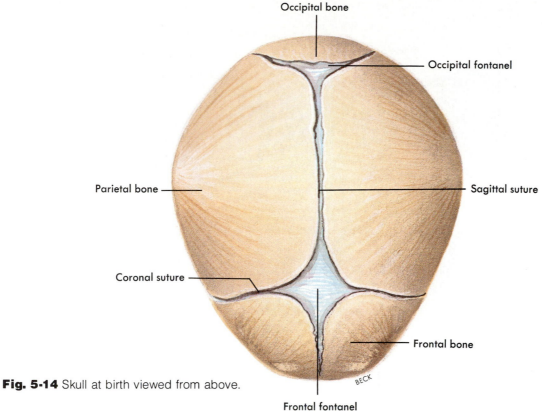

Fig. 5-14 Skull at birth viewed from above.

Occipital bone

Occipital fontanel

Parietal bone

Sagittal suture

Coronal suture

Frontal bone

Frontal fontanel

BECK

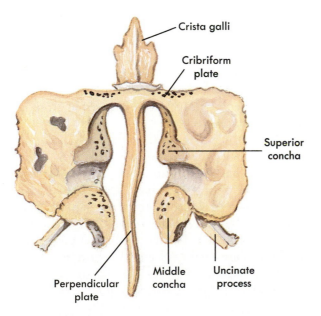

Crista galli

Cribriform plate

Superior concha

Perpendicular plate

Middle concha

Uncinate process

Fig. 5-15 Ethmoid bone viewed from behind.

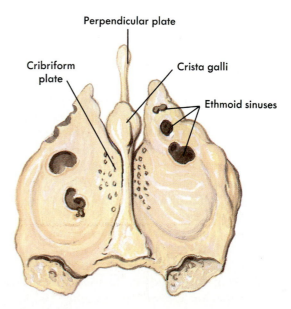

Perpendicular plate

Cribriform plate

Crista galli

Ethmoid sinuses

Fig. 5-16 Ethmoid bone viewed from above.

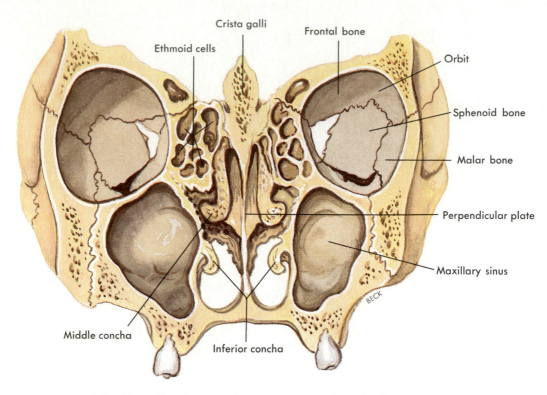

Crista galli

Ethmoid cells

Frontal bone

Orbit

Sphenoid bone

Malar bone

Perpendicular plate

Maxillary sinus

BECK

Middle concha

Inferior concha

Fig. 5-17 Skull in coronal section as seen from the front.

nasolacrimal (tear) duct, this bone is called the *lacrimal bone* (Fig. 5-12). It joins the maxilla, frontal bone, and ethmoid bone.

The two *palatine bones* join to each other in the midline like two L's facing each other. Their united horizontal portions form the posterior part of the hard palate (Fig. 5-10). The vertical portion of each palatine bone forms the lateral wall of the posterior part of each nasal cavity. The palatine bones articulate with the maxillae and the sphenoid bone.

There are two *inferior nasal conchae* (turbinates). Each one is scroll-like in shape and forms a kind of ledge projecting into the nasal cavity from its lateral wall. In each nasal cavity there are three of these ledges, formed, respectively, by the superior and middle conchae (which are projections of the ethmoid bone) and the in-

ferior concha (which is a separate bone). They are mucosa covered and divide each nasal cavity into three narrow, irregular channels, the *nasal meati*. The inferior nasal conchae form immovable joints with the ethmoid, lacrimal, maxillary, and palatine bones.

Two structures that enter into the formation of the nasal septum have already been mentioned—the perpendicular plate of the ethmoid bone and the septal cartilage. One other structure, the *vomer bone*, completes the septum posteriorly. It is usually described as being shaped like a ploughshare. It forms immovable joints with four bones: the sphenoid, ethmoid, palatine, and maxillae.

For a description of the *ear bones*, see Table 5-1, p. 117.

Special features of the skull include sutures,

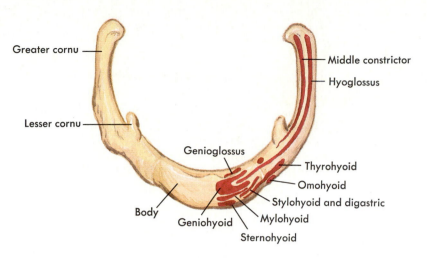

Fig. 5-18 Hyoid bone with muscle attachments.

fontanels, sinuses, orbits, nasal septum, and wormian bones and are described on p. 122.

Hyoid bone. The hyoid bone is a single bone in the neck—a part of the axial skeleton. Its U shape may be felt just above the larynx and below the mandible where it is suspended from the styloid processes of the temporal bones. Several muscles attach to the hyoid bone. Among them are an extrinsic tongue muscle (hyoglossus) and certain muscles of the floor of the mouth (mylohyoid, geniohyoid) (Fig. 5-18). The hyoid claims the distinction of being the only bone in the body that does not articulate with any other bone.

Vertebral column. The vertebral column constitutes the longitudinal axis of the skeleton. It is a flexible rather than a rigid column because it is segmented, that is, made up of 26 (typical in adult) separate bones called vertebrae, so joined to each other as to permit forward, backward, and sideways movement of the column. The head is balanced on top of this column, the ribs and viscera are suspended in front, the lower extremities are attached below, and the spinal cord is enclosed within. It is, indeed, the "backbone" of the body.

The 7 *cervical vertebrae* constitute the skeletal framework of the neck (Fig. 5-19). The next 12

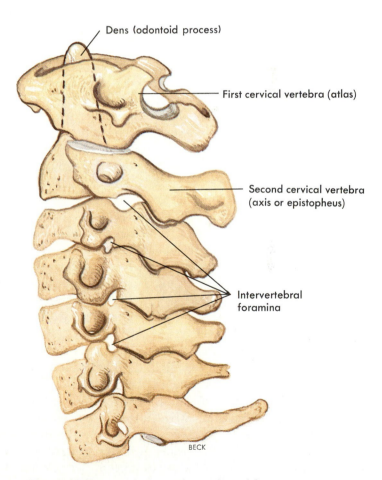

Fig. 5-19 The cervical vertebrae viewed from the side.

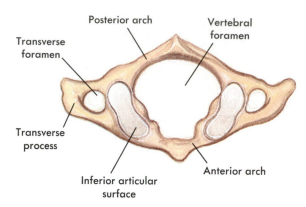

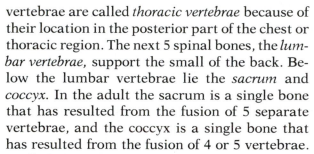

Fig. 5-20 First cervical vertebra (atlas) viewed from below.

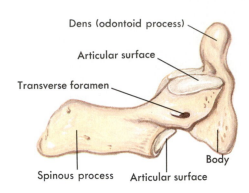

Fig. 5-21 Second cervical vertebra (axis) viewed from the side.

vertebrae are called *thoracic vertebrae* because of their location in the posterior part of the chest or thoracic region. The next 5 spinal bones, the *lumbar vertebrae,* support the small of the back. Below the lumbar vertebrae lie the *sacrum* and *coccyx.* In the adult the sacrum is a single bone that has resulted from the fusion of 5 separate vertebrae, and the coccyx is a single bone that has resulted from the fusion of 4 or 5 vertebrae.

All the vertebrae resemble each other in certain features and differ in others. For example, all except the first cervical vertebra have a flat, rounded mass placed anteriorly and centrally, known as the *body,* plus a sharp or blunt *spinous process* projecting inferiorly in the posterior midline and two transverse processes projecting laterally (Figs. 5-20 to 5-23). All but the sacrum and coccyx have a central opening, the *vertebral foramen.* An upward projection (the *dens*) from the body of the second cervical vertebra furnishes an axis for rotating the head. A long, blunt spinous process that can be felt at the back of the base of the neck characterizes the seventh cervical vertebra. Each thoracic vertebra has articular facets for the ribs. More detailed descriptions of separate vertebrae are given in Table 5-2. The vertebral column as a whole articulates with the

head, ribs, and hip bones. Individual vertebrae articulate with each other in joints between their bodies and between their articular processes. For a description of intervertebral joints, see Table 6-2, p. 145.

In order to increase the carrying strength of the vertebral column and to make balance possible in the upright position the vertebral column is curved. At birth there is a continuous posterior convexity from head to coccyx. Later, as the child learns to sit and stand, secondary posterior concavities necessary for balance develop in the cervical and lumbar regions. Not uncommonly, spinal curves deviate from the normal. For example, the lumbar curve frequently shows an exaggerated concavity *(lordosis),* whereas any of the regions may have a lateral curvature *(scoliosis).* The so-called hunchback is an exaggerated convexity in the thoracic region *(kyphosis).*

Sternum. The *medial part* of the anterior chest wall is supported by the *sternum,* a somewhat dagger-shaped bone consisting of three parts: the upper handle part, the *manubrium;* the middle blade part, the *body;* and a blunt cartilaginous lower tip, the *xiphoid process.* The latter ossifies during adult life. The manubrium articu-

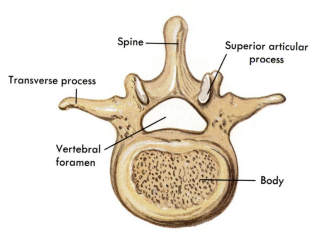

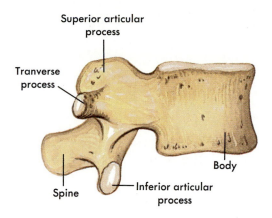

Fig. 5-22 Third lumbar vertebra viewed from above.

Fig. 5-23 Third lumbar vertebra viewed from the side.

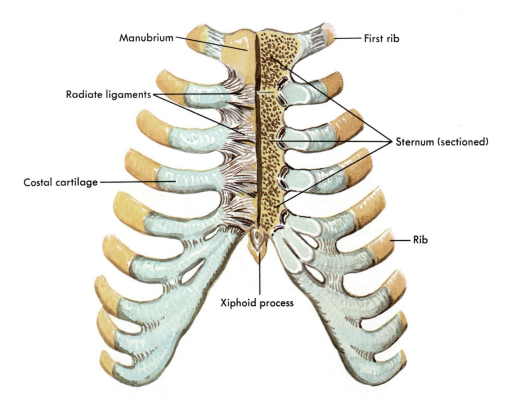

Fig. 5-24 The costal cartilages and their articulation with the sternum by radiate ligaments.

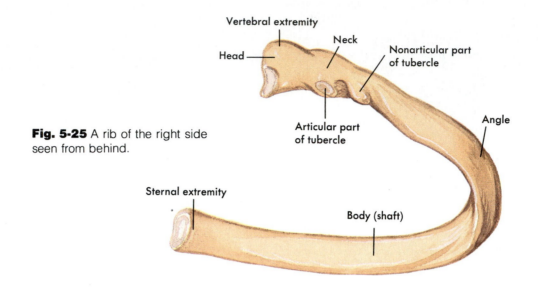

Fig. 5-25 A rib of the right side seen from behind.

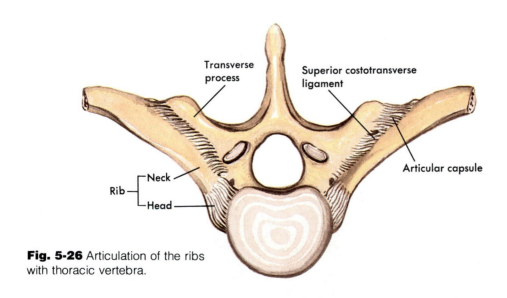

Fig. 5-26 Articulation of the ribs with thoracic vertebra.

lates with the clavicle and first rib, whereas the next nine ribs join the body of the sternum, either directly or indirectly, by means of the *costal cartilages* (Fig. 5-24).

Ribs. *Twelve pairs of ribs*, together with the vertebral column and sternum, form the bony cage known as the *thorax*. Each rib articulates with both the body and the transverse process of its corresponding thoracic vertebra. The head of each rib articulates with the body of the corresponding thoracic vertebra, and the tubercle of each rib articulates with the vertebra's transverse process (Fig. 5-25). In addition, the second through the ninth ribs articulate with the body of the vertebra above. From its vertebral attachment, each rib curves outward, then forward and downward (Fig. 5-26), a mechanical fact im-

portant for breathing. Anteriorly, each of the first seven ribs joins a costal cartilage that attaches it to the sternum. Each of the costal cartilages of the next three ribs, however, joins the cartilage of the rib above to be thus indirectly attached to the sternum. Because the eleventh and twelfth ribs do not attach even indirectly to the sternum, they are designated floating ribs.

Appendicular skeleton

Upper extremity. The upper extremity consists of the bones of the shoulder girdle, upper arm, lower arm, wrist, and hand. Two bones, the *clavicle* and *scapula*, compose the *shoulder girdle*. Contrary to appearances, this girdle forms only one bony joint with the trunk: the sternoclavicular joint between the sternum and clavicle. At

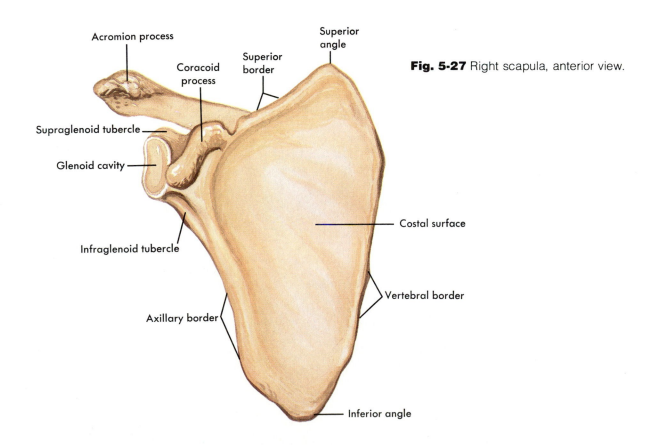

Fig. 5-27 Right scapula, anterior view.

its outer end the clavicle articulates with the scapula, which attaches to the ribs by muscles and tendons, not by a joint. All shoulder movements, therefore, involve the sternoclavicular joint. Various markings of the scapula are described in Table 5-2, p. 124 (see also Figs. 5-27 to 5-29).

The *humerus* or upper arm bone, like other long bones, consists of a shaft or diaphysis and two ends or epiphyses (Figs. 5-30 and 5-31). The upper epiphysis bears several identifying structures: the head, anatomical neck, greater and lesser tubercles, intertubercular groove, and surgical neck. On the diaphysis are found the deltoid tuberosity and the radial groove. The distal epiphysis has four projections—the medial and lateral epicondyles, the capitulum, and the trochlea—and two depressions—the olecranon and coronoid fossae. For descriptions of all of these markings, see Table 5-2. The humerus

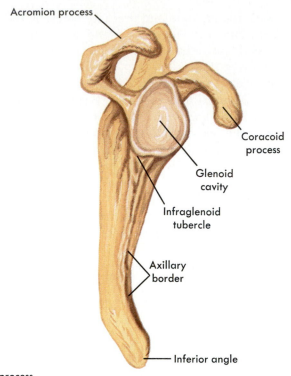

Fig. 5-28
Right scapula, lateral view.

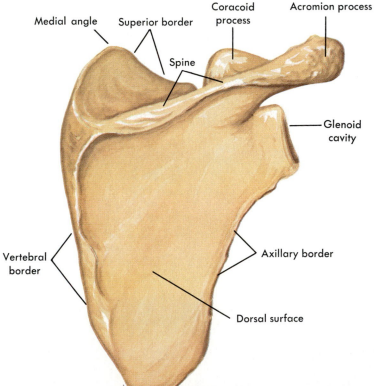

Fig. 5-29
Right scapula, posterior view.

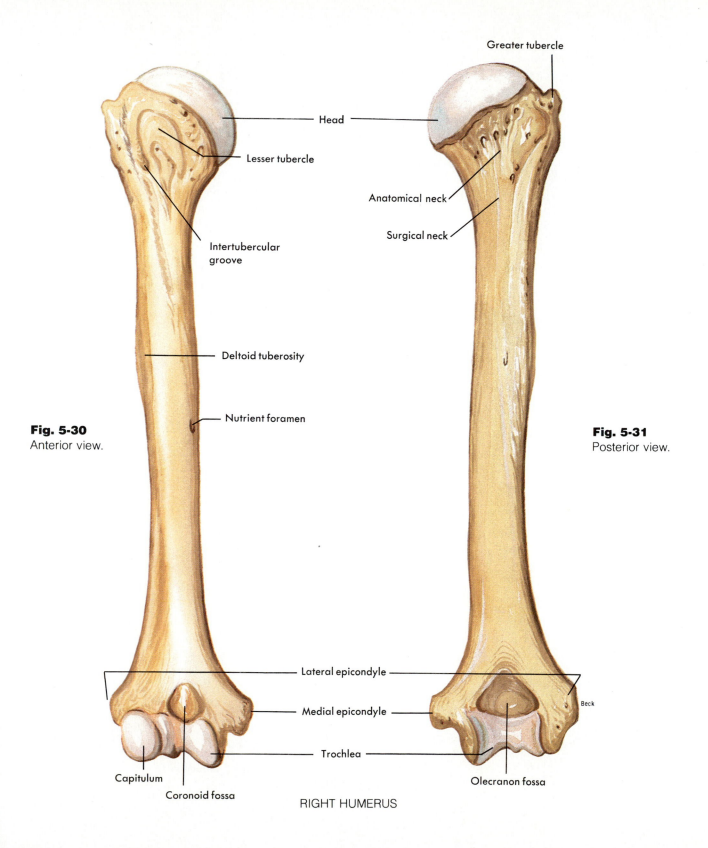

Greater tubercle

Head

Lesser tubercle

Anatomical neck

Surgical neck

Intertubercular
groove

Deltoid tuberosity

Nutrient foramen

Fig. 5-30
Anterior view.

Fig. 5-31
Posterior view.

Lateral epicondyle

Medial epicondyle

Trochlea

Beck

Capitulum

Coronoid fossa

Olecranon fossa

RIGHT HUMERUS

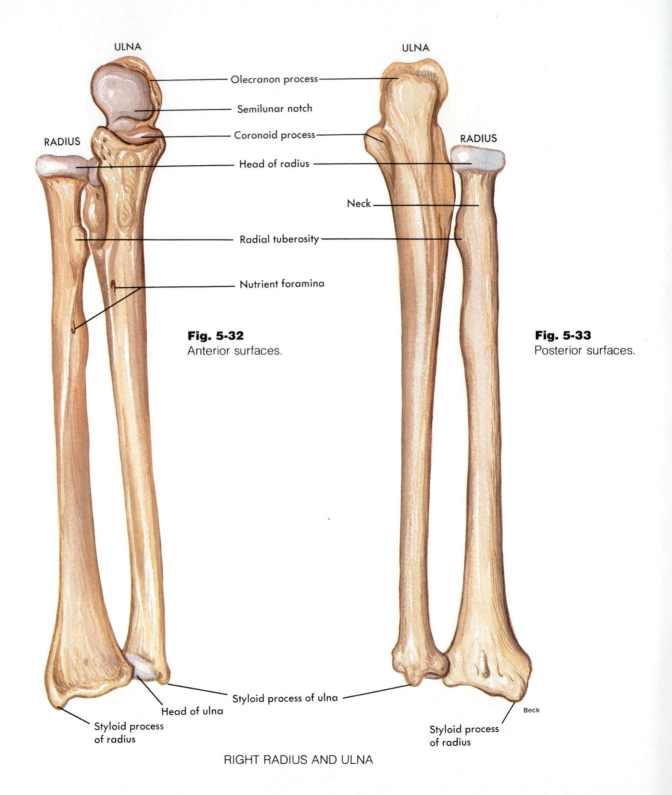

ULNA

Olecranon process

Semilunar notch

Coronoid process

RADIUS

Head of radius

Neck

Radial tuberosity

Nutrient foramina

Fig. 5-32
Anterior surfaces.

ULNA

RADIUS

Fig. 5-33
Posterior surfaces.

Styloid process of ulna

Head of ulna

Styloid process
of radius

Beck

Styloid process
of radius

RIGHT RADIUS AND ULNA

articulates proximally with the scapula and distally with both the radius and ulna.

Two bones form the framework for the lower arm: the *radius* on the thumb side and the *ulna* on the little finger side. At the proximal end of the ulna the olecranon process projects posteriorly and the coronoid process anteriorly. There are also two depressions:the semilunar notch on the anterior surface and the radial notch on the lateral surface. The distal end has two projections: a rounded head and a sharper styloid process. For more detailed identification of these markings, see Table 5-2 (p. 125). The ulna articulates proximally with the humerus and radius

and distally with a fibrocartilaginous disk but not with any of the carpal bones.

The radius has three projections: two at its proximal end, the head and radial tuberosity, and one at its distal end, the styloid process (Figs. 5-32 and 5-33). There are two proximal articulations: one with the capitulum of the humerus and the other with the radial notch of the ulna. The three distal articulations are with the scaphoid and lunate carpal bones and with the head of the ulna.

The eight *carpal bones* (Figs. 5-34 and 5-35) form what most people think of as the upper part of the hand but what, anatomically speaking, is

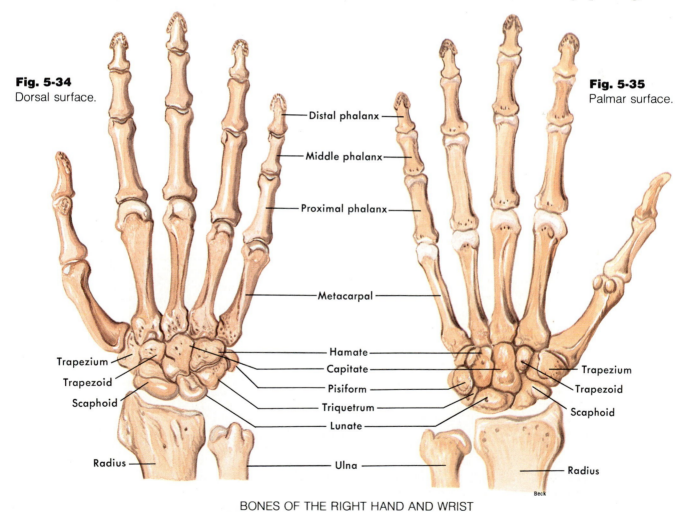

Fig. 5-34
Dorsal surface.

Fig. 5-35
Palmar surface.

Distal phalanx

Middle phalanx

Proximal phalanx

Metacarpal

Trapezium

Trapezoid

Scaphoid

Hamate
Capitate
Pisiform
Triquetrum
Lunate

Trapezium

Trapezoid

Scaphoid

Radius

Ulna

Radius

Beck

BONES OF THE RIGHT HAND AND WRIST

the wrist. Only one of these bones is evident from the outside, the *pisiform bone*, which projects posteriorly on the little finger side as a small rounded elevation. Ligaments bind the carpals closely and firmly together in two rows of four each: proximal row (from little finger toward thumb)—pisiform, triquetrum, lunate, and scaphoid bones; distal row—hamate, capitate, trapezoid, and trapezium bones. The joints between the carpals and radius permit wrist and hand movements.

Of the five *metacarpal bones* that form the framework of the hand, the thumb metacarpal forms the most freely movable joint with the carpals. This fact has great significance. Because of the wide range of movement possible between the thumb metacarpal and the trapezium, particularly the ability to oppose the thumb to the fingers, the human hand has much greater dexterity than the forepaw of any animal and has enabled man to manipulate his environment effectively. The heads of the metacarpals, prominent as the proximal knuckles of the hand, articulate with the phalanges.

Lower extremity. Bones of the hip, thigh, lower leg, ankle, and foot constitute the lower extremity. Strong ligaments bind each hip bone (*os coxae* or *os innominatum*) to the sacrum posteri-

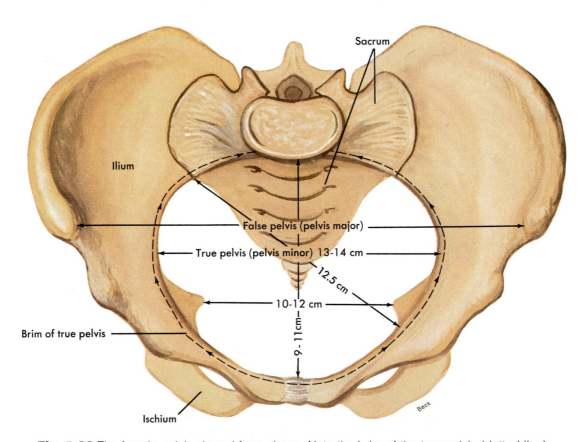

Fig. 5-36 The female pelvis viewed from above. Note the brim of the true pelvis (dotted line) that marks the boundary between the false pelvis (pelvis major) above and the true pelvis (pelvis minor) below it.

orly and to each other anteriorly to form the *pelvic girdle* (Figs. 5-36 to 5-38), a stable, circular base that supports the trunk and attaches the lower extremities to it. In early life, each innominate bone is made up of three separate bones. Later on, they fuse into a single, massive, irregular bone that is broader than any other bone in the body. The largest and uppermost of the three bones is the *ilium;* the strongest and lowermost, the *ischium;* and the anteriormost, the *pubis.* Numerous markings are present on the three bones. These are identified on pp. 125-126 (also see Fig. 5-39).

The two thigh bones or *femurs* have the dis-

tinction of being the longest and heaviest bones in the body. Several prominent markings characterize them. For example, three projections are conspicuous at each epiphysis: the head and greater and lesser trochanters proximally and the medial and lateral condyles and adductor tubercle distally (Fig. 5-40). Both condyles and the greater trochanter may be felt externally. For a description of the various femur markings, see p. 126.

The largest sesamoid bone in the body, and the one that is almost universally present, is the *patella* or kneecap, located in the tendon of the quadriceps femoris muscle as a projection to the

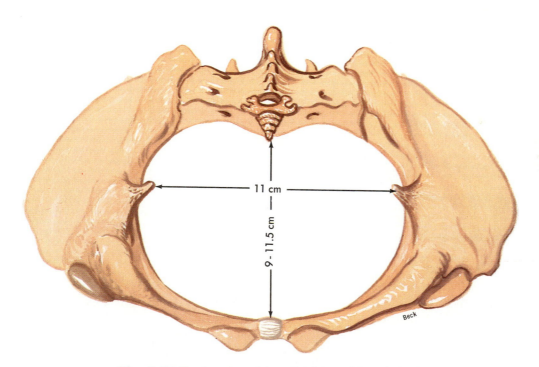

11 cm

9 - 11.5 cm

Beck

Fig. 5-37 The female pelvic outlet (viewed from below).

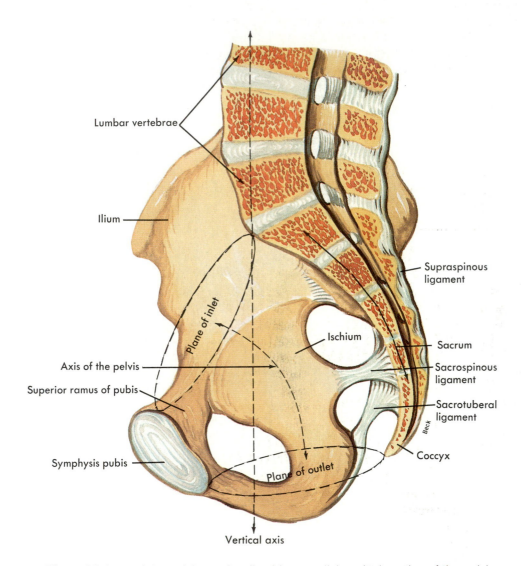

Lumbar vertebrae

Ilium

Supraspinous
ligament

Plane of inlet

Ischium

Axis of the pelvis

Sacrum

Sacrospinous
ligament

Superior ramus of pubis

Sacrotuberal
ligament

Beck

Symphysis pubis

Coccyx

Plane of outlet

Vertical axis

Fig. 5-38 Axes of the pelvis as visualized in a medial sagittal section of the pelvis.

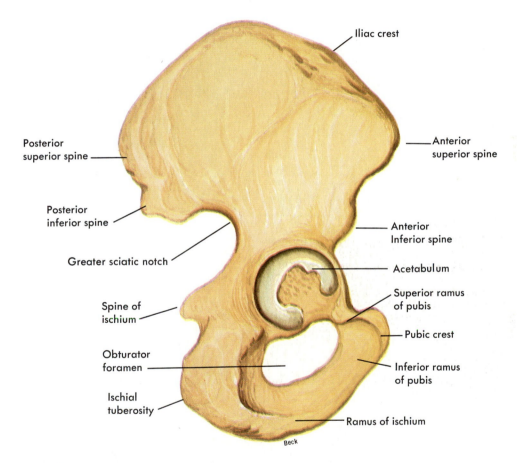

Iliac crest

Posterior
superior spine

Anterior
superior spine

Posterior
inferior spine

Greater sciatic notch

Anterior
Inferior spine

Acetabulum

Spine of
ischium

Superior ramus
of pubis

Pubic crest

Obturator
foramen

Inferior ramus
of pubis

Ischial
tuberosity

Ramus of ischium

Beck

Fig. 5-39 Right hip bone disarticulated from the skeleton viewed from the side with the
bone turned so as to look directly into the acetabulum.

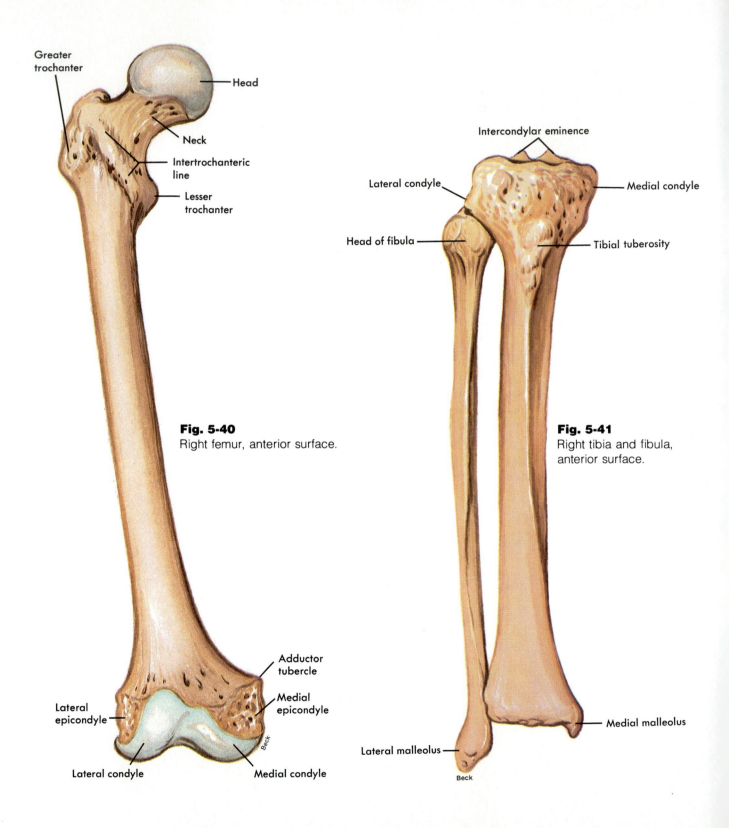

Fig. 5-40
Right femur, anterior surface.

Greater trochanter

Head

Neck

Intertrochanteric line

Lesser trochanter

Adductor tubercle

Medial epicondyle

Lateral epicondyle

Lateral condyle

Medial condyle

Beck

Fig. 5-41
Right tibia and fibula, anterior surface.

Intercondylar eminence

Lateral condyle

Medial condyle

Head of fibula

Tibial tuberosity

Medial malleolus

Lateral malleolus

Beck

underlying knee joint. Although some individuals have sesamoid bones in tendons of other muscles, lists of bone names do not include them because they are not always present, are not found in any particular tendons, and seemingly are of no importance. When the knee joint is extended, the patellar outline may be distinguished through the skin, but as the knee flexes, it sinks into the intercondylar notch of the femur and can no longer be delineated.

The *tibia* is the larger and stronger and the more medially and superficially located of the two lower leg bones. The *fibula* is smaller and more laterally and deeply placed. At its proximal end it articulates with the lateral condyle of the tibia. The proximal end of the tibia, in turn, articulates with the femur to form the knee joint, the largest and one of the most stable joints of the body. Distally the tibia articulates with the fibula and also with the talus. The latter fits into a boxlike socket (ankle joint) formed by the medial and lateral malleoli, projections of the tibia and fibula, respectively. For other tibial markings, see Table 5-2, p. 127, and Fig. 5-41.

Structure of the *foot* is similar to that of the hand with certain differences that adapt it for supporting weight. One example of this is the much greater solidity and the more limited mobility of the great toe compared to the thumb. Then, too, the foot bones are held together in such a way as to form springy lengthwise and crosswise arches. This is architecturally sound, since arches are known to furnish more supporting strength per given amount of structural material than any other type of construction. Hence the two-way arch construction makes a highly stable base. The longitudinal arch has an inner or medial portion and an outer or lateral portion, both of which are formed by the placement of tarsals and metatarsals. Specifically, some of the tarsals (calcaneus, talus, navicular, and cuneiform) and the first three metatarsals (starting with the great toe) form the medial longitudinal arch. The calcaneus and cuboid tarsals plus the fourth and fifth metatarsals shape the

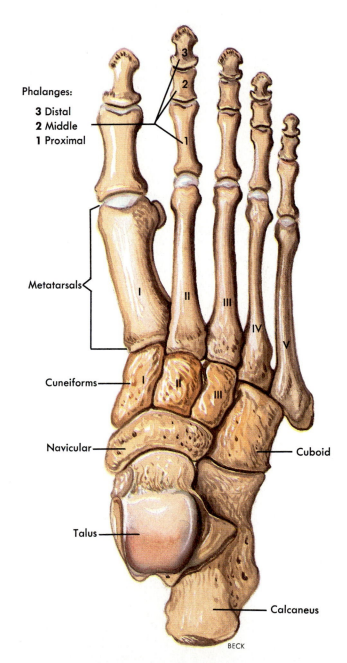

Phalanges:
3 Distal
2 Middle
1 Proximal

Metatarsals

Cuneiforms

Navicular

Talus

Cuboid

Calcaneus

BECK

Fig. 5-42 Bones of right foot viewed from above. Tarsal bones consist of cuneiforms, navicular, talus, cuboid, and calcaneus.

Fig. 5-43 Longitudinal arches of the foot. Medial formed by calcaneus, talus, navicular, cuneiforms, and three metatarsals; lateral formed by calcaneus, cuboid, and two lateral metatarsals.

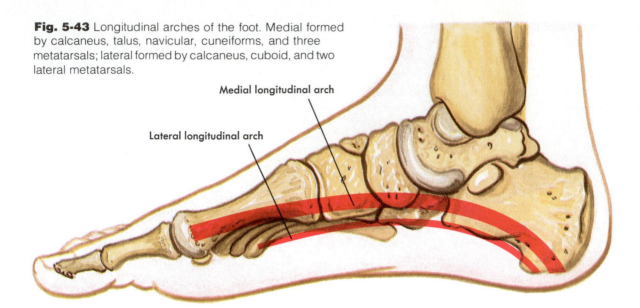

Medial longitudinal arch

Lateral longitudinal arch

Fig. 5-44 Transverse arch in the tarsal and metatarsal region of the right foot (phalanges removed).

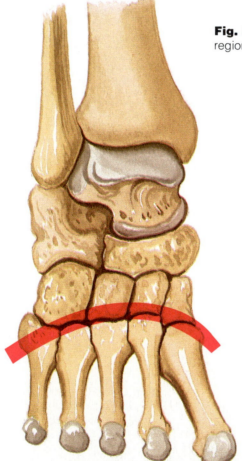

lateral longitudinal arch (Figs. 5-42 to 5-44). The transverse arch results from the relative placement of the distal row of tarsals and the five metatarsals. (See Table 5-2 for specific bones of different arches.) Strong ligaments and leg muscle tendons normally hold the foot bones firmly in their arched positions, but not infrequently these weaken, causing the arches to flatten, a condition aptly called fallen arches or flatfeet (Fig. 5-45). Look at Fig. 5-46 to see what high heels do to the position of the foot. They give a forward thrust to the body, which forces an undue amount of weight on the heads of the metatarsals. Normally the tarsals and metatarsals play the major role in the functioning of the foot as a supporting structure, with the phalanges relatively unimportant. The reverse is true for the hand. Here, manipulation is the main function rather than support. Consequently, the phalanges are all important, and the carpals and metacarpals are subsidiary.

Fig. 5-45 Flatfoot results when there is a weakening of tendons and ligaments attached to the tarsal bones. Downward pressure by the weight of the body gradually flattens out the normal arch of bones.

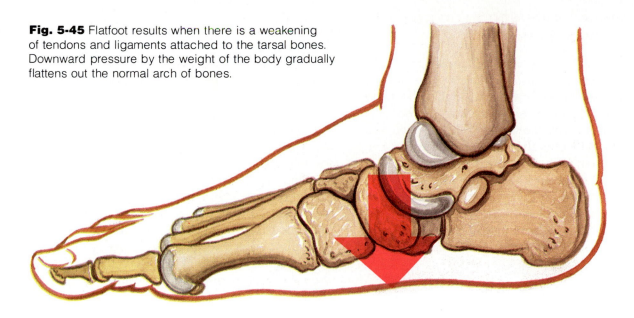

Fig. 5-46 High heels throw the weight forward, causing the heads of the metatarsals to bear most of the body's weight.

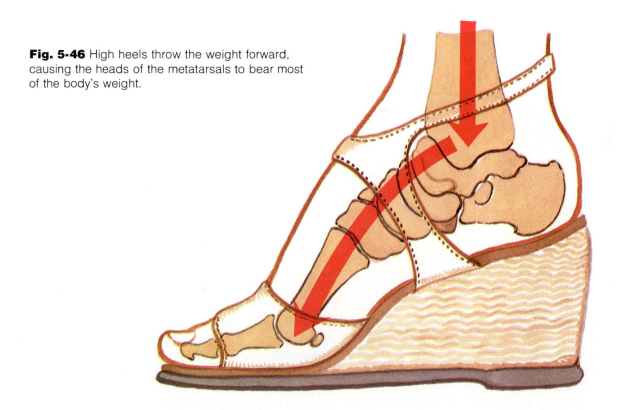

Differences between male and female skeletons

Both general and specific differences exist between male and female skeletons. The general difference is one of size and weight, the male skeleton being larger and heavier. The specific differences concern the shape of the pelvic bones and cavity. Whereas the male pelvis is deep and funnel shaped, with a narrow pubic arch (usually less than 90 degrees), the female pelvis, as Figs. 5-37 to 5-39 show, is shallow, broad, and flaring, with a wider pubic arch (usually greater than 90 degrees). The childbearing function obviously explains the necessity for these and certain other modifications of the female pelvis.

Age changes in skeleton

Changes in the skeleton as a whole from infancy to adulthood are mainly changes in the size of the bones and in the proportionate sizes between different bones. Changes in individual bones from young adulthood to old age, on the other hand, are mainly a matter of changes in the extent of calcification, in the texture, and in the contour of the margins and bone markings. Some of the major changes in the skeleton from infancy to young adulthood are as follows:

1 The head becomes proportionately smaller. Whereas the infant head is approximately one fourth the total height of the body, the adult head is only about one eighth the total height.

2 The thorax changes shape, roughly speaking, from round to elliptical.

3 The pelvis becomes relatively larger and in the female relatively wider.

4 The legs become proportionately longer and the trunk proportionately shorter.

5 The vertebral column develops two curves not present at birth—the cervical curve when the infant starts lifting up his head (at about 3 months of age) and the lumbar curve when the child begins standing (toward the end of the first year). Both of these secondary curves are con-cave posteriorly, whereas the primary thoracic and sacral curves are convex posteriorly.

6 The cranium shows several modifications. It grows rapidly during early childhood, enlarging its capacity from approximately 360 ml at birth to approximately 1,500 ml (about adult size) by 6 years of age. The fontanels close by about 1½ or 2 years of age, and the sutures begin to fuse in the twenties.

7 The facial bones also show several changes between infancy and adulthood. Unlike the cranial bones, their growth is slow during early childhood but rapid during the teens. Whereas the infant face compared with the entire skull bears the relationship of 1 : 8, the adult face bears the relationship of 1 : 2 to the adult skull. The sinuses are much larger in the adult. For example, at birth, only rudimentary maxillary and mastoid sinuses exist. The ethmoid and sphenoid sinuses start to appear at about 6 years of age and the frontal at about 7 years. All of the bony sinuses, but especially the frontal, grow rapidly during adolescence.

8 The epiphyses of the long bones are composed of cartilage at birth but become completely ossified (except for the thin layer of articular cartilage) by adulthood. Demonstration on x-ray films of epiphyseal cartilage between epiphyses and diaphysis indicates that skeletal growth has not ceased.

Changes in the skeleton continue to occur from adulthood to old age. Many of these changes stem from the fact that during this period of life bone is lost at the inner or endosteal surfaces and gained at the outer or periosteal surfaces of bone. Consequently, bone margins and projections do not look the same in old bones and young bones. Instead of clean-cut, distinct margins, old bones characteristically have indistinct, shaggy-appearing margins (marginal lipping and spurs)—a regrettable change because the restricted movements of old age stem partly from this piling up of bone around joint margins. Also, an increase in bone along various projections develops in old age, making ridges and processes more pronounced.

Table 5-1 Bones of skeleton

Axial skeleton (80 bones)			Bones that form upright axis on body—skull, hyoid, vertebral column, sternum, and ribs
Part of body	**Name of bone**	**Number**	**Identification**
Skull (28 bones)			
Cranium (8 bones)			Cranium forms floor for brain to rest on and helmetlike covering over it
	Frontal	1	Forehead bone; also forms most of roof of orbits (eye sockets) and anterior part of cranial floor
	Parietal	2	Prominent, bulging bones behind frontal bone; form topsides of cranial cavity
	Temporal	2	Form lower sides of cranium and part of cranial floor; contain middle and inner ear structures
	Occipital	1	Forms posterior part of cranial floor and walls
	Sphenoid	1	Keystone of cranial floor; forms its midportion; resembles bat with wings outstretched and legs extended downward posteriorly; lies behind and slightly above nose and throat; forms part of floor and sidewalls of orbit
	Ethmoid	1	Complicated irregular bone that helps make up anterior portion of cranial floor, medial wall of orbits, upper parts of nasal septum, and sidewalls and part of nasal roof; lies anterior to sphenoid and posterior to nasal bones
Face (14 bones)	Nasal	2	Small bones forming upper part of bridge of nose
	Maxillary	2	Upper jaw bones; form part of floor of orbit, anterior part of roof of mouth, and floor of nose and part of sidewalls of nose
	Zygomatic (malar)	2	Cheekbones; form part of floor and sidewall of orbit
	Mandible	1	Lower jawbone; largest, strongest bone of face
	Lacrimal	2	Thin bones about size and shape of fingernail; posterior and lateral to nasal bones in medial wall of orbit; help form sidewall of nasal cavity, often missing in dry skull
	Palatine	2	Form posterior part of hard palate, floor, and part of sidewalls of nasal cavity and floor of orbit
	Inferior conchae (turbinates)	2	Thin scroll of bone forming kind of shell along inner surface of sidewall of nasal cavity; lies above roof of mouth
	Vomer	1	Forms lower and posterior part of nasal septum; shaped like ploughshare
Ear bones (6 bones)	Malleus (hammer)	2	Tiny bones referred to as auditory ossicles in middle ear cavity in temporal bones; resemble, respectively, miniature hammer, anvil, and stirrup
	Incus (anvil)	2	
	Stapes (stirrup)	2	

Continued.

Table 5-1 Bones of skeleton—cont'd

Axial skeleton (cont'd)

Part of body	Name of bone	Number	Identification
Hyoid bone		1	U-shaped bone in neck between mandible and upper part of larynx; claims distinction as only bone in body not forming a joint with any other bone; suspended by ligaments from styloid processes of temporal bones
Vertebral column (26 bones)			Not actually a column but a flexible segmented rod shaped like an elongated letter S; forms axis of body; head balanced above, ribs and viscera suspended in front, and lower extremities attached below; encloses spinal cord
	Cervical vertebrae	7	First or upper 7 vertebrae
	Thoracic vertebrae	12	Next 12 vertebrae; 12 pairs of ribs attached to these
	Lumbar vertebrae	5	Next 5 vertebrae
	Sacrum	1	Five separate vertebrae until about 25 years of age; then fused to form 1 wedge-shaped bone
	Coccyx	1	Four or 5 separate vertebrae in child but fused into 1 in adult
Sternum and ribs (25 bones)			Sternum, ribs, and thoracic vertebrae together form bony cage known as *thorax;* ribs attach posteriorly to vertebrae, slant downward anteriorly to attach to sternum (see description of false ribs below)
	Sternum	1	Breastbone; flat dagger-shaped bone
	True ribs	7 pairs	Upper 7 pairs; fasten to sternum by costal cartilages
	False ribs	5 pairs	False ribs do not attach to sternum directly; upper 3 pairs of false ribs attach by means of costal cartilage of seventh ribs; last 2 pairs do not attach to sternum at all; therefore called *"floating"*

Appendicular skeleton (126 bones)

Bones that are appended to axial skeleton: upper and lower extremities, including shoulder and hip girdles

Part of body	Name of bone	Number	Identification
Upper extremities (including shoulder girdle) (64 bones)	Clavicle	2	Collar bones; shoulder girdle joined to axial skeleton by articulation of clavicles with sternum; scapula does not form joint with axial skeleton
	Scapula	2	Shoulder blades; scapulae and clavicles together comprise shoulder girdle
	Humerus	2	Long bone of upper arm
	Radius	2	Bone of thumb side of forearm
	Ulna	2	Bone of little finger side of forearm; longer than radius
	Carpals (scaphoid, lunate, triquetrum, pisiform, trapezium, trapezoid, capitate, and hamate)	16	Arranged in two rows at proximal end of hand (Figs. 5-30 and 5-35)
	Metacarpals	10	Long bones forming framework of palm of hand
	Phalanges	28	Miniature long bones of fingers, 3 in each finger, 2 in each thumb
Lower extremities (62 bones)	Ossa coxae or innominate bones	2	Large hip bones; with sacrum and coccyx, these 3 bones form basin-like pelvic cavity; lower extremities attached to axial skeleton by pelvic bones
	Femur	2	Thigh bone; largest strongest bone of body
	Patella	2	Kneecap; largest sesamoid bone of body*; embedded in tendon of quadriceps femoris muscle
	Tibia	2	Shin bone
	Fibula	2	Long, slender bone of lateral side of lower leg
	Tarsals (calcaneus, talus, navicular, first, second, and third cuneiforms, cuboid)	14	Bones that form heel and proximal or posterior half of foot (Fig. 5-42)
	Metatarsals	10	Long bones of feet
	Phalanges	28	Miniature long bones of toes; 2 in each great toe, 3 in other toes
Total		206*	

*An inconstant number of small, flat, round bones known as *sesamoid bones* (because of their resemblance to sesame seeds) is found in various tendons in which considerable pressure develops. Because the number of these bones varies greatly between individuals, only 2 of them, the patellae, have been counted among the 206 bones of the body. Generally, 2 of them can be found in each thumb (in flexor tendon near metacarpophalangeal and interphalangeal joints) and great toe plus several others in the upper and lower extremities. *Wormian bones,* the small islets of bone frequently found in some of the cranial sutures, have not been counted in this list of 206 bones either because of their variable occurrence.

Table 5-2 Identification of bone markings

Frontal

Marking	Description
■ Supraorbital margin	Arched ridge just below eyebrow
■ Frontal sinuses	Cavities inside bone just above supraorbital margin; lined with mucosa; contain air
Frontal tuberosities	Bulge above each orbit; most prominent part of forehead
Superciliary arches	Ridges caused by projection of frontal sinuses; eyebrows lie over these ridges
Supraorbital notch (sometimes foramen)	Notch or foramen in supraorbital margin slightly mesial to its midpoint; transmits supraorbital nerve and blood vessels
Glabella	Smooth area between superciliary ridges and above nose

Sphenoid

Marking	Description
■ Body	Hollow, cubelike central portion
■ Greater wings	Lateral projections from body; form part of outer wall of orbit
■ Lesser wings	Thin, triangular projections from upper part of sphenoid body; form posterior part of roof of orbit
■ Sella turcica (or *Turk's saddle*)	Saddle-shaped depression on upper surface of sphenoid body; contains pituitary gland
■ Sphenoid sinuses	Irregular air-filled mucosa-lined spaces within central part of sphenoid
Pterygoid processes	Downward projections on either side where body and greater wing unite; comparable to extended legs of bat if entire bone is likened to this animal; form part of lateral nasal wall
■ Optic foramen	Opening into orbit at root of lesser wing; transmits second cranial nerve
Superior orbital fissure	Slitlike opening into orbit; lateral to optic foramen; transmits third, fourth, and part of fifth cranial nerves
Foramen rotundum	Opening in greater wing that transmits maxillary division of fifth cranial nerve
Foramen ovale	Opening in a greater wing that transmits mandibular division of fifth cranial nerve

Temporal

Marking	Description
■ Mastoid process	Protuberance just behind ear
■ Mastoid air cells	Air-filled mucosa-lined spaces within mastoid process
■ External auditory meatus (or canal)	Opening into ear and tube extending into temporal bone
■ Zygomatic process	Projection that articulates with malar (or zygomatic) bone
■ Internal auditory meatus	Fairly large opening on posterior surface of petrous portion of bone; transmits eighth cranial nerve to inner ear and seventh cranial nerve on its way to facial structures
Squamous portion	Thin, flaring upper part of bone
Mastoid portion	Rough-surfaced lower part of bone posterior to external auditory meatus
Petrous portion	Wedge-shaped process that forms part of center section of cranial floor between sphenoid and occipital bones; name derived from Greek word for stone because of extreme hardness of this process; houses middle and inner ear structures
■ Mandibular fossa	Oval-shaped depression anterior to external auditory meatus; forms socket for condyle of mandible
■ Styloid process	Slender spike of bone extending downward and forward from undersurface of bone anterior to mastoid process; often broken off in dry skull; several neck muscles and ligaments attach to styloid process
Stylomastoid foramen	Opening between styloid and mastoid processes where facial nerve emerges from cranial cavity
Jugular fossa	Depression on undersurface of petrous portion; dilated beginning of internal jugular vein lodged here
Jugular foramen	Opening in suture between petrous portion and occipital bone; transmits lateral sinus and ninth, tenth, and eleventh cranial nerves
Carotid canal (or foramen)	Channel in petrous portion; best seen from undersurface of skull; transmits internal carotid artery

■ Those that seem particularly important because they are places of muscle attachments.

Occipital

Marking	Description
■ Foramen magnum	Hole through which spinal cord enters cranial cavity
■ Condyles	Convex, oval processes on either side of foramen magnum; articulate with depressions on first cervical vertebra
External occipital protuberance	Prominent projection on posterior surface in midline short distance above foramen magnum; can be felt as definite bump
Superior nuchal line	Curved ridge extending laterally from external occipital protuberance
Inferior nuchal line	Less well-defined ridge paralleling superior nuchal line short distance below it
Internal occipital protuberance	Projection in midline on inner surface of bone; grooves for lateral sinuses extend laterally from this process and one for sagittal sinus extends upward from it

Ethmoid

Marking	Description
■ Horizontal (cribriform) plate	Olfactory nerves pass through numerous holes in this plate
■ Crista galli	See Figs. 5-15 and 5-17; meninges attach to this process
■ Perpendicular plate	Forms upper part of nasal septum (Figs. 5-15 and 5-17)
■ Ethmoid sinuses	Honeycombed, mucosa-lined air spaces within lateral masses of bone (Figs. 5-16 and 5-17)
■ Superior and middle turbinates (conchae)	Help to form lateral walls of nose (Figs. 5-15 and 5-17)
Lateral masses	Compose sides of bone; contain many air spaces (ethmoid cells or sinuses); inner surface forms superior and middle conchae

Palatine

Horizontal plate	Joined to palatine processes of maxillae to complete part of hard palate

Mandible

Marking	Description
Body	Main part of bone; forms chin
Ramus	Process, one on either side, that projects upward from posterior part of body
■ Condyle (or head)	Part of each ramus that articulates with mandibular fossa of temporal bone
Neck	Constricted part just below condyles
■ Alveolar process	Teeth set into this arch
■ Mandibular foramen	Opening on inner surface of ramus; transmits nerves and vessels to lower teeth
■ Mental foramen	Opening on outer surface below space between two bicuspids; transmits terminal branches of nerves and vessels that enter bone through mandibular foramen; dentists inject anesthetics through these foramina
Coronoid process	Projection upward from anterior part of each ramus; temporal muscle inserts here
Angle	Juncture of posterior and inferior margins of ramus

Maxilla

Marking	Description
■ Alveolar process	Arch containing teeth
■ Maxillary sinus or antrum of Highmore	Large air-filled mucosa-lined cavity within body of each maxilla; largest of sinuses
■ Palatine process	Horizontal inward projection from alveolar process; forms anterior and larger part of hard palate
Infraorbital foramen	Hole on external surface just below orbit; transmits vessels and nerves
Lacrimal groove	Groove on inner surface; joined by similar groove on lacrimal bone to form canal housing nasolacrimal duct

Continued.

Table 5-2 Identification of bone markings—cont'd

Special features of skull

Marking	Description
■ Sutures (Fig. 5-8)	Immovable joints between skull bones
1 Sagittal	**1** Line of articulation between two parietal bones
2 Coronal	**2** Joint between parietal bones and frontal bone
3 Lambdoidal	**3** Joint between parietal bones and occipital bone
■ Fontanels (Figs. 5-13 and 5-14)	"Soft spots" where ossification incomplete at birth; allow some compression of skull during birth; also important in determining position of head before delivery; 6 such areas located at angles of parietal bones
1 Anterior (or frontal)	**1** At intersection of sagittal and coronal sutures (juncture of parietal bones and frontal bone); diamond shaped; largest of fontanels; usually closed by 1½ years of age
2 Posterior (or occipital)	**2** At intersection of sagittal and lambdoidal sutures (juncture of parietal bones and occipital bone); triangular in shape; usually closed by second month
3 Anterolateral (or sphenoid)	**3** At juncture of frontal, parietal, temporal, and sphenoid bones
4 Posterolateral (or mastoid)	**4** At juncture of parietal, occipital, and temporal bones; usually closed by second year

Marking	Description
■ Sinuses	
1 Air (or bony)	**1** Spaces or cavities within bones; those that communicate with nose called *paranasal sinuses* (frontal, sphenoidal, ethmoidal, and maxillary); mastoid cells communicate with middle ear rather than nose, therefore not included among paranasal sinuses
2 Blood	**2** Veins within cranial cavity (Figs. 14-19 and 14-20, p. 391)
■ Orbits (Fig. 5-12) formed by	
1 Frontal	**1** Roof of orbit
2 Ethmoid	**2** Medial wall
3 Lacrimal	**3** Medial wall
4 Sphenoid	**4** Lateral wall
5 Zygomatic	**5** Lateral wall
6 Maxillary	**6** Floor
7 Palatine	**7** Floor
■ Nasal septum (Fig. 5-11) formed by	Partition in midline of nasal cavity; separates cavity into right and left halves
1 Perpendicular plate of ethmoid bone	**1** Forms upper part of septum
2 Vomer bone	**2** Forms lower, posterior part
3 Cartilage	**3** Forms anterior part
■ Wormian bones	Small islands of bones within suture

■ Those that seem particularly important because they are places of muscle attachments.

Vertebral column

Marking	Description
General features	Anterior part of vertebrae (except first two cervical) consists of body; posterior part of neural arch, which, in turn, consists of 2 pedicles, 2 laminae, and 7 processes projecting from laminae
■ Thoracic vertebrae	
1 Body	1 Main part; flat, round mass located anteriorly; supporting or weight-bearing part of vertebra
2 Pedicles	2 Short projections extending posteriorly from body
3 Laminae	3 Posterior part of vertebra to which pedicles join and from which processes project
4 Neural arch	4 Formed by pedicles and laminae; protects spinal cord posteriorly; congenital absence of one or more neural arches known as *spina bifida* (cord may protrude right through skin)
5 Spinous process	5 Sharp process projecting inferiorly from laminae in midline
6 Transverse processes	6 Right and left lateral projections from laminae
7 Superior articulating processes	7 Project upward from laminae
8 Inferior articulating processes	8 Project downward from laminae; articulate with superior articulating processes of vertebrae below
9 Spinal foramen	9 Hole in center of vertebra formed by union of body, pedicles, and laminae; spinal foramina, when vertebrae superimposed one on other, form spinal cavity that houses spinal cord
■ Cervical vertebrae (Figs. 5-19 to 5-21)	
1 General features	1 Foramen in each transverse process for transmission of vertebral artery, vein, and plexus of nerves; short bifurcated spinous processes except on seventh vertebrae, where it is extra long and may be felt as protrusion when head bent forward; bodies of these vertebrae small, whereas spinal foramina large and triangular
2 Atlas	2 First cervical vertebra; lacks body and spinous process; superior articulating processes concave ovals that act as rockerlike cradles for condyles of occipital bone; named atlas because sup-

Marking	Description
	ports head as Atlas was thought to have supported world (Fig. 5-20)
3 Axis (epistropheus)	3 Second cervical vertebra, so named because atlas rotates about this bone in rotating movements of head; *dens,* or odontoid process, peglike projection upward from body of axis, forming pivot for rotation of atlas (Fig. 5-21)
■ Lumbar vertebrae (Figs. 5-22 and 5-23)	Strong, massive; superior articulating processes directed inward instead of upward; inferior articulating processes, outward instead of downward; short, blunt spinous process (Fig. 5-22)
■ Sacral promontory	Protuberance from anterior, upper border of sacrum into pelvis; of obstetrical importance because its size limits anteroposterior diameter of pelvic inlet
■ Intervertebral foramina	Opening between vertebrae through which spinal nerves emerge
■ Curves	Curves have great structural importance because increase carrying strength of vertebral column, make balance possible in upright position (if column were straight, weight of viscera would pull body forward), absorb jars from walking (straight column would transmit jars straight to head), and protect column from fracture
1 Primary	1 Column curves at birth from head to sacrum with convexity posteriorly; after child stands, convexity persists only in *thoracic* and *sacral* regions, which, therefore, are called primary curves
2 Secondary	2 Concavities in *cervical* and *lumbar* regions; cervical concavity results from infant's attempts to hold head erect (3 to 4 months); lumbar concavity, from balancing efforts in learning to walk (10 to 18 months)
3 Abnormal	3 *Kyphosis,* exaggerated convexity in thoracic region (hunchback); *lordosis,* exaggerated concavity in lumbar region, a very common condition; *scoliosis,* lateral curvature in any region

Continued.

Table 5-2 Identification of bone markings—cont'd

Sternum (Fig. 5-24)

Marking	Description
■ Body	Main central part of bone
■ Manubrium	Flaring, upper part
■ Xiphoid process	Projection of cartilage at lower border of bone

Ribs (Figs. 5-25 and 5-26)

Marking	Description
■ Head	Projection at posterior end of rib; articulates with corresponding thoracic vertebra and one above, except last three pairs, which join corresponding vertebrae only
Neck	Constricted portion just below head
Tubercle	Small knob just below neck; articulates with transverse process of corresponding thoracic vertebra; missing in lowest 3 ribs
Body or shaft	Main part of rib
■ Costal cartilage	Cartilage at sternal end of true ribs; attaches ribs (except floating ribs) to sternum

Scapula (Figs. 5-27 to 5-29)

Marking	Description
■ Borders 1 Superior 2 Vertebral 3 Axillary	1 Upper margin 2 Margin toward vertebral column 3 Lateral margin
■ Spine	Sharp ridge running diagonally across posterior surface of shoulder blade
■ Acromion process	Slightly flaring projection at lateral end of scapular spine; may be felt as tip of shoulder; articulates with clavicle
■ Coracoid process	Projection on anterior surface from upper border of bone; may be felt in groove between deltoid and pectoralis major muscles, about 1 inch below clavicle
■ Glenoid cavity	Arm socket

Humerus (Figs. 5-30 and 5-31)

Marking	Description
■ Head	Smooth, hemispherical enlargement at proximal end of humerus
Anatomical neck	Oblique groove just below head
Greater tubercle	Rounded projection lateral to head on anterior surface
Lesser tubercle	Prominent projection on anterior surface just below anatomical neck
Intertubercular	Deep groove between greater and lesser tubercles; long tendon of biceps muscle lodges here
Surgical neck	Region just below tubercles; so named because of its liability to fracture
Deltoid tuberosity	V-shaped, rough area about midway down shaft where deltoid muscle inserts
Radial groove	Groove running obliquely downward from deltoid tuberosity; lodges radial nerve
■ Epicondyles (medial and lateral)	Rough projections at both sides of distal end
■ Capitulum	Rounded knob below lateral epicondyle; articulates with radius; sometimes called radial head of humerus
■ Trochlea	Projection with deep depression through center similar to shape of pulley; articulates with ulna
■ Olecranon fossa	Depression on posterior surface just above trochlea; receives olecranon process of ulna when lower arm extends
■ Coronoid fossa	Depression on anterior surface above trochlea; receives coronoid process of ulna in flexion of lower arm

■ Those that seem particularly important because they are places of muscle attachments.

Ulna (Figs. 5-32 and 5-33)

Marking	Description
■ Olecranon process	Elbow
■ Coronoid process	Projection on anterior surface of proximal end of ulna; trochlea of humerus fits snugly between olecranon and coronoid processes
■ Semilunar notch	Curved notch between olecranon and coronoid process, into which trochlea fits
■ Radial notch	Curved notch lateral and inferior to semilunar notch; head of radius fits into this concavity
Head	Rounded process at distal end; does not articulate with wrist bones but with fibrocartilaginous disk
Styloid process	Sharp protuberance at distal end; can be seen from outside on posterior surface

Radius (Figs. 5-32 and 5-33)

Marking	Description
■ Head	Disk-shaped process forming proximal end of radius; articulates with capitulum of humerus and with radial notch of ulna
Radial tuberosity	Roughened projection on ulnar side, short distance below head; biceps muscle inserts here
Styloid process	Protuberance at distal end on lateral surface (with forearm supinated as in anatomical position)

Os coxae (Figs. 5-36 to 5-39)

Marking	Description
■ Ilium	Upper, flaring portion
■ Ischium	Lower, posterior portion
■ Pubic bone or pubis	Medial, anterior section
■ Acetabulum	Hip socket; formed by union of ilium, ischium, and pubis
■ Iliac crests	Upper, curving boundary of ilium
■ Iliac spines	
■ 1 Anterior superior	1 Prominent projection at anterior end of iliac crest; can be felt externally as "point" of hip
2 Anterior inferior	2 Less prominent projection short distance below anterior superior spine
3 Posterior superior	3 At posterior end of iliac crest
4 Posterior inferior	4 Just below posterior superior spine
Greater sciatic notch	Large notch on posterior surface of ilium just below posterior inferior spine
Gluteal lines	Three curved lines across outer surface of ilium—posterior, anterior, inferior, respectively
Iliopectineal line	Rounded ridge extending from pubic tubercle upward and backward toward sacrum
Iliac fossa	Large, smooth, concave inner surface of ilium above iliopectineal line
■ Ischial tuberosity	Large, rough, quadrilateral process forming inferior part of ischium; in erect sitting position body rests on these tuberosities
Ischial spine	Pointed projection just above tuberosity
■ Symphysis pubis	Cartilaginous, amphiarthrotic joint between pubic bones
Superior pubic ramus	Part of pubis lying between symphysis and acetabulum; forms upper part of obturator foramen
Inferior pubic ramus	Part extending down from symphysis; unites with ischium

Continued.

Table 5-2 Identification of bone markings—cont'd

Os coxae (cont'd)

Marking	Description
■ Pubic arch	Angle formed by two inferior rami
Pubic crest	Upper margin of superior ramus
Pubic tubercle	Rounded process at end of crest
■ Obturator foramen	Large hole in anterior surface of os coxa; formed by pubis and ischium; largest foramen in body
■ Pelvic brim (or inlet; Figs. 5-36 and 5-38)	Boundary of aperture leading into true pelvis; formed by pubic crests, iliopectineal lines, and sacral promontory; size and shape of this inlet has great obstetrical importance, since if any of its diameters too small, infant skull cannot enter true pelvis for natural birth
■ True (or lesser) pelvis	Space below pelvic brim; true "basin" with bone and muscle walls and muscle floor; pelvic organs located in this space
■ False (or greater) pelvis	Broad, shallow space above pelvic brim, or pelvic inlet; name "false pelvis" is misleading, since this space is actually part of abdominal cavity, not pelvic cavity
Pelvic outlet (Figs. 5-37 and 5-38)	Irregular circumference marking lower limits of true pelvis; bounded by tip of coccyx and two ischial tuberosities
Pelvic girdle (or bony pelvis)	Complete bony ring; composed of two hip bones (ossa coxae), sacrum, and coccyx; forms firm base by which trunk rests on thighs and for attachment of lower extremities to axial skeleton

Femur (Fig. 5-40)

Marking	Description
■ Head	Rounded, upper end of bone; fits into acetabulum
Neck	Constricted portion just below head
■ Greater trochanter	Protuberance located inferiorly and laterally to head
■ Lesser trochanter	Small protuberance located inferiorly and medially to greater trochanter
Linea aspera	Prominent ridge extending lengthwise along concave posterior surface
Gluteal tubercle	Rounded projection just below greater trochanter; rudimentary third trochanter
Supracondylar ridges	Two ridges formed by division of linea aspera at its lower end; medial supracondylar ridge extends inward to inner condyle, lateral ridge to outer condyle
■ Condyles	Large, rounded bulges at distal end of femur; one on medial and one on lateral surface
Adductor tubercle	Small projection just above inner condyle; marks termination of medial supracondylar ridge
Trochlea	Smooth depression between condyles on anterior surface; articulates with patella
Intercondyloid notch	Deep depression between condyles on posterior surface; cruciate ligaments that help bind femur to tibia lodge in this notch

■ Those that seem particularly important because they are places of muscle attachments.

Tibia (Fig. 5-41)

Marking	Description
■ Condyles	Bulging prominences at proximal end of tibia; upper surfaces concave for articulation with femur
Intercondylar eminence	Upward projection on articular surface between condyles
■ Crest	Sharp ridge on anterior surface
Tibial tuberosity	Projection in midline on anterior surface
Popliteal line	Ridge that spirals downward and inward on posterior surface of upper third of tibial shaft
■ Medial malleolus	Rounded downward projection at distal end of tibia; forms prominence on inner surface of ankle

Fibula (Fig. 5-41)

Marking	Description
■ Lateral malleolus	Rounded prominence at distal end of fibula; forms prominence on outer surface of ankle

Tarsals (Figs. 5-42 and 5-45)

■ Calcaneus	Heel bone
■ Talus	Uppermost of tarsals; articulates with tibia and fibula; boxed in by medial and lateral malleoli
■ Longitudinal arches **1** Medial **2** Lateral	Tarsals and metatarsals so arranged as to form arch from front to back of foot **1** Formed by calcaneus, talus, navicular, cuneiforms, and 3 medial metatarsals **2** Formed by calcaneus, cuboid, and 2 lateral metatarsals
■ Transverse (or metatarsal) arch	Metatarsals and distal row of tarsals (cuneiforms and cuboid) so articulated as to form arch across foot; bones kept in 2 arched positions by means of powerful ligaments in sole of foot and by muscles and tendons

Outline summary

Functions

A Furnishes supporting framework

B Affords protection

C Movement—bones constitute levers for muscle action

D Reservoir—bones serve as the major reservoir for calcium deposits and withdrawals, thereby playing essential part in maintaining blood calcium homeostasis

E Hemopoiesis—blood cell formation by red bone marrow, that is, myeloid tissue

Types of bones

A Long (femur)

B Short (carpal)

C Flat (parietal)

D Irregular (vertebrae)

Macroscopic structure

Long bones

A Diaphysis—hollow, shaftlike portion composed of thick compact or dense bone

B Epiphyses—extremities of long bones composed of spongy or cancellous bone; marrow fills spaces of cancellous bone (yellow marrow in most adult epiphyses but red marrow in proximal epiphyses of humerus and femur)

C Articular cartilage—thin layer of hyaline cartilage covering joint surfaces of epiphyses

D Periosteum—dense white fibrous membrane covering bone except at joint surfaces; firmly attached to underlying bone; muscle and tendons attached firmly to periosteum by interlacing fibers; periosteum contains blood vessels and bone-forming cells so essential for maintenance, growth, and repair of bones

E Medullary (marrow) cavity—hollow in diaphysis filled with yellow (fatty) marrow

F Endosteum—membrane lining medullary cavity

Short, flat, and irregular bones

Cancellous bone forms inside of these bones and compact bone forms outside; red marrow in spaces of cancellous bone inside a few irregular and flat bones, for example, vertebrae and sternum

Microscopic structure

Bone

A Mainly calcified matrix—cement substance impregnated with calcium salts and reinforced by collagenous fibers

B Lamellae—concentric cylindrical layers of calcified matrix enclosing haversian canal that contains blood vessel

C Haversian system—canal and surrounding lamellae

D Lacunae—microscopic spaces containing osteocytes (bone cells); lie between lamellae

E Canaliculi—microscopic canals radiating in all directions from lacunae, connecting them with haversian canals; routes by which tissue fluid reaches osteocytes

F Compact bone (or solid bone)—no empty spaces; lamellae fit closely together

G Cancellous bone (or spongy bone)—many spaces in matrix arranged mainly in trabeculae rather than lamellae

Cartilage

A Intercellular substance (matrix) predominates over cells

B Matrix—firm gel; no canals or blood vessels in cartilage matrix

Formation and growth of bone

A Formation

1 Skeleton preformed in hyaline cartilage and fibrous membranes; most cartilaginous or membranous structures changed into bone before birth but process not complete until about 25 years of age

2 Endochondral ossification—incompletely understood process that replaces hyaline cartilage "bones" with true bones

3 Intramembranous ossification—process that replaces fibrous membrane "bones" with true bones

B Growth

1 In length—by continual thickening of epiphyseal cartilage followed by ossification

2 In diameter—medullary cavity enlarged by osteoclasts destroying bone around it while new bone added around circumference by osteoblasts

C Opposing processes of bone tissue formation and destruction (resorption) go on concurrently throughout life

1 Bone formation exceeds resorption during growth years, from infancy through adolescence

2 Bone formation and resorption balance each other during young adulthood

3 After young adulthood (age 35 to 40 years), more bone resorbed at endosteal surface than formed at periosteal surface; net loss of bone tissue weakens bones, causing them to fracture more easily

Bone markings

A Depressions and openings
  1 Fossa—hollow or depression
  2 Sinus—cavity or spongelike air spaces within bone
  3 Foramen—hole
  4 Meatus—tube-shaped opening
B Projections or processes
  1 Those that fit into joints
    a Condyle—rounded projection entering into formation of joint
    b Head—rounded projection beyond narrow neck
  2 Those to which muscles attach
    a Trochanter—very large process
    b Crest—ridge
    c Spinous process or spine—sharp projection
    d Tuberosity—large, rounded projection
    e Tubercle—small, rounded projection

Divisions of skeleton

A Names and definitions
  1 Axial skeleton—bones that form the upright axis of the body plus ear bones
  2 Appendicular skeleton—bones appended to axial skeleton, that is, bones of upper and lower extremities
B Names and numbers of bones in axial skeleton (80 bones)
  1 Skull (28 bones)
    a Cranium (8 bones)—frontal, parietal (2), temporal (2), occipital, sphenoid, and ethmoid
    b Face (14 bones)—nasal (2), maxillary (2), zygomatic or malar (2), mandible, lacrimal (2), palatine (2), inferior conchae or turbinates (2), and vomer
    c Ear bones (6)—malleus (2), incus (2), and stapes (2)
  2 Hyoid
  3 Vertebral column (26 vertebrae)
    a Cervical (7)
    b Thoracic (12)
    c Lumbar (5)
    d Sacrum
    e Coccyx
  4 Sternum and ribs (25 bones)—sternum, true ribs (7 pairs), false ribs (5 pairs, 2 pairs of which are floating)
C Names and numbers of bones in appendicular skeleton (126 bones)
  1 Upper extremities (64 bones)—clavicle (2), scapula (2), humerus (2), radius (2), ulna (2), carpals (16), metacarpals (10), and phalanges (28)
  2 Lower extremities (62 bones)—ossa coxae (2), femur (2), patella (2), tibia (2), fibula (2), tarsals (14), metatarsals (10), and phalanges (28)

Description of bones

See Tables 5-1 and 5-2

Differences between male and female skeletons

A Male skeleton larger and heavier
B Male pelvis deep and funnel shaped with narrow pubic arch; female pelvis shallow, broad, and flaring with wider pubic arch and larger iliosacral notch

Age changes in skeleton

A From infancy to young adulthood—absolute and relative sizes of bones change
B From young adulthood to age—texture of bones changes, as do contour of bone margins and markings

Review questions

1 What general functions does the skeletal system perform?
2 Describe the microscopic structure of bone and cartilage.
3 Describe the structure of a long bone.
4 Explain the functions of the periosteum.
5 Describe the general plan of the skeleton.
6 Name the bones of the axial skeleton.
7 Name the primary and secondary curves of the spine. Describe each.
8 Name the five pairs of bony sinuses in the skull.
9 Name the bones that fuse to form the coccyx.
10 What is the true pelvis? The false pelvis? Name the boundary line between the true and false pelves.
11 Through what opening does the spinal cord enter the cranial cavity?
12 Explain the basic steps in the process of ossification according to the concept described in the text.
13 Compare osteoblasts and osteoclasts as to function.
14 Define or make an identifying statement about each of the following terms: condyle, crest, diaphysis, endosteum, epiphysis, foramen, fossa, haversian system, kyphosis, lordosis, medullary cavity, periosteum, scoliosis, sinus, spinous process, trochanter, trabeculae.
15 Describe some common age changes in the skeleton.

The skeletal system—articulations

Articulations, in the body, are joints between bones. They perform two seemingly contradictory functions. Articulations hold bones firmly bound to each other, and yet, they also permit movement between them. If you, like most of us, are fortunate enough to have normal, healthy joints, you may never have been aware of these structures and their importance. To say merely that joints are important is an understatement. The existence of joints between bones makes possible movements of body parts. Movements, in turn, make a major contribution to the maintenance of homeostasis and, therefore, to survival. But they also do something more. Movements provide a large measure of our enjoyment of life. Without joints between our bones, we could make no movements; our bodies would be rigid, immobile hulks. This chapter includes names and descriptions of different kinds of joints and descriptions of age changes and certain diseases of joints.

Kinds of joints

Currently a popular way of classifying joints is according to characteristic structural features. This classification identifies three kinds of joints: fibrous, cartilaginous, and synovial. Another classification divides joints according to the degree of movement they allow into these three types: synarthroses (no movement), amphiarthroses (slight movement), and diarthroses (free movement).

Fibrous joints

Fibrous joints, as the name suggests, are those in which the articular surfaces of two bones are connected by fibrous connective tissue that binds them closely and tightly to each other. Fibrous joints are subdivided into *sutures* and *syndesmoses*. Sutures occur only in the skull, and their structure permits no movement between the articulating bones. *Syndesmoses* unite bones by means of dense fibrous tissue, and this allows almost no movement at these joints. The distal tibiofibular joint is a syndesmosis.

Cartilaginous joints

Cartilaginous joints, as the name suggests, are those in which cartilage joins one bone to another. There are two types of cartilaginous joints, namely, *symphyses* (from the Greek for "a growing together") and *synchondroses* (from the Greek *syn*, together, and *chondros*, cartilage). Symphyses are located in the midline of the body. Examples are the symphysis pubis, the joint between the two pubic bones, and the joint between the bodies of the vertebrae. Examples of synchondroses are the joints between the ribs and sternum. These bones are bound to each other by large pieces of cartilage called costal cartilages. Only very slight movement can occur at symphyses and synchondroses.

Synovial joints

Synovial joints include by far the majority of the body's articulations—fortunately, because they are the most mobile of the three types of joints. They also have the most complex struc-ture. Refer to Fig. 6-1 to identify the following structural features of synovial joints:

1 *Joint capsule*—sleevelike extension of the periosteum of each of the articulating bones. The capsule completely encases the ends of the bones and binds them to each other.

2 *Synovial membrane*—moist, slippery membrane that lines the inner surface of the joint capsule. It attaches, as Fig. 6-1 shows, to the margins of the articular cartilage and secretes synovial fluid, which lubricates the inner joint surfaces.

3 *Articular cartilage*—hyaline cartilage that covers and cushions the articulating ends of the bones.

4 *Joint cavity*—small space between the articulating surfaces of the two bones of the joint. Because of this cavity with no tissue growing between the articulating surfaces of the bones, the bones are free to move against one another. Synovial joints, therefore, are diarthroses, or freely movable joints.

5 *Ligaments*—strong cords of dense white fibrous tissue at most synovial joints. These grow between the bones, lashing them even more firmly together than possible with the joint capsule alone.

Synovial, or diarthrotic, joints permit one or more of the following kinds of movement: flexion, extension, abduction, adduction, rotation, circumduction. Some of them permit special movements such as supination, pronation, inversion, eversion, protraction, and retraction.

Flexion. Flexion decreases the size of the angle between the anterior surfaces of articulated bones (exception: flexion of the knee and toe joints decreases the angle between the posterior surfaces of the articulated bones). Flexing movements are bending or folding movements. For example, bending the head forward is flexion of the joint between the occipital bone and the atlas, and bending the elbow is flexion of the elbow joint or of the lower arm (Fig. 6-2). Flexing movements of the arms and legs may be thought of as "withdrawing" movements.

Extension. Extension is the return from flexion. Whereas bending movements are flexions, straightening movements are extensions. Extension restores a part to its anatomical position from the flexed position. Continuation of extension beyond the anatomical position is called hyperextension. Examples include flexion of the head, bending it forward as in prayer; extension of the head, returning it to the upright anatomical position from the flexed position; and hyperextension of the head, stretching it backward from the upright position. Extension of the foot at the ankle joint is commonly referred to as plantar flexion, whereas flexion of the ankle joint is called dorsal flexion.

Abduction. Abduction moves a bone away from the median plane of the body. An example is moving the arms straight out to the sides.

Adduction. Adduction is the opposite of abduction. It moves the part toward the median plane of the body. Examples are bringing the arms back to the sides; moving the fingers toward the third finger; and moving the toes toward the second toe.

Rotation. Rotation is the pivoting of a bone on its own axis somewhat as a top turns on its axis. An example is holding the head in an upright position and turning it from one side to the other.

Circumduction. Circumduction causes the bone to describe the surface of a cone as it moves. The distal end of the bone describes a circle. It combines flexion, abduction, extension, and adduction in succession. Examples are dropping the head to one shoulder, then to the chest, to the

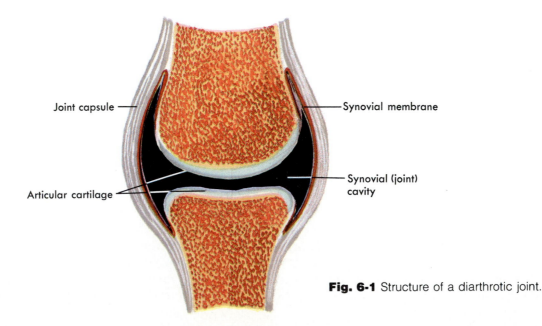

Joint capsule

Synovial membrane

Articular cartilage

Synovial (joint) cavity

Fig. 6-1 Structure of a diarthrotic joint.

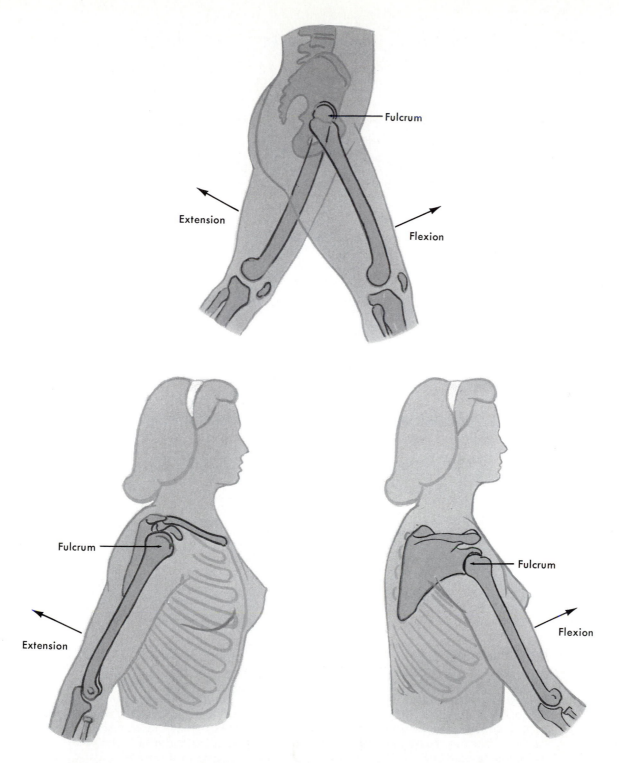

Fig. 6-2 Top, flexion and extension of the thigh at the hip joint. Lower figures show flexion and extension of the upper arm at the shoulder joint.

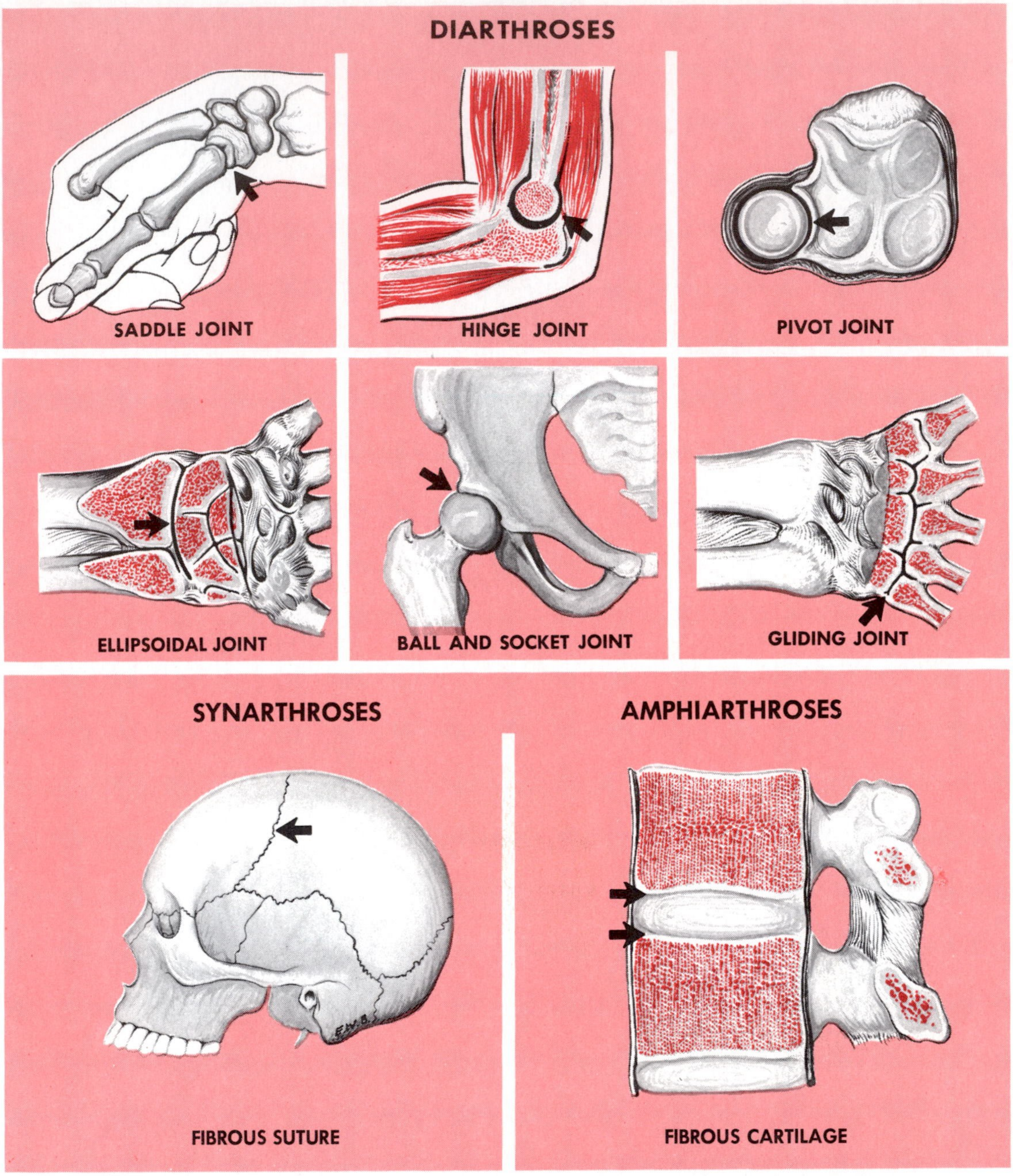

DIARTHROSES

SADDLE JOINT

HINGE JOINT

PIVOT JOINT

ELLIPSOIDAL JOINT

BALL AND SOCKET JOINT

GLIDING JOINT

SYNARTHROSES

FIBROUS SUTURE

AMPHIARTHROSES

FIBROUS CARTILAGE

Fig. 6-3 Typical joints. Saddle joint at base of thumb, between first metacarpal bone and a carpal bone (trapezium). Hinge joint at elbow between humerus and ulna and between humerus and radius. Pivot joint at elbow between radius and ulna. Ellipsoidal joint at wrist between carpal bones and radius and ulna. Ball and socket hip joint between head of femur and os coxae. Gliding joints at wrist between various carpal bones. Fibrous synarthroses, sutures between skull bones. Cartilaginous amphiarthroses between bodies of vertebrae.

other shoulder, and backward and describing a circle with the arms outstretched.

Special movements. *Supination* is a movement of the forearm that turns the palm forward, as it is in the anatomical position. *Pronation* is turning the forearm so as to bring the back of the hand forward. *Inversion* is a special movement of the ankle that turns the sole of the foot inward; *eversion* turns it outward. *Protraction* moves a part forward, such as sticking out the jaw. *Retraction* is the reverse of protraction.

Types (Fig. 6-3)

Because synovial joints differ somewhat in structure and in the kinds of movement they permit, they are subdivided into six types: ball and socket, hinge, pivot, ellipsoidal, saddle, and gliding.

Ball and socket joints are those in which a ball-shaped head of one bone fits into a concave socket of another bone. Examples are the shoulder and hip joints. Of all the joints in our bodies, ball and socket joints permit the widest range of movements, namely, flexion, extension, abduction, adduction, rotation, and circumduction.

Hinge joints permit only flexion and extension —movements that might be called hinged-door movements.

Pivot joints are those in which a small projection of one bone pivots in an arch of another bone, causing the first bone to rotate on its axis. There are two pivot joints in the body, one between the first two cervical vertebrae (atlas and axis) and the other between the proximal ends of the radius and ulna.

Ellipsoidal joints are those in which an oval-shaped condyle fits into an elliptical socket. The radius joins the carpal bones (scaphoid, lunate, and triquetrum) by means of an ellipsoidal joint, which permits the movements of flexion and extension of the hand in one axis and the movements of abduction and adduction in another axis. In short, it permits biaxial movement. Next to the ball and socket joints, ellipsoidal joints allow the widest range of movements.

Saddle joints, like ellipsoidal joints, permit biaxial movements. The shapes of the articulating ends of the bones, however, differ in these two types of joints. An oval-shaped projection fits into an elliptical socket in an elliptical joint. In a saddle joint, on the other hand, a saddle-shaped surface of one bone fits into a saddle-shaped surface of another bone. Only one pair of saddle joints exists in the body—one in each hand between the metacarpal bone of the thumb and the carpal bone named the trapezium (Fig. 5-34, p. 107, and Fig. 6-3). These are deservedly famous joints—at least to anatomists. They make possible the great mobility of the human thumbs. We can flex, extend, abduct, adduct, and circumduct them. But most important of all, we can oppose them, that is, we can move our thumbs to touch the tip of any one of our fingers. If we did not have a saddle joint at the base of each thumb, we could not oppose the thumb and so would be unable to do such simple things as picking up a pin or grasping a pencil between thumb and forefinger.

Gliding joints include most of the joints between both the carpal and tarsal bones and also all of the joints between the articular processes of the vertebrae. These joints allow only the simplest kind of motion, specifically, a little gliding back and forth or sideways.

Descriptions of representative joints

Shoulder joint (Figs. 6-4 and 6-5). The shoulder joint claims distinction as our most mobile joint. Several anatomical facts account for this. The head of the humerus is a large rounded projection (Fig. 5-30, p. 105), but it fits into a shallow socket, the glenoid cavity of the scapula (Fig. 5-27, p. 103). In addition, the joint capsule is loose fitting and allows for considerable movement. Several muscle tendons exert a stabilizing influence on the shoulder joint. So, too, do

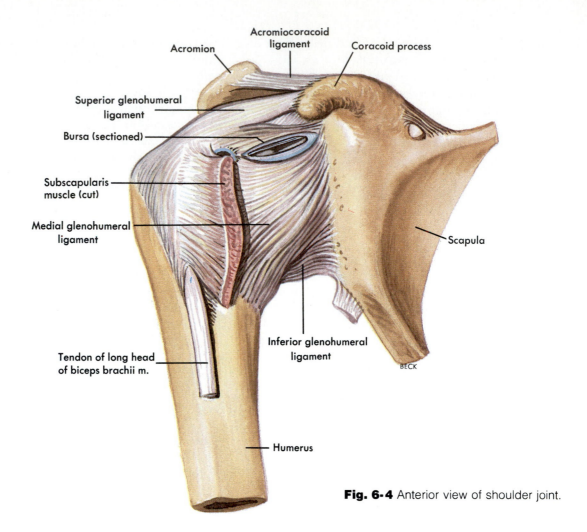

Acromion

Acromiocoracoid ligament

Coracoid process

Superior glenohumeral ligament

Bursa (sectioned)

Subscapularis muscle (cut)

Medial glenohumeral ligament

Scapula

Tendon of long head of biceps brachii m.

Inferior glenohumeral ligament

BECK

Humerus

Fig. 6-4 Anterior view of shoulder joint.

several bursae. The main one of these is the large subdeltoid (subacromial) bursa wedged between the superior surface of the joint capsule and the inferior surface of the deltoid muscle. All in all, however, the shoulder joint is a more mobile than stable joint, and dislocations of the humerus from the glenoid cavity occur rather frequently.

Hip joint (Figs. 6-6 and 6-7). The first characteristic to remember about the hip joint is stability; the second is mobility. The stability of the hip joint derives largely from the shape of the head of the femur and of the acetabulum, the socket of the hip bone into which the femur head fits. Turn to Fig. 5-38 and note the deep, cuplike shape of the acetabulum, and then observe the ball-like head of the femur in Fig. 5-40. Compare these with the shallow, almost saucer-shaped glenoid cavity (Fig. 5-27) and the head of the humerus (Fig. 5-30). A joint capsule and several ligaments hold the femur and hip bones together. The ilio-femoral ligament between the ilium and femur, for example, is said to be one of the strongest ligaments in the body. Other ligaments also con-

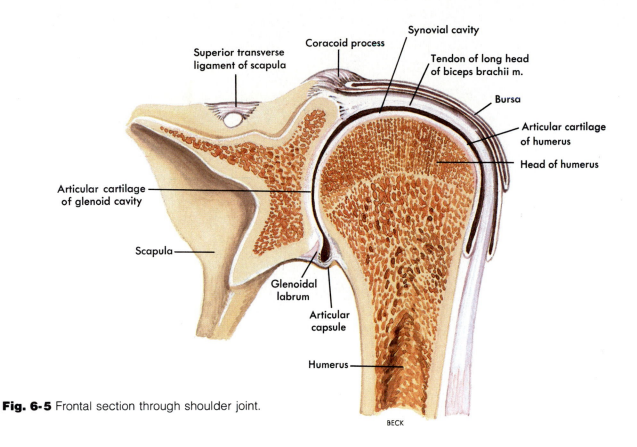

Superior transverse
ligament of scapula

Coracoid process

Synovial cavity

Tendon of long head
of biceps brachii m.

Bursa

Articular cartilage
of humerus

Head of humerus

Articular cartilage
of glenoid cavity

Scapula

Glenoidal
labrum

Articular
capsule

Humerus

BECK

Fig. 6-5 Frontal section through shoulder joint.

tribute to the stability of the joint by binding the femur firmly to the ischium and pubic bone. The mobility of the hip joint is somewhat limited by its structure, although it permits the same kinds of movement as does the shoulder joint—flexion, extension, abduction, adduction, circumduction, and rotation.

Knee joint. The knee joint is the largest and one of the most complex and most frequently injured joints in the body. The condyles of the femur articulate with the flat upper surface of the tibia. Although this is a precariously unstable ar-

rangement, there are counteracting forces supplied by a joint capsule, cartilages, and numerous ligaments and muscle tendons. Note, for example, in Fig. 6-8, the shape of the two cartilages labeled medial meniscus and lateral meniscus. They attach to the flat top of the tibia and, because of their concavity, form a kind of shallow socket for the condyles of the femur. Of the many ligaments that hold the femur bound to the tibia, four can be seen in Fig. 6-8. The anterior cruciate ligament attaches to the anterior part of the tibia, between its condyles, then crosses over

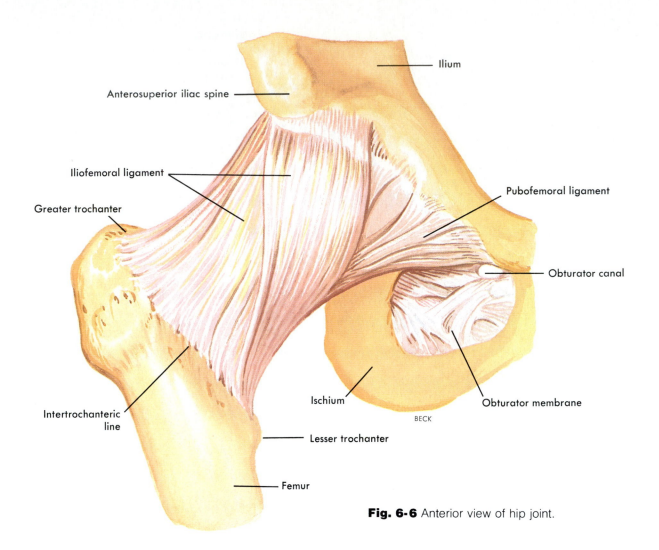

Anterosuperior iliac spine

Iliofemoral ligament

Greater trochanter

Intertrochanteric line

Femur

Lesser trochanter

Ischium

Ilium

Pubofemoral ligament

Obturator canal

Obturator membrane

BECK

Fig. 6-6 Anterior view of hip joint.

and backward and attaches to the posterior part of the lateral condyle. The posterior cruciate ligament attaches posteriorly to the tibia and lateral meniscus, then crosses over and attaches to the front part of the femur's medial condyle. The ligament of Wrisberg attaches posteriorly to the lateral meniscus and extends up and over to attach to the medial condyle behind the attachment of the posterior cruciate ligament (Figs. 6-8 and 6-9). The transverse ligament connects the anterior margins of the two menisci.

Strong ligaments, the fibular and tibial collateral ligaments, located at the sides of the knee joint can be seen in Figs. 6-9 and 6-10.

Compared to the hip joint, the knee joint is relatively unprotected by surrounding muscles. Consequently, the knee, more often than the hip, is injured by blows or sudden stops and turns. Athletes, for example, frequently tear a knee cartilage (one of the menisci).

A baker's dozen of bursae serve as pads around the knee joint: four in front, four located later-

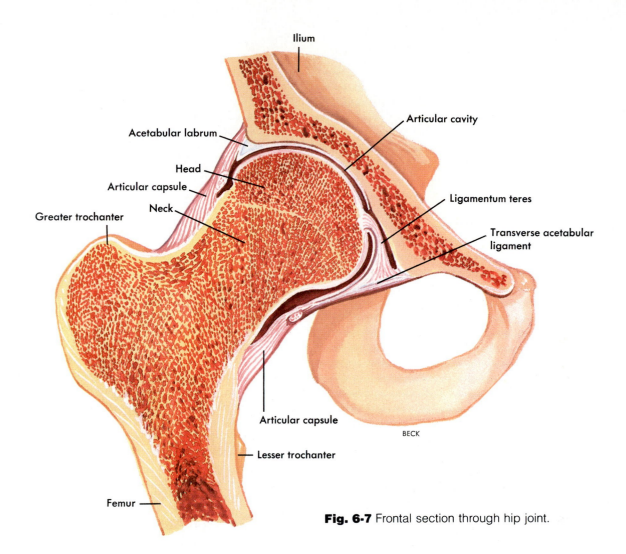

Ilium

Acetabular labrum

Head

Articular capsule

Neck

Greater trochanter

Articular cavity

Ligamentum teres

Transverse acetabular ligament

Articular capsule

Lesser trochanter

Femur

BECK

Fig. 6-7 Frontal section through hip joint.

ally, and five medially. Of these, the largest is the prepatellar bursa (Fig. 6-11) inserted in front of the patellar ligament, between it and the skin. The painful ailment called "housemaid's knee" is prepatellar bursitis.

The structure of the knee joint permits the hingelike movements of flexion and extension. Also, with the knee flexed, some internal and external rotation can occur. In most of our day-to-day activities—such ordinary ones as walking, going up and down stairs, and getting into and

out of chairs—our knees bear the brunt of the load; they are the main weight bearers. Injury or disease of them, therefore, can be badly crippling.

Vertebral joints. One vertebra connects to another by several joints—between their bodies, laminae, and articular, transverse, and spinous processes. These joints hold the vertebrae firmly together so they are not easily dislocated, but also these joints form a flexible column. Consider how many ways you can move the trunk of

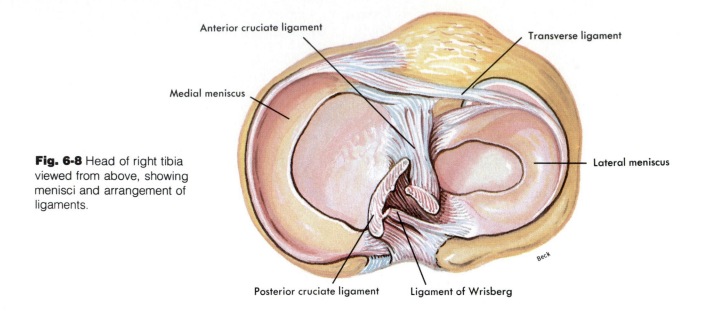

Anterior cruciate ligament

Transverse ligament

Medial meniscus

Lateral meniscus

Fig. 6-8 Head of right tibia viewed from above, showing menisci and arrangement of ligaments.

Posterior cruciate ligament

Ligament of Wrisberg

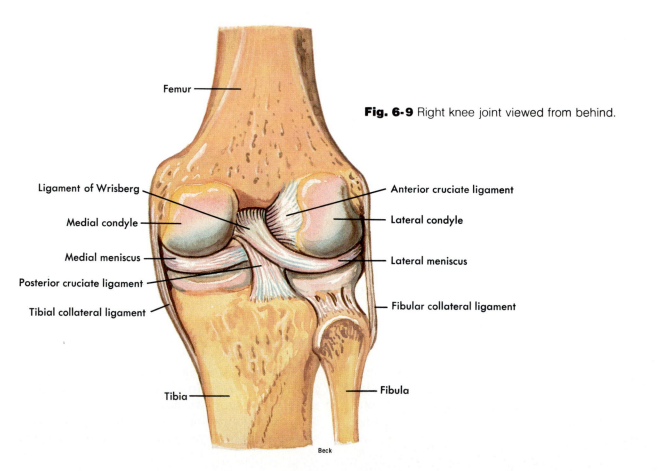

Femur

Fig. 6-9 Right knee joint viewed from behind.

Ligament of Wrisberg

Anterior cruciate ligament

Medial condyle

Lateral condyle

Medial meniscus

Lateral meniscus

Posterior cruciate ligament

Tibial collateral ligament

Fibular collateral ligament

Tibia

Fibula

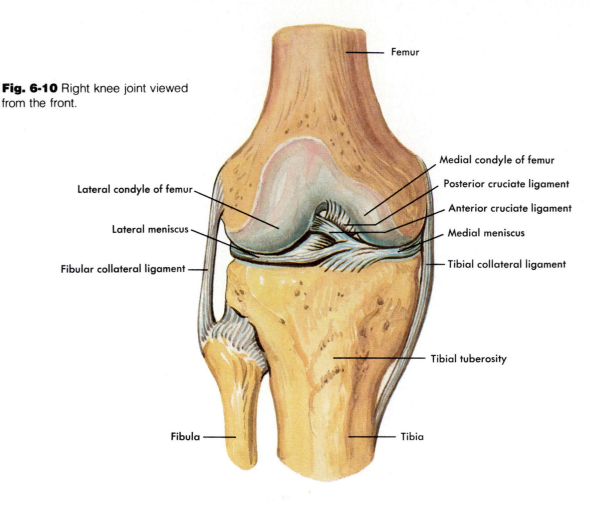

Fig. 6-10 Right knee joint viewed from the front.

Femur

Medial condyle of femur

Posterior cruciate ligament

Anterior cruciate ligament

Medial meniscus

Tibial collateral ligament

Lateral condyle of femur

Lateral meniscus

Fibular collateral ligament

Tibial tuberosity

Fibula

Tibia

your body. You can flex it forward or laterally, you can extend it, and you can circumduct or rotate it. The bodies of adjacent vertebrae are connected by intervertebral disks and strong ligaments. Fibrous tissue and fibrocartilage form a disk's outer rim (called the annulus fibrosus). Its central core (the nucleus pulposus), in contrast, consists of a pulpy, elastic substance. With age the nucleus loses some of its resiliency. It may then be suddenly compressed by exertion or trauma and pushed through the annulus, with fragments protruding into the spinal canal and pressing on spinal nerve routes or the spinal

cord itself. Severe pain results. In medical terminology, this is called a herniated disk; in popular language, it is a "slipped disk."

Identify in Fig. 6-12 the following ligaments that bind the vertebrae together. The *anterior longitudinal ligament*, a strong band of fibrous tissue, connects the anterior surfaces of the vertebral bodies from the atlas down to the sacrum. Connecting the posterior surfaces of the bodies is the *posterior longitudinal ligament*. The *ligamenta flava* bind the laminae of adjacent vertebrae firmly together. Spinous processes are connected by *interspinous ligaments*. In addition, the tips of

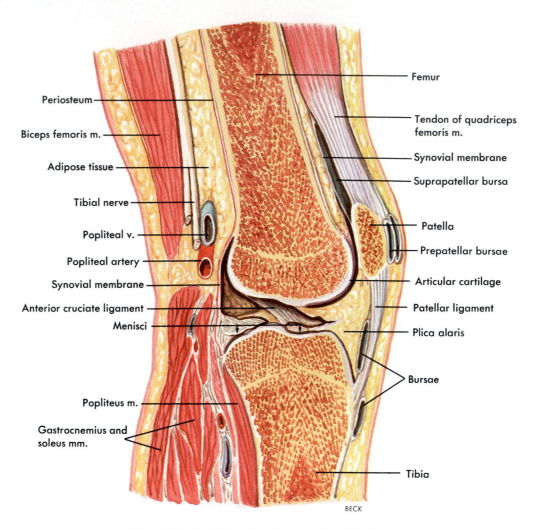

Periosteum

Biceps femoris m.

Adipose tissue

Tibial nerve

Popliteal v.

Popliteal artery

Synovial membrane

Anterior cruciate ligament

Menisci

Popliteus m.

Gastrocnemius and soleus mm.

Femur

Tendon of quadriceps femoris m.

Synovial membrane

Suprapatellar bursa

Patella

Prepatellar bursae

Articular cartilage

Patellar ligament

Plica alaris

Bursae

Tibia

BECK

Fig. 6-11 Sagittal section through knee joint.

the spinous processes of the cervical vertebrae are connected by the *ligamentum nuchae;* its extension, the *supraspinous ligament* connects the tips of the rest of the vertebrae down to the sacrum. And finally, *intertransverse ligaments* connect the transverse processes of adjacent vertebrae.

Table 6-1, p. 144, classifies joints according to structure and range of movement. Table 6-2,

p. 146, gives brief descriptions of most of the synovial joints of the body.

Joint age changes and diseases

By far the commonest age change in joints is a degenerative condition called *osteoarthritis.* Loss of articular cartilage with formation of bone at

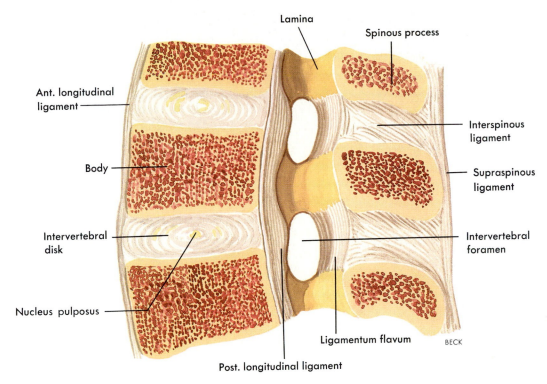

Lamina

Spinous process

Ant. longitudinal ligament

Body

Intervertebral disk

Nucleus pulposus

Interspinous ligament

Supraspinous ligament

Intervertebral foramen

Ligamentum flavum

BECK

Post. longitudinal ligament

Fig. 6-12 Sagittal section of two lumbar vertebrae and their ligaments.

the margin of the articular cartilage characterizes this condition. This bone enlarges and may deform the affected joints and interfere with movement at them. Involvement of the knee or hip joints may become a major disability. Through surgery, it is now possible to replace these joints with artificial joints.*

Rheumatoid disease is a painful, disabling condition involving inflammation of the synovial membranes (synovitis). It occurs most commonly between the ages of 40 and 60 years but may afflict people of any age. Granulation tissue (called pannus) forms on the articular cartilages of the affected joints. In time, this may erode not

only the cartilages but also bone and even ligaments and tendons in the area.

The disease known as *gout* is really a group of diseases characterized by higher than normal levels of uric acid in the blood (hyperuricemia), a condition generally unknown to the individual. The excess uric acid forms sodium urate crystals in the extracellular fluid. Most often the crystals first make their presence known by severe pain in a joint when they appear in its synovial fluid. The great toe joint between the first metatarsal and phalangeal bone is the one most frequently affected by gouty arthritis. Gout has a genetic basis and to some extent runs in families. Also, it has long been associated with the intemperate use of alcohol. An old saying declares that "In the young, wine goes to the head; in the aged, it goes to the feet."

*See Sonstegard, D. A., Matthews, L. S., and Kaufer, H.: The surgical replacement of the human knee joint, Sci. Am. **238:** 44-50, Jan., 1978.

Table 6-1 Classification of joints			
Joints	**Description**	**Movement**	**Example**
Fibrous			
Sutures	Thin layer of fibrous tissue (extension of periosteum); grows between articulating surfaces of bones; fibrous connection replaced by bone in older adults	None (synarthroses)	Sutures between certain skull bones
Syndesmoses	Same as sutures, but fibrous tissue not eventually replaced by bone	None (synarthroses)	Distal tibiofibular
Cartilaginous			
Symphyses	Fibrocartilage connects articulating surfaces of bones; joint capsule present, and sometimes a joint cavity	Slight (amphiarthroses)	Symphysis pubis; joints between bodies of vertebrae
Synchondroses	Hyaline cartilage connects epiphyses to diaphysis of growing bones	Ordinarily none	In all growing long bones
Synovial			
Ball and socket (spheroidal, endarthroses)	Ball-shaped head fits into concave socket	Widest range of all joints; triaxial	Shoulder joint and hip joint
Hinge (ginglymus)	Spool-shaped surface fits into concave surface	In one plane about single axis (uniaxial); like hinged-door movement, namely, flexion and extension	Elbow, knee, ankle, and interphalangeal joints
Pivot (trochoid)	Arch-shaped surface rotates about rounded or peglike pivot	Rotation; uniaxial	Between axis and atlas; between radius and ulna
Ellipsoidal (condyloid, ovoid)	Oval-shaped condyle fits into elliptical cavity	In two planes at right angles to each other—specifically, flexion, extension, abduction, and adduction; biaxial	Wrist joint (between radius and carpals)
Saddle	Saddle-shaped bone fits into socket that is concave-convex in opposite direction; modification of condyloid joint	Same kinds of movement as condyloid joint but freer; like rider in saddle; biaxial	Thumb, between first metacarpal and trapezium, is only saddle joint
Gliding (arthrodia)	Articulating surfaces; usually flat	Gliding—a nonaxial movement	Between carpals and between tarsals

Table 6-2 Description of individual joints

Name	Articulating bones	Type	Movements
Atlantoepistropheal	Anterior arch of atlas rotates about dens of axis (epistropheus)	Diarthrotic (pivot type)	Pivoting or partial rotation of head
Vertebral	Between bodies of vertebrae	Amphiarthrotic, cartilaginous	Slight movement between any two vertebrae but considerable motility for column as whole
	Between articular processes	Diarthrotic (gliding)	
Sternoclavicular	Medial end of clavicle with manubrium of sternum; only joint between upper extremity and trunk	Diarthrotic (gliding)	Gliding; weak joint that may be injured comparatively easily
Acromioclavicular	Distal end of clavicle with acromion of scapula	Diarthrotic (gliding)	Gliding; elevation, depression, protraction, and retraction
Thoracic	Heads of ribs with bodies of vertebrae	Diarthrotic (gliding)	Gliding
	Tubercles of ribs with transverse processes of vertebrae	Diarthrotic (gliding)	Gliding
Shoulder	Head of humerus in glenoid cavity of scapula	Diarthrotic (ball and socket type)	Flexion, extension, abduction, adduction, rotation, and circumduction of upper arm; one of most freely movable of joints
Elbow	Trochlea of humerus with semilunar notch of ulna; head of radius with capitulum of humerus	Diarthrotic (hinge type)	Flexion and extension
	Head of radius in radial notch of ulna	Diarthrotic (pivot type)	Supination and pronation of lower arm and hand; rotation of lower arm on upper as in using screwdriver
Wrist	Scaphoid, lunate, and triquetral bones articulate with radius and articular disk	Diarthrotic (condyloid)	Flexion, extension, abduction, and adduction of hand
Carpal	Between various carpals	Diarthrotic (gliding)	Gliding
Hand	Proximal end of first metacarpal with trapezium	Diarthrotic (saddle)	Flexion, extension, abduction, adduction, and circumduction of thumb and opposition to fingers; motility of this joint accounts for dexterity of human hand compared with animal forepaw

Continued.

Table 6-2 Description of individual joints—cont'd

Name	Articulating bones	Type	Movements
Hand—cont'd	Distal end of metacarpals with proximal end of phalanges	Diarthrotic (hinge)	Flexion, extension, limited abduction, and adduction of fingers
	Between phalanges	Diarthrotic (hinge)	Flexion and extension of finger sections
Sacroiliac	Between sacrum and two ilia	Diarthrotic (gliding); joint cavity mostly obliterated after middle life	None or slight, for example, during late months of pregnancy and during delivery
Symphysis pubis	Between two pubic bones	Synarthrotic (or amphiarthrotic), cartilaginous	Slight, particularly during pregnancy and delivery
Hip	Head of femur in acetabulum of os coxae	Diarthrotic (ball and socket)	Flexion, extension, abduction, adduction, rotation, and circumduction
Knee	Between distal end of femur and proximal end of tibia; largest joint in body	Diarthrotic (hinge type)	Flexion and extension; slight rotation of tibia
Tibiofibular	Head of fibula with lateral condyle of tibia	Diarthrotic (gliding type)	Gliding
Ankle	Distal ends of tibia and fibula with talus	Diarthrotic (hinge type)	Flexion (dorsiflexion) and extension (plantar flexion)
Foot	Between tarsals	Diarthrotic (gliding)	Gliding; inversion and eversion
	Between metatarsals and phalanges	Diarthrotic (hinge type)	Flexion, extension, slight abduction, and adduction
	Between phalanges	Diarthrotic (hinge type)	Flexion and extension

Meaning and functions

A Articulations are joints between bones
B Articulations hold bones together but also, in most cases, permit movement between them

Kinds of joints

A Fibrous joints
1 Fibrous tissue connects articular surfaces of bones
    a Sutures—immovable joints (synarthroses) between various skull bones
    b Syndesmoses—fibrous joints between a few bones other than skull bones, for example, between distal ends of tibia and fibula
B Cartilaginous joints
1 Cartilage connects articular surfaces of bones
    a Symphyses—slightly movable joints (amphiarthroses) in midline of body, for example, symphysis pubis
    b Synchondroses—temporary, essentially immovable joints in which epiphyseal cartilage connects diaphysis to epiphyses of growing bones
C Synovial joints
1 Most mobile and most numerous of joints
    a Main structural features—joint capsule, capsule lined by synovial membrane, articulating ends of bones covered by articular cartilage, joint cavity present between articular surfaces of bones; another name for synovial joints in diarthroses, or freely movable joints
    b Movements at synovial joints—one or more of the following: flexion, extension, abduction, adduction, rotation, and circumduction; special movements at some joints: supination, pronation, inversion, eversion, protraction, and retraction
    c Types of synovial joints—ball and socket, hinge, pivot, ellipsodial, saddle, and gliding

Descriptions of representative joints

See Table 6-2

Joint age changes and diseases

A Osteoarthritis—commonest age change in joints; a degenerative condition with loss of articular cartilage and formation at margin of articular cartilage, which enlarges and may deform joints and interfere with movements
B Rheumatoid disease—painful disabling inflammation of synovial membranes; pannus or granulation tissue forms on articular cartilages and may erode them and other joint structures including bones, ligaments, and tendons
C Gout—group of diseases characterized by hyperuricemia; sodium urate crystals form in extracellular fluid; their appearance in synovial fluid causes severe pain in affected joint; great toe joint between first metatarsal and phalangeal bone is most frequently affected

Review questions

1 Identify three kinds of joints according to their structure.
2 Identify three kinds of joints according to the degree of movement possible at them.
3 Identify six types of synovial joints and describe the kinds of movements possible at them.
4 Define and give an example of each of these movements: flexion, extension, abduction, adduction, circumduction and rotation.
5 Define: supination, pronation, inversion, eversion, protraction, retraction.
6 What joint makes possible much of the dexterity of the human hand? Describe it and the movements it permits.
7 Describe vertebral joints.
8 Why does the range of movement vary at different joints? What factors determine this?
9 What is the most common age change in joints?
10 Define: osteoarthritis, rheumatoid disease, gout.

chapter 7

Skeletal muscles

Man's survival depends in large part on his ability to adjust to the changing conditions of his environment. Movements constitute the major part of this adjustment. Whereas most of the systems of the body play some role in accomplishing movement, it is the skeletal and muscular systems acting together that actually produce movements. We have investigated the architectural plan of the skeleton and have seen how its firm supports and joint structures make movement possible. However, bones and joints cannot move themselves. They must be moved by something. Muscle tissue, because of its irritability, contractility, extensibility, and elasticity, is admirably suited to this function. Our subject, then, for this chapter is the 40% to 50% of our body weight that is skeletal muscle—those muscle masses that attach to bones and move them about, the "red meat" of the body. (Cardiac muscle will be discussed in Chapter 14, and information about smooth muscle appears in several chapters.)

In this chapter we shall try to discover how muscles move bones, and to do this we shall try to answer many other questions—how the structure of muscles adapts them to their function, how energy is made available for their work, and how muscle activity contributes to the health and survival of the whole body—to mention only a few.

General functions

If you have any doubts about the importance of muscle function to normal life, you have only to observe a person with extensive paralysis—a victim of advanced muscular dystrophy, for example. Any of us possessed of normal powers of movement can little imagine life with this matchless power lost. But cardinal as it is, movement is not the only contribution muscles make to healthy survival. They also perform two other essential functions: production of a large portion of body heat and maintenance of posture.

1 *Movement.* Skeletal muscle contractions produce movements either of the body as a whole (locomotion) or of its parts.

2 *Heat production.* Muscle cells, like all cells, produce heat by the process known as catabolism (discussed in Chapter 19). But because skeletal muscle cells are both highly active and numerous, they produce a major share of total body heat. Skeletal muscle contractions, therefore, constitute one of the most important parts of the mechanism for maintaining homeostasis of temperature.

3 *Posture.* The continued partial contraction of many skeletal muscles makes possible standing, sitting, and other maintained positions of the body.

Skeletal muscle cells
Microscopic structure

Look at Fig. 7-1, *A* and *B*. As you can see there, a skeletal muscle is composed of bundles of skeletal muscle fibers that generally extend the entire length of the muscle. They are called fibers, instead of cells, because of their threadlike shape (1 to 40 mm long but with a diameter of only 10 to 100 μm). Skeletal muscle fibers have many of the same structural parts as other cells. Several of them, however, bear different names in muscle fibers. For example, *sarcolemma* is the plasma membrane of a muscle fiber. *Sarcoplasm* is its cytoplasm. Muscle cells contain a network of tubules and sacs known as the *sarcoplasmic reticulum*—a structure analogous, but not identical, to the endoplasmic reticulum of other cells. Muscle fibers contain many mitochondria, and, unlike other cells, they have several nuclei.

Certain structures not found in other cells are present in skeletal muscle fibers. For instance, bundles of very fine fibers—*myofibrils*—extend lengthwise of skeletal muscle fibers and almost fill their sarcoplasm. Myofibrils, in turn, are made up of still finer fibers called thick and thin filaments. The red lines in Fig. 7-1, *D*, represent thick filaments, and the gray lines represent thin filaments. Find the label sarcomere in this drawing. Note that a *sarcomere* is a segment between two successive Z lines. Each myofibril consists of a lineup of several sarcomeres, each of which functions as a contractile unit. The A bands of the sarcomeres appear as relatively wide, dark stripes (cross striae) under the microscope, and they alternate with narrower, lighter colored stripes formed by the I bands (Fig. 7-1, *C*). Because of its cross striae, skeletal muscle is also called striated muscle. Electron microscopy of skeletal muscle (Fig. 7-2) has revolutionized our concept of both its structure and function.

Another structure unique to skeletal and cardiac muscle cells is the transverse tubular or *T system.* This name derives from the fact that this system consists of tubules that extend transversely into the sarcoplasm. T system tubules, as Fig. 7-3 shows, enter the sarcoplasm at the levels of the Z lines, that is, in the middle of the light I bands. Because invaginations of the sarcolemma form the T system tubules, they open to the exterior of the muscle fiber and may serve to carry interstitial fluid directly into each sarcomere.

The sarcoplasmic reticulum is also a system of tubules in a muscle fiber. It is separate from the T system and differs from it in that the tubules of the sarcoplasmic reticulum run parallel to muscle fibers and terminate in closed sacs at the ends of each sarcomere, that is, immediately above and below each Z line. Since T tubules constitute the Z lines, the sacs of the sarcoplasmic reticulum of one sarcomere lie just above a T tubule, whereas those of the next sarcomere lie just below it. This forms a triple-layered structure (a T tubule sandwiched between sacs of the sarcoplasmic reticulum) called a *triad*.

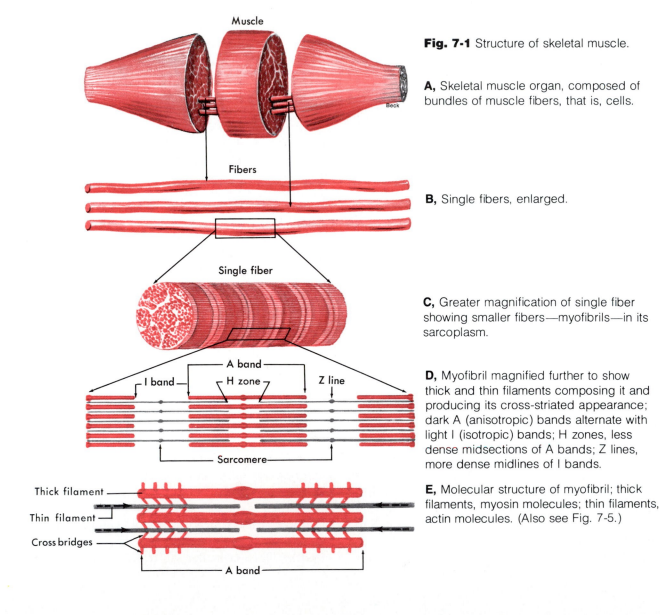

Fig. 7-1 Structure of skeletal muscle.

A, Skeletal muscle organ, composed of bundles of muscle fibers, that is, cells.

B, Single fibers, enlarged.

C, Greater magnification of single fiber showing smaller fibers—myofibrils—in its sarcoplasm.

D, Myofibril magnified further to show thick and thin filaments composing it and producing its cross-striated appearance; dark A (anisotropic) bands alternate with light I (isotropic) bands; H zones, less dense midsections of A bands; Z lines, more dense midlines of I bands.

E, Molecular structure of myofibril; thick filaments, myosin molecules; thin filaments, actin molecules. (Also see Fig. 7-5.)

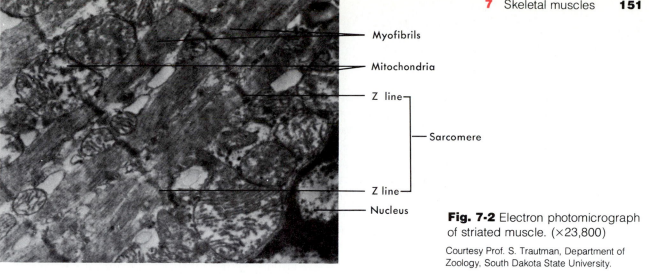

Myofibrils

Mitochondria

Z line ⎤

Sarcomere

Z line ⎦

Nucleus

Fig. 7-2 Electron photomicrograph of striated muscle. (×23,800)

Courtesy Prof. S. Trautman, Department of Zoology, South Dakota State University.

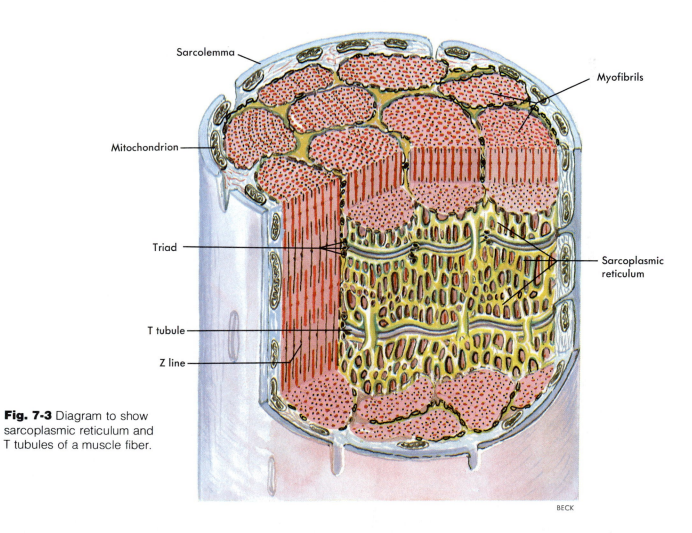

Sarcolemma

Myofibrils

Mitochondrion

Triad

Sarcoplasmic reticulum

T tubule

Z line

BECK

Fig. 7-3 Diagram to show sarcoplasmic reticulum and T tubules of a muscle fiber.

Molecular structure

In recent years, researchers have made many discoveries about the molecular structure of muscle fibers. Here are a few of them. They have learned that four proteins—myosin, actin, tropomyosin, and troponin—are found only in the sarcomere units of muscle fibers and that these protein compounds interact to produce muscle contraction. Each muscle fiber may contain a thousand or more subunits, about 1 μm in diameter, called *myofibrils*. Lying side by side in each myofibril are up to 2,500 actin and myosin *myofilaments*. Researchers have established that the *thick filaments* of a myofibril consist almost entirely of myosin molecules and that there are 300 or 400 of them per filament. They know the shape of myosin molecules. Each one is a thin rod with two rounded heads at one end. They know, too, that these myosin molecules are arranged lengthwise in the thick filaments and that some of their heads point toward one end of the filament and some point toward the other end. Thus the heads of the myosin molecules jut out from the surface of both ends of the thick filaments in projections that are called *cross bridges*. A schematic diagram of these appears in Fig. 7-1, *E*.

Thin filaments consist of a complex arrangement of three kinds of protein compounds, namely, actin, tropomyosin, and troponin.* Within a myofibril the thick and thin filaments alternate, as shown in Fig. 7-1, *D*. This arrangement is crucial for contraction. Another fact important for contraction is that the thin filaments attach to both Z lines of a sarcomere and that they extend in from the Z lines part way toward the center of the sarcomere. When the muscle fiber is relaxed, the thin filaments terminate at the outer edges of the H zones. In contrast, the thick myosin filaments do not attach to the Z lines and they extend only the length of the A bands of the sarcomeres.

*Murray, J. M., and Weber, A.: The cooperative action of muscle proteins, Sci. Am. **230**:59-71, Feb., 1974.

Functions

Skeletal muscle fibers specialize in the function of contraction. When nerve impulses arrive at a skeletal muscle fiber, they initiate impulse conduction over its sarcolemma and inward via its T tubules. This triggers the release of calcium ions from the sacs of the sarcoplasmic reticulum into the sarcoplasm. Here, calcium ions combine with the troponin molecules in the thin filaments of the myofibrils. In a resting muscle fiber, troponin prevents myosin from interacting with actin. Stated differently, troponin that is not bound to calcium prevents the cross bridges of the thick filaments from attaching to the actin molecules of the thin filaments. But calcium-bound troponin permits this action. Therefore myosin interacts with actin and pulls the thin filaments toward the center of each sarcomere. This shortens the sarcomeres and thereby shortens the myofibrils and the muscle fibers they compose. If sufficient numbers of fibers composing a skeletal muscle organ shorten, the muscle itself shortens. In a word, it contracts.

Relaxation of a muscle fiber is now believed to be brought about by a reversal of the contraction mechanism just described. The calcium-troponin combinations separate, calcium ions reenter the sacs of the sarcoplasmic reticulum, and troponin, now no longer bound to calcium, inhibits myosin-actin interaction. The sarcoplasmic reticulum is sometimes called the "relaxing factor" of muscle cells because it has a very strong affinity for calcium. Calcium ions freed by a nerve impulse to initiate contraction are available for only a few milliseconds before becoming firmly bound, once again, to the sarcoplasmic reticulum. This process of rapid binding of calcium ions helps to maintain muscle fibers in a relaxed state.

Summarizing, the release of calcium ions from the sacs of the sarcoplasmic reticulum turns on muscle contraction. The withdrawal of calcium ions back into the sarcoplasmic reticulum's sacs turns off muscle contraction and turns on muscle relaxation.

Muscle cells obey the all-or-none law when they contract. This means that they either contract with all the force possible under existing conditions or they do not contract at all. However, if conditions at the time of stimulation change, the force of the cell's contraction changes. Suppose, for example, that a particular muscle fiber receives an adequate oxygen supply at one time and an inadequate supply at another time. It will contract more forcefully with an adequate than with a deficient oxygen supply.

Energy sources for muscle contraction

The high-energy compound adenosine triphosphate (ATP) that is present in muscle fibers and all cells breaks down, releasing energy that does the work of contracting muscles. As Fig. 7-4 shows, nerve impulses arriving at a muscle fiber trigger both ATP breakdown and the release of calcium ions from the sacs of the sarcoplasmic reticulum into the sarcoplasm of the muscle fiber. The calcium ions combine with the troponin molecules of the thin filaments in myofibrils.

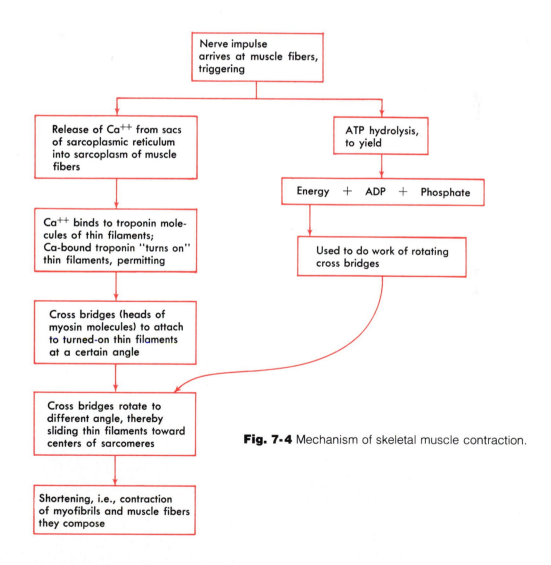

Fig. 7-4 Mechanism of skeletal muscle contraction.

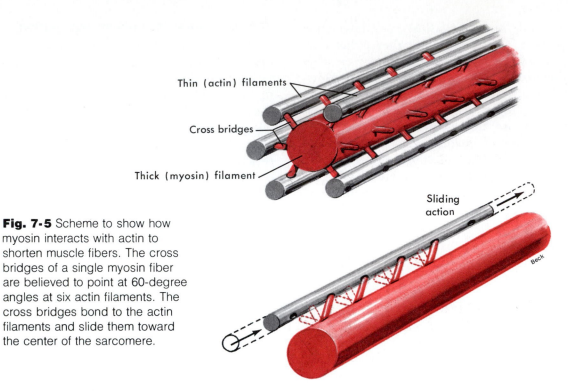

Fig. 7-5 Scheme to show how myosin interacts with actin to shorten muscle fibers. The cross bridges of a single myosin fiber are believed to point at 60-degree angles at six actin filaments. The cross bridges bond to the actin filaments and slide them toward the center of the sarcomere.

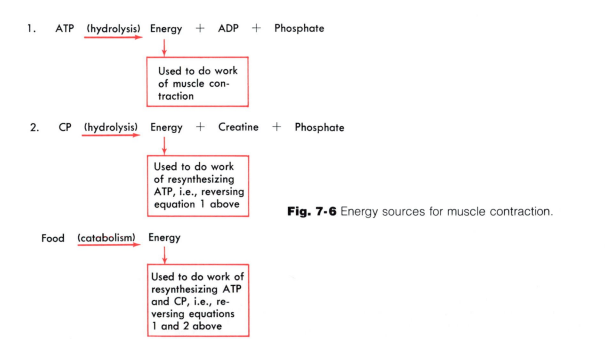

Fig. 7-6 Energy sources for muscle contraction.

Calcium-combined troponin acts in some way to "turn on" the thin filaments. What this means is that the actin molecules of the thin filaments become able to combine with the heads of the myosin molecules that make up the cross bridges of the thick filaments. The energy released from ATP breakdown then does the work of rotating the cross bridges to a different angle (Fig. 7-5). As the cross bridges rotate, they move the thin filaments to which they are attached in toward the centers of the sarcomeres. This necessarily shortens the sarcomeres and the myofibrils. Thus energy released from ATP breakdown does the work of muscle contraction.

Muscle fibers must continually resynthesize ATP because they can store only small amounts of it. Immediately after ATP breaks down, energy for its resynthesis is supplied by the breakdown of another high-energy compound, creatine phosphate (CP), which is also present in small amounts in muscle fibers. But ultimately, energy for both ATP and CP synthesis comes from the catabolism of foods (see Fig. 19-1 and discussion in Chapter 19).

Skeletal muscle organs
Structure
Size, shape, and fiber arrangement

The structures called skeletal muscles are organs. They consist mainly of skeletal muscle tissue plus important connective and nervous tissue components. Skeletal muscles vary considerably in size, shape, and arrangement of fibers. They range from extremely tiny strands such as the stapedius muscle of the middle ear to large masses such as the muscles of the thigh. Some skeletal muscles are broad in shape and some narrow. Some are long and tapering and some short and blunt. Some are triangular, some quadrilateral, and some irregular. Some form flat sheets and others bulky masses.

Arrangement of fibers varies in different muscles. In some muscles the fibers are parallel to the long axis of the muscle, in some they converge to a narrow attachment, and in some they are oblique and either pennate (like the feathers in an old-fashioned plume pen) or bipennate (double-feathered, as in the rectus femoris). Fibers may even be curved, as in the sphincters of the face, for example. The direction of the fibers composing a muscle is significant because of its relationship to function. For instance, a muscle with the bipennate fiber arrangement can produce the strongest contraction.

Connective tissue components

A fibrous connective tissue sheath (*epimysium*) envelops each muscle and extends into it as partitions between bundles of its fibers (*perimysium*) and between individual fibers (*endomysium*). Because all three of these structures are continuous with the fibrous structures that attach muscles to bones or other structures, muscles are most firmly harnessed to the structures they pull on during contraction. The epimysium, perimysium, and endomysium of a muscle, for example, may be continuous with fibrous tissue that extends from the muscle as a *tendon*, a strong tough cord continuous at its other end with the fibrous covering of bone (periosteum). Or the fibrous wrapping of a muscle may extend as a broad, flat sheet of connective tissue (*aponeurosis*) to attach it to adjacent structures, usually the fibrous wrappings of another muscle. So tough and strong are tendons and aponeuroses that they are not often torn, even by injuries forceful enough to break bones or tear muscles. They are, however, occasionally pulled away from bones.

Tube-shaped structures of fibrous connective tissue called *tendon sheaths* enclose certain tendons, notably those of the wrist and ankle. Like the bursae, tendon sheaths have a lining of synovial membrane. Its moist smooth surface enables the tendon to move easily, almost frictionlessly, in the tendon sheath.

You may recall that a continuous sheet of loose connective tissue known as the superficial fascia lies directly under the skin. Under this lies a layer of dense fibrous connective tissue, the *deep fascia*. Extensions of the deep fascia form the epimysium, perimysium, and endomysium of muscles and their attachments to bones and other structures and also enclose viscera, glands, blood vessels, and nerves.

Nerve supply

A nerve cell that transmits impulses to a skeletal muscle is called a *somatic motoneuron*. One such neuron plus the muscle cells in which its axon terminates constitutes a *motor unit* (Fig. 7-7). The single axon fiber of a motor unit divides, on entering the skeletal muscle, into a variable number of branches. Those of some motor units terminate in only a few muscle fibers, whereas others terminate in numerous fibers. Consequently, impulse conduction by one motor unit may stimulate only a half dozen or so muscle fibers to contract at one time, whereas conduction by another motor unit may activate a hundred or more fibers simultaneously. This fact bears a relationship to the function of the muscle as a whole. As a general rule, the fewer the number of fibers supplied by a skeletal muscle's individual motor units, the more precise the movements that muscle can produce. For example, in certain small muscles of the hand, each motor unit includes only a few muscle fibers, and these muscles produce precise finger movements. In contrast, motor units in large abdominal muscles that do not produce precise movements are reported to include more than a hundred muscle fibers each.

The area of contact between a nerve and muscle fiber is known as the *motor end-plate* or *neuromuscular junction* (Fig. 7-7, inset). When nerve impulses reach the ends of the axon fibers in a skeletal muscle, small vesicles in the axon terminals release a chemical—acetylcholine—into the neuromuscular junction. Diffusing swiftly across this microscopic trough, acetylcholine contacts the sarcolemma of the adjacent muscle fibers, stimulating the fiber to contract. In addition to the many motor nerve endings, there are also many sensory nerve endings in skeletal muscles.

Age changes

As a person grows old, his skeletal muscles undergo a process called *fibrosis*. Gradually, some of the skeletal muscle fibers degenerate, and fibrous connective tissue replaces them. With this loss of muscle fibers and increase in connective tissue comes waning muscular strength, a common finding in elderly people.* Other factors probably also contribute to decreasing muscular strength in advanced age.

Function

Several methods of study have been used to amass the present-day store of knowledge about how muscles function. They vary from the traditional and relatively simple procedures, such as observing and palpating muscles in action, manipulating dissected muscles to observe movements, or deducing movements from knowledge of muscle anatomy, to the newer more complicated method of electromyography (recording action potentials from contracting muscles). As a result, there is now a somewhat overwhelming amount of knowledge about muscle functions. So perhaps we can thread our way through this maze of detail more easily if we start with general principles and then go on to the details that seem to us most useful.

Basic principles

1 *Skeletal muscles contract only if stimulated.* They do not have the quality of automaticity inherent in cardiac and visceral muscle. Although nerve impulses are the natural stimuli for skeletal muscles, electrical and some other artificial stimuli such as heat or injury can also activate them. A skeletal muscle deprived of nerve impulses by whatever cause is a functionless mass.

*Shock, N. W.: The physiology of aging, Sci. Am. **206:**100-110, Jan., 1962.

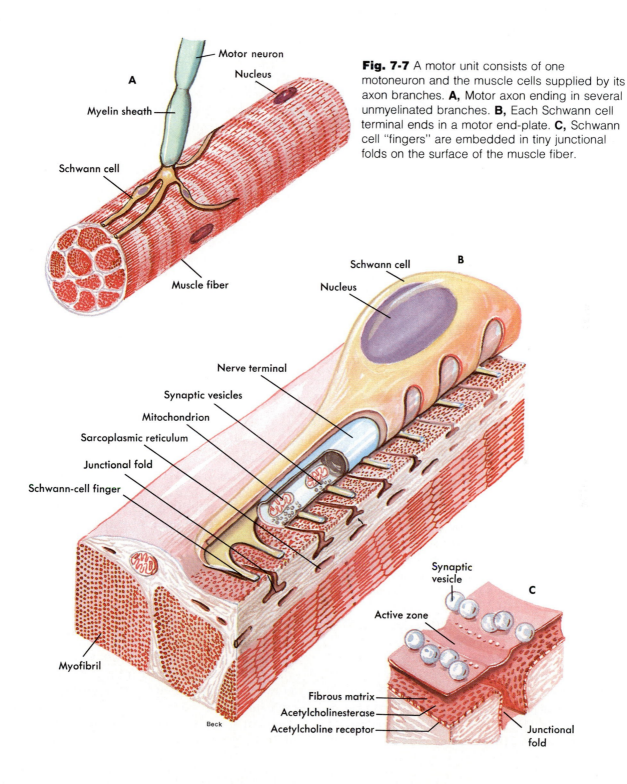

A

Motor neuron

Nucleus

Myelin sheath

Schwann cell

Muscle fiber

Fig. 7-7 A motor unit consists of one motoneuron and the muscle cells supplied by its axon branches. **A,** Motor axon ending in several unmyelinated branches. **B,** Each Schwann cell terminal ends in a motor end-plate. **C,** Schwann cell "fingers" are embedded in tiny junctional folds on the surface of the muscle fiber.

B

Schwann cell

Nucleus

Nerve terminal

Synaptic vesicles

Mitochondrion

Sarcoplasmic reticulum

Junctional fold

Schwann-cell finger

Myofibril

Beck

Synaptic vesicle

Active zone

C

Fibrous matrix

Acetylcholinesterase

Acetylcholine receptor

Junctional fold

One should, therefore, think of a skeletal muscle and its motor nerve as a physiological unit, always functioning together, either useless without the other.

2 *A skeletal muscle contraction may be any one of several types*. It may be a tonic contraction, an isotonic contraction, an isometric contraction, a twitch contraction, or a tetanic contraction. And there are also other types of contraction—called treppe, fibrillation, and convulsions.

a A *tonic contraction* (*tonus*, tone) is a continual, partial contraction. At any one moment a small number of the total fibers in a muscle contract, producing a tautness of the muscle rather than a recognizable contraction and movement. Different groups of fibers scattered throughout the muscle contract in relays. Tonic contraction, or tone, is characteristic of the muscles of normal individuals when they are awake. It is particularly important for maintaining posture. A striking illustration of this fact is the following: When a person loses consciousness, his muscles lose their tone and he collapses in a heap, unable to maintain a sitting or standing posture. Muscles with less tone than normal are described as flaccid muscles and those with more than normal tone are called spastic. Impulses over stretch reflex arcs (Fig. 8-5, p. 209) maintain tone. Specialized sensory (stretch) receptors called *muscle spindles* and *neurotendinous endorgans* are capable of detecting the degree of stretch in a muscle or at the junction of a muscle with its tendon. These receptors permit regulation of muscle tone at a reflex level. Blocking passage of impulses from these sensory endorgans by cutting the reflex arc will result in loss of muscle tonus.

b An *isotonic contraction* (*iso*, same; *tonic*, tone, pressure, or tension) is a contraction in which the tone or tension within a muscle remains the same but the length of the muscle changes. It shortens, producing movement.

c An *isometric contraction* is a contraction in which muscle length remains the same but in which muscle tension increases. You can observe isometric contraction by pushing your arms against a wall and feeling the tension increase in your arm muscles. Isometric contractions "tighten" a muscle, but they do not produce movements or do work. Isotonic contractions, on the other hand, both produce movements and do work. Although a majority of muscles can contract either isotonically or isometrically, most body movements are a mixture of the two. Can you explain how walking or running is an example of movement resulting from both isotonic and isometric muscle contraction?

d A *twitch contraction* is a quick, jerky contraction in response to a single stimulus. Fig. 7-8 shows a record of such a contraction. It reveals that the muscle does not shorten at the instant of stimulation, but rather a fraction of a second later, and that it reaches a peak of shortening and then gradually resumes its former length. These three phases of contraction are spoken of, respectively, as the *latent period*, the *contraction phase*, and the *relaxation phase*. The entire twitch usually lasts less than $1/10$ of a second. Twitch contractions rarely occur in the body.

e A *tetanic contraction* (*tetanus*) is a more sustained contraction than a twitch. It is produced by a series of stimuli bombarding the muscle in rapid succession. About 30 stimuli per second, for example, evoke a tetanic contraction by a frog gastrocnemius muscle, but the rate varies for different muscles and different conditions. Fig. 7-9 shows records of incomplete and complete tetanus. Normal movements are said to be produced by incomplete tetanic contractions.

f *Treppe (staircase phenomenon)* is a phenomenon in which increasingly stronger twitch contractions occur in response to constant-strength stimuli repeated at the rate of about once or twice a second. In other words, a muscle contracts more forcefully after it has contracted a few times than when it first contracts—a principle made practical use of by athletes when they warm up but one not yet satisfactorily explained. Presumably, it relates partly to the rise

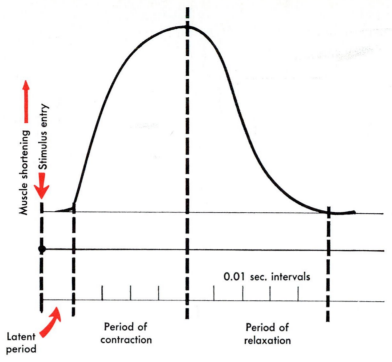

Fig. 7-8 The simple muscle twitch and its time components. (From Stacy, R. W., and Santolucito, J. A.: Modern college physiology, St. Louis, The C. V. Mosby Co.)

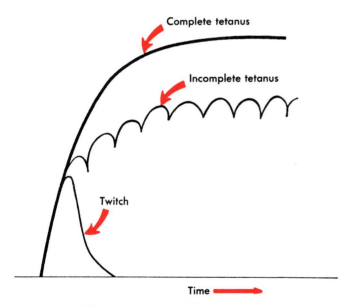

Fig. 7-9 Single twitch, incomplete tetanus, and complete tetanus of muscle. (From Stacy, R. W., and Santolucito, J. A.: Modern college physiology, St. Louis, The C. V. Mosby Co.)

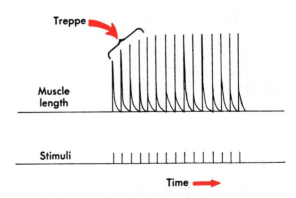

Fig. 7-10 Record of several successive muscle contractions, showing treppe occurring in the first few. (From Stacy, R. W., and Santolucito, J. A.: Modern college physiology, St. Louis, The C. V. Mosby Co.)

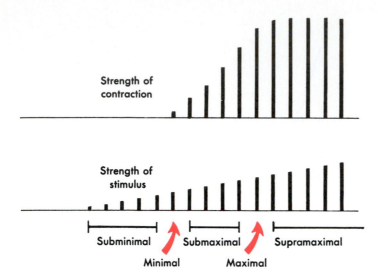

Strength of
contraction

Strength of
stimulus

Subminimal   Submaximal   Supramaximal

Minimal   Maximal

Fig. 7-11 Variation of strength of contraction of muscle with strength of stimulus. (From Stacy, R. W., and Santolucito, J. A.: Modern college physiology, St. Louis, The C. V. Mosby Co.)

in temperature of active muscles and partly to their accumulation of metabolic products. After the first few stimuli, muscle responds to a considerable number of successive stimuli with maximal contractions (Fig. 7-10). Eventually, it will respond with less and less strong contractions. The relaxation phase becomes shorter and finally disappears entirely. In other words, the muscle stays partially contracted—an abnormal state of prolonged contraction called *contracture.*

Repeated stimulation of muscle in time lessens its irritability and contractility and may result in muscle fatigue, a condition in which the muscle does not respond to the strongest stimuli. Complete muscle fatigue, however, very seldom occurs in the body but can be readily induced in an excised muscle.

g *Fibrillation* is an abnormal type of contraction in which individual fibers contract asynchronously, producing a flutter of the muscle but no effective movement. Fibrillation of the heart, for example, occurs fairly often.

h *Convulsions* are abnormal uncoordinated tetanic contractions of varying groups of muscles.

3 *Skeletal muscles* (organs) *contract according to the graded strength principle* (Fig. 7-11)—not according to the all-or-none principle, as do the individual muscle cells composing them. In other words, skeletal muscles contract with varying degrees of strength at different times—a fact of practical importance. (How else, for example, could we match the force of a movement to the demands of a task?)

Several generalizations may help explain the fact of graded strength contractions. The strength of the contraction of a skeletal muscle bears a direct relationship to the initial length of its fibers, to their metabolic condition, and to the number of them contracting. If a muscle is moderately stretched at the moment when contraction begins, the force of its contraction increases. This principle, established years ago, applies experimentally to heart muscle also (Starling's law of the heart, discussed in Chapter 15). Outstanding among metabolic conditions that influence contraction are oxygen and food supply. With adequate amounts of these essentials a muscle can contract with greater force than possible with deficient amounts. The greater the number of muscle fibers contracting simultane-

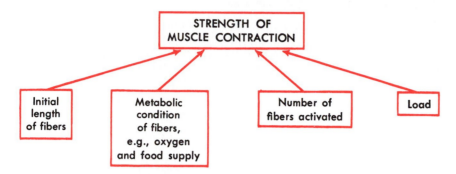

Fig. 7-12 Factors that influence the strength of muscle contraction.

ously, the stronger the contraction of a muscle. How large this number is depends on how many motor units are activated, and this, in turn, depends on the intensity and frequency of stimulation. In general, the more intense and the more frequent a stimulus, the more motor units and therefore the more fibers are activated and the stronger the contraction. Contraction strength also relates to previous contraction, the warm-up principle discussed on p. 158.

Another factor that influences the force of contraction is the size of the load imposed on the muscle. Within certain limits, the heavier the load, the stronger the contraction. Lift a pencil, for example, and then a heavy book and you can feel your arm muscles contract more strongly with the book.

The factors that influence muscle contraction are summarized in Fig. 7-12.

4 *Skeletal muscles produce movements by pulling on bones.* Most of our muscles span at least one joint and attach to both articulating bones. When they contract, therefore, their shortening puts a pull on both bones, and this pull moves one of the bones at the joint—draws it toward the other bone, much as a pull on marionette strings moves a puppet's parts. (In case you are wondering why both bones do not move, since both are pulled on by the contracting muscle, the reason is that one of them is normally stabilized by isometric contractions of other muscles or by certain features of its own that makes it less mobile.)

5 *Bones serve as levers, and joints serve as fulcrums of these levers.* (By definition, a *lever* is any rigid bar free to turn about a fixed point called its *fulcrum*.) A contracting muscle applies a pulling force on a bone lever at the point of the muscle's attachment to the bone. This causes the bone (referred to as the insertion bone) to move about its joint fulcrum. We have already noted that a skeletal muscle and its motor nerve act as a functional unit. Now we can add bones and joints to this unit and describe the physiological unit for movement as a neuromusculoskeletal unit. Disease or injury of any one of these parts of the unit—of nerve or muscle or bone or joint —can, as you might surmise, cause abnormal movements or complete loss of movement. Poliomyelitis, for example, and multiple sclerosis and hemiplegia all involve the neural part of the unit. In contrast, muscular dystrophy affects the

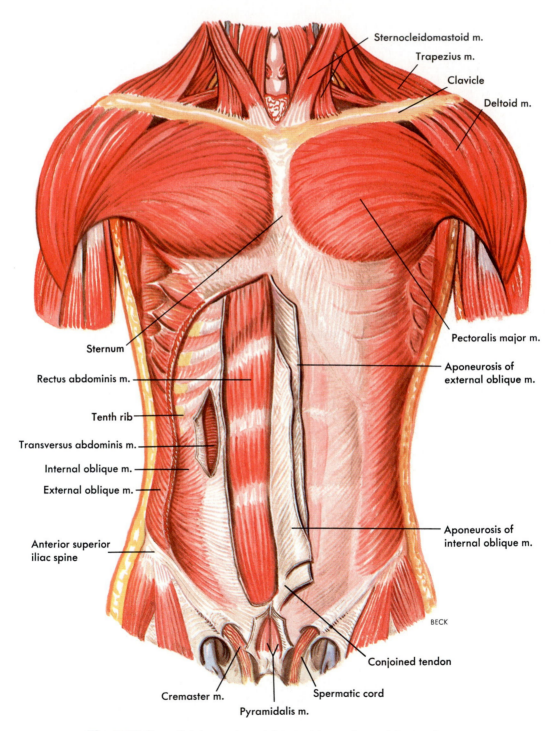

Sternocleidomastoid m.

Trapezius m.

Clavicle

Deltoid m.

Pectoralis major m.

Aponeurosis of
external oblique m.

Aponeurosis of
internal oblique m.

Sternum

Rectus abdominis m.

Tenth rib

Transversus abdominis m.

Internal oblique m.

External oblique m.

Anterior superior
iliac spine

Conjoined tendon

Spermatic cord

Cremaster m.

Pyramidalis m.

BECK

Fig. 7-13 Superficial muscles of the anterior surface of the trunk.

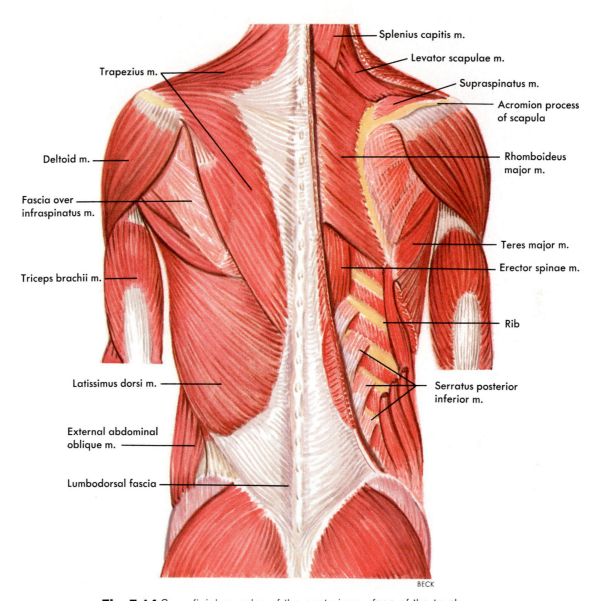

Trapezius m.

Deltoid m.

Fascia over
infraspinatus m.

Triceps brachii m.

Latissimus dorsi m.

External abdominal
oblique m.

Lumbodorsal fascia

Splenius capitis m.

Levator scapulae m.

Supraspinatus m.

Acromion process
of scapula

Rhomboideus
major m.

Teres major m.

Erector spinae m.

Rib

Serratus posterior
inferior m.

BECK

Fig. 7-14 Superficial muscles of the posterior surface of the trunk.

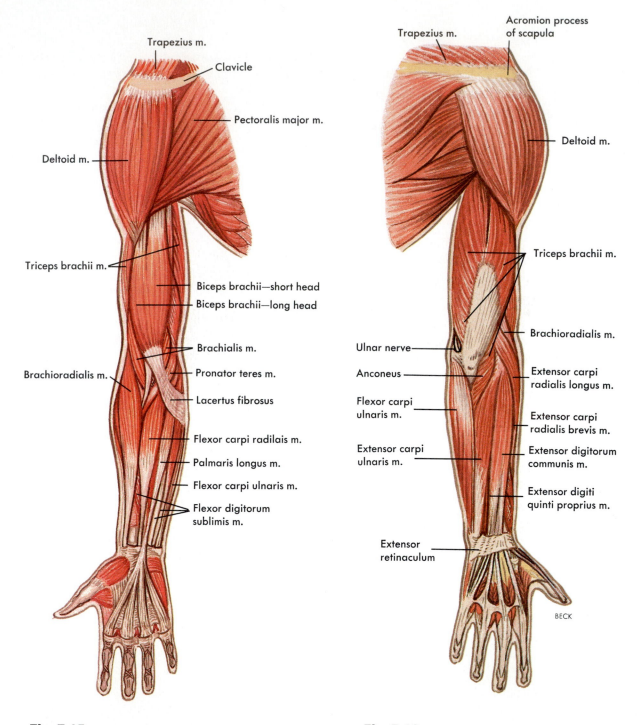

Fig. 7-15
Muscles of the flexor surface of the upper extremity.

Fig. 7-16
Muscles of the extensor surface of the upper extremity.

muscular part and arthritis the skeletal part.

6 *Muscles that move a part usually do not lie over that part.* In most cases the body of a muscle lies proximal to the part moved. Thus muscles that move the lower arm lie proximal to it, that is, in the upper arm. Applying the same principle, where would you expect muscles that move the hand to be located? Those that move the lower leg? Those that move the upper arm?

7 *Skeletal muscles almost always act in groups rather than singly.* In other words, most movements are produced by the coordinated action of several muscles. Some of the muscles in the group contract while others relax. To identify each muscle's special function in the group the following classification is used.

  a *prime movers*—muscle or muscles whose contraction actually produces the movement

  b *antagonists*—muscles that relax while the prime mover is contracting to produce movement (exception: contraction of the antagonist at the same time as the prime mover when some part of the body needs to be held rigid, such as the knee joint when standing*)

  c *synergists*—muscles that contract at the same time as the prime mover (may help the prime mover produce its movement or may stabilize a part—hold it steady—so that the prime mover produces a more effective movement)

Hints on how to deduce actions

To understand muscle actions, you need first to know certain anatomical facts such as which

*Antagonistic muscles have opposite actions and opposite locations. If the flexor lies anterior to the part, the extensor will be found posterior to it. For example, the pectoralis major, the flexor of the upper arm, is located on the anterior aspect of the chest, whereas the latissimus dorsi, the extensor of the upper arm, is located on the posterior aspect of the chest. The antagonist of a flexor muscle is obviously an extensor muscle and that of an abductor muscle, an adductor muscle. Some frequently used antagonists are listed in Table 7-2, p. 183.

bones muscles attach to and which joints they pull across. Then if you relate these structural facts to functional principles (for instance, those discussed in the preceding paragraphs), you may find your study of muscles more interesting and less difficult than you anticipate. Some specific suggestions for deducing muscle actions follow.

1 Start by making yourself familiar with the names, shapes, and general locations of the larger muscles, using Table 7-1, p. 182, as a guide.

2 Try to deduce which bones the two ends of a muscle attach to from your knowledge of the shape and general location of the muscle. For example, look carefully at the deltoid muscle as illustrated in Figs. 7-13 to 7-16. To what bones does it seem to attach? Check your deductions with Table 7-4, p. 185.

3 Next, make a guess as to which bone moves when the muscle shortens. (The bone moved by a muscle's contraction is its *insertion* bone; the bone that remains relatively stationary is its *origin* bone.) In many cases, you can tell by trying to move one bone and then another which one is the insertion bone. In some cases, either bone may function as the insertion. Although not all muscle attachments can be deduced as readily as those of the deltoid, they can all be learned more easily by using this deduction method than by relying on rote memory alone.

4 Deduce a muscle's actions by applying the principle that its insertion moves toward its origin. Check your conclusions with the text. Here, as in steps 2 and 3, the method of deduction is intended merely as a guide and is not adequate by itself for determining muscle actions.

5 To deduce which muscle produces a given action (instead of which action a given muscle produces, as in step 4), start by inferring the insertion bone (bone that moves during the action). The body and origin of the muscle will lie on one or more of the bones toward which the insertion moves—often a bone or bones proximal to the insertion bone. Couple these conclusions about origin and insertion with your knowledge of muscle names and locations to deduce the muscle that produces the action.

For example, if you wish to determine the prime mover for the action of raising the upper arms straight out to the sides, you infer that the muscle inserts on the humerus, since this is the bone that moves. It moves toward the shoulder, that is, the clavicle and scapula, so that probably the muscle has its origin on these bones. Because you know that the deltoid muscle fulfills these conditions, you conclude, and rightly so, that it is the muscle that raises the upper arms sidewise.

6 Do not try to learn too many details about muscle origins, insertions, and actions. Remember, it is better to start by learning a few important facts thoroughly than to half learn a mass of relatively unimportant details. Remember, too, that trying to learn too many minute facts may well result in your not retaining even the main facts.

Names
Reasons for names

Muscle names seem more logical and therefore easier to learn when one understands the reasons for the names. Each name describes one or more of the following features about the muscle.

1 *Its action*—as flexor, extensor, adductor, etc.

2 *Direction of its fibers*—as rectus or transversus

3 *Its location*—as tibialis or femoris

4 *Number of divisions composing a muscle*—as biceps, triceps, or quadriceps

5 *Its shape*—as deltoid (triangular) or quadratus (square)

6 *Its points of attachment*—as sternocleidomastoid

A good way to start the study of a muscle is by trying to find out what its name means.

Muscles grouped according to location

Just as names of people are learned by associating them with physical appearance, so the names of muscles should be learned by associating them with their appearance. As you learn each muscle name, study Figs. 7-13 to 7-36 to familiarize yourself with the muscle's size, shape, and general location. To help you in this task, the names of some of the major muscles are grouped according to their location in Table 7-1, p. 182.

Muscles grouped according to function

The following terms are used to designate muscles according to their main actions (see Table 7-2, p. 183, for examples).

flexors—decrease the angle of a joint (between the anterior surfaces of the bones except in the knee and toe joints)

extensors—return the part from flexion to normal anatomical position; increase the angle of a joint

abductors—move the bone away from midline

adductors—move the part toward the midline

rotators—cause a part to pivot on its axis

levators—raise a part

depressors—lower a part

sphincters—reduce the size of an opening

tensors—tense a part, that is, make it more rigid

supinators—turn the hand palm upward

pronators—turn the hand palm downward

Origins, insertions, functions, and innervations of representative muscles

Basic information about many muscles is given in Tables 7-3 to 7-15. Each table has a description of a group of muscles that move one part of the body. Muscles that, in our judgment, are the most important for beginning students of anatomy to know are preceded by a block, and the origins and insertions so judged are set in boldface type. Remember that the actions listed for each muscle are those for which it is a prime mover. Actually, a single muscle contracting alone rarely accomplishes a given action. Instead, muscles act in groups as prime movers, synergists, and antagonists (p. 165) to bring about movements. As you study the muscles de-

Text continued on p. 178.

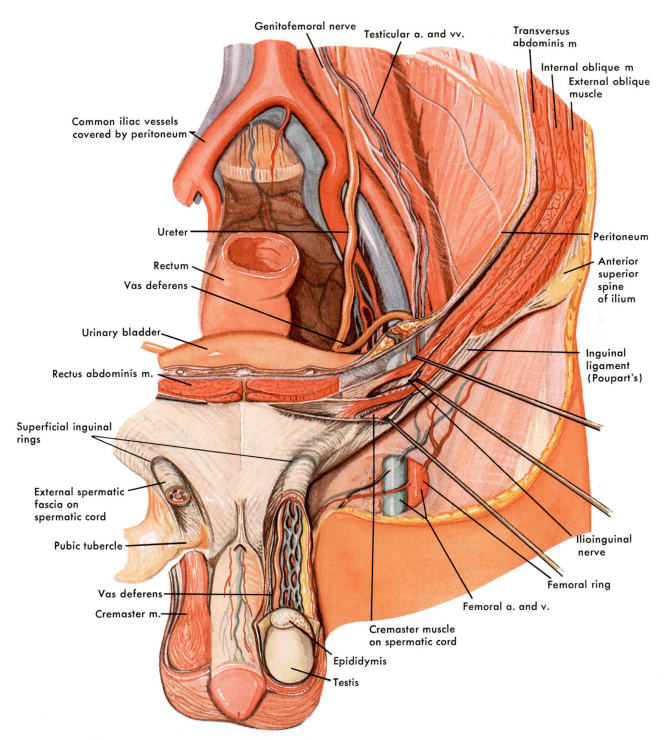

Genitofemoral nerve

Testicular a. and vv.

Transversus abdominis m

Internal oblique m

External oblique muscle

Common iliac vessels covered by peritoneum

Peritoneum

Anterior superior spine of ilium

Ureter

Rectum

Vas deferens

Urinary bladder

Inguinal ligament (Poupart's)

Rectus abdominis m.

Superficial inguinal rings

External spermatic fascia on spermatic cord

Pubic tubercle

Ilioinguinal nerve

Femoral ring

Vas deferens

Cremaster m.

Femoral a. and v.

Cremaster muscle on spermatic cord

Epididymis

Testis

Fig. 7-17 The anterolateral abdominal wall and inguinal area in the male. Note the placement of the superficial inguinal rings in the aponeurosis of the external oblique muscle.

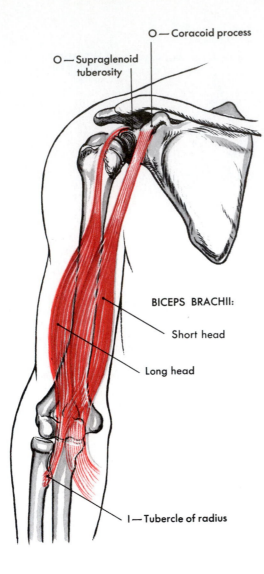

O — Coracoid process

O — Supraglenoid tuberosity

BICEPS BRACHII:

Short head

Long head

I — Tubercle of radius

Fig. 7-18
Biceps brachii muscle. *O*, Origin. *I*, Insertion.

Fig. 7-19
Triceps brachii muscle. *O*, Origin. *I*, Insertion.

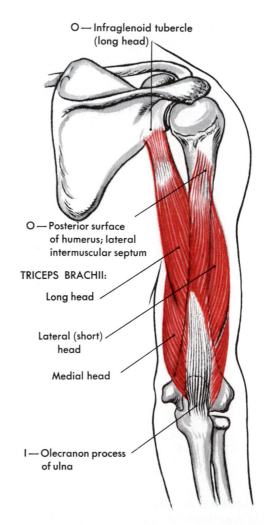

O — Infraglenoid tubercle (long head)

O — Posterior surface of humerus; lateral intermuscular septum

TRICEPS BRACHII:

Long head

Lateral (short) head

Medial head

I — Olecranon process of ulna

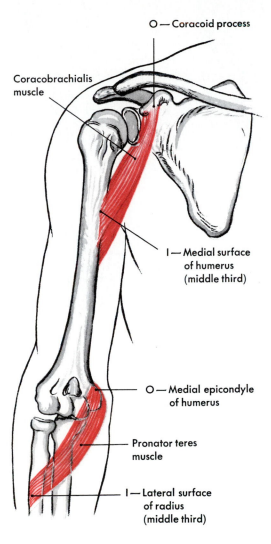

O — Coracoid process

Coracobrachialis muscle

I — Medial surface of humerus (middle third)

O — Medial epicondyle of humerus

Pronator teres muscle

I — Lateral surface of radius (middle third)

Fig. 7-20
Coracobrachialis and pronator teres muscles. *O*, Origin. *I*, Insertion.

Fig. 7-21
Brachialis muscle. *O*, Origin. *I*, Insertion.

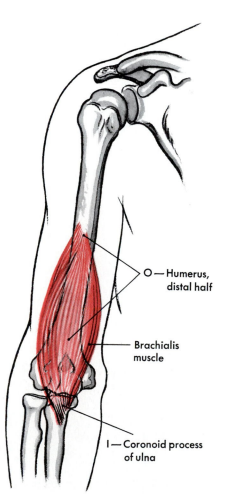

O — Humerus, distal half

Brachialis muscle

I — Coronoid process of ulna

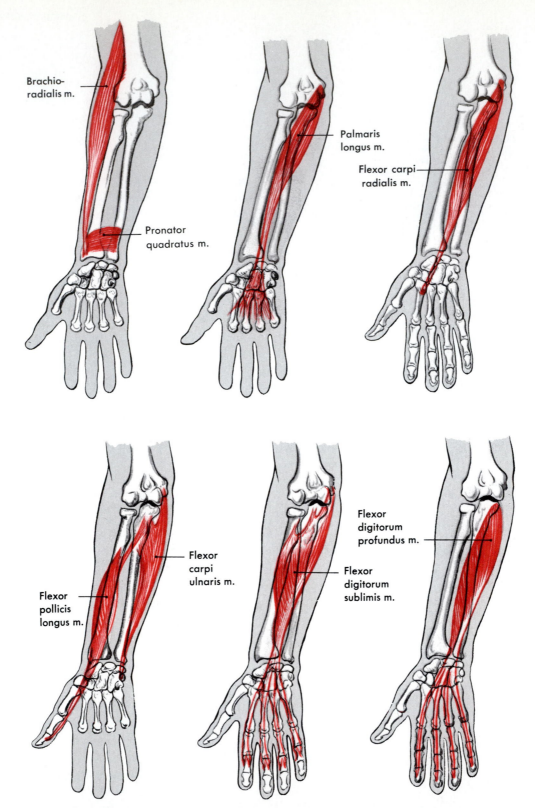

Fig. 7-22 Some muscles of the anterior (volar) aspect of the right forearm.

Brachio-radialis m.

Pronator quadratus m.

Palmaris longus m.

Flexor carpi radialis m.

Flexor pollicis longus m.

Flexor carpi ulnaris m.

Flexor digitorum profundus m.

Flexor digitorum sublimis m.

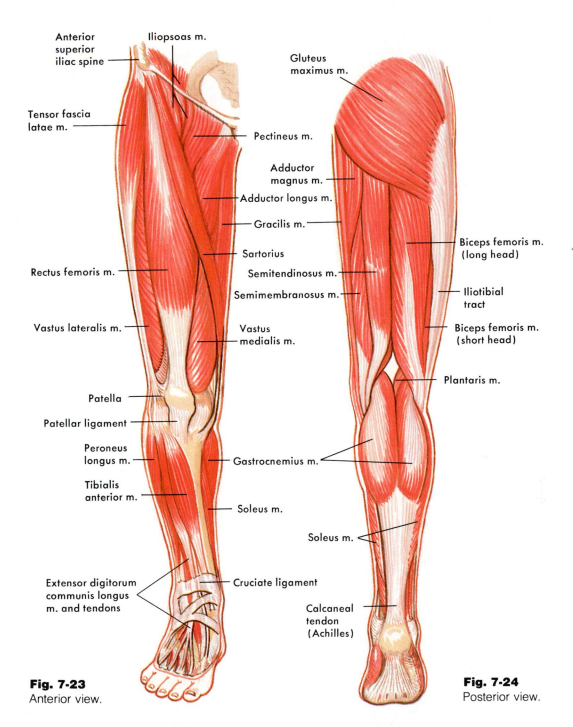

Anterior superior iliac spine

Iliopsoas m.

Tensor fascia latae m.

Pectineus m.

Adductor magnus m.

Adductor longus m.

Gracilis m.

Sartorius

Rectus femoris m.

Semitendinosus m.

Semimembranosus m.

Vastus lateralis m.

Vastus medialis m.

Patella

Patellar ligament

Peroneus longus m.

Gastrocnemius m.

Tibialis anterior m.

Soleus m.

Extensor digitorum communis longus m. and tendons

Cruciate ligament

Gluteus maximus m.

Biceps femoris m. (long head)

Iliotibial tract

Biceps femoris m. (short head)

Plantaris m.

Soleus m.

Calcaneal tendon (Achilles)

Fig. 7-23
Anterior view.

Fig. 7-24
Posterior view.

SUPERFICIAL MUSCLES OF THE RIGHT THIGH AND LEG

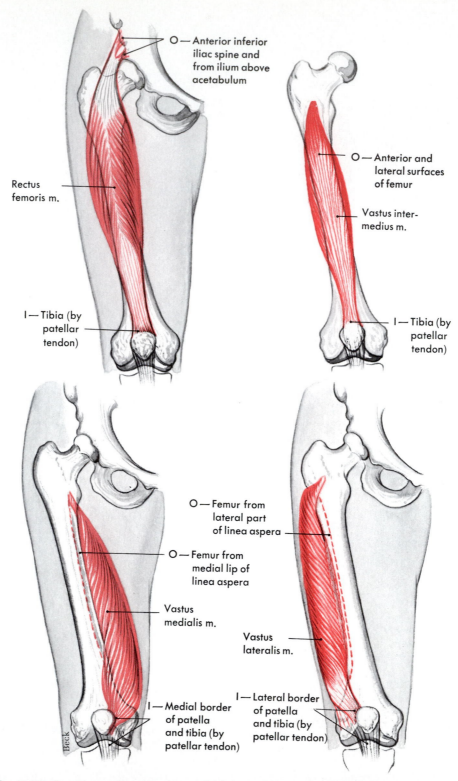

O — Anterior inferior iliac spine and from ilium above acetabulum

Rectus femoris m.

I — Tibia (by patellar tendon)

O — Anterior and lateral surfaces of femur

Vastus inter- medius m.

I — Tibia (by patellar tendon)

O — Femur from lateral part of linea aspera

O — Femur from medial lip of linea aspera

Vastus medialis m.

Vastus lateralis m.

I — Medial border of patella and tibia (by patellar tendon)

I — Lateral border of patella and tibia (by patellar tendon)

Beck

Fig. 7-25 Quadriceps femoris group of thigh muscles: rectus femoris, vastus intermedius, vastus medialis, and vastus lateralis.

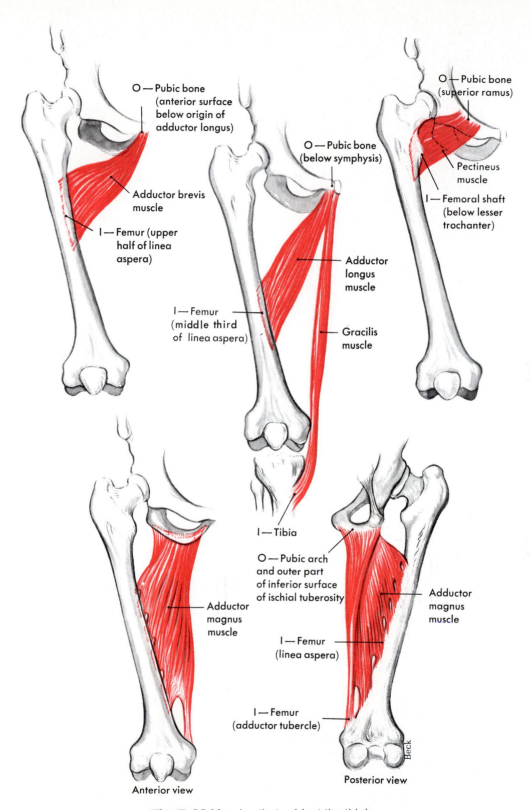

O—Pubic bone (anterior surface below origin of adductor longus)

Adductor brevis muscle

I—Femur (upper half of linea aspera)

O—Pubic bone (below symphysis)

Adductor longus muscle

I—Femur (middle third of linea aspera)

Gracilis muscle

O—Pubic bone (superior ramus)

Pectineus muscle

I—Femoral shaft (below lesser trochanter)

I—Tibia

O—Pubic arch and outer part of inferior surface of ischial tuberosity

Adductor magnus muscle

I—Femur (linea aspera)

Adductor magnus muscle

I—Femur (adductor tubercle)

Anterior view

Posterior view

Beck

Fig. 7-26 Muscles that adduct the thigh.

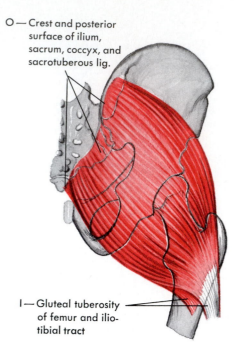

O — Crest and posterior surface of ilium, sacrum, coccyx, and sacrotuberous lig.

I — Gluteal tuberosity of femur and ilio-tibial tract

Fig. 7-27 Gluteus maximus muscle. *O,* Origin. *I,* Insertion.

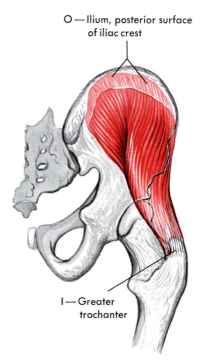

O — Ilium, posterior surface of iliac crest

I — Greater trochanter

O — Gluteal surface of ilium

I — Greater tro-chanter of femur

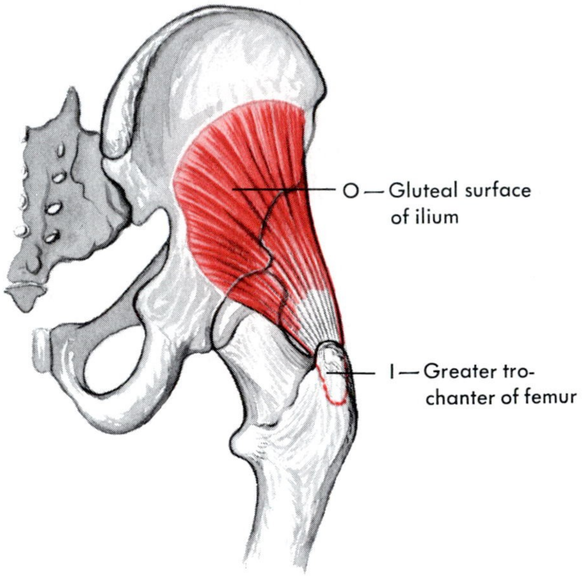

Fig. 7-28 Gluteus minimus muscle. *O,* Origin. *I,* Insertion.

Fig. 7-29 Gluteus medius muscle. *O,* Origin. *I,* Insertion.

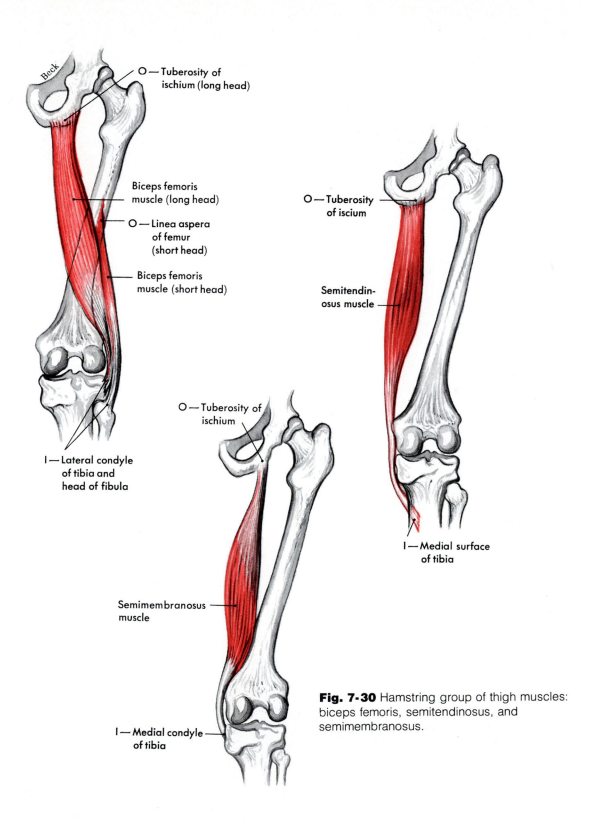

Beck

O — Tuberosity of ischium (long head)

Biceps femoris muscle (long head)

O — Linea aspera of femur (short head)

Biceps femoris muscle (short head)

I — Lateral condyle of tibia and head of fibula

O — Tuberosity of iscium

Semitendinosus muscle

I — Medial surface of tibia

O — Tuberosity of ischium

Semimembranosus muscle

I — Medial condyle of tibia

Fig. 7-30 Hamstring group of thigh muscles: biceps femoris, semitendinosus, and semimembranosus.

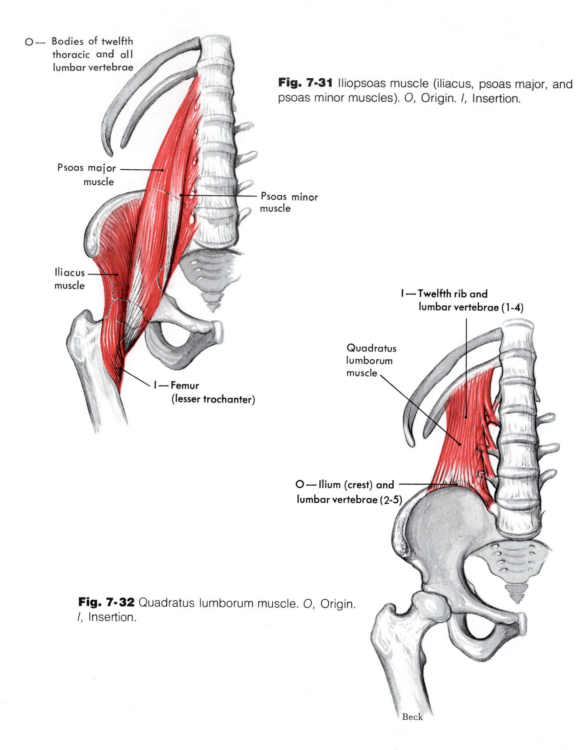

O— Bodies of twelfth thoracic and all lumbar vertebrae

Fig. 7-31 Iliopsoas muscle (iliacus, psoas major, and psoas minor muscles). *O,* Origin. *I,* Insertion.

Psoas major muscle

Psoas minor muscle

Iliacus muscle

I— Femur (lesser trochanter)

I— Twelfth rib and lumbar vertebrae (1-4)

Quadratus lumborum muscle

O— Ilium (crest) and lumbar vertebrae (2-5)

Fig. 7-32 Quadratus lumborum muscle. *O,* Origin. *I,* Insertion.

Beck

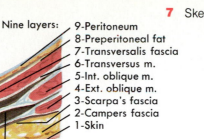

Nine layers:
9-Peritoneum
8-Preperitoneal fat
7-Transversalis fascia
6-Transversus m.
5-Int. oblique m.
4-Ext. oblique m.
3-Scarpa's fascia
2-Campers fascia
1-Skin

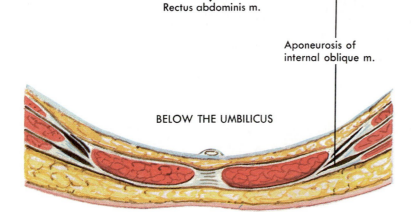

ABOVE THE UMBILICUS

Rectus abdominis m.

Aponeurosis of
internal oblique m.

BELOW THE UMBILICUS

Fig. 7-33 Horizontal section of the abdominal wall. The aponeurosis of the internal oblique muscle splits into two sections, one lying anterior and the other posterior to the rectus abdominis muscle, thereby forming an encasing sheath around this muscle, above in umbilicus. Below the umbilicus the aponeuroses of all the muscles pass anterior to the rectus.

Fig. 7-34 The diaphragm as seen from the front. Note the openings in the vertebral portion for the inferior vena cava, esophagus, and aorta.

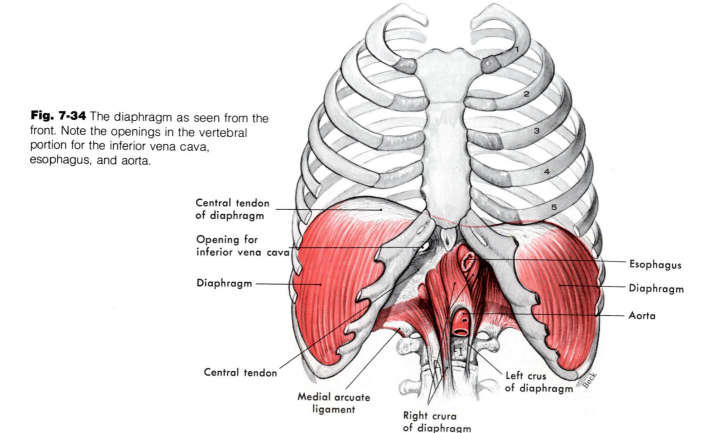

Central tendon of diaphragm

Opening for inferior vena cava

Diaphragm

Central tendon

Medial arcuate ligament

Right crura of diaphragm

Left crus of diaphragm

Esophagus

Diaphragm

Aorta

Beck

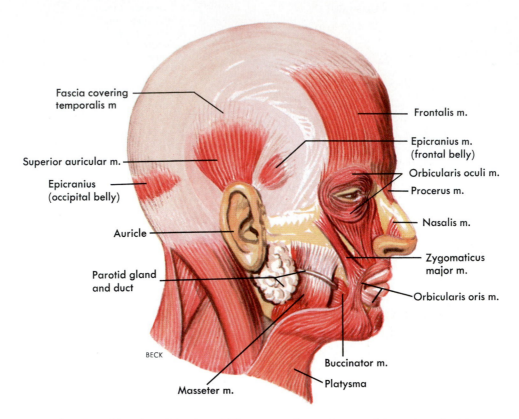

Fascia covering temporalis m

Frontalis m.

Epicranius m. (frontal belly)

Superior auricular m.

Orbicularis oculi m.

Epicranius (occipital belly)

Procerus m.

Nasalis m.

Auricle

Zygomaticus major m.

Parotid gland and duct

Orbicularis oris m.

BECK

Buccinator m.

Platysma

Masseter m.

Fig. 7-35 Muscles of the head. These muscles make possible various facial expressions.

scribed in Tables 7-3 to 7-15, try to follow the hints for deducing muscle actions given on p. 165.

Weak places in abdominal wall

There are several places in the abdominal wall where rupture (hernia) with protrusion of part of the intestine may occur. At these points the wall is weakened because of the presence of an interval or space in the abdominal aponeuroses. Any undue pressure on the abdominal viscera, therefore, can force a portion of the parietal peritoneum, and often a part of the intestine as well,

through these nonreinforced places. The weak places are (1) the inguinal canals, (2) the femoral rings, and (3) the umbilicus. Congenital defects or traumatic injury may permit abdominal viscera to project into the thorax (diaphragmatic hernia) or, on rare occasions, through the floor of the pelvis and some other areas.

Piercing the aponeuroses of the abdominal muscles are two canals, the *inguinal canals*, one on the right and the other on the left. They lie above, but parallel to, the inguinal ligaments (p. 167) and are about 5 cm long. In the male the spermatic cords extend through the canals into the scrotum (Fig. 7-17), whereas in the female the round ligaments of the uterus are in this location. The internal opening of each canal is a cir-

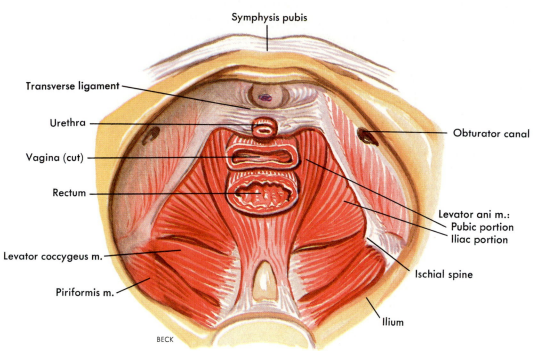

Symphysis pubis

Transverse ligament

Urethra

Vagina (cut)

Rectum

Levator coccygeus m.

Piriformis m.

Obturator canal

Levator ani m.:
Pubic portion
Iliac portion

Ischial spine

Ilium

BECK

Fig. 7-36
A, Female pelvic floor viewed from above.

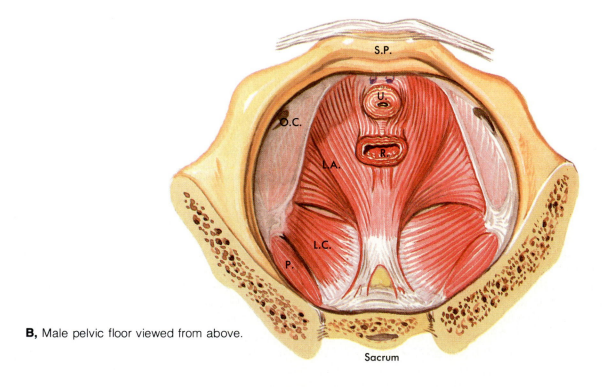

S.P.

U.

O.C.

R.

L.A.

L.C.

P.

Sacrum

B, Male pelvic floor viewed from above.

cular space in the aponeurosis of the transverse muscle known as the *internal (abdominal) inguinal ring*. The external openings called the *external (superficial) inguinal rings* are triangular spaces in the aponeuroses of the external oblique muscles. They are located inferiorly and mesially to the internal rings just above and lateral to the pubic crest. The upper surface of the inguinal ligament forms the floor of the canal as it passes obliquely downward and medially through the abdominal wall. The fact that the inguinal canal is larger in the male than in the female probably explains why inguinal hernia occurs more often in men than in women.

The *femoral (saphenous) rings* are openings in the fascia lata of the thigh below and lateral to the external inguinal rings and pubic tubercle (Fig. 7-17). The openings are covered by a thin membrane called the *cribriform fascia* that is perforated by the great saphenous vein. They have a diameter of about ½ inch and are usually somewhat larger in females, a fact that accounts for the greater prevalence of femoral hernia in women than in men.

Bursae

Definition

Bursae are small connective tissue sacs lined with synovial membrane and containing synovial fluid.

Locations

Bursae are located wherever pressure is exerted over moving parts, for example, between skin and bone, between tendons and bone, or between muscles or ligaments and bone. Some bursae that fairly frequently become inflamed (bursitis) are as follows: the subacromial bursa, between the head of the humerus and the acromion process and the deltoid muscle; the olecranon bursa, between the olecranon process and the skin; and the prepatellar bursa, between the patella and the skin. Inflammation of the prepatellar bursa is known as housemaid's knee, whereas olecranon bursitis is called student's elbow.

Function

Bursae act as cushions, relieving pressure between moving parts.

Posture

We have already discussed the major role muscles play in movement and heat production. We shall now turn our attention to a third way in which muscles serve the body as a whole—that of maintaining the posture of the body. Let us consider a few aspects of this important function.

Meaning

The term posture means simply position or alignment of body parts. "Good posture" means many things. It means body alignment that most favors function; it means position that requires the least muscular work to maintain, which puts the least strain on muscles, ligaments, and bones; it means keeping the body's center of gravity over its base. Good posture in the standing position, for example, means head and chest held high, chin, abdomen, and buttocks pulled in, knees bent slightly, and feet placed firmly on the ground about 6 inches apart.

How maintained

Since gravity pulls on the various parts of the body at all times, and since bones are too irregularly shaped to balance themselves on each other, the only way the body can be held upright is for muscles to exert a continual pull on bones in the opposite direction from gravity. Gravity tends to pull the head and trunk forward and downward; muscles (head and trunk extensors) must therefore pull backward and upward on them. Gravity pulls the lower jaw downward;

muscles must pull upward on it, etc. Muscles exert this pull against gravity by virtue of their property of tonicity. Because tonicity is absent during sleep, muscle pull does not then counteract the pull of gravity. Hence, for example, we cannot sleep standing up.

Many structures other than muscles and bones play a part in the maintenance of posture. The nervous system is responsible for the existence of muscle tone and also regulates and coordinates the amount of pull exerted by the individual muscles. The respiratory, digestive, circulatory, excretory, and endocrine systems all contribute something toward the ability of muscles to maintain posture. This is one of many examples of the important principle that all body functions are interdependent.

Importance to body as whole

The importance of posture can perhaps be best evaluated by considering some of the effects of poor posture. Poor posture throws more work on muscles to counteract the pull of gravity and therefore leads to fatigue more quickly than good posture. Poor posture puts more strain on ligaments. It puts abnormal strains on bones and may eventually produce deformities. It interferes with various functions such as respiration, heart action, and digestion. It probably is not going too far to say that it even detracts from one's feeling of self-confidence and joy. In support of this claim, consider our use of such expressions as "shoulders squared, head erect" to denote confidence and joy and "down-in-the-mouth," "long-faced," and "bowed down" to signify dejection and anxiety. The importance of posture to the body as a whole might be summed up in a single sentence: maximal health and good posture are reciprocally related, that is, each one depends on the other.

Tables 7-1 to 7-15 follow.

Table 7-1 Muscles grouped according to location

Location	Muscles	Figures illustrating
Neck	Sternocleidomastoid	7-13
Back	Trapezius	7-13 to 7-16
	Latissimus dorsi	7-14
Chest	Pectoralis major	7-13
	Serratus anterior	7-13
Abdominal wall	External oblique	7-13, 7-14
Shoulder	Deltoid	7-13 to 7-16
Upper arm	Biceps brachii	7-15, 7-18
	Triceps brachii	7-15, 7-16, 7-19
	Brachialis	7-15, 7-21
Forearm	Brachioradialis	7-15, 7-16, 7-22
	Pronator teres	7-15, 7-20
Buttocks	Gluteus maximus	7-24, 7-27
	Gluteus minimus	7-28
	Gluteus medius	7-29
	Tensor fascia latae	7-23
Thigh		
Anterior surface	Quadriceps femoris group	
	Rectus femoris	7-23, 7-25
	Vastus lateralis	7-23, 7-25
	Vastus medialis	7-23, 7-25
	Vastus intermedius	7-25
Medial surface	Gracilis	7-23, 7-24, 7-26
	Adductor group (brevis, longus, magnus)	7-23, 7-24, 7-26
Posterior surface	Hamstring group	
	Biceps femoris	7-24, 7-30
	Semitendinosus	7-24, 7-30
	Semimembranosus	7-24, 7-30
Leg		
Anterior surface	Tibialis anterior	7-23
Posterior surface	Gastrocnemius	7-24
	Soleus	7-24
Pelvic floor	Levator ani	7-36
	Levator coccygeus	7-36
	Rectococcygeus	7-36

Table 7-2 Muscles grouped according to function

Part moved	Example of flexor	Example of extensor	Example of abductor	Example of adductor
Head	Sternocleidomastoid	Semispinalis capitis		
Upper arm	Pectoralis major	Trapezius Latissimus dorsi	Deltoid	Pectoralis major with latissimus dorsi
Forearm	With forearm supinated: biceps brachii With forearm pronated: brachialis With semisupination or semipronation: brachioradialis	Triceps brachii		
Hand	Flexor carpi radialis and ulnaris Palmaris longus	Extensor carpi radialis, longus, and brevis Extensor carpi ulnaris	Flexor carpi radialis	Flexor carpi ulnaris
Thigh	Iliopsoas Rectus femoris (of quadriceps femoris group)	Gluteus maximus	Gluteus medius and gluteus minimus	Adductor group
Leg	Hamstrings	Quadriceps femoris group		

Foot	Tibialis anterior	Gastrocnemius
		
		Soleus
			Evertors
			    Peroneus longus
			    Peroneus brevis
				Invertor
				    Tibialis anterior

Trunk	Iliopsoas Rectus abdominis	Sacrospinalis		

Table 7-3 Muscles that move shoulder*

Muscle	Origin	Insertion	Function	Innervation
■ Trapezius	**Occipital bone†** (protuberance)	**Clavicle**	Raises or lowers shoulders and shrugs them	Spinal accessory, second, third, and fourth cervical nerves
	Vertebrae (cervical and thoracic)	**Scapula** (spine and acromion)	Extends head when occiput acts as insertion	
■ Pectoralis minor	**Ribs** (second to fifth)	**Scapula** (coracoid)	Pulls shoulder down and forward	Medial and lateral anterior thoracic nerves
■ Serratus anterior	**Ribs** (upper eight or nine)	**Scapula** (anterior surface, vertebral border)	Pulls shoulder forward; abducts and rotates it upward	Long thoracic nerve

*When trying to learn the origins and insertion of the muscles listed, refer frequently to illustrations of each muscle and to the skeleton. Also, when possible, feel each muscle on your own body.
†Origins and insertions judged to be most important for beginning students of anatomy to know are indicated by boldface type.
■ Muscles judged to be most important for beginning students to know.

Table 7-4 Muscles that move upper arm*

Muscle	Origin	Insertion	Function	Innervation
■ Pectoralis major	**Clavicle** (medial half)† **Sternum** **Costal cartilages of true ribs**	**Humerus** (greater tubercle)	Flexes upper arm Adducts upper arm anteriorly; draws it across chest	Medial and lateral anterior thoracic nerves
■ Latissimus dorsi	**Vertebrae** (spines of lower thoracic, lumbar, and sacral) **Ilium** (crest) Lumbodorsal fascia‡	**Humerus** (intertubercular groove)	Extends upper arm Adducts upper arm posteriorly	Thoracodorsal nerve
■ Deltoid	**Clavicle** **Scapula** (spine and acromion)	**Humerus** (lateral side about halfway down— deltoid tubercle)	Abducts upper arm Assists in flexion and extension of upper arm	Axillary nerve
Coracobrachialis	Scapula (coracoid process)	Humerus (middle third, medial surface)	Adduction; assists in flexion and medial rotation of arm	Musculocutaneous nerve
Supraspinatus	Scapula (supraspinous fossa)	Humerus (greater tubercle)	Assists in abducting arm	Suprascapular nerve
Teres major	Scapula (lower part, axillary border)	Humerus (upper part, anterior surface)	Assists in extension, adduction, and medial rotation of arm	Lower subscapular nerve
Teres minor	Scapula (axillary border)	Humerus (greater tubercle)	Rotates arm outward	Axillary nerve
Infraspinatus	Scapula (infraspinatus border)	Humerus (greater tubercle)	Rotates arm outward	Suprascapular nerve

*When trying to learn the origins and insertion of the muscles listed, refer frequently to illustrations of each muscle and to the skeleton. Also, when possible, feel each muscle on your own body.
†Origins and insertions judged to be most important for beginning students of anatomy to know are indicated by boldface type.
‡Lumbodorsal fascia—extension of aponeurosis of latissimus dorsi; fills in space between last rib and iliac crest.
■ Muscles judged to be most important for beginning students to know.

Table 7-5 Muscles that move lower arm*

Muscle	Origin	Insertion	Function	Innervation
■ Biceps brachii	**Scapula** (supra-glenoid tuberosity)† **Scapula** (coracoid)	**Radius** (tubercle at proximal end)	Flexes supinated forearm Supinates forearm and hand	Musculocuta-neous nerve
■ Brachialis	**Humerus** (distal half, anterior surface)	**Ulna** (front of coronoid process)	Flexes pronated forearm	Musculocuta-neous nerve
Brachioradialis	Humerus (above lateral epicondyle)	Radius (styloid process)	Flexes semipronated or semisupinated forearm; supinates forearm and hand	Radial nerve
■ Triceps brachii	**Scapula** (infraglenoid tuberosity) **Humerus** (posterior surface—lateral head above radial groove; medial head, below)	**Ulna** (olecranon process)	Extends lower arm	Radial nerve
Pronator teres	Humerus (medial epicondyle) Ulna (coronoid process)	Radius (middle third of lateral surface)	Pronates and flexes forearm	Median nerve
Pronator quadratus	Ulna (distal fourth, anterior surface)	Radius (distal fourth, anterior surface)	Pronates forearm	Median nerve
Supinator	Humerus (lateral epicondyle) Ulna (proximal fifth)	Radius (proximal third)	Supinates forearm	Radial nerve

*When trying to learn the origins and insertion of the muscles listed, refer frequently to illustrations of each muscle and to the skeleton. Also, when possible, feel each muscle on your own body.

†Origins and insertions judged to be most important for beginning students of anatomy to know are indicated by boldface type.

■ Muscles judged to be most important for beginning students to know.

Table 7-6 Muscles that move hand*

Muscle	Origin	Insertion	Function	Innervation
Flexor carpi radialis	Humerus (medial epicondyle)	Second metacarpal (base of)	Flexes hand Flexes forearm	Median nerve
Palmaris longus	Humerus (medial epicondyle)	Fascia of palm	Flexes hand	Median nerve
Flexor carpi ulnaris	Humerus (medial epicondyle) Ulna (proximal two thirds)	Pisiform bone Third, fourth, and fifth metacarpals	Flexes hand Adducts hand	Ulnar nerve
Extensor carpi radialis longus	Humerus (ridge above lateral epicondyle)	Second metacarpal (base of)	Extends hand Abducts hand (moves toward thumb side when hand supinated)	Radial nerve
Extensor carpi radialis brevis	Humerus (lateral epicondyle)	Second, third meta-carpals (bases of)	Extends hand	Radial nerve
Extensor carpi ulnaris	Humerus (lateral epicondyle) Ulna (proximal three fourths)	Fifth metacarpal (base of)	Extends hand Adducts hand (move toward little finger side when hand supinated)	Radial nerve

*When trying to learn the origins and insertion of the muscles listed, refer frequently to illustrations of each muscle and to the skeleton. Also, when possible, feel each muscle on your own body.

Table 7-7 Muscles that move thigh*

Muscle	Origin	Insertion	Function	Innervation
■ Iliopsoas (iliacus and psoas major)	**Ilium** (iliac fossa)† **Vertebrae** (bodies of twelfth thoracic to fifth lumbar)	**Femur** (small trochanter)	Flexes thigh Flexes trunk (when femur acts as origin)	Femoral and second to fourth lumbar nerves
■ Rectus femoris	**Ilium** (anterior, inferior spine)	**Tibia** (by way of patellar tendon)	Flexes thigh Extends lower leg	Femoral nerve
■ Gluteal group Maximus	**Ilium** (crest and posterior surface) Sacrum and coccyx (posterior surface) Sacrotuberous ligament	**Femur** (gluteal tuberosity) **Iliotibial tract‡**	Extends thigh—rotates outward	Inferior gluteal nerve
Medius	**Ilium** (lateral surface)	**Femur** (greater trochanter)	Abducts thigh—rotates outward; stabilizes pelvis on femur	Superior gluteal nerve
Minimus	**Ilium** (lateral surface)	**Femur** (greater trochanter)	Abducts thigh; stabilizes pelvis on femur Rotates thigh medially	Superior gluteal nerve
■ Tensor fasciae latae	**Ilium** (anterior part of crest)	**Tibia** (by way of **iliotibial tract**)	Abducts thigh Tightens iliotibial tract†	Superior gluteal nerve
Piriformis	Vertebrae (front of sacrum)	Femur (medial aspect of greater trochanter)	Rotates thigh outward Abducts thigh Extends thigh	First or second sacral nerves
■ Adductor group Brevis	**Pubic bone**	**Femur** (linea aspera)	Adducts thigh	Obturator nerve
Longus	**Pubic bone**	**Femur** (linea aspera)	Adducts thigh	Obturator nerve
Magnus	**Pubic bone**	**Femur** (linea aspera)	Adducts thigh	Obturator nerve
Gracilis	Pubic bone (just below symphysis)	Tibia (medial surface behind sartorius)	Adducts thigh and flexes and adducts leg	Obturator nerve

*When trying to learn the origins and insertion of the muscles listed, refer frequently to illustrations of each muscle and to the skeleton. Also, when possible, feel each muscle on your own body.
†Origins and insertions judged to be most important for beginning students of anatomy to know are indicated by boldface type.
‡The iliotibial tract is part of the fascia enveloping all the thigh muscles. It consists of a wide band of dense fibrous tissue attached to the iliac crest above and the lateral condyle of the tibia below. The upper part of the tract encloses fasciae latae muscle.
■ Muscles judged to be most important for beginning students to know.

Table 7-8 Muscles that move lower leg*

Muscle	Origin	Insertion	Function	Innervation
■ Quadriceps femoris group				
Rectus femoris	**Ilium** (anterior, inferior spine)†	**Tibia** (by way of patellar tendon)	Flexes thigh Extends leg	Femoral nerve
Vastus lateralis	**Femur** (linea aspera)	**Tibia** (by way of patellar tendon)	Extends leg	Femoral nerve
Vastus medialis	**Femur**	**Tibia** (by way of patellar tendon)	Extends leg	Femoral nerve
Vastus intermedius	**Femur** (anterior surface)	**Tibia** (by way of patellar tendon)	Extends leg	Femoral neve
■ Sartorius	**Os innominatum** (anterior, superior iliac spines)	**Tibia** (medial surface of upper end of shaft)	Adducts and flexes leg Permits crossing of legs tailor fashion	Femoral nerve
■ Hamstring group				
Biceps femoris	**Ischium** (tuberosity)	**Fibula** (head of)	Flexes leg	Hamstring nerve (branch of sciatic nerve)
	Femur (linea aspera)	**Tibia** (lateral condyle)	Extends thigh	Hamstring nerve
Semitendinosus	**Ischium** (tuberosity)	**Tibia** (proximal end, medial surface)	Extends thigh	Hamstring nerve
Semimembranosus	**Ischium** (tuberosity)	**Tibia** (medial condyle)	Extends thigh	Hamstring nerve

*When trying to learn the origins and insertion of the muscles listed, refer frequently to illustrations of each muscle and to the skeleton. Also, when possible, feel each muscle on your own body.
†Origins and insertions judged to be most important for beginning students of anatomy to know are indicated by boldface type.
■ Muscles judged to be most important for beginning students to know.

Table 7-9 Muscles that move foot*

Muscle	Origin	Insertion	Function	Innervation
■ Tibialis anterior	**Tibia** (lateral condyle of upper body)†	**Tarsal** (first cuneiform) Metatarsal (base of first)	Flexes foot Inverts foot	Common and deep peroneal nerves
■ Gastrocnemius	**Femur** (condyles)	**Tarsal** (calcaneus by way of Achilles tendon)	Extends foot Flexes lower leg	Tibial nerve (branch of sciatic nerve)
■ Soleus	**Tibia** (underneath gastrocnemius) **Fibula**	**Tarsal** (calcaneus by way of Achilles tendon)	Extends foot (plantar flexion)	Tibial nerve
Peroneus longus	Tibia (lateral condyle) Fibula (head and shaft)	First cuneiform Base of first metatarsal	Extends foot (plantar flexion) Everts foot	Common peroneal nerve
Peroneus brevis	Fibula (lower two thirds of lateral surface of shaft)	Fifth metatarsal (tubercle, dorsal surface)	Everts foot Flexes foot	Superficial peroneal nerve
Tibialis posterior	Tibia (posterior surface) Fibula (posterior surface)	Navicular bone Cuboid bone All three cuneiforms Second and fourth metatarsals	Extends foot (plantar flexion) Inverts foot	Tibial nerve
Peroneus tertius	Fibula (distal third)	Fourth and fifth metatarsals (bases of)	Flexes foot Everts foot	Deep peroneal nerve

*When trying to learn the origins and insertion of the muscles listed, refer frequently to illustrations of each muscle and to the skeleton. Also, when possible, feel each muscle on your own body.

†Origins and insertions judged to be most important for beginning students of anatomy to know are indicated by boldface type.

■ Muscles judged to be most important for beginning students to know.

Table 7-10 Muscles that move head*				
Muscle	**Origin**	**Insertion**	**Function**	**Innervation**
■ Sternocleidomastoid	**Sternum†** **Clavicle**	**Temporal bone** (mastoid process)	Flexes head (prayer muscle) One muscle alone, rotates head toward opposite side; spasm of this muscle alone or associated with trapezius called torticollis or wryneck	Accessory nerve
Semispinalis capitis	Vertebrae (transverse processes of upper six thoracic, articular processes of lower four cervical)	Occipital bone (between superior and inferior nuchal lines)	Extends head; bends it laterally	First five cervical nerves
Splenius capitis	Ligamentum nuchae Vertebrae (spinous processes of upper three or four thoracic)	Temporal bone (mastoid process) Occipital bone	Extends head Bends and rotates head toward same side as contracting muscle	Second, third, and fourth cervical nerves
Longissimus capitis	Vertebrae (transverse processes of upper six thoracic, articular processes of lower four cervical)	Temporal bone (mastoid process)	Extends head Bends and rotates head toward contracting side	

*When trying to learn the origins and insertion of the muscles listed, refer frequently to illustrations of each muscle and to the skeleton. Also, when possible, feel each muscle on your own body.
†Origins and insertions judged to be most important for beginning students of anatomy to know are indicated by boldface type.
■ Muscles judged to be most important for beginning students to know.

Table 7-11 Muscles that move abdominal wall*

Muscle	Origin	Insertion	Function	Innervation
■ External oblique	**Ribs** (lower eight)	**Ossa coxae** (iliac crest and pubis by way of inguinal ligament)† **Linea alba‡** by way of an aponeurosis§‖	Compresses abdomen Important postural function of all abdominal muscles is to pull front of pelvis upward, thereby flattening lumbar curve of spine; when these muscles lose their tone, common figure faults of protruding abdomen and lordosis develop	Lower seven intercostal nerves and iliohypogastric nerves
■ Internal oblique	**Ossa coxae** (iliac crest and inguinal ligament) **Lumbodorsal fascia**	**Ribs** (lower three) **Pubic bone** **Linea alba**	Same as external oblique	Last three intercostal nerves; iliohypogastric and ilioinguinal nerves
■ Transversalis	**Ribs** (lower six) **Ossa coxae** (iliac crest, inguinal ligament) **Lumbodorsal fascia**	**Pubic bone** **Linea alba**	Same as external oblique	Last five intercostal nerves; iliohypogastric and ilioinguinal nerves
■ Rectus abdominis	**Ossa coxae** (pubic bone and symphysis pubis)	**Ribs** (costal cartilage of fifth, sixth, and seventh ribs) Sternum (xiphoid process)	Same as external oblique; because abdominal muscles compress abdominal cavity, they aid in straining, defecation, forced expiration, childbirth, etc.; abdominal muscles are antagonists of diaphragm, relaxing as it contracts and vice versa Flexes trunk	Last six intercostal nerves

*When trying to learn the origins and insertion of the muscles listed, refer frequently to illustrations of each muscle and to the skeleton. Also, when possible, feel each muscle on your own body.

†Inguinal ligament (or Poupart's)—lower edge of aponeurosis of external oblique muscle, extending between the anterior superior iliac spine and the tubercle of the pubic bone. This edge is doubled under like a hem on material. The inguinal ligament forms the upper boundary of the femoral triangle, a large triangular area in the thigh; its other boundaries are the adductor longus muscle mesially and the sartorius muscle laterally.

‡Linea alba—literally, a white line; extends from xiphoid process to symphysis pubis; formed by fibers of aponeuroses of the right abdominal muscles interlacing with fibers of aponeuroses of the left abdominal muscles; comparable to a seam up the midline of the abdominal wall, anchoring its various layers. During pregnancy the linea alba becomes pigmented and is known as the linea niger.

§Aponeurosis—sheet of white fibrous tissue that attaches one muscle to another or attaches it to bone or other movable structures, for example, the right external oblique muscle attaches to the left external oblique muscle by means of an aponeurosis.

‖Origins and insertions judged to be most important for beginning students of anatomy to know are indicated by boldface type.

■ Muscles judged to be most important for beginning students to know.

Table 7-12 Muscles that move chest wall*

Muscle	Origin	Insertion	Function	Innervation
External intercostals	Rib (lower border; forward fibers)	Rib (upper border of rib below origin)	Elevate ribs	Intercostal nerves
Internal intercostals	Rib (inner surface, lower border; backward fibers)	Rib (upper border of rib below origin)	Probably depress ribs	Intercostal nerves
■ Diaphragm	**Lower circumference of thorax** (of rib cage)†	**Central tendon of diaphragm**	Enlarges thorax, causing inspiration	Phrenic nerves

*When trying to learn the origins and insertion of the muscles listed, refer frequently to illustrations of each muscle and to the skeleton. Also, when possible, feel each muscle on your own body.
†Origins and insertions judged to be most important for beginning students of anatomy to know are indicated by boldface type.
■ Muscles judged to be most important for beginning students to know.

Table 7-13 Muscles of pelvic floor*

Muscle	Origin	Insertion	Function	Innervation
Levator ani	Pubis (posterior surface) Ischium (spine)	Coccyx	Together form floor of pelvic cavity; support pelvic organs; if these muscles are badly torn at childbirth or become too relaxed, uterus or bladder may prolapse, that is, drop out	Pudendal nerve
Coccygeus (posterior continuation of levator ani)	Ischium (spine)	Coccyx Sacrum	Same as levator ani	Pudendal nerve

*When trying to learn the origins and insertion of the muscles listed, refer frequently to illustrations of each muscle and to the skeleton. Also, when possible, feel each muscle on your own body.

Table 7-14 Muscles that move trunk*

Muscle	Origin	Insertion	Function	Innervation
Sacrospinalis (erector spinae)			Extend spine; maintain erect posture of trunk Acting singly, abduct and rotate trunk	Posterior rami of first cervical to fifth lumbar spinal nerves
Lateral portion: Iliocostalis lumborum	Iliac crest, sacrum (posterior surface), and lumbar vertebrae (spinous processes)	Ribs, lower six		
Iliocostalis dorsi	Ribs, lower six	Ribs, upper six		
Iliocostalis cervicis	Ribs, upper six	Vertebrae, fourth to sixth cervical		
Medial portion: Longissimus dorsi	Same as iliocostalis lumborum	Vertebrae, thoracic ribs		
Longissimus cervicis	Vertebrae, upper six thoracic	Vertebrae, second to sixth cervical		
Longissimus capitis	Vertebrae, upper six thoracic and last four cervical	Temporal bone, mastoid process		
Quadratus lumborum (forms part of posterior abdominal wall)	Ilium (posterior part of crest) Vertebrae (lower three lumbar)	Ribs (twelfth) Vertebrae (transverse processes of first four lumbar)	Both muscles together extend spine One muscle alone abducts trunk toward side of contracting muscle	First three or four lumbar nerves
■ Iliopsoas	See muscles that move thigh, p. 188		Flexes trunk	

*When trying to learn the origins and insertion of the muscles listed, refer frequently to illustrations of each muscle and to the skeleton. Also, when possible, feel each muscle on your own body.
■ Muscles judged to be most important for beginning students to know.

Table 7-15 Muscles of facial expression and of mastication*

Muscle	Origin	Insertion	Function	Innervation
Muscles of facial expression				
Epicranius (occipito-frontalis)	Occipital bone	Tissues of eyebrows	Raises eyebrows, wrinkles forehead horizontally	Cranial nerve VII
Corrugator supercilii	Frontal bone (super-ciliary ridge)	Skin of eyebrow	Wrinkles forehead vertically	Cranial nerve VII
Orbicularis oculi	Encircles eyelid		Closes eye	Cranial nerve VII
Orbicularis oris	Encircles mouth		Draws lips together	Cranial nerve VII
Platysma	Fascia of upper part of deltoid and pectoralis major	Mandible (lower border) / Skin around corners of mouth	Draws corners of mouth down—pouting	Cranial nerve VII
Buccinator	Maxillae	Skin of sides of mouth	Permits smiling / Blowing, as in playing a trumpet	Cranial nerve VII
Muscles of mastication				
Masseter	Zygomatic arch	Mandible (external surface)	Closes jaw	Cranial nerve V
Temporal	Temporal bone	Mandible	Closes jaw	Cranial nerve V
Pterygoids (internal and external)	Undersurface of skull	Mandible (mesial surface)	Grate teeth	Cranial nerve V

*When trying to learn the origins and insertion of the muscles listed, refer frequently to illustrations of each muscle and to the skeleton. Also, when possible, feel each muscle on your own body.

Outline summary
General functions

A Movement—sometimes locomotion, sometimes movement within given area
B Posture
C Heat production

Skeletal muscle cells
Microscopic structure

A Muscle cells usually called muscle fibers, term descriptive of their long, narrow shape
B Sarcolemma—cell membrane of muscle fiber
C Sarcoplasm—cytoplasm of muscle fiber
D Sarcoplasmic reticulum—analogous but not identical to endoplasmic reticulum of cells other than muscle fibers
E Myofibrils—numerous fine fibers packed close together in sarcoplasm
F Cross striae
  1 Dark stripes called A bands; light H zone runs across midsection of each dark A band
  2 Light stripes called I bands; dark Z line extends across center of each light I band
G Sarcomere—section of myofibril extending from one Z line to next; each myofibril consists of several sarcomeres
H T system—transverse tubules that extend into sarcoplasm at levels of Z lines; formed by invaginations of sarcolemma
I Triad—triple-layered structure consisting of T tubule sandwiched between sacs of sarcoplasmic reticulum

Molecular structure

A Proteins
  1 Thick filaments—composed almost entirely of myosin molecules; heads of myosin molecules are cross bridges of thick filament
  2 Thin filaments—composed of actin, tropomyosin, and troponin molecules arranged in complex fashion; thin filaments attach to Z lines, extend from them in toward center of sarcomeres; thick and thin filaments alternate in myofibrils

Functions

A Contraction—cross bridges of thick filaments attach to thin filaments and pull them toward middle of each sarcomere (Fig. 7-5, p. 154)
B Muscle cells obey all-or-none law when they contract, that is, they either contract with all force possible under existing conditions or do not contract at all
C Sarcoplasmic reticulum—sometimes called "relaxing factor" of muscle cells because it has a very strong affinity for calcium

Energy sources for muscle contraction

See Fig. 7-6, p. 154

Skeletal muscle organs
Structure

A Size, shape, and fiber arrangement—wide variation in different muscles
B Connective tissue components
  1 Epimysium—fibrous connective tissue sheath that envelops each muscle
  2 Perimysium—extensions of epimysium, partitioning each muscle into bundles of fibers
  3 Endomysium—extensions of perimysium between individual muscle fibers
  4 Tendon—strong, tough cord continuous at one end with fibrous wrappings (epimysium, etc.) of muscle and at other end with fibrous covering of bone (periosteum)
  5 Aponeurosis—broad flat sheet of fibrous connective tissue continuous on one border with fibrous wrappings of muscle and at other border with fibrous coverings of some adjacent structure, usually another muscle
  6 Tendon sheaths—tubes of fibrous connective tissue that enclose certain tendons, notably those of wrist and ankle; synovial membrane lines tendon sheaths
  7 Deep fascia—layer of dense fibrous connective tissue underlying superficial fascia under skin; extensions of deep fascia form epimysium, etc. and also enclose viscera, glands, blood vessels, and nerves
C Nerve supply
One motoneuron, together with skeletal muscle fibers it supplies, constitutes *motor unit;* number of muscle fibers per motor unit varies; in general, more precise movements produced by muscle in which motor units include fewer muscle fibers
D Age changes
  1 Fibrosis—with advancing years, some skeletal muscle fibers degenerate and are replaced by fibrous connective tissue
  2 Decreased muscular strength, resulting in part from fibrosis

Function

A Basic principles
1 Skeletal muscles contract only if stimulated; natural stimulus is nerve impulses, artificial stimuli, for example, electrical, or injury
2 Skeletal muscle contractions of several types
  a Tonic contraction (tone, tonus)—continual, partial contractions produced by simultaneous activation of small group of motor units, followed by relaxation of their fibers and activation of another group of motor units; all healthy muscles exhibit tone when individuals are awake; specialized receptors called muscle spindles and neurotendinous end-organs detect degree of muscle stretch
  b Isotonic contraction—muscle shortens but its tension remains constant; isotonic contractions produce movements
  c Isometric contraction—muscle length remains unchanged but tension within muscle increases; isometric contractions "tighten" muscles but do not produce movements
  d Twitch contraction—quick, jerky contraction in response to single stimulus; consists of three phases—latent period, contraction phase, and relaxation phase; twitch contractions rare in normal body
  e Tetanic contraction (tetanus)—sustained smooth contraction produced by series of stimuli bombarding muscle in rapid succession; normal movements said to be produced by incomplete tetanic contractions
  f Treppe (staircase phenomenon)—series of increasingly stronger contractions in response to constant-strength stimuli applied at rate of 1 or 2 per second; contracture—incomplete relaxation after repeated stimulation; fatigue—failure of muscle to contract in response to strongest stimuli after repeated stimulation; true muscle fatigue seldom occurs in body
  g Fibrillation—abnormal contraction in which individual muscle fibers contract asynchronously, producing no effective movement
  h Convulsions—uncoordinated tetanic contractions of varying groups of muscles
3 Skeletal muscles contract according to graded-strength principle in contrast to individual muscle cells that compose them, which contract according to all-or-none law
4 Skeletal muscles produce movement by pulling on insertion bones across joints
5 Bones serve as levers and joints as fulcrums of these levers
6 Muscles that move part usually do not lie over that part but proximal to it
7 Skeletal muscles almost always act in groups rather than singly, that is, most movements produced by coordinated action of several muscles

B Hints of how to deduce actions
1 Deduce bones that muscle attaches to from illustrations of muscle
2 Make guess as to which bone moves (insertion)
3 Deduce movement muscle produces by applying principle that its insertion moves toward its origin

Names

A Reasons for names—muscle names describe one or more of following features about muscle
1 Its action
2 Direction of fibers
3 Its location
4 Number of divisions composing it
5 Its shape
6 Its points of attachment
B Muscles grouped according to location—see Table 7-1
C Muscles grouped according to function—see also Table 7-2
  1 Flexors—decrease angle of joint
  2 Extensors—return part from flexion to normal anatomical position
  3 Abductors—move bone away from midline of body
  4 Adductors—move bone toward midline of body
  5 Rotators—cause part to pivot on its axis
  6 Levators—raise part
  7 Depressors—lower part
  8 Sphincters—reduce size of opening
  9 Tensors—tense part or make it more rigid
  10 Supinators—turn hand palm upward
  11 Pronators—turn hand palm downward

Origins, insertions, functions, and innervations of representative muscles

See Tables 7-3 to 7-15, pp. 184-195

Weak places in abdominal wall

A Inguinal rings—right and left internal; right and left external
B Femoral rings—right and left
C Umbilicus
D Diaphragm

Bursae

A Definition—small connective tissue sacs lined with synovial membrane and containing synovial fluid

B Locations—wherever pressure exerted over moving parts

  1 Between skin and bone

  2 Between tendons and bone

  3 Between muscles or ligaments and bone

  4 Names of bursae that frequently become inflamed (bursitis)

    a Subacromial—between deltoid muscle and head of humerus and acromion process

    b Olecranon—between olecranon process and skin; inflammation called student's elbow

    c Prepatellar—between patella and skin; inflammation called housemaid's knee

C Function—act as cushion, relieving pressure between moving parts

Posture

A Meaning—position or alignment of body parts

B How maintained—by continual pull of muscles on bones in opposite direction from pull of gravity, that is, posture maintained by continued partial contraction of muscles, or muscle tone; therefore indirectly dependent on many other factors, for example, normal nervous, respiratory, and circulatory systems, health in general

C Importance to body as whole—essential for optimal functioning of most of body, including respiration, circulation, digestion, joint action; briefly, maximal health depends on good posture, good posture depends on health

Review questions

1 Differentiate between the three kinds of muscle tissue as to structure, location, and innervation.

2 Describe several physiological properties of muscle tissue.

3 What property is more highly developed in muscle than in any other tissue?

4 State a principle describing the usual relationship between a part moved and the location of muscles (insertion, body, and origin) moving the part.

5 Applying the principle stated in question 4, where would you expect muscles that move the head to be located? Name two or three muscles that fulfill these conditions.

6 Applying the principle stated in question 4, what part of the body do thigh muscles move? Name several muscles that fulfill these conditions.

7 What bone or bones serve as a lever in movements of the forearm? What structure constitutes the fulcrum for this lever?

8 Explain the meaning of the term neuromusculoskeletal unit.

9 Name the main muscles of the back, chest, abdomen, neck, shoulder, upper arm, lower arm, thigh, buttocks, leg, and pelvic floor.

10 Name the main muscles that flex, extend, abduct, and adduct the upper arm; that raise and lower the shoulder; that flex and extend the lower arm; that flex, extend, abduct, and adduct the thigh; that flex and extend the lower leg and thigh; that flex and extend the foot; that flex, extend, abduct, and adduct the head; that move the abdominal wall; that move the chest wall.

11 Discuss the chemical reactions thought to make available energy for muscle contraction.

12 What physiological reason can you give for athletes using a warm-up period before starting a game?

13 In general, where are bursae located? Give several specific locations.

14 Name several weak places in the abdominal wall where hernia may occur.

15 What and where are the inguinal canals? Of what clinical importance are they?

16 Good posture depends on tonicity of the antigravity muscles, particularly of those that hold the head and trunk erect and the abdominal wall pulled in. Name several muscles that perform these functions.

17 Define the following terms: aponeurosis, bursa, contraction, contracture, elasticity, extensibility, fibrillation, insertion, motor unit, origin, oxygen debt, tetanus, tone, treppe, twitch.

18 Explain briefly the current theory about the role of calcium ions in muscle contraction and relaxation.

Communication, control, and integration

chapter 8

Nervous system cells

If you want to understand the body, you need to remind yourself frequently of some principles stated in the first chapter of this book—briefly, that the body is made up of millions of smaller structures that carry on a host of different activities. But, and this is the important point, together all of these diverse activities accomplish the one big, all-encompassing function of the body—survival. Think about this for a moment. How can many parts of any kind—whether they be cells, people, or parts of a machine—be made to accomplish any one big function? How can many be made one? How can unification (integration) be achieved? Communication and control are the keys that unlock the secret of integration. Communication makes possible control, and control makes possible integration. Many familiar examples of this principle suggest themselves. A hospital, to name just one example, is a functional unit. Quite obviously the activities of the hundreds of individuals who make up a modern hospital must be organized, coordinated, and integrated. Communication between the individual workers is what makes possible this organization, coordination, and integration. Without communication, chaos would prevail and the hospital would not survive. In no time at all it would be utterly unable to carry on its one great function of giving care to patients.

Nerve impulses and chemicals—chiefly hormones, carbon dioxide, and certain ions—constitute the two kinds of "messages" or communications sent within the body. The nervous system and the endocrine system constitute its two main communication systems. The circulatory system serves as an assistant communication system by distributing hormones and other chemical messages.

Facts, theories, and questions about the nervous system are as abundant and complex as they are fascinating. We shall approach this large body of material by considering, in this chapter, the cells of the nervous systems and the mechanism of nerve impulse conduction. Chapter 9 presents information about the somatic nervous system, Chapter 10 discusses the autonomic nervous system, and Chapter 11 deals with the special senses. Discussion of the endocrine system appears in Chapter 12.

Cells

Two main kinds of cells compose nervous system structures—neurons and neuroglia. Although outnumbered by neuroglia by at least two to one, neurons are the "specialists" of the nervous system. They specialize in impulse conduction, the function that makes possible all other nervous system functions. Neuroglia perform the less specialized functions described below.

Neuroglia
Types

Histologists identify three major kinds of neuroglia, namely, astrocytes, oligodendroglia, and microglia. To see microscopic views of these cells, look at Fig. 8-1.

Structure and function

Astrocytes (from the Greek *astron*, star, and *kytos*, hollow, that is, a cell) are the neuroglia that have a starlike shape because of their many branching processes. They are the most numerous of all neuroglia. Large numbers of them occur in the brain and cord, located mainly between neurons and blood vessels, with the tiny footlike ends of their processes in contact with the vessels. By means of these contact regions, astrocytes are thought to function as part of the so-called blood-brain barrier, which restricts the movement of certain substances into the brain.

Oligodendroglia are smaller cells and have fewer processes than astrocytes. Some oligodendroglia lie clustered around nerve cell bodies, some are arranged in rows between nerve fibers in the brain and cord. They help hold nerve fibers together and also serve another and probably more important function—they produce the fatty myelin sheath that envelops nerve fibers located in the brain and cord.

Microglia are small, usually stationary cells. In inflamed or degenerating brain tissue, however, microglia enlarge, move about, and carry on phagocytosis. In other words, they engulf and destroy microbes and cellular debris.

Neurons
Types

Neurons are classified according to two different criteria—the direction in which they conduct impulses and the number of processes they have.

Classified according to the direction in which

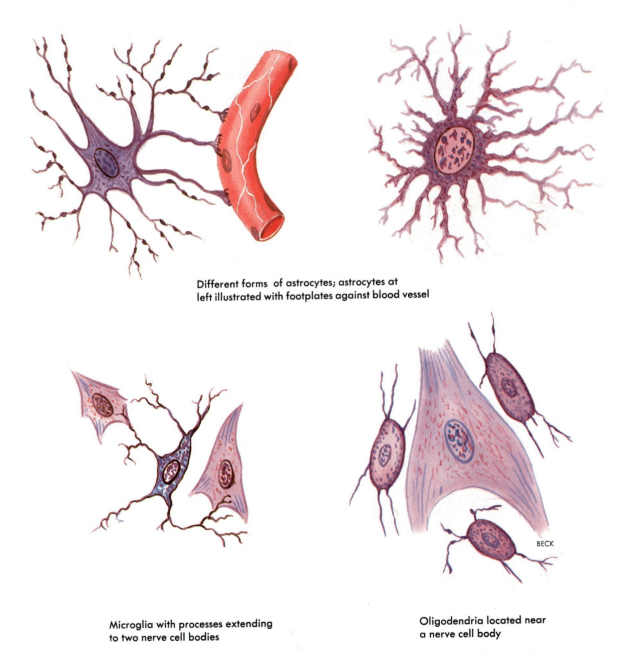

Different forms of astrocytes; astrocytes at
left illustrated with footplates against blood vessel

Microglia with processes extending
to two nerve cell bodies

Oligodendria located near
a nerve cell body

BECK

Fig. 8-1 Neuroglia, the special connecting, and supporting cells of the brain and cord.

they conduct impulses, there are three types of neurons: sensory, motor, and interneurons. *Sensory (afferent) neurons* transmit nerve impulses to spinal cord or brain. *Motoneurons (motor or efferent neurons)* transmit nerve impulses away from brain or spinal cord to or toward muscle or glandular tissue. *Interneurons* conduct impulses from sensory to motor neurons. Interneurons lie entirely within the central nervous system (brain and spinal cord).

Classified according to the number of their processes, there are also three types of neurons: multipolar, bipolar, and unipolar. *Multipolar neurons* have only one axon but several dendrites. Most of the neurons in the brain and spinal cord are multipolar. *Bipolar neurons* have only one axon and also only one dendrite and are the least numerous kind of neuron. They are found in the retina of the eye, for example, and in the spiral ganglion of the inner ear. *Unipolar neurons* originate in the embryo as bipolar neurons. But in the course of development, their two processes become fused into one for a short distance beyond the cell body. Then they separate into clearly distinguishable axon and dendrite. Sensory neurons, like the one shown in Fig. 8-2, are usually unipolar. How would you classify the interneuron shown in this figure? Is it unipolar, bipolar, or multipolar? How would you classify the motoneuron?

Structure

All neurons consist of a cell body (also called the *soma* or *perikaryon*) and at least two processes: one axon and one or more dendrites. Because dendrites and axons are threadlike extensions from a neuron's soma (its cell body), they are often called *nerve fibers.*

Clusters of neuron cell bodies have a slightly gray color. *Gray matter* in the brain and spinal cord, for example, consists largely of neuron cell bodies. In many respects the cell body, the largest part of a nerve cell, resembles other cells. It contains a nucleus, cytoplasm, and various organelles found in other cells, for example, mitochondria and a Golgi apparatus. Incidentally, Golgi first saw this apparatus in neurons. A neuron's cytoplasm extends through its cell body and its processes. A cytoplasmic membrane encloses the entire neuron.

Certain structures—dendrites, axons, neurofibrils, Nissl bodies, myelin sheath, and neurilemma—are found only in neurons. The following paragraphs describe them briefly.

Dendrites, as you can see in Fig. 8-2, branch extensively, like tiny trees. In fact, their name derives from the Greek word for tree. The distal ends of dendrites of sensory neurons are called *receptors* because they receive the stimuli that initiate conduction. Dendrites conduct impulses to the cell body of the neuron.

The *axon* of a neuron is a single process that extends out from the neuron cell body. Although a neuron has only one axon, it often has one or more side branches *(axon collaterals).* Moreover, axons terminate in many branched filaments. The endings of these filaments contain numerous vesicles and mitochondria.

Axons vary in both length and diameter. Some are 3 feet long or more. Some measure only a fraction of an inch. Axon diameters also vary considerably, from about 20 μm down to about 1 μm—a point of interest because axon diameter relates to velocity of impulse conduction. In general, the larger the diameter, the more rapid the conduction. A neuron's axon conducts impulses away from its cell body.

Neurofibrils are very fine fibers extending through dendrites, cell bodies, and axons. Electron micrographs seem to indicate that neurofibrils consist of bundles of thinner fibers, namely, microtubules and microfilaments, structures present also in the cytoplasm of other cells.

Nissl bodies consist of layers of small pieces of the endoplasmic reticulum with many ribosomes lying between them. Seen with the light microscope, Nissl bodies appear as rather large granules widely scattered through the cyto-

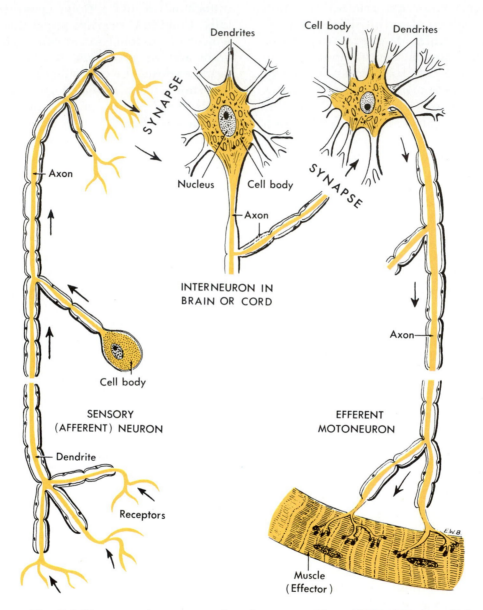

Fig. 8-2 Diagrammatic representation of structure of three kinds of neurons. Note that each neuron has three parts: a cell body and two types of extensions, dendrite(s) and an axon. Arrows indicate direction of impulse conduction. (See Fig. 8-3 for more details about coverings of processes.)

plasm of the neuron cell body but not present in the axon or the axon hillock (part of the neuron cell body from which the axon emerges). Nissl bodies specialize in protein synthesis, thereby providing the protein needs for maintaining and regenerating neuron processes and for renewing chemicals involved in the transmission of nerve impulses from one neuron to another.

The *myelin sheath,* a segmented wrapping around a nerve fiber, consists of myelin, a fatty substance. One segment of a myelin sheath extends from one node of Ranvier to the next (Fig. 8-3) and is produced by either a Schwann cell or an oligodendroglial cell. To form the myelin sheath the cell wraps itself around the nerve fiber in jelly roll fashion. Schwann cells form the myelin sheaths around peripheral nerve fibers, that is, fibers located outside of the brain and spinal cord. Oligodendroglia form the myelin sheaths around central nerve fibers, that is, those located in the brain and cord. Fibers that have a myelin sheath are called *myelinated fibers,* but those that have only a thin layer of myelin are called *unmyelinated.* Because of the high fat content of myelin, bundles of myelinated fibers have a creamy white color, and they make up the white matter of the nervous system. *White matter* in the brain and spinal cord is composed of tracts, and *tracts* consist of bundles of myelinated fibers (axons). White matter found outside the brain and cord consists of *nerves,* that is, cordlike structures composed of bundles of myelinated fibers.

The *neurilemma,* or sheath of Schwann, consists of a Schwann cell wrapped once around a nerve fiber to form a continuous sheath enclosing the segmented myelin sheaths of peripheral nerve fibers. Brain and cord fibers do not have a neurilemma. Because the neurilemma plays an essential part in the regeneration of cut or injured nerve fibers, peripheral nerve fibers regenerate but brain and cord fibers presumably do not.

Function

Neurons perform the specific function of conducting impulses and the general function of providing the body with its most rapid means of communication, control, and integration.

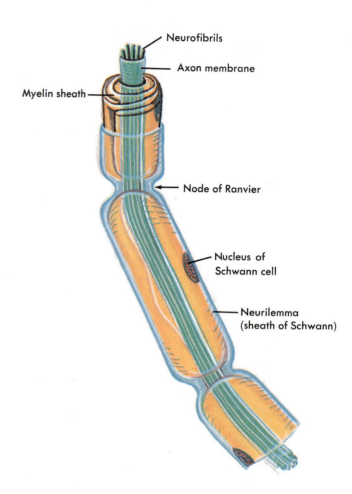

Fig. 8-3 Diagram of a nerve fiber and its coverings. Note these features of the myelin sheath: its concentric layers surround the nerve fiber—the axon in this figure—and it is segmented. One segment is the section of myelin sheath located between two successive nodes of Ranvier. The neurilemma surrounds the myelin sheath and is continuous, not segmented.

Nerve impulse conduction

Definitions

In order to understand the discussion of nerve impulse conduction, you will need to familiarize yourself with the meanings of the following terms.

potential difference—an electrical difference; an electrical gradient, to use another term. A potential difference is a difference between the amounts of electrical charge present at two points. A potential difference is a form of potential energy, a force that has the power to move positively charged ions down an electrical gradient, that is, from a point with a higher positive charge to a point with a lower positive charge. The magnitude of a potential difference is measured in volts or millivolts (mv).

polarized membrane—membrane whose outer and inner surfaces bear different amounts of electrical charges. In short, a potential difference exists across a polarized membrane.

depolarized membrane—membrane whose outer and inner surfaces bear equal amounts of electrical charges. A potential difference does not exist across a depolarized membrane.

resting potential—potential difference that exists across the membrane of a neuron when it is not conducting impulses—when it is in the resting state. The magnitude of the resting potential varies but usually ranges between 70 and 90 mv. The inner surface of the resting neuron's membrane is 70 to 90 mv negative to its outer surface.

action potential—potential difference that exists across the membrane of a neuron when it is conducting impulses—when it is active.

stimulus—change in the environment. Some common kinds of stimuli are changes in the pressure, temperature, or chemical composition of either the body's external or internal environment.

Mechanism of resting potential

The direct cause of the resting potential is the presence of a minute excess of positive ions outside a neuron's cell membrane and a minute excess of negative ions inside it (illustrated by the first diagram in Fig. 8-4). The following factors bring about this ionic imbalance. The so-called sodium pump actively transports positive sodium ions out of the intracellular fluid and into the extracellular fluid. This produces a higher concentration of sodium ions in the extracellular fluid than in the intracellular fluid. A potassium pump—postulated to be coupled to the sodium pump—also operates across cell membranes. It actively transports positive potassium ions into the cell but only about one third the number of positive sodium ions pumped out of the cell. Since a normal cell membrane is some 50 to 100 times more permeable to potassium ions than to sodium ions, more potassium ions diffuse back out of the cell than sodium ions diffuse in. Moreover, since a normal cell membrane is impermeable or relatively impermeable to the negative ions inside a cell, too few negative ions diffuse out of the cell to balance the number of positive potassium ions that have left it. This leaves a minute excess of negative ions inside the cells and produces a minute excess of positive ions outside it; this, in turn, creates a resting potential across a nonconducting neuron's cell membrane. In essence, the outward diffusion of potassium down its concentration gradient creates the resting potential, but this concentration gradient is established by the sodium-potassium pump. Anything that interrupts the active transport of sodium and potassium is soon followed by both the equilibration of ions and electric charges on the two sides of the neuron cell membrane and the depolarization of the membrane.

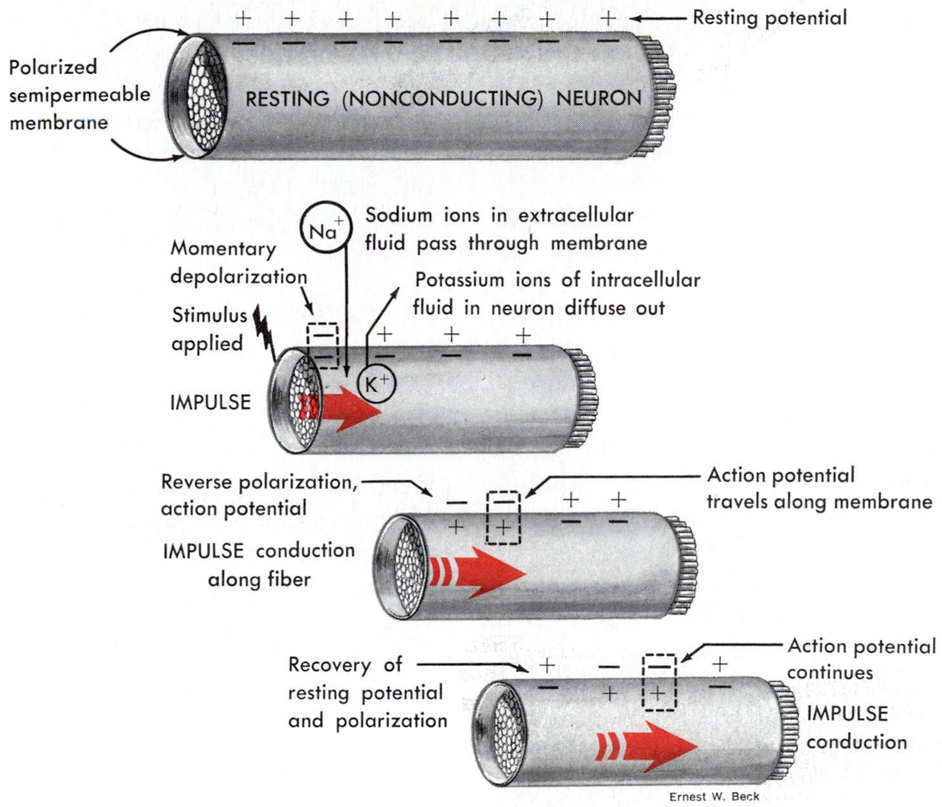

Fig. 8-4 Upper diagram represents polarized state of the membrane of a nerve fiber when it is not conducting impulses. Lower diagrams represent nerve impulse conduction—a self-propagating wave of negativity or action potential travels along membrane.

Mechanism of action potential (nerve impulse)

According to a widely accepted definition, the action potential or nerve impulse is a self-propagating wave of electrical negativity that travels along the surface of a neuron's cytoplasmic membrane. A step-by-step description of the mechanism that produces the action potential follows. Refer to Fig. 8-4 as you read each step.

1 When an adequate stimulus is applied to a neuron, it greatly increases the membrane's permeability to sodium ions at the point of stimulation.

2 Sodium ions rush into the cell at the stimulated point. Therefore at this point the excess of positive ions outside rapidly dwindles to zero. Also, the membrane potential decreases to zero at this point. In other words the stimulated point of the membrane is no longer polarized. It is *depolarized*, but only for an instant. As more sodium ions continue to stream into the cell, they almost instantaneously produce an excess of positive ions inside the cell and leave an excess of negative ions outside. In short, the influx of sodium ions reverses the resting potential and thereby changes it into an action potential. With a typical resting potential the inner surface of a neuron's membrane is, for example, 70 mv negative to its outer surface. In contrast, with a typical action potential the inner surface of the neuron's membrane is about 30 mv positive to its outer surface.

3 The stimulated negatively charged point of the membrane sets up a local current with the positive point adjacent to it, and this local current acts as a stimulus. Consequently, within a fraction of a second the adjacent point on the membrane becomes depolarized and its potential reverses from positive to negative. The action potential has thus moved from the point originally stimulated to the adjacent point on the membrane. As the cycle goes on repeating itself over and over again in rapid succession, the action potential travels point by point out of the full length of the neuron. Much as a wave of water moves in from sea to shore, so an action potential moves along a neuron's membrane from its point of stimulation (on dendrites or cell body) to the neuron's axon endings. From the facts stated in this paragraph came the definition of the *action potential* (or its synonym, *nerve impulse*) as a self-propagating wave of negativity that travels along the surface of a neuron's cytoplasmic (plasma) membrane.

4 By the time the action potential has moved from one point on the membrane to the next (a matter of thousandths of a second), the first point has repolarized—its resting potential has been restored. *Repolarization* results from the fact that the increased permeability to sodium induced by stimulation lasts only momentarily. It is quickly replaced by increased permeability to potassium, which therefore diffuses outward (because potassium concentration inside the cell is much greater than that outside the cell). Then the permeability of the membrane decreases, and its pumping activity again becomes effective. Once again, it actively transports sodium ions out of the cell and potassium ions into it. Once again—and in much less time than it takes to tell it—more positive ions accumulate on the outer than on the inner surface of the cytoplasmic membrane. The resting potential of the cell has, in other words, been restored. The outer surface of its membrane is again electrically positive to its inner surface. (Or, stated differently, the inner surface of the neuron's membrane again becomes electrically negative to its outer surface.)

Initiation of action potential

When a stimulus acts on a neuron, the mechanism just described operates to bring about a change in that neuron's membrane potential. Only if its potential reaches a certain critical level, known as the threshold of stimulation, is impulse conduction triggered. In short, an action potential is initiated by a change in a neuron's membrane potential from its resting level, for example, -70 mv, to its threshold of stimulation, for example, $+40$ mv. Once initiated, the action potential then rapidly propagates itself the full length of the neuron. Whether or not a stimulus initiates conduction depends on the intensity or "strength" of the stimulus. A stimulus just strong enough to change the membrane potential to its threshold is just strong enough to initiate impulse conduction and is called a *threshold stimulus* (or liminal stimulus). Any stimulus weaker than this is a *subthreshold stimulus* (or subliminal stimulus). It changes the membrane potential, but not enough to reach the threshold level, so it does not trigger impulse conduction. The effects of two or more subthreshold stimuli, however, can add together or "summate." By summation, they may decrease the membrane potential to its threshold level and thus trigger impulse conduction.

Speed of impulse conduction

How fast does a neuron conduct impulses? It all depends on the diameter of its axon, the thickness of its axon's myelin sheath, and the distance between the myelin's nodes of Ranvier. The larger an axon's diameter, the faster its conduction. Fibers with a large diameter (A fibers) conduct most rapidly. Impulses travel along them at a speed of about 100 meters per second, or more than 200 miles per hour. Fibers with a small diameter (C fibers) conduct most slowly, about 0.5 meter per second, or 1 mile per hour. Fibers with a diameter of intermediate size (B fibers) conduct at speeds between those of the large, fast A fibers and the small, slow C fibers.

Heavily myelinated fibers with a greater distance between successive nodes conduct, according to another principle, many times faster than do so-called unmyelinated fibers with a lesser distance between nodes. ("Unmyelinated" fibers are not really unmyelinated. They have, we know now, a thin layer of myelin around them.)

Impulse conduction routes

The route traveled by many nerve impulses is known as the *reflex arc*. Basically the reflex arc route consists of two or more neurons, arranged in series, that conduct impulses from the periphery to the central nervous system and back out to the periphery. Periphery, as used in the preceding sentence, means any place in the body outside of the central nervous system *(spinal cord and brain)*. Impulse conduction over a reflex arc begins in receptors and ends in effectors. Look at Fig. 8-2, and you will see that receptors are the beginnings—or in technical terms, the distal ends—of a sensory neuron's dendrite. The effector shown in this figure is striated or skele-

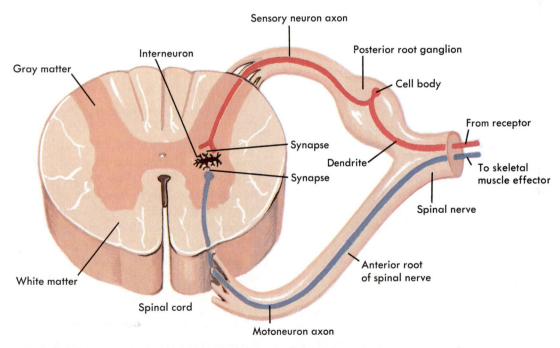

Fig. 8-5 Three-neuron ipsilateral reflex arc, consisting of a sensory neuron, an interneuron, and a motoneuron. Note the presence of two synapses in this arc: (1) between sensory neuron axon terminals and interneuron dendrites and (2) between interneuron axon terminals and motoneuron dendrites and cell bodies (located in anterior gray matter). Nerve impulses traversing such arcs produce many spinal reflexes. Example: withdrawing the hand from a hot object.

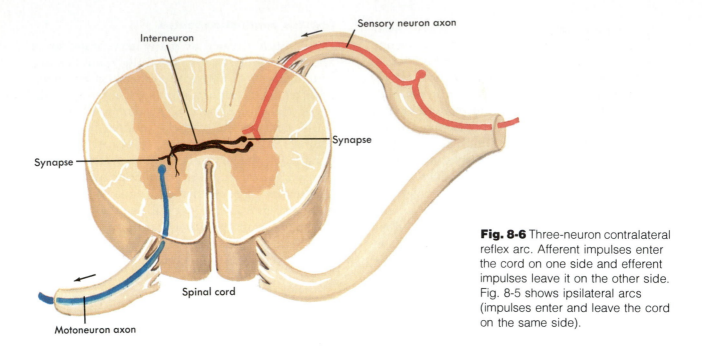

Interneuron

Sensory neuron axon

Synapse

Synapse

Spinal cord

Motoneuron axon

Fig. 8-6 Three-neuron contralateral reflex arc. Afferent impulses enter the cord on one side and efferent impulses leave it on the other side. Fig. 8-5 shows ipsilateral arcs (impulses enter and leave the cord on the same side).

tal muscle. The only other kinds of effectors are smooth (visceral) muscle and glandular tissues.

The simplest kind of reflex arc is called a two-neuron or a monosynaptic arc. As these names suggest, such an arc consists of two neurons and one synapse. A *synapse* is the place where a nerve impulse is transmitted from one neuron to another. All the synapses in reflex arcs whose motoneurons conduct impulses to skeletal muscles lie in either spinal cord or brain gray matter.

A three-neuron reflex arc consists of three kinds of neurons—sensory neurons and motoneurons with interneurons between them. As you can see in Fig. 8-5, there are two synapses in a three-neuron arc. The distal ends of both the sensory neuron and the motoneuron in this figure are located on the same side of the body—hence this is an *ipsilateral reflex arc*. Now examine Fig. 8-6 to find out how a contralateral arc differs from an ipsilateral arc.

Besides simple two-neuron and three-neuron arcs, intersegmental arcs (Fig. 8-7) and even more complex multineuron, multisynaptic arcs also exist. All impulses that start in receptors, however, do not invariably travel over a complete reflex arc and terminate in effectors. Many impulses fail to be conducted across synapses. Moreover, all impulses that terminate in effectors do not invariably start in receptors. Many of them, for example, are thought to originate in the brain.

Mechanism of conduction across synapses

A synapse is the place where nerve impulses are transmitted from one neuron, called the *presynaptic neuron*, to another nueron, called the

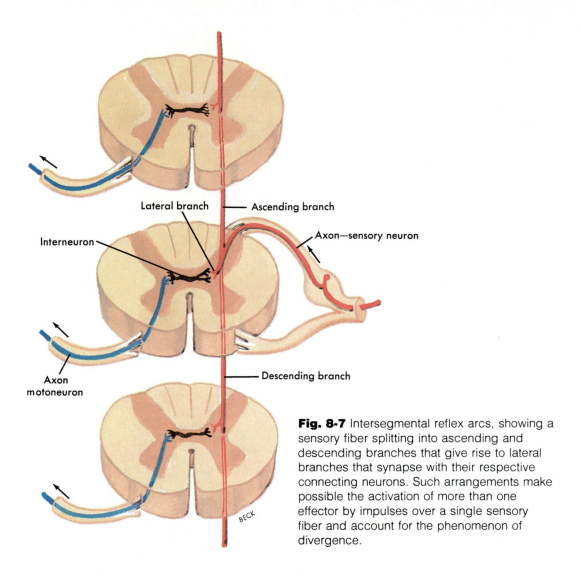

Lateral branch — Ascending branch

Interneuron

Axon—sensory neuron

Axon
motoneuron

Descending branch

Fig. 8-7 Intersegmental reflex arcs, showing a sensory fiber splitting into ascending and descending branches that give rise to lateral branches that synapse with their respective connecting neurons. Such arrangements make possible the activation of more than one effector by impulses over a single sensory fiber and account for the phenomenon of divergence.

BECK

postsynaptic neuron. Structurally a synapse consists of a synaptic knob, a synaptic cleft, and the cytoplasmic membrane of a postsynaptic neuron's dendrite or cell body. A *synaptic knob* is a tiny distention at the end of one of the fine filaments in which a presynaptic neuron's axon terminates. Fig. 8-8 shows several of these knobs under low magnification. Fig. 8-9 shows one synaptic knob under very high magnification.

Note especially, in Fig. 8-9, the mitochondria and numerous vesicles (small closed sacs) present in the synaptic knob. A *synaptic cleft* is the space between a synaptic knob and the cytoplasmic membrane of a postsynaptic neuron's dendrite or cell body. It is an incredibly narrow space measuring only 200 to 300 Å or about one-millionth of an inch in width! Identify the synaptic cleft in Fig. 8-9. Similar narrow spaces called

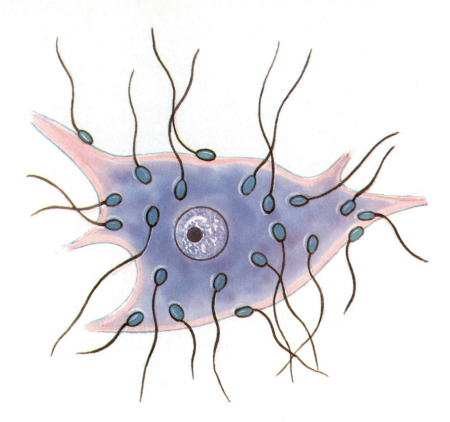

Fig. 8-8 Synaptic knobs (end feet or end buttons) on motoneuron cell body.

neuroeffector junctions separate the axon terminals of motoneurons from effector cells. Effector cells are either muscle or glandular cells. Where they are muscle cells, neuroeffector junctions are called *neuromuscular junctions* or *nerve-muscle synapses*. Where effectors are glandular cells, neuroeffector junctions are called *neuroglandular junctions*.

Across each synaptic cleft, opposite a synaptic knob, lie protein molecules embedded in the cytoplasmic membrane of the postsynaptic neuron. The crucial functions these protein molecules perform in synaptic conduction will be described subsequently.

An action potential (nerve impulse) that has traveled the length of a neuron stops at its axon terminals. Action potentials cannot cross synaptic clefts, miniscule barriers though they are. To initiate an action potential in a postsynaptic neuron a chemical mechanism operates. It consists of the following steps.

1 When an action potential reaches axon terminals, a chemical called a *neurotransmitter* is released from the synaptic knobs into the synaptic cleft. Each one of numerous small vesicles in a synaptic knob stores some 10,000 or so neurotransmitter molecules (Fig. 8-9). The vesicles move outward when an action potential arrives at a synaptic knob, and as they reach the knob's surface, they fuse with its surface membrane and form openings. Out of these openings, thousands of neurotransmitter molecules spurt forth into the synaptic cleft. (Each neuron releases only one specific neurotransmitter, but several different kinds of neurotransmitters are known to exist.)

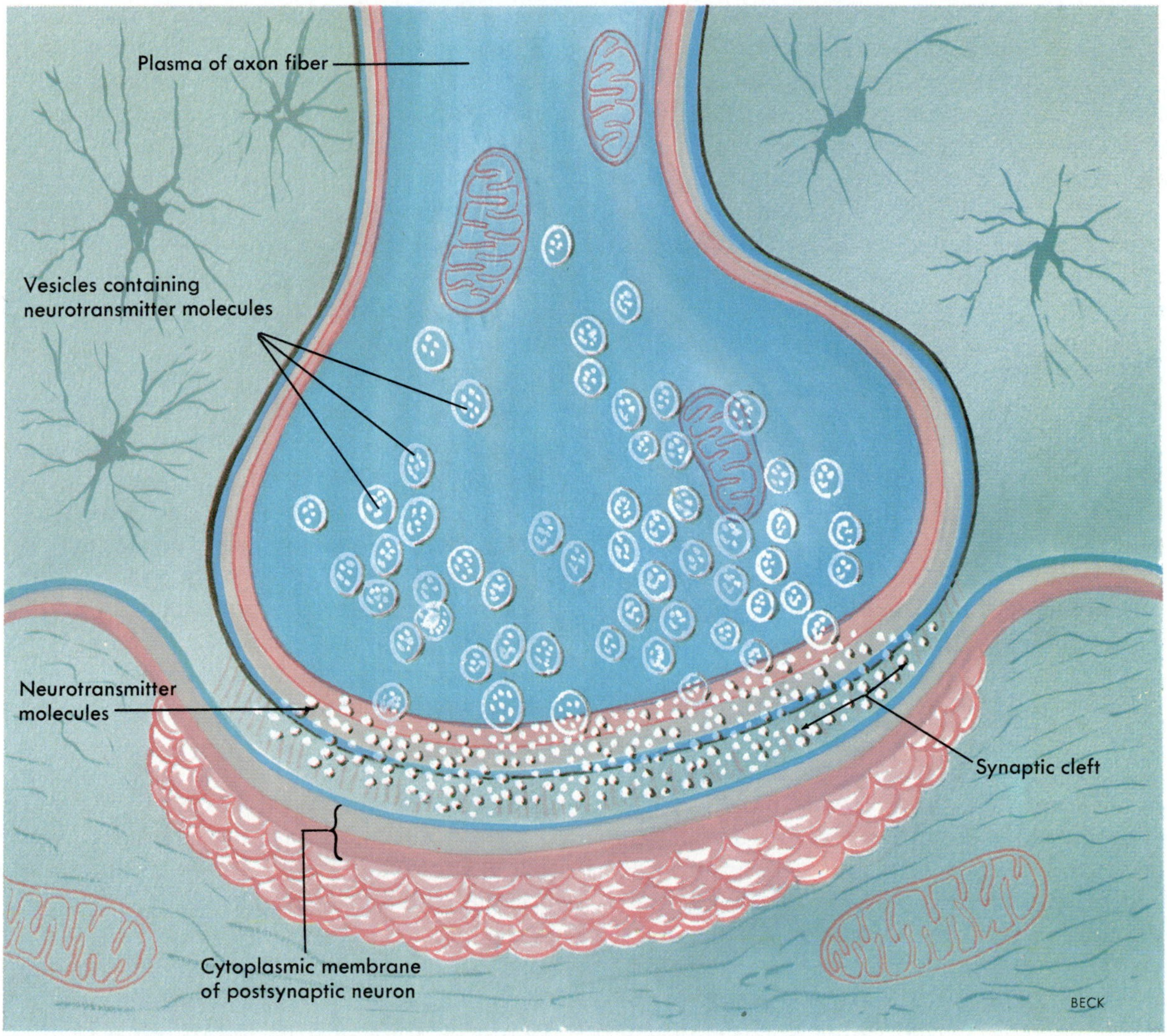

Fig. 8-9 Diagram of a synapse showing its components: a synaptic knob (axon terminal) of a presynaptic neuron, the cytoplasmic membrane of a postsynaptic neuron, and a synaptic cleft (space about one millionth of an inch wide) between the two neurons. On the arrival of an action potential at a synaptic knob, neurotransmitter molecules are released from vesicles in the knob into the synaptic cleft. The combining of neurotransmitter molecules with receptor molecules in the cytoplasmic membrane of the postsynaptic neuron initiates impulse conduction by it.

2 Neurotransmitter molecules diffuse rapidly across the microscopic width of the synaptic cleft and contact the postsynaptic neuron's cytoplasmic membrane. Here they bind to specific protein molecules, called *neurotransmitter receptors*. This leads to the opening of channels in the membrane through which sodium ions diffuse into and potassium ions diffuse out of the interior of the postsynaptic neuron. This flow of ions depolarizes the neuron and gives rise to an action potential—called the *excitatory postsynaptic potential*. In other words the flow of ions initiates nerve impulse transmission by the postsynaptic neuron. A chemical, a neurotransmitter, has thus bridged the synaptic cleft and accomplished conduction by the synapse.

3 Once impulse conduction by postsynaptic neurons is initiated, neurotransmitter activity is rapidly terminated by one or both of two mechanisms. Some neurotransmitter molecules that remain in the synaptic cleft move back into the synaptic knobs from which they came and others are metabolized into inactive compounds by specific enzymes (p. 216).

Research in recent years has revealed additional information about neurotransmitter actions. When certain neurotransmitters bind to their receptors in a postsynaptic neuron's cytoplasmic membrane, they activate an enzyme—adenyl cyclase—also present in the membrane. In fact, the neurotransmitter receptor and adenyl cyclase are different parts of one large protein molecule that extends all the way through the neuron's cytoplasmic membrane. The neurotransmitter receptor is the part of the molecule located at the outer surface of the membrane facing the synaptic cleft. The part of the molecule located at the inner surface of the membrane facing the cytoplasm of the postsynaptic neuron is adenyl cyclase. Adenyl cyclase, activated by the binding of a neurotransmitter to its receptor, catalyzes the conversion of ATP to cyclic AMP (adenosine monophosphate). Cyclic AMP then activates two enzymes called kinases. One kinase is located in the surface membrane of the postsynaptic neuron. When activated, it adds a phosphate group to another protein located in the membrane. This is postulated to change the shape or the position of the protein in such a way as to open channels in the membrane and thereby allow the inward diffusion of sodium ions, which initiates an action potential, that is, impulse conduction by the postsynaptic neuron. The other kinase activated by cyclic AMP is located in the cytoplasm of the postsynaptic neuron. When activated, this kinase is thought to move into the cell's nucleus. There, by altering gene activity, it leads to change in protein synthesis by the cell. This second effect of cyclic AMP occurs more slowly but lasts longer than the first impulse conduction.

Cyclic AMP has a descriptive nickname—the "second messenger." In postsynaptic neurons where cyclic AMP is formed, neurotransmitter molecules serve as the first messenger, a messenger from a presynaptic neuron that brings about formation of cyclic AMP by the postsynaptic neuron. The second messenger, cyclic AMP, then initiates both the fleeting electrical changes associated with impulse conduction and the longer lasting chemical changes associated with protein synthesis.

Many neurotransmitters, as just described, decrease the negativity of the postsynaptic membrane potentials. These are classified as *excitatory neurotransmitters*. Others, called *inhibitory neurotransmitters*, increase the negativity of postsynaptic membrane potentials. In most cases, a single postsynaptic neuron receives transmitters from many presynaptic axons. Moreover, some of these axons release excitatory and some release inhibitory transmitters into the synaptic cleft. Summation of their opposite influences produces one of three changes in the postsynaptic neuron, namely, facilitation, impulse conduction, or inhibition.

Facilitation is a decrease in the negativity of the postsynaptic neuron's membrane potential, changing it from the usual potential to what is called the *excitatory postsynaptic potential* (commonly abbreviated to EPSP). Suppose a postsynaptic neuron's resting potential is -70 mv and its threshold of stimulation is -50 mv, then -60 mv would be an example of an EPSP for that neuron. Note the relationship between the EPSP and the resting potential and threshold of stimulation. It lies somewhere between them. The negativity of the EPSP is always lower than the neuron's resting potential but higher than its threshold of stimulation. And because the EPSP is higher than the threshold of stimulation, it does not trigger impulse conduction by the postsynaptic neuron. It "facilitates" it, that is, makes it ready for a weaker subsequent stimulus to initiate conduction by it. An EPSP is produced in a postsynaptic neuron when the amount of excitatory transmitter released into a synaptic cleft by presynaptic axons slightly exceeds the amount of inhibitory transmitter released by other presynaptic axons into that cleft.

Impulse conduction by a postsynaptic neuron is initiated when the amount of excitatory transmitter exceeds the amount of inhibitory transmitter released by presynaptic axons sufficiently to decrease the negativity of the postsynaptic neuron's potential down to its threshold of stimulation, for example, from a resting potential of -70 mv (inner surface of neuron's membrane negative to its outer surface) to a threshold of stimulation of -50 mv. Instantaneously, when the potential reaches the critical threshold level, its negativity decreases explosively through 0 until the membrane's inner surface becomes about 30 mv positive to its outer surface. At this point, it becomes an action potential, and impulse conduction by the postsynaptic neuron begins.

Inhibition of postsynaptic neurons occurs when the amount of inhibitory transmitter released by presynaptic axons exceeds the amount of excitatory transmitter released by other presynaptic axons into the same synaptic cleft. Inhibition produces an *inhibitory postsynaptic potential* (IPSP) in the postsynaptic neuron. The IPSP is higher than the usual resting potential, for example, -80 mv instead of -70 mv. When

an IPSP exists in a neuron, it is said to be inhibited.

Summarizing, dozens of synaptic knobs from dozens of presynaptic axons synapse with any one postsynaptic neuron, most commonly with its dendrites or soma. Some knobs release excitatory transmitter, some release inhibitory transmitter. The algebraic sum of these antagonistic chemicals determines their effect on the postsynaptic neuron. Facilitation, impulse conduction, or inhibition of the postsynaptic neuron may result from the summation of excitatory and inhibitory transmitters in a synaptic cleft.

Whereas synapses are regions of contact between presynaptic axon terminals and postsynaptic neurons, neuroeffector junctions are regions of contact between a motoneuron's axon terminals and an effector—muscle or glandular cells. Transmitter substances released at neuroeffector junctions stimulate muscle cells to contract and glandular cells to secrete.

Neurotransmitters

Several different chemical compounds are now known to serve as neurotransmitters. A major one is a compound named acetylcholine (ACh). ACh is released at neuromuscular junctions with skeletal muscle cells, at some neuromuscular junctions with smooth muscle and cardiac muscle cells, and at some neuroglandular junctions. At other neuroglandular junctions and at some neuromuscular junctions with smooth muscle and cardiac muscle cells the compound that serves as the neurotransmitter is norepinephrine (NE). Various substances function as neurotransmitters at synapses in the brain and spinal cord. We shall discuss several of these in the next chapter.

At neuromuscular junctions with skeletal muscle cells, ACh binds to specific receptors in the muscle cell membrane. This is postulated to change the shape of the receptor molecules in a way that opens channels in the membrane and permits the influx of sodium ions, which initiates conduction by the muscle cell and results in its contraction. In short, ACh functions directly at skeletal muscle synapse without the formation of the second messenger, cyclic AMP. An enzyme, acetylcholinesterase, rapidly terminates ACh's stimulation of skeletal muscle by hydrolyzing ACh to acetate and choline.

Curare (plant poison used on arrows by South American Indians) binds to ACh receptors, but it does not induce muscle contraction. Because ACh cannot bind to receptors already bound to curare, muscle cells cannot be stimulated to contract—in other words, they are paralyzed by the curare. Present-day anesthesiologists use curare preparations and drugs with similar actions, for example, pancuronium (Pavulon), to produce skeletal muscle relaxation during surgery.

In the chronic disease myasthenia gravis, inadequate conduction by ACh occurs across neuromuscular junctions with skeletal muscles. This leads to severe muscular weakness, a characteristic of this disease.

Outline summary

Cells

Neuroglia

A Types—astrocytes, oligodendria, and microglia
B Structure and function
  1 Astrocytes—star shaped; numerous processes twine around neurons and attach them to blood vessels; postulated to function as part of the blood-brain barrier
  2 Oligodendroglia—fewer processes than other two types of neuroglia; support neurons; produce myelin sheath around nerve fibers in the brain and cord
  3 Microglia—small cells but enlarge and move about in inflamed brain tissue; carry on phagocytosis

Neurons

A Types
  1 Classified according to direction of impulse conduction
    a Sensory or afferent—conduct impulses to cord or brain
    b Motoneurons or efferent—conduct impulses away from brain or cord to or toward muscle or glandular tissue
    c Interneurons (internuncial or intercalated neurons)—conduct from sensory to mononeurons
  2 Classified according to number of processes
    a Multipolar—one axon and several dendrites
    b Bipolar—one axon and one dendrite
    c Unipolar—one process comes off neuron cell body but divides almost immediately into one axon and one dendrite

B Structure—identify parts of neurons in Figs. 8-2 and 8-3
  1 Soma—cell body of neuron
  2 Gray matter—clusters of neuron cell bodies
  3 Nerve fibers—dendrites or axons
  4 Dendrites—branching process of a neuron; conduct impulses to its cell body
  5 Receptors—distal ends of dendrites of sensory neurons
  6 Axon—single process of a neuron, but may have collateral branches; conducts impulses away from neuron cell body
  7 Neurofibrils—fine fibers extending through dendrites, cell bodies, and axons; seem to consist of bundles of thinner fibers, that is, microtubules and microfilaments
  8 Nissl bodies—layered fragments of the endoplasmic reticulum; appear as good-sized granules scattered through cytoplasm of neuron cell body; specialize in synthesizing proteins for maintaining and regenerating neuron processes and for renewing neurotransmitters
  9 Myelin sheath—segmented wrapping around a nerve fiber; segments separated by nodes of Ranvier; Schwann cells form myelin sheaths around peripheral nerve fibers; oligodendroglia form myelin sheaths around central nerve fibers, that is, those located in the brain and cord
  10 White matter—consists of myelinated fibers
  11 Tracts—bundles of myelinated fibers located in the brain and cord
  12 Nerves—bundles of myelinated fibers located outside the brain and cord
  13 Neurilemma or sheath of Schwann—continuous sheath enclosing myelin sheath; neurilem-

ma plays essential part in regenerating injured or cut nerve fibers; brain and cord fibers do not have neurilemma, so presumably do not regenerate

C Function—respond to stimulation by conducting impulses (body's most rapid means of communication)

Nerve impulse conduction
Definitions

A Potential difference—difference in electrical charge

B Polarized membrane—one whose outer surface bears different electrical charge than its inner surface

C Depolarized membrane—one with no potential difference between its outer and inner surfaces

D Resting potential—potential difference between outer and inner surfaces of membrane of nonconducting neuron; magnitude of resting potential ranges between 70 to 90 mv, with inside of membrane negative to outside

E Action potential—potential difference that exists across neuron membrane when it is conducting impulses

F Stimulus—change in the environment, for example, a change in pressure, temperature, or chemical composition

Mechanism of resting potential

A Sodium pump transports sodium ions out of neuron's intracellular fluid into surrounding extracellular fluid, whereas potassium pump transports about one third as many potassium ions into neuron's intracellular fluid as sodium pump transports out of it

B Because neuron cell membrane is much more permeable to potassium than to sodium, more potassium diffuses back out of the intracellular fluid than sodium diffuses back in

C Because neuron cell membrane is impermeable to some intracellular negative ions and only slightly permeable to others, the number of negative ions diffusing out of the cell is less than the number of positive (potassium) ions diffusing out; result—minute excess of negative ions inside neuron cell membrane and minute excess of positive ions outside it; this ionic imbalance across neuron cell membrane produces resting potential; steps 1 and 2 also contribute to minute excess of positive ions outside neuron cell membrane

Mechanism of action potential (nerve impulse)

A Definition—self-propagating wave of electrical negativity that travels along surface of neuron membrane

B Mechanism
1 Stimulus increases permeability of neuron membrane to sodium ions
2 Rapid inward diffusion of sodium ions causes inner surface of membrane to become positive to outer surface; this reversal of resting potential marks beginning of action potential or nerve impulse

C Speed of conduction
1 Large-diameter A fibers conduct at speed of about 100 meters per second (more than 200 miles per hour)
2 Small-diameter C fibers conduct at speed of about 0.5 meter per second (1 mile per hour)
3 Intermediate-diameter B fibers conduct at speeds intermediate between A and C fibers
4 Heavily myelinated fibers with greater distance between successive nodes of Ranvier conduct many times faster than thinly myelinated fibers

D Initiation of action potential
1 Stimulus, that is, change in the environment, initiates action potential by increasing neuron membrane permeability to sodium ions
2 Threshold (liminal) stimulus—just strong enough to decrease negativity of neuron membrane potential from level of resting potential down to critical level that initiates acting potential
3 Subthreshold (subliminal) stimulus—decreases membrane negativity below resting level but not down to threshold level

E Impulse conduction routes
1 Many, but by no means all, impulses are conducted over route known as reflex arc, that is, two or more neurons that conduct impulses from periphery to spinal cord or brain stem and back to periphery; impulse begins in receptors and ends in effectors; receptors—distal ends of sensory neurons; effectors—muscle or glandular cells
2 Two-neuron or monosynaptic reflex arc—simplest arc possible; consists of at least one sensory neuron, one synapse, and one motoneuron (synapse is contact region between axon terminals of one neuron and dendrites or cell body of another neuron)

3 Three-neuron arc—consists of at least one sensory neuron, synapse, interneuron, synapse, and motoneuron; two synapses in three-neuron arc

4 Complex, multisynaptic neural pathways also exist; many not clearly understood

Mechanism of conduction across synapses

A Definition of synapse—site where impulses are transmitted from a presynaptic to a postsynaptic neuron

B Structure of synapse—consists of

1 Synaptic knobs—tiny distentions at presynaptic neuron's axon terminals; numerous vesicles containing neurotransmitter molecules present in each synaptic knob

2 Synaptic cleft—microscopic space between a synaptic knob and a postsynaptic neuron's dendrite or cell body; neuroeffector junctions—microscopic spaces between axon terminals of motoneurons and effector cells; neuromuscular junctions (nerve-muscle synapses) and neuroglandular junctions are types of neuroeffector junctions

3 Postsynaptic cell membrane—protein molecules embedded in membrane function as neurotransmitter receptors and as enzymes

C Synaptic conduction mediated by chemicals

1 When action potential reaches axon terminals, neurotransmitter molecules released from synaptic knob vesicles into synaptic cleft

2 Neurotransmitter molecules diffuse across cleft, bind to specific receptors in postsynaptic neuron's membrane, opening channels in it, through which sodium ions diffuse into and potassium ions diffuse out of interior of neuron, thereby depolarizing it and giving rise to excitatory postsynaptic potential

3 Neurotransmitters are either excitatory or inhibitory; excitatory transmitters decrease the negativity of the postsynaptic potential; inhibitory transmitters increase postsynaptic negativity

4 Summation—dozens of presynaptic axons synapse with any one postsynaptic neuron; some of these presynaptic axons release excitatory transmitter, some release inhibitory transmitter into the same synaptic cleft; adding together, or summation, of these antagonistic chemicals determines effect on postsynaptic neuron (facilitation, inhibition, or impulse conduction)

5 Facilitation—decrease in the negativity of the postsynatpic neuron's membrane potential to a level below its resting potential but above its threshold of stimulation; this lowered potential called the excitatory postsynaptic potential (EPSP); result of summation with slight excess of excitatory transmitter in synaptic cleft; does not initiate impulse conduction by postsynaptic neuron

6 Inhibition—increase in the negativity of the postsynaptic neuron's membrane potential above its usual resting potential; this increased potential called the inhibitory postsynaptic potential (IPSP); result of summation with excess of inhibitory transmitter in synaptic cleft

7 Impulse conduction by postsynaptic neuron initiated when summation results in an excess of excitatory transmitter sufficient to decrease the negativity of the postsynaptic neuron's membrane potential to its threshold of stimulation; depolarization, followed by a reverse potential of about 30 mv; when inner surface of membrane is about 30 mv positive to outer surface, it has become an action potential (nerve impulse)

D Neurotransmitters

1 At neuroeffector junctions

a Acetylcholine (ACh)—known transmitter at all neuroeffector junctions with skeletal muscle cells and at some neuroeffector junctions with cardiac muscle, smooth muscle, and glandular cells

b Norepinephrine (NE)—known transmitter at some neuroeffector junctions with cardiac muscle, smooth muscle, and glandular cells

2 At synapses in central nervous system, several substances act as neurotransmitters (discussed in next chapter)

Review questions

1 In a word or two, what general function does the nervous system perform for the body?
2 Name another system that serves the same general function.
3 Compare neurons and neuroglia.
4 Differentiate between sensory neurons, motoneurons, and interneurons.
5 Differentiate between myelin sheath and neurilemma
6 Differentiate between white matter and gray matter, tracts and nerves.
7 Differentiate between polarized membrane and depolarized membrane.
8 Differentiate between action potential and resting potential.
9 The term receptor has one meaning when used in reference to sensory neurons and another meaning when used in reference to postsynaptic neurons. Differentiate between these two meanings.
10 What are effectors?
11 Define briefly: facilitation, inhibition, summation.
12 What substance serves as the neurotransmitter at neuromuscular junctions with skeletal muscle cells?
13 Name two substances that function at neuromuscular junctions with smooth muscle, cardiac muscle, and glandular cells.
14 Explain the rationale for giving a curare preparation before surgery.

chapter 9

The somatic nervous system

Divisions of nervous system

From the title of this chapter you might infer that your body has more than one nervous system. It does not. It has only one nervous system, but to discuss this highly complicated set of organs, we pretty clearly must divide it up some way. Nervous system organs may be separated by location into two divisions, namely, the central nervous system (CNS) and the peripheral nervous system (PNS). The CNS consists of the centrally located nervous system organs, that is, the brain and spinal cord. All other nervous system organs—cranial nerves, spinal nerves, autonomic nerves, and ganglia—compose the PNS. For our discussion, we shall separate the nervous system into two divisions based on the type of effectors they innervate. The names of these divisions are the somatic nervous system and the autonomic nervous system. The *somatic nervous system* consists of organs that innervate somatic effectors, that is, skeletal muscles. Included in the somatic nervous system are both organs of the central nervous system and most organs of the peripheral nervous system. Specifically the somatic nervous system consists of the brain, spinal cord, cranial nerves, and spinal nerves. The *autonomic nervous system*, to be discussed in the next chapter, consists of organs that innervate autonomic effectors, that is, smooth muscle, cardiac muscle, and glandular epithelial tissue.

This chapter presents basic information about the anatomy and physiology of the somatic nervous system. Our plan is to start with a description of the coverings of the brain and spinal cord. Then we shall discuss the spinal cord and its nerves and proceed upward through the lower parts of the brain to its highest part, the cerebrum. We shall also discuss brain and cord neurotransmitters and somatic sensory and motor pathways as well as some reflexes.

Brain and cord coverings

Because the brain and spinal cord are both delicate and vital, nature has provided them with two protective coverings. The outer covering consists of bone: cranial bones encase the brain and vertebrae encase the cord. The inner covering consists of membranes known as *meninges*. Three distinct layers compose the meninges: the dura mater, the arachnoid membrane, and the pia mater. Observe their respective locations in Figs. 9-1 and 9-2. The dura mater, made of strong white fibrous tissue, serves both as the outer layer of the meninges and also as the inner periosteum of the cranial bones. The arachnoid membrane, a delicate, cobwebby layer, lies between the dura mater and the pia mater or innermost layer of the meninges. The transparent pia mater adheres to the outer surface of the brain and cord and contains blood vessels.

Three extensions of the dura mater should be mentioned: the falx cerebri, falx cerebelli, and

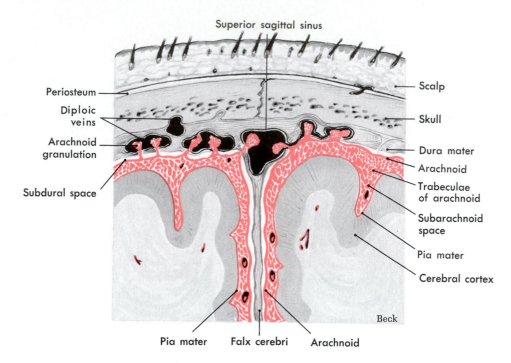

Superior sagittal sinus

Periosteum

Diploic
veins

Arachnoid
granulation

Subdural space

Scalp

Skull

Dura mater

Arachnoid

Trabeculae
of arachnoid

Subarachnoid
space

Pia mater

Cerebral cortex

Beck

Pia mater   Falx cerebri   Arachnoid

Fig. 9-1 Meninges of the brain as seen in coronal section through the skull.

tentorium cerebelli. The falx cerebri projects downward into the longitudinal fissure to form a kind of partition between the two cerebral hemispheres. The falx cerebelli separates the two cerebellar hemispheres. The tentorium cerebelli separates the cerebellum from the occipital lobe of the cerebrum. It takes its name from the fact that it forms a tentlike covering over the cerebellum.

Between the dura mater and the arachnoid membrane is a small space called the subdural space and between the arachnoid and the pia mater is another space, the subarachnoid space. Inflammation of the meninges is called meningitis. It most often involves the arachnoid and pia mater or the *leptomeninges*, as they are sometimes called.

The meninges of the cord continue on down inside the spinal cavity for some distance below

the end of the spinal cord. The pia mater forms a slender filament known as the filum terminale. At the level of the third segment of the sacrum, the filum terminale blends with the dura mater to form a fibrous cord that disappears in the periosteum of the coccyx. This extension of the meninges beyond the cord is convenient for performing lumbar punctures. It makes it possible to insert a needle between the third and fourth or fourth and fifth lumbar vertebrae into the subarachnoid space to withdraw cerebrospinal fluid without danger of injuring the spinal cord, which ends more than an inch above that point. The fourth lumbar vertebra can be easily located because it lies on a line with the iliac crest. Placing the patient on his side and arching his back by drawing the knees and chest together separates the vertebrae sufficiently to introduce the needle.

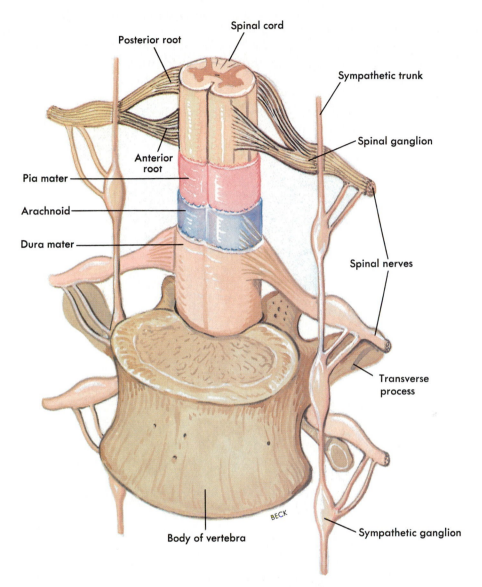

Fig. 9-2 Spinal cord showing meninges, formation of the spinal nerves, and relations to a vertebra and to the sympathetic trunk and ganglia.

Brain and cord fluid spaces

In addition to the bony and membranous coverings, nature has further fortified the brain and spinal cord against injury by providing a cushion of fluid both around them and within them. The fluid is called *cerebrospinal fluid*, and the spaces containing it are as follows:

1 The subarachnoid space around the brain
2 The subarachnoid space around the cord
3 The ventricles and aqueduct inside the brain
4 The central canal inside the cord

The ventricles are cavities or spaces inside the brain. They are four in number. Two of them, the lateral (or first and second) ventricles, are located one in each cerebral hemisphere. Note in Fig. 9-3 the shape of these ventricles—roughly like the hemispheres themselves. The third ventricle is little more than a lengthwise slit in the cerebrum beneath the midportion of the corpus callosum and longitudinal fissure. The fourth ventricle is a diamond-shaped space between the cerebellum posteriorly and the medulla and pons anteriorly. Actually, it is an expansion of the central canal of the cord after the cord en-

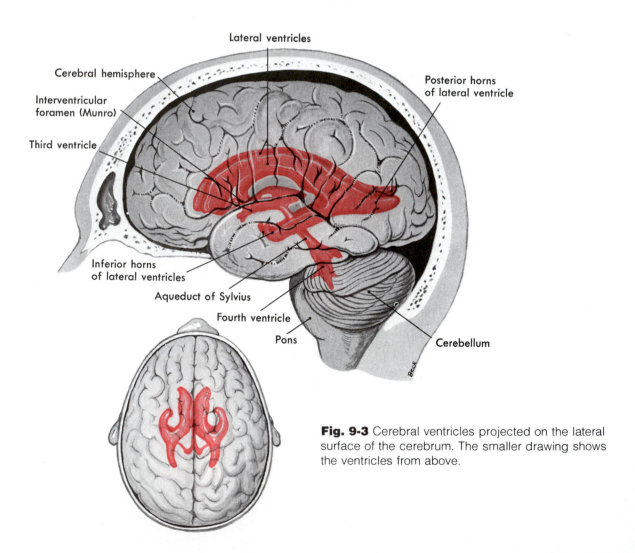

Lateral ventricles

Cerebral hemisphere

Posterior horns of lateral ventricle

Interventricular foramen (Munro)

Third ventricle

Inferior horns of lateral ventricles

Aqueduct of Sylvius

Fourth ventricle

Pons

Cerebellum

Fig. 9-3 Cerebral ventricles projected on the lateral surface of the cerebrum. The smaller drawing shows the ventricles from above.

ters the cranial cavity and becomes enlarged to form the medulla.

Formation, circulation, and functions of cerebrospinal fluid

Formation of cerebrospinal fluid occurs mainly by secretion by the choroid plexi. Choroid plexi are networks of capillaries that project from the pia mater into the lateral ventricles and into the roofs of the third and fourth ventricles. From each lateral ventricle the fluid seeps through an opening, the interventricular foramen (of Munro), into the third ventricle, then through a narrow channel, the cerebral aqueduct (or aqueduct of Sylvius), into the fourth ventricle, from which it circulates into the central canal of the cord. Openings in the roof of the fourth ventricle (the foramen of Magendie and foramina of Luschka) permit the flow of fluid into the subarachnoid space around the cord and then into the subarachnoid space around the brain. From the latter space, it is gradually absorbed into the venous blood of the brain. Thus cerebrospinal fluid "circulates" from blood in the choroid plexuses, through ventricles, central canal, and subarachnoid spaces, and back into the blood.

Occasionally, some condition interferes with this circuit. For example, a brain tumor may press against the cerebral aqueduct, shutting off the flow of fluid from the third to the fourth ventricle. In such an event the fluid accumulates within the lateral and third ventricles because it continues to form even though its drainage is blocked. This condition is known as internal hydrocephalus. If the fluid accumulates in the subarachnoid space around the brain, external hydrocephalus results. Subarachnoid hemorrhage, for example, may lead to formation of blood clots that block drainage of the cerebrospinal fluid from the subarachnoid space. With decreased drainage an increased amount of fluid, of course, remains in the space.

Withdrawal of some of the cerebrospinal fluid from the subarachnoid space in the lumbar region of the cord is known as a lumbar puncture.

The amount of cerebrospinal fluid in the average adult is about 140 ml (about 23 ml in the ventricles and 117 ml in the subarachnoid space of brain and cord).*

Cerebrospinal fluid serves as a protective cushion around and within the brain and cord. However, it is now known to function in other ways as well. For instance, changes in its carbon dioxide content affect neurons of the respiratory center in the medulla and thereby help control respiration.

Spinal cord

Structure

The spinal cord lies within the spinal cavity, extending from the foramen magnum to the lower border of the first lumbar vertebra (Fig. 9-4), a distance of 17 or 18 inches in the average body. The cord does not completely fill the spinal cavity—it also contains the meninges, spinal fluid, a cushion of adipose tissue, and blood vessels.

The spinal cord is an oval-shaped cylinder that tapers slightly from above downward and has two bulges, one in the cervical region and the other in the lumbar region. Two deep grooves, the anterior median fissure and the posterior median sulcus, just miss dividing the cord into separate symmetrical halves. The anterior fissure is the deeper and the wider of the two grooves—a useful factor to remember when you examine spinal cord diagrams. It enables you to tell at a glance which part of the cord is anterior and which is posterior. Gray matter composes the inner core of the cord. Although it looks like a flat letter H in cross-section views of the cord, it actually has three dimensions, since the gray matter extends the length of the cord. The

---

*Mountcastle, V. B., editor: Medical physiology, ed. 13, St. Louis, 1974, The C. V. Mosby Co., p. 1116.

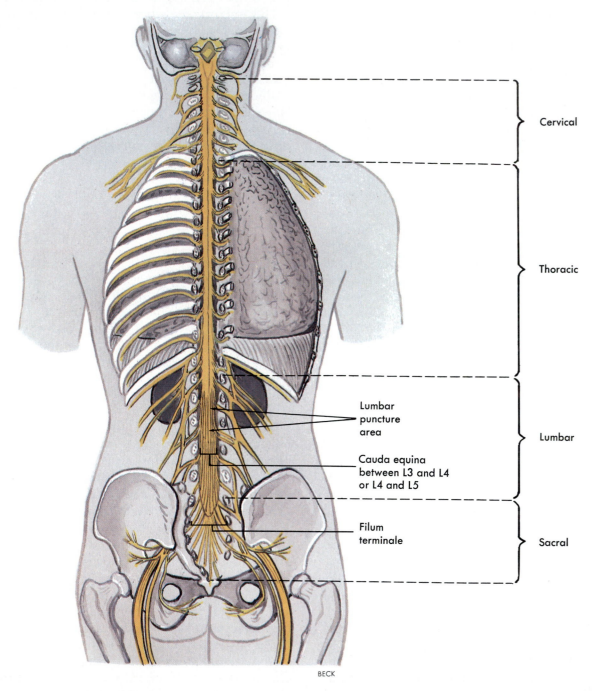

Cervical

Thoracic

Lumbar puncture area

Cauda equina between L3 and L4 or L4 and L5

Filum terminale

Lumbar

Sacral

BECK

Fig. 9-4 Relation of the spinal cord, part of the brain, and some of the spinal nerves to surrounding structures. (Adapted from Mettler, F. A.: Neuroanatomy, St. Louis, The C. V. Mosby Co.)

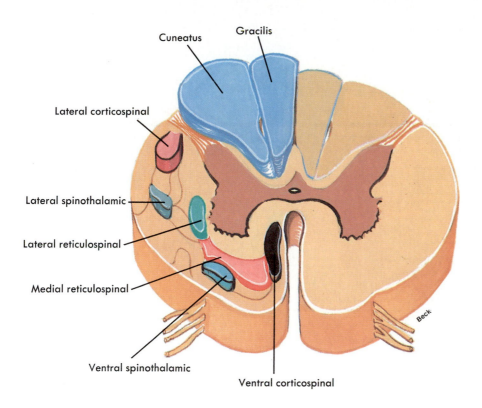

Cuneatus

Gracilis

Lateral corticospinal

Lateral spinothalamic

Lateral reticulospinal

Medial reticulospinal

Ventral spinothalamic

Ventral corticospinal

Beck

Fig. 9-5 Location in the spinal cord of some major projection tracts. Ascending or sensory tracts described in Table 9-1. Table 9-2 describes descending or motor tracts.

limbs of the **H** are called either anterior, posterior, or lateral horns of gray matter or anterior, posterior, or lateral gray columns. They consist primarily of cell bodies of interneurons and motoneurons.

White matter surrounding the gray matter is subdivided in each half of the cord into three columns (or funiculi): the anterior, posterior, and lateral white columns. Each white column or funiculus consists of a large bundle of nerve fibers (axons) divided into smaller bundles called tracts, as shown in Fig. 9-5. The names of most spinal cord tracts indicate the white column in which the tract is located, the structure in which the axons that make up the tract originate, and the structure in which they terminate. Examples: The lateral corticospinal tract is located in the lateral white column of the cord, and the axons that compose it originate from neuron cell bodies located in the cortex (of the cerebrum) and terminate in the spinal cord. The

ventral spinothalamic tract lies in the ventral, that is, anterior, white column, and the axons that compose it originate from neuron cell bodies located in the spinal cord and terminate in the thalamus.

Functions

The spinal cord performs sensory, motor, and reflex functions. The following paragraphs describe these functions briefly.

Spinal cord tracts serve as two-way conduction paths between peripheral nerves and the brain. Ascending tracts conduct impulses up the cord to the brain. Descending tracts conduct impulses down the cord from the brain. Bundles of axons compose all tracts. Tracts are both structural and functional organizations of these nerve fibers. They are structural organizations in that all of the axons of any one tract originate from neuron cell bodies located in the same structure and all of the axons terminate in the same struc-

Table 9-1 Major ascending tracts of spinal cord

Name	Function	Location	Origin*	Termination†
Lateral spinothalamic	Pain, temperature, and crude touch opposite side	Lateral white columns	Posterior gray column opposite side	Thalamus
Ventral spinothalamic	Crude touch, pain, and temperature	Anterior white columns	Posterior gray column opposite side	Thalamus
Fasciculi gracilis and cuneatus	Discriminating touch and pressure sensations, including vibration, stereognosis, and two-point discrimination; also conscious kinesthesia	Posterior white columns	Spinal ganglia same side	Medulla
Spinocerebellar	Unconscious kinesthesia	Lateral white columns	Posterior gray column	Cerebellum

*Location of cell bodies of neurons from which axons of tract arise.
†Structure in which axons of tract terminate.

Table 9-2 Major descending tracts of spinal cord

Name	Function	Location	Origin*	Termination†
Lateral corticospinal (or crossed pyramidal)	Voluntary movement, contraction of individual or small groups of muscles, particularly those moving hands, fingers, feet, and toes of opposite side	Lateral white columns	Motor areas cerebral cortex (mainly areas 4 and 6) opposite side from tract location in cord	Intermediate or anterior gray columns
Ventral corticospinal (direct pyramidal)	Same as lateral corticospinal except mainly muscles of same side	Lateral white columns	Motor cortex but on same side as tract location in cord	Intermediate or anterior gray columns
Lateral reticulospinal	Mainly facilitatory influence on motoneurons to skeletal muscles	Lateral white columns	Reticular formation, midbrain, pons, and medulla	Intermediate or anterior gray columns
Medial reticulospinal	Mainly inhibitory influence on motoneurons to skeletal muscles	Anterior white columns	Reticular formation, medulla mainly	Intermediate or anterior gray columns

*Location of cell bodies of neurons from which axons of tract arise.
†Structure in which axons of tract terminate.

ture. For example, all the fibers of the spinothalamic tract are axons originating from neuron cell bodies located in the spinal cord and terminating in the thalamus. Tracts are functional organizations in that all the axons that compose one tract serve one general function. For instance, fibers of the spinothalamic tracts serve a sensory function. They transmit impulses that produce our sensations of crude touch, pain, and temperature.

Because so many different tracts make up the white columns of the cord, we shall mention only a few that seem most important in man. Locate each tract in Fig. 9-5. Consult Tables 9-1 and 9-2 for a brief summary of information about tracts. Four important ascending or sensory tracts and their functions, stated very briefly, are as follows:

1 Lateral spinothalamic tracts—crude touch, pain, and temperature
2 Ventral spinothalamic tracts—crude touch, pain, and temperature
3 Fasciculi gracilis and cuneatus—discriminating touch and conscious kinesthesia
4 Spinocerebellar tracts—unconscious kinesthesia

Further discussion of the sensory neural pathways may be found on pp. 259-262.

Four important descending or motor tracts and their functions in brief are as follows:

1 Lateral corticospinal tracts—voluntary movement; contraction of individual or small groups of muscles, particularly those moving hands, fingers, feet, and toes on opposite side of body
2 Ventral corticospinal tracts—same as preceding except mainly muscles of same side of body
3 Lateral reticulospinal tracts—mainly facilitatory impulses to anterior horn motoneurons to skeletal muscles
4 Medial reticulospinal tracts—mainly inhibitory impulses to anterior horn motoneurons to skeletal muscles

Nuclei in spinal cord gray matter function as reflex centers for all spinal reflexes. The term *re-*

flex center means the center of a reflex arc or the place in the arc where incoming sensory impulses become outgoing motor impulses. They are structures that switch impulses from afferent to efferent neurons. In two-neuron arcs, reflex centers are merely synapses between sensory and motor neurons. In all other arcs, reflex centers consist of interneurons interposed between sensory and motor neurons.

A *reflex* is an action that results from impulse conduction over a reflex arc. A reflex, therefore, is a response to a stimulus. Usually only unwilled or involuntary responses are called reflexes. Of all reflexes—and there are many of them—probably the most familiar is the knee jerk or patellar reflex. Two-neuron arcs mediate this spinal cord reflex. The reflex centers of these arcs lie in the cord's gray columns, in the second, third, and fourth lumbar segments.

Spinal nerves

Structure

Thirty-one pairs of nerves are connected to the spinal cord. They have no special names but are merely numbered according to the level of the spinal column at which they emerge from the spinal cavity. Thus there are 8 cervical, 12 thoracic, 5 lumbar, and 5 sacral pairs and 1 coccygeal pair of spinal nerves. The first cervical nerves emerge from the cord in the space above the first cervical vertebra (between it and the occipital bone). The rest of the cervical and all of the thoracic nerves pass out of the spinal cavity horizontally through the intervertebral foramina of their respective vertebrae. For example, the second cervical nerves emerge through the foramina above the second cervical vertebra.

Lumbar, sacral, and coccygeal nerve roots, on the other hand, descend from their point of origin at the lower end of the cord (which terminates at the level of the first lumbar vertebra) before reaching the intervertebral foramina of their respective vertebrae, through which the nerves then emerge. This gives the lower end of the cord, with its attached spinal nerve roots, the appearance of a horse's tail. In fact, it bears the name *cauda equina* (Latin equivalent for horse's tail). (See Fig. 9-4.)

Each spinal nerve, instead of attaching directly to the cord, attaches indirectly by means of two short roots, anterior and posterior. The posterior roots are readily recognized by the presence of a swelling, the posterior root ganglion or *spinal ganglion*. The roots and ganglia lie, respectively, within the spinal cavity in the intervertebral foramina (Fig. 9-6). Fig. 9-7 shows the structure of a nerve in a cross-section view.

After each spinal nerve emerges from the spinal cavity, it divides into anterior, posterior, and white rami. Anterior and posterior rami contain fibers belonging to the voluntary nervous system. White rami contain fibers of the autonomic nervous system. The posterior rami subdivide into lesser nerves that extend into the muscles

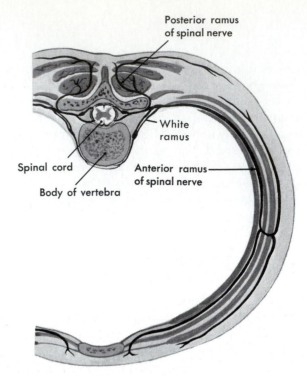

Fig. 9-6 Branchings of a spinal nerve. Note the anterior and posterior roots of the nerve within the vertebral foramen, and anterior and posterior rami outside of it.

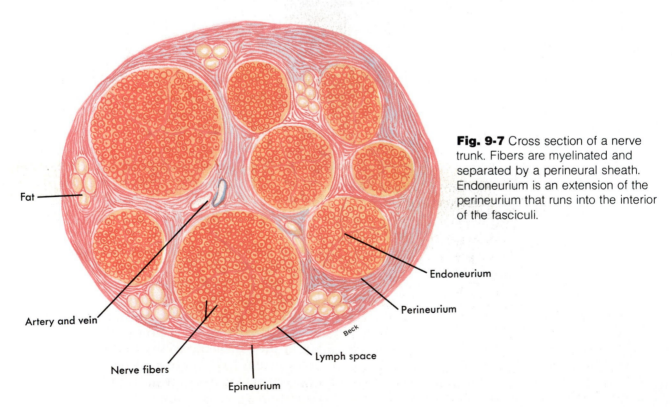

Fig. 9-7 Cross section of a nerve trunk. Fibers are myelinated and separated by a perineural sheath. Endoneurium is an extension of the perineurium that runs into the interior of the fasciculi.

and skin of the posterior surface of the head, neck, and trunk.

The anterior rami (except those of the thoracic nerves) subdivide and supply fibers to skeletal muscles and skin of the extremities and of anterior and lateral surfaces. Subdivisions of the anterior rami form complex networks or plexuses (Table 9-3). For example, fibers from the lower four cervical and first thoracic nerves intermix in such a way as to form a fairly defi-

nite, although apparently hopelessly confused, pattern called the *brachial plexus* (Fig. 9-8). Emerging from this plexus are smaller nerves bearing names descriptive of their locations, such as the median nerve, the musculocutaneous nerve, and the ulnar nerve. These nerves (each containing fibers from more than one spinal nerve) divide further into smaller and smaller branches, resulting ultimately in the complete innervation of the hand and most of the arm. The

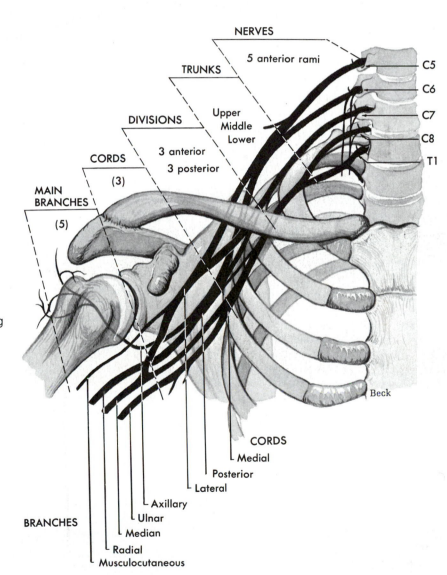

Fig. 9-8 Brachial plexus. Intermixing of fibers from the lower four cervical and first thoracic nerves.

NERVES

5 anterior rami

C5
C6
C7
C8
T1

TRUNKS

Upper
Middle
Lower

DIVISIONS

3 anterior
3 posterior

CORDS

(3)

MAIN
BRANCHES

(5)

Beck

CORDS
└ Medial
└ Posterior
└ Lateral

└ Axillary
└ Ulnar
└ Median
BRANCHES
└ Radial
└ Musculocutaneous

Table 9-3 Spinal nerves and peripheral branches

Spinal nerves	Plexuses formed from anterior rami	Spinal nerve branches from plexuses	Parts supplied
Cervical 1 2 3 4	Cervical plexus	Lesser occipital Great auricular Cutaneous nerve of neck Anterior supraclavicular Middle supraclavicular Posterior supraclavicular Branches to numerous neck muscles	Sensory to back of head, front of neck, and upper part of shoulder; motor to numerous neck muscles
		Phrenic (branches from cervical nerves before formation of plexus; most of its fibers from fourth cervical nerve)	Diaphragm
		Suprascapular and dorsoscapular	Superficial muscles* of scapula
		Thoracic nerves, medial and lateral branches	Pectoralis major and minor
		Long thoracic nerve	Serratus anterior
Cervical 5 6 7 8 Thoracic (or dorsal) 1	Brachial plexus	Thoracodorsal	Latissimus dorsi
		Subscapular	Subscapular and teres major muscles
		Axillary (circumflex)	Deltoid and teres minor muscles and skin over deltoid
		Musculocutaneous	Muscles of front of arm (biceps brachii, coracobrachialis, and brachialis) and skin on outer side of forearm
2 3 4 5 6 7 8 9 10 11 12	No plexus formed; branches run directly to intercostal muscles and skin of thorax	Ulnar	Flexor carpi ulnaris and part of flexor digitorum profundus; some of muscles of hand; sensory to medial side of hand, little finger, and medial half of fourth finger
		Median	Rest of muscles of front of forearm and hand; sensory to skin of palmar surface of thumb, index, and middle fingers
		Radial	Triceps muscle and muscles of back of forearm; sensory to skin of back of forearm and hand
		Medial cutaneous	Sensory to inner surface of arm and forearm

*Although nerves to muscles are considered motor, they do contain some sensory fibers that transmit proprioceptive impulses.

Spinal nerves	Plexus formed from anterior rami	Spinal nerve branches from plexuses	Parts supplied
		Iliohypogastric ⎫ Sometimes fused	Sensory to anterior abdominal wall
		Ilioinguinal ⎭	Sensory to anterior abdominal wall and external genitalia; motor to muscles of abdominal wall
		Genitofemoral	Sensory to skin of external genitalia and inguinal region
		Lateral cutaneous of thigh	Sensory to outer side of thigh
Lumbar 1 2 3 4 5 Sacral 1 2 3 4 5 Coccygeal 1	Lumbosacral plexus	Femoral	Motor to quadriceps, sartorius, and iliacus muscles; sensory to front of thigh and medial side of lower leg (saphenous nerve)
		Obturator	Motor to adductor muscles of thigh
		Tibial† (medial popliteal)	Motor to muscles of calf of leg; sensory to skin of calf of leg and sole of foot
		Common peroneal (lateral popliteal)	Motor to evertors and dorsiflexors of foot; sensory to lateral surface of leg and dorsal surface of foot
		Nerves to hamstring muscles	Motor to muscles of back of thigh
		Gluteal nerves, superior and inferior	Motor to buttock muscles and tensor fasciae latae
		Posterior cutaneous nerve	Sensory to skin of buttocks, posterior surface of thigh, and leg
		Pudendal nerve	Motor to perineal muscles; sensory to skin of perineum

†Sensory fibers from the tibial and peroneal nerves unite to form the *medial cutaneous* (or sural) *nerve* that supplies the calf of the leg and the lateral surface of the foot. In the thigh the tibial and common peroneal nerves are usually enclosed in a single sheath to form the *sciatic nerve,* the largest nerve in the body with its width of approximately ¾ of an inch. About two thirds of the way down the posterior part of the thigh, it divides into its component parts. Branches of the sciatic nerve extend into the hamstring muscles.

brachial plexus is located in the shoulder region from the neck to the axilla. It is of clinical significance, since it is sometimes stretched or torn at birth, causing paralysis and numbness of the baby's arm on that side. If untreated, it results in a withered arm. Branching from this plexus are several nerves to the skin and to voluntary muscles.

The right and left phrenic nerves, whose fibers come from the third and fourth or fourth and fifth cervical spinal nerves before formation of the brachial plexus, have considerable clinical interest, since they supply the diaphragm muscle. If the neck is broken in a way that severs or crushes the cord above this level, nerve impulses from the brain can, of course, no longer reach the phrenic nerves, and therefore the diaphragm stops contracting. Unless artificial respiration of some kind is provided, the patient dies of respiratory paralysis as a result of the broken neck. Poliomyelitis that attacks the cord between the third and fifth cervical segments also paralyzes the phrenic nerve and, therefore, the diaphragm.

Another spinal nerve plexus is the *lumbar plexus*, formed by the intermingling of fibers from the first four lumbar nerves. This network of nerves is located in the lumbar region of the back in the psoas muscle. The large femoral nerve is one of several nerves emerging from the lumbar plexus. It divides into many branches supplying the thigh and leg.

Fibers from the fourth and fifth lumbar nerves and the first, second, and third sacral nerves form the *sacral plexus*, located in the pelvic cavity on the anterior surface of the piriformis muscle. Among other nerves that emerge from the sacral plexus are the tibial and common peroneal nerves, which, in the thigh, form the largest nerve in the body, namely, the great sciatic nerve (Fig. 9-9). It pierces the buttocks and runs down the back of the thigh. Its many branches supply nearly all the skin of the leg, the posterior thigh muscles, and the leg and foot muscles. Sciatica or neuralgia of the sciatic nerve is a fairly common and very painful condition.

At first glance the distribution of spinal nerves does not appear to follow an ordered arrangement. But detailed mapping of the skin surface has revealed a close relationship between the source on the cord of each spinal nerve and the vertical position of the body it innervates (Figs. 9-10 and 9-11). Knowledge of the segmental arrangement of spinal nerves has proved useful to physicians. For instance, a neurologist can identify the site of spinal cord or nerve abnormality from the area of the body insensitive to a pinprick.

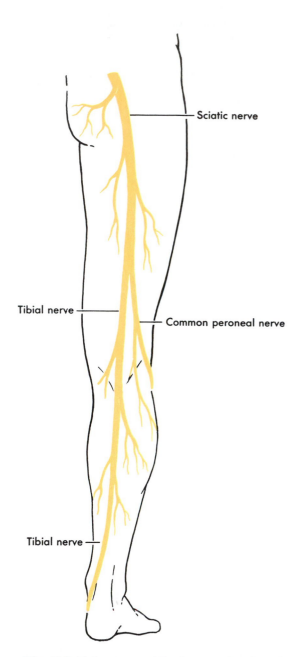

Fig. 9-9 Main nerves of the lower extremity.

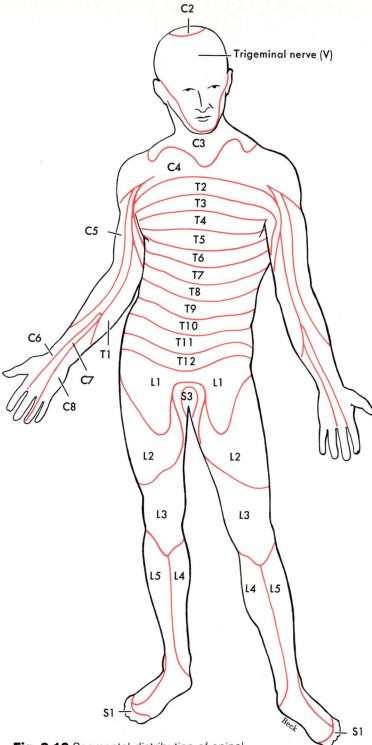

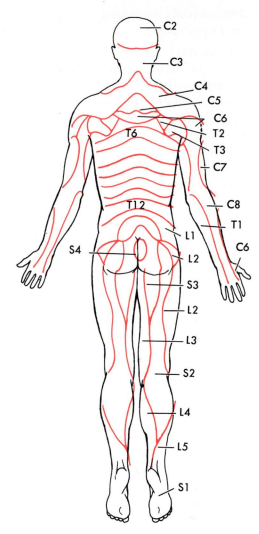

Fig. 9-11 Segmental distribution of spinal nerves to the back of the body. *C,* Cervical segments; *T,* thoracic segments; *L,* lumbar segments; *S,* sacral segments.

Fig. 9-10 Segmental distribution of spinal nerves to the front of the body. *C,* Cervical segments; *T,* thoracic segments; *L,* lumbar segments; *S,* sacral segments.

Microscopic structure and functions

Look back now at Fig. 8-5. Note the microscopic structure located in the posterior root ganglion—the cell body of a sensory neuron. Note, too, that the sensory neuron's dendrite lies in the spinal nerve leading to the ganglion. Its axon lies in the posterior root of the spinal nerve. Thus all of these structures—spinal nerves, posterior root ganglia, and posterior roots of spinal nerves—conduct impulses toward the cord. In short, they serve a sensory function. Next observe that the anterior root of the spinal nerve in Fig. 8-5 contains a motor axon that comes from a cell body located in the anterior horn of gray matter of the cord and that extends out to a skeletal muscle, a somatic effector. Thus the anterior gray column of the cord and the anterior roots of spinal nerves conduct impulses out of the cord. In short, they serve a motor function. Neurons whose dendrites and cell bodies lie in the anterior gray columns are known by three names—*anterior horn neurons, somatic motoneurons* and *lower motoneurons.* Summarizing: Spinal nerves are *mixed nerves,* that is, they contain both sensory dendrites and motor axons. Posterior root ganglia contain sensory cell bodies. Posterior roots contain sensory axons. Anterior gray columns of the cord contain dendrites and cell bodies of somatic motoneurons. Anterior roots of spinal nerves contain the axons of somatic motoneurons.

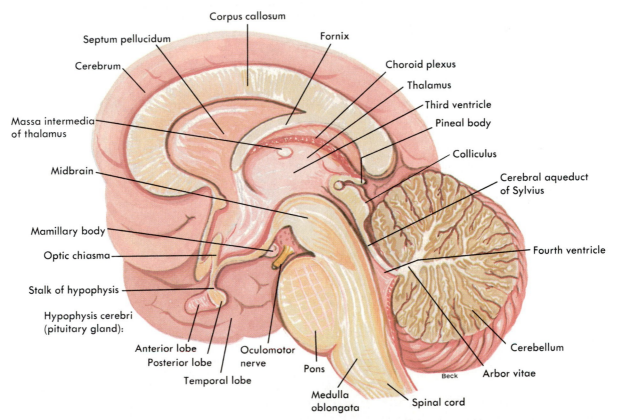

Fig. 9-12 Sagittal section through midline of the brain showing structures around the third ventricle.

Divisions and size of brain

The brain is one of the largest of adult organs. It consists of several billion neurons and presumably even more neuroglia. In most adults, it weighs about 3 pounds but generally is smaller in women than in men and in older people than in younger people. Neurons of the brain undergo mitosis only during the prenatal period and the first few months of postnatal life. Although they grow in size after that, they do not increase in number. Malnutrition during those crucial prenatal months of neuron multiplication is reported to hinder the process and result in fewer brain cells. The brain attains full size by about the eighteenth year but grows rapidly only during the first 9 years or so.

The brain has six major divisions: cerebrum, diencephalon, cerebellum, medulla oblongata, pons, and midbrain. The medulla, pons, and midbrain (mesencephalon) constitute the brain stem—an apt term, since, viewed from the side, they look like a stem for the rest of the brain (Fig. 9-12). Each division of the brain consists of gray matter (nuclei and centers) and white matter (tracts).

Brain stem

Structure

Three divisions of the brain make up the brain stem—so-called because of its resemblance to a stem (Fig. 9-12). The medulla oblongata forms the lowest part of the brain stem, the midbrain forms the uppermost part, and the pons lies between them, that is, above the medulla and below the midbrain.

Medulla

The medulla or bulb is the part of the brain that attaches to the spinal cord. It is, in fact, an enlarged extension of the cord located just above the foramen magnum. It measures only slightly more than an inch in length and is separated from the pons above by a horizontal groove. It is composed mainly of white matter (projection tracts) and reticular formation, a term that means the interlacement of gray and white matter present in the cord, brain stem, and diencephalon. Nuclei in the reticular formation of the medulla include such important centers as respiratory and vasomotor centers.

On each side of the lower posterior part of the medulla are two prominent nuclei, the nucleus gracilis and the nucleus cuneatus. Here, afferent fibers from the posterior white columns (fasciculi gracilis and cuneatus) of the cord synapse with neurons whose axons extend to the thalamus and cerebellum.

The pyramids (Fig. 9-13) are two bulges of white matter located on the anterior surface of the medulla formed by fibers of the pyramidal projection tracts.

The olive (Fig. 9-13) is an oval projection appearing one on each side of the anterior surface of the medulla. It contains the inferior olivary nucleus and two accessory olivary nuclei. Fibers from the cells of these nuclei run through the inferior cerebellar peduncles (restiform bodies) into the cerebellum. Motor nuclei of the ninth to the twelfth cranial nerves are also located in the medulla.

Pons

Just above the medulla lies the pons, composed, like the medulla, of white matter and a few nuclei. Fibers that run transversely across the pons and through the brachia pontis (middle cerebellar peduncles) into the cerebellum make up the external white matter of the pons and give it its bridgelike appearance. The reticular formation extends into the pons from the medulla. One important reticular nucleus in the pons is called the pneumotaxic center. (It functions in the control of respiration.) Nuclei of the fifth to eighth cranial nerves are located in the upper part of the pons.

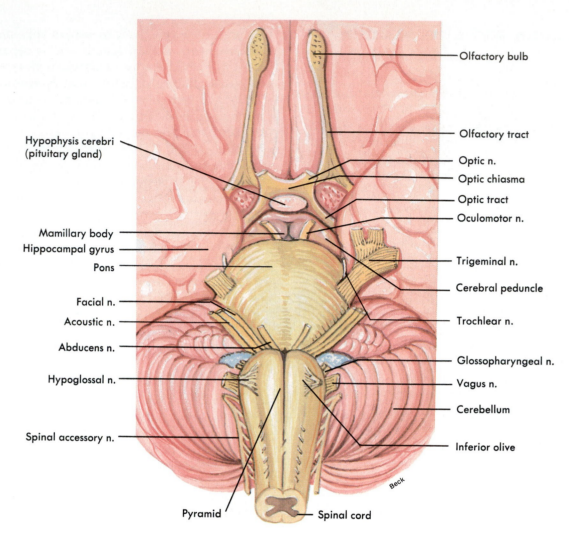

Hypophysis cerebri (pituitary gland)

Mamillary body

Hippocampal gyrus

Pons

Facial n.

Acoustic n.

Abducens n.

Hypoglossal n.

Spinal accessory n.

Olfactory bulb

Olfactory tract

Optic n.

Optic chiasma

Optic tract

Oculomotor n.

Trigeminal n.

Cerebral peduncle

Trochlear n.

Glossopharyngeal n.

Vagus n.

Cerebellum

Inferior olive

Pyramid

Spinal cord

Beck

Fig. 9-13 Ventral surface of the brain showing attachment of the cranial nerves.

Midbrain (mesencephalon)

The mesencephalon, or as it is more commonly called, the midbrain, lies below the inferior surface of the cerebrum and above the pons. It consists mainly of white matter with some internal gray matter around the cerebral aqueduct, the channel within the midbrain. The cerebral peduncles form the ventral part of the midbrain, and the corpora quadrigemina or colliculi form the dorsal part. The cerebral peduncles are two ropelike masses of white matter that extend divergently from the pons to the undersurface of the cerebral hemispheres (Fig. 9-13). In other words the cerebral peduncles are made up of tracts that constitute the main connection between the forebrain and hindbrain. Hence the midbrain, by function as well as by location, is well named.

The corpora quadrigemina consist of four rounded eminences, the two superior and the two inferior colliculi (Fig. 9-12), which form the dorsal part of the midbrain. Certain auditory reflex centers lie in the inferior colliculi and visual centers in the superior colliculi.

An important nucleus in the midbrain reticular formation is the red nucleus, a large gray mass ventral to the superior colliculi. Fibers from the cerebellum and the frontal lobe of the cerebral cortex end here, whereas fibers that extend into the rubrospinal tracts of the cord have their cells of origin here. Nuclei of the third and fourth cranial nerves and the anterior part of the nucleus of the fifth cranial nerve are located deep in the midbrain. Also, as mentioned in the preceding paragraph, nuclei for certain auditory and visual reflexes lie in the colliculi of the midbrain.

Functions

The brain stem, like the spinal cord, performs sensory, motor, and reflex functions. The spinothalamic tracts are important sensory tracts that pass through the brain stem on their way to the thalamus. The fasciculi cuneatus and gracilis and the spinoreticular tracts are sensory tracts whose axons terminate in the gray matter of the brain stem. Corticospinal and reticulospinal tracts are two of the major tracts present in the white matter of the brain stem.

Nuclei in the medulla contain a number of reflex centers. (The term *nucleus* used in reference to the nervous system means a cluster of neuron cell bodies located in the gray matter of the central nervous system.) Of first importance among reflex centers in the medulla are the cardiac, vasomotor, and respiratory centers. Because their functioning is essential for survival, they are called the vital centers. They serve as the centers for various reflexes controlling heart action, blood vessel diameter, and respiration. Because the medulla contains these centers, it is the most vital part of the entire brain—so vital, in fact, that injury or disease of the medulla often proves fatal. Blows at the base of the skull and bulbar poliomyelitis, for example, cause death if they interrupt impulse conduction by the vital respiratory centers. Other centers present in the medulla are those for various nonvital reflexes such as vomiting, coughing, sneezing, hiccuping, and swallowing.

The pons contains centers for reflexes mediated by the fifth, sixth, seventh, and eighth cranial nerves. (See Table 9-4 for functions.) In addition, the pons contains the pneumotaxic centers that help regulate respiration.

The midbrain, like the pons, contains reflex centers for certain cranial nerve reflexes, for example, pupillary reflexes and eye movements, mediated by the third and fourth cranial nerves, respectively.

Cranial nerves

Twelve pairs of nerves arise from the undersurface of the brain (Fig. 9-13), mostly from the brain stem. (This is our reason for discussing them here following the discussion of the brain stem.) After leaving the cranial cavity by way of small foramina in the skull, they extend to their respective destinations. Both names and numbers identify the cranial nerves. Their names suggest their distribution or function. Their numbers indicate the order in which they emerge from front to back. Some cranial nerves consist of both afferent and efferent fibers—in short, they are mixed nerves. On the other hand, some cranial nerves consist only of afferent fibers and some mainly of efferent fibers. Cell bodies of the efferent fibers lie in the various nuclei of the brain stem. Cell bodies of the afferent fibers, with few exceptions, are located in the ganglia outside the brain stem, for example, the trigeminal (gasserian) ganglion of the fifth cranial nerve. The first, second, and eighth cranial nerves are purely afferent. Information about cranial nerves is summarized in Tables 9-4 and 9-5.

Table 9-4 Cranial nerves

Nerve*	Sensory fibers†			Motor fibers†		Functions‡
	Receptors	Cell bodies	Termination	Cell bodies	Termination	
I Olfactory	Nasal mucosa	Nasal mucosa	Olfactory bulbs (new relay of neurons to olfactory cortex)			Sense of smell
II Optic	Retina	Retina	Nucleus in thalamus (lateral geniculate body); some fibers terminate in superior colliculus of midbrain			Vision
III Oculomotor	External eye muscles except superior oblique and lateral rectus	?	?	Midbrain (oculomotor nucleus and Edinger-Westphal nucleus)	External eye muscles except superior oblique and lateral rectus; fibers from Edinger-Westphal nucleus terminate in ciliary ganglion and then to ciliary and iris muscles	Eye movements, regulation of size of pupil, accommodation, proprioception (muscle sense)
IV Trochlear	Superior oblique	?	?	Midbrain	Superior oblique muscle of eye	Eye movements, proprioception
V Trigeminal	Skin and mucosa of head, teeth	Gasserian ganglion	Pons (sensory nucleus)	Pons (motor nucleus)	Muscles of mastication	Sensations of head and face, chewing movements, muscle sense
VI Abducens	Lateral rectus			Pons	Lateral rectus muscle of eye	Abduction of eye, proprioception
VII Facial	Taste buds of anterior two thirds of tongue	Geniculate ganglion	Medulla (nucleus solitarius)	Pons	Superficial muscles of face and scalp	Facial expressions, secretion of saliva, taste
VIII Acoustic 1 Vestibular branch	Semicircular canals and vestibule (utricle and saccule)	Vestibular ganglion	Pons and medulla (vestibular nuclei)			Balance or equilibrium sense

2 Cochlear or auditory branch	Organ of Corti in cochlear duct	Pons and medulla (cochlear nuclei)	Spiral ganglion			Hearing
IX Glossopharyngeal	Pharynx; taste buds and other receptors of posterior one third of tongue	Medulla (nucleus solitarius)	Jugular and petrous ganglia	Medulla (nucleus ambiguus)	Muscles of pharynx	Taste and other sensations of tongue, **swallowing movements, secretion of saliva,** aid in reflex control of blood pressure and respiration
	Carotid sinus and carotid body	Medulla (respiratory and vasomotor centers)	Jugular and petrous ganglia	Medulla at junction of pons (nucleus salivatorius)	Otic ganglion and then to parotid gland	
X Vagus	Pharynx, larynx, carotid body, and thoracic and abdominal viscera	Medulla (nucleus solitarius), pons (nucleus of fifth cranial nerve)	Jugular and nodose ganglia	Medulla (dorsal motor nucleus)	Ganglia of vagal plexus and then to muscles of pharynx, larynx, and thoracic and abdominal viscera	Sensations and **movements** or organs supplied; for example, **slows heart, increases peristalsis, and contracts muscles for voice production**
XI Spinal accessory	?	?	?	Medulla (dorsal motor nucleus of vagus and nucleus ambiguus)	Muscles of thoracic and abdominal viscera and pharynx and larynx	Shoulder movements, turning movements of head, movements of viscera, voice productions, proprioception?
				Anterior gray column of first five or six cervical segments of spinal cord	Trapezius and sternocleidomastoid muscle	
XII Hypoglossal	?	?	?	Medulla (hypoglossal nucleus)	Muscles of tongue	Tongue movements, proprioception?

*The first letters of the words in the following sentence are the first letters of the names of the cranial nerves. Many generations of anatomy students have used this sentence as an aid to memorizing these names. It is "On Old Olympus Tiny Tops, A Finn and German Viewed Some Hops." (There are several slightly differing versions of this mnemonic.)

†Italics indicate sensory fibers and functions. Boldface type indicates motor fibers and functions.

‡An aid for remembering the general function of each cranial nerve is the following 12-word saying: "Some say marry money but my brothers say bad business marry money." Words beginning with S indicate sensory function. Words beginning with M indicate motor function. Words beginning with B indicate both sensory and motor functions. For example, the first, second, and eighth cranial nerves perform sensory functions.

Table 9-5 Cranial nerves contrasted with spinal nerves

	Cranial nerves	Spinal nerves
Origin	Base of brain	Spinal cord
Distribution	Mainly to head and neck	Skin, skeletal muscles, joints, blood vessels, sweat glands, and mucosa except of head and neck
Structure	Some composed of sensory fibers only; some of both motor axons and sensory dendrites; some motor fibers belong to somatic nervous system, some to autonomic	All of them composed of both sensory dendrites and motor axons; some of latter, somatic, some autonomic
Function	Vision, hearing, sense of smell, sense of taste, eye movements, etc.	Sensations, movements, and sweat secretion

First (olfactory)

The olfactory nerves are composed of axons of neurons whose dendrites and cell bodies lie in the nasal mucosa, high up along the septum and superior conchae (turbinates). Axons of these neurons form about 20 small fibers that pierce each cribriform plate and terminate in the olfactory bulbs, where they synapse with olfactory neurons II, whose axons compose the olfactory tracts. Summarizing:

Olfactory neurons I
  Dendrites⎫
  Cell body⎬   In nasal mucosa
  Axons   In small fibers that extend
                  through cribriform plate
                  to olfactory bulb

Olfactory neurons II
  Dendrites⎫
  Cell body⎬   In olfactory bulb
  Axons   In olfactory tracts

Second (optic)

Axons from the third and innermost layer of neurons of the retina compose the second cranial nerves. After entering the cranial cavity through the optic foramina the two optic nerves unite to form the *optic chiasma,* in which some of the fibers of each nerve cross to the opposite side and continue in the *optic tract* of that side (see Fig. 11-22). Thus each optic nerve contains fibers only from the retina of the same side, whereas each optic tract has fibers in it from both retinae, a fact of importance in interpreting certain visual disorders. Most of the optic tract fibers terminate in the thalamus (in the portion known as the lateral geniculate body). From here a new relay of fibers runs to the visual area of the occipital lobe cortex. A few optic tract fibers terminate in the superior colliculi of the midbrain, where they synapse with motor fibers to the external eye muscles (third, fourth, and sixth cranial nerves).

Third (oculomotor)

Fibers of the third cranial nerve originate from cells in the oculomotor nucleus in the ventral part of the midbrain and extend to the various external eye muscles, with the exception of the superior oblique and the lateral rectus. Autonomic fibers whose cells lie in a nucleus of the midbrain are also contained in the oculomotor nerves. These fibers terminate in the ciliary ganglion, where they synapse with cells whose postganglionic fibers supply the intrinsic eye

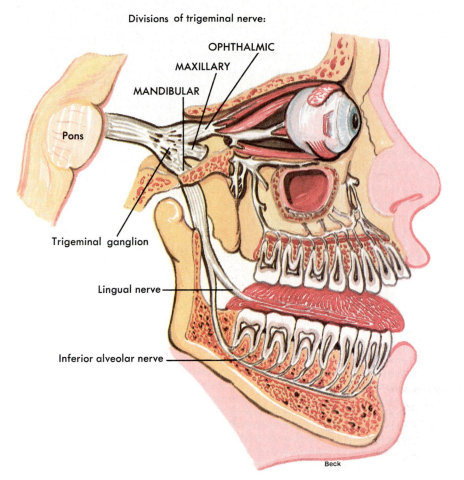

Divisions of trigeminal nerve:

OPHTHALMIC

MAXILLARY

MANDIBULAR

Pons

Trigeminal ganglion

Lingual nerve

Inferior alveolar nerve

Beck

Fig. 9-14 Trigeminal (fifth cranial) nerve and its three main divisions.

muscles (ciliary and iris). Still a third group of fibers is found in the third cranial nerves, namely, sensory fibers from proprioceptors in the eye muscles.

Fourth (trochlear)

Motor fibers of the fourth cranial nerve have their origin in cells in the midbrain, from which they extend to the superior oblique muscles of the eye. Afferent fibers from proprioceptors in these muscles are also contained in the trochlear nerves.

Fifth (trigeminal, trifacial)

Three sensory branches (ophthalmic, maxillary, and mandibular nerves) carry afferent impulses from the skin and mucosa of the head and from the teeth to cell bodies in the trigeminal or gasserian ganglion (a swelling on the nerve, lodged in the petrous portion of the temporal bone) (Fig. 9-14). Fibers extend from the ganglion to the main sensory nucleus of the fifth cranial nerve situated in the pons. A smaller motor root of the trigeminal nerve originates in the trifacial motor nucleus located in the pons just medial to the sensory nucleus. Fibers run from the motor root to the muscles of mastication by way of the mandibular nerve.

Neuralgia of the trigeminal nerve, known as tic douloureaux, is an extremely painful condition that can be relieved by removing the gasserian (or trigeminal) ganglion, the large ganglion on the posterior root of the nerve (Fig. 9-14) containing the cell bodies of the nerve's afferent fibers. After such an operation the patient's face, scalp, teeth, and conjunctiva on the side treated show anesthesia. Special care, such as wearing protective goggles and irrigating the eye frequently, is therefore prescribed. The patient is instructed also to visit the dentist regularly, since he can no longer experience a toothache as a warning of diseased teeth.

Sixth (abducens)

The sixth cranial nerve is a motor nerve with fibers originating from a nucleus in the pons in the floor of the fourth ventricle and extending to the lateral rectus muscles of the eyes. It contains also some afferent fibers from proprioceptors in the lateral rectus muscles.

Seventh (facial)

The motor fibers of the seventh cranial nerve arise from a nucleus in the lower part of the pons, from which they extend by way of several branches to the superficial muscles of the face and scalp and to the submaxillary and sublingual glands. Sensory fibers from the taste buds of the anterior two thirds of the tongue run in the facial nerve to cell bodies in the geniculate ganglion, a small swelling on the facial nerve, where it passes through a canal in the temporal bone. From the ganglion, fibers extend to the nucleus solitarius in the medulla.

Eighth (vestibulocochlear)

The eighth cranial nerve has two distinct divisions: the vestibular nerve and the cochlear nerve. Both are sensory. Fibers from the semicircular canals run to the vestibular ganglion (in the internal auditory meatus), where their cell bodies are located and from which fibers extend to the vestibular nuclei in the pons and medulla. Together, these fibers constitute the *vestibular nerve*. Some of its fibers run to the cerebellum. The vestibular nerve transmits impulses that result in sensations of balance or imbalance. The *cochlear nerve* consists of fibers starting in the organ of Corti in the cochlea, have their cell bodies in the spiral ganglion in the cochlea, and terminate in the cochlear nuclei located between the medulla and pons. Conduction by the cochlear nerve results in sensations of hearing.

Ninth (glossopharyngeal)

Both sensory and motor fibers compose the ninth cranial nerve. This nerve supplies fibers not only to the tongue and pharynx, as its name implies, but also to other structures, for example, to the carotid sinus. The latter plays an important part in the control of blood pressure. Sensory fibers, with their receptors in the pharynx and posterior third of the tongue, have their cell bodies in the jugular (superior) and petrous (inferior) ganglia, located, respectively, in the jugular foramen and the petrous portion of the temporal bone. From these, fibers extend to the nucleus solitarius in the medulla. The motor fibers of the ninth cranial nerve originate in cells in the nucleus ambiguus in the medulla and run to muscles of the pharynx. There are also secretory fibers in this nerve, with cells of origin in the nucleus salivatorius (at the junction of the pons and medulla). These fibers run to the otic ganglion, from which postganglionic fibers extend to the parotid gland.

Tenth (vagus)

The tenth cranial nerve is widely distributed and contains both sensory and motor fibers. Its sensory fibers supply the pharynx, larynx, trachea, heart, carotid body, lungs, bronchi, esophagus, stomach, small intestine, and gallbladder. Cell bodies for these sensory dendrites lie in the jugular and nodose ganglia, located, respectively, in the jugular foramen and just inferior to it on the trunk of the nerve. Centrally the sensory axons terminate in the medulla (in the nucleus solitarius) and in the pons (in the nucleus of the trigeminal nerve). Motor fibers of the vagus originate in cells in the medulla (in the dorsal motor nucleus of the vagus) and extend to various autonomic ganglia in the vagal plexus, from which postganglionic fibers run to muscles of the pharynx, larynx, and thoracic and abdominal viscera.

Eleventh (accessory)

The eleventh cranial nerve is a motor nerve. Some of its fibers originate in cells in the medulla (in the dorsal motor nucleus of the vagus and in the nucleus ambiguus) and pass by way of vagal branches to thoracic and abdominal viscera. The rest of the fibers have their cells of origin in the anterior gray column of the first five or six segments of the cervical spinal cord and extend through the spinal root of the accessory nerve to the trapezius and sternocleidomastoid muscles.

Twelfth (hypoglossal)

Motor fibers with cell bodies in the medulla (in the hypoglossal nucleus) compose the twelfth cranial nerve. They supply the muscles of the tongue. According to some anatomists, this nerve also contains sensory fibers from proprioceptors in the tongue.

■ ■ ■

The main facts about the distribution and function of each of the cranial nerve pairs are summarized in Table 9-4.

Severe head injuries often damage one or more of the cranial nerves, producing symptoms analogous to the functions of the nerve affected. For example, injury of the sixth cranial nerve causes the eye to turn in, because of paralysis of the abducting muscle of the eye, whereas injury of the eighth cranial nerve produces deafness.

Injury to the facial nerve results in a poker-faced expression and a drooping of the corner of the mouth from paralysis of the facial muscles.

Cerebellum

Structure

The cerebellum, the second largest part of the brain, is located just below the posterior portion of the cerebrum and is partially covered by it. A transverse fissure separates the cerebellum from the cerebrum. These two parts of the brain have several characteristics in common. For instance, gray matter makes up their outer portions and white matter predominates in their interiors. Turn to Fig. 9-12 to observe the arbor vitae, that is, the internal white matter of the cerebellum. Note its distinctive pattern, similar to the veins of a leaf. Note, too, that the surfaces of both the cerebellum and the cerebrum have numerous grooves (sulci) and convolutions (gyri). The convolutions of the cerebellum, however, are much more slender and less prominent than those of the cerebrum. The cerebellum has two large lateral masses, the cerebellar hemispheres, and a central section called the vermis because in shape it resembles a worm coiled on itself. (For a detailed description of the several subdivisions of the cerebellum, consult a textbook of neuroanatomy.)

The internal white matter of the cerebellum is composed of some short and some long tracts. The short association tracts conduct impulses from neuron cell bodies located in the cerebellar cortex to neurons whose dendrites and cell bodies compose nuclei located in the interior of the cerebellum. The longer tracts conduct impulses to and from the cerebellum. Fibers of the longer tracts enter or leave the cerebellum by way of its three pairs of penducles, as follows:

1 Inferior cerebellar peduncles (or restiform bodies)—composed chiefly of tracts into the cerebellum from the medulla and cord (no-

tably, spinocerebellar, vestibulocerebellar, and reticulocerebellar tracts)

2 Middle cerebellar peduncles (or brachia pontis)—composed almost entirely of tracts into the cerebellum from the pons, that is, pontocerebellar tracts

3 Superior cerebellar peduncles (or brachia conjunctivum cerebelli)—composed principally of tracts from dentate nuclei through the red nucleus of the midbrain to the thalamus

An important pair of cerebellar nuclei are the dentate nuclei, one of which lies in each hemisphere. Tracts connect the nuclei with motor areas of the cerebral cortex (the dentatorubrothalamic tracts to the thalamus and thalamocortical tracts to the cortex). By means of these tracts, cerebellar impulses influence the motor cortex. Impulses also travel the reverse direction. Corticopontine and pontocerebellar tracts enable the motor cortex to influence the cerebellum.

Functions

The cerebellum performs three general functions, all of which have to do with the control of skeletal muscles. It acts with the cerebral cortex to produce skilled movements by coordinating the activities of groups of muscles. It controls skeletal muscles so as to maintain equilibrium. It helps control posture. It functions below the level of consciousness to make movements smooth instead of jerky, steady instead of trembling, and efficient and coordinated instead of ineffective, awkward, and uncoordinated (asynergic).

There have been many theories about cerebellar functions. One theory, based on comparative anatomy studies and substantiated by experimental methods, regards the cerebellum as three organs, each with a somewhat different function: synergic control of muscle action, excitation and inhibition of postural reflexes, and maintenance of equilibrium.

Synergic control of muscle action, which is ascribed to the neocerebellum (superior vermis and hemispheres), is closely associated with cerebral motor activity. Normal muscle action, you will recall, involves groups of muscles, the various members of which function together as a unit. In any given action, for example, the prime mover contracts and the antagonist relaxes but then contracts weakly at the proper moment to act as a brake, checking the action of the prime mover. Also, the synergists contract to assist the prime mover, and the fixation muscles of the neighboring joint contract. Through such harmonious, coordinated group action, normal movements are smooth, steady, and precise as to force, rate, and extent. Achievement of such movements results from cerebellar activity added to cerebral activity. Impulses from the cerebrum may start the action, but those from the cerebellum synergize or coordinate the contractions and relaxations of the various muscles once they have begun. Some physiologists consider this the main, if not the sole, function of the cerebellum.

One part of the cerebellum is thought to be concerned with both exciting and inhibiting postural reflexes.

Part of the cerebellum presumably discharges impulses important to the maintenance of equilibrium. Afferent impulses from the labyrinth of the ear reach the cerebellum. Here, connections are made with the proper efferent fibers for contraction of the necessary muscles for equilibrium.

Cerebellar disease (abscess, hemorrhage, tumors, trauma, etc.) produces certain characteristic symptoms, among which ataxia (muscle incoordination), hypotonia, tremors, and disturbances of gait and equilibrium predominate. One example of ataxia is overshooting a mark or stopping before reaching it when trying to touch a given point on the body (finger-to-nose test). Drawling, scanning, or singsong speech are also examples of ataxia. Tremors are particularly pronounced toward the end of the movements and with the exertion of effort. Disturbances of gait and equilibrium vary, depending on the muscle groups involved, but the walk is often characterized by staggering or lurching and by a clumsy manner of raising the foot too high and bringing it down with a clap. Paralysis does not result from loss of cerebellar function.

Diencephalon

The diencephalon is the part of the brain located between the cerebrum and the mesencephalon (midbrain). Although the diencephalon consists of several structures located around the third ventricle, the main ones are the thalamus, the hypothalamus, and the neurohypophysis (posterior pituitary gland), associated with the hypothalamus.

Structure

The right thalamus is a rounded mass of gray matter about ½ inch wide and 1½ inches long, bulging into the right lateral wall of the third ventricle. The left thalamus is a similar mass in the left lateral wall. Each thalamus consists of numerous nuclei. They are arranged in groups named the anterior, lateral, intralaminar, midline (or medial), and posterior groups of nuclei. Two important structures in the posterior group of nuclei are the medial and lateral geniculate bodies. Large numbers of axons conduct impulses into the thalamus from the cord, brain stem, cerebellum, basal ganglia, and various parts of the cerebrum. These axons terminate in thalamic nuclei, where they synapse with neurons whose axons conduct impulses out of the thalamus to virtually all areas of the cerebral cortex. Thus the thalamus serves as the major relay station for sensory impulses on their way to the cerebral cortex.

The hypothalamus consists of several structures that lie beneath the thalamus and form the third ventricle's floor and the lower part of its sidewall. Prominent among the structures composing the hypothalamus are the supraoptic nuclei, the paraventricular nuclei, the stalk of the hypophysis (pituitary gland), the neurohypophy-sis (the posterior lobe of the pituitary gland), and the mamillary bodies. Identify as many of these as you can in Figs. 9-12 and 9-13. The supraoptic nuclei consist of gray matter located just above and on either side of the optic chiasma. The paraventricular nuclei of the hypothalamus are so named because of their location close to the wall of the third ventricle. The midportion of the hypothalamus consists of the stalk and the posterior lobe of the pituitary gland (neurohypophysis). The posterior part of the hypothalamus consists mainly of the mamillary bodies, in which are located the mamillary nuclei.

Functions

The *thalamus* performs the following functions:

1 Plays two parts in the mechanism responsible for sensations
  a Impulses from appropriate receptors, on reaching the thalamus, produce conscious recognition of the cruder, less critical sensations of pain, temperature, and touch
  b Neurons whose dendrites and cell bodies lie in certain nuclei of the thalamus relay all kinds of sensory impulses, except possibly olfactory, to the cerebrum
2 Plays a part in the mechanism responsible for emotions by associating sensory impulses with feelings of pleasantness and unpleasantness
3 Plays a part in the arousal or alerting mechanism
4 Plays a part in mechanisms that produce complex reflex movements

The *hypothalamus* is a small but functionally mighty area of the brain. It weighs little more than ¼ ounce, yet it performs many functions of the greatest importance both for survival and for the enjoyment of life. For instance, it functions as a link between the psyche (mind) and the soma (body). It also links the nervous system to the endocrine system. Certain areas of the hypothalamus function as pleasure centers or reward centers for the primary drives such as eating, drinking, and mating. The following paragraphs give a brief summary of hypothalamic functions.

1 The hypothalamus functions as a higher autonomic center or, rather, as several higher autonomic centers. By this we mean that axons of neurons whose dendrites and cell bodies lie in nuclei of the hypothalamus extend in tracts from the hypothalamus to both parasympathetic and sympathetic centers in the brain stem and cord. (Fig. 10-2 indicates these tracts in blue.) Thus impulses from the hypothalamus can simultaneously or successively stimulate or inhibit few or many lower autonomic centers. In other words the hypothalamus serves as a regulator and coordinator of autonomic activities. It helps control and integrate the responses made by visceral effectors all over the body.

2 The hypothalamus functions as the major relay station between the cerebral cortex and lower autonomic centers. Tracts conduct impulses from various centers in the cortex to the hypothalamus (also shown in blue in Fig. 10-2). Then, via numerous synapses in the hypothalamus, these impulses are relayed to other tracts that conduct them on down to autonomic centers in the brain stem and cord and also to spinal cord somatic centers (anterior horn motoneurons). Thus the hypothalamus functions as the link between the cerebral cortex and lower centers—hence between the psyche and the soma. It provides a crucial part of the route by which emotions can express themselves in changed bodily functions. It is the all-important relay station in the neural pathways that makes possible the mind's influence over the body—sometimes, unfortunately, even to the profound degree of producing "psychosomatic disease."

3 Neurons in the supraoptic and paraventricular nuclei of the hypothalamus synthesize the hormones secreted by the posterior pituitary gland (neurohypophysis). Because one of these hormones affects the volume of urine excreted, the hypothalamus plays an indirect but essential role in maintaining water balance (see Fig. 21-9, p, 578).

4 Some of the neurons in the hypothalamus function as endocrine glands. Their axons secrete chemicals, called releasing hormones, into blood, which circulates to the anterior pituitary gland. Releasing hormones control the release of certain anterior pituitary hormones—specifically growth hormone and hormones that control hormone secretion by the sex glands, the thyroid gland, and the adrenal cortex (discussed in Chapter 11). Thus indirectly the hypothalamus helps control the functioning of every cell in the body (see pp. 327-329).

The Nobel Prize in medicine for 1977 was shared by three individuals. Two of them, Andrew W. Schally and Roger C. L. Guillemin, received their awards for identifying, isolating, and synthesizing a protein hormone, somatostatin, secreted by the hypothalamus. Somatostatin has been shown to switch off the production of somatotropin (growth hormone) by the pituitary gland.

5 The hypothalamus plays an essential role in maintaining the waking state. Presumably it functions as part of an arousal or alerting mechanism. Clinical evidence of this is that somnolence characterizes some hypothalamic disorders.

6 The hypothalamus functions as a crucial part of the mechanism for regulating appetite and therefore the amount of food intake. Experimental and clinical findings seem to indicate the presence of a "feeding or appetite center" in the lateral part of the hypothalamus and a "satiety center" located medially. For example, an animal with an experimental lesion in the ventromedial nucleus of the hypothalamus will consume tremendous amounts of food. Similarly, a human being with a tumor in this region of the hypothalamus may eat insatiably and and gain an enormous amount of weight.

7 The hypothalamus functions as a crucial

part of the mechanism for maintaining normal body temperature. Hypothalamic neurons whose fibers connect with autonomic centers for vasoconstriction, dilation, and sweating and with somatic centers for shivering constitute heat-regulating centers. Marked elevation of body temperature frequently characterizes injuries or other abnormalities of the hypothalamus.

Cerebrum

Structure

The cerebrum is the largest and most superiorly located division of the brain. A deep groove, the longitudinal fissure, divides the cerebrum into two halves, the right and left cerebral hemispheres. These halves, however, are not completely separate organs. A structure composed of white matter (tracts) and known as the corpus callosum joins them medially (Fig. 9-16). Prominent fissures, in addition to the longitudinal fissure already named, include the central sulcus (fissure of Rolando), the lateral fissure (of Sylvius), and the parieto-occipital fissure. These deep grooves subdivide each cerebral hemisphere into four lobes. Each lobe bears the name of the bone that lies over it: frontal lobe, parietal lobe, temporal lobe, and occipital lobe (Fig. 9-15). A fifth lobe, the insula (island of Reil), lies hidden from view in the lateral fissure. To see it, one must dissect the brain. The central fissure separates the frontal lobe from the parietal lobe. The lateral fissure separates the temporal lobe, which lies below the fissure, from the frontal and parietal lobes, which lie above it. The parieto-occipital fissure separates the occipital lobe from the two parietal lobes.

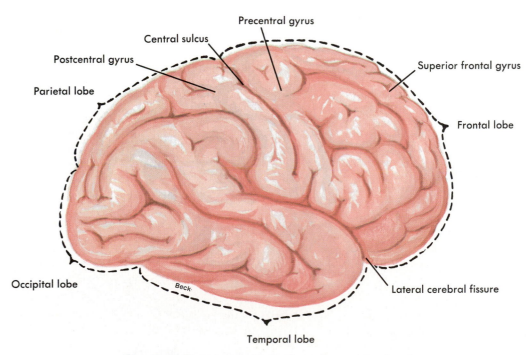

Fig. 9-15 Right hemisphere of cerebrum, lateral surface.

Each hemisphere of the cerebrum consists of external gray matter, internal white matter, and islands of internal gray matter. The cerebral cortex is the thin surface layer of the cerebrum. Gray matter only 2 to 4 mm (roughly 1/12 to 1/6 inch) thick composes it. But despite its thinness the cerebral cortex consists of six layers, each one a dense network of millions of axon terminals synapsing with millions of dendrites and cell bodies of other neurons. Under the cortex lies the white matter that composes the bulk of the cerebrum's interior. Presumably, when early anatomists observed the cerebrum's outer darker layer, it reminded them of tree bark—hence their choice of the name cortex (Latin for bark) for it.

Provided one uses a little imagination, the surface of the cerebrum looks like a group of small sausages. Each "sausage" represents a convolution or gyrus. Between adjacent gyri lie sulci or fissures. Identify the following structures in Figs. 9-15 and 9-16: precentral gyrus, postcentral gyrus, cuneus, cingulate gyrus, hippocampal gyrus, uncus. These are some of the structures we shall refer to later in our discussion of the physiology of the brain.

Cerebral tracts lie interior to the cortex and are composed of great numbers of nerve fibers (axons). In one part of the interior of the cerebrum a group of sensory and motor projection tracts forms a large irregular mass of white matter known as the internal capsule. It lies between the thalamus and the basal ganglia (Figs. 9-18 and 9-22). Some tracts are short, extending from one convolution to another in the same hemisphere. These are called association tracts.

Basal ganglia (or cerebral nuclei*) are islands of gray matter deep inside each cerebral hemisphere. Authorities agree that the most important basal ganglia are the caudate nucleus, the

*When applied to the nervous system, the term nucleus means an area of gray matter in the brain or cord (composed mainly, as is all gray matter, of neuron cell bodies and dendrites). Such a cluster located outside the brain and cord is called a ganglion. So cerebral nuclei is a more accurate (but less common) name than basal ganglia.

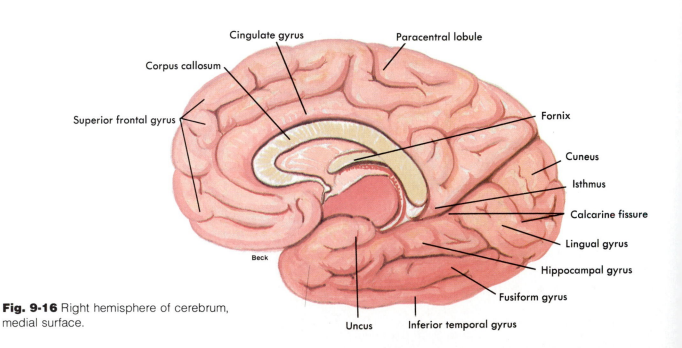

Fig. 9-16 Right hemisphere of cerebrum, medial surface.

Cingulate gyrus

Corpus callosum

Paracentral lobule

Superior frontal gyrus

Fornix

Cuneus

Isthmus

Calcarine fissure

Lingual gyrus

Hippocampal gyrus

Fusiform gyrus

Beck

Uncus

Inferior temporal gyrus

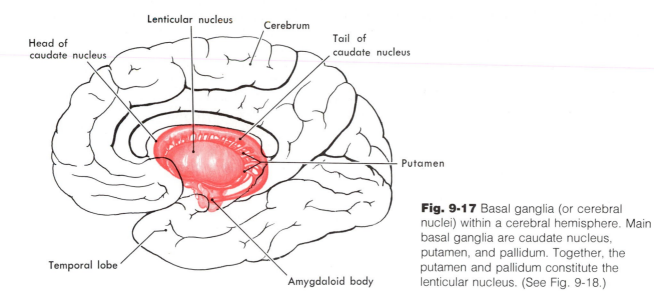

Fig. 9-17 Basal ganglia (or cerebral nuclei) within a cerebral hemisphere. Main basal ganglia are caudate nucleus, putamen, and pallidum. Together, the putamen and pallidum constitute the lenticular nucleus. (See Fig. 9-18.)

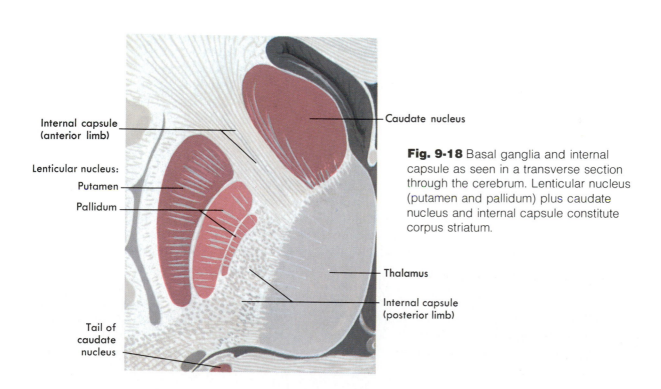

Fig. 9-18 Basal ganglia and internal capsule as seen in a transverse section through the cerebrum. Lenticular nucleus (putamen and pallidum) plus caudate nucleus and internal capsule constitute corpus striatum.

putamen, and the globus pallidus (pallidum). They differ somewhat, however, as to what other structures they include as basal ganglia. Look now at Fig. 9-17 to observe the size, shape, and location of the caudate nucleus and putamen. The amygdaloid body shown in this figure is also classified as one of the basal ganglia by some authorities. Next, look at Fig. 9-18 and note the location of the pallidum. The pallidum and putamen together are sometimes called the lenticular nucleus. The pallidum and putamen, plus two other structures, namely, the caudate nucleus and internal capsule, on the other hand, are together commonly referred to as the corpus striatum. The term means "striped body." You can see the striped appearance of this internal region of the cerebrum in Fig. 9-18.

Functions

The cerebrum above all other bodily structures deserves the title "the most important human organ." Why? Because it can claim two distinctions not shared by any other organ. The cerebrum is the top executive of the world's most complex organization—the human body. In addition, the cerebrum provides the functions that endow us with our uniquely human qualities. During the past decade or so, research scientists in various fields—neurophysiology, neurosurgery, neuropsychiatry, and others—have added mountains of information to our knowledge about the brain. But questions come faster than answers, and clear, complete understandings of the brain's mechanisms still elude us. Perhaps they forever will. Perhaps the capacity of the human brain falls short of the ability to understand its own complexity. We shall start our discussion of the cerebrum with some generalizations about its functions and follow these with more specific information about each generalization.

1 Cerebral activity goes on as long as life itself. Only when life ceases (or moments before) does the cerebrum cease its functioning. Only then do all of its neurons stop conducting impulses. Proof of this has come from records of brain electrical potentials known as electroencephalograms, or EEGs, or "brain waves."

2 The cerebrum performs three kinds of functions: sensory functions, motor functions, and a group of activities less easily named and even less easily defined or explained. "Integrative functions" is one name for them. What most of us think of as mental activities are part, but not all, of the cerebrum's integrative functions.

3 The right and left hemispheres of the cerebrum specialize in different functions. Also, certain regions of each hemisphere play key roles in particular functions.

Electroencephalograms

Electroencephalograms (EEGs) are records of the cerebrum's electrical activity. These records are usually made from a number of electrodes placed on different regions of the scalp, and they consist of waves—brain waves, as they are called. Four types of brain waves are recognized based on the frequency and amplitude of the waves. Frequency, or the number of wave cycles per second, is usually referred to as Hertz (Hz, from Hertz, German physicist). Amplitude means voltage. Listed in order of their frequency from the fastest to the slowest, brain wave names are beta, alpha, theta, and delta. *Beta waves* have a frequency of over 13 Hz and a relatively low voltage. *Alpha waves* have a frequency of 8 to 13 Hz and a relatively high voltage. *Theta waves* have both a relatively low frequency—4 to 7 Hz—and a low voltage. *Delta waves* have the slowest frequency—less than 4 Hz—but a high voltage. Brain waves vary in different regions of the brain, in different states of awareness, and in abnormal conditions of the cerebrum.

Fast, low-voltage beta waves characterize EEGs recorded from the frontal and central regions of the cerebrum when an individual is awake, has his eyes open, and is alert and attentive. They predominate when the cerebrum is busiest, when it is engaged with sensory stimulation or mental activities. In short, beta waves are "busy waves." Alpha waves, in contrast, are "relaxed waves." These moderately fast, relatively high-voltage waves dominate EEGs recorded from the parietal, occipital, and posterior parts of the temporal lobes when an individual is awake but has his eyes closed and is in a relaxed, nonattentive state—when his cerebrum is idling, so to speak. When drowsiness descends, moderately slow, low-voltage theta waves appear. Theta waves are "drowsy waves." "Deep sleep waves," on the other hand, are delta waves. These slowest brain waves characterize the deep sleep from which one is not easily aroused. For this reason, deep sleep is referred to as slow-wave sleep (SWS).

Physicians use electroencephalograms to help localize areas of brain dysfunction, to identify altered states of consciousness, and often to establish death. Two flat EEG recordings (no brain waves) taken 24 hours apart in conjunction with no spontaneous respiration and total absence of somatic reflexes are criteria accepted as evidence of death.

Sensory functions

Many of the nervous system's gross structures function to produce sensations—nerves, ganglia, and sensory pathways ascending through the spinal cord, brain stem, thalamus, and cerebrum. Impulse conduction by neurons in the cerebral cortex makes possible complex, discriminative sensations. Most essential for normal sensations are the somatic sensory, visual, and auditory areas of the cerebral cortex (Fig. 9-19). These regions of the cortex do more than just register separate and simple sensations. They compare and evaluate them. They integrate them into perceptions of wholes. Suppose, for example, that someone blindfolded you and then put an ice cube in your hand. You would, of course, sense something cold touching your hand. But also, you would probably know that it was an ice cube because you would sense a total impression compounded of many sensations such as temperature, shape, size, weight, texture, and movement and position of your hand and arm. Somatic sensory pathways are discussed on p. 259.

Somatic motor functions

Mechanisms that control voluntary movements are extremely complex and imperfectly understood. It is known, however, that for normal movements to take place, many parts of the nervous system—including certain areas of the cerebral cortex—must function. The precentral gyrus, that is, the most posterior gyrus of the frontal lobe, area 4, Fig. 9-19, constitutes the primary motor area. However, the gyrus immediately anterior to the precentral gyrus also contains motoneurons. So, too, do many other regions, including even the somatic sensory areas. Neurons in the precentral gyrus are said to control individual muscles, especially those that produce movements of distal joints (wrist, hand, finger, ankle, foot, and toe movements). Neurons in the gyrus just anterior to the precentral gyrus are thought to activate groups of muscles simultaneously. Motor pathways descending from the cerebrum through the brain stem and spinal cord are discussed on p. 263.

Integrative functions

"Integrative functions of the cerebrum" is, to say the least, a nebulous phrase, and the neural processes it designates are even more obscure. In general, they consist of all events that take place in the cerebrum between its reception of sensory impulses and its sending out of motor impulses. Integrative functions of the cerebrum include consciousness and mental activities of all kinds. Consciousness, memory, use of language, and emotions are the integrative cerebral functions that we shall discuss briefly.

Consciousness may be defined as a state of

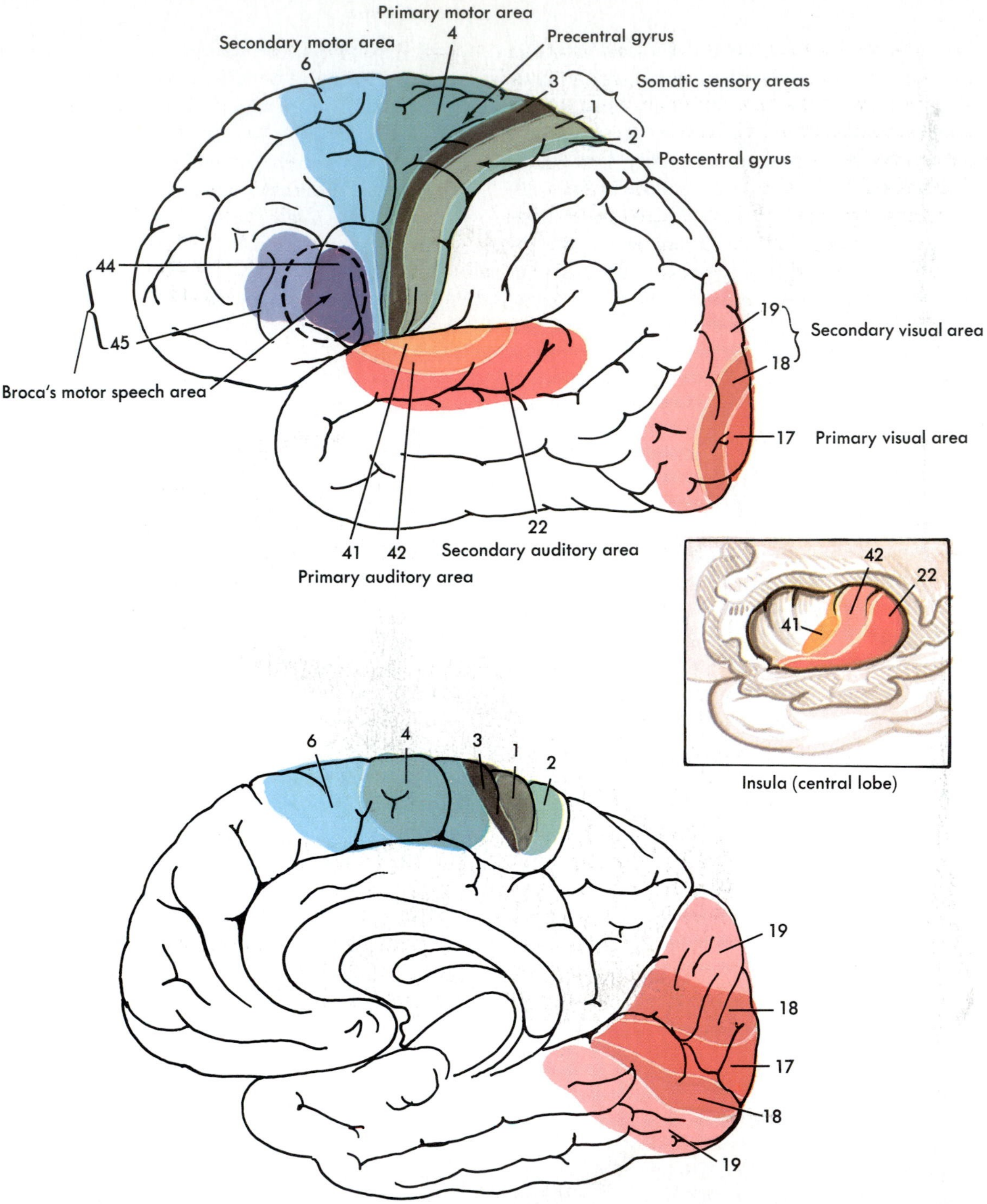

Fig. 9-19 Map of human cortex. Identity of each numbered area is determined by structural differences in neurons that compose it. (Adapted from Brodmann, K.: Feinere Anatomie des Grosshirns. In Handbuch der Neurologie, Berlin, 1910, Springer-Verlag.)

awareness of one's self, one's environment, and other beings. Very little is known about the neural mechanisms that produce consciousness. One fact known, however, is that consciousness depends on excitation of cortical neurons by impulses conducted to them by a relay of neurons known as the reticular activating system. The *reticular activating system* consists of centers in the brain stem reticular formation that receive impulses from the cord and relay them to the thalamus and from the thalamus to all parts of the cerebral cortex. Both direct spinal reticular tracts and collateral fibers from the specialized sensory tracts (spinothalamic, lemniscal, auditory, and visual) relay impulses over the reticular activating system to the cortex. Without continual excitation of cortical neurons by reticular activating impulses, an individual is uncon-

scious and cannot be aroused. Here, then, are two accepted concepts about the reticular activating system: it functions as the arousal or alerting system for the cerebral cortex, and its functioning is crucial for maintaining consciousness. Drugs known to depress the reticular activating system decrease alertness and induce sleep. Barbiturates, for example, act this way. Amphetamine, on the other hand, a drug known to have a stimulating effect on the cerebrum and to enhance alertness and produce wakefulness, probably acts by stimulating the reticular activating system.

Certain variations in the levels or states of consciousness are normal. All of us, for example, experience different levels of wakefulness. At times, we are highly alert and attentive. At other times, we are relaxed and nonattentive. All of us

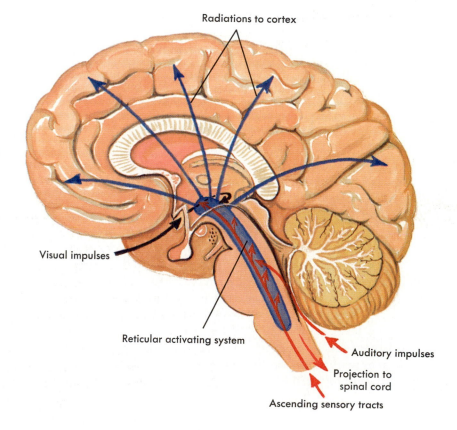

Fig. 9-20 Reticular activating system. Consists of centers in the brain stem reticular formation plus fibers that conduct to the centers from below and fibers that conduct from the centers to widespread areas of the cerebral cortex. Functioning of the reticular activating system is essential for consciousness.

Radiations to cortex

Visual impulses

Reticular activating system

Auditory impulses

Projection to spinal cord

Ascending sensory tracts

also experience different levels of sleep. Two of the best known stages are those called slow-wave sleep (SWS) and rapid eye movement (REM) sleep. SWS sleep takes it name from the slow frequency, high-voltage delta waves that identify it. It is almost entirely a dreamless sleep. REM sleep, on the other hand, is associated with dreaming.

In addition to the various normal states of consciousness, altered states of consciousness (ASC) also occur under certain conditions. Anesthetic drugs produce an ASC, namely anesthesia. Disease or injury of the brain may produce the type of ASC called coma. And lysergic acid diethylamide (LSD), a "mind-altering" drug, induces a type of ASC known in the drug culture as a "trip." Peoples of Eastern cultures have long been familiar with an ASC called *yoga*, or *meditation*. Yoga is a waking state but differs markedly in certain respects from the usual waking state. According to an accepted definition, yoga is a "higher" or "expanded" level of consciousness. This higher consciousness is accompanied, almost paradoxically, by a high degree of both relaxation and alertness. With training in meditation techniques and practice, an individual can enter the meditative state at will and remain in it for an extended period. Undoubtedly the most widely practiced form of meditation in this country is transcendental meditation (TM) as taught by Maharishi Mahesh Yogi. Certain physiological changes have been reported to occur during meditation. We shall describe these in Chapter 10.

Memory is one of our major mental activities. The mechanisms responsible for memory, like those for consciousness, have not yet yielded their secrets. Many investigators have tried and are now trying to search out answers to explain memory. What have they learned so far? One fact they have established is that the cerebral cortex functions in memory. But they cannot yet explain how impulse conduction by cortical neurons can possibly produce memories. They know that memory does not reside in any one part of the cortex. Apparently several parts of it store memories. In the 1950s, Dr. Wilder Penfield, a noted Canadian neurosurgeon, electrically stimulated the temporal lobes of patients undergoing brain surgery. They responded, much to his surprise, by recalling in the most minute detail songs and events from their past. They seemed more to be reliving than remembering their experiences. Other investigators have accumulated evidence of memory storage in the occipital and parietal lobes.

More recently, researchers have reported with some certainty that the cerebrum's limbic system—the "emotional brain" as it is called—plays a key role in memory. Various observations support this view. To mention one, when the hippocampus (part of the limbic system) is removed, the patient loses his ability to recall new information. Another idea now widely accepted is that protein synthesis constitutes a crucial part of the mechanism for long-term memory. Whether or not these proteins serve as human "memory molecules" has not yet been proved. Studies done on rats suggest that the process of memory may also involve an increase in the number of synapses between neurons in the brain.

Language functions consist of the ability to speak and write words and the ability to understand spoken and written words. Certain areas in the frontal, parietal, and temporal lobes serve as speech centers—as crucial areas, that is, for language functions. The left cerebral hemisphere contains these areas in about 90% of the population; in the remaining 10%, either the right hemisphere or both hemispheres contain them. Lesions in speech centers give rise to language defects called *aphasias*. For example, with damage to an area in the inferior gyrus of the frontal lobe (Broca's area, Fig. 9-19), a person becomes unable to articulate words but can still make vocal sounds and understand words he hears and reads.

Emotions—both the subjective experiencing and objective expression of them—involve functioning of the cerebrum's limbic system. The name limbic, which derives from the Latin word for border or fringe, suggests the shape of the cortical structures that make up the system. They lie on the medial surface of the cerebrum and form a curving border around the corpus callosum, the structure that connects the two cerebral hemispheres. Look now at Fig. 9-16. Here you can identify most of the structures of the limbic system. They are the cingulate gyrus, the isthmus, the hippocampal gyrus, the uncus, and the hippocampus (the extension of the hippocampal gyrus that protrudes into the floor of the inferior horn of the lateral ventricle). These limbic system structures have primary connections with various other parts of the brain, notably the septum, the amygdala (the tail of the caudate nucleus, one of the basal ganglia), and the hypothalamus. Some physiologists, therefore, include these connected structures as parts of the limbic system.

The limbic system (or to use its more descriptive name, the emotional brain) functions in some way to make us experience many kinds of emotions—anger, fear, sexual feelings, pleasure, and sorrow, for example. To bring about the normal expression of emotions, parts of the cerebral cortex other than the limbic system must also function. Considerable evidence exists that limbic activity without the modulating influence of the other cortical areas may bring on the attacks of abnormal, uncontrollable rage suffered periodically by some unfortunate individuals.

Hemispheric specialization and functional localization within a hemisphere

Each hemisphere of the human cerebrum specializes in certain functions. For example, as already noted, the left hemisphere specializes in language functions—it does the talking, so to speak. The left hemisphere also appears to dominate the control of certain kinds of hand movements, notably skilled and gesturing movements. Most people use their right hands for performing skilled movements, and the left side of the cerebrum controls the muscles on the right side that execute these movements. The next time you are with a group of people who are talking, observe their gestures. The chances are about nine to one that they will gesture mostly with their right hands—indicative of left cerebral control.

Evidence that the right hemisphere of the cerebrum specializes in certain functions has rather recently been reported. As of now, it seems that one of the right hemisphere's specialties is the perception of certain kinds of auditory material. For instance, some studies have shown that the right hemisphere perceives nonspeech sounds such as melodies, coughing, crying, and laughing better than does the left hemisphere. The right hemisphere may also function better at tactual perception and for perceiving and visualizing spatial relationships.

Certain areas of the cortex in each hemisphere of the cerebrum engage predominantly in one particular function. What that function is depends on what structures the cortical area receives impulses from or sends impulses to. For example, the postcentral gyrus (areas 3, 1, and 2, in Brodmann's map) functions mainly as a general somatic sensory area. It receives impulses from receptors activated by heat, cold, and touch stimuli. The precentral gyrus (area 4) on the other hand, functions chiefly as the somatic motor area. Impulses from neurons in this area descend over motor tracts and eventually stimulate somatic effectors, the skeletal muscles. The transverse gyrus of the temporal lobe (areas 41 and 42 in Fig. 9-19) serves as the primary auditory area. The primary visual area lies in the occipital lobe (area 17). It is important to remember that no part of the brain functions alone. Many structures of the central nervous system must function in order for any one part of the brain to function.

Neurotransmitters in cord and brain

Neurotransmitters* are chemicals by which neurons talk to one another. At literally billions of synapses in the central nervous system, axons of presynaptic neurons release neurotransmitters that then act on postsynaptic neurons to facilitate, stimulate, or inhibit them. Acetylcholine is the only compound so far established as a neurotransmitter at spinal cord synapses. Because it increases the resting potential of postsynaptic neurons, it functions as an excitatory neurotransmitter. Many other substances also presumably function in this way—glutamic acid, for example. In contrast, glycine and gamma-aminobutyric acid (GABA) appear to act as inhibitory neurotransmitters at some spinal cord synapses. But for most synapses in both the cord and brain the neurotransmitters are still unknown. It is known, however, that the axons of certain neurons in the brain release catecholamines, that is, norepinephrine (NE), dopamine (DA), and serotonin (abbreviated 5-HT for 5-hydroxytryptamine). For instance, NE is released by axons that originate from cell bodies in the locus ceruleus and that terminate at synapses in the cerebellum, cerebral cortex, and hypothalamus. (Locus ceruleus, literally a blue place, is a small region in the brain stem in the upper part of the floor of the fourth ventricle.) DA is known to be released from axons originating from neuron cell bodies in the midbrain (in the substantia nigra, so called because of the deep pigmentation of this area) and terminating in the caudate nucleus and putamen. Here in these basal ganglia, DA serves as a neurotransmitter. It is normally present in them in high amounts. A marked deficiency in the basal ganglia DA content occurs, however, in the condition called parkinsonism or Parkinson's disease. The reason for the DA deficiency is that the neurons in the substantia nigra that synthesize DA have degenerated. Since the late 1960s, L-dopa, a precursor of dopamine, has been used to treat parkinsonism—often with dramatic relief of symptoms (tremors, rigidity, and other abnormalities of movements). In contrast to the DA deficiency of Parkinson's disease, excess DA release in certain parts of the brain, according to one postulate, may bring on schizophrenia, a major mental disease.

5-HT is a neurotransmitter in the limbic system of the cerebrum. Axons that release 5-HT originate from neuron cell bodies located in the brain stem's midline nuclei (raphe nuclei). In cats, excess 5-HT has been shown to lead to excess sleep and lessened sexual activity.

Norepinephrine, dopamine, and serotonin all serve as excitatory neurotransmitters. They bind to specific receptors in the membranes of postsynaptic neurons. This event initiates impulse conduction by the postsynaptic neurons and also leads to cyclic AMP formation in their cytoplasm.

In 1975 John Hughes and Hans W. Kosterlitz of the University of Aberdeen submitted evidence of an exciting new kind of brain and cord

---

*Much of our present knowledge about chemical transmitters stems from years of research by Sweden's Dr. Ulf S. von Euler, the United States' Dr. Julius Axelrod, and England's Sir Bernard Katz. For their work, these eminent scientists shared the 1970 Nobel Prize in medicine and physiology. Their most significant findings include the following:

1 Identification of norepinephrine (NE) as the major transmitter in some brain synapses (by Dr. von Euler)
2 Identification of the enzyme catechol-*O*-methyl transferase (COMT) and discovery that it inactivates norepinephrine (by Dr. Axelrod)
3 Discovery that conducting cholinergic fibers rapidly release numerous packets of acetylcholine into synapses (by Sir Bernard Katz)

The above knowledge led to other discoveries and to valuable applications. For example, researchers later learned that severe psychic depression occurs when a deficit of NE exists in certain brain synapses. This finding led to the development of antidepressant drugs. Certain ones of these inhibit COMT. Because inhibited COMT does not inactivate NE, the amount of active NE in brain synapses increases, and this relieves the individual's depression. The graphic names "psychic energizers" and "mood elevators" refer to antidepressant drugs, now valuable weapons in medicine's arsenal for combating mental disease.

neurotransmitter—exciting because it seemed to function as a natural painkiller! Somewhat later these scientists isolated two peptides, consisting of only five amino acids each, from the brains of pigs. They gave them a name suggestive of their source—*enkephalins*—from the Greek for "in the head." (These peptides are further identified as methionine-enkephalin and isoleucine-enkephalin. These names emphasize the only amino acid difference between the two compounds. One has methionine where the other has isoleucine; their other four amino acids are identical.)

Axon terminals that release enkephalins have been found to be concentrated in the spinal cord in the substantia gelatinosa (approximately the most posterior part of the posterior horn of gray matter). In the brain, enkephalin-releasing axons are concentrated in the central part of the thalamus and in the amygdala (part of the cerebrum's limbic system, so-called emotional brain). All of these structures serve as part of a pathway for pain perception. Enkephalins act as inhibitory neurotransmitters in the pathway, thereby decreasing not only the perception of pain but also its emotional component. Based on experimental data, some neurobiologists have postulated that enkephalins bind to the same receptors in neuron membranes that morphine does. Enkephalins, in short, might be considered internal opiates—the body's own natural painkillers.

Recent investigations have revealed other natural peptides besides enkephalins that also suppress pain. These other peptides came from animal pituitary glands, however, not from their brains. *Endorphin* (from "endogenous morphine") is the name coined to designate these pituitary peptide pain killers. One of them, beta-endorphin, consists of a known sequence of 31 amino acids, the first 5 of which are identical to the amino acids that compose methionine-enkephalin. Moreover, beta-endorphin's amino acid sequence is identical to part of a peptide hormone from the pituitary gland. Beta-lipopro-tein is its name. It consists of 91 amino acids and induces fat metabolism. Whether the pituitary peptides, endorphin and beta-lipoprotein, serve as precursors for the brain peptides, enkephalins, is not known.

Somatic sensory pathways

1 Sensory neural pathways consist of relays of sensory neurons that conduct impulses from any part of the body to the spinal cord or brain stem and from these lower levels of the central nervous system up to its highest level, the cerebral cortex. Most impulses that reach the cortex have traveled over a relay of at least three sensory neurons. Although they are usually called first-order, second-order, and third-order sensory neurons, we shall shorten these names to sensory neurons I, II, and III. Some important general principles about sensory neural pathways from the periphery—any part of the body outside of the central nervous system—to the cerebral cortex follow.

a Sensory neurons I of the relay conduct from the periphery to the central nervous system. If the receptors of these neurons lie in regions supplied by spinal nerves, their dendrites lie in a spinal nerve and their axons terminate in gray matter of the cord or brain stem. Where are their cell bodies located? (Confirm or find your answer in Fig. 8-5.) If receptors of sensory neurons I lie in regions supplied by cranial nerves, their dendrites lie in a cranial nerve, their cell bodies lie in cranial nerve ganglia, and their axons terminate in gray matter of the brain stem. In either case, sensory neuron I axon terminals synapse with sensory neuron II dendrites or cell bodies (Fig. 9-21).

b Sensory neurons II conduct from the cord or brain stem up to the thalamus. Their

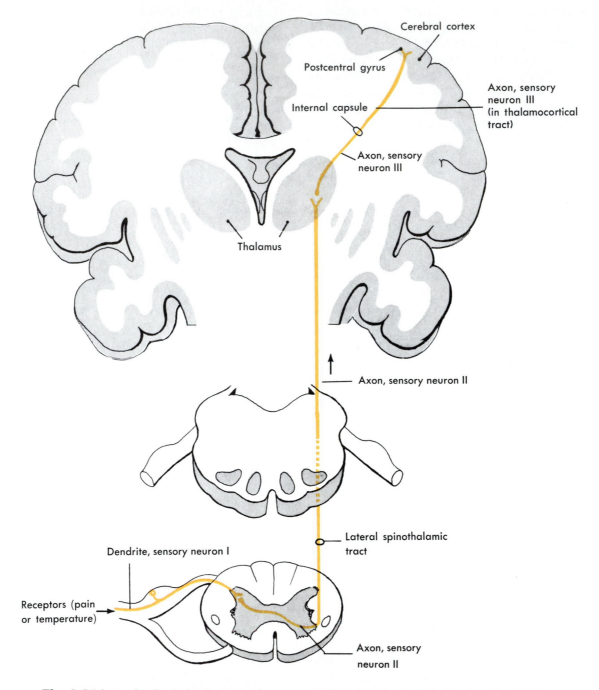

Cerebral cortex

Postcentral gyrus

Internal capsule

Axon, sensory
neuron III
(in thalamocortical
tract)

Axon, sensory
neuron III

Thalamus

Axon, sensory neuron II

Lateral spinothalamic
tract

Dendrite, sensory neuron I

Receptors (pain
or temperature)

Axon, sensory
neuron II

Fig. 9-21 Lateral spinothalamic tract relays sensory impulses from crude touch, pain, and temperature receptors up the cord to the thalamus. Thalamocortical tract fibers relay them to the somatic sensory area of the cortex (postcentral gyrus).

dendrites and cell bodies are located in cord or brain stem gray matter. Their axons ascend in ascending tracts up the cord, through the brain stem, and terminate in the thalamus. Here they synapse with sensory neuron III dendrites or cell bodies (Fig. 9-21).

c Sensory neurons III conduct from the thalamus to the postcentral gyrus of the parietal lobe, the somatosensory area. Bundles of axons of sensory neurons III form thalamocortical tracts. They extend through the portion of cerebral white matter known as the *internal capsule* to the cerebral cortex (Fig. 9-21).

2 For the most part, sensory pathways to the the cerebral cortex are crossed pathways. This means that each side of the brain registers sensations from the opposite side of the body. Look again at Fig. 9-21. The axon that decussates in this sensory pathway is part of which sensory neuron—I? II? or III? Usually it is the axon of sensory neuron II that decussates at some level in its ascent to the thalamus.

3 The neural pathway for sensations of crude touch, pain, and temperature is called the spinothalamic pathway.

4 Two neural pathways conduct impulses that produce sensations of touch and pressure, namely, the medial lemniscal system and the spinothalamic pathway. The *medial lemniscal system* consists of the tracts that make up the posterior white columns of the cord (the fasciculi cuneatus and gracilis) plus the *medial lemniscus*, a flat band of white fibers extending through the medulla, pons, and midbrain. (Derivation of the name lemniscus may interest you. It comes from the Greek word *lemniskos* meaning "woolen band." Apparently, to some early anatomist, the medial lemniscus looked like a band of woolen material running through the brain stem.)

The fibers of the medial lemniscus, like those of the spinothalamic tracts, are axons of sensory neurons II. They originate from cell bodies in the medulla, decussate, and then extend upward to terminate in the thalamus on the opposite side. The function of the medial lemniscal system is to transmit impulses that produce our more discriminating touch and pressure sensations, including stereognosis (awareness of an object's size, shape, and texture), precise localization, two-point discrimination, weight discrimination, and sense of vibrations.

Crude touch and pressure sensations are functions of the spinothalamic, not the medial lemniscal, pathway. Knowing that something touches the skin is a crude touch sensation, whereas knowing its precise location, size, shape, or texture involves discriminating touch sensations.

5 The neural pathway for *kinesthesia* (sense of movement and position of body parts) is also the medial lemniscal system. See Table 9-6 for a brief summary of the sensory neural pathways named in items 2, 3, and 4.

6 The first part of the cerebral cortex reached by sensory impulses conducted up spinothalamic and medial lemniscal pathways is the postcentral gyrus of the parietal lobe—the somatosensory area or areas 3, 1, and 2 in Brodman's map of the cortex (Fig. 9-19). It functions to produce our general senses—not only the common ones such as pain, heat, cold, touch, and pressure, but also those less familiar senses such as stereognosis, kinesthesia, vibratory sense, and two-point and weight discrimination. General sensations of the right side of the body are predominantly experienced by the left cerebral hemisphere's somatosensory area. General sensations of the left side of the body are predominantly experienced by the right hemisphere's somatosensory area. A second somatosensory area is located posterior to the central sulcus along the upper margin of the fissure of Sylvius.

Table 9-6 Sensory neural pathways

Neurons	Gross structures in which neuron parts are located
Pain, temperature, and crude touch	
Sensory neuron I	
Receptors	In skin, mucosa, muscles, tendons, viscera
Dendrite	In spinal nerve and branch of spinal nerve
Cell body	In spinal ganglion, on posterior root of spinal nerve
Axon	In posterior root of spinal nerve; terminates in posterior gray column of cord
Sensory neuron II	
Dendrite	Posterior gray column
Cell body	Posterior gray column
Axon	Decussates and ascends in lateral spinothalamic tract (Figs. 9-5 and 9-21); terminates in thalamus
Sensory neuron III	
Dendrite	Thalamus
Cell body	Thalamus
Axon	Thalamus via thalamocortical tract in internal capsule to general sensory area of cerebral cortex, that is, postcentral gyrus in parietal lobe
Discriminating touch (two-point discrimination, vibrations), deep touch, and pressure and kinesthesia	
Sensory neuron I	Same as sensory neuron I for pain, temperature, and crude touch stimuli, except that axon extends up cord in posterior white columns (fasciculi gracilis and cuneatus, Fig. 9-5) to nucleus gracilis or cuneatus in medulla instead of terminating in posterior gray columns of cord
Sensory neuron II	
Dendrite	In nucleus gracilis or cuneatus of medulla
Cell body	In nucleus gracilis or cuneatus of medulla
Axon	Decussates and ascends in medial lemniscus (broad band of fibers extending up through medulla and midbrain) and terminates in thalamus
Sensory neuron III	Same as sensory neuron III for pain, temperature, and crude touch stimuli

Somatic motor pathways

Somatic motor pathways consist of motoneurons that conduct impulses from the central nervous system to somatic effectors, that is, skeletal muscles. Some motor pathways are extremely complex and not at all clearly defined. Others, notably spinal cord reflex arcs, are simple and well established. You read about these in Chapter 8. Look back now at Fig. 8-5. From this diagram you can derive a cardinal principle about somatic motor pathways—the *principle of the final common path*. It is this: Only one final common path, namely, the anterior horn motoneuron, conducts impulses to skeletal muscles. Anterior horn motoneuron axons are the only ones that terminate in skeletal muscle cells. This principle of the final common path to skeletal muscles has important practical implications. For example, it means that any condition that makes anterior horn motoneurons unable to conduct impulses also makes skeletal muscle cells supplied by these neurons unable to contract. They cannot be willed to contract nor can they contract reflexly. They are, in short, paralyzed. Most famous of the diseases that produce paralysis by destroying anterior horn motoneurons is poliomyelitis. Numerous somatic motor paths conduct impulses from motor areas of the cerebrum down to anterior horn motoneurons at all levels of the cord.

Two methods are used to classify them—one based on the location of their fibers in the medulla and the other on their influence on the lower motoneurons. The first method divides them into pyramidal (Table 9-7) and extrapyramidal tracts. The second classifies them as facilitatory and inhibitory tracts.

Pyramidal tracts are those whose fibers come together in the medulla to form the pyramids, hence their name. Because axons composing the pyramidal tracts originate from neuron cell bodies located in the cerebral cortex, they also bear another name—*corticospinal tracts*. About three fourths of their fibers decussate (cross over from one side to the other) in the medulla. After decussating, they extend down the cord in the crossed corticospinal tract located on the opposite side of the cord in the lateral white column. About one fourth of the corticospinal fibers do not decussate. Instead, they extend down the same side of the cord as the cerebral area from which they came. One pair of uncrossed tracts lies in the ventral white columns of the cord, namely, the ventral corticospinal tracts. The other uncrossed corticospinal tracts form part of the lateral corticospinal tracts (Fig. 9-5). About 60% of corticospinal fibers are axons that arise from neuron cell bodies in the precentral (frontal lobe) region of the cortex.* In Fig. 9-19, these are the areas numbered 4 and 6. About 40% of corticospinal fibers originate from neuron cell bodies located in postcentral areas of the cortex, areas classified as sensory; now, more accurately, they are often called sensorimotor areas.

Relatively few corticospinal tract fibers synapse directly with anterior horn motoneurons. Most of them synapse with interneurons, which in turn synapse with anterior horn motoneurons. All corticospinal fibers conduct impulses that facilitate, that is, lower the resting negativity of anterior horn motoneurons. The effects of facilitatory impulses acting rapidly on any one neuron add up or summate. Each impulse, in other words, decreases the neuron's resting potential a little bit more. If sufficient numbers of impulses impinge rapidly enough on a neuron, its negativity decreases to threshold level. And at that moment the neuron starts conducting impulses. In short, it is stimulated. Stimulation of anterior horn motoneurons by corticospinal tract impulses results in stimulation of individual muscle groups (mainly of the hands and feet). Precise control of their contractions is, in short, the function of the cortico-

---

*Mountcastle, V. B., editor: Medical physiology, ed. 13, St. Louis, 1974, The C. V. Mosby Co., p. 748.

Table 9-7 Pyramidal path from cerebral cortex

Microscopic structures	Macroscopic structures in which neurons located
Upper motor neuron (Betz cells)	
Dendrite	Motor area of cerebral cortex
Cell body	Motor area of cerebral cortex
Axon	Motor area of cerebral cortex; descends in corticospinal (pyramidal) tract through cerebrum and brain stem; decussates in medulla and continues descent in lateral corticospinal (crossed pyramidal) tract in lateral white column; or may descend uncrossed in ventral, or direct, pyramidal tract in anterior white column and either decussate or not prior to terminating in anterior gray column
Lower motor neuron	
Dendrite	Anterior gray column
Cell body	Anterior gray column
Axon	Anterior gray column to anterior root of spinal nerve to spinal nerve and branches; terminates in somatic effector, that is, skeletal muscle

spinal tracts. Without stimulation of anterior horn motoneurons by impulses over corticospinal fibers, willed movements cannot occur. This means that paralysis results whenever pyramidal corticospinal tract conduction is interrupted. For instance, the paralysis that so often follows cerebral vascular accidents ("strokes") comes from pyramidal neuron injury—sometimes of their cell bodies in the motor areas, sometimes of their axons in the internal capsule (Fig. 9-22).

Extrapyramidal tracts are much more complex than pyramidal tracts. They consist of all motor tracts from the brain to the spinal cord anterior horn motoneurons except the corticospinal (pyramidal) tracts. Within the brain, extrapyramidal tracts consist of numerous but as yet incompletely worked-out relays of motoneurons between motor areas of the cortex, basal ganglia, thalamus, cerebellum, and brain stem. In the cord, some of the most important extrapyramidal tracts are the reticulospinal tracts.

Fibers of the *reticulospinal tracts* originate from cell bodies in the reticular formation of the brain stem and terminate in gray matter of the spinal cord, where they synapse with interneurons that synapse with lower (anterior horn) motoneurons. Some reticulospinal tracts function as facilitatory tracts, others as inhibitory tracts. Impulses over facilitatory tracts tend to decrease the lower motoneuron's resting potential. Impulses over inhibitory tracts tend to increase its negativity. Summation of these opposing influences determines the lower motoneuron's response. It initiates impulse conduction only when facilitatory impulses exceed inhibitory impulses sufficiently to decrease its negativity to its threshold level.

Conduction by extrapyramidal tracts plays a crucial part in producing our larger, more automatic movements because extrapyramidal impulses bring about contractions of groups of muscles in sequence or simultaneously. Such muscle action occurs, for example, in swimming and walking and, in fact, in all normal voluntary movements.

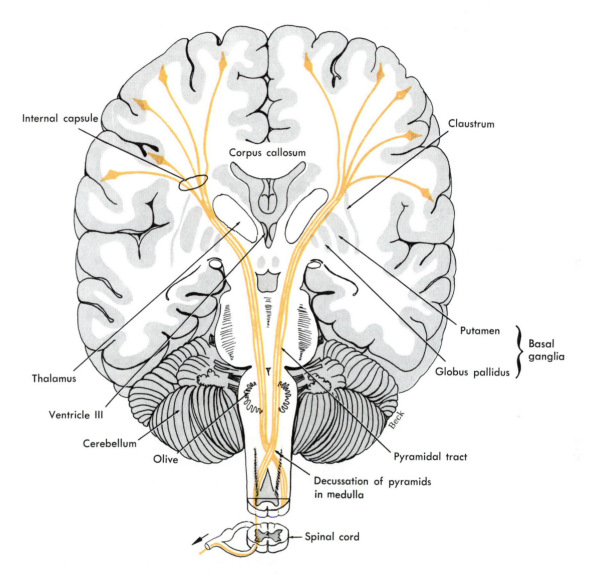

Fig. 9-22 Crossed corticospinal (pyramidal) tracts. Axons that compose pyramidal tracts come from neuron cell bodies in the cerebral cortex. After they descend through the internal capsule of the cerebrum and the white matter of the brain stem, about three fourths of the fibers decussate—cross over from one side to the other—in the medulla, as shown here. Then they continue downward in the lateral corticospinal tract (Fig. 9-5) on the opposite side of the cord. Each crossed corticospinal tract, therefore, conducts motor impulses from one side of the brain to interneurons or anterior horn motoneurons on the opposite side of the cord. Therefore impulses from one side of the cerebrum cause movements of the opposite side of the body.

Conduction by extrapyramidal tracts plays an important part in our emotional expressions. For instance, most of us smile automatically at things that amuse us and frown at things that irritate us. And it is extrapyramidal, not pyramidal, impulses that produce the smiles or frowns.

Axons of many different neurons converge on, that is, synapse, with each anterior horn motoneuron. Hence many impulses from diverse sources—some facilitatory and some inhibitory—continually bombard this final common path to skeletal muscles. Together the added or summated effect of these opposing influences determines lower motoneuron functioning. Facilitatory impulses reach these cells via sensory neurons (whose axons, you will recall, lie in the posterior roots of spinal nerves), pyramidal (corticospinal) tracts, and extrapyramidal facilitatory reticulospinal tracts. According to recent evidence, impulses over facilitatory reticulospinal fibers facilitate the lower motoneurons that supply extensor muscles. And at the same time, they reciprocally inhibit the lower motoneurons that supply flexor muscles. Hence facilitatory reticulospinal impulses tend to increase the tone of extensor muscles and decrease the tone of flexor muscles.

Inhibitory impulses reach lower motoneurons mainly via inhibitory reticulospinal fibers that originate from cell bodies located in the *bulbar inhibitory area* in the medulla. They inhibit the lower motoneurons to extensor muscles (and reciprocally stimulate those to flexor muscles). Hence inhibitory reticulospinal impulses tend to decrease extensor muscle tone and increase flexor muscle tone—opposite effects from facilitatory reticulospinal impulses.

Brain stem inhibitory and facilitatory areas are influenced directly or indirectly by impulses from various higher motor centers. Some experimental data* suggest that the basal ganglia and

*Evarts, E. V.: Brain mechanisms in movement, Sci. Am. **209**:96-103, July, 1973.

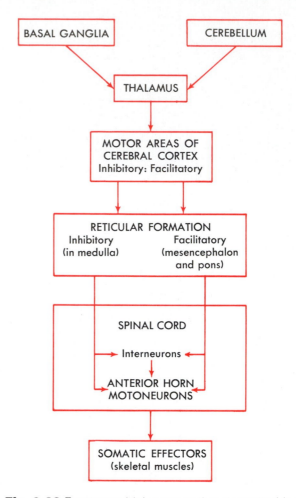

Fig. 9-23 Extrapyramidal motor paths, suggested by recent experimental studies.

cerebellum sends impulses to the thalamus, which, in turn, relays them to motor areas of the cortex (Fig. 9-23). This concept, in other words, sees the basal ganglia and cerebellum functioning as the highest command centers for movements and relaying their orders through the cerebral motor cortex to skeletal muscles. If this is true, it implies that the function of the cerebral motor cortex may be to refine skeletal muscle actions rather than to will them. The traditional view is quite different. It sees the cerebral motor cortex functioning as the highest command center for movements and relaying some of its orders through the basal ganglia and cerebellum. If this is true, it implies that the cerebral motor cortex wills skeletal muscle actions and that the basal ganglia and cerebellum refine them. The truth is that no one yet knows the truth. Physiologists still cannot tell us positively and precisely what part of the brain initiates the commands that lead to movements, nor can they describe with assurance the neural pathways through the brain from higher to lower motor centers. Whatever these pathways may be, the ratio of facilitatory and inhibitory impulses converging on lower motoneurons (anterior horn motoneurons) is such as to maintain normal muscle tone. In other words, facilitatory impulses normally somewhat exceed inhibitory impulses. But disease sometimes alters this ratio. Parkinson's disease and "strokes," for example, may interrupt transmission by inhibitory extrapyramidal paths from basal ganglia to bulbar inhibitory centers. Facilitatory impulses then predominate, and excess muscle tone (rigidity or spasticity) develops. Injury of *upper motoneurons* (those whose axons lie in either pyramidal or extrapyramidal tracts) produces symptoms frequently referred to as "pyramidal signs," notably a spastic type of paralysis, exaggerated deep reflexes, and a positive Babinski reflex (p. 269). Actually, pyramidal signs result from interruption of both pyramidal and extrapyramidal pathways. The paralysis stems from interruption of pyramidal tracts, whereas the spasticity (rigidity) and exaggerated reflexes come from interruption of inhibitory extrapyramidal pathways.

Injury of lower motoneurons produces symptoms different from those of upper motoneuron injury. Anterior horn cells or lower motoneurons, you will recall, constitute the final common path by which impulses reach skeletal muscles. This means that if they are injured, impulses can no longer reach the skeletal muscles they supply. And this, in turn, results in the ab-

sence of all reflex and willed movements produced by contraction of the muscles involved. Unused, the muscles soon lose their normal tone and become soft and flabby (flaccid). In short, absence of reflexes and flaccid paralysis are the chief "lower motoneuron signs."

Reflexes

Definition

The action that results from a nerve impulse passing over a reflex arc is called a *reflex*. In other words, a reflex is a response to a stimulus. It may or may not be conscious. Usually the term is used to mean only involuntary responses rather than those directly willed, that is, involving cerebral cortex activity.

A reflex consists either of muscle contraction or glandular secretion. *Somatic reflexes* are contractions of skeletal muscles. Impulse conduction over somatic reflex arcs—arcs whose motoneurons are somatic motoneurons, that is, anterior horn neurons or lower motoneurons—produces somatic reflexes. *Autonomic* (or *visceral*) *reflexes* consist either of contractions of smooth or cardiac muscle or secretion by glands; they are mediated by impulse conduction over autonomic reflex arcs, the motoneurons of which are autonomic neurons (discussed in the next chapter). The following paragraphs describe only somatic reflexes.

Some somatic reflexes of clinical importance

Clinical interest in reflexes stems from the fact that they deviate from normal in certain diseases. So the testing of reflexes is a valuable diagnostic aid. Physicians frequently test the following reflexes: knee jerk, ankle jerk, Babinski reflex, corneal reflex, and abdominal reflex.

The *knee jerk* or patellar reflex is an extension of the lower leg in response to tapping of the patellar tendon. The tap stretches both the tendon and its muscles, the quadriceps femoris, and thereby stimulates muscle spindles (receptors) in the muscle and initiates conduction over the following two-neuron reflex arc:

1 *Sensory neurons*
   a Dendrites—in femoral and second, third, and fourth lumbar nerves
   b Cell bodies—second, third, and fourth lumbar ganglia
   c Axons—in posterior roots of second, third, and fourth lumbar nerves; terminate in these segments of the spinal cord; synapse directly with lower motoneurons
2 *Reflex center*—synapses in anterior gray column between axons of sensory neurons and dendrites and cell bodies of lower motoneurons

3 *Motoneurons*
   a Dendrites and cell bodies—in spinal cord anterior gray column
   b Axons—in anterior roots of second, third, and fourth lumbar spinal nerves and femoral nerves; terminate in quadriceps femoris muscle

The knee jerk can be classified in various ways as follows:

1 As a *spinal cord reflex*—because the center of the reflex arc (which transmits the impulses that activate the muscles producing the knee jerk) lies in the spinal cord gray matter
2 As a *segmental reflex*—because impulses that mediate it enter and leave the same segment of the cord

Table 9-8 Correlation of microscopic and macroscopic structures of a three-neuron cord reflex arc*

Microscopic structures	Macroscopic structures in which neurons located
Sensory neuron	
Receptor	In skin or mucosa
Dendrite	In spinal nerve and branches
Cell body	In spinal ganglion on posterior root of spinal nerve
Axon	Posterior root of spinal nerve; terminates in posterior gray column of cord
Interneuron	
Dendrite	Posterior gray column
Cell body	Posterior gray column
Axon	Central gray matter of cord, extending into anterior gray column
Motor neuron	
Dendrite	Anterior gray column
Cell body	Anterior gray column
Axon	Anterior gray column, extending into anterior root of spinal nerve, spinal nerve, and its branches
Effector	In skeletal muscles

*See Figs. 8-5 and 8-6.

3 As an *ipsilateral reflex*—because the impulses that mediate it come from and go to the same side of the body
4 As a *stretch reflex*, or *myotatic reflex* (from the Greek *mys*, muscle, and *tasis*, stretching)—because of the kind of stimulation used to evoke it
5 As an *extensor reflex*—because produced by extensors of lower leg (muscles located on an anterior surface of thigh, which extend the lower leg)
6 As a *tendon reflex*—because tapping of a tendon is the stimulus that elicits it
7 *Deep reflex*—because of the deep location (in tendon and muscle) of the receptors stimulated to produce this reflex (*superficial reflexes*—those elicited by stimulation of receptors located in the skin or mucosa)

When a physician tests a patient's reflexes, he interprets the test results on the basis of what he knows about the reflex arcs that must function to produce normal reflexes.

To illustrate, suppose that a patient has been diagnosed as having poliomyelitis. In examining him the physician finds that he cannot elicit the knee jerk when he taps the patient's patellar tendon. He knows that the poliomyelitis virus attacks anterior horn motoneurons. He also knows the information previously related about which cord segments contain the reflex centers for the knee jerk. On the basis of this knowledge, therefore, he deduces that in this patient the poliomyelitis virus has damaged the second, third, and fourth lumbar segments of the spinal cord. Do you think that this patient's leg would be paralyzed, that he would be unable to move it voluntarily? What neurons would not be able to function that must function to produce voluntary contractions?

Ankle jerk or Achilles reflex is an extension (plantar flexion) of the foot in response to tapping of the Achilles tendon. Like the knee jerk, it is a tendon reflex and a deep reflex mediated by two-neuron spinal arcs, but the centers for the ankle jerk lie in the first and second sacral segments of the cord.

The *Babinski reflex* is an extension of the great toe, with or without fanning of the other toes, in response to stimulation of the outer margin of the sole of the foot. Normal infants, up until they are about 1½ years old, show this positive Babinski reflex. By about this time, corticospinal fibers have become fully myelinated and the Babinski reflex becomes suppressed. Just why this is so is not clear. But at any rate it is, and a positive Babinski reflex after this age is abnormal. From then on the normal response to stimulation of the outer edge of the sole is the *plantar reflex*. It consists of a curling under of all the toes (plantar flexion) plus a slight turning in and flexion of the anterior part of the foot. A positive Babinski reflex is one of the pyramidal signs (p. 267) and is interpreted to mean destruction of pyramidal tract (corticospinal) fibers.

The *corneal reflex* is winking in response to touching the cornea. It is mediated by reflex arcs with sensory fibers in the ophthalmic branch of the fifth cranial nerve, centers in the pons, and motor fibers in the seventh cranial nerve.

The *abdominal reflex* is drawing in of the abdominal wall in response to stroking the side of the abdomen. It is mediated by arcs with sensory and motor fibers in the ninth to twelfth thoracic spinal nerves and centers in these segments of the cord. It is classified as a superficial reflex. A decrease in this reflex or its absence occurs in lesions involving pyramidal tract upper motoneurons.

Outline summary

Divisions of nervous system

A Classified by location
  1 Central nervous system (CNS)—brain and spinal cord
  2 Peripheral nervous system (PNS)—cranial nerves and ganglia, spinal nerves and ganglia, autonomic nerves and ganglia
B Classified by effectors innervated
  1 Somatic nervous system—innervates somatic effectors, that is, skeletal muscles; includes brain, cord, cranial nerves, and spinal nerves
  2 Autonomic nervous system—innervates visceral (autonomic) effectors, that is, smooth muscle, cardiac muscle, and glandular tissue; organs of autonomic system are autonomic nerves and ganglia

Brain and cord coverings

A Bony—vertebrae around cord; cranial bones around brain
B Membranous—called meninges and consist of three layers
  1 Dura mater—white fibrous tissue outer layer
  2 Arachnoid membrane—cobwebby middle layer
  3 Pia mater—transparent; adherent to outer surface of cord and brain; contains blood vessels, therefore, nutritive layer

Brain and cord fluid spaces

A Names
  1 Subarachnoid space around cord
  2 Subarachnoid space around brain
  3 Central canal inside cord
  4 Ventricles and cerebral aqueduct inside brain—four cavities within brain:
    a First and second (lateral) ventricles—large cavities, one in each cerebral hemisphere
    b Third ventricle—vertical slit in cerebrum beneath corpus callosum and longitudinal fissure
    c Fourth ventricle—diamond-shaped space between cerebellum and medulla and pons; expansion of central canal of cord
B Formation, circulation, and function of cerebrospinal fluid
  1 Formed by plasma filtering from network of capillaries (choroid plexus) in each ventricle
  2 Circulates from lateral ventricles to third ventricle, cerebral aqueduct, fourth ventricle, central canal of cord, subarachnoid space of cord and brain; venous sinuses
  3 Function—protection of brain and cord

Spinal cord

Structure

A Size, shape, and location—about 1½ feet long, smaller around than spinal cavity in which it is located
B Sulci
  1 Two deep grooves, anterior median fissure and posterior median sulcus, incompletely divide cord into right and left symmetrical halves; anterior median fissure deeper and wider than posterior groove
C Gray matter—shaped like three-dimensional letter H
D White matter—present in columns (funiculi) anterior, lateral, and posterior; composed of numerous projection tracts

Functions

A Sensory tracts conduct up to brain from peripheral nerves; motor tracts conduct down from brain to peripheral nerves; see Tables 9-1 and 9-2 for names of spinal cord tracts
B Gray matter nuclei function as reflex centers for all spinal reflexes

Spinal nerves

Structure

A Number of spinal nerves—31 pairs (8 cervical, 12 thoracic, 5 lumbar, 5 sacral, and 1 coccygeal)
B Origin—originate by anterior and posterior roots from cord, emerge through intervertebral foramina; spinal ganglion on each posterior root
C Distribution—branches distributed to skin, mucosa, skeletal muscles; branches form plexuses, such as brachial plexus, from which nerves emerge to supply various parts
D Microscopic structure—consist of sensory dendrites and motor axons, that is, are mixed nerves; also contain autonomic postganglionic fibers, see Table 9-3 for peripheral branches

Functions

Conduct impulses both to cord from periphery and from cord to periphery

Divisions and size of brain

A Divisions—cerebrum, diencephalon, cerebellum, medulla, pons, and midbrain

B Size—adult brain weighs about 3 pounds; consists of billions of neurons and an even larger number of neuroglia

Brain stem

A Structure—made up of three divisions

  1 Medulla—part formed by enlargement of cord as it enters cranial cavity through foramen magnum; consists of white matter (ascending and descending tracts) and of mixture of gray and white matter (reticular formation)

  2 Pons—lies just above medulla; ascending and descending tracts make up white matter of pons; contains small amount of gray matter (nuclei)

  3 Midbrain—lies just above pons, below diencephalon and cerebrum; contains ascending and descending tracts and few nuclei; cerebral peduncles connect pons to cerebrum (consist of tracts); corpora quadrigemina—two superior and two inferior colliculi—are rounded eminences on dorsal surface of midbrain; red nucleus lies in gray matter of midbrain; cerebral aqueduct is fluid space in midbrain

B Functions

  1 Medulla—two-way conduction between cord and brain; cardiac, vasomotor, and respiratory centers in medulla are vital in control of heartbeat, blood pressure, and respiration; centers for reflexes of vomiting, coughing, and hiccuping also in medulla

  2 Pons—two-way conduction between cord and brain; centers for cranial nerves V through VIII in pons

  3 Midbrain—two-way conduction between cord and brain; centers for cranial nerves III and IV (pupillary reflexes and eye movements) in midbrain

Cranial nerves

A Structure—see Table 9-4

B Functions—see Table 9-4

Cerebellum

A Structure—center section, the vermis, lies between two hemispheres of cerebellum; surface is grooved with sulci and has slightly raised, slender convolutions; internal white matter in leaflike pattern; tracts in cerebellum are inferior, middle, and superior cerebellar peduncles; dentate nucleus in cerebellum

B Functions—synergic control of skeletal muscles; mediates postural and equilibrium reflexes

Diencephalon

A Structure—thalamus and hypothalamus are two major parts of diencephalon; thalamus—large rounded mass of gray matter, one in each hemisphere of cerebrum, lateral to third ventricle; hypothalamus—includes gray matter around optic chiasma, pituitary stalk, posterior lobe of pituitary gland, mamillary bodies and adjacent regions; prominent nuclei in hypothalamus are supraoptic, paraventricular, and mamillary nuclei; tracts connect diencephalon with cerebral cortex, basal ganglia, brain stem, and cord

B Functions

  1 Thalamus—relays sensory impulses from cord to cerebral cortex; registers crude sensations of pain, temperature, and touch; emotions of pleasantness or unpleasantness associated with sensations; part of pathway for arousal or alerting and for complex reflex movements

  2 Hypothalamus—functions as higher center for both parasympathetic and sympathetic divisions of autonomic system, regulating and coordinating them and thereby integrating responses by visceral effectors; functions as link between psyche and soma by relaying impulses from cerebral cortex to autonomic centers; supraoptic and paraventricular nuclei of hypothalamus synthesize posterior pituitary hormones; many hypothalamic neurons synthesize and secrete hormones that regulate hormone secretion by anterior pituitary gland; functions as part of pathways for arousal and alerting and for regulating appetite and temperature

Cerebrum

A Structure

  1 Hemispheres, fissures, and lobes—longitudinal fissure divides cerebrum into two hemispheres, connected only by corpus callosum; each cerebral hemisphere divided by fissures into five lobes: frontal, parietal, temporal, occipital, and island of Reil (insula)

  2 Cerebral cortex—outer layer of gray matter arranged in ridges called convolutions or gyri

3 Cerebral tracts—bundles of axons compose white matter in interior of cerebrum; ascending projection tracts transmit impulses toward or to brain; descending projection tracts transmit impulses down from brain to cord; commissural tracts transmit from one hemisphere to other; association tracts transmit from one convolution to another in same hemisphere

4 Basal ganglia (or cerebral nuclei)—masses of gray matter embedded deep inside white matter in interior of cerebrum; caudate, putamen, and pallidum; putamen and pallidum constitute lenticular nucleus

B Functions

1 Generalizations

a Cerebral activity goes on as long as life itself; electroencephalograms are records of cerebrum's electrical activity

b Cerebrum performs sensory, motor, and integrative functions

c Right and left cerebral hemispheres specialize in different functions; also certain regions of each hemisphere play key roles in particular functions

d Certain chemical transmitters are released only at synapses in brain and cord

2 Electroencephalograms (EEGs)

a Definition—EEGs are records of cerebrum's electrical activity

b Brain waves—alpha waves are moderately fast, moderately high-voltage waves; dominant in relaxed, nonattentive state; beta waves are fast, low-voltage waves, dominant in attentive waking state; delta waves are very slow, high-voltage waves, characterize deep sleep; theta waves are moderately slow, low-voltage waves, appear as drowsiness descends; absence of brain waves ("flat EEG") generally considered best criterion of death; brain wave patterns vary in altered states of consciousness

3 Sensory functions of cerebrum—comparison, evaluation, and integration of sensations to form total perceptions

4 Somatic motor functions of cerebrum—control voluntary (skeletal muscle) movements

5 Integrative functions of the cerebrum

a Consciousness—state of awareness of one's self, one's environment, and other beings; depends on excitation ("arousal" or "alerting") of cortical neurons by impulses conducted to them via neurons of reticular activating system; normal variations in degree of level of consciousness—waking, dream or rapid eye movement (REM) sleep, and deep, dreamless sleep; also several kinds of altered states of consciousness (ASC), for example, anesthesia, coma, and yoga or meditative state

b Memory—function of cerebral cortex neurons but mechanisms not definitely known; evidence that several parts of cortex, for example, temporal, parietal, occipital lobes, store memories and that limbic system ("emotional brain") and protein synthesis play key roles in memory; also, protein synthesis appears to be crucial for long-term memory

c Speech functions—the use of language (speaking and writing) and the understanding of language (spoken and written) depend on widespread integrated cortical processes involving many parts of cortex but certain regions in frontal, parietal, and temporal lobes called speech centers serve as focal points for integration of speech processes; speech defects (aphasias) of different kinds result from lesions of different speech centers

d Emotions—both subjective experience and objective expression of emotions involve functioning of cerebrum's limbic system, that is, part of the cortex located on medial surface of cerebrum that forms a border around the corpus callosum; limbic system made up of cingulate gyrus, isthmus, hippocampal gyrus, hippocampus, and uncus and tracts connecting these structures to several parts of brain, notably the septum, amygdala, and hypothalamus; for normal expression of emotions, other parts of the cortex must modulate limbic system activity

6 Hemispheric specialization and functional localization within a hemisphere

a Left hemisphere specializes in, that is, dominates, functions of producing and understanding speech sounds and also dominates control of skilled movements and gesturing movements of right hand

b Right cerebral hemisphere specializes in perception of nonspeech sounds, for example, melodies, coughing, crying, laughing, and in locating objects in space; right hemisphere may also specialize in tactual perception and perceptions of spatial relationships

c Certain areas of cortex in each hemisphere play dominant role in a particular function, for example, primary somatic sensory area (postcentral gyrus of parietal lobe) crucial for experiencing general sensations (heat, cold, and touch); primary somatic motor area (precentral gyrus of frontal lobe) dominates control of somatic effectors (skeletal muscles); primary auditory area (transverse gyrus of temporal lobe) crucial for auditory sensations; primary visual area (area 17 of occipital lobe) crucial for vision; important to remember that any one function requires functioning of many parts of the nervous system

Neurotransmitters in cord and brain

A In cord
   1 Acetylcholine—known to be excitatory neurotransmitter released at some cord synapses
   2 Glutamic acid—some evidence suggests that this is excitatory neurotransmitter at some cord synapses
   3 Glycine and gamma-aminobutyric acid—some evidence that this is inhibitory neurotransmitter in cord
   4 Enkephalins—inhibitory neurotransmitter at synapses in substantia gelatinosa; from axons of sensory neurons I in pain pathway
B In brain
   1 Norepinephrine (NE)—excitatory neurotransmitter at some synapses in cerebral cortex, cerebellum, and hypothalamus; released by axons from cell bodies in locus ceruleus
   2 Dopamine (DA)—excitatory neurotransmitter at some synapses in caudate nucleus and putamen; released by axons from cell bodies in substantia nigra; DA deficiency in parkinsonism
   3 Serotonin (5-HT)—excitatory neurotransmitter in limbic system; released from axons from cell bodies in raphe nuclei of brain stem
   4 Enkephalins—inhibitory neurotransmitters in thalamus and amygdala (of limbic system) at some synapses that are part of pain perception pathways; function as natural painkillers

Somatic sensory pathways

A Sensory pathways from periphery to cerebral cortex consist of three-neuron relays as follows
   1 Sensory neuron I conducts from periphery to cord or to brain stem
   2 Sensory neuron II conducts from cord to brain stem to thalamus
   3 Sensory neuron III conducts from thalamus to general sensory area of cerebral cortex (areas 3, 1, and 2)
B Most sensory neuron II axons decussate; so one side of brain registers mainly sensations for opposite side of body
C Pain and temperature—spinothalamic tracts to thalamus
D Touch and pressure
   1 Discriminating touch and pressure (stereognosis, precise localization, and vibratory sense)—medial lemniscal system (fasciculi cuneatus and gracilis to medulla; medial lemniscus to thalamus)
   2 Crude touch and pressure—spinothalamic tracts
E Conscious proprioception or kinesthesia—medial lemniscal system to thalamus (see also Table 9-6)
F Impulses conducted up spinothalamic and medial lemniscal pathways reach somatic sensory area of cerebral cortex (postcentral gyrus or areas 3, 1, and 2 on Brodmann's map of cortex), the primary area for general sensations
G Several secondary somatic sensory areas exist in cerebral cortex

Somatic motor pathways

A Principle of final common path—motoneurons in anterior gray horns of cord constitute final common path for impulses to skeletal muscles; are only neurons transmitting impulses into skeletal muscles
B Motor pathways from cerebral cortex to anterior horn cells classified according to place fibers enter cord
   1 Pyramidal (or corticospinal) tracts—dendrites and cells in cortex; axons descend through internal capsule, pyramids of medulla; enter cord from pyramids; descend and synapse directly with anterior horn cells or indirectly via internuncial neurons; impulses via pyramidal tracts essential for voluntary contractions of individual muscles to produce small discrete movements; also help maintain muscle tone
   2 Extrapyramidal tracts—all pathways between motor cortex and anterior horn cells except pyramidal tracts; upper extrapyramidal tracts relay impulses between cortex, basal ganglia, thalamus, and brain stem; reticulospinal tracts main lower extrapyramidal tracts; impulses via

extrapyramidal tracts essential for large, automatic movements; also for facial expressions and movements accompanying many emotions
C Motor pathways from cerebral cortex to anterior horn cells also classified according to influence on anterior horn cells
1 Facilitatory tracts—have facilitating or stimulating effect on anterior horn cells; all pyramidal tracts and some extrapyramidal tracts facilitatory, notably facilitatory reticulospinal fibers
2 Inhibitory tracts—have inhibiting effect on anterior horn cells; inhibitory reticulospinal fibers main ones
D Principle of convergence—axons of many neurons synapse with each anterior horn motoneuron
E Ratio of facilitatory and inhibitory impulses impinging on anterior horn cells controls activity; normally, slight predominance of facilitatory impulses maintains muscle tone

Reflexes

A Definition—action resulting from conduction over reflex arc; reflex is response (either muscle contraction or glandular secretion) to stimulus; term reflex usually used to mean only involuntary responses
B Some reflexes of clinical importance (see also pp. 267-269)
1 Knee jerk
2 Ankle jerk
3 Babinski reflex
4 Corneal reflex
5 Abdominal reflex

1 What term means the membranous coverings of the brain and cord? What three layers compose this covering?
2 Where does a physician do a lumbar puncture? Why?
3 What are the cavities inside the brain called? How many are there? What do they contain?
4 Describe the circulation of cerebrospinal fluid.
5 What function does the cerebrospinal fluid serve?
6 Explain the system for naming individual tracts.
7 Based on the explanation you gave in question 6, what is a spinothalamic tract?
8 Name several ascending or sensory tracts in the spinal cord.
9 Name two descending or motor tracts in the spinal cord.
10 Name the tracts that transmit impulses from each of the following types of receptors to the brain and state in which column of the cord each tract is located: pain and temperature receptors, crude touch receptors, proprioceptors (name two tracts for these), discriminating touch receptors (two tracts).
11 Describe the spinal cord's general functions.
12 Explain what the term reflex center means.
13 Describe the medulla's general functions.
14 Which cranial nerves transmit impulses that result in vision? In eye movements?
15 Which cranial nerves transmit impulses that result in hearing? In taste sensations?
16 Which cranial nerve serves as the great sensory nerve of the head?
17 Digoxin (Lanoxin), a drug that stimulates the vagus nerve, has been administered to a patient. What effect, if any, would you expect it to have on the patient's pulse rate?
18 Describe the general functions of the cerebellum.
19 Describe the general functions of the thalamus.
20 Describe the general functions of the hypothalamus.
21 Describe the general functions of the cerebrum.
22 Define the term electroencephalogram. What abbreviation stands for this term?
23 Identify the following kinds of brain waves according to their frequency, voltage, and the level of consciousness in which they predominate: alpha, beta, delta, theta.
24 What general functions does the cerebral cortex perform?

25 Define consciousness. Name the normal states or levels of consciousness.
26 Explain briefly what is meant by the arousal or alerting mechanism.
27 Name some altered states of consciousness.
28 Compare yoga (the meditative state) with the usual waking state.
29 Describe some of the current ideas about memory.
30 What part of the brain is called the "emotional brain"?
31 Locate the dendrite, cell body, and axon of sensory neurons I, II, and III.
32 Explain each of the following: lower motoneuron, upper motoneuron, and final common path.
33 Compare pyramidal tract and extrapyramidal tract functions.

chapter 10

The autonomic nervous system

The body has only one nervous system, but this one system has two major subdivisions: the somatic nervous system and the autonomic nervous system. The many structures that make up these two divisions function together as a single unit. What function do they jointly achieve? They regulate and integrate the multitudinous activities of the body's working parts—its somatic and visceral effectors—so as to normally maintain homeostasis.

Definitions

One way to define the autonomic nervous system is by the type of neurons that compose it. Using this criterion the autonomic nervous system is the part of the nervous system composed of autonomic or visceral motoneurons (neurons that conduct impulses from the central nervous system to visceral effectors). *Visceral effectors* are cardiac muscle, smooth muscle, and glandular epithelial tissues. Most of the organs that contain these tissues are listed in Table 10-1. Read through the list and ask yourself this question about each organ: Does it seem to you to function on its own, or do you have to will its action? Each of these visceral effector organs functions automatically. We do not have to will them to function—in fact, most of us cannot voluntarily control these organs. This leads to another definition for the autonomic nervous system: it is the part of the nervous system that regulates the automatic functions of the body.

Even though the autonomic nervous system consists, by definition, only of motoneurons (specifically, those that conduct impulses from the brain stem or cord to visceral effectors), nevertheless sensory neurons also take part in autonomic functioning. The autonomic nervous system functions on the reflex arc principle just as the somatic nervous system does. But any sensory neuron can theoretically function in both autonomic and somatic reflex arcs. For example, stimulation of cold receptors in the skin can initiate both an autonomic reflex (vasoconstriction of skin vessels) and a somatic reflex (shivering). In short, any one sensory neuron can function in both somatic and autonomic arcs, whereas any one motoneuron can function in either a somatic arc or an autonomic arc but not in both.

Structure

Two anatomically and physiologically separate divisions compose the autonomic nervous system: the sympathetic (or thoracolumbar) division and the parasympathetic (or craniosacral) division. Both divisions consist of autonomic ganglia and nerves. Autonomic ganglia contain synapses and cell bodies of autonomic neurons. Autonomic nerves contain axons of autonomic neurons. There are two kinds of autonomic neurons, namely, preganglionic and postganglionic. These names suggest the direction of impulse conduction by the neurons. Preganglionic neurons conduct impulses before they reach a ganglion. In other words, preganglionic neurons conduct from the central nervous system to an autonomic neuron. Postganglionic neurons, on the other hand, conduct impulses after they reach a ganglion, that is, from a ganglion to a visceral effector.

Sympathetic system

Sympathetic ganglia lie lateral to the anterior surface of the spinal column. Because short fibers extend between the sympathetic ganglia, connecting them to each other, they look a little like two chains of beads (one chain on each side of the spinal column from the level of the second cervical vertebra to the coccyx) and are often re-

Table 10-1 Visceral effector tissues and organs

Cardiac muscle	Smooth muscle	Glandular epithelium
Heart	Blood vessels	Sweat glands
	Bronchial tubes	Lacrimal glands
	Stomach	Digestive glands (salivary, gastric, pancreas, liver)
	Gallbladder	
	Intestines	Adrenal medulla
	Urinary bladder	
	Spleen	
	Eye (iris and ciliary muscles)	
	Hair follicles	

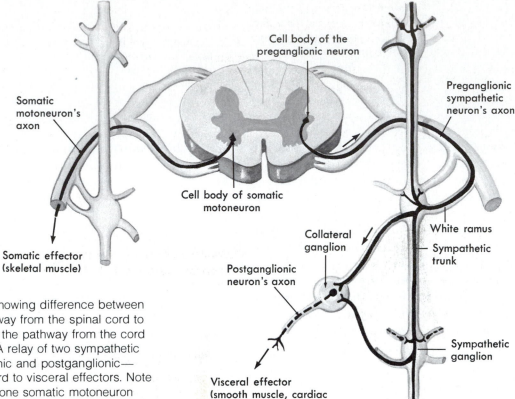

Cell body of the
preganglionic neuron

Somatic
motoneuron's
axon

Preganglionic
sympathetic
neuron's axon

Cell body of somatic
motoneuron

Somatic effector
(skeletal muscle)

Collateral
ganglion

White ramus

Sympathetic
trunk

Postganglionic
neuron's axon

Sympathetic
ganglion

Visceral effector
(smooth muscle, cardiac
muscle, glands)

Fig. 10-1 Diagram showing difference between the sympathetic pathway from the spinal cord to visceral effectors and the pathway from the cord to somatic effectors. A relay of two sympathetic neurons—preganglionic and postganglionic— conducts from the cord to visceral effectors. Note that, in contrast, only one somatic motoneuron (anterior horn neuron) conducts from the cord to somatic effectors with no intervening synapses. Parasympathetic impulses also travel over a relay of two neurons (parasympathetic preganglionic and postganglionic) to reach visceral effectors from the central nervous system. Note the location of the sympathetic preganglionic neuron's cell body and axon. Where in this figure do you find a sympathetic postganglionic neuron's cell body and axon located?

Sympathetic ganglia—usually there are 22 sympathetic ganglia: 3 cervical, 11 thoracic, 4 lumbar, and 4 sacral in each sympathetic chain. *Collateral sympathetic ganglia*—located a short distance from spinal cord. Examples include *celiac ganglia (solar plexus)*—two fairly large, flat ganglia located on each side of the celiac artery just below the diaphragm; *superior mesenteric ganglion*—small ganglion located near the beginning of the superior mesenteric artery; and *inferior mesenteric ganglion*—small ganglion located close to the beginning of the inferior mesenteric artery.

ferred to as the "sympathetic chain ganglia."* Short fibers connect each sympathetic ganglion with the spinal cord.

Sympathetic preganglionic neurons have their cell bodies in the lateral gray columns of the thoracic and first three or four lumbar segments of the cord. Look now at the right side of Fig. 10-1. Note that the axon of the sympathetic preganglionic neuron shown there leaves the cord in an anterior root of a spinal nerve. It next enters the spinal nerve but soon leaves it by way of a short branch (white ramus) and goes to and through a sympathetic chain ganglion to terminate in a collateral ganglion. Here it synapses with several postganglionic neurons whose axons terminate in visceral effectors.

Other preganglionic axons send branches up

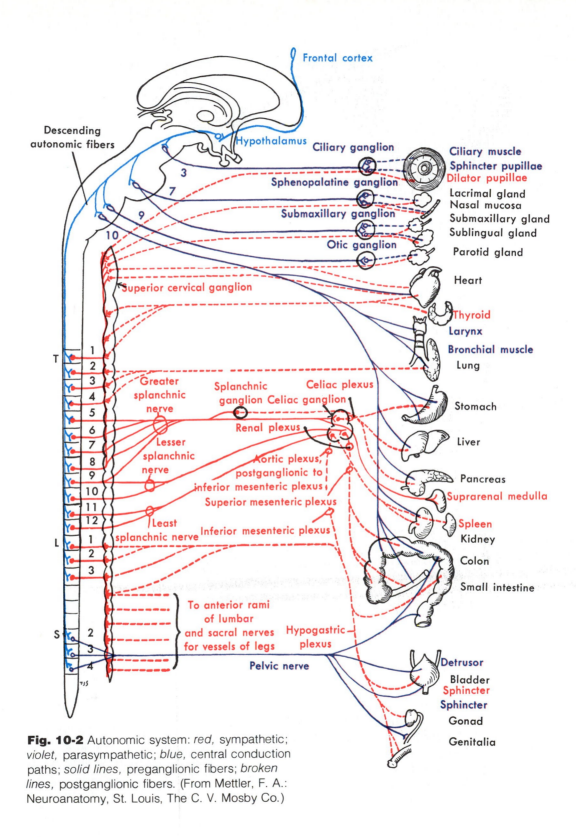

Fig. 10-2 Autonomic system: *red,* sympathetic; *violet,* parasympathetic; *blue,* central conduction paths; *solid lines,* preganglionic fibers; *broken lines,* postganglionic fibers. (From Mettler, F. A.: Neuroanatomy, St. Louis, The C. V. Mosby Co.)

and down the sympathetic chain to terminate in ganglia above and below their point of origin. Still others extend through the sympathetic ganglia, out through the splanchnic nerves,* and terminate in the collateral ganglia.

No matter which course a sympathetic preganglionic axon follows, it synapses with many postganglionic neurons, and these frequently terminate in widely separated organs. This anatomical fact partially explains a well-known physiological principle—sympathetic responses are usually widespread, involving many organs and not just one.

Sympathetic postganglionic neurons have their dendrites and cell bodies in the sympathetic chain ganglia or in collateral ganglia. Their axons are distributed by both spinal nerves and separate autonomic nerves. They reach the spinal nerves via small filaments (gray rami) that connect the sympathetic ganglia with the spinal nerves. They then travel in the spinal nerves to blood vessels, sweat glands, and arrector hair muscles all over the body.

The course of postganglionic axons through autonomic nerves is somewhat more complex. These nerves form complicated plexuses before fibers are finally distributed to their respective destinations. For example, postganglionic fibers from the celiac and superior mesenteric ganglia pass through the celiac plexus before reaching the abdominal viscera, those from the inferior mesenteric ganglion pass through the hypogastric plexus on their way to the lower abdominal and pelvic viscera, and those from the cervical ganglion pass through the cardiac nerves and the cardiac plexus at the base of the heart and are then distributed to the heart.

*The three splanchnic nerves constitute the main branches from the thoracic sympathetic trunk. Since their fibers synapse in the celiac and other collateral ganglia with postganglionic neurons to abdominal structures, they form the main routes for sympathetic stimulation of abdominal viscera.

Parasympathetic system

Parasympathetic preganglionic neurons have their cell bodies in nuclei in the brain stem or in the lateral gray columns of the sacral cord. Their axons are contained in cranial nerves III, VII, IX, X, and XI and in some pelvic nerves. They extend a considerable distance before synapsing with postganglionic neurons. For example, axons arising from cell bodies in the vagus nuclei (located in the medulla) travel in the vagus nerve for a distance of a foot or more before reaching their terminal ganglia in the chest and abdomen (Fig. 10-2 and Table 10-2).

Parasympathetic postganglionic neurons have their dendrites and cell bodies in the outlying parasympathetic ganglia and send short axons into the nearby structures. Each parasympathetic preganglionic neuron synapses, therefore, only with postganglionic neurons to a single effector. For this reason, parasympathetic stimulation frequently involves response by only one organ, as contrasted with sympathetic responses, which, as noted before, usually involve numerous organs.

General principles

Table 10-2 compares the locations of sympathetic and parasympathetic preganglionic and postganglionic neurons. Read this carefully and try to absorb the information it summarizes. Then turn your attention to the left side of Fig. 10-1. Note that one neuron—the somatic motoneuron—conducts from the cord (or brain stem) to somatic effectors. Contrast that with the conduction pathway to visceral effectors shown on the right side of this figure. It consists of a two-neuron relay, preganglionic and postganglionic neurons. The conduction path to visceral effectors, therefore, contains synapses in autonomic ganglia located outside of the central nervous system. The conduction pathway to somatic effectors from the central nervous system, in con-

Table 10-2 Sympathetic and parasympathetic neurons compared as to location

	Sympathetic	Parasympathetic
Preganglionic neurons		
Dendrites and cell bodies	In lateral gray columns of thoracic and first four lumbar segments of spinal cord	In nuclei of brain stem and cord (in lateral gray columns of sacral segments of cord)
Axons	In anterior roots of spinal nerves, to spinal nerves (thoracic and first four lumbar), to white rami, and then in any of three following pathways: (1) ganglia; (2) through white rami to and through sympathetic ganglia, then up or down sympathetic trunk before synapsing in a sympathetic ganglion; (3) through white rami to and through sympathetic ganglia, to and through splanchnic nerves to collateral ganglia (celiac, superior, and inferior mesenteric ganglia)	From brain stem nuclei through cranial nerve III to ciliary ganglion From nuclei in pons: through cranial nerve VII to sphenopalatine or submaxillary ganglion From nuclei in medulla through cranial nerve IX to otic ganglion or through cranial nerves X and XI to cardiac and celiac ganglia, respectively
Postganglionic neurons		
Dendrites and cell bodies	In sympathetic and collateral ganglia	In parasympathetic ganglia (for example, ciliary, sphenopalatine, submaxillary, otic, cardiac, celiac) located in or near visceral effector organs
Axons	In autonomic nerves and plexuses that innervate thoracic and abdominal viscera and blood vessels in these cavities In gray rami to spinal nerves, to smooth muscle of skin blood vessels and hair follicles, and to sweat glands	In short nerves to various visceral effector organs

trast, contains no synapses. These facts have practical value. Investigators have been able to positively identify acetylcholine as the neurotransmitter released at these external autonomic synapses. To identify neurotransmitters at synapses buried deep in central gray matter is technically so difficult that as yet it has been accomplished for relatively few locations.

Both parasympathetic and sympathetic postganglionic axons supply many visceral effector organs. All visceral effector organs are innervated by sympathetic postganglionic axons, and some are supplied by only sympathetic postganglionic fibers, that is, they receive no parasympathetic innervation. These anatomical facts are referred to as the principles of dual autonomic innervation and single autonomic innervation, respectively. Terminals of autonomic axons, like those of all axons, release chemicals that transmit impulses across synapses and neuroeffector junctions. Autonomic axons fall into

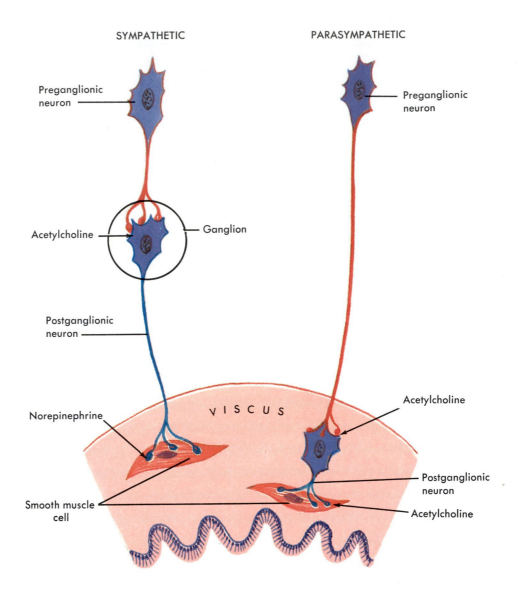

SYMPATHETIC   PARASYMPATHETIC

Preganglionic
neuron

Preganglionic
neuron

Acetylcholine   Ganglion

Postganglionic
neuron

Norepinephrine

VISCUS

Acetylcholine

Smooth muscle
cell

Postganglionic
neuron

Acetylcholine

Fig. 10-3 Cholinergic fibers release acetylcholine (ACh). Adrenergic fibers release norepinephrine (NE). Note the three kinds of cholinergic fibers shown in the diagram: axons of both sympathetic and parasympathetic preganglionic neurons and of which postganglionic neurons? Which postganglionic neurons would you classify as adrenergic? Almost immediately after its release, ACh is inactivated by the enzyme acetylcholinesterase (AChE). The enzyme catechol-O-methyl transferase (COMT) inactivates NE.

two classifications based on the chemical transmitters that they release. Some release acetylcholine and so are classified as cholinergic fibers. Others release norepinephrine and so are classified as adrenergic fibers.

Here are some facts worth remembering about autonomic neurotransmitters:

1 The only autonomic fibers that release norepinephrine—that are adrenergic—are sympathetic postganglionic axons, but even a few of these are not adrenergic but cholinergic, that is, they release acetylcholine. (Cholinergic sympathetic postganglionic fibers are those to sweat glands, to external genitalia, and to smooth muscle of the blood vessels in skeletal muscles.)

2 All preganglionic axons, both sympathetic and parasympathetic, release acetylcholine. Also, all or nearly all parasympathetic postganglionic fibers are cholinergic.

3 Acetylcholine combines with two kinds of receptors in the membranes of postganglionic neurons. One kind is named nicotinic receptors, because nicotine also combines with them. Curare blocks these receptors so that acetylcholine cannot combine with them and impulse conduction across the synapses is prevented. The other type of receptors that acetylcholine combines with are called muscarinic receptors. The drug atropine blocks these receptors.

4 Acetylcholine's action is terminated by its being hydrolyzed by the enzyme acetylcholinesterase.

5 Norepinephrine combines with two kinds of receptors, namely, alpha and beta receptors.

6 Norepinephrine's action is terminated in two ways—by its being broken down by the enzyme catechol-O-methyl transferase (COMT) and by its being transported back into axon endings, where it may be broken down by another enzyme, monoamine oxidase.

Functions
Sympathetic division

The sympathetic division of the autonomic nervous system functions chiefly as an emergency system. Under stress conditions, from either physical or emotional causes, sympathetic impulses increase greatly to most visceral effectors. In fact, one of the very first steps in the body's complex defense mechanism against stress is a sudden and marked increase in sympathetic activity. This brings about a group of responses that all go on at the same time. Together they make the body ready to expend maximum energy and to engage in maximum physical exertion—as, for example, in running or fighting. Walter B. Cannon coined a descriptive and now famous phrase—the "fight or flight" reaction—as his name for this group of sympathetic responses. Read the middle column of Table 10-3, and you will find many of the "fight or flight" physiological changes. Some particularly important changes for maximum energy expenditure are the faster, stronger heartbeat, the dilated blood vessels in skeletal muscles, the dilated bronchi, the increased blood sugar levels from stimulated glycogenolysis, and the increased secretion of epinephrine. Epinephrine is a hormone secreted by the adrenal medulla. It reinforces and prolongs the effects of the neurotransmitter norepinephrine released by most sympathetic postganglionic fibers.

Sympathetic impulses usually dominate the control of most visceral effectors in times of stress, but not always. Curiously enough, parasympathetic impulses frequently become excessive to some effectors at such times. For instance, one of the first symptoms of emotional stress in many individuals is that they feel hungry and want to eat more than usual. Presumably this is partly caused by increased parasympathetic impulses to the smooth muscle of the stomach. This stimulates increased gastric contractions, which in turn may cause the feeling of hunger. The most famous disease of para-

Table 10-3 Autonomic functions

Visceral effectors	Sympathetic impulses	Parasympathetic impulses
Cardiac muscle	Stimulate pacemaker, accelerate rate and strength of heartbeat	Inhibit pacemaker; decrease rate and strength of heartbeat
Smooth muscle of blood vessels		
Skin blood vessels	Stimulate; constrict	No parasympathetic fibers
Skeletal muscle blood vessels	Inhibit; dilate	No parasympathetic fibers
Coronary blood vessels	Inhibit; dilate	No parasympathetic fibers
Blood vessels in cerebrum and abdominal viscera	Stimulate; constrict	Inhibit; dilate; no parasympathetic fibers to some viscera
Blood vessels in external genitalia	Inhibit; dilate	Stimulate; constrict
Smooth muscle of hollow organs and sphincters		
Bronchi	Inhibit; dilate	Stimulate; constrict
Digestive tract, except sphincters	Inhibit; decrease peristalsis	Stimulate; increase peristalsis
Sphincters of digestive tract	Stimulate; close	Inhibit; open
Urinary bladder	Inhibit; relax	Stimulate; contract
Urinary sphincters	Stimulate; close	Inhibit; open
Eye		
Iris	Stimulate radial muscle; dilate pupil	Stimulate circular muscle; constrict pupil
Ciliary	Inhibit; accommodate for far vision	Stimulate; accommodate for near vision
Hairs (pilomotor muscles)	Stimulate; produce "goose pimples" (piloerection)	No parasympathetic fibers
Glands		
Sweat	Stimulate; increase sweat	No parasympathetic fibers
Digestive (salivary, gastric, etc.)	Inhibit; decrease secretion of saliva; not known for others	Stimulate; increase secretion of saliva
Pancreas, including islets	Inhibit; decrease secretion	Stimulate; increase secretion of pancreatic juice and insulin
Liver	Stimulate glycogenolysis; increase blood sugar level	No parasympathetic fibers
Adrenal medulla	Stimulate; increase epinephrine secretion	No parasympathetic fibers

sympathetic excess is peptic ulcer. In some individuals (presumably those born with certain genes), chronic stress leads to excessive parasympathetic stimulation of gastric hydrochloric acid glands, with eventual development of peptic ulcer.

Parasympathetic division

The parasympathetic division functions under normal, everyday conditions as the main regulator of many visceral effectors (heart, digestive tract smooth muscle, glands that secrete digestive juices, and endocrine gland cells that secrete insulin). Ordinarily these visceral effectors receive more impulses over cholinergic parasympathetic fibers than they do over adrenergic fibers. Acetylcholine is the neurotransmitter of the parasympathetic systems. It stimulates the secretion of digestive juices and insulin. It also stimulates the smooth muscle of the digestive tract and thereby increases peristalsis. In short, the parasympathetic system functions to promote digestion and elimination.

Autonomic nervous system as a whole

The autonomic nervous system as a whole functions to regulate visceral effectors in ways that tend to maintain or quickly restore homeostasis. Many autonomic fibers are tonically active, that is, they continually conduct impulses. Doubly innervated organs continually receive both sympathetic and parasympathetic impulses, and these tend to influence their functioning in opposite or antagonistic ways—a fact that we might call the principle of autonomic antagonism. For example, the heart continually receives sympathetic impulses that tend to make it beat faster and parasympathetic impulses that tend to slow it down. The ratio between the two antagonistic forces determines the actual heartbeat.

The name autonomic nervous system is misleading. It wrongly implies independence from the rest of the nervous system. Quite the contrary is true. The autonomic nervous system is far from being autonomous either anatomically or physiologically. Axons of many central nervous system neurons synapse with sympathetic and parasympathetic preganglionic neurons and thereby exert a major influence over their activity. Neurons located at higher and higher levels of the central nervous system function as a hierarchy in their control of autonomic activity. Lowest ranking in this hierarchy are the so-called lower autonomic centers in the spinal cord and brain stem. Here, axons from neurons located at higher levels synapse with preganglionic neurons. Highest ranking in the hierarchy of autonomic control, we now know, are certain areas of the cerebral cortex itself. Fig. 10-4 indicates the central nervous system hierarchy that regulates autonomic neuron conduction and thereby regulates the functioning of the visceral effectors they supply. Neurons in certain areas (called autonomic centers) of the cerebral cortex send impulses that either stimulate or inhibit autonomic centers in the limbic system.* Limbic system autonomic centers then relay impulses to centers in the hypothalamus, which in turn relay impulses to lower autonomic centers—to either parasympathetic centers in the brain stem or sacral segments of the spinal cord, or to sympathetic centers in the thoracolumbar segments of the cord, or to both parasympathetic and sympathetic centers. Finally, these lowest autonomic centers stimulate or inhibit conduction by parasympathetic or sympathetic preganglionic and postganglionic neurons and thereby increase or decrease visceral effector activities.

You may be wondering why the name autonomic system was ever chosen in the first place if the system is really not autonomous. Originally the term seemed appropriate. The autonomic system seemed to be self-regulating and independent of the rest of the nervous system. Common observations furnished abundant evidence of its independence from cerebral control, from direct control by the will, that is. But later, even

---

*See p. 257. An older name for essentially the same structures now called the limbic system was rhinencephalon (literally "nosebrain"), a name that refers to its role in olfaction.

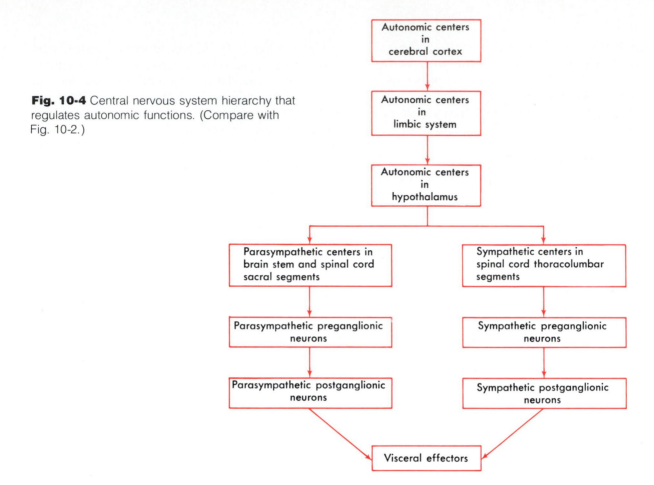

Fig. 10-4 Central nervous system hierarchy that regulates autonomic functions. (Compare with Fig. 10-2.)

this was found to be not entirely true. Some rare and startling exceptions were discovered. I have seen one such exception—a man who sat in a brightly lighted amphitheater in front of a class of medical students and made his pupils change from small, constricted dots (normal response to bright light) to widely dilated circles. This same man also willed gooseflesh to appear on his arms by contracting the smooth muscle of the hairs.

"Fight or flight" response and physiological changes in meditation

Wallace and Benson, in their major study of the physiology of meditation,* noted that a spe-

*Wallace, R. K., and Benson, H.: The physiology of meditation, Sci. Am. **226:**85-90, Feb., 1972.

cific set of physiological changes—an integrated response"—occurred during meditation. They observed, too, that this integrated response to meditation was a hypometabolic state and that it represented quiescence of the sympathetic nervous system. In other words, it seemed to be an integrated response that was essentially opposite to the integrated "fight or flight" response. They found that the integrated response to meditation consisted mainly of the following physiological changes: a dramatic decrease in oxygen consumption and respiratory rate and volume, a moderate slowing of the heart, and a low, almost unchanging blood pressure. The integrated "fight or flight" response, in contrast, includes opposite physiological changes: a marked increase in oxygen consumption and respiratory rate and volume, an accelerated heart rate, and an increased blood pressure. Interestingly, the integrated response to meditation also included several physiological changes that had previously been shown to accompany a highly relaxed state of mind—intensification of slow (8 to 9 cycles per second) alpha waves in electroencephalograms of the frontal and central brain regions, a precipitous decrease in the concentration of lactate in the blood, and a rapid and marked increase in the electrical resistance of the skin. The integrated "fight or flight" response, on the other hand, is accompanied by a high degree of tenseness and often anxiety.

Autonomic learning and biofeedback

Only a few years ago, physiologists generally took for granted that learning occurred only in the somatic nervous system and that it could not take place in the autonomic nervous system. They knew, however, that conditioning could occur in the autonomic system. Pavlov had long since proved this with his dogs who salivated at the sound of a bell. But this classical or pavlovian conditioning of a visceral response was not considered true autonomic learning. True learning resulted from a process called operant conditioning, which differed somewhat from the pavlovian method. Briefly, operant conditioning consists of presenting a stimulus and a choice of several responses. When the desired response is made, it is immediately reinforced by giving a reward of some kind.

That operant conditioning, or true learning, can occur in the autonomic system is now no longer doubted. By the late 1960s, hard-to-refute evidence was indicating that the epithet "stupid autonomic nervous system" was no longer appropriate. Respected scientists in the United States and abroad were demonstrating that both animals and humans could learn to make a specific visceral response to a specific stimulus, providing they could be informed that they were making the desired visceral response and were rewarded for it. A number of so-called biofeedback instruments were designed to inform subjects when they made a desired visceral response. Various kinds of rewards were used to reinforce the response.

For example, one of the early experiments in human autonomic learning was conducted at Harvard Medical School by David Shapiro and his colleagues. In this experiment a group of young men learned to increase their blood pressure by about 4 mm Hg. Without their knowledge, their blood pressure was monitored; each time it happened to increase a light flashed (the biofeedback). After a certain number of flashes, they were rewarded with the sight of a nude pin-up of a pretty girl.

At the Menninger Foundation in Topeka, Kansas, patients who suffered migraine headache attacks learned to dilate blood vessels in their hands. A biofeedback instrument that detected slight temperature changes was attached to their hands and emitted a high sound each time they happened to dilate their hand blood vessels. The reward for these patients was a lessening of the migraine pain—presumably because of the shunting of blood away from the head to the hands. (Migraine headaches initially involve distention of head blood vessels.) Ninety of the first 100 subjects trained to dilate their hand blood vessels learned to control their headaches in this way. One volunteer was reported to have raised her hand temperature 10° in 2 minutes![*]

Recently a good deal of interest has been focused on the spectacular feat of brain-wave training, using electroencephalograms for the biofeedback.[†]

---

[*]Ferguson, Marilyn: The brain revolution, New York, 1973, Taplinger Publishing Co.
[†]Trotter, R. J.: Listen to your head, Sci. News **100**:314-316, Nov., 1971.

Outline summary

Definition

Nerves and ganglia that contain motoneurons to visceral effectors (smooth muscle, cardiac muscle, and glands); part of nervous system that regulates automatic functions of body

Structure

A Sympathetic division
  1 Sympathetic division consists of two chains of ganglia (one on either side of backbone) and fibers that connect ganglia with each other and with thoracic and lumbar segments of cord; other fibers extend from sympathetic ganglia out to visceral effectors
  2 Sympathetic preganglionic neurons—cell bodies in lateral gray columns of thoracic and first three or four lumbar segments of cord; axons leave cord in anterior roots of spinal nerves, leave spinal nerve by way of white ramus, and then follow one of three paths: to terminate in sympathetic ganglia; or to and through sympathetic ganglia, then up or down sympathetic trunk to terminate in a higher or lower sympathetic ganglion; or to and through sympathetic ganglia, through splanchnic nerves to terminate in collateral ganglia
  3 Sympathetic postganglionic neurons—cell bodies in sympathetic chain or collateral ganglia (celiac, superior, and inferior mesenteric); axons in autonomic nerves and plexuses or in gray rami and spinal nerves
B Parasympathetic division
  1 Parasympathetic preganglionic neurons—cell bodies in nuclei of brain stem and of sacral segments of cord; axons from brain stem nuclei go through cranial nerve III, terminate in ciliary ganglion; from nuclei in pons through cranial nerve VII, terminate in sphenopalatine or submaxillary ganglion; from nuclei in medulla through cranial nerve IX to otic ganglion or through cranial nerves X and XI to cardiac and celiac ganglia
  2 Parasympathetic postganglionic neurons—cell bodies in ganglia on or near organs innervated; axons lie in short nerves extending into effector

General principles

A Contrast between autonomic and somatic conduction paths from central nervous system to effectors
  1 Autonomic path—two-neuron relay consisting of preganglionic neuron (from cord to brain stem to autonomic ganglion), synapse, and postganglionic neuron (from autonomic ganglion to visceral effector)
  2 Somatic path—somatic motoneuron from cord or brain stem to somatic effector
  3 Double autonomic innervation, that is, both sympathetic and parasympathetic fibers, to some visceral effectors but single innervation (sympathetic fibers only) to some visceral effectors; all somatic effectors receive single innervation, that is, by somatic motoneurons
B Autonomic neurotransmitters
  1 Acetylcholine—released by all preganglionic fibers, presumably by all postganglionic fibers of parasympathetic division and by postganglionic fibers of sympathetic division to sweat glands, external genitalia, and smooth muscle of skeletal muscle blood vessels
  2 Norepinephrine—released by all postganglionic fibers of sympathetic division except those to sweat glands, external genitalia, and smooth muscle

Functions

A Sympathetic division—chiefly an emergency system that greatly increases conduction to most visceral effectors during stress and prepares body for maximum energy expenditure and physical exertion by producing "fight or flight" reaction, a group of sympathetic responses including increased epinephrine secretion by adrenal medulla (increases and prolongs effects of norepinephrine)
B Parasympathetic division—regulates many visceral effectors under normal, everyday conditions; in general, its neurotransmitter, acetylcholine, stimulates digestive juice secretion, insulin secretion, and contraction of smooth muscle of digestive tract, thereby increasing peristalsis
C Autonomic system does not function autonomously or independently of the central nervous system; impulses from centers of spinal cord, brain stem, and cerebral cortex regulate conduction by autonomic neurons that make up conduction path from central nervous system to visceral effectors
D "Fight or flight" response and physiological changes in meditation—essentially opposite integrated responses
E Automatic learning and biofeedback—operant conditioning of the autonomic nervous system is possible; humans can learn specific visceral responses to specific stimuli if feedback rewarding correct responses is provided

Review questions

1 Contrast the meanings of the terms visceral effectors and somatic effectors.
2 Differentiate between a preganglionic neuron and a postganglionic neuron.
3 Compare sympathetic and parasympathetic preganglionic neurons as to locations of their cell bodies and axons.
4 Compare sympathetic and parasympathetic postganglionic neurons as to the locations of their cell bodies and axons.
5 All preganglionic neurons release the same neurotransmitter. Name it.
6 Only one kind of autonomic fiber releases norepinephrine. Which kind?
7 Are all sympathetic postganglionic fibers adrenergic? If not, which ones are cholinergic?
8 Acetylcholine combines with two kinds of receptors. Name them.
9 Name the enzyme that terminates the action of acetylcholine.
10 Norepinephrine combines with two kinds of receptors. Name them.
11 Describe two mechanisms by which the action of norepinephrine is normally terminated.
12 Compare parasympathetic and sympathetic functions.
13 Explain why the name autonomic nervous system is misleading.
14 Classify the following structures as somatic effectors, visceral effectors, or neither: adrenal glands, biceps femoris muscle, heart, iris, skin.
15 Which of the following would indicate an increase in sympathetic impulses and which might indicate an increase in parasympathetic impulses to visceral effectors: constipation, dilated pupils, dry mouth, goose pimples, "I'm always hungry and eat too much when I am upset," rapid heartbeat?
16 What is the limbic system and what general function does it perform?
17 Compare the "fight or flight" reaction with the physiological changes of meditation.
18 Can learning in the autonomic nervous system occur? If so, explain the method for learning visceral responses.

chapter 11

Sense organs

The body has millions of sense organs. All of its receptors, that is, the distal ends of dendrites of all its sensory neurons, are its sense organs. This chapter deals briefly with the receptors for general sensations, then discusses more fully some special sense organs, the eye, ear, and olfactory and gustatory sense organs.

Classifications

Charles S. Sherrington, an English physiologist, divided receptors into exteroceptors, visceroceptors, and proprioceptors. *Exteroceptors* are located roughly at the surface of the body, in the skin, mucosa, eye, and ear. They are stimulated by changes in the external environment. Visceroceptors and proprioceptors are both located internally. *Visceroceptors* are found, for example, in the walls of blood vessels, stomach, intestines, and various other organs. *Proprioceptors* are located in muscles, tendons, joints, and the internal ear.

Another classification divides receptors into superficial, deep, visceral, and special receptors. *Superficial receptors* are located in the skin and mucosa; stimulation of them initiates sensations of touch, pressure, warmth, cold, and pain. *Deep receptors* are located in muscles, tendons, and joints; stimulation of them initiates sensations of position, vibration, deep pressure, and deep pain. *Visceral receptors* are located in the viscera; stimulation of them initiates such sensations as hunger, nausea, and visceral pain. *Special receptors* are located in the eye, ear, mouth, and nose; stimulation of them initiates what we call our

Fig. 11-1 Free nerve endings.

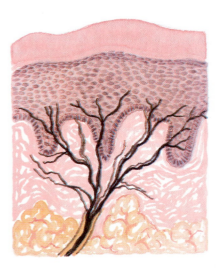

A, In dermis of skin.

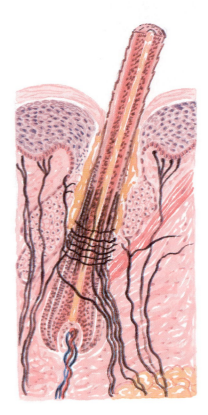

B, Surrounding, in linear and circular fashion, the root of a hair follicle.

C, In the cornea.

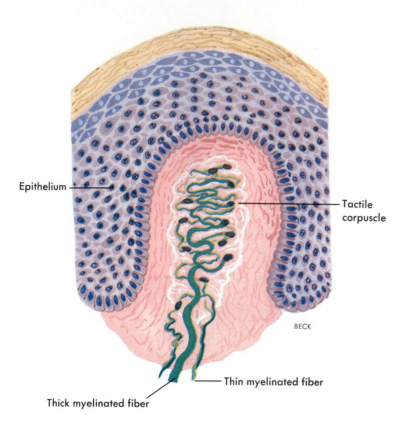

Epithelium

Tactile corpuscle

BECK

Thin myelinated fiber

Thick myelinated fiber

Fig. 11-2 Meissner's corpuscle (tactile corpuscle) in skin papilla, an encapsulated nerve ending found in hairless portions of skin. Meissner's corpuscles mediate sensation of touch.

special senses—vision, hearing, and the senses of equilibrium, taste, and smell.

Structure and functions

Receptors differ in structure considerably, from the very highly complex to the very simple. The simplest of all receptors are merely the free endings of sensory dendrites in the skin that give rise to sensations of touch, pain, and temperature (Fig. 11-1). Somewhat more complex skin receptors are those named Meissner's corpuscles (Fig. 11-2), Ruffini's corpuscles (Fig. 11-3), pacinian corpuscles (Fig. 11-4), and Krause's end-

bulbs (Fig. 11-5). Table 11-1 shows the sensations these receptors mediate.

According to the principle of specificity of receptors, specific kinds of receptors mediate specific sensations because they are sensitive to specific kinds of stimuli. This principle, however, is now questioned by some investigators. They present data that support the concepts that one kind of receptor may mediate more than one kind of sensation and that one kind of sensation may be mediated by more than one kind of receptor. Generally accepted ideas about which types of receptors mediate which sensations are included in Table 11-1.

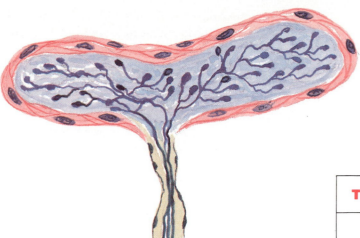

Fig. 11-3 Ruffini corpuscle, a skin receptor that probably mediates touch, rather than heat, as formerly thought.

Table 11-1 Skin receptors	
Receptors	**Sensations**
Free nerve endings	Pain, touch, temperature
Meissner's corpuscles	Touch
Ruffini's corpuscles	Touch, probably, and not heat as formerly thought
Pacinian corpuscles	Pressure
Krause's end-bulbs	Touch(?), cold(?)

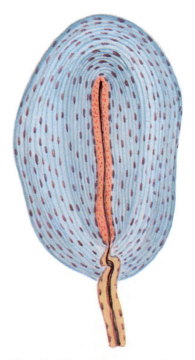

Fig. 11-4 Pacinian corpuscle, an encapsulated nerve ending widely distributed in subcutaneous tissue; mediates sensation of pressure.

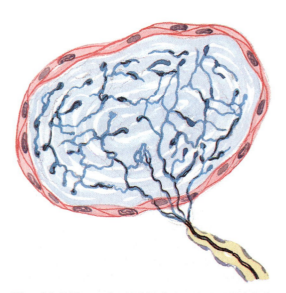

Fig. 11-5 Krause's end-bulb, an encapsulated nerve ending; may mediate sensation of cold, but evidence indicates that it is not the only type of receptor for cold.

Pain receptors

Because stimulation of pain receptors may give warning of potentially harmful environmental changes, pain receptors are also called *nociceptors* (from the Latin *noceo*, to injure). Any type of stimulus, provided that it be sufficiently intense, seems to be adequate for stimulating nociceptors in the skin and mucosa. In contrast, only marked changes in pressure and certain chemicals can stimulate nociceptors located in the viscera. Sometimes this knowledge proves useful. For example, it enables a physician to cauterize the uterine cervix without giving an anesthetic but with assurance that the patient will not suffer pain from the intense heat. On the other hand, he knows that if the intestine becomes markedly distended (as it sometimes does following surgery), the patient will experience pain. So, too, will the individual whose heart becomes ischemic because of coronary occlusion. Presumably the resulting cellular oxygen deficiency leads to the formation or accumulation of chemicals that stimulate nociceptors in the heart.

Types of pain

Pain is classified according to the location of the pain receptors stimulated as either somatic or visceral.

Somatic pain may be *superficial,* as when it arises from stimulation of skin receptors, or it may be *deep,* as when it results from stimulation of receptors in the skeletal muscles, fascia, tendons, and joints.

Visceral pain results from stimulation of receptors located in the viscera. Impulses are conducted from these receptors to the cord primarily by sensory fibers in sympathetic nerves and only rarely in parasympathetic nerves.

The cerebrum does not always interpret the source of pain accurately. Sometimes it erroneously refers the pain to a surface area instead of to the region in which the stimulated receptors actually are located. This phenomenon is called *referred pain.* It occurs only as a result of stimulation of pain receptors located in deep structures (notably, viscera, joints, and skeletal muscles)—never from stimulation of skin receptors. In other words, deep somatic pain and visceral pain may be referred but not superficial somatic pain.

Pain originating in the viscera and other deep structures is generally interpreted as coming from the skin area whose sensory fibers enter the same segment of the spinal cord as the sensory fibers from the deep structure. For example, sensory fibers from the heart enter the first to fourth thoracic segments. And so do sensory fibers from the skin areas over the heart and on the inner surface of the left arm. Pain originating in the heart is referred to those skin areas, but the reason for this is not clear.

Eye
Anatomy

Coats of eyeball

Approximately five sixths of the eyeball lies recessed in the orbit, protected by this bony socket. Only the small anterior surface of the eyeball is exposed. Three layers of tissues or coats compose the eyeball. From the outside in they are the sclera, the choroid, and the retina. Both the sclera and the choroid coats consist of an anterior and a posterior portion.

Tough white fibrous tissue fashions the *sclera.* Deep within the anterior part of the sclera at its junction with the cornea lies a ring-shaped venous sinus, the *canal of Schlemm* (Figs. 11-6 and 11-9 to 11-11). The anterior portion of the sclera is called the *cornea* and lies over the colored part of the eye (iris). The cornea is transparent, whereas the rest of the sclera is white and opaque, a fact that explains why the visible anterior surface of the sclera is usually spoken of as the "white" of the eye. No blood vessels are found in the cornea, in the aqueous and vitreous humors, or in the lens.

The middle or *choroid coat* of the eye contains

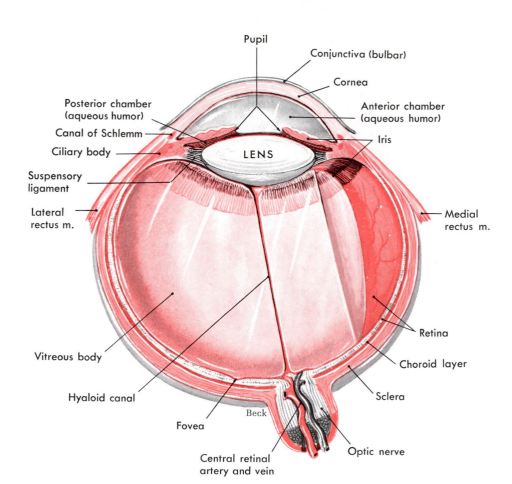

Pupil

Conjunctiva (bulbar)

Cornea

Posterior chamber
(aqueous humor)

Anterior chamber
(aqueous humor)

Canal of Schlemm

Iris

Ciliary body

LENS

Suspensory
ligament

Lateral
rectus m.

Medial
rectus m.

Vitreous body

Retina

Choroid layer

Hyaloid canal

Sclera

Fovea

Beck

Central retinal
artery and vein

Optic nerve

Fig. 11-6 Horizontal section through left eyeball.

a great many blood vessels and a large amount of pigment. Its anterior portion is modified into three separate structures: the ciliary body, the suspensory ligament, and the iris. (See Fig. 11-6.)

The *ciliary body* is formed by a thickening of the choroid and fits like a collar into the area between the anterior margin of the retina and the posterior margin of the iris. The small *ciliary muscle*, composed of both radial and circular smooth muscle fibers, lies in the anterior part of the ciliary body. Attached to the ciliary body is the *suspensory ligament*, which blends with the elastic capsule of the *lens* and holds it suspended in place.

The *iris* or colored part of the eye consists of circular and radial smooth muscle fibers arranged so as to form a doughnut-shaped struc-

ture (the hole in the middle is called the *pupil*). The iris attaches to the ciliary body.

The *retina* is the incomplete innermost coat of the eyeball—incomplete in that it has no anterior portion. It consists mainly of nervous tissue and contains three layers of neurons (Fig. 11-7). Named in the order in which they conduct impulses, they are photoreceptor neurons, bipolar neurons, and ganglion neurons. The distal ends of the dendrites of the photoreceptor neurons have been given names descriptive of their shapes. Because some look like tiny rods and others like cones, they are called *rods* and *cones*, respectively. They constitute our visual receptors, structures highly specialized for stimulation by light rays (discussed on p. 307). They differ as to numbers, distribution, and function.

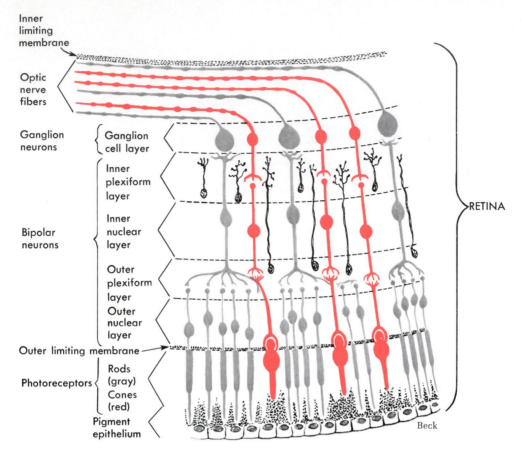

Inner
limiting
membrane

Optic
nerve
fibers

Ganglion
neurons

Ganglion
cell layer

Inner
plexiform
layer

Bipolar
neurons

Inner
nuclear
layer

Outer
plexiform
layer

Outer
nuclear
layer

Outer limiting membrane

Photoreceptors

Rods
(gray)
Cones
(red)

Pigment
epithelium

RETINA

Beck

Fig. 11-7 Layers that compose the retina. The inner limiting
membrane lies nearest the inside of the eyeball. It adheres to the
vitreous humor. The pigment epithelium lies farthest from the inside of
the eyeball. It adheres to the choroid coat. Note relay of three neurons
in the retina: photoreceptor, bipolar, and ganglion neurons (named in
order of impulse transmission). Light rays pass through the vitreous
humor and various layers of retina to stimulate rods and cones, the
receptors of the photoreceptor neurons.

Cones are less numerous than rods and are most
densely concentrated in the *fovea centralis,* a
small depression in the center of a yellowish
area, the *macula lutea,* found near the center
of the retina. They become less and less dense
from the fovea outward. Rods, on the other
hand, are absent entirely from the fovea and
macula and increase in density toward the periphery of retina. How these anatomical facts
relate to rod and cone functions is revealed on
p. 307.

All the axons of ganglion neurons extend back
to a small circular area in the posterior part of
the eyeball known as the *optic disk* or papilla.
This part of the sclera contains perforations
through which the fibers emerge from the eyeball as the *optic nerves.* The optic disk is also
called the *blind spot* because light rays striking
this area cannot be seen, since it contains no
rods or cones, only nerve fibers.

For an outline summary of the coats of the eyeball see Table 11-2.

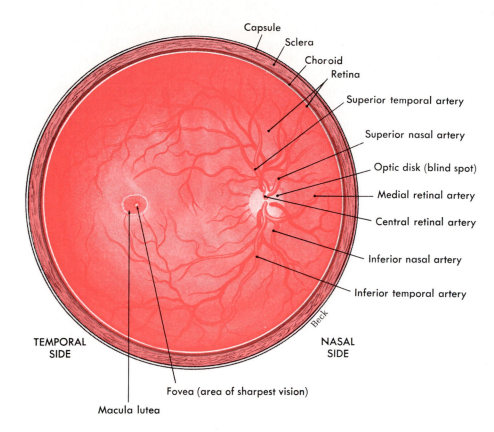

Fig. 11-8 Right eyeground (fundus) showing vessels, optic disk, macula lutea, and layers of the eyeball.

Table 11-2 Coats of the eyeball

Location	Posterior portion	Anterior portion	Characteristics
Outer coat (sclera)	Sclera proper	Cornea	Protective fibrous coat; cornea transparent; rest of coat white and opaque
Middle coat (choroid)	Choroid proper	Ciliary body; suspensory ligament; iris (pupil is hole in iris); lens suspended in suspensory ligament	Vascular, pigmented coat
Inner coat (retina)	Retina	No anterior portion	Nervous tissue; rods and cones (receptors for second cranial nerve) located in retina

Cavities and humors

The eyeball is not a solid sphere but contains a large interior cavity that is divided into two cavities, anterior and posterior.

The *anterior cavity* has two subdivisions known as the *anterior* and *posterior chambers*. As Fig. 11-6 shows, the entire anterior cavity lies in front of the lens. The posterior chamber of the anterior cavity consists of the small space directly posterior to the iris but anterior to the lens. And the anterior chamber of the anterior cavity is the space anterior to the iris but posterior to the cornea. *Aqueous humor* fills both chambers of the anterior cavity. This substance is clear and watery and often leaks out when the eye is injured.

The *posterior cavity* of the eyeball is considerably larger than the anterior, since it occupies all the space posterior to the lens, suspensory ligament, and ciliary body. It contains *vitreous humor*, a substance with a consistency comparable to soft gelatin. This semisolid material helps maintain sufficient intraocular pressure to prevent the eyeball from collapsing. (An obliterated artery, the hyaloid canal, runs through the vitreous humor between the lens and optic disk.)

An outline summary of the cavities of the eye is included in Table 11-3.

Still not established is the mechanism by which aqueous humor forms. It comes from blood in capillaries (located mainly in the ciliary body). Presumably the ciliary body actively secretes aqueous humor into the posterior chamber. But also, passive filtration from capillary blood may contribute to aqueous humor formation. From the posterior chamber, aqueous humor moves from the area between the iris and the lens through the pupil into the anterior chamber (Fig. 11-9). From here, it drains into the canal of Schlemm and moves on into small veins. Normally, aqueous humor drains out of the anterior chamber at the same rate at which it enters the posterior chamber, so the amount of aqueous humor in the eye remains relatively constant—and so, too, does intraocular pressure. But sometimes something happens to upset this balance, and intraocular pressure increases above the normal level of about 20 to 25 mm Hg pressure. The individual then has the eye disease known as glaucoma (Figs. 11-10 and 11-11). Either excess formation or, more often, decreased absorption is seen as an immediate cause of this condition, but underlying causes are unknown.

Table 11-3 Cavities of the eye

Cavity	Divisions	Location	Contents
Anterior	Anterior chamber	Anterior to iris and posterior to cornea	Aqueous humor
	Posterior chamber	Posterior to iris and anterior to lens	Aqueous humor
Posterior	None	Posterior to lens	Vitreous humor

Fig. 11-9 Aqueous humor (heavy arrows) is believed to be formed mainly by secretion by the ciliary body into the posterior chamber. It passes into the anterior chamber through the pupil, from which it is drained away by the ring-shaped canal of Schlemm, and finally into the anterior ciliary veins. Small arrows indicate pressure of the aqueous humor.

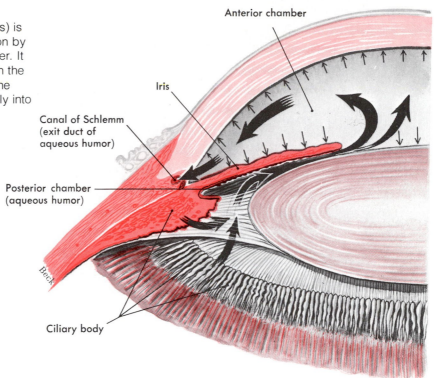

Anterior chamber

Iris

Canal of Schlemm (exit duct of aqueous humor)

Posterior chamber (aqueous humor)

Ciliary body

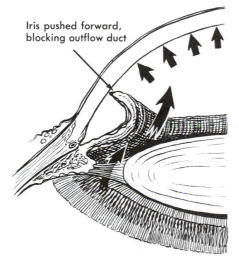

Iris pushed forward, blocking outflow duct

Fig. 11-10

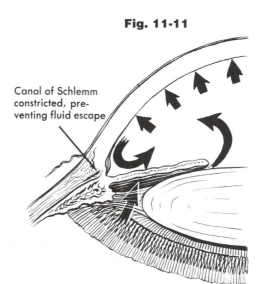

Fig. 11-11

Canal of Schlemm constricted, preventing fluid escape

Figs. 11-10 and **11-11** Acute glaucoma. When pressure of the aqueous humor in the anterior chamber becomes extreme, the iris is pushed forward, causing it to press on and block the canal of Schlemm from draining the fluid. In most instances, proper eyedrop medication will dilate the duct, permitting excess fluid escape. When medication fails, a new outflow duct can be created surgically.

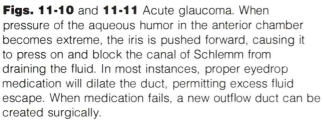

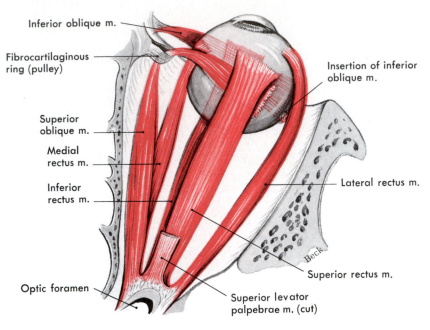

Inferior oblique m.

Fibrocartilaginous ring (pulley)

Superior oblique m.

Medial rectus m.

Inferior rectus m.

Optic foramen

Insertion of inferior oblique m.

Lateral rectus m.

Superior rectus m.

Superior levator palpebrae m. (cut)

Fig. 11-12 Muscles that move the right eye as viewed from above.

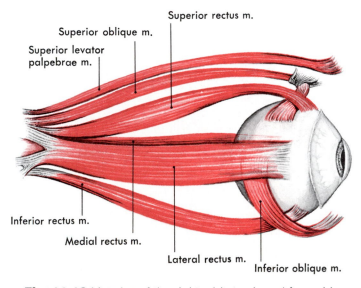

Superior rectus m.

Superior oblique m.

Superior levator palpebrae m.

Inferior rectus m.

Medial rectus m.

Lateral rectus m.

Inferior oblique m.

Fig. 11-13 Muscles of the right orbit as viewed from side.

Table 11-4 Eye muscles

	Extrinsic muscles	Intrinsic muscles
Names	Superior rectus Inferior rectus Lateral rectus Mesial rectus Superior oblique Inferior oblique	Iris Ciliary muscle
Kind of muscle	Voluntary (striated, skeletal)	Involuntary (smooth, visceral)
Location	Attached to eyeball and bones of orbit	Modified anterior portion of choroid coat of eyeball; iris, doughnut-shaped sphincter muscle; pupil, hole in center of iris
Functions	Eye movements	Iris regulates size of pupil—therefore amount of light entering eye; ciliary muscle controls shape of lens (accommodation)—therefore its refractive power
Innervation	Somatic fibers of third, fourth, and sixth cranial nerves	Autonomic fibers of third and fourth cranial nerves

Muscles

Eye muscles are of two types: extrinsic and intrinsic.

Extrinsic eye muscles are those that attach to the outside of the eyeball and to the bones of the orbit. They move the eyeball in any desired direction and are, of course, voluntary muscles. Four of them are straight muscles, and two are oblique. Their names describe their positions on the eyeball. They are the superior, inferior, mesial, and lateral rectus muscles and superior and inferior oblique muscles (Figs. 11-12 and 11-13).

Intrinsic eye muscles are those located within the eye. Their names are the iris and the ciliary muscles, and they are involuntary. Incidentally, the eye is the only organ in the body in which both voluntary and involuntary muscles are found. The iris regulates the size of the pupil. The ciliary muscle controls the shape of the lens. As the ciliary muscle contracts, it releases the suspensory ligament from the backward pull usually exerted on it. And this allows the elastic lens, suspended in the ligament, to bulge or become more convex—a necessary accommodation for near vision (pp. 306-307). Some essential facts about eye muscles are summarized in Table 11-4.

Accessory structures

Accessory structures of the eye include the eyebrows, eyelashes, eyelids, and lacrimal apparatus.

Eyebrows and eyelashes. The eyebrows and eyelashes serve a cosmetic purpose and give some protection against the entrance of foreign objects into the eyes. Small glands located at the base of the lashes secrete a lubricating fluid. They frequently become infected forming a *sty.*

Eyelids. The eyelids or palpebrae consist mainly of voluntary muscle and skin, with a border of thick connective tissue at the free edge of each

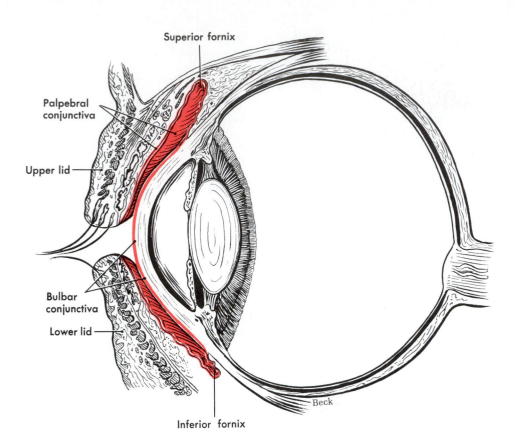

Superior fornix

Palpebral
conjunctiva

Upper lid

Bulbar
conjunctiva

Lower lid

Beck

Inferior fornix

Fig. 11-14 Longitudinal section through eyeball and eyelids showing the conjunctiva, a lining of mucous membrane. Conjunctiva covering the cornea is called bulbar, and that lining the posterior surface of the eyelids is called palpebral.

lid known as the tarsal plate. One can feel the tarsal plate as a ridge when turning back the eyelid to remove a foreign object. Mucous membrane called conjunctiva lines each lid (Fig. 11-14). It continues over the surface of the eyeball, where it is modified to give transparency. Inflammation of the conjunctiva (conjunctivitis) is a fairly common infection. Because it produces a pinkish discoloration of the eye's surface, it is called pinkeye.

The opening between the eyelids bears the technical name of palpebral fissure. The height of the fissure determines the apparent size of the eyes. In other words, if the eyelids are habitually held widely opened, the eyes appear large, although actually there is very little difference in size between eyeballs of different adults. Eyes

appear small if the upper eyelids droop. Plastic surgeons can correct this common aging change with an operation called blepharoplasty. The upper and lower eyelids join, forming an angle or corner known as a *canthus*, the inner canthus being the mesial corner of the eye and the outer canthus the lateral corner.

Lacrimal apparatus. The lacrimal apparatus consists of the structures that secrete tears and drain them from the surface of the eyeball. They are the lacrimal glands, lacrimal ducts, lacrimal sacs, and nasolacrimal ducts (Fig. 11-15).

The *lacrimal glands*, comparable in size and shape to a small almond, are located in a depression of the frontal bone at the upper outer margin of each orbit. Approximately a dozen small ducts lead from each gland, draining the tears

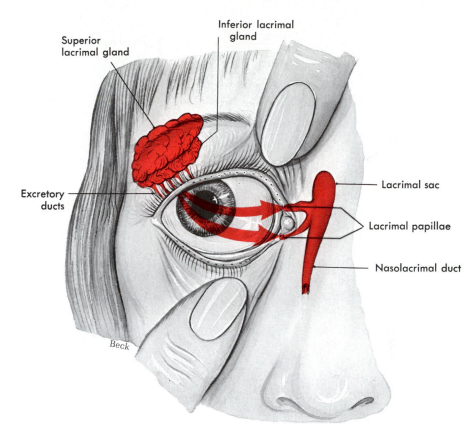

Superior
lacrimal gland

Inferior lacrimal
gland

Excretory
ducts

Lacrimal sac

Lacrimal papillae

Nasolacrimal duct

Beck

Fig. 11-15 Lacrimal apparatus. Arrows indicate direction of drainage from the excretory ducts of the lacrimal glands across the eye to the nasolacrimal duct.

onto the conjunctiva at the upper outer corner of the eye.

The *lacrimal canals* are small channels, one above and the other below each *caruncle* (small red body at inner canthus). They empty into the lacrimal sacs. The openings into the canals are called *punctae* and can be seen as two small dots at the inner canthus of the eye. The *lacrimal sacs* are located in a groove in the lacrimal bone. The *nasolacrimal ducts* are small tubes that extend from the lacrimal sac into the inferior meatus of the nose. All the tear ducts are lined with mucous membrane, an extension of the mucosa that lines the nose. When this membrane becomes inflamed and swollen, the nasolacrimal ducts become plugged, causing the tears to overflow from the eyes instead of draining into the

nose as they do normally. Hence when we have a common cold, "watering" eyes add to our discomfort.

Physiology of vision

In order for vision to occur, the following conditions must be fulfilled: an image must be formed on the retina to stimulate its receptors (rods and cones) and the resulting nerve impulses must be conducted to the visual areas of the cerebral cortex.

Formation of retinal image

Four processes focus light rays so that they form a clear image on the retina: *refraction* of the light rays, *accommodation* of the lens, *constric-*

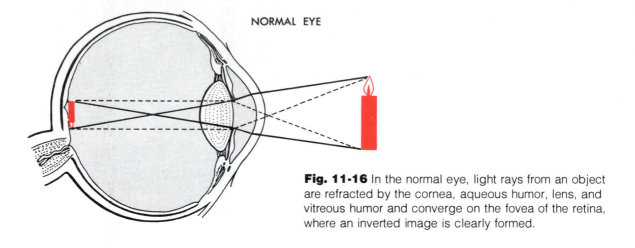

NORMAL EYE

Fig. 11-16 In the normal eye, light rays from an object are refracted by the cornea, aqueous humor, lens, and vitreous humor and converge on the fovea of the retina, where an inverted image is clearly formed.

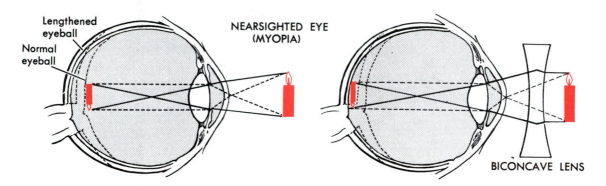

Lengthened eyeball

Normal eyeball

NEARSIGHTED EYE (MYOPIA)

BICONCAVE LENS

Fig. 11-17 Nearsighted or myopic eye focuses the image in front of the retina. This may occur when the eyeball is too long or the lens is too thick. Correction is by concave lens.

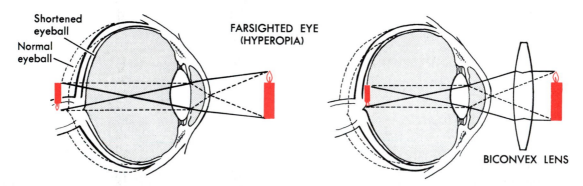

Shortened eyeball

Normal eyeball

FARSIGHTED EYE (HYPEROPIA)

BICONVEX LENS

Fig. 11-18 The farsighted or hyperopic eye could only focus the image at a hypothetical distance behind the retina. This may occur when the eyeball is too short or the lens is too thin. Correction is by convex lens.

tion of the pupil, and *convergence* of the eyes.

Refraction of light rays. Refraction means the deflection or bending of light rays. It is produced by light rays passing obliquely from one transparent medium into another of different optical density; the more convex the surface of the medium, the greater its refractive power. The refracting media of the eye are the cornea, aqueous humor, lens, and vitreous humor. Light rays are bent or refracted at the anterior surface of the cornea as they pass from the rarer air into the denser cornea, at the anterior surface of the lens as they pass from the aqueous humor into the denser lens, and at the posterior surface of the lens as they pass from the lens into the rarer vitreous humor.

When an individual goes to an ophthalmologist for an eye examination, the doctor does a "refraction." In other words, by various specially designed methods, he measures the refractory or light-bending power of that person's eyes.

In a relaxed normal (emmetropic) eye the four refracting media together bend light rays sufficiently to bring to a focus on the retina the parallel rays reflected from an object 20 or more feet away (Fig. 11-16). Of course a normal eye can also focus objects located much nearer than 20 feet from the eye. This is accomplished by a mechanism known as accommodation (discussed on pp. 306-307). Many eyes, however, show *errors of refraction*, that is, they are not able to focus the rays on the retina under the stated conditions. Some common errors of refraction are *nearsightedness* (myopia), *farsightedness* (hypermetropia), and *astigmatism*.

The nearsighted (myopic) eye sees distant objects as blurred images because it focuses rays from the object at a point in front of the retina (Fig. 11-17). According to one theory, this occurs because the eyeball is too long and the distance too great from the lens to the retina in the myopic eye. Concave glasses, by lessening refraction, can give clear distant vision to nearsighted individuals. Presumably, opposite conditions exist in the farsighted (hyperopic) eye (Fig. 11-18).

Astigmatism is a more complicated condition in which the curvature of the cornea or of the lens is uneven, causing horizontal and vertical rays to be focused at two different points on the retina (Figs. 11-19 and 11-20). Instead of the

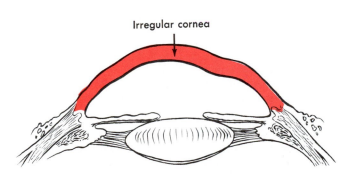

Fig. 11-19 Astigmatism results most frequently from an irregular cornea. The cornea may be only slightly flattened horizontally, vertically, or diagonally to produce distortion of vision. The compensating shape of the lens largely nullifies the irregularities of the cornea.

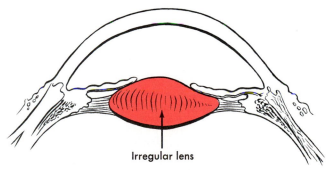

Fig. 11-20 An irregular lens causes light to be bent in such a way that it does not focus the image on the sharpest area of vision on the retina. An astigmatic eye with this defect causes distorted or blurred vision.

curvature of the cornea being a section of a sphere, it is more like that of a teaspoon, with horizontal and vertical arcs uneven. Suitable glasses correct the refraction of such an eye.

Visual acuity or the ability to distinguish form and outline clearly is indicated by a fraction that compares the distance at which an individual sees an object (usually letters of a definite size and shape) clearly with the distance at which the normal eye would see the object. Thus if he sees clearly at 20 feet an object that the normal eye would be able to see clearly at 20 feet, his visual acuity is said to be 20/20 or normal. But if he sees an object clearly at 20 feet that the normal eye sees clearly at 30 feet, his visual acuity is 20/30 or two thirds of normal.

As people grow older, they tend to become far-sighted because lenses lose their elasticity and therefore their ability to bulge and to accommodate for near vision. This condition is called *presbyopia*.

Accommodation of lens. Accommodation for near vision necessitates three changes: increase in the curvature of the lens, constriction of the pupils, and convergence of the two eyes. Light rays from objects 20 or more feet away are practically parallel. The normal eye, as previously noted, refracts such rays sufficiently to focus them clearly on the retina. But light rays from nearer objects are divergent rather than parallel. So obviously they must be bent more acutely to bring them to a focus on the retina. Accommodation of the lens or, in other words, an increase in its curvature takes place to achieve this greater refraction. (It is a physical fact that the greater the convexity of a lens, the greater its refractive power.) Most observers accept Helmholtz' theory about the mechanism that produces accommodation of the lens. According to his theory, the ciliary muscle contracts, pulling the ciliary body and choroid forward toward the lens. This releases the tension on the suspensory ligament and therefore on the lens, which, being elastic, immediately bulges. For near vision, then, the ciliary muscle is contracted and the lens is bulging, whereas for far vision the ciliary muscle is relaxed and the lens is comparatively flat. Continual use of the eyes for near work produces eyestrain because of the prolonged contraction of the ciliary muscle. Some of the strain can be avoided by looking into the distance at intervals while doing close work.

Constriction of pupil. The muscles of the iris play an important part in the formation of clear retinal images. Part of the accommodation mechanism consists of contraction of the circular fibers of the iris, which constricts the pupil. This prevents divergent rays from the object from entering the eye through the periphery of the cornea and lens. Such peripheral rays could not be refracted sufficiently to be brought to a focus on the retina and therefore would cause a blurred image. Constriction of the pupil for near vision is called the *near reflex* of the pupil and occurs simultaneously with accommodation of the lens in near vision. The pupil constricts also in bright light (*photopupil reflex* or *pupillary light reflex*) to protect the retina from too intense or too sudden stimulation.

Convergence of eyes. Single binocular vision (seeing only one object instead of two when both eyes are used) occurs when light rays from an object fall on corresponding points of the two retinas. The foveas and all points lying equidistant and in the same direction from the foveas are corresponding points. Whenever the eyeballs move in unison, either with the visual axes parallel (for far objects) or converging on a common point (for near objects), light rays strike corresponding points of the two retinas. Convergence is the movement of the two eyeballs inward so that their visual axes come together or converge at the object viewed. The nearer the object, the greater the degree of convergence necessary to maintain single vision. A simple procedure serves to demonstrate the fact that single binocular vision results from stimulation of corresponding points of the two retinas. Gently press one eyeball out of line while viewing an object. Instead of one object, you will then see two. In order to achieve unified movement of the two eyeballs, a functional balance between the antagonistic extrinsic muscles must exist. For clear distant vision, the muscles must hold the visual axes of the two eyes parallel. For clear near vision, they must converge them. These conditions cannot be met if, for example, the internal rectus muscle of one eye should contract more forcefully than its antagonist, the external rectus muscle. That eye would then be pulled in toward the nose. The movement of its visual axis would not coordinate with that of the other eye. Light rays from an object would then fall on noncorresponding points of the two retinas, and the object would be seen double (diplopia). Sometimes the individual can overcome the deviation of the visual axes by neuromuscular effort (extra conduction to and contraction of the weak muscle) and thereby achieve single vision, but only at the expense of muscular and nervous strain. The condition in which the imbalance of the eye muscles can be overcome by extra effort on the part of the weak muscle is called *heterophoria* (*esophoria* if the internal rectus muscle is stronger and pulls the eye nasalward and *exophoria* if the external rectus muscle is stronger and pulls the eye temporalward). *Strabismus* (cross-eye or squint) is an exaggerated esophoria that cannot be overcome by neuromuscular effort. An individual with strabismus usually does not have double vision, as you might expect, because he learns to suppress one of the images.

Stimulation of retina

Rods are known to contain *rhodopsin* (visual purple), a pigmented compound. As shown in Fig. 11-21, it forms by a protein, opsin, combining with retinal, a derivative of vitamin A. Rhodopsin is highly light sensitive, so that when light rays strike a rod, its rhodopsin rapidly breaks down (also shown in Fig. 11-21). In some way, this chemical change initiates impulse conduction by the rod. Then, if the rod is exposed to darkness for a short time, rhodopsin re-forms from the opsin and retinal and is ready to func-

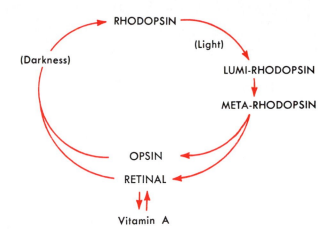

Fig. 11-21 Rhodopsin cycle.

tion again. *Cones* also contain photosensitive chemicals—iodopsin and possibly others. Presumably the cone compounds are less sensitive to light than rhodopsin. Brighter light seems necessary for their breakdown. Cones, therefore, are considered to be the receptors responsible for daylight and color visions. Rods, on the other hand, are believed to be the receptors for night vision because their rhodopsin quickly becomes almost depleted in bright light because of rapid breakdown but slow regeneration. This explains why you cannot see for a little while after you go from a bright light to darkness. But when rhodopsin has had time to re-form, the rods again start functioning and dark adaptation has occurred. Or, as we say, we "can see in the dark once we get used to it." Night blindness occurs in marked vitamin A deficiency. Why? Fig. 11-21 contains a clue. The fovea contains the greatest concentration of cones and is, therefore, the point of clearest vision in good light. For this reason, when we want to see an object clearly in the daytime, we look directly at it so as to focus the image on the fovea. But in dim light or darkness, we see an object better if we look slightly to the side of it, thereby focusing the image nearer the periphery of the retina, where rods are more plentiful.

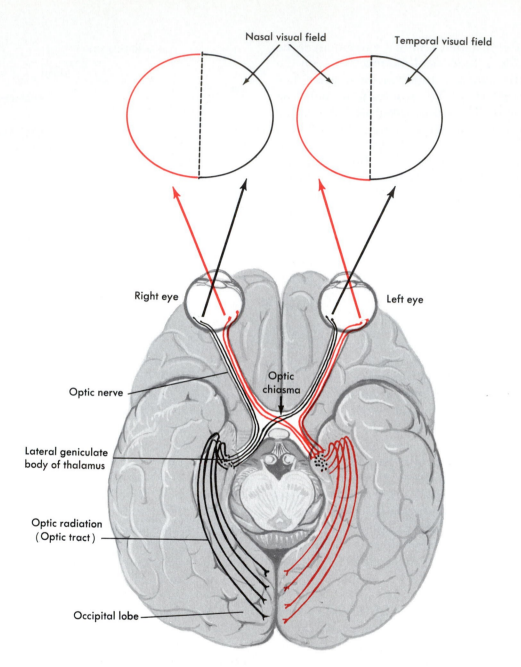

Fig. 11-22 Visual pathways. Note structures that compose each pathway: optic nerve, optic chiasma, lateral geniculate body of thalamus, optic radiations, and visual cortex of occipital lobe. Fibers from nasal portion of each retina cross over to opposite side at optic chiasma, hence terminate in lateral geniculate body of opposite side. Location of a lesion in the visual pathway determines the resulting visual defect. Examples: Destruction of an optic nerve produces permanent blindness in the same eye. Pressure on the optic chiasma—by a pituitary tumor, for instance—produces bitemporal hemianopia, or more simply, blindness in both temporal visual fields. Why? Because it destroys fibers from nasal sides of both retinas.

Conduction to visual area

Fibers that conduct impulses from the rods and cones reach the visual cortex in the occipital lobes via the optic nerves, optic chiasma, optic tracts, and optic radiations. Look closely at Fig. 11-22. Notice that each optic nerve contains fibers from only one retina but that the optic chiasma contains fibers from the nasal portions of both retinas. Each optic tract also contains fibers from both retinas. These anatomical facts explain certain peculiar visual abnormalities that sometimes occur. Suppose a man's right optic tract were injured so that it could not conduct impulses. He would be totally blind in neither eye but partially blind in both eyes. Specifically, he would be blind in his right nasal and left temporal visual fields. Here are the reasons. The right optic tract contains fibers from the right retina's temporal area, the area that sees the right nasal visual field. And, in addition, the right optic tract contains fibers from the left retina's nasal area, the area that sees the left temporal visual field.

Auditory apparatus
Anatomy

Ears, auditory nerves, and auditory areas of the temporal lobes of the cerebrum compose the auditory apparatus. Each ear has three parts: external, middle, and internal (Fig. 11-23).

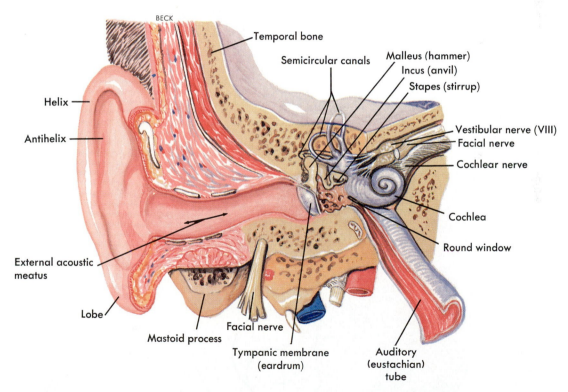

Fig. 11-23 Components of the ear. External ear consists of auricle (pinna), external acoustic meatus (ear canal), and tympanic membrane (eardrum). Middle ear (tympanic cavity) includes malleus (hammer), incus (anvil), and stapes (stirrup). Internal ear contains semicircular canals, vestibule, and cochlea.

External ear

The external ear has two divisions: the flap or modified trumpet on the side of the head called the *auricle* or *pinna* and the tube leading from the auricle into the temporal bone and named the *external acoustic meatus (ear canal)*. This canal is about 1¼ inches long and takes, in general, an inward, forward, and downward direction, although the first portion of the tube slants upward and then curves downward. Because of this curve in the auditory canal, the auricle should be pulled up and back to straighten the tube when medications are to be dropped into the ear. Modified sweat glands in the auditory canal secrete *cerumen* (waxlike substance), which occasionally becomes impacted and may cause pain and deafness. The *tympanic membrane* (eardrum) stretches across the inner end of the auditory canal, separating it from the middle ear.

Middle ear

The middle ear (tympanic cavity), a tiny epithelial-lined cavity hollowed out of the temporal bone, contains the three auditory ossicles: the malleus, incus, and stapes (Fig. 11-24). The names of these very small bones describe their shapes (hammer, anvil, and stirrup). The "handle" of the malleus is attached to the inner surface of the tympanic membrane, whereas the "head" attaches to the incus, which, in turn, attaches to the stapes. There are several openings into the middle ear cavity: one from the external auditory meatus, covered over with the tympanic membrane; two into the internal ear, the fenestra ovalis (oval window), into which the stapes fits, and the fenestra rotunda (round window), which is covered by a membrane; and one into the eustachian tube.

Posteriorly the middle ear cavity is continuous with a number of mastoid air spaces in the tem-

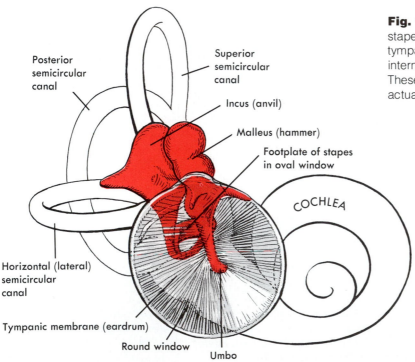

Posterior semicircular canal

Superior semicircular canal

Incus (anvil)

Malleus (hammer)

Footplate of stapes in oval window

COCHLEA

Horizontal (lateral) semicircular canal

Tympanic membrane (eardrum)

Round window

Umbo

Fig. 11-24 Ossicles—malleus, incus, and stapes—of the middle ear as seen through the tympanic membrane and in relation to the internal ear's semicircular canals and cochlea. These structures are shown 4½ times their actual size.

poral bone. The clinical importance of these middle ear openings is that they provide routes for infection to travel. Head colds, for example, especially in children, may lead to middle ear or mastoid infections via the nasopharynx–eustachian tube–middle ear–mastoid path.

The *eustachian* or *auditory tube* is composed partly of bone and partly of cartilage and fibrous tissue and is lined with mucosa. It extends downward, forward, and inward from the middle ear cavity to the nasopharynx (the part of the throat behind the nose).

Thus the eustachian tube provides the path by which throat infections may invade the middle ear. But the eustachian tube also serves a useful function. It makes possible equalization of pressure against inner and outer surfaces of the tympanic membrane and therefore prevents membrane rupture and the discomfort that marked pressure differences produce. The way the eustachian tube equalizes tympanic membrane pressures is this. When one swallows or yawns, air spreads rapidly through the open tube. Atmospheric pressure then presses against the inner surface of the tympanic membrane. And since atmospheric pressure is continually exerted against its outer surface, the pressures are equal. You might test this mechanism sometime when you are ascending or descending in an airplane—start chewing gum to increase your swallowing and observe whether this relieves the discomfort in your ears.

Internal ear

The internal ear is also called the labyrinth because of its complicated shape. It consists of two main parts, a bony labyrinth and, inside this, a membranous labyrinth. The bony labyrinth consists of three parts: vestibule, cochlea, and semicircular canals. The membranous labyrinth, as Fig. 11-25 shows, consists of the utricle and saccule inside the vestibule, the cochlear duct inside the cochlea, and the membranous semicircular canals inside the bony ones.

Table 11-5 Divisions of internal ear (labyrinth)

Bony labyrinth (part of temporal bone)	Membranous labyrinth (inside bony labyrinth)
1 Vestibule—central section of bony labyrinth; oval and round windows are openings of middle ear into vestibule; bony semicircular canals also open into vestibule	1 Utricle—one of the two parts of the membranous labyrinth contained in bony vestibule; utricle contains fluid called endolymph and an equilibrium sense organ, the macula; macula senses both head positions and movements and acceleration and deceleration of the body; vestibular nerve (branch of eighth cranial nerve) supplies macula 2 Saccule—the other part of the membranous labyrinth within bony vestibule; contains endolymph and macula
2 Cochlea—spiraling bony tube that resembles a snail shell in shape	3 Cochlear duct—membranous tube that forms shelf across interior of bony cochlea; contains endolymph and organ of Corti, the sense organ for hearing; cochlear nerve (branch of eighth cranial nerve) supplies organ of Corti; cochlear duct separated from bony cochlea by scala vestibuli and scala tympani, spaces that contain perilymph
3 Bony semicircular canals—three of these semicircular-shaped canals; each lies at approximately right angle to the others	4 Membranous semicircular canals—separated from bony semicircular canals by perilymph; contain endolymph and the crista, an equilibrium sense organ that senses head movements; vestibular nerve supplies crista

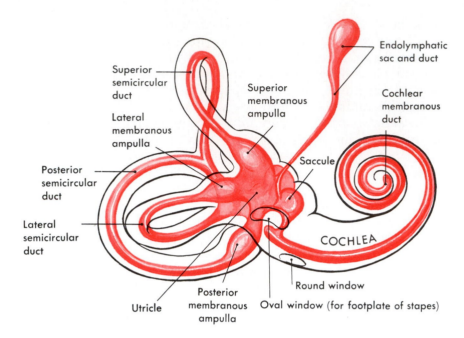

Fig. 11-25 Membranous labyrinth (red) of the internal ear shown in relation to bony labyrinth.

Superior semicircular duct

Lateral membranous ampulla

Posterior semicircular duct

Lateral semicircular duct

Superior membranous ampulla

Endolymphatic sac and duct

Cochlear membranous duct

Saccule

COCHLEA

Utricle

Posterior membranous ampulla

Round window

Oval window (for footplate of stapes)

Vestibule, utricle, and saccule. The vestibule constitutes the central section of the bony labyrinth. Into it open both the oval and round windows from the middle ear as well as the three semicircular canals of the internal ear. The utricle and saccule have membranous walls and are suspended within the vestibule. They are separated from the bony walls of the vestibule by fluid (perilymph), and both utricle and saccule contain a fluid called endolymph.

Located within the utricle (and also within the saccule) lies a small structure called the *macula*. It consists mainly of hair cells and a gelatinous membrane that contains *otoliths* (tiny ear "stones," that is, small particles of calcium carbonate). A few delicate hairs protrude from the hair cells and are embedded in the gelatinous membrane. Receptors for the vestibular branch of the eighth cranial nerve contact the hair cells of the macula located in the utricle. Changing the position of the head produces a change in the amount of pressure on the gelatinous membrane and causes the otoliths to pull on the hair cells. This stimulates the adjacent receptors of the

vestibular nerve. Its fibers conduct impulses to the brain that produce a sense of the position of the head and also a sensation of a change in the pull of gravity, for example, a sensation of acceleration. In addition, stimulation of the macula in the utricle evokes *righting reflexes*, muscular responses to restore the body and its parts to their normal position when they have been displaced. (Impulses from proprioceptors and from the eyes also activate righting reflexes. Interruption of the vestibular or visual or proprioceptive impulses that initiate these reflexes may cause disturbances of equilibrium, nausea, vomiting, and other symptoms.)

Cochlea and cochlear duct. The word cochlea, which means snail, describes the outer appearance of this part of the bony labyrinth. When sectioned, the cochlea resembles a tube wound spirally around a cone-shaped core of bone, the *modiolus*. The modiolus houses the spiral ganglion, which consists of cell bodies of the first sensory neurons in the auditory relay. Inside the cochlea lies the membranous *cochlear duct*—the only part of the internal ear concerned with

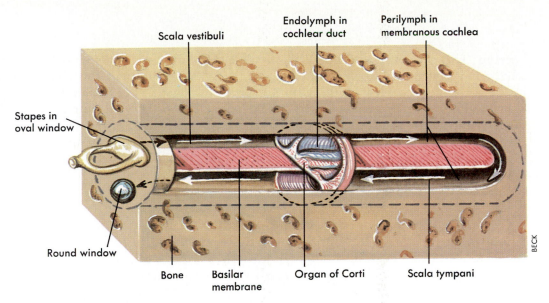

Scala vestibuli

Endolymph in cochlear duct

Perilymph in membranous cochlea

Stapes in oval window

Round window

Bone

Basilar membrane

Organ of Corti

Scala tympani

BECK

Fig. 11-26 Diagram of the bony and membranous cochlea, uncoiled. Note end-organ of Corti projecting into endolymph contained in cochlear duct (membranous cochlea). Perilymph indicated above endolymph occupies the scala vestibuli. That in lower compartment lies in the scala tympani (see also Fig. 11-27).

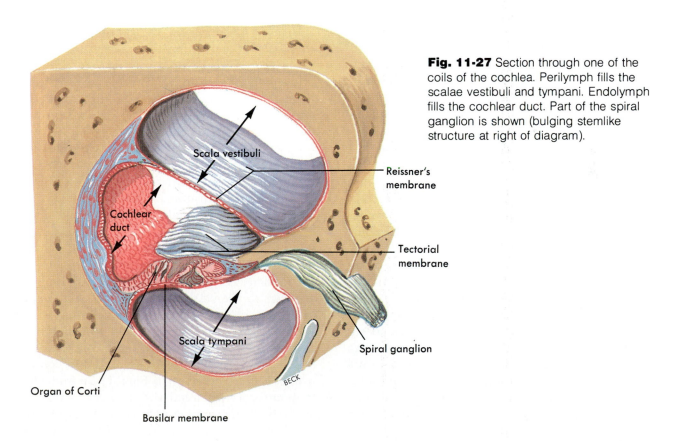

Scala vestibuli

Reissner's membrane

Cochlear duct

Tectorial membrane

Spiral ganglion

Scala tympani

Organ of Corti

Basilar membrane

BECK

Fig. 11-27 Section through one of the coils of the cochlea. Perilymph fills the scalae vestibuli and tympani. Endolymph fills the cochlear duct. Part of the spiral ganglion is shown (bulging stemlike structure at right of diagram).

hearing. This structure is shaped like a somewhat triangular tube. It forms a shelf across the inside of the bony cochlea, dividing it into upper and lower sections all along its winding course (Figs. 11-26 and 11-27). The upper section (above the cochlear duct, that is) is called the *scala vestibuli*, whereas the lower section below the cochlear duct is the *scala tympani*. The roof of the cochlear duct is known as *Reissner's membrane* or the vestibular membrane. *Basilar membrane* is the name given the floor of the cochlear duct. It is supported by bony and fibrous projections from the wall of the cochlea. Perilymph fills the scala vestibuli and scala tympani, and endolymph fills the cochlear duct.

The hearing sense organ, the *organ of Corti*, rests on the basilar membrane throughout the whole length of the cochlear duct. The structure of the organ of Corti resembles that of the equilibrium sense organ, that is, the macula in the utricle. It consists of supporting cells plus the important *hair cells* that project into the endolymph and are topped by an adherent gelatinous membrane called the *tectorial membrane*. Dendrites of the sensory neurons whose cells lie in the spiral ganglion in the modiolus have their beginnings around the bases of the hair cells of the organ of Corti. Axons of these neurons extend in the cochlear nerve (a branch of the eighth cranial nerve) to the brain. They conduct impulses that produce the sensation of hearing.

Semicircular canals. Three semicircular canals, each in a plane approximately at right angles to the others, are found in each temporal bone. Within the bony semicircular canals and separated from them by perilymph are the membranous semicircular canals. Each contains endolymph and connects with the utricle, one of the membranous sacs inside the bony vestibule. Near its junction with the utricle the canal enlarges into an *ampulla*. Some of the receptors for the vestibular branch of the eighth cranial nerve lie in each ampulla. Like all receptors for both vestibular and auditory branches of this nerve, these receptors, too, lie in contact with hair cells in a supporting structure. Here in the ampulla the hair cells and supporting structure together are named the *crista ampullaris*, whereas in the utricle and saccule they are called the macula and in the cochlear duct, the organ of Corti. Sitting atop the crista is a gelatinous structure called the *cupula*.

The crista with its vestibular nerve endings presumably functions as the end-organ for sensations of head movements, whereas the macula in the utricle serves as the end-organ for sensations of head positions. Hence both the crista and the macula of the utricle function as end-organs for the sense of equilibrium.

The function of the macula in the saccule is not known. Some evidence, however, suggests that it serves as a receptor for vibratory stimuli.

Physiology

Hearing

Hearing results from stimulation of the auditory area of the cerebral cortex (temporal lobe, Fig. 9-19, p. 254). Before reaching this area of the brain, however, sound waves must be projected through air, bone, and fluid to stimulate nerve endings and set up impulse conduction over nerve fibers.

Sound waves in the air enter the external auditory canal, probably without much aid from the pinna in collecting and reflecting them because of its smallness in man. At the inner end of the canal, they strike against the tympanic membrane, setting it in vibration. Vibrations of the tympanic membrane move the malleus, whose handle attaches to the membrane. The head of the malleus attaches to the incus, and the incus attaches to the stapes. So when the malleus vibrates, it moves the incus, which moves the stapes against the oval window into which it fits so precisely. At this point, fluid conduction of sound waves begins. To understand this, you will probably need to refer to Figs. 11-26 and 11-27 frequently as you read the next few sentences. When the stapes moves against the oval window, pressure is exerted inward into the perilymph in the scala vestibuli of the cochlea. This starts a "ripple" in the perilymph that is transmitted through Reissner's membrane (the roof of the cochlear duct) to endolymph inside the duct and then to the organ of Corti and to the basilar membrane that supports the organ of Corti and forms the floor of the cochlear duct. From the basilar membrane the ripple is next transmitted to and through the perilymph in the scala tympani and finally expends itself against the round window—like an ocean wave expending itself as it breaks against the shore, although on a much reduced scale. Dendrites (of neurons whose cell bodies lie in the spiral ganglion and whose axons make up the cochlear nerve) terminate around the bases of the hair cells of the organ of Corti, and the tectorial membrane adheres to their upper surfaces. The movement of the hair cells against the adherent tectorial membrane somehow stimulates these dendrites and initiates impulse conduction by the cochlear nerve to the brain stem. Before reaching the auditory area of the temporal lobe, impulses pass through "relay stations" in nuclei in the medulla, pons, midbrain, and thalamus.

Equilibrium

In addition to hearing, the internal ear aids in the maintenance of equilibrium. It does this by sensing head movements and positions and also by sensing acceleration or deceleration of the body (see pp. 312 and 314).

Olfactory sense organs

The receptors for the fibers of the olfactory (first) cranial nerves lie in the mucosa of the upper part of the nasal cavity. Their location here explains the necessity for sniffing or drawing air forcefully up into the nose in order to smell delicate odors. The olfactory sense organ consists of hair cells and is relatively simple compared with the complex visual and auditory organs. Where-

as the olfactory receptors are extremely sensitive, that is, stimulated by even very slight odors, they are also easily fatigued—a fact that explains why odors that are at first very noticeable are not sensed at all after a short time.

Gustatory sense organs

The receptors for the taste nerve fibers (in branches of the seventh and ninth cranial nerves) are known as *taste buds*. Most of them are located on the tongue and the roof of the mouth. Taste buds are exteroceptors of the type called chemoreceptors, for the obvious reason that chemicals stimulate them. Only four kinds of taste sensations—sweet, sour, bitter, and salty—result from stimulation of taste buds. All the other flavors we sense result from a combination of taste bud and olfactory receptor stimulation. In other words the myriads of tastes recognized are not tastes alone but tastes plus odors. For this reason a cold that interferes with the stimulation of the olfactory receptors by odors from foods in the mouth markedly dulls one's taste sensations.

Outline summary

Classifications

A Sherrington's classification
  1 Exteroreceptors—in skin, mucosa, eye, and ear; stimulated by changes in external environment
  2 Visceroreceptors—in walls of viscera and blood vessels; stimulated by changes in internal environment
  3 Proprioceptors—in muscles, tendons, joints, and internal ear

B Another classification
  1 Superficial receptors—in skin and mucosa; stimulation gives rise to sensations of temperature, pain, touch, and pressure
  2 Deep receptors—in muscles, tendons, and joints; stimulation gives rise to sensations of kinesthesia (position and movement), vibrations, deep pressure, and deep pain
  3 Visceral receptors—in viscera; stimulation gives rise to sensations such as hunger, nausea, and visceral pain
  4 Special receptors—in eyes, ears, mouth, and nose; stimulation gives rise to sensations of vision, hearing, taste, and smell

Pain receptors

A In skin and mucosa, stimulated by any kind of intense stimuli

B In viscera, stimulated by marked changes in pressure and by certain chemicals, for example, some produced in tissues with inadequate oxygen supply

Types of pain

A Superficial somatic—from stimulation of pain receptors in skin and mucosa

B Deep somatic—from stimulation of deep pain receptors

C Visceral—from stimulation of visceral pain receptors

D Referred—from stimulation of visceral receptors but interpreted as coming from surface regions

Eye

Anatomy

A Coats of eyeball
  1 Outer coat (sclera)
  2 Middle coat (choroid)
  3 Inner coat (retina)
  4 For outline summary, see Table 11-2

B Cavities and humors
  1 Anterior cavity
  2 Posterior cavity
  3 For outline summary, see Table 11-3

C Muscles
  1 Extrinsic
    a Attach to outside of eyeball and to bones of orbit
    b Voluntary muscles; move eyeball in desired directions
    c Four straight (rectus) muscles—superior, inferior, lateral, and mesial; two oblique muscles—superior and inferior
  2 Intrinsic
    a Within eyeball; named iris and ciliary muscles
    b Involuntary muscles
    c Iris regulates size of pupil
    d Ciliary muscle controls shape of lens, making possible accommodation for near and far objects
  3 For outline summary, see Table 11-4

D Accessory structures
  1 Eyebrows and eyelashes—protective and cosmetic
  2 Eyelids
    a Lined with mucous membrane that continues over surface of eyeball; called conjunctiva
    b Opening between eyelids called palpebral fissure
    c Corners where upper and lower eyelids join called canthus, mesial and lateral
  3 Lacrimal apparatus—lacrimal glands, lacrimal canals, lacrimal sacs, and nasolacrimal ducts

Physiology of vision

A Fulfillment of following conditions results in conscious experience known as vision: formation of retinal image, stimulation of retina, and conduction to visual area

B Formation of retinal image
  1 Accomplished by four processes:
    a Refraction or bending of light rays as they pass through eye
    b Accommodation or bulging of lens—normally occurs if object viewed lies nearer than 20 feet from eye
    c Constriction of pupil; occurs simultaneously with accommodation for near objects and also in bright light
    d Convergence of eyes for near objects so light rays from object fall on corresponding points of two retinas; necessary for single binocular vision

C Stimulation of retina
Accomplished by light rays producing photochemical change in rods and cones
D Conduction to visual area
Fibers that conduct impulses from rods and cones reach visual cortex in occipital lobes via optic nerves, optic chiasma, optic tracts, and optic radiations

Auditory apparatus
Anatomy

A External ear
  1 Auricle or pinna
  2 External acoustic meatus (ear canal)
B Middle ear
  1 Separated from external ear by tympanic membrane
  2 Contains auditory ossicles (malleus, incus, and stapes) and openings from external acoustic meatus, internal ear, eustachian tube, and mastoid sinuses
  3 Eustachian tube, collapsible tube, lined with mucosa, extending from nasopharynx to middle ear
    a Equalizes pressure on both sides of eardrum
    b Open when yawning or swallowing
C Internal ear
  1 Consists of bony and membranous portions, latter contained within former
  2 Bony labyrinth has three divisions—vestibule, cochlea, and semicircular canals
  3 Membranous cochlear duct contains receptors for cochlear branch of eighth cranial nerve (sense of hearing)
  4 Utricle and membranous semicircular canals contain receptors for vestibular branch of eighth cranial nerve (sense of equilibrium)

Physiology

A Hearing—results from stimulation of auditory area of temporal lobes by impulses over cochlear nerves, which are stimulated by sound waves being projected through air, bone, and fluid before reaching auditory receptors (organ of Corti in cochlear duct)
B Equilibrium—stimulation of receptors (crista in semicircular canal and macula in utricle) leads to sense of equilibrium; also initiates righting reflexes essential for balance

Olfactory sense organs

A Receptors for olfactory (first) cranial nerve located in nasal mucosa high along septum
B Receptors very sensitive but easily fatigued

Gustatory sense organs

A Receptors for taste nerve fibers (in branches of seventh and ninth cranial nerves) called taste buds located on tongue and roof of mouth
B Four kinds of taste—sweet, sour, salty, and bitter
C All other tastes result from fusion of two or more of these tastes or from olfactory stimulation

Review questions

  1 What two general functions do sense organs perform?
  2 Explain briefly the principle of specificity of receptors.
  3 Describe briefly one theory about the mechanism of referred pain.
  4 Explain briefly the mechanism for accommodation for near vision.
  5 Define briefly the term *refraction*. Name the refractory media of the eye.
  6 Concave glasses are prescribed for nearsighted vision. On what principle is this based?
  7 What is the name of the receptors for vision in dim light? For bright light?
  8 Distinguish between exteroceptors, proprioceptors, and visceroceptors.
  9 Describe the main features of middle ear structure.
10 Name the parts of the bony and membranous labyrinths and describe the relationship of membranous labyrinth parts to those of the bony labyrinth.
11 In what ear structure(s) is the hearing sense organ located? The equilibrium sense organs?
12 What is the name of the hearing sense organ? Of the equilibrium sense organs?

chapter 12

The endocrine system

Meaning

The endocrine system and the nervous system both function to achieve and maintain homeostasis or stability of our internal environment. Working alone or in concert as a single *neuroendocrine system*, they perform the same general functions for the body: communication, integration, and control. But they accomplish these general functions through different kinds of mechanisms and with somewhat different types of results. The mechanism of the nervous system consists of nerve impulses conducted by neurons from one specific structure to another. In contrast, the mechanism of the endocrine system consists of circulation of minute quantities of specific chemical messengers that, as you undoubtedly know, are called *hormones*. Hormones (from the Greek root *hormaein*, "to excite") are secreted from endocrine gland cells directly into the blood where they circulate to all parts of the body. Nerve impulses and hormones do not exert exactly the same kind of control, nor do they control precisely the same structures and functions. This is fortunate. It makes for better timing and more precision of control. For instance, nerve impulses produce rapid, short-lasting responses, whereas hormones produce slower and generally longer lasting responses. Nerve impulses control directly only two kinds of cells: muscle and gland cells. Although certain hormones such as thyroid and growth hormone have widespread physiologic activity, most hormones, despite their widespread distribution by the blood to every body tissue, are highly specific in their action. Cells that are acted on and

Table 12-1 Names and locations of endocrine glands

Name	Location
Pituitary gland (hypophysis cerebri) and hypothalamus	Cranial cavity
Anterior lobe (adenohypophysis)	
Intermediate lobe (pars intermedia)	
Posterior lobe (neurohypophysis)	
Pineal gland (epiphysis)	Cranial cavity
Thyroid gland	Neck
Parathyroid glands	Neck
Thymus	Mediastinum
Adrenal glands	Abdominal cavity (retroperitoneal)
Adrenal cortex	
Adrenal medulla	
Islands of Langerhans	Abdominal cavity (pancreas)
Gastric and intestinal mucosa	Abdominal cavity
Ovaries	Pelvic cavity
Graafian follicle	
Corpus luteum	
Testes (interstitial cells)	Scrotum
Placenta	Pregnant uterus

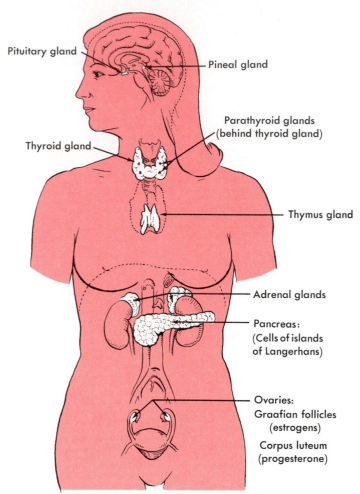

Pituitary gland

Pineal gland

Parathyroid glands
(behind thyroid gland)

Thyroid gland

Thymus gland

Adrenal glands

Pancreas:
(Cells of islands
of Langerhans)

Ovaries:
Graafian follicles
(estrogens)

Corpus luteum
(progesterone)

Fig. 12-1 Location of the endocrine glands in the female. Dotted line around thymus gland indicates maximum size at puberty.

respond to a particular hormone are referred to descriptively as *target organ cells*.

Endocrine glands secrete hormones directly into the blood. Because they have no duct system, they are often called "ductless glands." Exocrine or duct glands, in contrast, release their secretions into ducts. Although the list of endocrine glands and hormone-secreting cell types is growing, the names and locations of the main endocrine glands are given in Table 12-1 and Fig. 12-1.

Endocrine gland cells synthesize hormones by the process of anabolism. Hormones are either protein compounds, or compounds derived from proteins, or steroid compounds. All have relatively large molecules. For instance, human growth hormone consists of a chain of 188 amino acids and has a molecular weight of 20,500.* Compared with a carbon dioxide molecule, molecular weight 44, the growth hormone molecule is indeed a giant. (Carbon dioxide is one of several so-called "regulatory chemicals." These compounds resemble hormones in two respects —they are released into the blood from cells, and they exert profound influence on various structures and functions. They differ from hormones, however, in that they are neither proteins nor steroids but are much smaller, inorganic molecules or ions, they are present in the blood in much larger amounts than are hormones, and they are not products of anabolism. Instead, some regulatory chemicals, including carbon dioxide, are products of catabolism.)

Exaggerating the importance of endocrine glands is almost impossible. Hormones are the main regulators of metabolism, of growth and development, of reproduction, and of stress responses. They play roles of the utmost importance in maintaining homeostasis—fluid and electrolyte balance, acid-base balance, and energy balance, for example. Excesses or deficiencies of hormones make the difference between normalcy and all sorts of abnormalities such as dwarfism, gigantism, and sterility—and even the difference between life and death in some instances.

Prostaglandins (tissue hormones)

The prostaglandins (PGs) are a unique group of biologic compounds (20-carbon fatty acids that have a 5-carbon ring) that serve important

*Mountcastle, V. B.: Medical physiology, ed. 13, St. Louis, 1974, The C. V. Mosby Co., p. 1614.

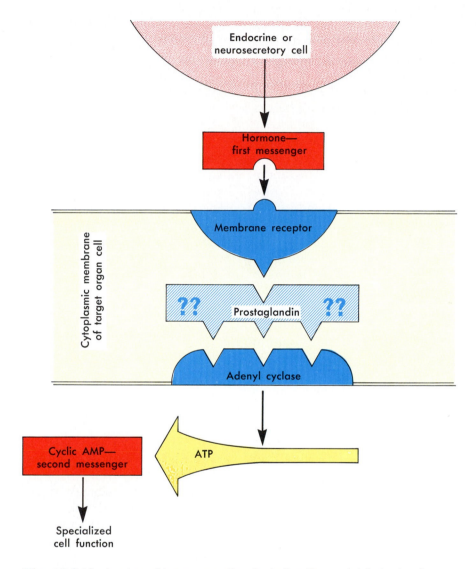

Fig. 12-2 Mechanism of hormone action, including the postulated role of prostaglandins. Hormone acts as "first messenger," delivering its message to a membrane receptor in the target organ cell. Prostaglandins appear to regulate hormone activity by influencing adenyl cyclase and cyclic AMP activity within the cell. Cyclic AMP serves as the "second messenger."

and widespread integrative functions in the body but do not meet the definition of a typical hormone. Although they may be secreted directly into the bloodstream like hormones in general, they are rapidly metabolized and inactivated, so that circulating levels are extremely low. The term "tissue hormone" is appropriate because in many instances the substance is produced in a tissue and diffuses only a short distance to act on cells within that tissue. Whereas typical hormones integrate activities of widely separated organs, typical prostaglandins integrate activities of neighboring cells.

Three classes of prostaglandins, prostaglandin A (PGA), prostaglandin E (PGE), and prostaglandin F (PGF), have been isolated and identified from a wide variety of tissues, including the seminal vesicles, kidneys, lungs, iris, brain, and thymus. As a group, the prostaglandins have diverse physiologic effects and are among the most varied and potent of any naturally occurring biologic compound. They are intimately involved in overall endocrine regulation by influencing adenyl cyclase and cyclic AMP activity within the cell (Fig. 12-2). Specific biologic effects depend on the class of prostaglandin.

Intra-arterial infusion of prostaglandins in group A (PGA) results in an immediate fall in blood pressure accompanied by an increase in regional blood flow to several areas, including the coronary and renal systems. Just how PGA infusion reduces blood pressure remains unclear. The mechanism probably involves relaxation of smooth muscle fibers in the walls of muscular arteries and arterioles.

PGE compounds play an important role in a number of systemic vascular, metabolic, and gastrointestinal functions. Vascular effects of PGE_1 and PGE_2 include regulation of both red blood cell "deformability" and platelet aggregation tendencies (see Chapter 13, p. 359). PGE_1 also plays a role in certain systemic manifestations of inflammation such as fever. Evidence now suggests that aspirin and many other common anti-inflammatory agents function, at least

in part, by inhibition of PGE synthesis. PGE_1 is also important in the regulation of hydrochloric acid secretion in the stomach and the prevention of gastric ulcer.

Prostaglandins of the F class play an especially important role in the reproductive system. $PGF_{2\alpha}$ causes uterine muscle to contract and can be used (often with PGE_2) to induce labor and accelerate delivery. PGF compounds also affect intestinal motility and are required for normal peristalsis.

Information and understanding about basic biological and cellular processes gained as a result of prostaglandin research may soon lead to powerful new tools in medicine. The potential therapeutic use of compounds found in almost every body tissue and capable of regulating hormone activity on the cellular level has been described as the most revolutionary development in medicine since the advent of antibiotics. They may eventually play an important role in the treatment of such diverse diseases as hypertension, coronary thrombosis, asthma, and ulcers.

How hormones act

What a hormone does to its target cells to cause them to respond in particular ways has been the subject of much conjecture and research. Dr. Earl W. Sutherland isolated a chemical, adenosine-3′,5′-monophosphate (cyclic AMP), and suggested the part it might play in hormone action. For his brilliant work, Dr. Sutherland received the 1971 Nobel Prize in medicine and physiology. He conceived what is known as the "second messenger hypothesis" of hormone action. According to this concept a hormone acts as a "first messenger," that is, it delivers its chemical message from the cells of an endocrine gland to highly specific membrane receptor sites on cells of a target organ. On reaching a target cell the hormone reacts with an enzyme, adenyl cyclase, present in the cell's plasma membrane, causing it to act on ATP molecules present inside the cell. Adenyl cyclase catalyzes the conversion of ATP to cyclic AMP. Cyclic AMP serves as the "second messenger," delivering information inside the cell that causes the cell to respond by performing its specialized function. For example, cyclic AMP causes thyroid cells to respond by secreting thyroid hormones. Current research involving a number of tropic hormones suggests that other intermediate messengers, especially the prostaglandins, increase the activation of adenyl cyclase and therefore serve to regulate the levels of cyclic AMP and ultimately target cell function.

Summarizing, hormones serve as first messengers, providing communication between endocrine and target gland cells. Cyclic AMP then acts as the second messenger, providing communication within a hormone's target cells. Fig. 12-2 summarizes the mechanism of hormone action, including the postulated role of prostaglandins.

Pituitary gland (hypophysis cerebri)
Size, location, and component glands

The pituitary gland is truly a small but mighty structure. At its largest diameter, it measures only 1.2 to 1.5 cm (about ½ inch). By weight, it is even less impressive—only about 0.5 gm, $\frac{1}{60}$ ounce! And yet, so crucial are the functions of the anterior lobe of this gland that it is called the "master gland."

The pituitary body has a protected location. It lies in the sella turcica and is covered over by an extension of the dura mater known as the pituitary diaphragm. The deepest part of the sella turcica (saddle-shaped depression in the sphenoid bone) is called the pituitary fossa, since the pituitary body lies in it. A stemlike portion of the gland, the infundibulum or pituitary stalk, juts up through a tiny perforation in the pituitary diaphragm and attaches the gland to the undersurface of the brain. More specifically, the

stalk attaches the pituitary body to the hypothalamus.

Although the pituitary looks like just one gland, it actually consists of two separate glands—the adenohypophysis, or anterior pituitary gland, and the neurohypophysis, or posterior pituitary gland. The anterior portion of the gland is itself partially divided by a narrow cleft and a connective tissue remnant into the larger *pars anterior* and small *pars intermedia* (Fig. 12-7). The anterior and posterior divisions of the pituitary gland develop from different embryonic structures, have different microscopic structure, and secrete different hormones. The anterior pituitary or adenohypophysis develops as an upward projection from the embryo's pharynx, whereas the posterior pituitary or neurohypophysis develops as a downward projection from the brain. Microscopic differences between the two glands are suggested by their names—*adeno* means gland and *neuro* means nervous. The adenohypophysis has the microscopic structure of an endocrine gland, whereas the neurohypophysis has the structure of nervous tissue. Hormones secreted by the adenohypophysis serve very different functions from those released from the neurohypophysis.

Anterior pituitary gland (adenohypophysis)

The anterior pituitary gland (pars anterior and pars intermedia) is composed of irregular clumps or nests of secretory cells supported by fine connective tissue fibers and surrounded by a rich vascular network. Three types of cells can be identified according to their affinity (or lack of affinity) for certain types of stains or dyes. About half of all cells present show little affinity for any type of stain or dye and are appropriately called *chromophobes* (from the Greek *chroma*, color, and the Latin *phobia*, fear of). Of the remaining cells, about 40% are called *acidophils* (from the Greek *philia*, love of) because of their affinity for acid dyes and 10% are *basophils*,

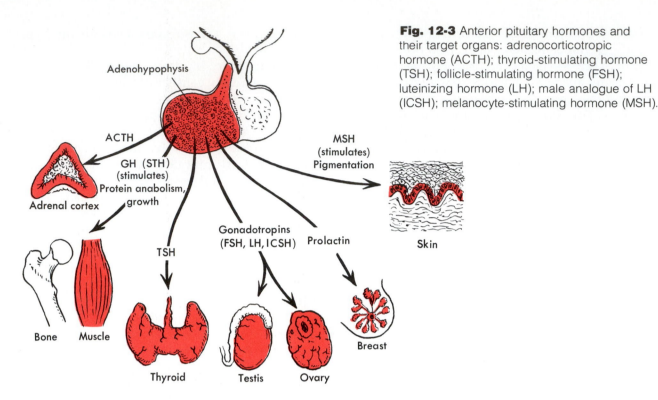

Fig. 12-3 Anterior pituitary hormones and their target organs: adrenocorticotropic hormone (ACTH); thyroid-stimulating hormone (TSH); follicle-stimulating hormone (FSH); luteinizing hormone (LH); male analogue of LH (ICSH); melanocyte-stimulating hormone (MSH).

which stain easily with basic dyes. Acidophils secrete growth hormone (GH; also called somatotropin [STH]) and prolactin (also called luteotropic hormone). Basophils secrete the other five hormones of the anterior lobe: thyrotropin (TH; also called thyroid-stimulating hormone [TSH]), adrenocorticotropin (ACTH), two gonadotropins (follicle-stimulating hormone [FSH] and luteinizing hormone [LH, or ICSH in the male]), and melanocyte-stimulating hormone (MSH). MSH is secreted by basophils in the pars intermedia of the anterior pituitary gland. Fig. 12-3 summarizes the anterior pituitary hormones and their target organs.

Growth hormone

Growth hormone or somatotropin (from the Greek *soma*, body, and *trope*, turning) is thought to promote bodily growth indirectly by accelerating amino acid transport into cells—evidence: blood amino acid content decreases within hours after administration of growth hormone to a fasting animal. With the faster entrance of amino acid into cells, anabolism of amino acids to form tissue protein also accelerates. This in turn tends to promote cellular growth. Growth hormone stimulates growth of both bone and soft tissues. If the anterior pituitary gland secretes an excess of growth hormone during the growth years, that is, before closure of the epiphyseal cartilages, bones grow more rapidly than normal and *gigantism* results (Fig. 12-4). Undersecretion of the growth hormone produces *dwarfism* when it occurs during the years of skeletal growth. If oversecretion of growth hormone occurs after the individual is fully grown, the condition known as *acromegaly* develops. Characteristic of this disease are enlargement of the bones of the hands, feet, jaws, and cheeks and an increase, too, in their overlying soft tissues (Fig. 12-5).

In addition to its stimulating effect on protein

Courtesy Dr. Edmund E. Beard, Cleveland, Ohio.

Fig. 12-4 A pituitary giant and dwarf contrasted with normal-sized men. Excessive secretion of growth hormone* by the anterior lobe of the pituitary gland during the early years of life produces giants of this type, whereas deficient secretion of this substance produces well-formed dwarfs.

*The structure of the human growth hormone molecule was identified by Dr. C. H. Li, Professor of Biochemistry at the University of California, in 1966. It consists of a chain of 188 amino acids with one loop of 93 subunits and another of 6 subunits. In January of 1971, Dr. Li announced that his laboratory had succeeded in synthesizing human growth hormone.

anabolism, growth hormone also influences fat metabolism and thereby affects carbohydrate (glucose) metabolism. Briefly, growth hormone tends to accelerate both the mobilization of fats from adipose tissues and their catabolism by other tissues. In other words, it tends to cause cells to shift from glucose catabolism to fat catabolism for their energy supply. Less glucose, therefore, leaves the blood to enter cells and more remains in the blood. Thus growth hormone tends to increase blood's concentration of glucose. In a word, it tends to have a hyperglycemic effect. Insulin produces opposite effects. It increases carbohydrate metabolism, thereby tending to decrease the blood concentration of

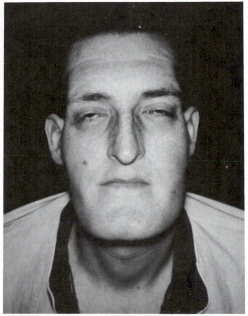

Fig. 12-5 Acromegaly. Note large head, exaggerated forward projection of jaw, and protrusion of frontal bone. (From Rimoin, D. L.: N. Engl. J. Med. **272:**923, 1965.)

glucose—to have a hypoglycemic effect. Where as adequate amounts of insulin prevent diabetes mellitus, long-continued excess amounts of growth hormone produce diabetes. In short, growth hormone and insulin function as antagonists. Or, as more commonly stated, growth hormone has an anti-insulin effect, a fact of considerable clinical importance. Summarizing very briefly, growth hormone helps regulate metabolism in the following ways:

1 It promotes protein anabolism (synthesis of tissue proteins), so is essential for normal growth and for tissue repair and healing.

2 It promotes fat mobilization and catabolism and decreases glucose catabolism. By decreasing glucose catabolism an excess of growth hormone may in time produce hyperglycemia and diabetes.

Prolactin (lactogenic or luteotropic hormone)

Acidophils in the anterior pituitary gland secrete two hormones: growth hormone and prolactin. Another name for prolactin—lactogenic hormone—suggests that it "generates," that is, initiates, milk secretion. It stimulates the mammary glands to start secreting soon after delivery of an infant. During pregnancy, it helps promote the breast development that makes possible milk secretion after delivery. In addition to its effect on milk secretion after childbirth, prolactin plays a supportive role (with luteinizing hormone) in maintaining the corpus luteum late in the postovulatory or premenstrual phase of each menstrual cycle. Because of this supportive role, prolactin is also called luteotropic hormone (LTH).

Tropic hormones

Basophil cells in the anterior pituitary gland secrete four tropic hormones—hormones that have a stimulating effect on other endocrine glands. Names of the tropic hormones are thyrotropin or thyroid-stimulating hormone, adrenocorticotropic hormone, and the two primary gonadotropins—follicle-stimulating hormone and luteinizing hormone.

All tropic hormones perform the same general function. Each one stimulates one other endocrine gland—stimulates it both to grow and to secrete its hormone at a faster rate. It seems to affect only this one structure as if, like a bullet, it had been aimed at a target, hence the name "target gland."

Individual tropic hormones perform the following functions:

1 *Thyrotropin* (thyroid-stimulating hormone, TSH) promotes and maintains growth and development of its target gland, the thyroid, and stimulates it to secrete thyroid hormone.

2 *Adrenocorticotropin* (ACTH) promotes and maintains normal growth and development of the adrenal cortex and stimulates it to secrete cortisol and other glucocorticoids.

3 *Follicle-stimulating hormone* (FSH) stimulates primary graafian follicles to start growing and to continue developing to maturity, that is, to the point of ovulation. FSH also stimulates follicle cells to secrete estrogens, one type of female sex hormones. In the male, FSH stimulates development of the seminiferous tubules and maintains spermatogenesis by them.

4 *Luteinizing hormone* (LH in the female, ICSH in the male). The name of this hormone suggests one of its functions, that of stimulating formation of the corpus luteum. Prior to this, LH acts with FSH to bring about complete maturation of the follicle; LH then produces ovulation and stimulates formation of the corpus luteum in the ruptured follicle. Finally, LH in human females stimulates the corpus luteum to secrete progesterone and estrogens. The male pituitary gland also secretes LH, but it is called *interstitial cell–stimulating hormone* (ICSH) because it stimulates interstitial cells in the testes to develop and secrete testosterone.

■ ■ ■

During childhood the anterior pituitary gland secretes insignificant amounts of gonadotropins.

Then it steps up their production gradually a few years before puberty. Just before puberty begins, presumably the anterior pituitary's secretion of these hormones takes a sudden and marked spurt. As a result, the blood concentration of gonadotropins increases markedly. This first high blood concentration of gonadotropins serves as the stimulus that brings on the first menstrual period and the many other changes that signal the beginning of puberty.

Melanocyte-stimulating hormone

Basophils in the pars intermedia of the anterior pituitary gland secrete the hormone generally believed to play the most important role in maintaining human pigmentation—melanocyte-stimulating hormone (MSH) or *intermedin*. In some species, including man, administration of MSH will cause a rapid increase in the synthesis and dispersion of melanin (pigment) granules in skin cells called melanocytes. The resulting hyperpigmentation resembles that which may occur when blood levels of other "darkening hormones" such as estrogen or progesterone become elevated. High estrogen and progesterone levels in pregnancy often cause darkening of the face ("mask of pregnancy") as well as the areolae, nipples, and, to a lesser extent, the genitalia. Another anterior pituitary hormone (namely, ACTH) will produce increased pigmentation of the skin. Structurally, also, the two molecules resemble each other. Several of the same amino acids occur in the same sequence in both MSH and ACTH.

Control of secretion

Neurons in certain parts of the hypothalamus synthesize chemicals that their axons secrete into blood in a complex of small veins, the *pituitary portal system*. Via this system the neurosecretions travel from the hypothalamus to the anterior lobe of the pituitary gland. There they stimulate the gland to release various hormones —hence *"releasing hormones"* is the other name for neurosecretions of the hypothalamus. Here are their abbreviated names, full names, and functions.

1 GRH, or growth hormone–releasing hormone—stimulates anterior pituitary gland to release, that is, to secrete, growth hormone.*

2 CRF, or corticotropin-releasing factor—stimulates anterior pituitary secretion of ACTH.

3 TRH, or thyrotropin-releasing hormone—stimulates anterior pituitary secretion of TSH (thyroid-stimulating hormone).

4 FSH-RH, or follicle-stimulating hormone–releasing hormone—stimulates anterior pituitary secretion of FSH.

5 LH-RH, or luteinizing hormone–releasing hormone—stimulates anterior pituitary secretion of LH.

6 PIF, or prolactin-inhibitory factor—hypothalamic substance that *inhibits* anterior pituitary secretion of prolactin except during pregnancy and after delivery. However, even if a woman is not pregnant or nursing, low levels of prolactin are secreted to help maintain the corpus luteum during the late stages of each menstrual cycle (luteotropic effects of prolactin). Evidence suggests that prolactin-releasing factor (PRF), also in hypothalamus extracts, acts with PIF in a dual control system to regulate prolactin blood levels.

Negative feedback mechanisms control the pituitary gland's secretion of tropic hormones. Such mechanisms operate on the following principles. A high blood level of a tropic hormone stimulates its target gland to increase its hormone secretion. This results in a high blood level of the target gland's hormone, which feeds back by way of the circulating blood to directly and indirectly inhibit secretion of the tropic hormone by the anterior pituitary gland. Fig. 12-6 illustrates the negative feedback mechanism that directly inhibits pituitary secretion of ACTH. Indirect inhibition of ACTH occurs as fol-

---

*Current evidence indicates that the hypothalamus also secretes somatostatin, which, as the name suggests, inhibits somatotropin secretion.

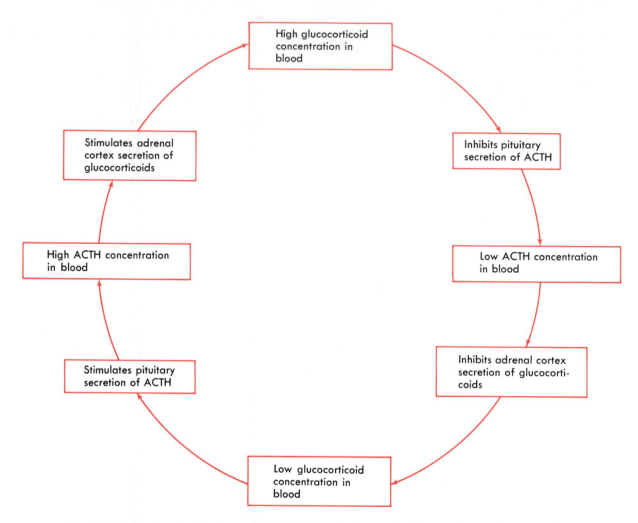

Fig. 12-6 Negative feedback control of ACTH and glucocorticoid secretion—a homeostatic mechanism that tends to keep the blood level of glucocorticoids within a narrow range.

lows. A high blood glucocorticoid concentration inhibits the hypothalamus from secreting corticotropin-releasing factor. Result? A low CRF concentration in pituitary portal vein blood inhibits pituitary secretion of ACTH. Similar mechanisms regulate the anterior lobe's secretion of thyrotropin and the thyroid gland's secretion of thyroid hormone. Also, a similar mechanism controls the anterior lobe's secretion of FSH and the ovary's secretion of estrogens.

Clinical facts furnish interesting evidence about feedback control of hormone secretion by the anterior pituitary gland and its target glands. For instance, if a person has his pituitary gland removed (hypophysectomy) surgically or by radiation, he must be given hormone replacement therapy for the rest of his life. If not, he will develop thyroid, adrenocortical, and gonadotropic deficiencies—deficiencies, that is, of the anterior pituitary's target gland hormones. Another well-known clinical fact is that estrogen deficiency develops in women between 40 and 50 years of age. By then the ovaries seem to have tired of producing hormones and ovulating each month. Or, at any rate, they no longer respond to FSH stimulation, so estrogen deficiency develops, brings about the menopause, and persists after the menopause. What, therefore, would you deduce is true of the blood concentration of FSH after the menopause? Apply the principle implied in the preceding paragraph that a low concentration of a target gland hormone stimulates tropic hormone secretion by the anterior pituitary gland.

Before leaving the subject of control of pituitary secretion, we want to call attention to another concept about the hypothalamus. It most likely functions as an important part of the body's complex machinery for coping with stress situations. For example, in severe pain or intense emotions, the cerebral cortex—especially its limbic lobe—is thought to send impulses to the hypothalamus. They stimulate it to secrete its releasing hormones into the pituitary portal veins. Circulating quickly to the anterior pituitary gland, they stimulate it to secrete more of its hormones. These, in turn, stimulate increased activity by the pituitary's target structures. In essence, what the hypothalamus does through its releasing hormones is to translate nerve impulses into hormone secretion by endocrine glands. Thus the hypothalamus links the nervous system to the endocrine system. It integrates the activities of these two great integrating systems—particularly, it seems, in times of stress. When healthy survival is threatened, the hypothalamus, via its releasing factors, can take over the command of the anterior pituitary gland. By so doing, it indirectly controls all of the pituitary's target glands—the thyroid, the adrenal cortex, and the gonads. Finally, by means of the hormones these glands secrete, the hypothalamus can dictate the functioning of literally every cell in our bodies.

These facts have tremendous implications. They mean that the cerebral cortex can do more than just receive impulses from all parts of the body and send out impulses to muscles and glands. They mean that the cerebral cortex— and therefore our thoughts and emotions—can, by way of the hypothalamus, influence the functioning of all our billions of cells. In short, the brain has two-way contact with every tissue of the body. Thus the state of the body can and does influence mental processes and, conversely, mental processes can and do influence the functioning of the body. In short, both somatopsychic and psychosomatic relationships exist between the body and the brain.

Posterior pituitary gland (neurohypophysis)

The posterior lobe of the pituitary gland serves as a storage area for the release of two hormones —one known as antidiuretic hormone (ADH) and the other called oxytocin. But strangely enough, cells of the posterior lobe, called *pituicytes*, do not themselves make these hormones. Instead, neurons in two areas of the hypothalamus called the supraoptic and paraventricular nuclei synthesize them (Fig. 12-7). From the cell bodies of these neurons in the hypothalamus the hormones pass down along axons (in the hypothalamohypophyseal tract) into the posterior lobe of the pituitary gland. Instead of the chemical "releasing factors" that triggered secretion of hormones from the anterior pituitary gland, release of both ADH and oxytocin into the blood is controlled by nervous stimulation.

Antidiuretic hormone

Antidiuresis means literally "against the production of a large urine volume." And this is exactly what ADH does. It prevents the formation of a large volume of urine. The one function ADH performs for the body is to decrease water loss from it. In other words, ADH is a water-retaining hormone. It performs this function by acting on cells of the distal and collecting tubules of the kidney to make them more permeable to water.

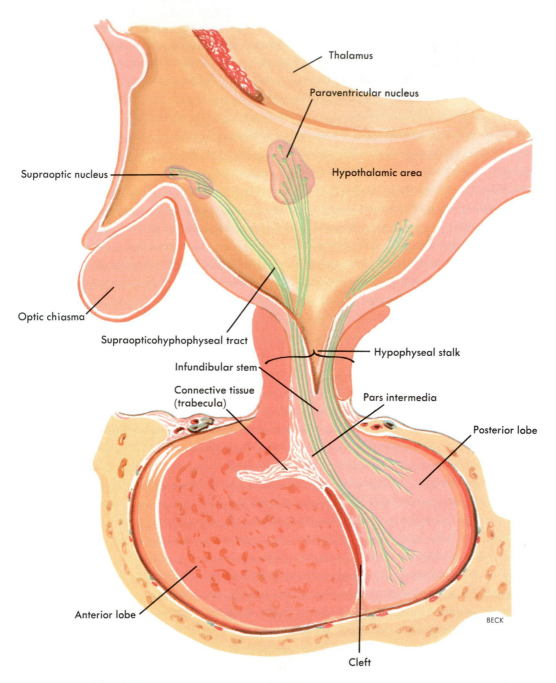

Fig. 12-7 Nerve tracts from hypothalamus to posterior lobe of pituitary gland.

This causes faster reabsorption of water from tubular urine into blood. And this, in turn, automatically produces antidiuresis (smaller urine volume). In cases of dehydration or water deprivation the osmolarity of plasma will increase as levels of solutes in the blood become more and more concentrated. As plasma osmolarity increases, specialized *osmoreceptors* near the supraoptic nucleus respond by stimulating ADH release from the posterior pituitary gland. The resulting increase in water reabsorption will then expand the plasma volume and reduce its osmolarity by diluting the blood solutes. Water is conserved, and urine volume decreases.

Marked diuresis, that is, abnormally large urine volume, occurs if ADH secretion is inadequate, such as occurs in the disease *diabetes insipidus*. A preparation of ADH used to treat this condition is called vasopressin (Pitressin). (The term "vasopressin" is sometimes used as another name for ADH. It can be misleading. Although large [pharmacological] doses of ADH will result in vasoconstriction of blood vessels, normal [physiological] levels produce little, if any, vasoconstrictor effects in humans.)

Oxytocin

Oxytocin has two actions: it stimulates powerful contractions by the pregnant uterus and it causes milk ejection from the lactating breast. Under the influence of this hormone from the posterior lobe of the pituitary gland, alveoli (cells that synthesize milk) release the milk into the ducts of the breast. This is a highly important function of oxytocin because milk cannot be removed by suckling unless it has first been ejected into ducts. However, it was oxytocin's other action—its stimulating effect on contractions of the pregnant uterus—that inspired its name. The term means "swift childbirth" (from the Greek *oxys*, swift, and *tokos*, childbirth). Whether or not oxytocin takes part in initiating labor is still an unsettled question. Commercial preparations of oxytocin are given to stimulate uterine contractions after delivery

of an infant in order to lessen the danger of uterine hemorrhage.

Control of secretion

Details of the mechanism that controls the secretion of ADH by the posterior lobe of the pituitary gland have not been established. However, review of the effects of dehydration and water deprivation on ADH levels illustrate the dominant roles of two important control factors, namely, the osmotic pressure of the extracellular fluid and its total volume. Without going into a discussion of evidence or details, the general principles of control of ADH secretion are as follows.

1 An increase in the osmotic pressure of extracellular fluid stimulates ADH secretion. This leads to decreased urine output and tends, therefore, to increase the volume of extracellular fluid, which in turn tends to decrease its osmotic pressure back toward normal. Opposite effects result from a decrease in the osmotic pressure of extracellular fluid.

2 A decrease in the total volume of extracellular fluid acts in some way to stimulate ADH secretion and thereby to decrease urine output and increase extracellular volume back toward normal.

3 Stress, whether induced by physical or emotional factors, brings about an increased secretion of ADH. Presumably, it acts in some way on the hypothalamus to stimulate ADH secretion.

■ ■ ■

About all that is known about the mechanism that controls the release of oxytocin from the posterior lobe of the pituitary gland is that stimulation of the nipples by the infant's nursing initiates sensory impulses that eventually reach the paraventricular nucleus of the hypothalamus, stimulating it to synthesize more oxytocin. The posterior lobe of the pituitary gland, in turn, increases its release of oxytocin. In addition to causing ejection of milk into the ducts of the breast, oxytocin also stimulates increased pro-

Table 12-2 Production, control, and effects of pituitary hormones

Pituitary gland (hypophysis cerebri)	Hormone	Source (cell type or location)	Control mechanism	Effect
Anterior pituitary gland (adenohypophysis)	Growth hormone (GH, somatotropin [STH])	Acidophils	GRH (growth hormone–releasing hormone) from hypothalamus	Promotes body growth, protein anabolism, and mobilization and catabolism of fats; decreases glucose catabolism; increases blood glucose levels
Pars anterior	Prolactin (lactogenic or luteotropic hormone [LTH])	Acidophils	Prolactin inhibitory factor (PIF); prolactin-releasing factor (PRF) from hypothalamus and high blood levels of oxytocin	Stimulates milk secretion and development of secretory alveoli; helps maintain corpus luteum
	Thyrotropin (TH); thyroid-stimulating hormone (TSH)	Basophils	Thyrotropin-releasing hormone (TRH) from hypothalamus	Growth and maintenance of thyroid gland and stimulation of thyroid hormone secretion
	Adrenocorticotropin (ACTH)	Basophils	Corticotropin-releasing factor (CRF) from hypothalamus	Growth and maintenance of adrenal cortex and stimulation of cortisol and other glucocorticoid secretions
	Follicle-stimulating hormone (FSH)	Basophils	Follicle-stimulating hormone–releasing hormone (FSH-RH) from hypothalamus	In female—stimulates follicle growth and maturation and estrogen secretion
				In male—stimulates development of seminiferous tubules and maintains spermatogenesis
	Luteinizing hormone (LH in female, ICSH in male)	Basophils	Luteinizing hormone–releasing hormone (LH-RH) from hypothalamus	In female (LH)—induces ovulation and stimulates formation of corpus luteum and progesterone secretion
				In male (ICSH)—stimulates interstitial cell secretion of testosterone
Pars intermedia	Melanocyte-stimulating hormone (MSH); intermedin	Basophils	Unknown in humans	May cause darkening of skin by increasing melanin production
Posterior pituitary gland (neurohypophysis)	ADH (vasopressin)	Hypothalamus, mainly supraoptic nucleus	Osmoreceptors in hypothalamus stimulated by increase in blood osmotic pressure, decrease in extracellular fluid volume, and stress	Decreased urine output
	Oxytocin	Hypothalamus, paraventricular nucleus	Nervous stimulation of hypothalamus caused by stimulation of nipples (nursing)	Contraction of uterine smooth muscle and ejection of milk into lactiferous ducts

duction of prolactin. Lactation is therefore controlled by two hormones: prolactin from the anterior pituitary gland, which stimulates milk production by the secretory alveoli, and oxytocin from the posterior pituitary gland, which results in ejection of milk from alveoli into the ducts. Table 12-2 summarizes the production, control, and effects of the pituitary hormones.

Thyroid gland

Location and structure

Two fairly large lateral lobes and a connecting portion, the isthmus, constitute the thyroid gland. Weight of the gland in the adult is somewhat variable but is usually about 30 gm. It is located in the neck just below the larynx. The isthmus lies across the anterior surface of the upper part of the trachea. In many individuals a thin pyramidal-shaped extension of thyroid tissue can be found arising from the upper surface of the connecting isthmus (Fig. 12-8). Thyroid tissue is made up of numerous structural units called *follicles*. Each follicle is a small closed sac consisting of an outer layer of simple cuboidal epithelium surrounding a central cavity. Colloid, composed largely of an iodine-containing protein known as thyroglobulin, fills these tiny sacs. Thyroid cells have a unique ability to actively take up and concentrate blood iodide (I^-). Once in the follicular cells I^- is changed to iodine (I) and used in the synthesis of the iodine-containing thyroid hormone.

Thyroid hormone and calcitonin

The thyroid gland secretes thyroid hormone and calcitonin. Actually, thyroid hormone consists of two hormones. Thyroxine is the name of the main one, and triiodothyronine is the name

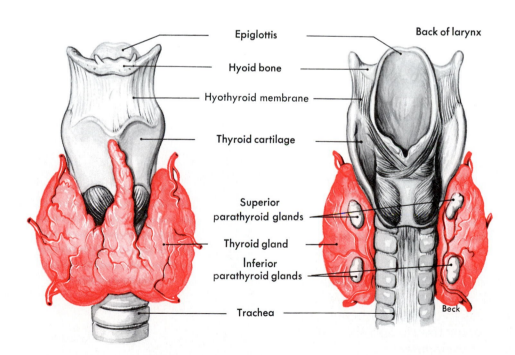

Epiglottis

Back of larynx

Hyoid bone

Hyothyroid membrane

Thyroid cartilage

Superior parathyroid glands

Thyroid gland

Inferior parathyroid glands

Trachea

Beck

Fig. 12-8 Thyroid and parathyroid glands. Note their relations to each other and to the larynx (voice box) and trachea.

of the less abundant one. One molecule of thyroxine contains four atoms of iodine, and one molecule of triiodothyronine, as its name suggests, contains three iodine atoms. After synthesizing its hormones the thyroid gland stores considerable amounts of them before secreting them. (Other endocrine glands do not store up their hormones.) As a preliminary to storage, thyroxine and triiodothyronine combine with a globulin in the thyroid cell to form a compound called thyroglobulin. Then thyroglobulin is stored in the colloid material in the follicles of the gland. Later the two hormones are released from thyroglobulin and secreted into the blood as thyroxine and triiodothyronine. Almost immediately, however, they combine with a blood protein. They travel in the bloodstream in this protein-bound form. But in the tissue capillaries, they are released from the protein and enter tissue cells as thyroxine and triiodothyronine.

The iodine in the protein-bound thyroxine and triiodothyronine is called protein-bound iodine (PBI). The amount of PBI can be measured by laboratory procedure and, in fact, is widely used as a test of thyroid functioning. (With a normal thyroid gland, the PBI is about 4 to 8 μg per 100 ml of plasma.)

The main physiological actions of thyroid hormone are to help regulate the metabolic rate and the processes of growth and tissue differentiation. Thyroid hormone increases the metabolic rate. How do we know this? Because oxygen consumption increases following thyroid administration. Like pituitary somatotropin, thyroid hormone stimulates growth, but unlike somatotropin, it also influences tissue differentiation and development. For example, cretins (individuals with thyroid deficiency) not only are dwarfed but also may be mentally retarded because the brain fails to develop normally. Their bones and many other tissues also show an abnormal pattern of development.

Convincing evidence indicates that the thyroid gland secretes *calcitonin*. Calcitonin acts quickly to decrease blood's calcium concentration. Presumably, it produces this effect by either or both of these actions: inhibiting bone breakdown with calcium release into blood or promoting calcium deposition in bone. The function of calcitonin is to help maintain blood calcium homeostasis and prevent harmful hypercalcemia. Parathyroid hormone serves as an antagonist to calcitonin (p. 335).

Effects of hypersecretion and hyposecretion

Hypersecretion of thyroid hormone produces the disease *exophthalmic goiter* (Graves' disease, Basedow's disease, and several other names) (Fig. 12-9), characterized by an elevated PBI level, an increased metabolism (+30 or more), an increased appetite, loss of weight, increased nervous irritability, and exophthalmos. Marked

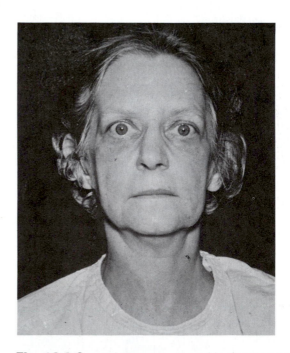

Fig. 12-9 Graves' disease, caused by hypersecretion by the thyroid gland.

Courtesy Dr. William McKendree Jefferies, Case Western Reserve University School of Medicine, Cleveland, Ohio.

edema of the fatty tissue behind the eye, attributed to the high thyrotropic hormone content in blood, produces the exophthalmos.

Hyposecretion during the formative years leads to malformed dwarfism or *cretinism*, a condition characterized by a low metabolic rate, retarded growth and sexual development, and often, too, retarded mental development. Later in life, deficient thyroid secretion produces the disease *myxedema* (Fig. 12-10). The low metabolic rate that characterizes myxedema leads to lessened mental and physical vigor, a gain in weight, loss of hair, and a thickening of the skin from an accumulation of fluid in the subcutaneous tissues. Because of a high mucoprotein content, this fluid is viscous. Therefore it gives firmness to the skin, and the skin does not pit when pressed, as it does in some other types of edema.

Fig. 12-10 Myxedema, a condition produced by hyposecretion by the thyroid gland during the adult years.
Courtesy Dr. Edmund E. Beard, Cleveland, Ohio.

Parathyroid glands
Location and structure

The parathyroid glands are small round bodies attached to the posterior surfaces of the lateral lobes of the thyroid gland (Fig. 12-8). Usually there are four or five, but sometimes there are fewer and sometimes more of these glands.

Parathyroid hormone

The parathyroid glands secrete *parathyroid hormone*. Its chief function is to help maintain homeostasis of blood calcium concentration by promoting calcium absorption into the blood and thereby tending to prevent hypocalcemia. It acts as follows.

1 Parathyroid hormone acts on intestine, bones, and kidney tubules to accelerate calcium absorption from them into the blood. Hence parathyroid hormone tends to increase the blood concentration of calcium. Its primary action on bone is to stimulate bone breakdown or resorption. This releases calcium and phosphate, which diffuse into the blood.

2 Parathyroid hormone acts on kidney tubules to accelerate their excretion of phosphates from the blood into the urine. Note, therefore, that parathyroid hormone has opposite effects on the kidney's handling of calcium and phosphate. It accelerates calcium reabsorption but phosphate excretion by the tubules. Consequently, parathyroid hormone tends to increase the blood concentration of calcium and to decrease the blood concentration of phosphate.

The maintenance of calcium homeostasis is highly important for healthy survival. Normal neuromuscular irritability, blood clotting, cell membrane permeability, and also normal functioning of certain enzymes all depend on the blood concentration of calcium being maintained at a normal level. Neuromuscular irritability is inversely related to blood calcium concentration. In other words, neuromuscular irritability increases when the blood concentration of calcium decreases. Suppose, for example, that a parathyroid hormone deficiency develops and causes *hypocalcemia* (lower than normal blood concentration of calcium). The hypocalcemia increases neuromuscular irritability—sometimes so much that it produces muscle spasms and convulsions, a condition called tetany.

Parathyroid hormone excess produces hypercalcemia or a higher than normal blood concentration of calcium. Sometimes it causes a bone disease with the long name of osteitis fibrosa cystica. Bone mass decreases (as a result of increased bone destruction followed by fibrous tissue replacement), decalcification occurs, and cystlike cavities appear in the bone.

Adrenal glands
Location and structure

The adrenal glands are located atop the kidneys, fitting like a cap over these organs. The outer portion of the gland is called the *cortex* and the inner substance the *medulla*. Although the adrenal cortex and adrenal medulla are structural parts of one organ, they function as two separate endocrine glands.

Adrenal cortex

Three different zones or layers of cells make up the adrenal cortex (Fig. 12-11). Starting with the zone directly under the outer capsule of the gland, their names are zona glomerulosa, zona fasciculata, and zona reticularis. The outer zone of adrenal cells secretes hormones called mineralocorticoids. The middle zone secretes glucocorticoids, and the innermost zone secretes small amounts of both glucocorticoids and sex hormones. We shall now discuss briefly the functions of these three kinds of adrenal cortical hormones.

Glucocorticoids

The chief glucocorticoids secreted by the zona fasciculata of the adrenal cortex are cortisol* (also called hydrocortisone) and corticosterone. Glucocorticoids affect literally every cell in the body. Although the precise primary actions of glucocorticoids remain unknown, their most outstanding effects are as follows.

1 Glucocorticoids tend to accelerate the breakdown of proteins to amino acids in all

*Steroid compounds have the following nucleus:

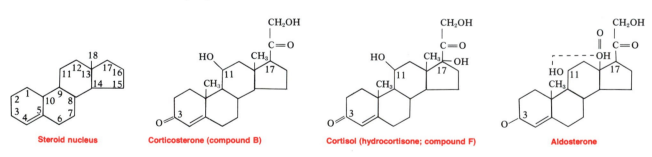

Steroid nucleus   Corticosterone (compound B)   Cortisol (hydrocortisone; compound F)   Aldosterone

Corticosterone (compound B) may be the parent substance of other corticoids.

Compound E or cortisone (chemical name, 17-hydroxy-11-dehydrocorticosterone, signifying that the molecule is the same as corticosterone with —OH instead of —H on C17, and H on C11).

DOC (11-desoxycorticosterone—corticosterone molecules without any oxygen at C11).

Relation of molecular structure to function:
1 Oxygen at C11 produces glucocorticoid effects described above and on pp. 338-339.
2 OH at C17 (in addition to oxygen at C11) enhances glucocorticoid effects.
3 Aldehyde group at C18 (as in aldosterone) produces marked salt-retaining effect.

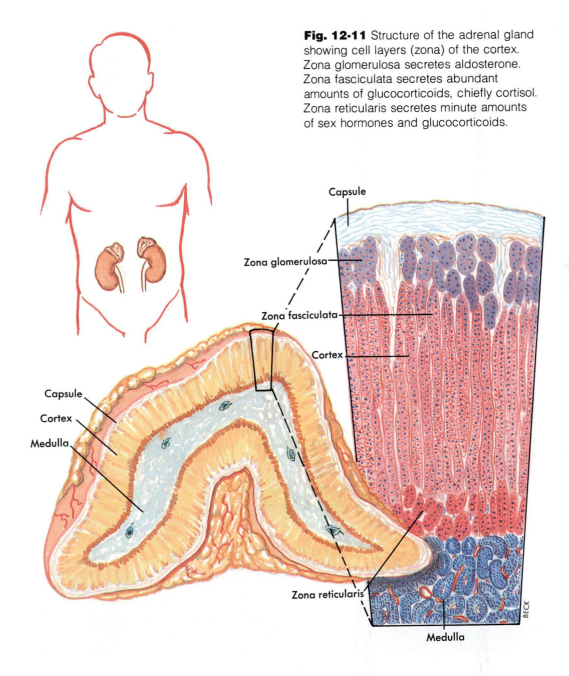

Fig. 12-11 Structure of the adrenal gland showing cell layers (zona) of the cortex. Zona glomerulosa secretes aldosterone. Zona fasciculata secretes abundant amounts of glucocorticoids, chiefly cortisol. Zona reticularis secretes minute amounts of sex hormones and glucocorticoids.

Capsule

Zona glomerulosa

Zona fasciculata

Cortex

Capsule

Cortex

Medulla

Zona reticularis

Medulla

cells except liver cells. These "mobilized" amino acids move out of tissue cells into blood and circulate to liver cells. Here they are changed to glucose by a process that consists of a series of chemical reactions and is called gluconeogenesis. A prolonged, high blood concentration of glucocorticoids in the blood, therefore, results in a net loss of tissue proteins, that is, a negative nitrogen balance or "tissue wasting," and a higher than normal blood glucose concentration, that is, hyperglycemia—a common symptom in diabetes mellitus. Summarizing, we might describe glucocorticoids as protein-mobilizing (or protein-catabolic), gluconeogenic, and hyperglycemic (or diabetogenic) hormones.

2 Glucocorticoids tend to accelerate both the mobilization of fats from adipose cells and the catabolism of fats by almost all kinds of cells. In other words, glucocorticoids tend to cause cells to "shift" for their energy supply from their usual carbohydrate catabolism to fat catabolism. But also the fats mobilized by glucocorticoids may be used by liver cells for gluconeogenesis. Chronic excess of glucocorticoids, as in Cushing's syndrome (Fig. 12-12), results in a redistribution of bony fat. It apparently accelerates fat mobilization from the arms and legs and paradoxically promotes fat deposition in the face ("moon face"), shoulders ("buffalo hump"), trunk, and abdomen.

Fig. 12-12 Cushing's syndrome, the result of chronic excess glucocorticoids.

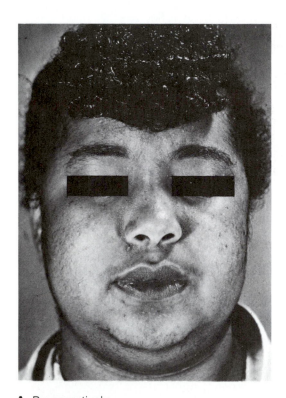

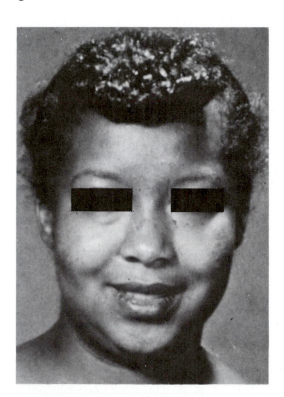

A, Preoperatively.

B, Six months postoperatively.

Courtesy Dr. William McKendree Jefferies, Case Western Reserve University School of Medicine, Cleveland, Ohio.

3 Glucocorticoids are essential for maintaining a normal blood pressure. Without adequate amounts of glucocorticoids in the blood, the hormones norepinephrine and epinephrine cannot produce their vasoconstricting effect on blood vessels, and blood pressure falls precariously.

4 Glucocorticoid secretion is known to increase during stress, particularly in stress produced by anxiety or severe injury. What is not known, after more than 30 years of study and debate, is how or even whether the increased blood glucocorticoid concentration helps the body cope successfully with factors that threaten its healthy survival.

5 A high blood concentration of glucocorticoids rather quickly causes a marked decrease in the number of eosinophils in blood (eosinopenia) and marked atrophy (decrease in size) of lymphatic tissues, particularly the thymus gland and lymph nodes. This, in turn, leads to a decrease in the number of lymphocytes and plasma cells in blood. Because of the decreased number of lymphocytes and plasma cells, antibody formation also decreases. Antibody formation is an important part of both immunity and allergy.

6 Normal physiological amounts of glucocorticoids act with epinephrine, a hormone secreted by the medullary portion of the adrenal glands, to bring about a normal recovery from injury produced by many kinds of inflammatory agents. How they accomplish this anti-inflammatory effect is still unsettled. Pharmacological amounts of glucocorticoids have been used for many years to relieve the symptoms of rheumatoid arthritis and some other inflammatory conditions.

Mineralocorticoids

Mineralocorticoids, as their name suggests, play an important part in regulating mineral salt (electrolyte) metabolism. In the human, aldosterone is the only physiologically important mineralocorticoid. Its primary general function seems to be to maintain homeostasis

of blood sodium concentration. It does this through its action on the distal renal tubule cells. Aldosterone stimulates them to increase their reabsorption of sodium ions from tubule urine back into blood. Because the tubule cells excrete either a potassium or a hydrogen ion in exchange for each sodium ion they reabsorb, aldosterone helps maintain a normal blood potassium concentration and a normal pH.

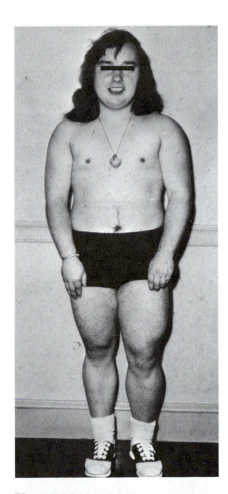

Fig. 12-13 Virilizing tumor of the adrenal cortex of a young girl. The tumor secretes excess androgens, thereby producing masculinizing adrenogenital syndrome.

Courtesy Dr. William McKendree Jefferies, Case Western Reserve University School of Medicine, Cleveland, Ohio.

Moreover, as each positive sodium ion is reabsorbed, a negative ion (bicarbonate or chloride) follows along, drawn by the attraction force between ions that bear opposite electrical charges. Also, the reabsorption of electrolytes causes net diffusion of water back into blood. Briefly, then, because of its primary sodium-reabsorbing effect on kidney tubules, aldosterone tends to produce sodium and water retention but potassium and hydrogen ion loss.

Sex hormones

The normal adrenal cortex in both sexes secretes small but physiologically significant amounts of male hormones (androgens) and insignificant amounts of female hormones (estrogens). The androgens secreted by the cortex do not have strong masculinizing properties, except for testosterone, but the cortex secretes only trace amounts of this. New information suggests that the small amounts of androgens secreted by the female cortex probably support sexual behavior. Tumors of the adrenal cortex that secrete large amounts of androgens are known as virilizing tumors. The excessive androgens produce masculinizing effects such as those evidenced in Fig. 12-13. Extreme androgen excess in a woman may cause a beard to grow.

Control of secretion

Glucocorticoids. Mechanisms that control secretion of glucocorticoids are discussed on pp. 327-329.

Mineralocorticoids (aldosterone mainly). Aldosterone secretion is largely controlled by the renin-angiotensin mechanism and by blood potassium concentration. The renin-angiotensin mechanism operates as indicated in Fig. 12-14. When blood pressure in the afferent arterioles of the kidney decreases below a certain level, it acts as a stimulus to the juxtaglomerular apparatus,*

---

*From the Latin *juxta*, near to. Cells located in afferent arterioles near their entry into the glomeruli constitute the juxtaglomerular apparatus.

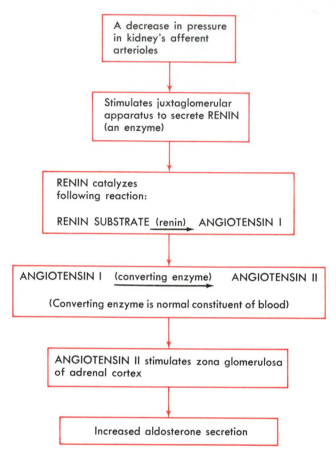

Fig. 12-14 Renin-angiotensin mechanism for regulating aldosterone secretion.

causing its cells to secrete renin into the blood and interstitial fluid. Renin is an enzyme. It catalyzes reactions that change a compound called renin substrate—a normal constituent of blood—into angiotensin I. Angiotensin I is immediately converted to angiotensin II by another enzyme normally present in blood and aptly named "converting enzyme." Angiotensin II stimulates the zona glomerulosa of the adrenal cortex to increase its secretion of aldosterone.

Blood potassium concentration also helps regulate aldosterone secretion. Specifically, a high blood potassium concentration stimulates aldosterone secretion and a low blood potassium concentration inhibits it.

Adrenal medulla

Epinephrine and norepinephrine

The adrenal medulla secretes two catecholamines—about 80% epinephrine and the rest norepinephrine. These hormones affect smooth muscle, cardiac muscle, and glands the way sympathetic stimulation does. They serve to increase and prolong sympathetic effects.

Control of secretion

Increased epinephrine secretion by the adrenal medulla is one of the body's first responses to stress. Impulses from the hypothalamus stimulate sympathetic preganglionic neurons, which synapse with cells of the adrenal medulla, stimulating them to increase their output of epinephrine.

Islands of Langerhans

The relationship between exocrine (digestive) and endocrine (hormone) secreting glandular cells in the pancreas is discussed in Chapter 18 (Fig. 18-17). Clusters of endocrine secreting cells called *islands of Langerhans* are responsible for hormone production in this dual-purpose gland.

The beta cells of the islands of Langerhans secrete insulin, and the alpha cells secrete glucagon. Insulin tends to accelerate the movement of glucose, amino acids, and fatty acids out of blood and through the cytoplasmic membranes of cells into their cytoplasm. Hence insulin tends to lower the blood concentrations of these food compounds and to promote their metabolism. The importance of insulin as a key regulator of cellular metabolic activity is discussed in Chapter 19.

Glucagon, the hormone secreted by the alpha cells of the islands of Langerhans, tends to increase blood glucose concentration. Chapter 19 describes the mechanism by which glucagon produces this opposite effect from insulin. It also describes the mechanisms that regulate insulin and glucagon secretion.

Current research findings of some investigators[*] have demonstrated a new polypeptide hormone secreted by the pancreas. The active component of the new hormone is a polypeptide residue containing 36 amino acids. Until the new hormone is named, scientists are referring to it simply as "pancreatic polypeptide" (PP).

There is a definite "distribution pattern" of hormone-producing cell types in the islands of Langerhans. Insulin-secreting beta cells tend to be concentrated in the central portion of each islet. Glandular cells somewhat different in appearance from the glucagon-secreting alpha cells, but found in association with them in a type of halo around the periphery of each clump of islet tissue, are now known to secrete pancreatic polypeptide and have been designated as PP cells.

Secretion of pancreatic polypeptide is elevated in diabetes and is influenced by blood levels of a number of compounds, including certain proteins and lipids. The hormone increases both gastric secretions and the production of glucagon. Available evidence seems to suggest that PP plays a variety of important functions in digestion of foodstuffs, distribution of nutrients to the tissues, and cellular metabolism.

Ovaries

The ovaries produce two kinds of steroid female hormones: estrogens (chiefly estradiol and estrone) and progesterone. The chief endocrine glands that secrete estrogens are microscopic structures located in the ovary and named ovarian or graafian follicles. Another endocrine gland in the ovary secretes mostly progesterone. Its name is the corpus luteum, and it has a claim to distinction. It is a temporary structure, replaced once a month during a woman's reproductive years, except in months when she is pregnant. In Chapter 25, we shall discuss the

[*] Dr. R. L. Hazelwood of the University of Houston, Houston, Texas, and others.

functions of estrogens and progesterone and the control of their secretion.

Testes

The interstitial cells of the testes secrete testosterone, a steroid hormone classed as an androgen, that is, a substance that promotes "maleness." Chapter 24 presents details about testosterone's functions and the control of its secretion.

Pineal gland (pineal body or epiphysis cerebri)

The pineal gland has long been a mystery organ. Even now its function in the human being remains a controversial matter. A small cone-shaped structure about 1 cm long, it is located in the cranial cavity behind the midbrain and the third ventricle (Fig. 9-13).

Melatonin, the primary hormone secreted by the pineal gland, appears to inhibit luteinizing hormone secretion and ovarian function. The pineal gland may also secrete a second hormone called *adrenoglomerulotropin* because of its ability to stimulate aldosterone secretion by the adrenal's zona glomerulosa.

Thymus

There is no longer any doubt that the thymus functions as an endocrine gland. Once relegated to a position of being a vestigial structure of little importance, the thymus is now considered the primary organ of the lymphatic system. As a component of the overall body immune system, the endocrine function of the thymus is not only important but essential. The anatomy of the thymus is discussed in Chapter 16.

The hormone *thymosin* has been isolated from thymus tissue and is considered responsible for its endocrine activity. Thymosin is actually a family of biologically active peptides that together play a critical role in the maturation and development of the immune system. In an individual with impaired thymic function, injection of thymosin will increase the population of specialized lymphocytes called "T cells" and activate the immune system (discussed in Chapter 26).

Suppression of the immune system sometimes occurs in certain disease states and in patients who are undergoing massive chemotherapy or radiotherapy for the treatment of cancer. Such individuals are said to be "immunosuppressed" and are extremely susceptible to infections. Thymosin may prove useful as an activator of the immune system in such patients.

Gastric and intestinal mucosa

The mucosal lining of the gastrointestinal tract, like the pancreas, contains cells that produce both endocrine and exocrine secretions. Gastrointestinal hormones such as gastrin, cholecystokinin-pancreozymin, and secretin play important regulatory roles in controlling and coordinating the secretory and motor activities involved in the digestive process. Chapter 18 describes the hormonal control of digestion in the stomach and small intestine.

Placenta

The placenta functions as a temporary endocrine gland. During pregnancy, it produces chorionic gonadotropins—so-called because they are secreted by cells of the chorion, the outermost fetal membrane. In addition to gonadotropins, the placenta also produces estrogens and progesterone. During pregnancy the kidneys excrete large amounts of gonadotropins in the urine. This fact, discovered more than a half century ago by Aschheim and Zondek, led to the development of the now familiar pregnancy tests.

Outline summary

Meaning

A Composed of glands that pour secretions into blood instead of into ducts

B General functions and importance of neuroendocrine system

  1 Communication, integration, and control—hormones produce slower and longer lasting responses in regulation of metabolism, reproduction, and responses to stress

  2 Nervous system (nerve impulses) produce rapid, short-lasting responses

  3 Cells that are acted on and respond to hormones are called target organ cells

  4 Names and locations of endocrine glands—see Table 12-1

Prostaglandins (tissue hormones)

A Unique group of widespread and potent biological compounds (20-carbon fatty acids with 5-carbon ring)

B Regulate endocrine activity at the cellular level by influencing adenyl cyclase and cyclic AMP activity —see Fig. 12-2

C Three classes of prostaglandins

  1 Prostaglandin A (PGA)—may help in regulation of blood pressure

  2 Prostaglandin E (PGE)—vascular effects involving red blood cell "deformability" and platelet aggregation; effect on hydrochloric acid secretion in stomach

  3 Prostaglandin F (PGF)—important in reproductive function

D May someday be used in treatment of such diverse diseases as hypertension, coronary thrombosis, asthma, and ulcers

Pituitary gland (hypophysis cerebri)

A Location—in sella turcica of sphenoid bone

B Consists of two endocrine glands: anterior lobe (adenohypophysis) and posterior lobe (neurohypophysis)

C Anterior portion divided into the larger pars anterior and small pars intermedia

D Table 12-2 summarizes the production, control, and effects of the pituitary hormones

Anterior pituitary gland (adenohypophysis)

A Cell types and functions

  1 Chromophobes—undifferentiated cells that do not stain well

  2 Chromophils

    a Acidophils—stain with acid dyes; secrete growth hormone (GH) and prolactin

    b Basophils secrete thyrotropin (TSH), adrenocorticotropin (ACTH), follicle-stimulating hormone (FSH), luteinizing hormone (LH), and melanocyte-stimulating hormone (MSH)

B Pars anterior secretes all hormones of the anterior lobe except melanocyte-stimulating hormone (MSH), which the pars intermedia secretes

C Growth hormone (somatotropin or somatotropic hormone)

  1 Accelerates protein anabolism, so promotes growth

    a Disorders caused by excess growth hormone— gigantism and acromegaly

    b Disorders caused by deficient growth hormone—dwarfism and pituitary cachexia

  2 Tends to accelerate fat mobilization from adipose cells and fat catabolism by other cells, thereby decreasing glucose catabolism and tending to increase blood glucose concentration (hyperglycemic effect); prolonged excess may produce diabetes

D Prolactin (lactogenic hormone or luteotropic hormone [LTH])

  1 During pregnancy, promotes breast development

  2 After delivery, initiates milk secretion

  3 Helps in maintaining corpus luteum

E Four tropic hormones (hormones that have a stimulating effect on other endocrine glands) are secreted by basophil cells of anterior pituitary gland

  1 Thyrotropin—thyroid-stimulating hormone (TSH)

    a Promotes growth and development of thyroid gland

    b Stimulates thyroid gland to secrete thyroid hormone

  2 Adrenocorticotropin (ACTH)

    a Promotes growth and development of adrenal cortex

    b Stimulates adrenal cortex to secrete cortisol and other glucocorticoids

  3 Follicle-stimulating hormone (FSH)

    a Stimulates primary graafian follicle to start growing and to develop to maturity b Stimulates follicle cells to secrete estrogens

c In male, stimulates development of seminiferous tubules and maintains spermatogenesis by them

4 Luteinizing hormone (LH)

a Acts with FSH to cause complete maturation of follicle

b Brings about ovulation

c Stimulates formation of corpus luteum (luteinizing effect)

d Stimulates corpus luteum to secrete progesterone and estrogens

e In male, LH called interstitial cell–stimulating hormone (ICSH); stimulates interstitial cells in testis to develop and secrete testosterone

f Just before puberty, presumably sudden marked increase in secretion of gonadotropins (FSH, LH) initiates first menses

F Melanocyte-stimulating hormone (MSH) or intermedin—tends to produce increased pigmentation of skin

G Control of secretion

1 Releasing hormones—neurosecretions produced in hypothalamus that reach anterior pituitary gland via blood in pituitary portal system to regulate hormone production

a Growth hormone–releasing hormone (GRH)

b Corticotropin-releasing factor (CRF)

c Thyrotropin-releasing hormone (TRH)

d Follicle-stimulating hormone–releasing hormone (FSH-RH)

e Luteinizing hormone–releasing hormone (LH-RH)

f Prolactin-inhibitory factor (PIF)

g Prolactin-releasing factor (PRF)

2 Negative feedback mechanisms operate between target glands and anterior lobe of pituitary gland

a High blood concentration of target gland hormone inhibits its pituitary secretion of tropic hormones either directly or indirectly, that is, via hypothalamic releasing hormones; conversely, low blood concentration of target gland hormones stimulates pituitary secretion of tropic hormones

b High blood concentration of tropic hormone stimulates target gland secretion of its hormone

3 Stress acts in some way to stimulate the hypothalamus to secrete releasing hormones, which in turn stimulate the anterior pituitary gland to increase its secretion of hormones

Posterior pituitary gland (neurohypophysis)

A Cell type—pituicytes do not secrete either of the hormones released from the posterior pituitary gland

B Releases antidiuretic hormone (ADH)—synthesized by neurons in supraoptic nucleus of hypothalamus—and oxytocin—synthesized by neurons in the paraventricular nucleus of hypothalamus

C Normal (physiological) levels of ADH (vasopressin or Pitressin) will stimulate water reabsorption by distal and collecting tubules and in large (pharmacological) doses will stimulate smooth muscle of blood vessels

D Oxytocin stimulates contractions of pregnant uterus and release of milk by lactating breast

E Control of secretion

1 ADH secretion—details of mechanism controlling secretion of this hormone not established but general principles follow

a Increased extracellular fluid osmotic pressure leads to decreased ADH secretion

b Decreased extracellular fluid volume leads to increased ADH secretion

c Stress leads to increased ADH secretion

2 Oxytocin—details of mechanism controlling oxytocin secretion not established but is known that stimulation of nipples by suckling leads to increased oxytocin secretion

Thyroid gland

A Location and structure

1 Located in neck just below larynx

2 Two lateral lobes connected by isthmus

B Thyroid hormone—stimulates rate of oxygen consumption (metabolic rate) of all cells and thereby helps regulate physical and mental development, development of sexual maturity, and numerous other processes

C Calcitonin—decreases blood calcium concentration, presumably by inhibiting bone breakdown with calcium release into blood—and/or increasing calcium deposition in bone

D Effects of hypersecretion and hyposecretion

1 Hypersecretion produces exophthalmic goiter

2 Hyposecretion in early life produces malformed dwarfism or cretinism; in later life, myxedema

Parathyroid glands

A Location and structure

1 Attached to posterior surfaces of thyroid gland

2 Small round bodies, usually four or five in number

B Parathyroid hormone
 1 Increases blood calcium by stimulating bone breakdown, releasing calcium and phosphate into blood from bone
 2 Increases blood calcium by accelerating calcium absorption from intestine and kidney tubules
 3 Accelerates kidney tubule excretion of phosphates from blood into urine
C Hypersecretion causes decrease in bone mass with replacement by fibrous tissue
D Hyposecretion produces hypocalcemia and tetany and death in few hours

Adrenal glands

A Location and structure
 1 Located atop kidneys
 2 Outer portion of gland called cortex and inner portion called medulla

Adrenal cortex

A Zones or layers (from outside in)
 1 Zona glomerulosa—secretes mineralocorticoids
 2 Zona fasciculata—secretes glucocorticoids
 3 Zona reticularis—secretes small amounts of glucocorticoids and sex hormones
B Glucocorticoids (mainly cortisol, smaller amounts of corticosterone)
 1 Tend to accelerate tissue protein mobilization; mobilized amino acids circulate to liver cells, where they are changed to glucose (process of gluconeogenesis)
 2 Tend to accelerate fat mobilization and catabolism, that is, tend to cause shift to fat utilization from usual carbohydrate utilization
 3 Necessary for norepinephrine's vasoconstricting effect on blood vessels and therefore essential for maintaining normal blood pressure
 4 Blood concentration of glucocorticoids increases during stress; value of this still controversial
 5 High blood concentration of glucocorticoids causes eosinopenia and marked atrophy of thymus gland and other lymphatic tissues; this atrophy leads to a decrease in the number of lymphocytes and plasma cells and decreased antibody formation, which decreases immunity and allergic reactions
 6 Glucocorticoids plus epinephrine promote normal recovery from injury by inflammatory agents (anti-inflammatory effect)
C Mineralocorticoids (mainly aldosterone)
 1 Accelerate renal tubule reabsorption of sodium ions and excretion of potassium ions (or hydrogen ions)

 2 Increased renal tubule reabsorption of bicarbonate ions (or chloride ions) and water result from increased sodium reabsorption
D Sex hormones—in both sexes, amount of male hormones secreted by adrenal cortex is physiologically significant, but amount of female hormones is insignificant
E Control of secretion
 1 Glucocorticoids (see outline, p. 344)
 2 Mineralocorticoids (aldosterone)
  a Renin-angiotensin mechanism, an important regulator of aldosterone secretion; see Fig. 12-14
  b Blood potassium concentration also helps regulate aldosterone secretion; high blood potassium concentration stimulates and low blood potassium concentration inhibits aldosterone secretion

Adrenal medulla

A Hormones—epinephrine mainly; some norepinephrine
B Functions of epinephrine—affects visceral effectors (smooth muscle, cardiac muscle, and glands) in same way as sympathetic stimulation of these structures; epinephrine from adrenal glands intensifies and prolongs sympathetic effects
C Control of secretion—stress acts in some way to stimulate hypothalamus, which sends impulses to adrenal medulla via preganglionic sympathetic neurons, stimulating medulla to increase its secretion of epinephrine

Islands of Langerhans

A Hormones
 1 Insulin (beta cells)
 2 Glucagon (alpha cells)
 3 Pancreatic polypeptide (PP cells)
B Functions
 1 Insulin
  a Promotes glucose transport into cells, thereby increasing glucose utilization (catabolism) and glycogenesis and decreasing blood glucose levels
  b Promotes fatty acid transport into cells and fat anabolism (lipogenesis or fat deposition) in them
  c Promotes amino acid transport into cells and protein anabolism
 2 Glucagon—accelerates liver glycogenolysis, so tends to increase blood glucose levels; in short, is insulin antagonist
 3 Pancreatic polypeptide

    a Stimulus for secretion includes blood levels of certain proteins and lipids
    b Increases gastric secretions
    c Increases production of glucagon
    d Blood levels increase in diabetes
    e May function in digestion of foodstuffs, distribution of nutrients to the tissues, and cellular metabolism

Ovaries

A Hormones
    1 Ovarian (graafian) follicles secrete estrogens
    2 Corpus luteum secretes progesterone
B Functions—see Chapter 25

Testes

A Hormones—androgens, most important of which is testosterone
B Functions—see Chapter 24

Pineal gland (pineal body or epiphysis cerebri)

A Hormones—melatonin and adrenoglomerulotropin
B Functions
    1 Melatonin—inhibits luteinizing hormone secretion and ovarian function
    2 Adrenoglomerulotropin—stimulates aldosterone secretion by zona glomerulosa of adrenal cortex

Thymus

A Anatomy—see Chapter 16
B Hormone—thymosin (family of biologically active peptides)
C Functions
    1 Needed for maturation and development of immune system
    2 Increases numbers of T cells in lymphocyte population
    3 Promotes an increase in many types of lymphocyte functions—see Chapter 26

Gastric and intestinal mucosa

A Hormones—gastrin, cholecystokinin-pancreozymin, and secretin
B Functions—see Chapter 18

Placenta

A Temporary endocrine gland
B Secretes estrogens, progesterone, and chorionic gonadotropin
C Helps maintain progestational state of endometrium

Review questions

1 Compare the activity of the nervous and endocrine systems in communication, integration, and control of body functions. In what ways are the two systems alike? How do they differ?
2 Define the terms hormone and target organ. Give examples of hormone–target organ interaction.
3 Name the endocrine glands and locate each one.
4 Discuss the difference between regular hormones, "regulatory chemicals," and prostaglandins.
5 What is the "second messenger" hypothesis of hormone action? Using a diagram, summarize the mechanism of hormone action, including the postulated role of prostaglandins.
6 List the three classes of prostaglandins.
7 Name the two subdivisions of the adenohypophysis and list the hormones produced in each area.
8 Discuss and identify by staining tendency and relative percentages the cell types present in the anterior pituitary gland.
9 List the hormones secreted by the acidophils and basophils of the anterior pituitary gland.
10 Explain the relationship of growth hormone to the following clinical conditions: gigantism, acromegaly, dwarfism.
11 What effect does growth hormone have on blood glucose concentration? Fat mobilization and catabolism? Protein anabolism?
12 Discuss the effect of prolactin on breast development and milk secretion. Why is the name "luteotropic hormone" (LTH) sometimes used in place of prolactin?
13 List the four tropic hormones secreted by the basophils of the anterior pituitary gland. Which of the tropic hormones are also called gonadotropins?
14 What is the role of the hypophyseal portal system in regulation of hormone secretion from the anterior pituitary gland?

15 List the "releasing hormones" produced by neurons in the hypothalamus and give their functions.

16 Explain how a "negative feedback system" can regulate hormone secretion. Give an example to illustrate such a control mechanism.

17 Describe a mechanism by which thoughts and emotions can influence literally all body functions.

18 Identify the areas in the hypothalamus that secrete ADH and oxytocin.

19 If chemical "releasing factors" are responsible for initiating secretion of hormones from the anterior pituitary, what triggers the release of ADH and oxytocin from the posterior pituitary gland?

20 Discuss the synthesis and storage of thyroxine. How is it transported in the blood? What effects would result from hypersecretion? Hyposecretion?

21 Discuss the functions of parathyroid hormone.

22 List the hormones produced by each "zone" of the adrenal cortex.

23 Tumors of which zone or layer of the adrenal cortex would produce masculinizing effects? Why?

24 Discuss the normal function of hormones produced by the adrenal medulla.

25 Identify the hormones produced by each of the cell types of the islands of Langerhans.

26 Compare the physiological effects of insulin and glucagon.

27 Discuss the endocrine function of the thymus gland.

28 Name the hormone or hormones that help control each of the following: blood sugar level, blood calcium level, blood sodium level, blood potassium level. Explain the mechanisms involved.

29 What hormone enhances and prolongs sympathetic effects?

30 Describe the mechanisms that bring about increased secretion of epinephrine, ACTH, ADH, and glucocorticoids during stress.

Transportation

chapter 13

Blood

This unit deals with transportation, one of the body's vital functions. Homeostasis of the internal environment—and therefore survival itself—depends on continual transportation to and from body cells. This chapter discusses the major transportation fluid, blood. Chapters 14 and 15 consider the major transportation system, the cardiovascular system, and Chapter 16 has for its subject a supplementary transportation system, the lymphatic system.

Blood is much more than the simple liquid it seems to be. Not only does it consist of a fluid but also of cells—billions and billions of them. The fluid portion of blood, that is, the *plasma*, is one of the three major body fluids (interstitial and intracellular fluids are the other two). The blood cells are suspended in the plasma. The term *formed elements* is used to designate the various kinds of blood cells collectively.

Blood is a complex transport medium that performs vital pickup and delivery services for the body. It picks up food and oxygen from the digestive and respiratory systems and delivers them to cells all over the body. It picks up wastes from cells and delivers them to excretory organs. It picks up hormones from endocrine glands and delivers them to the so-called target cells. It transports enzymes, buffers, and various other biochemical substances that serve important functions. And finally, blood serves as the keystone of the body's heat-regulating mechanism. Certain physical properties of the blood make it especially effective in this role. Its high specific heat and conductivity enable this unique fluid to both absorb large quantities of heat without an appreciable increase in its own temperature and to transfer this absorbed heat from the core of the body to its surface, where it can be more readily dissipated. Functionally the blood may be said to consist of ingenious, efficient, and precisely regulated transport and regulatory systems. Several chapters of this book discuss these systems.

Blood volume

How much blood does an adult body contain? We can answer that question only with some generalizations. The total blood volume varies markedly in different individuals. Age, body type, sex, and the method of measurement are all determinants. *Direct* measurement of total blood volume can be accomplished only by complete removal of all blood from an experimental animal. In humans, *indirect methods* of measurement that employ "tagging" of red blood cells or plasma components with radioisotopes, such as radiophosphorus (^{32}P) or radiochromium (^{51}Cr), can be used without endangering the subject. Normal values obtained will vary with the method and "tag" that is used. The principle is simply to introduce a known amount of radioisotope into the circulation, allow the material to distribute itself uniformly in the blood, and then analyze its concentration in a representative blood sample.

One of the chief variables influencing normal blood volume is the amount of body fat. Blood volume per kilogram of body weight varies inversely with the amount of excess body fat. This means that the less fat there is in your body, the more blood you have per kilogram of your body weight. Radiochromium studies indicate that healthy men average about 71 ml of blood per kilogram of body weight.

Blood cells

Three main kinds of blood cells are recognized: red blood cells (erythrocytes), white blood cells (leukocytes), and platelets (thrombocytes). Leukocytes are further divided as shown in the following classifications of blood cells:

1 Red blood cells or erythrocytes
2 White blood cells or leukocytes
   a Granular leukocytes: basophils, neutrophils, and eosinophils
   b Nongranular leukocytes: lymphocytes and monocytes
3 Platelets (thrombocytes)

Approximately 55% of total blood volume is plasma volume, and the rest is blood cell volume. Example: A man who weighs 70 kg (154 pounds) would have an average of about 5,000 ml of blood. Of this, approximately 2,750 ml would be plasma and 2,250 ml would be blood cells. Now we shall take a rather detailed look at blood cells and then consider blood plasma.

Erythrocytes

Appearance, size, shape, and number

Facts about normal red blood cell size, shape, and numbers hold more than academic interest. They are also clinically important. For example, an increase in red blood cell size characterizes one type of anemia, and a decrease in red blood cell size characterizes another type. Red blood cells are extremely small. More than 3,000 of them could be placed side by side in a 1-inch space, since they measure only about 7 μm in diameter. A normal mature red blood cell has no nucleus. Just before the cell reaches maturity and enters the bloodstream from the bone mar-

row, the nucleus is extruded, with the result that the cell caves in on both sides. So, as you can see in Fig. 13-1, normal mature red blood cells are shaped like tiny biconcave disks.

The depression on each flat surface of the cell results in a thin center and thicker edges. This unique shape of the red blood cell gives it a very large surface area relative to its volume. The

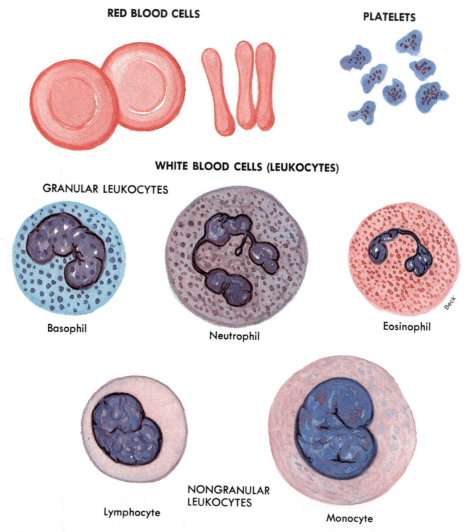

RED BLOOD CELLS

PLATELETS

WHITE BLOOD CELLS (LEUKOCYTES)

GRANULAR LEUKOCYTES

Basophil

Neutrophil

Eosinophil

Lymphocyte

NONGRANULAR LEUKOCYTES

Monocyte

Fig. 13-1 Human blood cells. There are close to 30 trillion red blood cells in the adult. Each cubic millimeter of blood contains from 4.5 to 5.5 million red blood cells and an average total of 7,500 white blood cells.

sides of the cell can actually move in and out, permitting it to undergo "deformity" or change in shape without injury as it moves through the very narrow capillary vessels. This ability to change shape is necessary for the survival of red blood cells, which are under almost constant mechanical shearing and bursting strains as they pass through the capillary system. In addition, the degree of cell deformity possible influences the speed of blood flow in the microcirculation.

One of the most important factors that determines the degree of red blood cell deformability other than physical shape is the blood level of certain prostaglandins (Chapter 12). Prostaglandin PGE_1 increases red blood cell deformability, whereas prostaglandin PGE_2 decreases it.

Red blood cells are the most numerous of the formed elements in blood. In males, red blood cell counts average about 5,500,000 per cubic millimeter (mm^3) of blood, and in females 4,800,000 per mm^3.

Structure and functions

Considered together, the total surface area of all the red blood cells in an adult is enormous! It provides an area larger than a football field for the exchange of respiratory gases between hemoglobin and interstitial fluid—a beautiful example of the familiar principle that structure determines function (specifically, the transport and exchange of oxygen and carbon dioxide). Packed within one tiny red blood cell are an estimated 200 to 300 million molecules of the complex compound *hemoglobin*. One hemoglobin molecule consists of a protein molecule (globin) combined with four molecules of a pigmented compound (heme). Because each molecule of heme contains one atom of iron, one hemoglobin molecule contains four iron atoms. This structural fact enables one hemoglobin molecule to unite with four oxygen molecules to form oxyhemoglobin (a reversible reaction). Hemoglobin can also combine with carbon dioxide to form carbaminohemoglobin (also reversible). But in

this reaction the structure of the globin part of the hemoglobin molecule rather than of its heme part makes the combining possible.

A man's blood usually contains more hemoglobin than a woman's. In most normal men, 100 ml of blood contains 14 to 16 gm of hemoglobin—and the closer to 16 the more "joie de vivre," in Stewart Alsop's words.* The normal hemoglobin content of a woman's blood is a little less—specifically, in the range of 12 to 14 gm per 100 ml. Any adult who has a hemoglobin content of less than 12 gm per 100 ml blood is diagnosed as having *anemia* (from the Greek *a-*, not, and *haima*, blood). In addition, the term may be used to describe a reduction in the number or volume of functional red blood cells in a given unit of whole blood. Anemias are classified according to the size and hemoglobin content of red blood cells. Such a morphological or structural classification of anemia would include the following types:

1 *Normocytic*—red blood cells are reduced in number but are normal in size.
2 *Macrocytic*—although reduced in number, those red blood cells that are produced are larger in size than normal.
3 *Microcytic*—the majority of red blood cells are smaller than normal.

The amount of hemoglobin (pigment) that a red blood cell contains will influence the intensity of its color. If the hemoglobin content of a red blood cell is high, the color will be more intense—such a cell is described as *hyperchromic* (from the Greek *hyper*, above, over, and *chroma*, color). Low hemoglobin levels result in *hypochromic* red blood cells. Red blood cells that have an average (normal) color and hemoglobin content are said to be *normochromic*. Example: An injury resulting in hemorrhage with rapid loss of whole blood would cause a reduction in circulating red blood cells and hemoglobin but would not change the average size or hemo-

*Alsop, S.: Stay of execution, New York, 1973, J. B. Lippincott Co., p. 30.

globin content of those cells that remained. The anemia that would result from such a rapid loss of blood would be classified as *normocytic normochromic*.

Formation (erythropoiesis)

The entire process of red blood cell formation is called erythropoiesis (from the Greek *poiesis*, production). Initially, erythrocytes are formed in the red bone marrow from nucleated cells known as hemocytoblasts or stem cells. These stem cells divide by mitosis and go through several stages of development. As development proceeds, the nucleus becomes smaller and eventually disappears. Newly formed red blood cells, at the time they leave the bone marrow and enter the blood, contain hemoglobin and a reticulum in their cytoplasm. For this reason they are called reticulocytes.

Frequently a physician needs information about the rate of erythropoiesis to help him make a diagnosis or prescribe treatment. A *reticulocyte count* gives this information. Approximately 0.5% to 1.5% of the red blood cells in normal blood are reticulocytes. A reticulocyte count of less than 0.5% of the red blood cell count usually indicates a slowdown in the process of red blood cell formation. Conversely, a reticulocyte count higher than 1.5% usually indicates an acceleration of red blood cell formation —as occurs, for example, following treatment of anemia.

Reticulocytes are almost 20% larger than mature red blood cells and are coated with a type of globulin that makes them more adhesive than adult corpuscles. This adhesive quality may help in retaining them within the bone marrow until maturity. Once in the circulation, reticulocytes pass through small blood vessels much more slowly than do mature red blood cells. After leaving the bone marrow the "life span" of a circulating reticulocyte rarely exceeds 30 hours.

Destruction

The life span of a red blood cell circulating in the bloodstream averages about 105 to 120 days. They break apart, that is, undergo fragmentation, in the capillaries. Reticuloendothelial cells in the lining of blood vessels, particularly in the liver, spleen, and bone marrow, then phagocytose the red blood cell fragments. In the process, iron is released from hemoglobin and the pigment bilirubin is formed. Both are transported to the liver, where iron is put in temporary storage and bilirubin is excreted in the bile. Eventually the bone marrow uses most of the iron over again for new red blood cell synthesis, and the liver excretes the bile pigments in the bile.

Erythrocyte homeostatic mechanism

Red blood cells are formed and destroyed at a breathtaking rate. Normally, every minute of every day of our adult lives, over 100 million red blood cells are formed to replace an equal number destroyed during that brief time. Since in health the number of red blood cells remains relatively constant at about 4.5 to 5.5 million per mm^3 of blood, efficient homeostatic mechanisms must operate to balance the number of cells formed against the number destroyed. The exact mechanism responsible for this constancy is not known. It is known, however, that the rate of red blood cell production soon speeds up if either the number of red blood cells decreases appreciably or tissue hypoxia (oxygen deficiency) develops. Either of these conditions acts in some way to stimulate the kidneys (and perhaps some other structures) to increase their secretion of a hormone named *erythropoietin*. The resulting increased blood concentration of erythropoietin stimulates bone marrow to accelerate its production of red blood cells (Fig. 13-2). Under maximum stimulation the bone marrow can increase red blood cell production about seven times. The name erythropoietin makes it easy to remember its function—erythropoietin stimulates erythropoiesis (process of red blood cell formation).

Note that for the red blood cell homeostatic mechanism to succeed in maintaining a normal number of red blood cells, the bone marrow

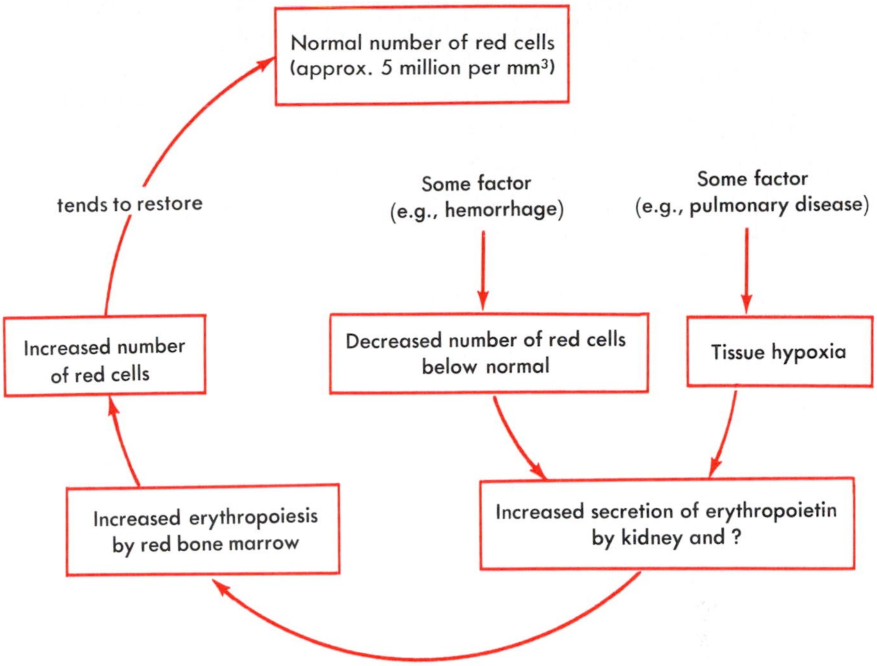

Fig. 13-2 Postulated red blood cell homeostatic mechanism—a negative feedback mechanism. A decrease in the number of circulating red blood cells "feeds back" to the red blood cell–forming structure (red bone marrow) to cause an increased rate of red blood cell formation, which in turn tends to increase the number of red blood cells sufficiently to restore their normal number.

must function adequately. To do this the blood must supply it with adequate amounts of several substances with which to form the new red blood cells—vitamin B_{12}, iron, and amino acids, for example, and also copper and cobalt to serve as catalysts. In addition, the gastric mucosa must provide some unidentified intrinsic factor necessary for absorption of vitamin B_{12} (called extrinsic factor because it derives from external sources in foods and is not synthesized by the body; vitamin B_{12} is also called antianemic principle).

Pernicious anemia develops when the gastric mucosa fails to produce the intrinsic factor needed for adequate vitamin B_{12} absorption. The marrow then produces fewer but larger red blood cells than normal (macrocytic anemia). Many of these cells are immature with overly fragile membranes, a fact that leads to their more rapid destruction.

A clinical example of failure of red blood cell homeostasis seen in recent years is anemia caused by bone marrow injury by x-ray or gam-

ma ray radiations. Other factors may also cause marrow damage. To help diagnose this condition a sample of marrow is removed, for example, from the sternum by means of a sternal puncture, and is studied microscopically for abnormalities. When damaged marrow can no longer keep red blood cell production apace with destruction, red blood cell homeostasis is not maintained. Instead, the red blood cell count falls below normal. Anemia (fewer red blood cells than normal) develops whenever the rates of red blood cell formation and destruction become unequal. Therefore either a decrease in red blood cell formation (as in pernicious anemia or bone marrow injury by radiation) or an increase in red blood cell destruction (as in infections and malignancies) can lead to anemia.

The number of red blood cells is determined by the *"red blood cell count"* or is estimated by the hematocrit. The *hematocrit* is the volume percentage of red blood cells in whole blood. To be more specific, a hematocrit of 47 means that in every 100 ml of whole blood there are 47 ml of blood cells and 53 ml of fluid (plasma). Normally the average hematocrit for a man is about 45 (±7, normal range) and for a woman about 42 (±5). Healthy individuals who live and work in high altitudes often have elevated red blood cell counts and hematocrit values. The condition is called *physiological polycythemia* (from the Greek *polys*, many, *kytos*, cell, and *haima*, blood). Can you explain the stimulus for increased erythropoiesis in these individuals?

Leukocytes

Appearance, size, shape, and number

There are five types of leukocytes classified according to the presence or absence of granules in their cytoplasm. Granular leukocytes consist of neutrophils, eosinophils, and basophils; nongranular leukocytes are lymphocytes and monocytes.

Consult Fig. 13-1. Note particularly the differences in color of the cytoplasmic granules and in the shapes of the nuclei of the granular leukocytes. Neutrophils take their name from the fact that their cytoplasmic granules stain a very light purple with neutral dyes. The granules in these cells are small and numerous and tend to give the cytoplasm a "coarse" appearance. Because their nuclei have two, three, or more lobes, neutrophils are also called *polymorphonuclear leukocytes* or, to avoid that tongue twister, simply "polys." Cytoplasmic granules in eosinophils are large and numerous and stain orange with acid dyes such as eosin. Their nuclei have two oval lobes. The relatively large but sparse cytoplasmic granules in basophils stain a dark purple with basic dyes. Their nuclei are roughly S shaped.

Most lymphocytes in a stained blood smear are about 8 μm in diameter—slightly larger than erythrocytes. In a typical lymphocyte a relatively large spherical nucleus with a slight indentation on one side is surrounded by a thin layer of homogeneous cytoplasm. An average monocyte—the largest of all leukocytes—measures about 15 to 20 μm in diameter. Monocytes have a kidney-shaped nucleus surrounded by abundant quantities of pale grayish or steel blue colored cytoplasm.

A cubic millimeter of normal blood usually contains about 5,000 to 9,000 leukocytes, with different percentages of each type. Because these numbers change in certain abnormal conditions, they have clinical significance. In acute appendicitis, for example, the percentage of neutrophils increases and so, too, does the total white blood cell count. In fact, these characteristic changes may be the deciding points for surgery.

The procedure in which the different types of leukocytes are counted and their percentage of the total white blood cell count is computed is known as a *differential count*. In other words, a differential count is a percentage count of white blood cells. The different kinds of white blood cells and a normal differential count are listed in Table 13-1. A decrease in the number of white blood cells is *leukopenia*. An increase in the number of white blood cells is *leukocytosis*. (*Leukemia* is a malignant disease characterized by a marked increase in the number of white blood cells.)

Functions

White blood cells function as part of the body's defense against microorganisms (Chapter 26). All leukocytes are motile cells. This characteristic enables them to move out of capillaries by squeezing through the intercellular spaces of the capillary wall—a process called *diapedesis* (from the Greek *dia*, through, and *pedesis*, a moving) and to migrate by ameboid movement toward microorganisms or other injurious particles that may have invaded the tissues. Neutrophils, monocytes, lymphocytes, and basophils are highly motile, whereas eosinophils are sluggishly so. Once in the tissue spaces the most important function of neutrophils and monocytes is to phagocytose—ingest and digest microbes or other injurious particles. At any one time, large numbers of white blood cells are moving about

Table 13-1 White blood cells (leukocytes)

Class	Differential count*	
	Normal range (%)	Typical normal (%)
Those with granular cytoplasm and irregular nuclei		
Neutrophils (neutral staining)	65 to 75	65
Eosinophils (acid staining)	2 to 5	3
Basophils (basic staining)	½ to 1	1
Those with nongranular cytoplasm and regular nuclei		
Lymphocytes (large and small)	20 to 25	25
Monocytes	3 to 8	6
Total		100

*In any differential count the sum of the percentages of the different kinds of leukocytes must, of course, total 100%.

in the tissues performing, or ready to perform, their protective functions. Red blood cells and platelets, in contrast, perform their functions within the blood vessels, not in the tissues.

You may recall that reticuloendothelial cells also perform the function of phagocytosis. In general, however, they do this work within more localized areas than do white blood cells. Reticuloendothelial cells, since they do not enter the bloodstream, cannot be transported by it to any part of the body, as white blood cells can.

Lymphocytes play a dominant and vital role in defending the body against microscopic invaders—notably bacteria, fungi, and viruses. Immunity to infectious diseases results largely from lymphocyte activities. Chapter 26 discusses the immune functions of lymphocytes.

Eosinophils are weak phagocytes and show only very limited motility. As a result, they do not play an important role in the body's defense against the usual types of infectious microorganisms. It has been suggested that eosinophils function to detoxify proteins and other harmful products that accumulate as a result of allergic reactions or cellular injury caused by infection with parasites such as hookworm.

The primary function of basophils in the circulating blood is almost totally unknown. Heparin, a highly active anticoagulant, can be obtained from the granules of basophils and may play a role in the prevention of intravascular coagulation.

Formation

Neutrophils, eosinophils, basophils, and a few lymphocytes and monocytes originate, as do erythrocytes, in red bone marrow (myeloid tissue). Most lymphocytes and monocytes derive from hemocytoblasts in lymphatic tissue. Although many lymphocytes are found in bone marrow, presumably most were formed in lymphatic tissues and carried to the bone marrow by the bloodstream.

Myeloid tissue (bone marrow) and lymphatic tissue together constitute the hemopoietic or blood cell–forming tissues of the body. Red bone marrow is myeloid tissue that is actually producing blood cells. Its red color comes from the red blood cells it contains. Yellow marrow, on the other hand, is yellow because it stores considerable fat. It is not active in the business of blood cell formation as long as it remains yellow. Sometimes, however, it becomes active and red in color when an extreme and prolonged need for red blood cell production occurs.

Destruction and life span

The life span of white blood cells is not known. Some evidence seems to indicate that granular leukocytes may live 3 days or less, whereas other evidence suggests that they may live about 12 days. Some of them are probably destroyed by phagocytosis and some by microorganisms. Contrary to the assumption accepted by many hematologists for many years, we now know that

a significant fraction of small lymphocytes survive a long time, 100 to 200 days or more.

Platelets

Appearance, size, shape, and number

To compare platelets with other blood cells as to appearance and size, see Fig. 13-1. In circulating blood, platelets are small, colorless bodies that usually appear as irregular spindles or oval disks about 2 to 4 μm in diameter.

Three important physical properties of platelets, namely, agglutination, adhesiveness, and aggregation, make attempts at classification on the basis of size or shape in dry blood smears all but impossible. As soon as blood is removed

Table 13-2 Blood cells

Cells	Number	Function	Formation (hemopoiesis)	Destruction
Red blood cells (erythrocytes)	4.5 to 5.5 million/mm³ (total of approximately 30 trillion in adult body)	Transport oxygen and carbon dioxide	Red marrow of bones (myeloid tissue)	Reticuloendothelial cells in lining of blood vessels in liver, spleen, and bone marrow phagocytose old red blood cells; live about 105 to 120 days in bloodstream
White blood cells (leukocytes)	Usually about 5,000 to 9,000/mm³	Play important part in producing immunity, for example, phagocytosis by neutrophils and monocytes; antibody formation by lymphocytes and their descendants, plasma cells; production of *heparin* by basophils; detoxification by eosinophils	Granular leukocytes in red marrow; before birth and for a few months after, some lymphocytes are formed in thymus gland; later, most lymphocytes and monocytes formed in lymph nodes and other lymphatic tissues	Not known definitely; probably some destroyed by phagocytosis
Platelets (thrombocytes)	150,000 to 350,000/mm³	Initiate blood clotting and hemostasis	Red marrow, lungs, and spleen	Unknown

from a vessel, the platelets adhere to each other and to every surface they contact; in so doing, they assume a variety of shapes and irregular forms.

Platelet counts in adults average about 250,000 per mm³ of blood. A range of 150,000 to 350,000 per mm³ is considered normal. Newborn infants often show reduced counts, but these rise gradually to reach normal adult values at about 3 months of age. There are no differences between the sexes in platelet count.

Functions

Platelets play an important role in both hemostasis (from the Greek *stasis*, a standing) and blood clotting. The two, although interrelated, are separate and distinct functions. Hemostasis refers to the stoppage of blood *flow* and may occur as an end result of any one of several body defense mechanisms. The role that platelets play in the operation of the blood clotting mechanism is discussed on p. 361.

Within 1 to 5 seconds after injury to a blood capillary, platelets will adhere to the damaged lining of the vessel and to each other to form a hemostatic platelet plug that helps to stop the flow of blood into the tissues. At least one of the prostaglandins (PGE₂) and certain prostaglandin-like substances called *thromboxanes*, which are both found in platelets, play roles in hemostasis and blood clotting. When released, these substances affect both local blood flow (by vasoconstriction) and platelet aggregation at the site of injury. If the injury is extensive, the blood clotting mechanism is activated to assist in hemostasis.

Formation and life span

Platelets are formed in the red bone marrow, lungs, and, to some extent, in the spleen by fragmentation of very large (40 to 170 μm) cells known as megakaryocytes. Platelets have a short life span, an average of about 10 days. A summary of the basic facts about blood cells is given in Table 13-2.

Blood types (or blood groups)

The term blood type refers to the type of antigens* (called *agglutinogens*) present on red blood cell membranes. Antigens A, B, and Rh are the most important blood antigens as far as transfusions and newborn survival are concerned. Many other antigens have also been identified, but they are less important clinically and are too complex to discuss here. Every person's blood belongs to one of the four AB blood groups and, in addition, is either Rh positive or Rh negative. Blood types are named according to the antigens present on red blood cell membranes. Here, then, are the four AB blood types:

1 *Type A*—antigen A on red blood cells
2 *Type B*—antigen B on red blood cells
3 *Type AB*—both antigen A and antigen B on red blood cells
4 *Type O*—neither antigen A nor antigen B on red blood cells

The term *Rh-positive blood* means that Rh antigen is present on its red blood cells. *Rh-negative blood*, on the other hand, is blood whose red cells have no Rh antigen present on them.

Blood plasma may or may not contain antibodies that can react with red blood cell antigens A, B, and Rh. An important principle about this is that plasma never contains antibodies against the antigens present on its own red blood cells—for obvious reasons. If it did, the antibody would react with the antigen and thereby destroy the red blood cells. But (and this is an equally important principle) plasma does contain antibodies against antigen A or antigen B if they are *not* present on its red blood cells. Applying these two principles: In type A blood, antigen A is present on its red blood cells; therefore its plasma contains no anti-A antibodies but does contain anti-B antibodies. In type B blood, antigen B is present on its red blood cells; therefore its plasma contains no anti-B antibodies

---

*Antigen—substance capable of stimulating formation of other substances called antibodies that can combine with the antigen, for example, to agglutinate or clump it.

but does contain anti-A antibodies. What antigens do you deduce are present on the red blood cells of type AB blood? What antibodies, if any, does its plasma contain?*

No blood normally contains anti-Rh antibodies. However, anti-Rh antibodies can appear in the blood of an Rh-negative person provided Rh-positive red blood cells have at some time entered his bloodstream. One way this can happen is by giving an Rh-negative person a transfusion of Rh-positive blood. In a short time, his body makes anti-Rh antibodies, and these remain in his blood. There is one other way in which Rh-positive red blood cells can enter the bloodstream of an Rh-negative individual—but this can happen only to a woman. If she becomes pregnant, and if her mate is Rh positive, and if the fetus, too, is Rh positive, some of the red blood cells of the fetus may find their way into her blood from the fetal blood capillaries (via the placenta). These Rh-positive red blood cells then stimulate her body to form anti-Rh antibodies. Briefly, the only people who can ever have anti-Rh antibodies in their plasma are Rh-negative men or women who have been transfused with Rh-positive blood or Rh-negative women who have carried an Rh-positive fetus.

Anti-A, anti-B, and anti-Rh antibodies are agglutinins. *Agglutinins* are chemicals that agglutinate cells, that is, make them stick together in clumps. The danger in giving a blood transfusion is that antibodies present in the plasma of the person receiving the transfusion (the recipient's plasma) may agglutinate the donor's red blood cells. If, for example, type B blood were used for transfusing a person who had type A blood, the anti-B antibodies in the recipient's blood would agglutinate the donor's type B red blood cells. These clumped cells are potentially lethal. They can plug vital small vessels and cause the recipient's death.

Type O blood is referred to as *universal donor*

blood, a term that implies that it can safely be given to any recipient. This, however, is not true because the recipient's plasma may contain agglutinins other than anti-A and anti-B antibodies. For this reason the recipient's and the donor's blood—even if it is type O—should be cross-matched, that is, mixed and observed for agglutination of the donor's red blood cells.

Universal recipient (type AB) blood contains neither anti-A nor anti-B antibodies, so it cannot agglutinate type A or type B donor's red blood cells. This does not mean, however, that any type of donor blood may be safely given to an individual who has type AB blood without first cross matching. Other agglutinins may be present in the so-called universal recipient blood and clump unidentified antigens (agglutinogens) in the donor's blood.

Blood plasma

Plasma is the liquid part of blood—whole blood minus its cells, in other words. In the laboratory, whole blood is centrifuged to form plasma. This consists of a rapid whirling process that hurls the blood cells to the bottom of the centrifuge tube. A clear, straw-colored fluid—blood plasma—lies above the cells. Plasma consists of 90% water and 10% solutes.* By far the largest quantity of these solutes are proteins; normally, they constitute about 6% to 8% of the plasma. Other solutes present in much smaller amounts in plasma are food substances (principally glucose, amino acids, and lipids), compounds formed by metabolism (for example, urea, uric acid, creatinine, and lactic acid),

---

*Type AB blood—antigens A and B present on red blood cells; no anti-A nor anti-B antibodies in plasma.

*Some of the solutes present in blood plasma are true solutes or crystalloids. Others are colloids. *Crystalloids* are solute particles less than 1 nm in diameter (ions, glucose, and other small molecules). *Colloids* are solute particles from 1 to about 100 nm in diameter (for example, proteins of all types). Blood solutes also may be classified as *electrolytes* (molecules that ionize in solution) or *nonelectrolytes*—examples: inorganic salts and proteins are electrolytes; glucose and lipids are nonelectrolytes.

respiratory gases (oxygen and carbon dioxide), and regulatory substances (hormones, enzymes, and certain other substances).

The proteins in blood plasma consist of three main kinds of compounds: albumins, globulins, and fibrinogen. Electrophoretic methods of measuring the amounts of these compounds indicate that 100 ml of plasma contains a total of approximately 6 to 8 gm of protein. Albumins constitute about 55% of this total, globulins about 38%, and fibrinogen about 7%.

Plasma proteins are crucially important substances. Fibrinogen, for instance, and an albumin named prothrombin play key roles in the blood-clotting mechanism. Globulins function as essential components of the immunity mechanism—circulating antibodies (immune bodies) are modified gamma globulins. All plasma proteins contribute to the maintenance of normal blood viscosity, blood osmotic pressure, and blood volume. Therefore plasma proteins play an essential part in maintaining normal circulation. Synthesis of plasma proteins takes place in liver cells. They form all kinds of plasma proteins except some of the gamma globulins—specifically, the circulating antibodies synthesized by plasma cells.

Blood coagulation

Purpose

The purpose of blood coagulation is obvious—to plug up ruptured vessels so as to stop bleeding and prevent loss of a vital body fluid.

Mechanism

Because of the function of coagulation, the mechanism for producing it must be swift and sure when needed, such as when a vessel is cut or ruptured. Equally important, however, coagulation needs to be prevented from happening when it is not needed because clots can plug up vessels that must stay open if cells are to receive blood's life-sustaining cargo of oxygen.

What makes blood coagulate? Over a period of many years a host of investigators have searched for the answer to this question. They have tried to find out what events compose the coagulation mechanism and what sets it in operation. They have succeeded in gathering an abundance of relevant information. But still the questions about this complicated and important process outnumber the answers. We shall discuss some of the answers and not delve too far into the questions.

The blood-coagulation mechanism presumably consists of a series of chemical reactions that take place in a definite and rapid sequence. The trigger that starts the first of these changes, that initiates blood clotting, is the appearance of a "rough" spot in the lining of a blood vessel—examples: a cut edge of a vessel, or most common of all, plaques (patchlike deposits) of a cholesterol-lipid substance. Within a matter of 1 or 2 seconds, clumps of platelets adhere to any portion of a blood vessel that loses its normal, perfectly smooth quality and release a variety of substances. Some accelerate coagulation* and others, notably serotonin (5-HT or 5-hydroxytryptamine) and thromboxane constrict blood vessels. After the platelets release these substances a series of chemical reactions takes place in rapid-fire succession. Although knowledge about these reactions is still tentative, this much is known—they cannot take place in the absence of calcium ions and certain phospholipids, and they result in the conversion of prothrombin to thrombin. Prothrombin and fibrinogen are the proteins involved in clotting that are made in the liver and are normally present in blood plasma.† Prothrombin is converted to

---

*VII—serum prothrombin conversion accelerator, or proconvertin; VIII—antihemophilic factor (AHF); IX—plasma thromboplastin component (PTC or Christmas factor); X—Stuart factor; XI—plasma thromboplastin antecedent (PTA); XII—Hageman factor.

†Normal prothrombin content of plasma = 10 to 15 mg per 100 ml. Normal fibrinogen content = 350 mg per 100 ml of plasma.

thrombin, and thrombin then converts fibrinogen to fibrin. We can summarize these changes by the following simple equations:

$$\text{Prothrombin} \xrightarrow{\text{(platelet factors, Ca}^{++}\text{ tissue factors)}} \text{Thrombin}$$

$$\text{Fibrinogen} \xrightarrow{\text{(thrombin)}} \text{Fibrin}$$

Prothrombin is normally present in plasma but is an inactive substance. Thrombin, on the other hand, is an active protein (an enzyme) and is not one of the proteins usually present in plasma. It must be formed from the normal plasma protein prothrombin. As thrombin forms, it catalyzes reactions that involve another plasma protein. Thrombin accelerates the conversion of the soluble plasma protein fibrinogen to insoluble fibrin. Fibrin appears in blood as fine threads all tangled together. Blood cells catch in the entanglement, and because most of the cells are red blood cells, clotted blood has a red color. The pale yellowish liquid left after a clot forms is *blood serum*. How do you think serum differs from plasma? What is plasma? To check your answers, see Fig. 13-3.

Liver cells synthesize both prothrombin and fibrinogen, as they do almost all other plasma proteins. In order for the liver to synthesize prothrombin at a normal rate, blood must contain an adequate amount of vitamin K. Vitamin K is absorbed into the blood from the intestine. Some foods contain this vitamin, but it is also synthesized in the intestine by certain bacteria (not present for a time in newborn infants). Because vitamin K is fat soluble, its absorption

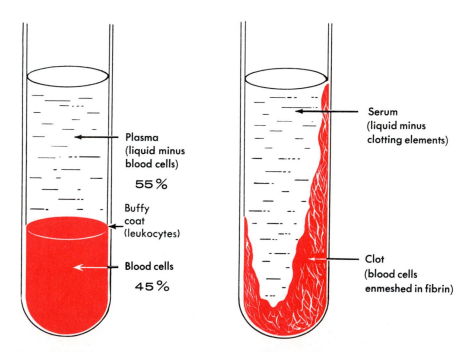

Plasma (liquid minus blood cells) **55%**

Buffy coat (leukocytes)

Blood cells **45%**

Serum (liquid minus clotting elements)

Clot (blood cells enmeshed in fibrin)

Fig. 13-3 Difference between blood plasma and blood serum. Plasma is whole blood minus cells. Serum is whole blood minus the clotting elements. Plasma is prepared by centrifuging blood. Serum is prepared by clotting blood.

requires bile. If, therefore, the bile ducts become obstructed and bile cannot enter the intestine, a vitamin K deficiency develops. The liver cannot then produce prothrombin at its normal rate, and the blood's prothrombin concentration soon falls below normal. A prothrombin deficiency gives rise to a bleeding tendency. As a preoperative safeguard, therefore, patients with obstructive jaundice are generally given some kind of vitamin K preparation.

Factors that oppose clotting

Although blood clotting probably goes on continuously and concurrently with clot dissolution (fibrinolysis), several factors operate to oppose clot formation in intact vessels. Most important by far is the perfectly smooth surface of the normal endothelial lining of blood vessels. Platelets do not adhere to it; consequently, they do not disintegrate and release platelet factors into the blood and, therefore, the blood-clotting mechanism does not get started in normal blood vessels. As an additional deterrent to clotting, blood contains certain substances called *antithrombins*. The name suggests their function—they oppose or inactivate thrombin. Thus antithrombins prevent thrombin from converting fibrinogen to fibrin. *Heparin*, a natural constituent of blood, acts as an antithrombin. It was first prepared from liver (hence its name), but various other organs also contain heparin. Its normal concentration in blood is too low to have much effect in keeping blood fluid. However, injections of heparin are used to prevent clots from forming in vessels. Coumarin compounds impair the liver's utilization of vitamin K and thereby slow its synthesis of prothrombin and factors VII, IX, and X. Indirectly, therefore, coumarin compounds retard coagulation.

Factors that hasten clotting

Two conditions particularly favor thrombus formation: a rough spot in the endothelium (blood vessel lining) and abnormally slow blood flow. Atherosclerosis, for example, is associated with an increased tendency toward thrombosis because of endothelial rough spots in the form of plaques of accumulated cholesterol-lipid material. Immobility, on the other hand, may lead to thrombosis because blood flow slows down as movements decrease. Incidentally, this fact is one of the major reasons why physicians insist that bed patients must either move or be moved frequently. Presumably, sluggish blood flow allows thromboplastin to accumulate sufficiently to reach a concentration adequate for clotting.

Once started, a clot tends to grow. Platelets enmeshed in the fibrin threads disintegrate, releasing more thromboplastin, which in turn causes more clotting, which enmeshes more platelets, and so on, in a vicious circle. Clot-retarding substances, available in recent years, have proved valuable for retarding this process.

Clot dissolution

Fibrinolysis is the physiological mechanism that dissolves clots. Newer evidence indicates that the two opposing processes of clot formation and fibrinolysis go on continuously. Dr. George Fulton of Boston University has presented one bit of dramatic evidence. He took micromovies that show tiny blood vessels rupturing under apparently normal circumstances and clots forming to plug them. Blood contains an enzyme, fibrinolysin, that catalyzes the hydrolysis of fibrin, causing it to dissolve. Many other factors, however, presumably also take part in clot dissolution, for instance, substances that activate profibrinolysin (inactive form of fibrinolysin). Streptokinase, an enzyme from certain streptococci, can act this way and so can cause clot dissolution and even hemorrhage.

Clinical methods of hastening clotting

One way of treating excessive bleeding is to speed up the blood-clotting mechanism. The principle involved is apparent—to increase any of the substances essential for clotting. Application of this principle is accomplished in the following ways:

1 By applying a rough surface such as gauze, by applying heat, or by gently squeezing the tissues around a cut vessel. Each of these procedures causes more platelets to disintegrate and release more platelet factors. This, in turn, accelerates the first of the clotting reactions.

2 By applying purified thrombin (in the form of sprays or impregnated gelatin sponges that can be left in a wound). Which stage of the clotting mechanism does this accelerate?

3 By applying fibrin foam, films, etc.

Outline summary

Blood

A Primary function—transportation of various substances to and from body cells; exchange of materials between respiratory, digestive, and excretory organs and blood and between blood and cells

B Secondary functions—contributes to all bodily functions, for example
1 Cellular metabolism
2 Homeostasis of fluid volume
3 Homeostasis of pH
4 Homeostasis of temperature
5 Defense against microorganisms

Blood volume

A Measurement—by direct and indirect methods using radioisotopes

B Effect of body fat—blood volume per kilogram of body weight varies inversely with the amount of excess body fat

C Average volume—70 kg adult male averages 71 ml/per kg of blood—about 5,000 ml total

Blood cells

A Three main kinds—red blood cells (erythrocytes), white blood cells (leukocytes), and platelets thrombocytes); see Fig. 13-1

B Whole blood consists of approximately 55% fluid (blood plasma) and 45% blood cells

Erythrocytes (red blood cells)

A Appearance, size, shape, and number—biconcave disks about 7 μm in diameter with large surface area relative to volume
1 Biconcave shape permits "deformity" to occur without injury
  a Prostaglandin (PGE_1) increases deformability; PGE_2 decreases it
  b Numbers of RBCs—male: 5,500,000 per mm^3; female, 4,800,000 per mm^3

B Structure and functions—millions of molecules of hemoglobin inside each red blood cell makes possible red blood cell functions of oxygen and carbon dioxide transport
1 Anemia—classification according to size and hemoglobin content
  a Size classification—normocytic, macrocytic, and microcytic
  b Classification by hemoglobin content—normochromic, hyperchromic, and hypochromic

C Formation (erythropoiesis)—by myeloid tissue (red bone marrow)

D Destruction—by fragmentation in capillaries; reticuloendothelial cells phagocytose red blood cell fragments and break down hemoglobin to yield iron-containing pigment and bile pigments; bone marrow reuses most of iron for new red blood cell synthesis; liver excretes bile pigments; life span of red cells about 120 days according to studies made with radioactive isotopes

E Erythrocyte homeostatic mechanism—stimulus thought to be tissue hypoxia; for description of mechanism, see Fig. 13-2

F Functions—see Table 13-2

Leukocytes (white blood cells)

A Appearance and size—vary; some relatively large cells, for example, monocytes, and some small cells, for example, small lymphocytes; nuclei vary from spherical to S shaped to lobulated; cytoplasm of neutrophils, eosinophils, and basophils contains granules that take neutral, acid, and basic stains, respectively

B Functions—defense or protection, for example, by carrying on phagocytosis

C Formation—in myeloid tissue (red bone marrow) for granular leukocytes and probably monocytes; lymphocytes formed in lymphatic tissue, that is, mainly in lymph nodes and spleen

D Destruction and life span—some destroyed by phagocytosis; life span unknown

E Numbers—about 5,000 to 10,000 leukocytes per mm^3 of blood; for differential count, see Table 13-1

F Functions—see Table 13-2

Platelets

A Appearance and size—platelets are small fragments of cells

B Functions—blood coagulation and hemostasis

C Prostaglandin (PGE_1) and thromboxane—role in hemostasis related to vasoconstriction and increase in platelet aggregation at site of injury

D Formation and life span—formed in red bone marrow, lungs, and spleen by fragmentation of large cells (megakaryocytes); life span about 10 days

E Functions—see Table 13-2

Blood types (or blood groups)

A Names—indicate type of antigen agglutinogen on red blood cell membranes

B Every person's blood belongs to one of the four AB blood groups (type A, type B, type AB, or type O) and, in addition, is either Rh positive or Rh negative

C Plasma does not normally contain antibodies against antigens present on its red blood cells but does normally contain antibodies against A or B antigens not present on its red blood cells

D No blood normally contains anti-Rh antibodies; anti-Rh antibodies appear in blood only of an Rh-negative person and only after Rh-positive red blood cells have entered the bloodstream, for example, by transfusion or by carrying an Rh-positive fetus

E Universal donor: type O; universal recipient: type AB but cross matching should be done before use for safety

Blood plasma

A Liquid part of blood or whole blood minus its cells; constitutes about 55% of total blood volume

B Composition
1 About 90% water and 10% solutes
2 Most solutes crystalloids but some colloids
3 Most solutes electrolytes but some nonelectrolytes
4 Solutes include foods, wastes, gases, hormones, enzymes, vitamins, and antibodies and other proteins

C Plasma proteins
1 Types—albumins, globulins, and fibrinogen
2 Total plasma proteins—6 to 8 gm per 100 ml of plasma
3 Albumins—about 55% of total plasma proteins
4 Globulins—about 38% of total plasma proteins
5 Fibrinogen—about 7% of total plasma proteins
6 Functions—contribute to blood viscosity, osmotic pressure, and volume; prothrombin (an albumin) and fibrinogen play key roles in blood clotting; modified gamma globulins are circulating antibodies, essential for immunity
7 All plasma proteins except circulating antibodies are synthesized in liver cells

Blood coagulation

A Purpose—to plug up ruptured vessels and thus prevent fatal hemorrhage

B Mechanism—complicated process, still incompletely understood
1 Triggered by appearance of rough spot in blood vessel lining

2 Clumps of platelets adhere to rough spot; they release "platelet factors" that initiate a series of rapidly occurring chemical reactions

3 Last two clotting reactions convert prothrombin to thrombin, an enzyme that catalyzes conversion of soluble fibrinogen to insoluble fibrin

4 Clot consists of tangled mass of fibrin threads and blood cells; blood serum is fluid left after clot forms

C Factors that oppose clotting

   1 Smoothness of endothelium that lines blood vessels prevents platelets' adherence and consequent disintegration; some platelet disintegration continuously despite this preventive

   2 Blood normally contains certain anticoagulants, for example, antithrombins, substances that inactivate thrombin so that it cannot catalyze fibrin formation

D Factors that hasten clotting

   1 Endothelial "rough" spots, for example, cholesterol-lipid plaques in atherosclerosis

   2 Sluggish blood flow

E Pharmaceutical preparations that retard clotting—commercial heparin, bishydroxycoumarin (Dicumarol), citrates (for transfusion blood)

F Clot dissolution—process called fibrinolysis; presumably this and opposing process (clot formation) go on all the time

G Clinical methods of hastening clotting

   1 Apply rough surfaces to wound to stimulate platelets and tissues to liberate more thromboplastin

   2 Apply purified thrombin

   3 Apply fibrin foam, film, etc.

Review questions

1 Compare different kinds of blood cells as to appearance and size; functions; formation, destruction, and life span; number per cubic millimeter of blood.

2 What is the effect of body fat on blood volume?

3 Explain the difference between direct and indirect methods used to measure blood volume.

4 Explain the effect of prostaglandins PGE_1 and PGE_2 on red blood cell "deformability."

5 Classify the various types of anemia according to size of red blood cell and hemoglobin content.

6 Describe the role of platelets in hemostasis and blood clotting.

7 A person suffering from a severe parasitic infestation, such as hookworms, would have an elevation in what leukocyte cell type?

8 "Graft-rejection cells" is a nickname for which kind of blood cells? Why?

9 What type of cells secrete "circulating" antibodies?

10 What is the function of circulating antibodies? Of cellular antibodies?

11 Since plasma cells synthesize and secrete circulating antibodies, what organelles would you expect to be prominent or abundant in plasma cells? (Review p. 37 if you cannot answer this question.)

12 Suppose your doctor has told you that you have a lymphocyte count of 2,000 per mm^3. Would you think this was normal or abnormal?

13 Why might a surgeon give antilymphocyte serum to a patient who had had a kidney transplant?

14 Explain what the term hematocrit means. Is it normally less than 50? More than 55?

15 What triggers blood clotting?

16 Write two equations to show the basic chemical reactions that produce a blood clot.

17 Why and how does a vitamin K deficiency affect blood clotting?

18 Explain some principles and methods by which blood clotting may be hastened.

19 Explain what these terms mean: type AB blood, Rh-negative blood.

20 Suppose you have type B blood. Which two of the following kinds of blood might be used for transfusions for you: type A? type B? type AB? type O?

21 Which of the following infants would be most likely to have erythroblastosis fetalis: fourth child of Rh-positive mother and Rh-negative father? first child of Rh-negative mother and Rh-positive father? second child of Rh-negative mother and Rh-positive father? Explain your reasoning.

22 The cells of what organ synthesize most plasma proteins? What is the normal plasma protein concentration?

23 What are some of the functions served by plasma proteins?

chapter 14

Anatomy of the cardiovascular system

The cardiovascular system is sometimes called the blood-vascular or simply the circulatory system. It consists of the heart, which is a muscular pumping device, and a closed system of vessels called arteries, veins, and capillaries. As the name implies, blood contained in the circulatory system is pumped by the heart around a closed circle or circuit of vessels as it passes again and again through the various "circulations" of the body (discussed on p. 394).

As in the adult, survival of the developing embryo depends on the circulation of blood to maintain homeostasis and a favorable cellular environment. In response to this need the cardiovascular system makes its appearance early in development and reaches a functional state long before any other major organ system. Incredible as it seems, the primitive heart begins to beat regularly early in the fourth week following fertilization!

Heart
Location, size, and position

The human heart is a four-chambered muscular organ, shaped and sized roughly like a man's closed fist. It lies in the mediastinum just behind the body of the sternum between the points of attachment of the second through the sixth ribs. Approximately two thirds of its mass is to the left of the midline of the body and one third to the right.

Posteriorly the heart rests on the bodies of the

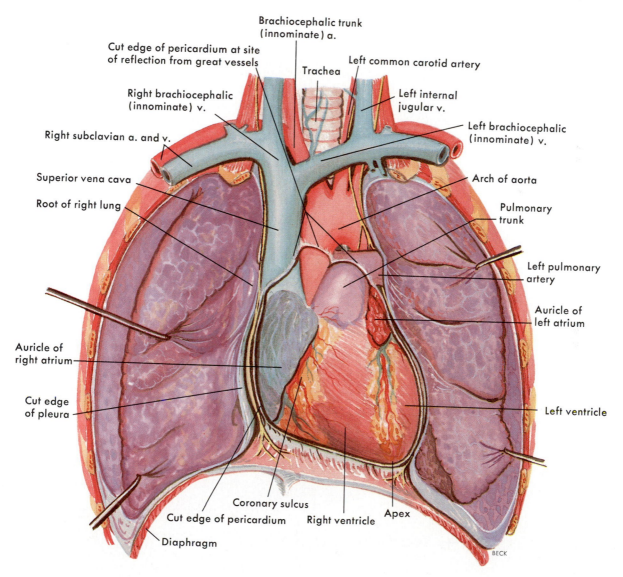

Brachiocephalic trunk (innominate) a.

Cut edge of pericardium at site of reflection from great vessels

Trachea

Left common carotid artery

Right brachiocephalic (innominate) v.

Left internal jugular v.

Right subclavian a. and v.

Left brachiocephalic (innominate) v.

Superior vena cava

Arch of aorta

Root of right lung

Pulmonary trunk

Left pulmonary artery

Auricle of left atrium

Auricle of right atrium

Cut edge of pleura

Left ventricle

Coronary sulcus

Cut edge of pericardium

Right ventricle

Apex

Diaphragm

BECK

Fig. 14-1 Anatomic relationship of the heart to other structures in the thoracic cavity.

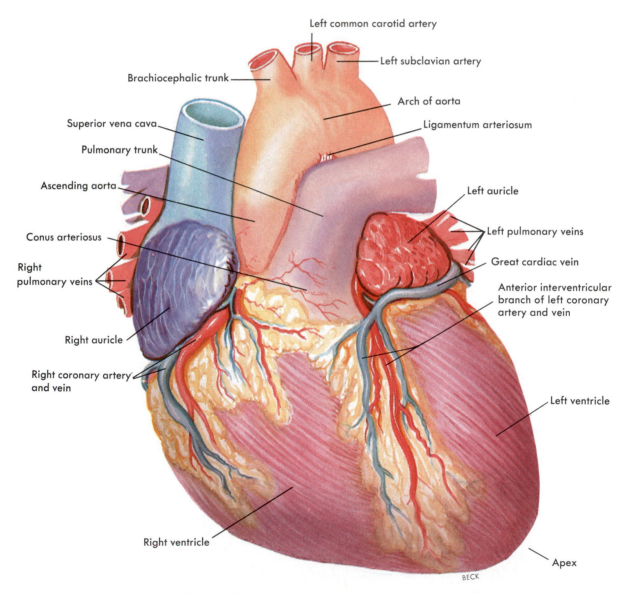

Left common carotid artery

Left subclavian artery

Brachiocephalic trunk

Arch of aorta

Superior vena cava

Ligamentum arteriosum

Pulmonary trunk

Ascending aorta

Left auricle

Conus arteriosus

Left pulmonary veins

Right
pulmonary veins

Great cardiac vein

Anterior interventricular
branch of left coronary
artery and vein

Right auricle

Right coronary artery
and vein

Left ventricle

Right ventricle

Apex

BECK

Fig. 14-2 Anterior view of the heart and great vessels.

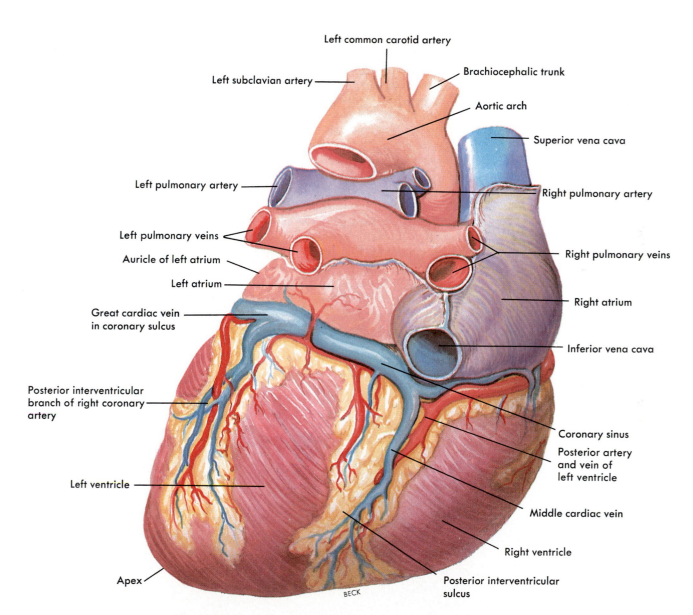

Left common carotid artery

Left subclavian artery

Brachiocephalic trunk

Aortic arch

Superior vena cava

Left pulmonary artery

Right pulmonary artery

Left pulmonary veins

Auricle of left atrium

Left atrium

Right pulmonary veins

Right atrium

Great cardiac vein in coronary sulcus

Inferior vena cava

Posterior interventricular branch of right coronary artery

Coronary sinus

Posterior artery and vein of left ventricle

Left ventricle

Middle cardiac vein

Right ventricle

Apex

Posterior interventricular sulcus

BECK

Fig. 14-3 Posterior view of the heart and great vessels.

fifth to the eighth thoracic vertebrae. Because of its placement between the sternum in front and the bodies of the thoracic vertebrae behind, it can be compressed by application of pressure to the lower portion of the body of the sternum using the heel of the hand. Rhythmic compression of the heart in this way can maintain blood flow in cases of cardiac arrest and, if combined with effective artificial respiration, the resulting procedure, called *cardiopulmonary resuscitation (CPR)*, can be lifesaving. The anatomical rela-

tionship of the heart to other structures in the thoracic cavity is shown in Fig. 14-1.

The lower border of the heart, which forms a blunt point known as the *apex*, lies on the diaphragm, pointing toward the left. To count the apical beat, one must place a stethoscope directly over the apex, that is, in the space between the fifth and sixth ribs (fifth intercostal space) on a line with the midpoint of the left clavicle.

The upper border of the heart, that is, its base, lies just below the second rib. The boundaries,

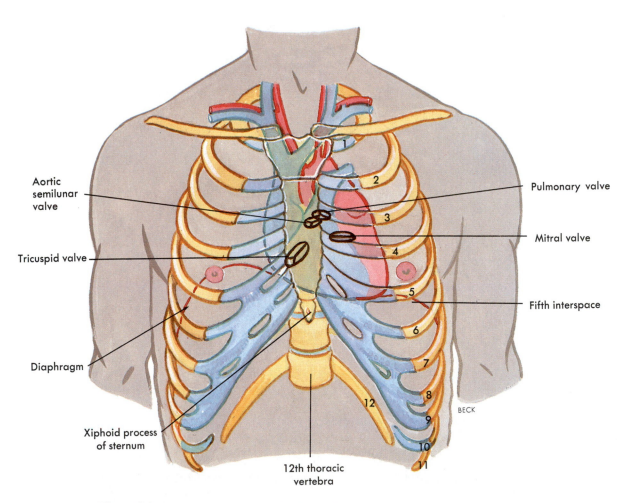

Fig. 14-4 Thoracic projection of the boundaries and valves of the heart on the chest wall.

which, of course, indicate its size, have considerable clinical importance, since a marked increase in heart size accompanies certain types of heart disease. Therefore when diagnosing heart disorders, the doctor charts the boundaries of the heart. The "normal" boundaries of the heart are, however, influenced by such factors as age, body build, and state of contraction.

At birth the heart is said to be transverse in type and appears large in proportion to the diameter of the chest cavity. In the infant, it is $1/130$ of the total body weight compared to about $1/300$ in the adult. Between puberty and 25 years of age the heart attains its adult shape and weight—about 310 gm is average for the male and 255 gm for the female.

In the adult the shape of the heart tends to resemble that of the chest. In tall, thin individuals the heart is frequently described as elongated, whereas in short, stocky individuals it has greater width and is described as transverse. In individuals of average height and weight it is neither long nor transverse but somewhat oblique and intermediate between the two. Its approximate dimensions are length, 12 cm; width, 9 cm; and depth, 6 cm. Fig. 14-2 shows the heart and great vessels in an anterior view and Fig. 14-3 shows a posterior view.

Surface projection

Fig. 14-4 shows several of the surface relationships between the adult heart and the thoracic cage. It is important to remember, however, that considerable variation within the normal range makes a precise "surface projection" outline of the boundaries or valves of the heart on the chest wall difficult.

Covering

Structure

The heart has its own special covering, a loose-fitting inextensible sac called the *pericardium*. The pericardium consists of two parts: a fibrous portion and a serous portion (Fig. 14-5). The sac itself is made of tough white fibrous tissue but is lined with smooth, moist serous membrane—the parietal layer of the serous pericardium. The same kind of membrane covers the entire outer surface of the heart. This covering layer is known as the visceral layer of the serous pericardium or as the *epicardium*. The fibrous sac attaches to the large blood vessels emerging from the top of the heart but not to the heart itself. Therefore it fits loosely around the heart, with a slight space between the visceral layer adhering to the heart and the parietal layer adhering to the inside of the fibrous sac. This space is called the *pericardial space*. It contains a few drops of lubricating fluid secreted by the serous membrane and called *pericardial fluid*.

The structure of the pericardium can be summarized in outline form as follows:

1 *Fibrous pericardium*—tough, loose-fitting and inelastic sac around the heart
2 *Serous pericardium*—consisting of two layers
  a *Parietal layer*—lining inside of the fibrous pericardium
  b *Visceral layer (epicardium)*—adhering to the outside of the heart; between visceral and parietal layers is a potential space, the pericardial space, which contains a few drops of pericardial fluid

Function

The fibrous pericardial sac with its smooth, well-lubricated lining provides protection against friction. The heart moves easily in this loose-fitting jacket with no danger of irritation from friction between the two surfaces as long as the serous pericardium remains normal. If, however, it becomes inflamed (pericarditis) and too much pericardial fluid, fibrin, or pus develops in the pericardial space, the visceral and parietal layers may stick together. In addition, inability of the fibrous pericardium to stretch as fluid accumulates in the pericardial space results in a rapid increase in pressure around the heart. If the resulting compression of the heart, called *cardiac tamponade*, seriously hampers

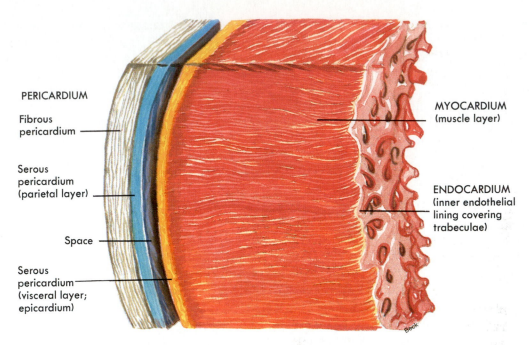

PERICARDIUM

Fibrous
pericardium

Serous
pericardium
(parietal layer)

Space

Serous
pericardium
(visceral layer;
epicardium)

MYOCARDIUM
(muscle layer)

ENDOCARDIUM
(inner endothelial
lining covering
trabeculae)

Fig. 14-5 Section of the heart wall showing the components of the outer pericardium (heart sac), muscle layer (myocardium), and inner lining (endocardium).

pumping efficiency, it sometimes becomes necessary to remove the fibrous pericardium with its lining of parietal serous membrane in order for the heart to continue functioning.

Structure

Wall

Three distinct layers of tissue make up the heart wall (Fig. 14-5) in both the atria and the ventricles.

1 The outer layer of the heart wall is called the *epicardium*—the visceral layer of the serous pericardium already described.
2 The bulk of the heart wall is the thick, contractile, middle layer of specially constructed and arranged cardiac muscle cells called the *myocardium*. The minute structure of cardiac muscle has been described in Chap-

ter 3. Look at the electron micrograph of cardiac muscle (Fig. 3-7, p. 61), and note the large number of mitochondria and the arrangement of the contractile components. As you can see, many of the same microscopic structural parts found in skeletal muscle cells and described in Chapter 6 can also be identified in cardiac muscle. On its inner surface the myocardium is raised into ridgelike projections, the papillary muscles (Figs. 14-6 and 14-7).
3 The lining of the interior of the myocardial wall is a delicate layer of endothelial* tissue known as the *endocardium*.

*Endothelial tissue resembles simple squamous epithelial tissue in that it consists of a single layer of flat cells. It differs from epithelial tissue in that it arises from the mesoderm layer of the embryo, whereas epithelial tissue arises from the ectoderm.

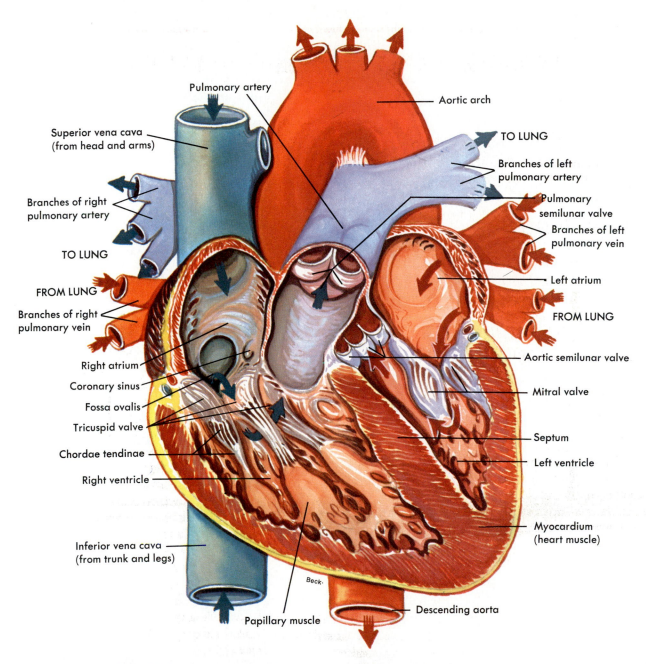

Pulmonary artery

Superior vena cava
(from head and arms)

Branches of right
pulmonary artery

TO LUNG

FROM LUNG

Branches of right
pulmonary vein

Right atrium

Coronary sinus

Fossa ovalis

Tricuspid valve

Chordae tendinae

Right ventricle

Inferior vena cava
(from trunk and legs)

Beck.

Papillary muscle

Aortic arch

TO LUNG

Branches of left
pulmonary artery

Pulmonary
semilunar valve

Branches of left
pulmonary vein

Left atrium

FROM LUNG

Aortic semilunar valve

Mitral valve

Septum

Left ventricle

Myocardium
(heart muscle)

Descending aorta

Fig. 14-6 Frontal section of the heart showing the four chambers, valves, openings, and major vessels. Arrows indicate direction of blood flow. Black arrows represent unoxygenated blood and red arrows oxygenated blood. The two branches of the right pulmonary vein extend from the right lung behind the heart to enter the left atrium.

Cavities

The interior of the heart is divided into four chambers, two upper and two lower. The upper cavities are named *atria** and the lower ones *ventricles* (Fig. 14-6). Of these, the ventricles are considerably larger and thicker walled than the atria because they carry a heavier pumping burden than the atria. Also, the left ventricle has thicker walls than the right for the same reason. It has to pump blood through all the vessels of the body, except those to and from the lungs, whereas the right ventricle sends blood only through the lungs.

Valves and openings

The heart valves are mechanical devices that permit the flow of blood in one direction only. Four sets of valves are of importance to the normal functioning of the heart (Figs. 14-6 to 14-

*The atria are sometimes called auricles. Strictly speaking, the latter term means the earlike flaps protruding from the atria. The terms atria and auricle should not be used synonymously.

8). Two of these, the cuspid (atrioventricular) valves, are located in the heart, guarding the openings between the atria and ventricles (atrioventricular orifices). The other two, the semilunar valves, are located inside the pulmonary artery and the great aorta just as they arise from the right and left ventricles, respectively.

The cuspid valve guarding the right atrioventricular orifice consists of three flaps of endocardium anchored to the papillary muscles of the right ventricle by several cordlike structures called *chordae tendineae*. Because this valve has three flaps, it is appropriately named the *tricuspid valve*. The valve that guards the left atrioventricular orifice is similar in structure to the tricuspid except that it has only two flaps and is, therefore, called the *bicuspid* or, more commonly, the *mitral valve*. (An easy way to remember which valve is on the right and which is on the left is this: The names whose first letters come nearest together in the alphabet go together—thus, L and M for *l*eft side, *m*itral valve, and R and T for *r*ight side, *t*ricuspid valve.)

The construction of both cuspid valves allows

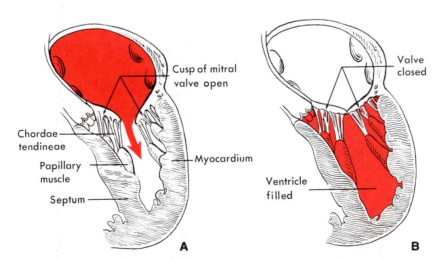

Fig. 14-7 Action of the cuspid (atrioventricular) valves. **A,** When the valves are open, blood passes freely from the atria to the ventricles. **B,** Filling of the ventricles closes the valves and prevents a backflow of blood into the atria when the ventricles contract.

blood to flow from the atria into the ventricles but prevents it from flowing back up into the atria from the ventricles. Ventricular contraction forces the blood in the ventricles hard against the cuspid valves, closing them and thereby ensuring the movement of the blood upward into the pulmonary artery and aorta as the ventricles contract (Fig. 14-7).

The *semilunar valves* consist of half-moon–shaped flaps growing out from the lining of the pulmonary artery and great aorta. When these valves are closed, as in Fig. 14-8, blood fills the spaces between the flaps and the vessel wall. Each flap then looks like a tiny filled bucket. Inflowing blood smooths the flaps against the blood vessel walls, collapsing the buckets and thereby opening the valves. Closure of the semilunar valves, as of the cuspid valves, simultaneously prevents backflow and ensures forward flow of blood in places where there would otherwise be considerable backflow. Whereas the cuspid valves prevent blood from flowing back up into the atria from the ventricles, the semilunar valves prevent it from flowing back down into the ventricles from the aorta and pulmonary artery.

Any one of the four valves may lose its ability to close tightly. Such a condition is known as *valvular insufficiency* or, because it permits blood to "leak" back into the part of the heart from which it came, "leakage of the heart." Leakage of blood through a diseased or malformed heart valve often creates turbulence and abnormal sounds called *murmurs*. Identification of murmurs may contribute important information to the recognition and diagnosis of heart disease.

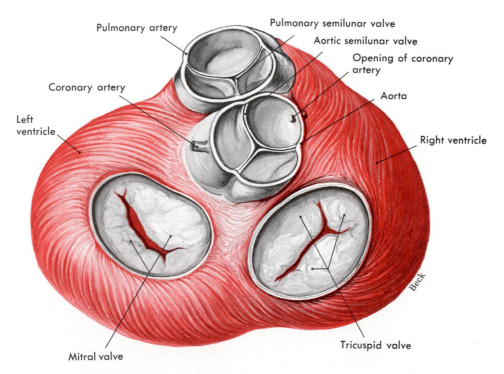

Fig. 14-8 Valves of the heart viewed from above. The atria are removed to show the mitral and tricuspid valves.

Mitral stenosis is an abnormality in which the left atrioventricular orifice becomes narrowed by scar tissue that forms as a result of disease. This hinders the passage of blood from the left atrium to the left ventricle and leads to circulatory failure. But fortunately the marvel of open-heart surgery has made possible the correction of many valvular defects.

Blood supply

Myocardial cells receive blood by way of two small vessels, the right and left coronary arteries. Since the openings into these vitally important vessels lie behind flaps of the aortic semilunar valve, they come off of the aorta at its very beginning and are its first branches. Both right and left coronary arteries have two main branches, as shown in Table 14-1.

More than a half million Americans die every year from coronary disease and another 3.5 million or more are estimated to suffer some degree of incapacitation from this great killer.* Knowledge about the distribution of coronary artery branches, therefore, has the utmost practical importance. Here, then, are some principles about the heart's own blood supply that seem worth noting. Both ventricles receive their blood supply from branches of both the right and left coronary arteries. Each atrium, in contrast, re-

*Effler, D. B.: Surgery for coronary disease, Sci. Am. **219:**36-43, Oct., 1968.

Table 14-1 Coronary arteries

Right coronary artery	Left coronary artery
Divides into two main branches: 1 Posterior descending artery—sends branches to both ventricles 2 Marginal artery—sends branches to right ventricle and right atrium	Divides into two main branches: 1 Anterior descending artery—sends branches to ventricles 2 Circumflex artery—sends branches to left ventricle and left atrium

ceives blood only from a small branch of the corresponding coronary artery (Table 14-1). The most abundant blood supply of all goes to the myocardium of the left ventricle—an appropriate fact, since the left ventricle does the most work and so needs the most oxygen and nutrients delivered to it. The right coronary artery is dominant in about 50% of all hearts and the left coronary artery in about 20%, and in about 30% neither right nor left coronary artery dominates.

Another fact about the heart's own blood supply—and one of life-and-death importance—is the fact that only a few anastomoses exist between the larger branches of the coronary arteries. An *anastomosis* consists of one or more branches from the proximal part of an artery to a more distal part of itself or of another artery. Thus anastomoses provide detours that arterial blood can travel if the main route becomes obstructed. In short, they provide collateral circulation to a part. This explains why the scarcity of anastomoses between larger coronary arteries looms so large as a threat to life. If, for example, a blood clot plugs one of the larger coronary artery branches, as it frequently does in coronary thrombosis or embolism, too little or no blood can reach some of the heart muscle cells. They become ischemic, in other words. Deprived of oxygen, they release too little energy for their own survival. Myocardial infarction (death of ischemic heart muscle cells) soon results. There is another anatomical fact, however, that brightens the picture somewhat—many anastomoses do exist between very small arterial vessels in the heart, and, given time, new ones develop and provide collateral circulation to ischemic areas. In recent years, several surgical procedures have been devised to aid this process.*

After blood has passed through capillary beds in the myocardium, it enters a series of coronary veins before draining into the right atrium

*Effler, D. B.: Surgery for coronary disease, Sci. Am. **219:**36-43, Oct., 1968.

through a common venous channel called the *coronary sinus*. Several veins that collect blood from a small area of the right ventricle do not end in the coronary sinus but instead drain directly into the right atrium. As a rule of thumb, the coronary veins follow a course that closely parallels that of the coronary arteries.

Conduction system

Four structures—the sinoatrial node, atrioventricular node, atrioventricular bundle, and Purkinje fibers—compose the conduction system of the heart. Each of these structures consists of cardiac muscle modified enough in structure to differ in function from ordinary cardiac muscle. The main specialty of ordinary cardiac muscle is contraction. In this, it is like all muscle, and, like all muscle, ordinary cardiac muscle can also conduct impulses. But conduction alone is the specialty of the modified cardiac muscle that composes the conduction system structures.

Sinoatrial node. The sinoatrial node (SA node, Keith-Flack node, or pacemaker) consists of hundreds of cells located in the right atrial wall near the opening of the superior vena cava. (See Fig. 15-1, p. 400.)

Atrioventricular node. The atrioventricular node (AV node or node of Tawara), a small mass of special cardiac muscle tissue, lies in the right atrium along the lower part of the interatrial septum.

Atrioventricular bundle and Purkinje fibers. The atrioventricular bundle (AV bundle or bundle of His) is a bundle of special cardiac muscle fibers that originate in the AV node and that extend by two branches down the two sides of the interventricular septum. From there, they continue as the Purkinje fibers. The latter extend out to the papillary muscles and lateral walls of the ventricles. The functioning of the conduction system of the heart will be discussed in Chapter 15.

Nerve supply

Both divisions of the autonomic nervous system send fibers to the heart. Sympathetic fibers (contained in the middle, superior, and inferior cardiac nerves) and parasympathetic fibers (in branches of the vagus nerve) combine to form *cardiac plexuses* located close to the arch of the aorta. From the cardiac plexuses, fibers accompany the right and left coronary arteries to enter the heart. Here, most of the fibers terminate in the SA node, but some end in the AV node and in the atrial myocardium. Sympathetic nerves to the heart are also called accelerator nerves. Vagus fibers to the heart serve as inhibitory or depressor nerves.

Blood vessels
Kinds

There are three kinds of blood vessels: arteries, veins, and capillaries. By definition an *artery* is a vessel that carries blood away from the heart. All arteries except the pulmonary artery and its branches carry oxygenated blood. Small arteries are called *arterioles*.

A *vein*, on the other hand, is a vessel that carries blood toward the heart. All of the veins except the pulmonary veins contain deoxygenated blood. Small veins are called *venules*. Both arteries and veins are macroscopic structures.

Capillaries are microscopic vessels that carry blood from small arteries to small veins, that is, from arterioles to venules. They represented the "missing link" in the proof of circulation for many years—from the time William Harvey first declared that blood circulated from the heart through arteries to veins and back to the heart until the time that microscopes made it possible to find these connecting vessels between arteries and veins. Many people rejected Harvey's theory of circulation on the basis that there was no possible way for blood to get from arteries to veins. The discovery of the capillaries formed the final proof that the blood actually does circulate from the heart into arteries to arterioles, to capillaries, to venules, to veins, and back to the heart.

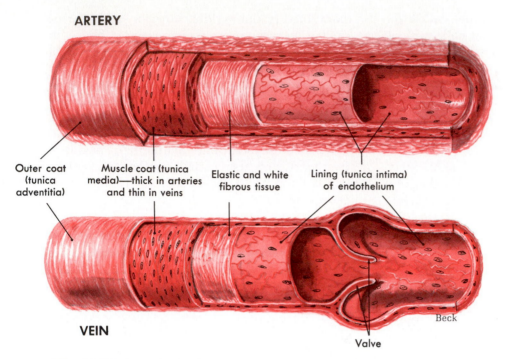

ARTERY

Outer coat (tunica adventitia)

Muscle coat (tunica media)—thick in arteries and thin in veins

Elastic and white fibrous tissue

Lining (tunica intima) of endothelium

Beck

VEIN

Valve

Fig. 14-9 Schematic drawings of an artery and vein showing comparative thicknesses of the three coats: outer coat (tunica adventitia), muscle coat (tunica media), and lining of endothelium (tunica intima). Note that the muscle and outer coats are much thinner in veins than in arteries and that veins have valves.

Fig. 14-10 The walls of capillaries consist of only a single layer of endothelial cells. These thin, flattened cells permit the rapid movement of substances between blood and interstitial fluid. Note that capillaries have no smooth muscle layer, elastic fibers, or surrounding adventitia.

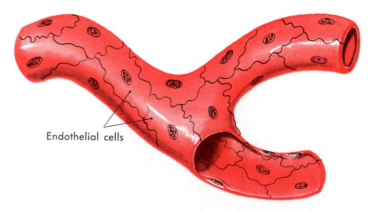

Endothelial cells

Table 14-2 Structure of blood vessels			
	Arteries	**Veins**	**Capillaries**
■ **Coats**	Outer coat (tunica adventitia or externa) of white fibrous tissue; causes artery to stand open instead of collapsing when cut (see Fig. 14-9) Muscle coat (tunica media) of smooth muscle, elastic, and some white fibrous tissues; this coat permits constriction and dilation Lining (tunica intima) of endothelium	Same three coats but thinner and fewer elastic fibers; veins collapse when cut; semilunar valves present at intervals	Only lining coat present; therefore walls only one cell thick

■ **Blood supply**
Endothelial lining cells supplied by blood flowing through vessels; exchange of oxygen, etc., between cells of middle coat and blood by diffusion; outer coat supplied by tiny vessels known as *vasa vasorum* or "vessels of vessels"

■ **Nerve supply**
Smooth muscle cells of tunica media innervated by autonomic fibers

■ **Abnormalities**
Atherosclerosis—hardening of walls of arteries (arteriosclerosis) characterized by lipid deposits in tunica intima
Aneurysm—saclike dilation of artery wall
Varicose veins—stretching of walls, particularly around semilunar valves
Phlebitis—inflammation of vein; "milk leg," phlebitis of femoral vein of women after childbirth

Structure

Consult Figs. 14-9 and 14-10 and Table 14-2 for the structure of the blood vessels.

The three coats or layers of the blood vessels are considered homologous with the layers of the heart wall—the outer tunica adventitia with the epicardium, the tunica media with the myocardium, and the tunica intima with the endocardium.

Functions

The capillaries, although seemingly the most insignificant of the three kinds of blood vessels because of their diminutive size, nevertheless are the most important vessels functionally. Since the prime function of blood is to transport essential materials to and from the cells and since the actual delivery and collection of these substances take place in the capillaries, the capillaries must be regarded as the most important blood vessels functionally. Arteries serve merely as "distributors," carrying the blood to the arterioles. Arterioles, too, serve as distributors, carrying blood from arteries to capillaries. But in addition, arterioles perform another function, one that is of great importance for maintaining normal blood pressure and circulation. They serve as resistance vessels (discussed in Chapter 15). Veins function both as collectors and as reservoir vessels. Not only do they return blood from the capillaries to the heart, but they also can accommodate varying amounts of blood. This reservoir function of veins, which we shall discuss later, plays an important part in maintaining normal circulation. The heart acts as a "pump," keeping the blood moving through this circuit of vessels—arteries, arterioles, capillaries, venules, and veins. In short, the entire

Table 14-3 Main arteries

Artery	Branches (only largest ones named)
Ascending aorta	Coronary arteries (two, to myocardium)
Aortic arch	Innominate (or brachiocephalic) artery Left subclavian Left common carotid
Innominate	Right subclavian Right common carotid
Subclavian (right and left)	Vertebral* Axillary (continuation of subclavian)
Axillary	Brachial (continuation of axillary)
Brachial	Radial Ulnar
Radial and ulnar	Palmar arches (superficial and deep arterial arches in hand formed by anastomosis of branches of radial and ulnar arteries; numerous branches to hand and fingers)
Common carotid (right and left)	Internal carotid (brain, eye, forehead, and nose)* External carotid (thyroid, tongue, tonsils, ear, etc.)
Descending thoracic aorta	Visceral branches to pericardium, bronchi, esophagus, mediastinum Parietal branches to chest muscles, mammary glands, and diaphragm
Descending abdominal aorta	Visceral branches: 1 Celiac axis (or artery), which branches into gastric, hepatic, and splenic arteries (stomach, liver, and spleen) 2 Right and left suprarenal arteries (suprarenal glands) 3 Superior mesenteric artery (small intestine) 4 Right and left renal arteries (kidneys) 5 Right and left spermatic (or ovarian) arteries (testes or ovaries) 6 Inferior mesenteric artery (large intestine) Parietal branches to lower surface of diaphragm, muscles and skin of back, spinal cord, and meninges Right and left common iliac arteries—abdominal aorta terminates in these vessels in an inverted Y formation

*The right and left vertebral arteries extend from their origin as branches of the subclavian arteries up the neck, through foramina in the transverse processes of the cervical vertebrae, and through the foramen magnum into the cranial cavity and unite on the undersurface of the brain stem to form the *basilar artery,* which shortly branches into the right and left *posterior cerebral arteries.* The internal carotid arteries enter the cranial cavity in the midpart of the cranial floor, where they become known as the *anterior cerebral arteries.* Small vessels, the *communicating arteries,* join the anterior and posterior cerebral arteries in such a way as to form an arterial circle (the *circle of Willis*) at the base of the brain, a good example of arterial anastomosis (Fig. 14-14).

Artery	Branches (only largest ones named)
Right and left common iliac	Internal iliac or hypogastric (pelvic wall and viscera) External iliac (to leg)
External iliac (right and left)	Femoral (continuation of external iliac after it leaves abdominal cavity)
Femoral	Popliteal (continuation of femoral)
Popliteal	Anterior tibial Posterior tibial
Anterior and posterior tibial	Plantar arch (arterial arch in sole of foot formed by anastomosis of terminal branches of anterior and posterior tibial arteries; small arteries lead from arch to toes)

circulatory mechanism pivots around one essential, that of keeping the capillaries supplied with an amount of blood adequate to the changing needs of the cells. All the factors governing circulation operate to this one end.

Although capillaries are very tiny (on the average, only 1 mm long or about $1/25$ inch), they are so numerous as to be incomprehensible. Someone has calculated that if these microscopic tubes were joined end to end, they would extend 62,000 miles, despite the fact that it takes 25 of them to reach a single inch! According to one estimate, 1 cubic inch of muscle tissue contains over 1.5 million of these important little vessels. None of the billions of cells composing the body lies very far removed from a capillary. The reason for this lavish distribution of capillaries is, of course, apparent in view of their function of keeping the cells supplied with vital materials and rid of injurious wastes.

Main blood vessels

Systemic circulation
Arteries

Locate the arteries listed in Table 14-3 (see also Figs. 14-11 to 14-14). You may find it easier to learn the names of blood vessels and the relation of the vessels to each other from diagrams than from descriptions.

As you learn the names of the main arteries, keep in mind that these are only the major pipelines distributing blood from the heart to the various organs and that in each organ the main artery resembles a tree trunk in that it gives off numerous branches that continue to branch and rebranch, forming ever smaller vessels (arterioles), which also branch, forming microscopic vessels, the capillaries. In other words, most arteries eventually ramify into capillaries. Arteries of this type are called *end-arteries*. Important organs or areas of the body supplied by end-arteries are subject to serious damage or death in occlusive arterial disease. As an example, permanent blindness results when the central artery of the retina, an end-artery, is occluded. Therefore occlusive arterial disease, such as atherosclerosis, is of great concern in clinical medicine when it affects important organs having an end-arterial blood supply.

A few arteries open into other branches of the same or other arteries. Such a communication is termed an *arterial anastomosis*. Anastomoses, we have already noted, fulfill an important protective function in that they provide detour routes for blood to travel in the event of obstruction of

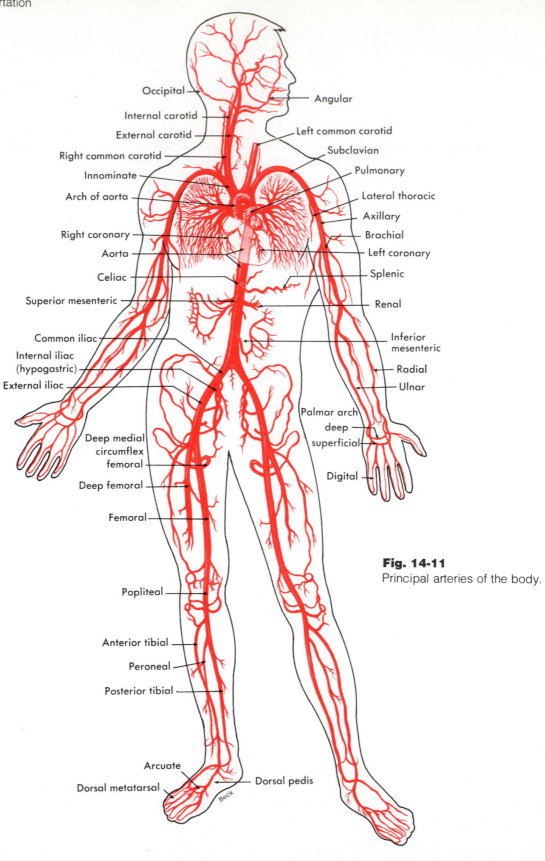

Fig. 14-11
Principal arteries of the body.

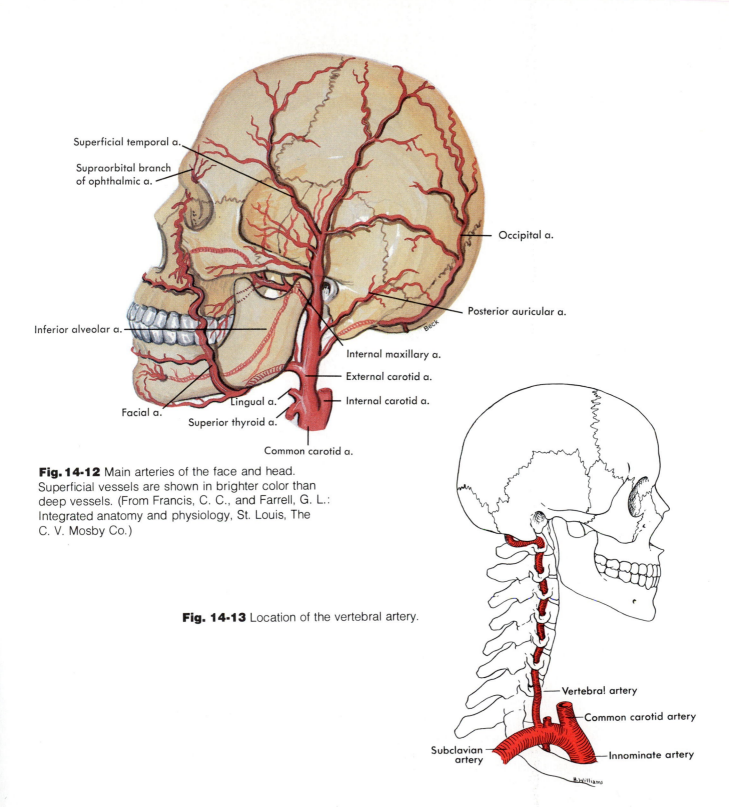

Superficial temporal a.

Supraorbital branch of ophthalmic a.

Occipital a.

Posterior auricular a.

Inferior alveolar a.

Internal maxillary a.

External carotid a.

Internal carotid a.

Lingual a.

Facial a.

Superior thyroid a.

Common carotid a.

Fig. 14-12 Main arteries of the face and head. Superficial vessels are shown in brighter color than deep vessels. (From Francis, C. C., and Farrell, G. L.: Integrated anatomy and physiology, St. Louis, The C. V. Mosby Co.)

Fig. 14-13 Location of the vertebral artery.

Vertebral artery

Common carotid artery

Subclavian artery

Innominate artery

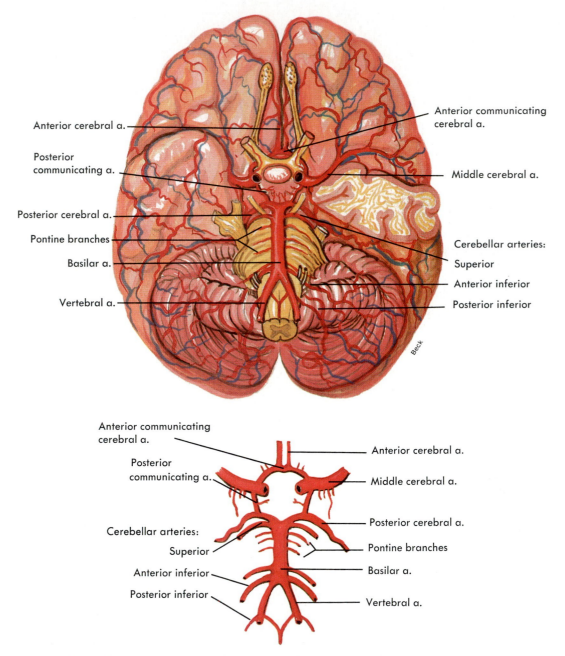

Fig. 14-14 Arteries at base of the brain. Those arteries that compose the circle of Willis are the two anterior cerebral arteries joined to each other by the anterior communicating cerebral artery and to the posterior cerebral arteries by the posterior communicating arteries.

a main artery. The incidence of arterial anastomoses increases as distance from the heart increases, and smaller arterial branches tend to anastomose more often than larger vessels. Examples of arterial anastomoses are the palmar and plantar arches. Other examples are found around several joints as well as in other locations.

Veins

Several facts should be borne in mind while learning the names of veins.

1 Veins are the ultimate extensions of capillaries, just as capillaries are the eventual extensions of arteries. Whereas arteries branch into vessels of decreasing size to form arterioles and eventually capillaries, capillaries unite into vessels of increasing size to form venules and eventually veins.

2 Many of the main arteries have corresponding veins bearing the same name and located alongside or near the arteries. These veins, like the arteries, lie in deep, well-protected areas, for the most part close along the bones—example: femoral artery and femoral vein, both located along the femur bone.

3 Veins found in the deep parts of the body are called *deep veins* in contradistinction to *superficial veins*, which lie near the surface. The latter are the veins that can be seen through the skin.

4 The large veins of the cranial cavity, formed by the dura mater, are not called veins but *sinuses*. They should not be confused with the bony sinuses of the skull.

5 Veins communicate (anastomose) with each other in the same way as arteries. Such venous anastomoses provide for collateral return blood flow in cases of venous obstruction.

6 Venous blood from the head, neck, upper extremities, and thoracic cavity, with the exception of the lungs, drains into the superior vena cava. Blood from the lower extremities and abdomen enters the inferior vena cava.

The following list identifies the major veins. Locate each one as named on Figs. 14-15 to 14-20.

Veins of upper extremities (Figs. 14-15 and 14-16)

Deep
Palmar (volar) arch (also superficial)
Radial (partially deep, partially superficial)
Ulnar (partially deep, partially superficial)
Brachial
Axillary (continuation of brachial)
Subclavian (continuation of axillary)

Superficial
Veins of hand from dorsal and volar venous arches, which, together with complicated network of superficial veins of lower arm, finally pour their blood into two large veins—cephalic (thumb side) and basilic (little finger side); these two veins empty into deep axillary vein

Veins of lower extremities (Figs. 14-15, 14-17, and 14-18)

Deep
Plantar arch
Anterior tibial
Posterior tibial
Popliteal
Femoral
External iliac

Superficial
Dorsal venous arch of foot
Great (or internal or long) saphenous
Small (or external or short) saphenous
(Great saphenous terminates in femoral vein in groin; small saphenous terminates in popliteal vein)

Veins of head and neck (Figs. 14-19 and 14-20)

Deep (in cranial cavity)
Longitudinal (or sagittal) sinus
Inferior sagittal and straight sinus
Numerous small sinuses
Right and left transverse (or lateral) sinuses
Internal jugular veins, right and left (in neck); continuations of transverse sinuses
Innominate veins, right and left; formed by union of subclavian and internal jugulars

Superficial
External jugular veins, right and left (in neck); receive blood from small superficial veins of face, scalp, and neck; terminate in subclavian veins (small emissary veins connect veins of scalp and face with blood si-

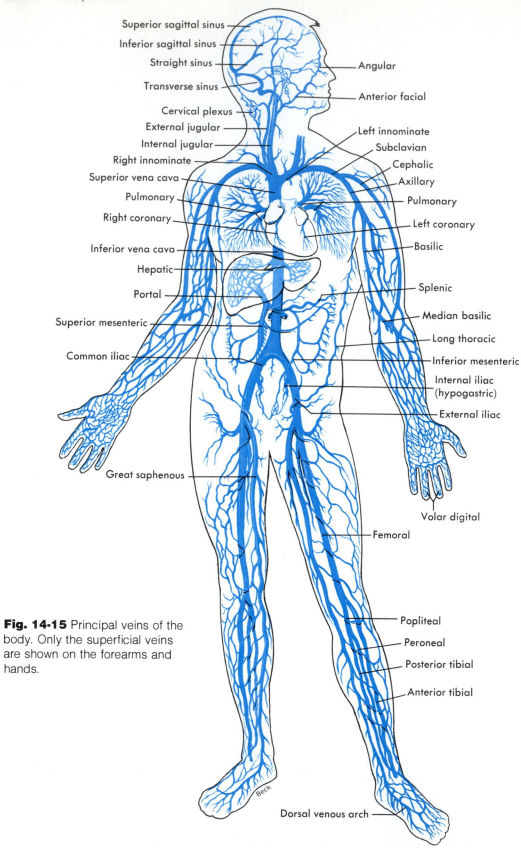

Superior sagittal sinus
Inferior sagittal sinus
Straight sinus
Transverse sinus
Cervical plexus
External jugular
Internal jugular
Right innominate
Superior vena cava
Pulmonary
Right coronary
Inferior vena cava
Hepatic
Portal
Superior mesenteric
Common iliac
Great saphenous

Angular
Anterior facial
Left innominate
Subclavian
Cephalic
Axillary
Pulmonary
Left coronary
Basilic
Splenic
Median basilic
Long thoracic
Inferior mesenteric
Internal iliac (hypogastric)
External iliac
Volar digital
Femoral
Popliteal
Peroneal
Posterior tibial
Anterior tibial
Dorsal venous arch

Beck

Fig. 14-15 Principal veins of the body. Only the superficial veins are shown on the forearms and hands.

nuses of cranial cavity, a fact of clinical interest as a possible avenue for infections to enter cranial cavity)

Veins of abdominal organs (Figs. 14-15 and 14-21)

Spermatic (or ovarian)
Renal
Hepatic   Drain into inferior
Suprarenal   vena cava

Left spermatic and left suprarenal veins usually drain into left renal vein instead of into inferior vena cava; for return of blood from abdominal digestive organs, see subsequent discussion of portal circulation; also Fig. 14-21

Veins of thoracic organs

Several small veins, such as bronchial, esophageal, pericardial, etc., return blood from chest organs (except lungs) directly into superior vena cava or azygos vein; azygos vein lies to right of spinal column and extends from inferior vena cava (at level of first or second lumbar vertebra) through diaphragm to terminal part of superior vena cava; hemiazygos vein lies to left of spinal column, extending from lumbar level of inferior vena cava through diaphragm to terminate in azygos vein; accessory hemiazygos vein connects some of superior intercostal veins with azygos or hemiazygos vein

Correlations. Middle ear infections sometimes cause infection of the transverse sinuses with the formation of a thrombus. In such cases the internal jugular vein may be ligated to prevent the development of a fatal cardiac or pulmonary embolism.

Intravenous injections are perhaps most often given into the median basilic vein at the bend of the elbow. Blood that is to be used for various laboratory tests is also usually removed from this vein. In an infant, however, the longitudinal sinus is more often punctured (through the anterior fontanel) because the superficial arm veins are too tiny for the insertion of a needle.

Portal circulation

Veins from the spleen, stomach, pancreas, gallbladder, and intestines do not pour their blood directly into the inferior vena cava as do the veins from other abdominal organs. Instead,

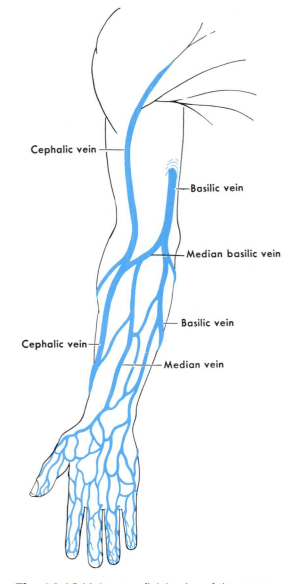

Fig. 14-16 Main superficial veins of the upper extremity, anterior view. The median basilic vein is commonly used for removing blood or giving intravenous infusions.

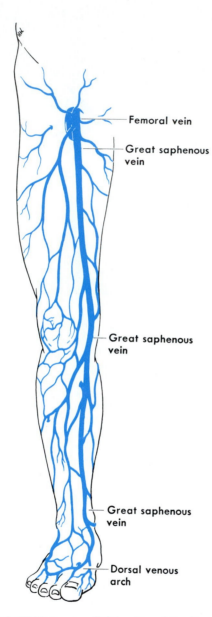

Femoral vein

Great saphenous
vein

Great saphenous
vein

Great saphenous
vein

Dorsal venous
arch

Fig. 14-17 Main superficial veins of the lower
extremity, anterior view.

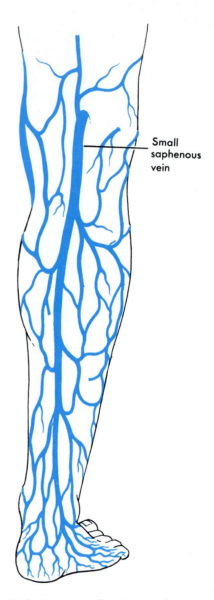

Small
saphenous
vein

Fig. 14-18 Main superficial veins of the lower
extremity, posterior view.

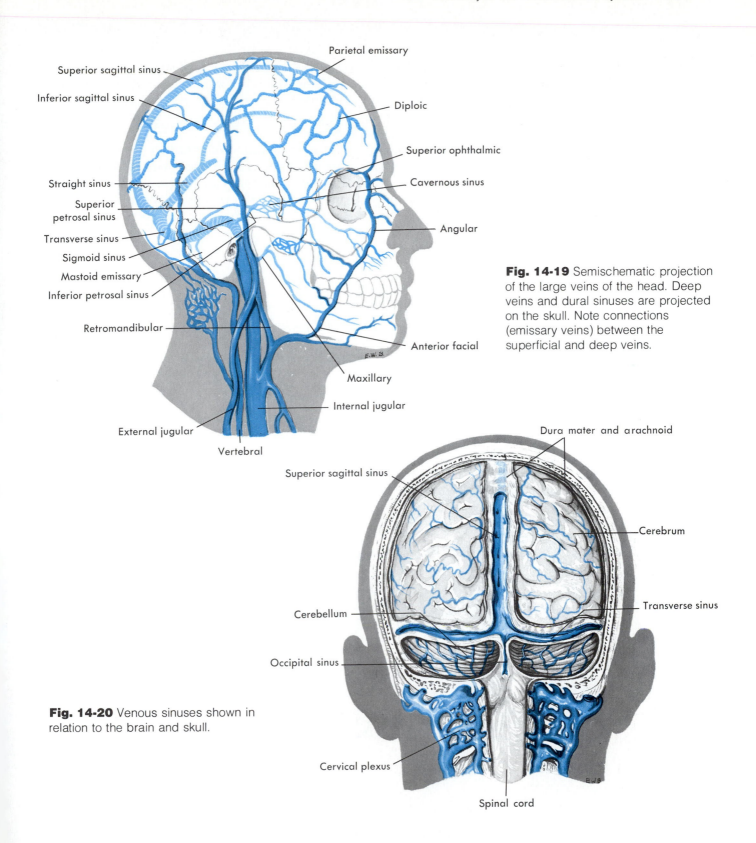

Superior sagittal sinus

Inferior sagittal sinus

Straight sinus

Superior petrosal sinus

Transverse sinus

Sigmoid sinus

Mastoid emissary

Inferior petrosal sinus

Retromandibular

External jugular

Vertebral

Maxillary

Internal jugular

Anterior facial

Angular

Cavernous sinus

Superior ophthalmic

Diploic

Parietal emissary

Fig. 14-19 Semischematic projection of the large veins of the head. Deep veins and dural sinuses are projected on the skull. Note connections (emissary veins) between the superficial and deep veins.

Dura mater and arachnoid

Superior sagittal sinus

Cerebrum

Cerebellum

Transverse sinus

Occipital sinus

Cervical plexus

Spinal cord

Fig. 14-20 Venous sinuses shown in relation to the brain and skull.

they send their blood to the liver by means of the portal vein. Here the blood mingles with the arterial blood in the capillaries and is eventually drained from the liver by the hepatic veins that join the inferior vena cava. The reason for this detouring of the blood through the liver before it returns to the heart will be discussed in the chapter on the digestive system.

Fig. 14-21 shows the plan of the portal system. In most individuals the portal vein is formed by the union of the splenic and superior mesenteric veins, but blood from the gastric, pancreatic, and inferior mesenteric veins drains into the splenic vein before it merges with the superior mesenteric vein.

If either portal circulation or venous return from the liver is interfered with (as they often are in certain types of liver or heart disease), venous drainage from most of the other abdominal organs is necessarily obstructed also. The

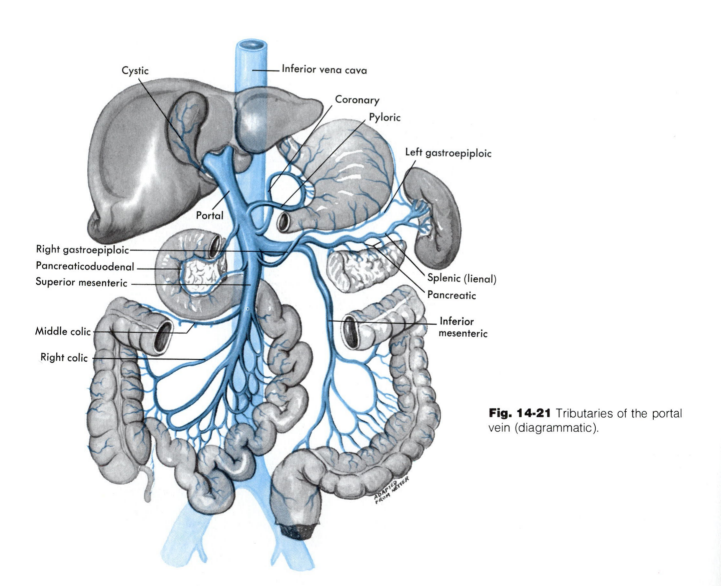

Fig. 14-21 Tributaries of the portal vein (diagrammatic).

accompanying increased capillary pressure accounts at least in part for the occurrence of ascites ("dropsy" of abdominal cavity) under these conditions.

Fetal circulation

Circulation in the body before birth necessarily differs from circulation after birth for one main reason—because fetal blood secures oxygen and food from maternal blood instead of

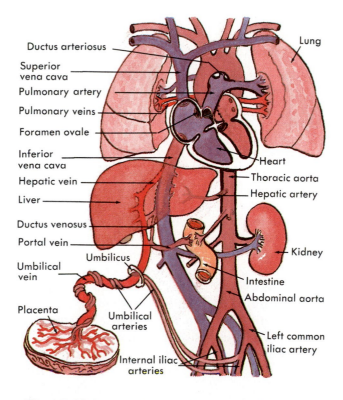

Ductus arteriosus

Superior vena cava

Pulmonary artery

Pulmonary veins

Foramen ovale

Inferior vena cava

Hepatic vein

Liver

Ductus venosus

Portal vein

Umbilicus

Umbilical vein

Placenta

Umbilical arteries

Internal iliac arteries

Lung

Heart

Thoracic aorta

Hepatic artery

Kidney

Intestine

Abdominal aorta

Left common iliac artery

Fig. 14-22 Scheme to show the plan of fetal circulation. Note the following essential features: (1) two umbilical arteries, extensions of the internal iliac arteries, carry blood to (2) the placenta, which is attached to the uterine wall; (3) one umbilical vein returns blood, rich in oxygen and food, from the placenta; (4) the ductus venosus, a small vessel that connects the umbilical vein with the inferior vena cava; (5) the foramen ovale, an opening in the septum between the right and left atria; and (6) the ductus arteriosus, a small vessel that connects the pulmonary artery with the thoracic aorta.

from its own lungs and digestive organs, respectively. Obviously, then, there must be additional blood vessels in the fetus to carry the fetal blood into close approximation with the maternal blood and to return it to the fetal body. These structures are the two *umbilical arteries*, the *umbilical vein*, and the *ductus venosus*. Also, there must be some structure to function as the lungs and digestive organs do postnatally, that is, a place where an interchange of gases, foods, and wastes between the fetal and maternal blood can take place. This structure is the *placenta*. The exchange of substances occurs without any actual mixing of maternal and fetal bloods, since each flows in its own capillaries.

In addition to the placenta and umbilical vessels, three structures located within the fetus' own body play an important part in fetal circulation. One of them (ductus venosus) serves as a detour by which most of the blood returning from the placenta bypasses the fetal liver. The other two (foramen ovale and ductus arteriosus) provide detours by which blood bypasses the lungs. A brief description of each of the six structures necessary for fetal circulation follows (also see Fig. 14-22).

1 The *two umbilical arteries* are extensions of the internal iliac (hypogastric) arteries and carry fetal blood to the placenta.

2 The *placenta* is a structure attached to the uterine wall. Exchange of oxygen and other substances between maternal and fetal blood takes place in the placenta.

3 The *umbilical vein* returns oxygenated blood from the placenta, enters the fetal body through the umbilicus, extends up to the undersurface of the liver where it gives off two or three branches to the liver, and then continues on as the ductus venosus. Two umbilical arteries and the umbilical vein together constitute the *umbilical cord* and are shed at birth along with the placenta.

4 The *ductus venosus* is a continuation of the umbilical vein along the undersurface of the liver and drains into the inferior vena cava.

Most of the blood returning from the placenta bypasses the liver. Only a relatively small amount enters the liver by way of the branches from the umbilical vein into the liver.

5 The *foramen ovale* is an opening in the septum between the right and left atria. A valve at the opening of the inferior vena cava into the right atrium directs most of the blood through the foramen ovale into the left atrium so that it bypasses the fetal lungs. A small percentage of the blood leaves the right atrium for the right ventricle and pulmonary artery. But even most of this does not flow on into the lungs. Still another detour, the ductus arteriosus, diverts it.

6 The *ductus arteriosus* is a small vessel connecting the pulmonary artery with the descending thoracic aorta. It therefore enables another portion of the blood to detour into the systemic circulation without going through the lungs.

Almost all fetal blood is a mixture of oxygenated and deoxygenated blood. Examine Fig. 14-22 carefully to determine why this is so. What happens to the oxygenated blood returned from the placenta via the umbilical vein? Note that it flows into the inferior vena cava.

Changes in the vascular system at birth

Since the six structures that serve fetal circulation are no longer needed after birth, several changes take place. As soon as the umbilical cord is cut, the two umbilical arteries, the placenta, and the umbilical vein obviously no longer function. The placenta is shed from the mother's body as the afterbirth, with part of the umbilical vessels attached. The sections of these vessels remaining in the infant's body eventually become fibrous cords that remain throughout life (the umbilical vein becomes the round ligament of the liver). The ductus venosus, no longer needed to bypass blood around the liver, eventually becomes the ligamentum venosum of the

liver. The foramen ovale normally becomes functionally closed soon after a newborn takes his first breath and full circulation through his lungs becomes established. Complete structural closure, however, usually requires 9 months or more. Eventually the foramen ovale becomes a mere depression (fossa ovalis) in the wall of the right atrial septum. The ductus arteriosus contracts as soon as respiration is established. Eventually, it also turns into a fibrous cord.

Circulation

Definitions

The term circulation of blood suggests its meaning, namely, blood flow through vessels arranged to form a circuit or circular pattern. Blood flow from the heart (left ventricle) through blood vessels to all parts of the body and back to the heart (to the right atrium) is spoken of as *systemic circulation*. The left ventricle pumps blood into the ascending aorta. From here it flows into arteries that carry it into the various tissues and organs of the body. Within each structure, blood moves, as indicated in Fig. 14-23, from arteries to arterioles to capillaries. Here the vital two-way exchange of substances occurs between blood and cells. Blood flows next out of each organ by way of its venules and then its veins to drain eventually into the inferior or superior vena cava. These two great veins of the body return venous blood to the heart (to the right atrium) to complete systemic circulation. But the blood has not quite come full circle back to its starting point, the left ventricle. To do this and start on its way again, it must first flow through another circuit, the *pulmonary circulation*. Observe in Fig. 14-23 that venous blood moves from the right atrium to the right ventricle to the pulmonary artery to lung arterioles and capillaries. Here, exchange of gases between blood and air takes place, converting venous blood to arterial blood. This oxygenated blood then flows on through lung venules into four pul-

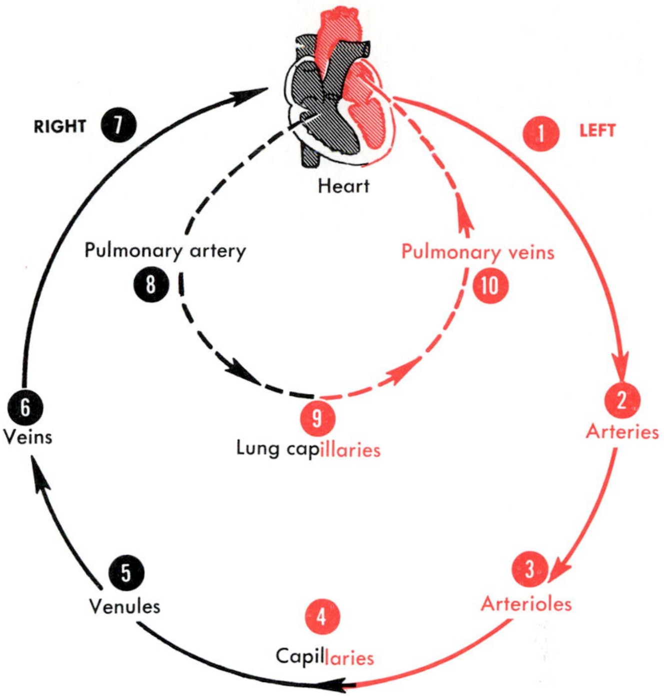

Fig. 14-23 Relationship of systemic and pulmonary circulation. As indicated by the numbers, blood circulates from the left side (ventricle) of the heart to arteries, to arterioles, to capillaries, to venules, to veins, to the right side of the heart (atrium to ventricle), to the lungs, and back to the left side of the heart, thereby completing a circuit. Refer to this diagram when tracing the circulation of blood to or from any part of the body.

monary veins and returns to the left atrium of the heart. From the left atrium it enters the left ventricle to be pumped again through the systemic circulation.

How to trace

In order to list the vessels through which blood flows in reaching a designated part of the body or in returning to the heart from a part, one must remember the following:

1 That blood always flows in this direction— from *left ventricle* of heart to *arteries*, to *arterioles*, to *capillaries* of each body part, to *venules*, to *veins*, to *right atrium*, *right ventricle*, *pulmonary artery*, *lung capillaries*, *pulmonary veins*, *left atrium*, and back to left ventricle (Fig. 14-23)

2 That when blood is in capillaries of abdominal digestive organs, it must flow through portal system (Fig. 14-21) before returning to heart

3 Names of main arteries and veins of body

For example, suppose glucose were instilled into the rectum. To reach the cells of the right little finger the vessels through which it would pass after absorption from the intestinal mucosa into capillaries would be as follows: *capillaries* into venules of large intestine into inferior mesenteric *vein*, splenic vein, portal vein, capillaries of liver, hepatic veins, inferior vena cava, *right atrium* of heart, *right ventricle*, *pulmonary artery*, *pulmonary capillaries*, *pulmonary veins*, *left atrium*, *left ventricle*, *ascending aorta*, aortic arch, innominate artery, right subclavian artery, right axillary artery, right brachial artery, right ulnar artery, arteries of palmar arch, arterioles, and *capillaries* of right little finger.

NOTE: The structures italicized show the direction of blood flow as described in points **1** and **2** and illustrated in Fig. 14-23. Follow this course of circulation first on Fig. 14-21 and then on Figs. 14-11 and 14-15. Answer review question 17 at the end of this chapter using the plan outlined.

Outline summary

Heart

A Four-chambered muscular organ

B Lies in mediastinum with apex on diaphragm, two thirds of its bulk to left of midline of body and one third to right

C Apical beat may be counted by placing stethoscope in fifth intercostal space on line with left midclavicular point

Covering

A Structure

   1 Loose-fitting, inextensible sac (fibrous pericardium) around heart, lined with serous pericardium (parietal layer), which also covers outer surface of heart (visceral layer or epicardium)

   2 Small space between parietal and visceral layers of serous pericardium contains few drops of pericardial fluid

B Function—protection against friction

Structure

A Wall

   1 Myocardium—name of muscular wall

   2 Endocardium—lining

   3 Pericardium—covering

B Cavities

   1 Upper two—atria

   2 Lower two—ventricles

C Valves and openings

   1 Openings between atria and ventricles—atrioventricular orifices, guarded by cuspid valves, tricuspid on right and mitral or bicuspid on left; valves consist of three parts; flaps, chordae tendineae, and papillary muscle

   2 Opening from right ventricle into pulmonary artery guarded by semilunar valves

   3 Opening from left ventricle into great aorta guarded by semilunar valves

D Blood supply—from coronary arteries; branch from ascending aorta behind semilunar valves

   1 Left ventricle receives blood via both major branches of left coronary artery and from one branch of right coronary artery

   2 Right ventricle receives blood via both major branches of right coronary artery and from one branch of left coronary artery

   3 Each atrium receives blood only from one branch of its respective coronary artery

   4 Usually only few anastomoses between larger branches of coronary arteries, so that occlusion of one of these produces areas of myocardial infarction; if not fatal, anastomoses between smaller vessels may grow and provide collateral circulation

E Conduction system

   1 Sinoatrial node (SA node; pacemaker of heart)—small mass of modified cardiac muscle at junction of superior vena cava and right atrium; inherent rhythmicity of SA node impulses set basic rate of heartbeat; ratio of sympathetic/parasympathetic impulses to node per minute and blood concentrations of epinephrine and thyroid hormone act on node to modify its activity and alter heart rate

   2 Atrioventricular node (AV node)—small mass of modified cardiac muscle in septum between two atria

   3 Atrioventricular bundle (AV bundle, bundle of His)—special cardiac muscle fibers originating in AV node and extending down interventricular septum

   4 Purkinje fibers—extension of bundle of His fibers out into walls of ventricles

F Nerve supply

   1 Sympathetic fibers (in cardiac nerves) and parasympathetic fibers (in vagus) form cardiac plexuses

   2 Fibers from plexuses terminate mainly in SA node

   3 Sympathetic impulses tend to accelerate and strengthen heartbeat

   4 Vagal impulses slow heartbeat

Blood vessels

A Kinds

   1 Arteries—vessels that carry blood away from heart; all except pulmonary artery carry oxygenated blood

   2 Veins—vessels that carry blood toward heart; all except pulmonary veins carry deoxygenated blood

   3 Capillaries—microscopic vessels that carry blood from small arteries (arterioles) to small veins (venules)

B Structure—see Table 14-2

C Functions

   1 Arteries and arterioles—carry blood away from heart to capillaries

   2 Capillaries—deliver materials to cells (by way of tissue fluid) and collect substances from them; vital function of entire circulatory system

3 Veins and venules—carry blood from capillaries back to heart

Main blood vessels

A Systemic circulation
1 Arteries—see p. 383, Fig. 14-11, and Table 14-3
2 Veins—see pp. 387 and 389, Figs. 14-15 to 14-21
B Portal circulation—see p. 389 and Fig. 14-21
C Fetal circulation—see p. 393 and Fig. 14-22

Circulation

A Definitions
1 Circulation—blood flow through closed circuit of vessels
2 Systemic circulation—blood flow from left ventricle into aorta, other arteries, arterioles, capillaries, venules, and veins of all parts of body to right atrium of heart
3 Pulmonary circulation—blood flow from right ventricle to pulmonary artery to lung arterioles, capillaries, and venules, to pulmonary veins, to left atrium

Review questions

1 Discuss the size, position, and location of the heart in the thoracic cavity.
2 How is compression of the heart in cardiopulmonary resuscitation effected?
3 Describe the pericardium, differentiating between the fibrous and serous portions.
4 The fibrous pericardium is tough and inelastic. Why is an understanding of this anatomical fact important in clinical medicine?
5 Exactly where is pericardial fluid found? Explain its function.
6 Describe the heart's own blood supply. Explain why occlusion of a large coronary artery branch has serious consequences.
7 Why would you expect large numbers of mitochondria to be present in cardiac muscle?

8 Identify, locate, and describe the functions of each of the following structures: SA node, AV node, bundle of His.
9 Compare arteries, veins, and capillaries as to structure and functions.
10 Differentiate between systemic, pulmonary, and portal circulations.
11 Explain why occlusion of an end-artery is more serious than occlusion of other small arteries.
12 Discuss the location, formation, and importance of the circle of Willis.
13 Name and locate the chambers and valves of the heart.
14 Trace the flow of blood through the heart.
15 Discuss the relationship between the layers of the blood vessels and the heart.
16 Explain the relationship between a heart "murmur" and valvular insufficiency.
17 Starting with the left ventricle of the heart, list the vessels through which blood would flow in reaching the small intestine; the large intestine; the liver (two ways); the spleen; the stomach; the kidneys; the suprarenal glands; the ovaries or testes; the anterior part of the base of the brain; the little finger of the right hand. List the vessels through which the blood returns from these parts to the right atrium of the heart (see Figs. 14-11, 14-15, and 14-23).
18 Briefly define the following terms: aneurysm, atherosclerosis, chordae tendineae, coronary sinus, endocardium, foramen ovale, mitral stenosis, myocardium, phlebitis, Purkinje fibers.
19 List the tributaries of the hepatic portal vein.
20 Why is the internal jugular vein sometimes ligated as a result of middle ear infections?
21 Give the general location of the following veins: longitudinal sinus, internal jugular vein, inferior vena cava, basilic vein, coronary sinus, great saphenous vein, innominate vein, hepatic portal vein.

chapter 15

Physiology of the cardiovascular system

Functions and importance of control mechanisms

The vital role of the cardiovascular system in maintaining homeostasis depends on the continuous and controlled movement of blood through the thousands of miles of capillaries that permeate every tissue and reach every cell in the body. It is in the microscopic capillaries that blood performs its ultimate transport function. Nutrients and other essential materials pass from capillary blood into fluids surrounding the cells as waste products are removed. Blood must not only be kept moving through its closed circuit of vessels by the pumping activity of the heart, but it must be directed and delivered to those capillary beds surrounding cells that need it most. Blood flow to cells at rest is minimal. In contrast, blood is shunted to the digestive tract following a meal or to skeletal muscles during exercise. The thousands of miles of capillaries could hold far more than the body's total blood volume if it were evenly distributed. Regulation of blood pressure and flow must, therefore, change in response to cellular activity.

Numerous control mechanisms help to regulate and integrate the diverse functions and component parts of the cardiovascular system in order to supply blood to specific body areas according to need. These mechanisms ensure a constant *milieu interieur*, that is, a constant internal environment surrounding each body cell regardless of differing demands for nutrients or production of waste products. This chapter presents information about several of the control mechanisms that regulate the pumping activity of the heart as well as the smooth and directed flow of blood through the complex channels of the circulation.

Physiology of the heart
Conduction system

The anatomy of four structures that compose the conduction system of the heart—the sino-atrial node, atrioventricular node, AV bundle, and Purkinje system—was discussed in Chapter 14. Each of these structures consists of cardiac muscle modified enough in structure to differ in function from ordinary cardiac muscle. The main specialty of ordinary cardiac muscle is contraction. In this, it is like all muscle, and like all muscle, ordinary cardiac muscle can also conduct impulses. But the conduction system structures (Fig. 15-1) are more highly specialized, both structurally and functionally, than ordinary cardiac muscle tissue. They are not contractile. Instead, they permit only generation or rapid conduction of an action potential through the heart.

The normal cardiac impulse that initiates mechanical contraction of the heart arises in the SA node (or pacemaker), located just below the atrial epicardium at its junction with the superior vena cava (Fig. 15-1). Specialized pacemaker cells in the node possess an *intrinsic rhythm*. This means that without any stimulation by nerve impulses from the brain and cord, they themselves initiate impulses at regular intervals. Even if the heart is removed from the body and placed in a nutrient solution, completely separated from all nervous and hormonal control, it will continue to beat!

Each impulse generated at the SA node travels swiftly throughout the muscle fibers of both atria. Thus stimulated, the atria begin to contract. As the action potential enters the AV node from the right atrium, its conduction slows markedly, thus allowing for complete contraction of both atrial chambers before the impulse reaches the ventricles. After passing slowly through the AV node, conduction velocity increases as the impulse is relayed through the AV bundle (bundle of His) into the ventricles. Here, right and left branches of the bundle fibers, and the Purkinje fibers in which they terminate, conduct the impulses throughout the muscle of both ventricles, stimulating them to contract almost simultaneously.

Thus the SA node initiates each heartbeat and sets its pace—it is the heart's own natural pacemaker. Normally the SA node will "discharge" or "fire" at an intrinsic rhythmic rate of 70 to 75 beats per minute. However, if for any reason the SA node loses its ability to generate an impulse, pacemaker activity will shift to another excitable component of the conduction system, such as the AV node or the Purkinje fibers. Pacemakers other than the SA node are called abnormal or *ectopic pacemakers*. Although ectopic pacemakers fire rhythmically, their rate of dis-

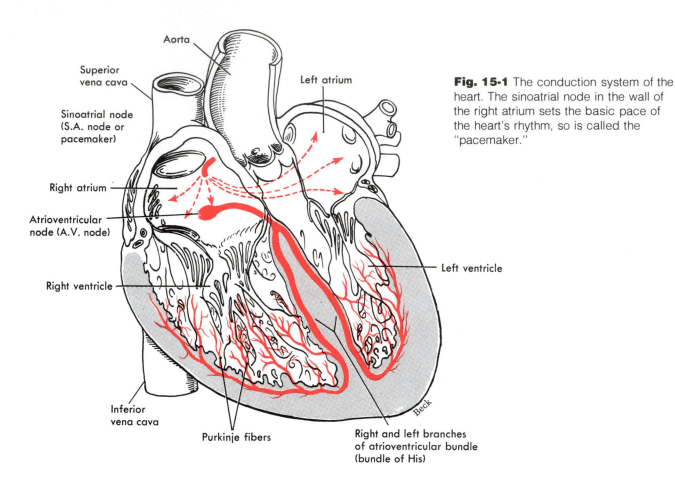

Aorta

Superior vena cava

Sinoatrial node (S.A. node or pacemaker)

Left atrium

Right atrium

Atrioventricular node (A.V. node)

Right ventricle

Left ventricle

Inferior vena cava

Purkinje fibers

Right and left branches of atrioventricular bundle (bundle of His)

Beck

Fig. 15-1 The conduction system of the heart. The sinoatrial node in the wall of the right atrium sets the basic pace of the heart's rhythm, so is called the "pacemaker."

charge is generally much slower than that of the SA node. For example, a pulse of 40 to 60 beats per minute would result if the AV node were forced to assume pacemaker activity.

Today, everyone has heard about artificial pacemakers, devices that electrically stimulate the heart at a set rhythm (continuously discharging type) or those that fire only when the heart rate decreases below a preset minimum (demand pacemakers). They do an excellent job of maintaining a steady heart rate and of keeping many individuals with damaged hearts alive for many years. Over 110,000 people currently have permanently implanted cardiac pacemakers. Nevertheless, these devices must be judged inferior to the heart's own natural pacemaker.

Why? Because they cannot speed up the heartbeat (as is necessary, for example, to make strenuous physical activity possible), nor can they slow it down again when the need has passed. The SA node, influenced as it is by autonomic impulses and hormones, can produce these changes. Discharging an average of 75 times each minute, this truly remarkable bit of specialized tissue will generate well over 2 billion action potentials in an average lifetime of some 70 years!

Electrocardiogram

Impulse conduction generates tiny electrical currents in the heart that spread through surrounding tissues to the surface of the body. This

Fig. 15-2

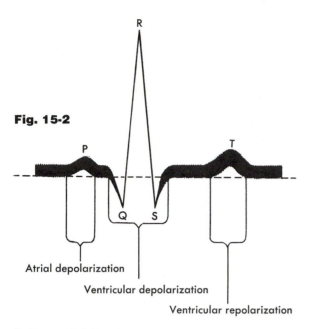

A, Normal ECG deflections. Depolarization and repolarization.

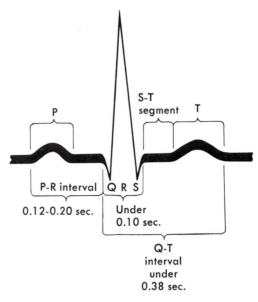

B, Principal ECG intervals between P, QRS, and T waves.

fact has great clinical importance. Why? Because from the skin, visible records of heart conduction can be made with an instrument called an electrocardiograph. Skilled interpretation of these records may sometimes make the difference between life and death.

The electrocardiogram (ECG) is a graphic record of the heart's action currents, its conduction of impulses. It is not a record of the heart's contractions but of the electrical events that precede them. Because electrocardiography is far too complex a science for us to attempt to explain here,* we shall describe only briefly normal ECG deflection waves and the intervals between them. As shown in Fig. 15-2, *A*, the normal ECG is composed of a P wave, a QRS complex, and a T wave. (The letters do not stand for any words but were chosen arbitrarily.) Briefly, the P wave represents depolarization of the

*For more information, see Conover, M. H., and Zalis, E. G.: Understanding electrocardiography: physiological and interpretive concepts, ed. 2, St. Louis, 1976, The C. V. Mosby Co.

atria, that is, the passage of a depolarization current from the SA node through the musculature of both atria, the QRS complex represents depolarization of the ventricles, and the T wave reflects repolarization (relaxation) of the ventricles. The principal ECG *intervals* between P, QRS, and T waves are shown in Fig. 15-2, *B*. Measurement of these intervals can provide valuable information concerning the rate of conduction of an action potential through the heart. Can you predict which interval would be lengthened if the passage of an action potential were slowed between atria and ventricles?

Control of heart rate
Autonomic nervous system control

Although the SA node normally initiates each heartbeat, the rate it sets is not an unalterable one. Various factors can and do change the rate of the heartbeat. One major modifier of SA node activity—and therefore of the heart rate—is the ratio of sympathetic and parasympathetic impulses conducted to the node per minute.

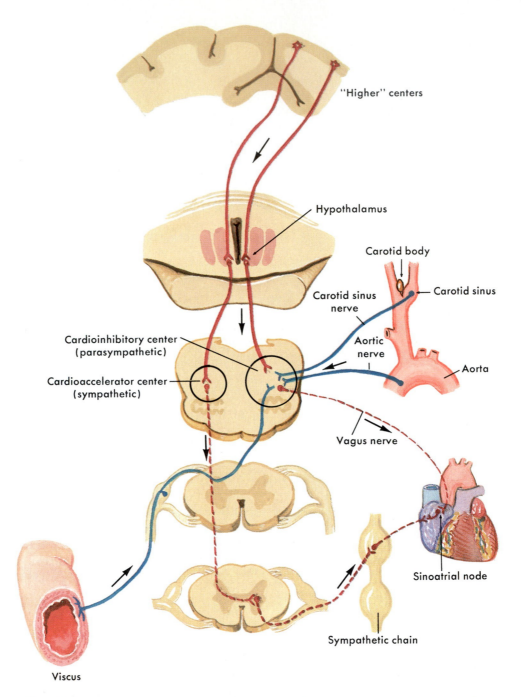

"Higher" centers

Hypothalamus

Carotid body

Carotid sinus nerve

Carotid sinus

Cardioinhibitory center (parasympathetic)

Aortic nerve

Carotid sinus

Cardioaccelerator center (sympathetic)

Aorta

Vagus nerve

Sinoatrial node

Sympathetic chain

Viscus

Fig. 15-3 Diagram of neural paths for controlling the heart rate. Note the following afferent paths to the cardioinhibitory and cardioaccelerator centers in the medulla: carotid sinus nerve, aortic nerve, and tracts from higher autonomic centers through the hypothalamus. What nerve constitutes the efferent path from the cardioinhibitory center to the heart? Do parasympathetic or sympathetic fibers compose the efferent path from the cardioaccelerator center to the heart?

Branches of the two divisions of the autonomic nervous system that reach the heart were discussed in Chapter 14. Autonomic control of heart rate is the result of opposing influences between parasympathetic (chiefly vagus) and sympathetic (accelerator nerve) stimulation. The results of parasympathetic stimulation on the heart are inhibitory and are mediated by vagal release of acetylcholine, whereas sympathetic (stimulatory) effects result from the release of norepinephrine.

Cardioinhibitory center. Parasympathetic impulses from a specialized area in the medulla called the cardioinhibitory center reach the SA node through the vagus. Normally the heart is under the restraint of vagal inhibition, as released acetylcholine decreases the rate of SA node firing. The vagus is said to act as a "brake" on the heart.

Cardioaccelerator center. Sympathetic impulses originate in the cardioaccelerator center of the medulla and reach the heart via sympathetic fibers (contained in the middle, superior, and inferior cardiac nerves). Norepinephrine released as a result of sympathetic stimulation increases both heart rate and strength of cardiac muscle contraction. Fig. 15-3 illustrates the control of cardioinhibitory and cardioaccelerator center activity or outflow from the medulla.

Cardiac pressoreflexes. Receptors sensitive to changes in pressure (baroreceptors) are located in two cardiovascular regions; they send afferent nerve fibers to the medullary control centers. Specifically, these stretch receptors are located in the aorta and carotid sinus and constitute one of the most important heart rate control mechanisms because of their effect on the cardioinhibitory and accelerator centers—and therefore on parasympathetic and sympathetic outflow.

Carotid sinus reflex. The carotid sinus is a small dilation at the beginning of the internal carotid artery just above the bifurcation of the common carotid artery to form the internal and external carotid arteries. The sinus lies just under the sternocleidomastoid muscle at the level of the upper margin of the thyroid cartilage. Sensory (afferent) fibers from carotid sinus baroreceptors run through the carotid sinus nerve (of Hering) and on through the glossopharyngeal (or ninth cranial) nerve to the cardioinhibitory center. Stimulation of these stretch receptors results in a reflex slowing of the heart.

Aortic reflex. Sensory (afferent) fibers also extend from baroreceptors located in the wall of the arch of the aorta through the aortic nerve and then through the vagus (tenth cranial) nerve to terminate in the cardioinhibitory center of the medulla.

■ ■ ■

If blood pressure within the aorta or carotid sinus increases suddenly, it stimulates the aortic or carotid baroreceptors, as shown in Fig. 15-3. The result of such stimulation is a reflex slowing of the heart. On the other hand, a decrease in aortic or carotid blood pressure usually initiates reflex acceleration of the heart. The lower blood pressure decreases baroreceptors' stimulation. Hence the cardioinhibitory center receives fewer stimulating impulses and the cardioaccelerator center fewer inhibitory impulses. Net result? The heart beats faster. Details of pressoreflex activity will also be included later in the chapter as part of a mechanism that tends to maintain or restore homeostasis of arterial blood pressure.

Miscellaneous factors that influence heart rate

Included in the miscellaneous category are such important factors as emotions, exercise, hormones, blood temperature, pain, and stimulation of various exteroceptors. Anxiety, fear, and anger often make the heart beat faster. Grief, in contrast, tends to slow it. Presumably, emotions produce changes in the heart rate through the influence of impulses from the "higher centers" in the cerebrum via the hypothalamus. Such impulses can influence activity of either the cardioinhibitory or the cardioaccelerator center.

In exercise the heart normally accelerates. The mechanism is not definitely known. But it is thought to include impulses from the cerebrum through the hypothalamus to cardiac centers. Epinephrine is the hormone most noted as a cardiac accelerator.

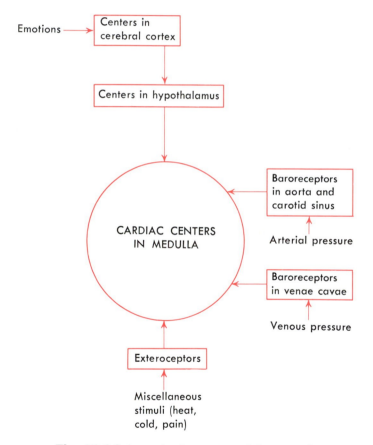

Fig. 15-4 Scheme to show some of the many factors that influence activity of the cardiac centers in the medulla. Impulses from various receptors are conducted by sensory fibers that terminate in synapses with neurons in cardiac centers. Motor fibers from the centers relay impulses to sympathetic and parasympathetic neurons, which transmit them to the heart. Also streaming into the cardiac centers are impulses from the hypothalamus, presumably part of the pathway by which emotions influence heart rate.

Increased blood temperature or stimulation of skin heat receptors tends to increase the heart rate, and decreased blood temperature or stimulation of skin cold receptors tends to slow it. Sudden intense stimulation of pain receptors in such visceral structures as the gallbladder, ureters, or intestines can result in such slowing of the heart that fainting may result. Fig. 15-4 shows the major factors that influence activity of the cardioinhibitory and cardioaccelerator centers.

Cardiac cycle

The term cardiac cycle means a complete heartbeat consisting of contraction (systole) and relaxation (diastole) of both atria plus contraction and relaxation of both ventricles. The two atria contract simultaneously. Then, as they relax, the two ventricles contract and relax, instead of the entire heart contracting as a unit. This gives a kind of milking action to the movements of the heart. The atria remain relaxed during part of the ventricular relaxation and then start the cycle over again. The cycle as a whole is often divided into a number of time intervals for discussion and study. The following list relates several of the important events of the cycle to time, and Fig. 15-5 is a composite intended to graphically illustrate and integrate changes in pressure gradients in the left atrium, left ventricle, and aorta with ECG and heart sound recordings. Aortic blood flow and changes in ventricular volume are also shown.

1 *Atrial systole.* The contracting force of the atria completes the emptying of blood out of the atria into the ventricles. Atrioventricular (AV or cuspid) valves are necessarily open during this phase; the ventricles are relaxed and filling with blood. The semilunar (SL) valves are closed so that blood does not flow on out into the pulmonary artery or aorta. This period of the cycle begins with the P wave of the ECG. Passage of the electrical wave of depolarization is then followed almost immediately by actual contraction of the atrial musculature.

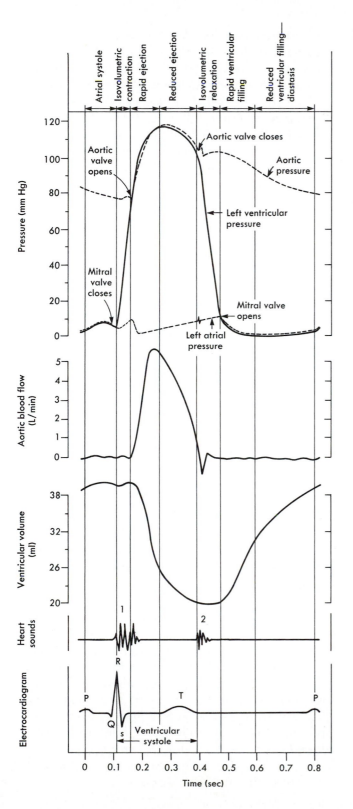

2 *Isovolumetric contraction. Iso* is a combining form denoting equality or uniformity. During the brief period of isovolumetric contraction, that is, between the start of ventricular systole and the opening of the SL valves, ventricular volume remains constant or uniform as the pressure increases rapidly. The onset of ventricular systole coincides with the R wave of the ECG and the appearance of the first heart sound.

3 *Ejection.* The SL valves open and blood is ejected from the heart when the pressure gradient in the ventricles exceeds the pressure in the pulmonary artery and aorta. An initial, shorter phase called *rapid ejection* is characterized by a marked increase in ventricular and aortic pressure and in aortic blood flow. The T wave of the ECG appears during the later, longer phase of *reduced ejection* (characterized by a less abrupt decrease in ventricular volume).

It is important to note that a considerable quantity of blood, called the *residual volume,* normally remains in the ventricles at the end of the ejection period. In heart failure the residual volume remaining in the ventricles may greatly exceed that ejected during systole.

4 *Isovolumetric relaxation.* Ventricular diastole or relaxation begins with this period of the cardiac cycle. It is the period between closure of the SL valves and opening of the AV valves. At the end of ventricular ejection the SL valves will close so that blood cannot reenter the ventricular chambers from the great vessels. The AV valves will not open until the pressure in the atrial chambers increases above that in the relaxing ventricles. The result is a dramatic fall in

Fig. 15-5 Left atrial, aortic, and left ventricular pressure pulses correlated in time with aortic flow, ventricular volume, heart sounds, venous pulse, and electrocardiogram for a complete cardiac cycle in the dog. (Modified from Berne, R. M., and Levy, M. N.: Cardiovascular physiology, ed. 3, St. Louis, 1977, The C. V. Mosby Co.)

intraventricular pressure but no change in volume. Both sets of valves are closed, and the ventricles are relaxing. The second heart sound is heard during this period of the cycle.

5 *Rapid ventricular filling.* Return of venous blood increases intra-atrial pressure until the AV valves are forced open and blood rushes into the relaxing ventricles. The rapid influx lasts about 0.1 second and results in a dramatic increase in ventricular volume.

6 *Reduced ventricular filling (diastasis).* The term diastasis is often used to describe a later, longer period of slow ventricular filling at the end of ventricular diastole. The abrupt inflow of blood that occurred immediately after opening of the AV valves is followed by a slow but continuous flow of venous blood into the atria and then through the open AV valves into the ventricles. Diastasis lasts about 0.2 second and is characterized by a gradual increase in ventricular pressure and volume.

Heart sounds during cycle. The heart makes certain typical sounds during each cycle that are described as sounding like lubb-dupp through a stethoscope. The first or systolic sound is believed to be caused primarily by the contraction of the ventricles and also by vibrations of the closing AV or cuspid valves. It is longer and lower than the second or diastolic sound, which is short, sharp, and thought to be caused by vibrations of the closing SL valves.

Both of these sounds have clinical significance, since they give information about the valves of the heart. Any variation from normal in the sounds indicates imperfect functioning of the valves. *Heart murmur* is one type of abnormal sound frequently heard and may signify incomplete closing of the valves (valvular insufficiency) or stenosis (constriction or narrowing) of them.

Control of circulation
Functions of control mechanisms

Circulation is, of course, a vital function. It constitutes the only means by which cells can receive materials needed for their survival and can have their wastes removed. Not only is circulation necessary, but circulation of different volumes of blood per minute is also essential for healthy survival. More active cells need more blood per minute than less active cells. The reason underlying this principle is obvious. The more work cells do, the more energy they use and the more oxygen and food they need to supply this energy. Only arterial blood can deliver these energy suppliers. So the more active any part of the body is, the greater the volume of blood circulated to it per minute must be. This requires that circulation control mechanisms accomplish two functions: maintain circulation (keep blood flowing) and vary the volume and distribution of the blood circulated. The greater the activity of any part of the body, the greater the volume of blood it needs circulating through it. Therefore as any structure increases its activity, an increased volume of blood must be distributed to it—must be shifted from the less active to the more active tissues.

To achieve these two ends, a great many factors must operate together as one smooth-running although complex machine. Incidentally, this is an important physiological principle that you have no doubt observed by now—that every body function depends on many other functions. A constellation of separate processes or mechanisms act as a single integrated mechanism. Together, they perform some one large function. For example, many mechanisms together accomplish the large function we call circulation.

Primary principle of circulation

Blood circulates for the same reason that any fluid flows—whether it be water in a river, in a garden hose, or in hospital tubing or blood in vessels. A fluid flows because a pressure gradient exists between different parts of its bed. This primary fluid flow principle derives from Newton's first and second laws of motion. In essence, these laws state the following principles:

1 A fluid does not flow when the pressure is the same in all parts of it.

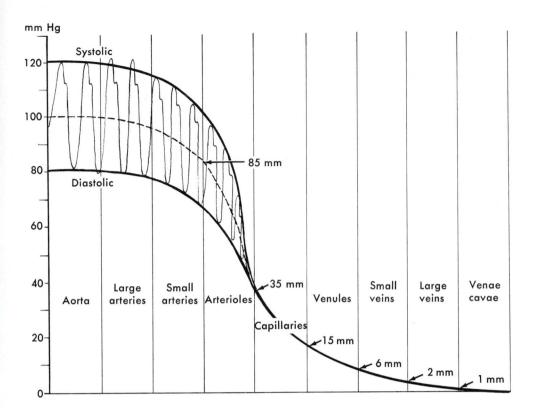

Fig. 15-6 Blood pressure gradient. Dotted line indicates the average or mean systolic pressure in arteries.

2 A fluid flows only when its pressure is higher in one area than in another, and it flows always from its higher pressure area toward its lower pressure area.

In brief, then, the primary principle about circulation is this: blood circulates from the left ventricle to the right atrium of the heart because a blood pressure gradient exists between these two structures. By blood pressure gradient, we mean the difference between the blood pressure in one structure and the blood pressure in another. For example, a typical normal blood pressure in the aorta, as the left ventricle contracts pumping blood into it, is 120 mm Hg, and as the left ventricle relaxes, it decreases to 80 mm Hg. The mean or average blood pressure, therefore, in the aorta in this instance is 100 mm Hg.

Fig. 15-6 shows the systolic and diastolic pressures in the arterial system and illustrates the progressive fall in pressure to 0 mm Hg by the time blood reaches the venae cavae and right atrium. The progressive fall in pressure as blood passes through the circulatory system is directly related to resistance. Resistance to blood flow in the aorta is almost zero. Although the pumping action of the heart causes fluctuations in aortic blood pressure (systolic 120 mm Hg; diastolic 80 mm Hg), the mean pressure remains almost constant, dropping perhaps only 1 or 2 mm Hg. The greatest drop in pressure (about 50 mm Hg) occurs across the arterioles because they present the greatest resistance to blood flow (Fig. 15-6).

The symbol $P_1 - P_2$ is often used to stand for a pressure gradient, with P_1 the symbol for the higher pressure and P_2 the symbol for the lower pressure. For example, blood enters the arterioles at 85 mm Hg and leaves at 35 mm Hg. Which is P_1? P_2? What is the blood pressure gradient? It would cause blood to flow from the arterioles into capillaries.

Control of arterial blood pressure

The primary determinant of arterial blood pressure is the volume of blood in the arteries. A direct relationship exists between arterial

ARTERIAL BLOOD PRESSURE

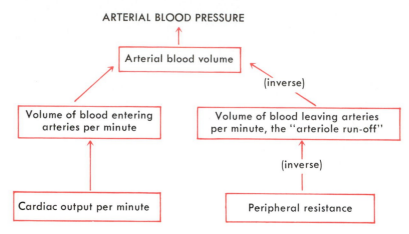

Fig. 15-7 Scheme to show how cardiac minute output and peripheral resistance affect arterial blood pressure. If cardiac minute output increases, the amount of blood entering the arteries increases and tends to increase the volume of blood in the arteries. The resulting increase in arterial volume increases arterial blood pressure. If peripheral resistance increases, it tends to increase arterial blood volume and pressure. Why? Because the increased resistance tends to decrease the amount of blood leaving the arteries, which tends to increase the amount of blood left in them. The increase in arterial volume increases arterial blood pressure.

blood volume and arterial pressure. This means that an increase in arterial blood volume tends to increase arterial pressure, and, conversely, a decrease in arterial volume tends to decrease arterial pressure.

Many factors together indirectly determine arterial pressure through their influence on arterial volume. Two of the most important are (1) cardiac output and (2) peripheral resistance (Fig. 15-7).

Cardiac output

The cardiac output (CO) is determined by both the volume of blood pumped out of the ventricles by each beat (stroke volume) and by the heart rate. Contraction of the heart is called *systole.* Therefore the volume of blood pumped by one contraction is known as *systolic discharge. Stroke volume* means the same thing, the amount of blood pumped by one stroke (contraction) of the ventricle.

Stroke volume

Stroke volume reflects the force or strength of ventricular contraction—the stronger the contraction, the greater the stroke volume tends to be. CO can be computed by the following simple equation:

$$\text{Stroke volume} \times \text{Heart rate} = \text{CO}$$

In practice, however, computing the CO is far from simple. It requires introducing a catheter into the right side of the heart (cardiac catheterization) and working a computation known as Fick's formula.*

Since the heart's rate and stroke volume determine its output, anything that changes either the rate of the heartbeat or its stroke volume tends to change CO, arterial blood volume, and blood pressure in the same direction. In other words,

*Mountcastle, V. B., editor: Medical physiology, ed. 13, St. Louis, 1974, The C. V. Mosby Co., pp. 906-907.

anything that makes the heart beat faster or anything that makes it beat stronger (increases its stroke volume) tends to increase CO and, therefore, arterial blood volume and pressure. Conversely, anything that causes the heart to beat more slowly or more weakly tends to decrease CO, arterial volume, and blood pressure. But do not overlook the word *tend* in the preceding sentences. A change in heart rate or stroke volume does not always change the heart's output, or the amount of blood in the arteries, or the blood pressure. To see whether this is true, do the following simple arithmetic, using the simple formula for computing CO. Assume a normal rate of 72 beats per minute and a normal stroke volume of 70 ml. Next, suppose the rate drops to 60 and the stroke volume increases to 100. Does the decrease in heart rate actually cause a decrease in CO in this case? Clearly not—the CO increases. Do you think it is valid, however, to say that a slower rate *tends* to decrease the heart's output? By itself, without any change in any other factor, would not a slowing of the heartbeat cause CO volume, arterial volume, and blood pressure to fall?

Mechanical, neural, and chemical factors regulate the strength of the heartbeat and therefore its stroke volume. The mechanical factor that helps determine stroke volume is the length of myocardial fibers at the beginning of ventricular contraction.

Many years ago Starling described a principle later made famous as Starling's law of the heart. In this principle he stated the factor he had observed as the main regulator of heartbeat strength in experiments performed on denervated animal hearts. Starling's law of the heart, in essence, is this: Within limits, the longer or more stretched the heart fibers at the beginning of contraction, the stronger is their contraction.

The factor determining how stretched the animal hearts were at the beginning of contractions was, as you might deduce, the amount of blood in the hearts at the end of diastole. The more blood returned to the hearts per minute, the more stretched were their fibers, the stronger were their contractions, and the larger was the volume of blood they ejected with each contraction. If, however, too much blood stretched the hearts too far, beyond a certain critical point, they seemed to lose their elasticity. They then contracted less vigorously—much as a rubber band, stretched too much, rebounds with less force.

Physiologists have long accepted as fact that Starling's law of the heart operates in animals under experimental conditions. But they have questioned its validity and importance in the intact human body. Today the prevailing opinion seems to be that Starling's law of the heart operates as a major regulator of stroke volume under ordinary conditions. Operation of Starling's law of the heart ensures that increased amounts of blood returned to the heart will be pumped out of it. It automatically adjusts CO to venous return under usual conditions.

Heart rate

Control mechanisms that influence heart rate have already been discussed. The pressoreflexes constitute the dominant heart rate control mechanism, although various other factors also influence heart rate (Figs. 15-3 and 15-4). Fig. 15-8 shows the relationship between homeostasis of arterial blood pressure and activation of cardiac pressoreflexes.

Peripheral resistance

Peripheral resistance helps determine arterial blood pressure. Specifically, arterial blood pressure tends to vary directly with peripheral resistance. Peripheral resistance means the resistance to blood flow imposed by the force of friction between blood and the walls of its vessels. Friction develops partly because of a characteristic of blood—its viscosity or stickiness—and partly from the small diameter of arterioles and capillaries. The resistance offered by arterioles, in particular, accounts for almost half of the total resistance in the systemic circulation. The

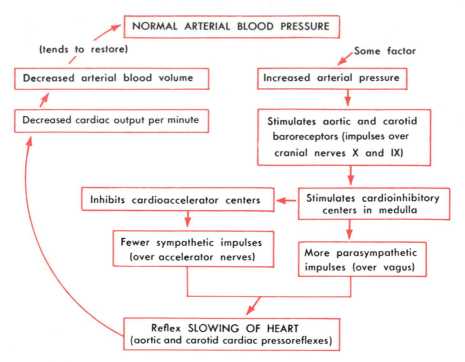

Fig. 15-8 Aortic and carotid cardiac pressoreflexes, a mechanism that tends to maintain or restore homeostasis of arterial blood pressure by regulating the rate of the heartbeat. Note that by this mechanism an increase in arterial blood pressure leads to reflex slowing of the heart and tends to lower blood pressure. The converse is also true. A decrease in blood pressure leads to reflex acceleration of the heart and tends to raise the pressure upward toward normal.

muscular coat that arterioles are vested with allows them to constrict or dilate and thus change the amount of resistance to blood flow. Peripheral resistance helps determine arterial pressure by controlling the rate of "arteriole runoff," the amount of blood that runs out of the arteries into the arterioles. The greater the resistance, the less the arteriole runoff or outflow tends to be—and, therefore, the more blood left in the arteries and the higher the arterial pressure tends to be.

Blood viscosity stems mainly from the red blood cells but also partly from the protein molecules present in blood. An increase in either blood protein concentration or red blood cell count tends to increase viscosity, and a decrease

in either tends to decrease it. Under normal circumstances, blood viscosity changes very little. But under certain abnormal conditions, such as marked anemia or hemorrhage, a decrease in blood viscosity may be the crucial factor lowering peripheral resistance and arterial pressure even to the point of circulatory failure.

Vasomotor control mechanism

Blood distribution patterns, as well as blood pressure, can be influenced by factors that control changes in the diameter of arterioles. Such factors might be said to constitute the *vasomotor control mechanism*. Like most physiological control mechanisms, it consists of many parts. An area in the medulla called the *vasomotor* or *vaso-*

constrictor center will, when stimulated, initiate an impulse outflow via sympathetic fibers that ends in the smooth muscle surrounding resistance vessels, arterioles, venules, and veins of the "blood reservoirs," causing their constriction. Thus the vasomotor control mechanism plays a role both in the maintenance of the general blood pressure and in the distribution of blood to areas of special need.

The main blood reservoirs are the venous plexuses and sinuses in the skin and abdominal organs (especially in the liver and spleen). In other words, blood reservoirs are the venous networks in most parts of the body—all but those in the skeletal muscles, heart, and brain. The term reservoir is apt, since these vessels serve as storage depots for blood. It can quickly be moved out of them and "shifted" to heart and skeletal muscles when increased activity demands. A change in either arterial blood's oxygen or carbon dioxide content sets a chemical vasomotor control mechanism in operation. A change in arterial blood pressure initiates a *vasomotor pressoreflex.*

Vasomotor pressoreflexes (Fig. 15-9). A sudden increase in arterial blood pressure stimulates aortic and carotid baroreceptors—the same ones that initiate cardiac reflexes. Not only does this lead to stimulation of cardioinhibitory centers but also to inhibition of vasoconstrictor centers. More impulses per second go out to the heart over parasympathetic vagal fibers and fewer over sympathetic fibers to blood vessels. As a result, the heartbeat slows, and arterioles and the venules of the blood reservoirs dilate. Since sympathetic vasoconstrictor impulses predominate at normal arterial pressures, inhibition of these is considered the major mechanism of vasodilation.

A decrease in arterial pressure causes the aortic and carotid baroreceptors to send more impulses to the medulla's vasoconstrictor centers, thereby stimulating them. These centers then send more impulses via the sympathetic fibers to stimulate vascular smooth muscle and cause vasoconstriction. This squeezes more blood out of the blood reservoirs, increasing the amount of venous return to the heart. Eventually, this extra blood is redistributed to more active structures such as skeletal muscles and heart because their arterioles become dilated largely from the operation of a local mechanism (p. 415). Thus the vasoconstrictor pressoreflex and the local vasodilating mechanism together serve as an important device for shifting blood from reservoirs to structures that need it more. It is an especially valuable mechanism during exercise.

Vasomotor chemoreflexes (Fig. 15-10). Chemoreceptors located in the aortic and carotid bodies are particularly sensitive to a deficiency of blood oxygen (hypoxia) and somewhat less sensitive to excess blood carbon dioxide (hypercapnia) and to decreased arterial blood pH. When one or more of these conditions stimulates the chemoreceptors, their fibers transmit more impulses to the medulla's vasoconstrictor centers, and vasoconstriction of arterioles and venous reservoirs soon follows. This stress mechanism functions as an emergency device when severe hypoxia or hypercapnia endangers survival.

Medullary ischemic reflex (Fig. 15-11). The medullary ischemic reflex mechanism is said to exert the most powerful control of all on small blood vessels. When the blood supply to the medulla becomes inadequate (ischemic), its neurons suffer from both oxygen deficiency and carbon dioxide excess. But, presumably, it is the latter, the hypercapnia, that intensely and directly stimulates the vasoconstrictor centers to bring about marked arteriole and venous constriction. If the oxygen supply to the medulla decreases below a certain level, its neurons, of course, cannot function and the medullary ischemic reflex cannot operate.

Vasomotor control by higher brain centers (Fig. 15-12). Impulses from centers in the cerebral cortex and in the hypothalamus are believed to be transmitted to the vasomotor centers in the medulla and to thereby help control vasoconstric-

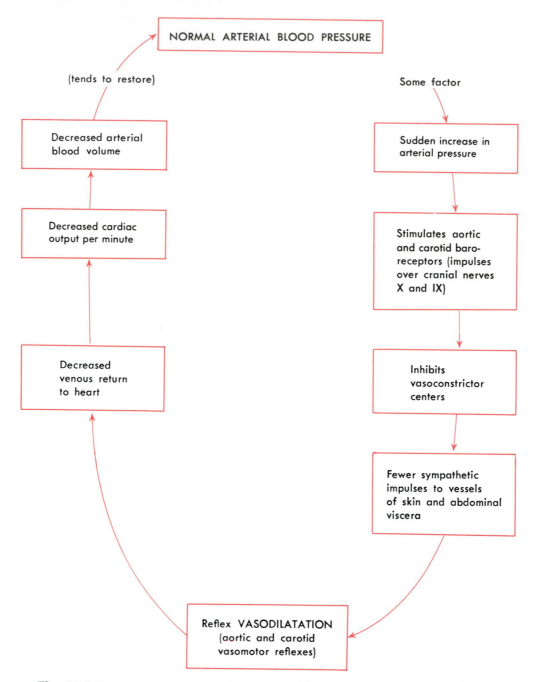

NORMAL ARTERIAL BLOOD PRESSURE

(tends to restore)

Some factor

Decreased arterial blood volume

Sudden increase in arterial pressure

Decreased cardiac output per minute

Stimulates aortic and carotid baro-receptors (impulses over cranial nerves X and IX)

Decreased venous return to heart

Inhibits vasoconstrictor centers

Fewer sympathetic impulses to vessels of skin and abdominal viscera

Reflex VASODILATATION (aortic and carotid vasomotor reflexes)

Fig. 15-9 The aortic and carotid vasomotor pressoreflex mechanism shown here is set in operation when some factor causes a sudden increase in arterial blood pressure. This mechanism and the aortic and carotid cardiac pressoreflexes (Fig. 15-8) operate simultaneously to maintain or restore homeostasis of arterial blood pressure.

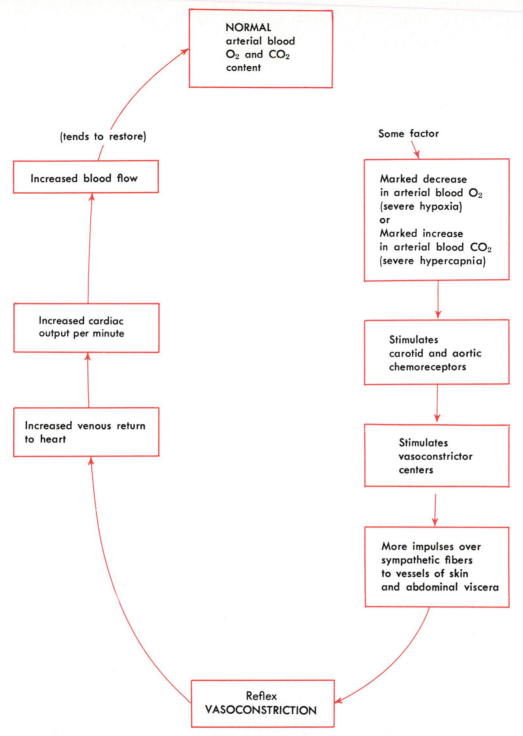

Fig. 15-10 The vasomotor chemoreflex shown here does not function under normal conditions. It operates as a response to the stress of severe hypoxia or hypercapnia and tends to restore normal blood oxygen and carbon dioxide levels. Because this mechanism brings about reflex vasoconstriction, it also tends to increase peripheral resistance and arterial blood pressure. (Also see Fig. 15-11.)

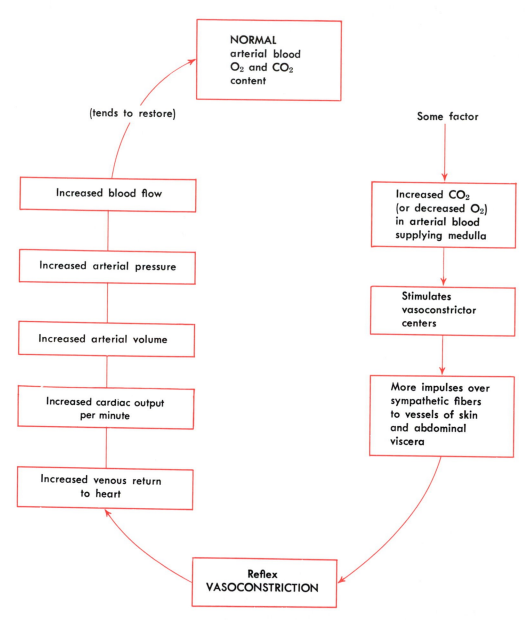

Fig. 15-11 Medullary ischemic reflex, a homeostatic mechanism that tends to restore homeostasis of blood oxygen and carbon dioxide content.

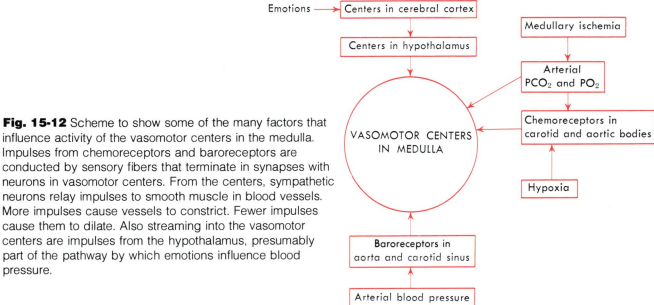

Fig. 15-12 Scheme to show some of the many factors that influence activity of the vasomotor centers in the medulla. Impulses from chemoreceptors and baroreceptors are conducted by sensory fibers that terminate in synapses with neurons in vasomotor centers. From the centers, sympathetic neurons relay impulses to smooth muscle in blood vessels. More impulses cause vessels to constrict. Fewer impulses cause them to dilate. Also streaming into the vasomotor centers are impulses from the hypothalamus, presumably part of the pathway by which emotions influence blood pressure.

tion and dilation. One evidence supporting this view, for example, is the fact that vasoconstriction and a rise in arterial blood pressure characteristically accompany emotions of intense fear or anger. Also, laboratory experiments on animals in which stimulation of the posterior or lateral parts of the hypothalamus leads to vasoconstriction support the belief that higher brain centers influence the vasomotor centers in the medulla.

Local control of arterioles

Some kind of local mechanism operates to produce vasodilation in localized areas. Although the mechanism is not clear, it is known to function in times of increased tissue activity. For example, it probably accounts for the increased blood flow into skeletal muscles during exercise. It also operates in ischemic tissues, serving as a homeostatic mechanism that tends to restore normal blood flow. Norepinephrine, histamine, lactic acid, and other locally produced substances have been suggested as the

stimuli that activate the local vasodilator mechanism. Local vasodilation is also referred to as *reactive hyperemia.*

Summarizing briefly, the volume of blood circulating through the body per minute is determined by the magnitude of both the blood pressure gradient and the peripheral resistance (Figs. 15-13 to 15-16).

A nineteenth century physiologist and physicist, Poiseuille, described the relation between these three factors—pressure gradient, resistance, and volume of fluid flow per minute—with a mathematical equation known as *Poiseuille's law.* In general, but with certain modifications, it applies to blood circulation. We can state it in a simplified form as follows: The volume of blood circulated per minute is directly related to mean arterial pressure minus central venous pressure and inversely related to resistance:

Volume of blood circulated per minute =

$$\frac{\text{Mean arterial pressure} - \text{Central venous pressure}}{\text{Resistance}}$$

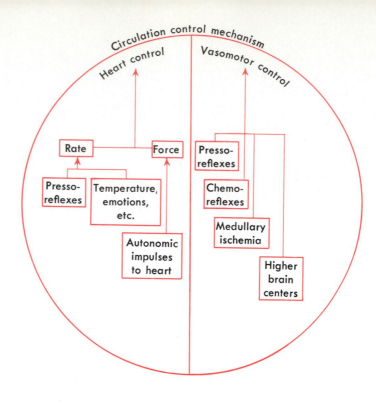

Fig. 15-13 Scheme to suggest some of the many factors that influence heart action and blood vessel diameter and thereby control circulation. (See also Figs. 15-14 and 15-15.)

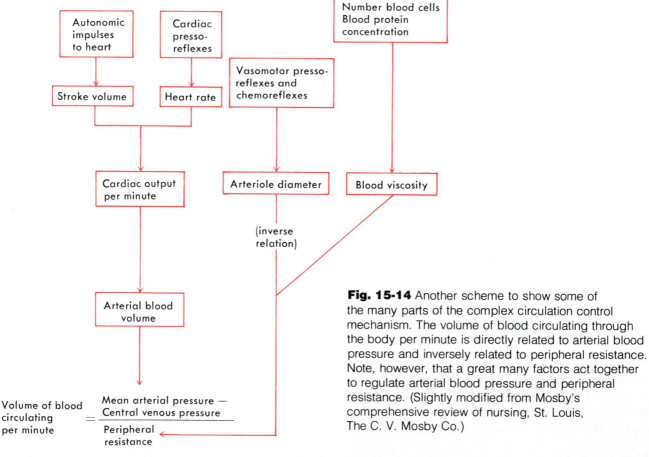

Fig. 15-14 Another scheme to show some of the many parts of the complex circulation control mechanism. The volume of blood circulating through the body per minute is directly related to arterial blood pressure and inversely related to peripheral resistance. Note, however, that a great many factors act together to regulate arterial blood pressure and peripheral resistance. (Slightly modified from Mosby's comprehensive review of nursing, St. Louis, The C. V. Mosby Co.)

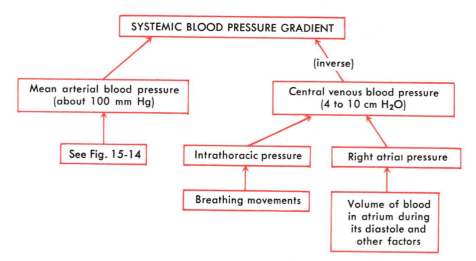

Fig. 15-15 Factors that determine the systemic blood pressure gradient.

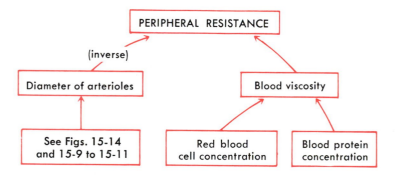

Fig. 15-16 The main determinants of peripheral resistance.

The preceding statement and equation need qualifying with regard to the influence of peripheral resistance on circulation. For instance, according to the equation, an increase in peripheral resistance would tend to decrease blood flow. (Why? Increasing peripheral resistance increases the denominator of the fraction in the preceding equation. Increasing the denominator of any fraction necessarily does what to its value?) Increased peripheral resistance, however, has a secondary action that opposes its primary tendency to decrease blood flow. An in-crease in peripheral resistance hinders or decreases arteriole runoff. This, of course, tends to increase the volume of blood left in the arteries and so tends to increase arterial pressure. Note also that increasing arterial pressure tends to increase the value of the fraction in Poiseuille's equation. Therefore it tends to increase circulation. In short, to say unequivocally what the effect of an increased peripheral resistance will be on circulation is impossible. It depends also on arterial blood pressure—whether it increases, decreases, or stays the same when peripheral resistance increases. The clinical condition arteriosclerosis with hypertension (high blood pressure) illustrates this point. Both peripheral resistance and arterial pressure are increased in this condition. If resistance were to increase more than arterial pressure, circulation, that is, volume of blood flow per minute, would decrease. But if arterial pressure increases proportionately to resistance, circulation remains normal.

Important factors influencing venous return to heart

Two important factors that promote the return of venous blood to the heart are respirations and skeletal muscle contractions. Both produce their facilitating effect on venous return by increasing the pressure gradient between peripheral veins and venae cavae.

The process of inspiration increases the pressure gradient between peripheral and central veins by decreasing central venous pressure and also by increasing peripheral venous pressure. Each time the diaphragm contracts, the thoracic cavity necessarily becomes larger and the abdominal cavity smaller. Therefore the pressures in the thoracic cavity, in the thoracic portion of the vena cava, and in the atria decrease, and those in the abdominal cavity and the abdominal veins increase. Deeper respirations intensify these effects and therefore tend to increase venous return to the heart more than do normal respirations. This is part of the reason why the principle is true that increased respirations and increased circulation tend to go hand in hand.

Skeletal muscle contractions serve as "booster pumps" for the heart. They promote venous return in the following way. As each skeletal muscle contracts, it squeezes the soft veins scattered through its interior, thereby milking the blood in them upward or toward the heart. The closing of the semilunar valves present in veins prevents blood from falling back as the muscle relaxes. Their flaps catch the blood as gravity pulls backward on it (Fig. 15-17). The net effect of skeletal

Fig. 15-17 Diagram showing the action of venous valves. **A,** External view of vein showing dilation at site of the valve. **B,** Interior of vein with the semilunar flaps in open position, permitting flow of blood through the valve. **C,** Valve flaps approximating each other, occluding the cavity and preventing the backflow of blood.

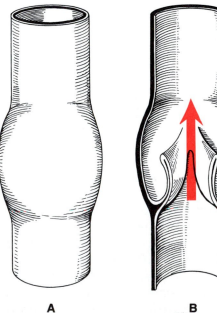
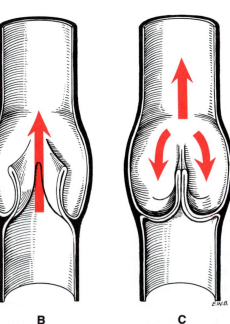

A   B   C

muscle contraction plus venous valvular action, therefore, is to move venous blood toward the heart, to increase the venous return.

The value of skeletal muscle contractions in moving blood through veins is illustrated by a common experience. Who has not noticed how much more uncomfortable and tiring standing still is than walking? After several minutes of standing quietly the feet and legs feel "full" and swollen. Blood has accumulated in the veins because the skeletal muscles are not contracting and squeezing it upward. The repeated contractions of the muscles when walking, on the other hand, keep the blood moving in the veins and prevent the discomfort of distended veins.

Blood pressure

Measurement of arterial blood pressure clinically

Blood pressure is measured with the aid of an apparatus known as a sphygmomanometer, which makes it possible to measure the amount of air pressure equal to the blood pressure in an artery. The measurement is made in terms of how many millimeters high the air pressure raises a column of mercury in a glass tube.

The sphygmomanometer consists of a rubber cuff attached by a rubber tube to a compressible bulb and by another tube to a column of mercury that is marked off in millimeters. The cuff is wrapped around the arm over the brachial artery, and air is pumped into the cuff by means of the bulb. In this way, air pressure is exerted against the outside of the artery. Air is added until the air pressure exceeds the blood pressure within the artery or, in other words, until it compresses the artery. At this time, no pulse can be heard through a stethoscope placed over the brachial artery at the bend of the elbow along the inner margin of the biceps muscle. By slowly releasing the air in the cuff the air pressure is decreased until it approximately equals the blood pressure within the artery. At this point the ves-

sel opens slightly and a small spurt of blood comes through, producing the first sound, one with a rather sharp, taplike quality. This is followed by increasingly louder sounds that suddenly change. They become more muffled, then disappear altogether. Nurses and physicians train themselves to hear these different sounds and simultaneously to read the column of mercury, since the first taplike sound represents the *systolic blood pressure*. Systolic pressure is the force with which the blood is pushing against the artery walls when the ventricles are contracting. The lowest point at which the sounds can be heard, just before they disappear, is approximately equal to the *diastolic pressure* or the force of the blood when the ventricles are relaxed. Systolic pressure gives valuable information about the force of the left ventricular contraction, and diastolic pressure gives valuable information about the resistance of the blood vessels. Clinically, diastolic pressure is considered more important than systolic pressure because it indicates the pressure or strain to which blood vessel walls are constantly subjected. It also reflects the condition of the peripheral vessels, since diastolic pressure rises or falls with the peripheral resistance. If, for instance, arteries are sclerosed, peripheral resistance and diastolic pressure both increase.

Blood in the arteries of the average adult exerts a pressure equal to that required to raise a column of mercury about 120 mm (or a column of water over 5 feet) high in a glass tube during systole of the ventricles and 80 mm high during their diastole. For the sake of brevity, this is expressed as a blood pressure of 120 over 80 (120/80). The first or upper figure represents systolic pressure and the second diastolic pressure. From the figures just given, we observe that blood pressure fluctuates considerably during each heartbeat. During ventricular systole the force is great enough to raise the mercury column 40 mm higher than during ventricular diastole. This difference between systolic and diastolic pressure is called *pulse pressure*. It characteristically increases in arteriosclerosis mainly because systolic pressure increases more than diastolic pressure does. Pulse pressure increases even more markedly in aortic valve insufficiency because of both a rise in systolic and a fall in diastolic pressure.

Relation to arterial and venous bleeding

Because blood exerts a comparatively high pressure in arteries and a very low pressure in veins, it gushes forth with considerable force from a cut artery but seeps in a slow, steady stream from a vein. As we have just seen, each ventricular contraction raises arterial blood pressure to the systolic level, and each ventricular relaxation lowers it to the diastolic level. As the ventricles contract, then, the blood spurts forth forcefully from the increased pressure in the artery, but as the ventricles relax, the flow ebbs to almost nothing because of the fall in pressure. In other words, blood escapes from an artery in spurts because of the alternate raising and lowering of arterial blood pressure but flows slowly and steadily from a vein because of the low, practically constant pressure. A uniform instead of a pulsating pressure exists in the capillaries and veins. Why? Because the arterial walls, being elastic, continue to squeeze the blood forward while the ventricles are in diastole. Therefore blood enters capillaries and veins under a steady pressure.

Velocity of blood

The speed with which blood flows, that is, distance per minute, through its vessels is governed in part by the physical principle that when a liquid flows from an area of one cross-section size to an area of larger size, its velocity slows in the area with the larger cross section. For example, a narrow river whose bed widens flows more slowly through the wide section than through the narrow. In terms of the blood vascular system, the total cross-section area of all arterioles together is greater than that of the arteries. Therefore blood flows more slowly through arterioles than through arteries. Likewise, the total cross-section area of all capillaries together is greater than that of all arterioles and, therefore, capillary flow is slower than arteriole flow. Venule cross-section area, on the other hand, is smaller than capillary cross-section area. Therefore the blood velocity increases in venules and again in veins, which have a still smaller cross-section area. In short, the most rapid blood flow takes place in arteries and the slowest in capillaries. Can you think of a valuable effect stemming from the fact that blood flows most slowly through the capillaries?

Pulse
Definition

Pulse is defined as the alternate expansion and recoil of an artery.

Cause

Two factors are responsible for the existence of a pulse that can be felt:

1 Intermittent injections of blood from the heart into the aorta, which alternately increase and decrease the pressure in that vessel. If blood poured steadily out of the heart into the aorta, the pressure there would remain constant and there would be no pulse.

2 The elasticity of the arterial walls, which makes it possible for them to expand with each injection of blood and then recoil. If the vessels were fashioned from rigid material such as glass, there would still be an alternate raising and lowering of pressure within them with each systole and diastole of the ventricles, but the walls could not expand and recoil and, therefore, no pulse could be felt.

Pulse wave

Each ventricular systole starts a new pulse that proceeds as a wave of expansion throughout the arteries and is known as the pulse wave. It gradually dissipates as it travels, disappearing entirely in the capillaries. The pulse felt in the radial artery at the wrist does not coincide with the contraction of the ventricles. It follows each contraction by an appreciable interval (the length of time required for the pulse wave to travel from an aorta to the radial artery). The farther from the heart the pulse is taken, therefore, the longer that interval is.

Any nurse has only to think of the number of times she or he has counted pulses to become aware of the diagnostic importance of the pulse. It reveals important information about the cardiovascular system, about heart action, blood vessels, and circulation.

Where pulse can be felt

In general the pulse can be felt wherever an artery lies near the surface and over a bone or other firm background. Some of the specific locations where the pulse is most easily felt are as follows:

1 *Radial artery*—at wrist
2 *Temporal artery*—in front of ear or above and to outer side of eye
3 *Common carotid artery*—along anterior edge of sternocleidomastoid muscle at level of lower margin of thyroid cartilage
4 *Facial artery*—at lower margin of lower jawbone on a line with corners of mouth and in groove in mandible about one third of way forward from angle
5 *Brachial artery*—at bend of elbow along inner margin of biceps muscle
6 *Posterior tibial artery*—behind the medial malleolus (inner "ankle bone")
7 *Dorsalis pedis artery*—on the dorsum (upper surface) of the foot

NOTE: The so-called pressure points or points at which pressure may be applied to stop arterial bleeding and the points where the pulse may be felt are related. To be more specific, both are found where an artery lies near the surface and near a bone that can act as a firm background

for pressure. There are six important pressure points:

1 *Temporal artery*—in front of ear
2 *Facial artery*—same place as pulse is taken
3 *Common carotid artery*—point where pulse is taken, with pressure back against spinal column
4 *Subclavian artery*—behind mesial third of clavicle, pressing against first rib
5 *Brachial artery*—few inches above elbow on inside of arm, pressing against humerus
6 *Femoral artery*—in middle of groin, where artery passes over pelvic bone; pulse can also be felt here

In trying to stop arterial bleeding by pressure, one must always remember to apply the pressure at the pressure point that lies between the bleeding part and the heart. Why? Because blood flows from the heart through the arteries to the part. Pressure between the heart and bleeding point, therefore, cuts off the source of the blood flow to that point.

Venous pulse

A pulse exists in the large veins only. It is most prominent in the veins near the heart because of changes in venous blood pressure brought about by alternate contraction and relaxation of the atria of the heart. Venous pulse does not have as great clinical significance as arterial pulse and so is less often measured.

Outline summary

Functions and importance of control mechanisms

A Movement of blood through its closed circuit of vessels by the heart

B Regulation of blood pressure and flow in response to changing cellular needs

Physiology of the heart

A Conduction system (Fig. 15-1)
  1 Conduction system structures—SA node, AV node, AV bundle, and Purkinje system (Chapter 14)
  2 Specialized, both structurally and functionally, for generation or conduction of action potential
    a SA node (pacemaker)—inherent rhythmicity sets basic heart rate
    b Ectopic pacemakers—conductive structures other than SA node; discharge rate is less than SA node
  3 Artificial pacemakers
    a Continuously discharging type—stimulates heart at a set rhythm
    b Demand pacemakers—fire when rate decreases below a preset minimum
B Electrocardiogram (ECG) (Fig. 15-2)—record of the heart's action currents (impulse conduction); composed of P wave, QRS complex, and T wave; interpretation complex but valuable aid for diagnosing certain heart disorders, especially disorders of heart's rhythm, for example, fibrillation; ECG intervals useful in interpretation of conduction velocity
C Control of heart rate (Fig. 15-3)
  1 Autonomic nervous system control is result of opposing influences between parasympathetic (vagus) and sympathetic (accelerator nerve) stimulation
    a Parasympathetic stimulation is inhibitory (vagal release of acetylcholine)
    b Sympathetic effects are stimulatory (norepinephrine)
  2 Cardioinhibitory center—source (medulla) of parasympathetic impulses that reach SA node via vagus (vagus is "brake" on heart)
  3 Cardioaccelerator center—source (medulla) of sympathetic stimuli that reach heart via middle, superior, and inferior cardiac nerves

  4 Cardiac pressoreflexes—stretch receptors that send impulses to the cardoinhibitory and accelerator centers and thereby influence parasympathetic and sympathetic outflow
    a Carotid sinus reflex—stimulation (by high blood pressure) causes reflex showing of heart via stimulation of cardoinhibitory center
    b Aortic reflex—stimulation (by high blood pressure) causes reflex slowing of heart
    c Decrease in aortic or carotid blood pressure initiates reflex acceleration of heart
  5 Miscellaneous factors that influence heart rate include emotions, exercise, hormones, blood temperature, pain, and stimulation of various exteroceptors
D Cardiac cycle (Fig. 15-5)
  1 Nature—consists of systole and diastole of atria and ventricles; atria contract and as they relax, ventricles contract
  2 Time required for cycle—about 0.8 second or from 70 to 80 times per minute
  3 Events of cycle
    a Atrial systole—AV valves open and SL valves closed; ventricles relaxed; preceded by P wave
    b Isovolumetric contraction—between start of ventricular systole and opening of SL valves; ventricular volume remains constant; onset coincides with R wave of ECG; first heart sound is heard
    c Ejection—initial, shorter, rapid ejection followed by longer phase of reduced ejection
      (1) Residual volume is blood remaining in ventricles following ejection phase—amount increases in heart failure
    d Isovolumetric relaxation—period between closure of SL valves and opening of AV valves; ventricles are relaxing; second heart sound heard during this period
    e Rapid ventricular filling—rapid influx of blood into ventricles after opening of AV valves
    f Reduced ventricular filling or diastasis—characterized by a gradual increase in ventricular pressure and volume
  4 Heart sounds during cycle—lubb from contraction of ventricles and closure of cuspid valves; dupp from closure of SL valves

Control of circulation

A Functions of control mechanisms
  1 Maintain circulation
  2 Vary circulation; increase blood flow per minute when activity increases and decrease blood flow when activity decreases
B Primary principle of circulation—blood circulates because blood pressure gradient exists within its vessels (Fig. 15-6)
C Control of arterial blood pressure
  1 Primary determinant of arterial pressure is volume of blood in arteries
  2 Many factors influence arterial volume and thereby influence arterial pressure; two of the most important are cardiac output (CO) and peripheral resistance (Fig. 15-7)
  3 CO is determined by both stroke volume and heart rate
    a Stroke volume reflects force or strength of ventricular contraction
      (1) Stroke volume × heart rate equals cardiac output
      (2) Stroke volume under ordinary conditions is determined by Starling's law of the heart
    b Heart rate is regulated by pressoreflexes and by many miscellaneous factors; increased arterial pressure in aorta or carotid sinus tends to produce reflex slowing of heart, whereas increased right atrial pressure tends to produce reflex cardiac acceleration (Fig. 15-8)
  4 Peripheral resistance determined mainly by blood viscosity and arteriole diameter; in general, less blood viscosity, less peripheral resistance, but smaller diameter of arterioles, greater peripheral resistance
  5 Blood viscosity determined by concentration of blood proteins and blood cells and directly related to both
  6 Vasomotor or vasoconstrictor control mechanism plays important role in control of changes in diameter of arterioles; center located in medulla; outflow is via sympathetic fibers to smooth muscle of blood vessels in "reservoir" areas
  7 Arteriole diameter regulated mainly by pressoreflexes and chemoreflexes; in general, increase in arterial pressure produces reflex dilation of arterioles, whereas hypoxia and hypercapnia cause constriction of arterioles in blood reservoir organs but dilation of them in local structures, notably in skeletal muscles, heart, and brain
  8 Volume of blood circulating per minute determined by blood pressure gradient and peripheral resistance; according to Poiseuille's law, directly related to pressure gradient and inversely related to peripheral resistance
D Important factors influencing venous return to heart
  1 Respirations—the deeper the respirations, the greater the venous return tends to be
  2 Skeletal muscle contractions serve as "booster pumps" that tend to increase venous return

Blood pressure

A How arterial blood pressure measured clinically
  1 Sphygmomanometer
  2 Systolic pressure normal range about 120 to 140 mm Hg and diastolic pressure about 80 to 90 mm Hg
B Relation to arterial and venous bleeding
  1 Arterial bleeding in spurts because of difference in amounts of systolic and diastolic pressures
  2 Venous bleeding—slow and steady because of low, practically constant venous pressure

Velocity of blood

A Speed with which blood flows
B Most rapid in arteries and slowest in capillaries

Pulse

A Definition—alternate expansion and recoil of artery
B Cause—intermittent injections of blood from heart into aorta with each ventricular contraction; pulse can be felt because of elasticity of arterial walls
C Pulse wave—pulse starts at beginning of aorta and proceeds as wave of expansion throughout arteries
D Where pulse can be felt—radial, temporal, common carotid, facial, brachial, femoral, and popliteal arteries; where near surface and over firm background, such as bone; pressure points, points where bleeding can be stopped by pressure, roughly related to places where pulse can be felt
E Venous pulse—in large veins only; caused by changes in venous pressure brought about by alternate contraction and relaxation of atria

Review questions

1 Identify, locate, and describe the function of each of the following structures: SA node, AV node, AV bundle, and Purkinje fibers.

2 Compare the intrinsic rhythm of the SA node with other components of the heart's conduction system. What is an ectopic pacemaker? Artificial pacemaker?

3 What does an electrocardiogram measure and record? List the normal ECG deflection waves and intervals. What do the various ECG waves represent?

4 Discuss and compare the effects of sympathetic and parasympathetic stimulation on heart rate. What effect would vagal stimulation have on heart rate?

5 Locate and describe the function of the cardioinhibitory and cardioaccelerator centers.

6 Explain the mechanism of action of the cardiac pressoreflexes on heart rate.

7 List and give the effect of several "miscellaneous" factors such as grief or pain on heart rate.

8 What is meant by the term cardiac cycle?

9 List the "periods" of the cardiac cycle and briefly describe the events that occur in each. Refer to Fig. 15-5 as you prepare your answer.

10 What is meant by the term residual volume as it applies to the heart?

11 Describe and explain the origin of the heart sounds.

12 State in your own words the basic principle of fluid flow.

13 What blood vessels present the greatest resistance to blood flow?

14 What is the primary determinant of arterial blood pressure?

15 List the two most important factors that indirectly determine arterial pressure through their influence on arterial volume.

16 How is cardiac output determined?

17 What two factors determine blood viscosity? What does viscosity mean? Give an example of a condition in which blood viscosity decreases. Explain its effect on circulation.

18 What mechanisms control peripheral resistance? Cite an example of the operation of one or more parts of this mechanism to increase resistance; to decrease it.
19 What is Starling's law of the heart?
20 What is arteriole runoff? What is the relationship of arteriole runoff to peripheral resistance?
21 What are the components of the vasomotor control mechanism?
22 Explain the mechanism of action of the medullary ischemic reflex.
23 State in your own words Poiseuille's law. Give an example of increased circulation to illustrate application of this law. Give an example of decreased circulation to illustrate application of this law.
24 What effect, if any, would a respiratory stimulant drug have on circulation? Explain why it would or would not affect circulation.
25 Describe the measurement of arterial blood pressure clinically.
26 List the so-called pressure points at which pressure can be applied to stop arterial bleeding.

chapter 16

The lymphatic system

Definitions
Lymphatic system

The lymphatic system is actually a specialized component of the circulatory system, since it consists of a moving fluid (lymph) derived from the blood and tissue fluid and a group of vessels (lymphatics) that return the lymph to the blood by a roundabout route. In addition to lymph and the lymphatic vessels, the system includes lymph nodes located along the paths of the collecting vessels (Fig. 16-1), isolated nodules of lymphatic tissue such as Peyer's patches in the intestinal wall, and specialized lymphatic organs such as the tonsils, thymus, and spleen.

Although it serves a unique transport function by returning tissue fluid, proteins, fats, and other substances to the general circulation, lymph flow differs from the true "circulation" of blood seen in the cardiovascular system. The lymphatic vessels do not, like vessels in the blood vascular system, form a closed ring or circuit but instead begin blindly in the intercellular spaces of the soft tissues of the body.

Lymph and interstitial fluid (tissue fluid)

Lymph is the clear, watery-appearing fluid found in the lymphatic vessels. Interstitial fluid, which fills the spaces between the cells, is not the clear watery fluid it seems to be. Recent studies show that it is a complex and "organized" material. In some tissues, it is part of a semifluid ground substance. In others, it is the bound water in a gelatinous ground substance. Interstitial fluid and blood plasma together constitute the extracellular fluid or, in the words of

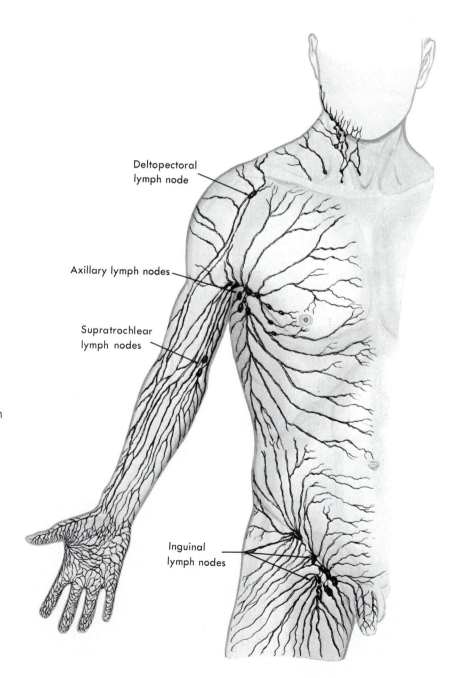

Deltopectoral
lymph node

Axillary lymph nodes

Supratrochlear
lymph nodes

Inguinal
lymph nodes

Fig. 16-1 Superficial lymphatics of
the upper extremity, anterior surface,
and superficial lymphatics and lymph
nodes of the front of the trunk.

Claude Bernard, the "internal environment of the body"—the fluid environment of cells in contrast to the atmosphere or external environment of the body.

Both lymph and interstitial fluid closely resemble blood plasma in composition. The main difference is that they contain a lower percentage of proteins than does plasma. Lymph is almost identical in chemical composition to interstitial fluid when comparisons are made between the two fluids taken from the same area of the body. However, the average concentration of protein (4 gm%) in lymph taken from the thoracic duct (Fig. 16-3) is about twice that found in most interstitial fluid samples. The elevated protein level of thoracic duct lymph (a mixture of lymph from all areas of the body) results from protein-rich lymph flowing into the duct from the liver and small intestine. A little over one half of the 2,500 to 2,800 ml total daily lymph flow through the thoracic duct is derived from these two organs.

Lymphatics
Formation and distribution

Lymphatic vessels originate as microscopic blind-end vessels called *lymphatic capillaries*. (Those originating in the villi of the small intestine are called *lacteals*.) The wall of the lymphatic capillary consists of a single layer of flattened endothelial cells. Each blindly ending capillary is attached or fixed to surrounding cells by tiny connective tissue filaments. Networks of lymphatic capillaries, which branch and anastomose freely, are located in the intercellular spaces and are widely distributed throughout the body. As a rule of thumb, lymphatic and blood capillary networks lie side by side but are always independent of each other.

As twigs of a tree join to form branches and branches join to form larger branches, and large branches join to form the tree trunk, so do lymphatic capillaries merge, forming slightly

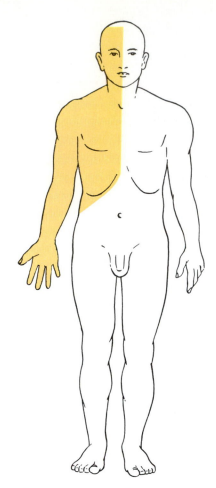

Fig. 16-2 Lymph drainage. The right lymphatic ducts drain lymph from the part of the body indicated by the stippled area. Lymph from all the rest of the body enters the general circulation by way of the thoracic duct.

larger lymphatics that join other lymphatics to form still larger vessels, which merge to form the main lymphatic trunks: the *right lymphatic ducts* and the *thoracic duct*. Lymph from the entire body, except the upper right quadrant (Fig. 16-2), drains eventually into the thoracic duct, which drains into the left subclavian vein at the point where it joins the left internal jugular vein. Lymph from the upper right quadrant of the body empties into the right lymphatic duct (or, more commonly, into three collecting ducts) and then into the right subclavian vein. Since most

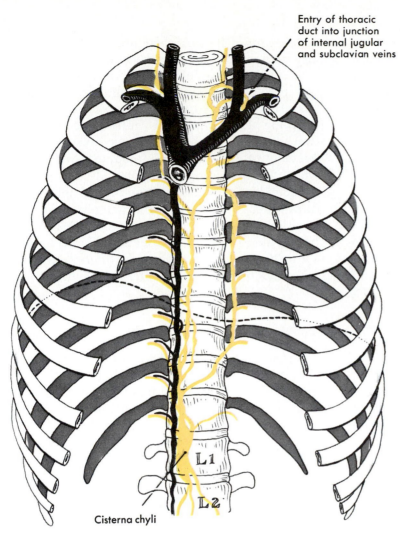

Entry of thoracic
duct into junction
of internal jugular
and subclavian veins

Fig. 16-3 Position of the cisterna chyli and the thoracic duct and its tributaries and the entry of the duct into the junction of the internal jugular and subclavian veins to form the innominate veins.

L 1

L 2

Cisterna chyli

of the lymph of the body returns to the bloodstream by way of the thoracic duct, this vessel is considerably larger than the other main lymph channels, the right lymphatic ducts, but is much smaller than the large veins, which it resembles in structure. It has a diameter about the size of a goose quill and a length of 15 to 18 inches. It originates as a dilated structure, the *cisterna chyli*, in the lumbar region of the abdominal cavity and ascends by a flexuous course to the root of the neck, where it joins the subclavian vein as just described (Fig. 16-3).

Structure

Lymphatics resemble veins in structure with these exceptions:

1 Lymphatics have thinner walls.
2 Lymphatics contain more valves.
3 Lymphatics contain lymph nodes located at certain intervals along their course.

The lymphatic capillary wall is formed by a single layer of large but very thin and flat endothelial cells. In the past, many investigators described what they believed were rather large and naturally occurring openings ("stomata") between adjacent and endothelial cells in the lymphatic capillary wall. Recent and more sophisticated injection techniques that are now being used to prepare lymphatic vessels for study have shown that such openings are in reality only preparation artifacts. Although very small intercellular openings (clefts) do exist between adjacent endothelial cells in the lymphatic capillary, there is no direct and open communication between the vessel lumen and the surrounding tissue spaces.

As lymph flows from the thin-walled capillaries into vessels with a larger diameter (0.2 to 0.3 mm), the walls become thicker and exhibit the three coats or layers typical of arteries and veins (see Table 14-2). Interlacing elastic fibers and several strata of circular smooth muscle bundles are found in both the tunica media and the tunica adventitia of the large lymphatic vessel wall. Boundaries between layers or coats are less distinct in the thinner lymphatic vessel walls than in arteries or veins.

Semilunar valves are extremely numerous in lymphatics of all sizes and give the vessels a somewhat varicose appearance. Valves are present every few millimeters in large lymphatics and are even more numerous in the smaller vessels. Formed from folds of the tunica intima, each valve projects into the vessel lumen in a slightly expanded area circled by bundles of smooth muscle fibers.

Experimental evidence suggests that most lymph vessels have the capacity for repair or regeneration when damaged. Formation of new lymphatic vessels occurs by extension of solid cellular cores or sprouts, formed by mitotic division of endothelial cells in existing vessels, which later become "canalized."

Functions

The lymphatics play a critical role in a number of interrelated homeostatic mechanisms. The high degree of permeability of the lymphatic capillary wall permits large molecular weight substances and even particulate matter, which cannot be absorbed into a blood capillary, to be removed from the interstitial spaces. Proteins that accumulate in the tissue spaces can return to blood only via lymphatics. This fact has great clinical importance. For instance, if anything blocks lymphatic return, blood protein concentration and blood osmotic pressure soon fall below normal and fluid imbalance and death will result (discussed in Chapter 21).

Lacteals (lymphatics in the villi of the small intestine) serve an important function in the absorption of fats and other nutrients. The milky lymph found in lacteals after digestion contains 1% to 2% fat and is called *chyle*. Interstitial fluid has a much lower lipid content than chyle.

Lymph circulation

Water and solutes continually filter out of capillary blood into the interstitial fluid. To balance this outflow, fluid continually reenters blood from the interstitial fluid. Experimental studies have shown that only about 40% of the fluid that filters out of blood capillaries returns to them by osmosis. The remaining 60% returns to the blood by way of the lymphatics. Also, newer evidence has disproved the old idea that healthy capillaries do not "leak" proteins. In truth, each day about 50% of the total blood proteins leak out of the capillaries into the tissue fluid and return to the blood by way of the lymphatic vessels. For more details about fluid exchange between blood and interstitial fluid, see Chapter 21. From lymphatic capillaries, lymph flows through progressively larger lymphatic vessels to eventually reenter blood at the junction of the internal jugular and subclavian veins (Fig. 16-3).

The "lymphatic pump"

Although there is no muscular pumping organ connected with the lymphatic vessels to force lymph onward as the heart does blood, still lymph moves slowly and steadily along in its vessels. Lymph flows through the thoracic duct and reenters the general circulation at the rate of about 125 ml per hour. This occurs despite the fact that most of the flow is uphill. It moves through the system in the right direction because of the large number of valves that permit fluid flow only in the central direction. What mechanisms establish the pressure gradient required by the basic law of fluid flow? Two of the same mechanisms that contribute to the blood pressure gradient also establish a lymph pressure gradient. These are breathing movements and skeletal muscle contractions.

The mechanism of inspiration, resulting from the descent of the diaphragm, causes intra-abdominal pressure to increase as intrathoracic pressure decreases. This simultaneously causes pressure to increase in the abdominal portion of the thoracic duct and to decrease in the thoracic portion. In other words, the process of inspiring establishes a pressure gradient in the thoracic duct that causes lymph to flow upward through it.

Contracting skeletal muscles also exert pressure on the lymphatics to push the lymph for-

ward. During exercise, lymph flow may increase as much as 10- to 15-fold. In addition, segmental contraction of the walls of the lymphatics themselves will result in lymph being pumped from one valved segment to the next.

Other pressure-generating factors that can compress the lymphatics will also contribute to the effectiveness of the "lymphatic pump." Examples of such factors include arterial pulsations, postural changes, and passive compression of the body soft tissues.

Lymph nodes
Structure

Lymph nodes or glands, as some people call them, are oval-shaped or bean-shaped structures. Some are as small as a pinhead and others as large as a lima bean. As shown in Fig. 16-4, lymph moves into the nodes via several afferent lymphatic vessels. Here it moves slowly through sinus channels lined with phagocytic reticuloendothelial cells and emerges usually by one ef-

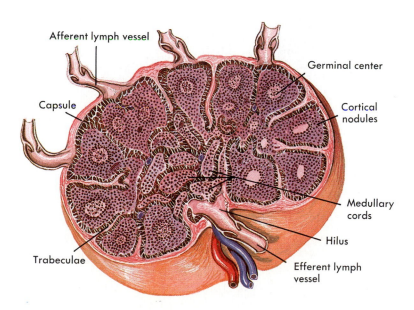

Fig. 16-4 Structure of a lymph node. Several afferent valved lymphatics bring lymph to the node. An efferent lymphatic leaves the node at the hilus. Note that the artery and vein enter and leave at the hilus.

Afferent lymph vessel

Germinal center

Capsule

Cortical nodules

Medullary cords

Hilus

Trabeculae

Efferent lymph vessel

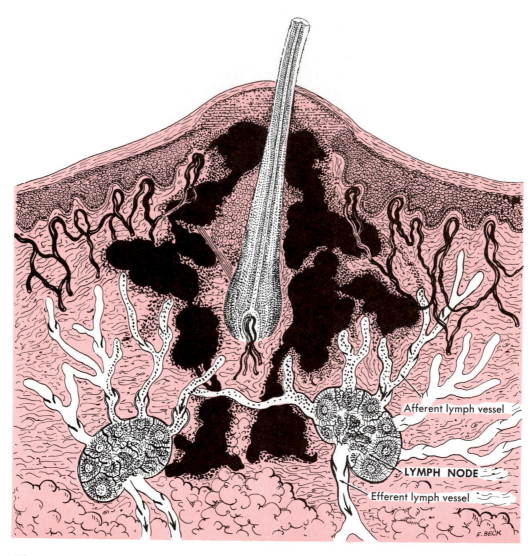

Fig. 16-5 Diagrammatic representation of a skin section in which an infection surrounds a hair follicle. The black areas represent dead and dying cells (pus). Black dots around the black areas represent bacteria. Leukocytes phagocytose many bacteria in tissue spaces. Others may enter the lymph nodes by way of afferent lymphatics. The nodes filter out those bacteria; here reticuloendothelial cells usually destroy them all by phagocytosis.

ferent vessel. Lymphatic tissue, densely packed with lymphocytes, composes the substance of the node.

Locations

With the exception of comparatively few single nodes, most of the lymph nodes occur in groups or clusters in certain areas. The group locations of greatest clinical importance are as follows:

1 *Submental and submaxillary groups* in the floor of the mouth—lymph from the nose, lips, and teeth drains through these nodes.
2 *Superficial cervical glands* in the neck along the sternocleidomastoid muscle—these nodes drain lymph from the head (which has already passed through other nodes) and neck.
3 *Superficial cubital* or *supratrochlear nodes* located just above the bend of the elbow—lymph from the forearm passes through these nodes.
4 *Axillary nodes* (20 to 30 large nodes clustered deep within the underarm and upper chest regions)—lymph from the arm and upper part of the thoracic wall, including the breast, drains through these nodes.
5 *Inguinal nodes* in the groin—lymph from the leg and external genitals drains through these.

Functions

Lymph nodes perform two unrelated functions: defense and hemopoiesis.

1 *Defense functions: filtration and phagocytosis.* The structure of the sinus channels within lymph nodes slows the lymph flow through them. This gives the reticuloendothelial cells that line the channels time to remove microorganisms and other injurious particles—cancer cells and soot, for example—from the lymph and phagocytose them (Fig. 16-5). Sometimes, however, such hordes of microorganisms enter the nodes that the phagocytes cannot destroy enough of them to prevent their injuring the node. An infection of the node, adenitis, then results. Also, because cancer cells often break away from a malignant tumor and enter lymphatics, they travel to the lymph nodes, where they may set up new growths. This may leave too few channels for lymph to return to the blood. For example, if tumors block axillary node channels, fluid accumulates in the interstitial spaces of the arm, causing the arm to become markedly swollen.

2 *Hemopoiesis.* The lymphatic tissue of lymph nodes forms lymphocytes and monocytes, the nongranular white blood cells, and plasma cells.

Lymphatic drainage of the breast

Cancer of the breast is one of the most common forms of malignancy in women. Unfortunately, cancerous cells from a single "primary" tumor in the breast often spread to other areas of the body through the lymphatic system. An understanding of the lymphatic drainage of the breast is, therefore, of particular importance in the diagnosis and treatment of this very common type of malignancy. Refer to Fig. 16-6 as you study the lymphatic drainage of the breast.

Location and distribution of lymphatics

The breast or mammary gland is drained by two sets of lymphatic vessels:

1 Lymphatics that originate in and drain the skin over the breast with the exception of the areola and nipple
2 Lymphatics that originate in and drain the substance of the breast itself as well as the skin of the areola and nipple

Superficial vessels that drain lymph from the skin and surface areas of the breast converge to form a diffuse *cutaneous lymphatic plexus*. Communication between the cutaneous plexus and large lymphatics that drain the secretory tissue and ducts of the breast occurs in the *subareolar plexus (plexus of Sappey)* located under the areola

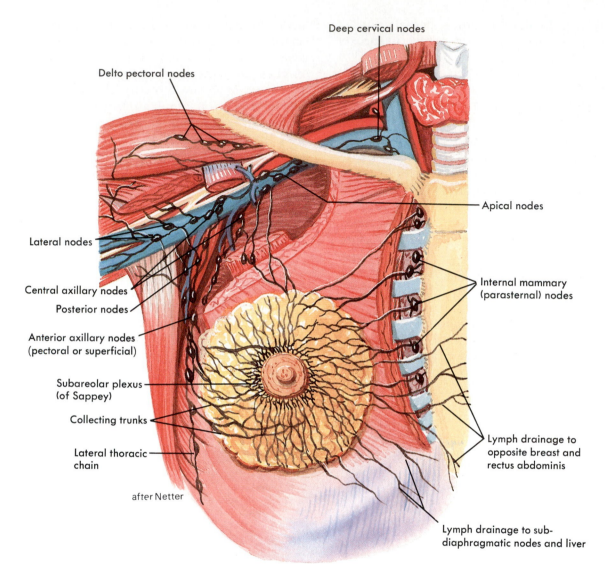

Deep cervical nodes

Delto pectoral nodes

Apical nodes

Lateral nodes

Internal mammary (parasternal) nodes

Central axillary nodes

Posterior nodes

Anterior axillary nodes (pectoral or superficial)

Subareolar plexus (of Sappey)

Collecting trunks

Lateral thoracic chain

Lymph drainage to opposite breast and rectus abdominis

after Netter

Lymph drainage to sub-diaphragmatic nodes and liver

Fig. 16-6 Lymphatic drainage of the breast. Note the extensive communication (anastomoses) between superficial and deep lymphatic vessels.

surrounding the nipple. Anastomoses also occur between superficial lymphatics from both breasts across the middle line. Such communication can result in the spread of cancerous cells to previously healthy tissue in the other breast.

Both superficial and deep lymphatic vessels also communicate with lymphatics in the fascia of the pectoralis major muscle. Removal of a wide area of deep fascia is therefore required in surgical treatment of advanced or diffuse breast malignancy (radical mastectomy). In addition, cancer cells from a breast tumor sometimes reach the abdominal cavity because of lymphatic communication through the upper part of the linea alba.

Lymph nodes

Over 85% of the lymph from the breast enters the axillary lymph nodes. Most of the remainder enters the parasternal nodes along the lateral edges of the sternum.

Lymph nodes in the axilla are divided into five sets and named according to their placement:
1 Anterior set
2 Posterior set
3 Lateral set
4 Central set
5 Apical (infraclavicular) set

The anterior set consists of several very large nodes that are in actual physical contact with an extension of breast tissue called the *axillary tail of Spence*. Because of the physical contact between these nodes and breast tissue, cancerous cells may spread by both lymphatic extension and contiguity of tissue. Other nodes in the axilla will enlarge and swell after being "seeded" with malignant cells as lymph from a cancerous breast flows through them.

Thymus

Location, appearance, and size

Intensive study and experimentation has identified the thymus as the primary central organ of the lymphatic system. It is a single unpaired organ consisting of two pyramidal-shaped lobes with delicate and finely lobulated surfaces. The thymus is located in the mediastinum, extending up into the neck as far as the lower edge of the thyroid gland and inferiorly as far as the fourth costal cartilage. Its size relative to the rest of the body is largest in a child about 2 years old. Its absolute size is largest at puberty, when its weight ranges between 35 and 40 gm. From then on, it gradually atrophies until in great old age, it may be largely replaced by fat, weigh less

than 10 gm, and be barely recognizable. The thymus is pinkish grey in color early in childhood but with advancing age becomes yellowish as lymphatic tissue is replaced by fat—a process called *involution*.

Structure

The pyramidal-shaped lobes of the thymus are subdivided into small (1 to 2 mm) lobules by connective tissue septa that extend inward from a covering fibrous capsule. Each lobule is composed of a dense cellular cortex and an inner, less dense medulla. Both cortex and medulla are composed of lymphocytes in an epithelial framework quite different from the supporting connective tissue seen in other lymphoid organs.

Function

One of the body's best-kept secrets has been the function of the thymus. Before 1961 there were no significant clues as to its role. Then a young Briton, Dr. Jacques F. A. P. Miller, removed the thymus glands from newborn mice. His findings proved startling and crucial. Almost like a chain reaction, further investigations followed and led to at least a partial uncovering of the thymus' long-held secret. It now seems clear that this small structure (it weighs at most a little over an ounce) plays a critical part in the body's defenses against infections—in its vital immunity mechanism (Chapter 26).

In the mouse, and presumably in man as well, the thymus does two things. First, it serves as the source of lymphocytes before birth. (Actually the fetal bone marrow forms lymphocytes, which then "seed" the thymus.) Many lymphocytes leave the thymus and circulate to the spleen, lymph nodes, and other lymphatic tissues. Soon after birth the thymus is postulated to start secreting a hormone that enables lymphocytes to develop into plasma cells. Since plasma cells synthesize antibodies against foreign proteins, the thymus functions as part of the immune mechanism. It probably completes its essential work early in childhood.

Spleen

Location

The spleen is located in the left hypochondrium directly below the diaphragm, above the left kidney and descending colon, and behind the fundus of the stomach.

Structure

As Fig. 16-7 shows, the spleen is roughly ovoid in shape. Its size varies greatly in different individuals and in the same individual at different times. For example, it hypertrophies during infectious diseases and atrophies in old age. Within the spleen, lymphocytes, monocytes, and neutrophils crowd the numerous spaces formed by a meshwork of interlacing fibers.

Functions

The spleen has long puzzled physiologists who have ascribed many and sundry functions to it. According to present-day knowledge, it performs several functions: defense, hemopoiesis, and red blood cell and platelet destruction; it also serves as a reservoir for blood.

1 *Defense.* As blood passes through the sinusoids of the spleen, reticuloendothelial cells (macrophages) lining these venous spaces remove microorganisms from the blood and destroy them by phagocytosis. Therefore the spleen plays a part in the body's defense against microorganisms.

2 *Hemopoiesis.* Nongranular leukocytes, that is, monocytes and lymphocytes, and plasma

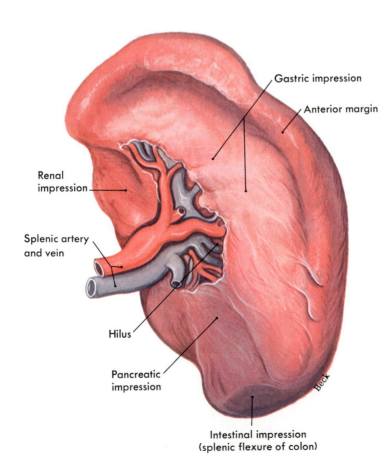

Gastric impression

Anterior margin

Renal impression

Splenic artery and vein

Hilus

Pancreatic impression

Intestinal impression (splenic flexure of colon)

Fig. 16-7 Spleen, medial aspect. Arrangement of the vessels at the hilus is highly variable.

cells are formed in the spleen. Before birth, red blood cells are also formed in the spleen, but after birth the spleen is said to form red blood cells only in extreme hemolytic anemia.

3 *Red blood cell and platelet destruction.* Macrophages lining the spleen's sinusoids remove worn-out red blood cells and imperfect platelets from the blood and destroy them by phagocytosis. They also break apart the hemoglobin molecules from the destroyed red blood cells and salvage their iron and globin content by returning them to the bloodstream for storage in bone marrow and liver.

4 *Blood reservoir.* The pulp of the spleen and its venous sinuses store considerable blood. Its normal volume of about 350 ml is said to decrease about 200 ml in less than a minute's time following sympathetic stimulation that produces marked constriction of its smooth capsule. This "self-transfusion" occurs, for example, as a response to the stress imposed by hemorrhage.

Although the spleen's functions make it a most useful organ, it is not a vital one. Dr. Charles Austin Doan in 1933 took the daring step of performing the first splenectomy. He removed the spleen from a 4-year-old girl who was dying of hemolytic anemia. Presumably, he justified his radical treatment on the basis of what was then merely conjecture, that is, that the spleen destroys red blood cells. The child recovered, and Dr. Doan's operation proved to be a landmark. It created a great upsurge of interest in the spleen and led to many investigations of it.

Outline summary

Definition

A Lymphatic system—part of circulatory system—consists of lymph, interstitial (tissue) fluid, lymphatics, lymph nodes, isolated nodules of lymphatic tissue, tonsils, thymus, and spleen

B Lymph and interstitial fluid (tissue fluid)
  1 Lymph—clear, watery fluid found in lymphatic vessels
  2 Interstitial fluid (tissue fluid)—complex and "organized" material that fills spaces between cells; interstitial fluid and blood together constitute extracellular fluid
  3 Composition of lymph and interstitial fluid—similar to that of blood plasma; main difference is that plasma contains higher concentration of proteins; lymph in thoracic duct has about twice as high protein concentration as most interstitial fluid

Lymphatics

A Formation and distribution
  1 Start as blind-end lymphatic capillaries in tissue spaces
  2 Widely distributed throughout body
  3 Two or more main lymphatic ducts—thoracic duct, which drains into left subclavian vein at junction of internal jugular and subclavian, and one or more right lymphatic ducts, which drain into right subclavian vein
  4 Lacteals—lymphatics originating in intestinal villi; after fatty meal contain milky lymph called chyle

B Structure
  1 Similar to veins except thinner walled
  2 Contain more valves and contain lymph nodes located at intervals

C Functions
  1 Return water and proteins from interstitial fluid to blood from which they came
  2 Lacteals absorb fats and other nutrients

Lymph circulation

A Direction of flow—water and solutes filter out of capillary blood into interstitial fluid, enter lymphatics and move through them to return to blood at junction of internal jugular and subclavian veins

B The lymphatic pump
  1 Lymph flow averages 125 ml per hour
  2 Mechanisms that contribute to effectiveness of "lymphatic pump"
    a Breathing movements
    b Skeletal muscle contractions
    c Arterial pulsations
    d Contraction of lymphatic walls

Lymph nodes

A Structure
  1 Lymphatic tissue, separated into compartments by fibrous partitions

2 Afferent lymphatics enter each node and efferent lymphatics leave each node

B Locations—usually in clusters (see pp. 433 and 435)

C Functions

1 Defense functions—filter out injurious substances and phagocytose them

2 Hemopoiesis—formation of lymphocytes and monocytes

Lymphatic drainage of the breast

A Location and distribution of lymphatics

1 Two sets of lymphatic vessels—superficial lymphatics drain skin of breast except of areola and nipple; deep lymphatics drain substance of breast and skin of areola and nipple

2 Subareolar plexus (plexus of Sappey)—under areola

3 Lymphatic anastomoses

a Between superficial and deep lymphatics

b Between superficial lymphatics of both breasts across middle line

c Between superficial lymphatics and lymphatics in fascia of pectoralis major

d Between lymphatics of breast and abdominal cavity through linea alba

B Lymph nodes

1 About 85% of lymph from breast to axillary nodes—some to parasternal nodes

2 Five sets of axillary nodes

3 Physical contact between anterior axillary nodes and axillary tail of Spence

Thymus

A Location—mediastinum, extends into lower neck

B Size—relatively largest in comparison to body size at about 2 years of age; absolutely largest at puberty, after which it gradually atrophies; almost disappears by advanced old age

C Function—forms lymphocytes before birth; postulated to secrete hormone, starting soon after birth, that permits lymphocytes to develop into plasma cells and secrete antibodies; hence thymus serves as part of body's defense against microbes and other foreign proteins

Spleen

A Location—left hypochondrium

B Structure

1 Similar to lymph nodes, ovoid in shape

2 Size varies

3 Contains numerous venous blood spaces that serve as blood reservoir

C Functions

1 Defense—protection by phagocytosis by reticuloendothelial cells and antibody formation by some lymphocytes

2 Hemopoiesis of nongranular leukocytes (monocytes and lymphocytes) and of red blood cells before birth; spleen also forms plasma cells

3 Red blood cell and platelet destruction—reticuloendothelial cells phagocytose these cells

4 Blood reservoir

Review questions

1 List the anatomical components of the lymphatic system.

2 Why is the term circulation more appropriate in describing the movement of blood than of lymph?

3 Lymph from what body areas enters the general circulation by way of the thoracic duct? The right lymphatic ducts?

4 How do interstitial fluid and lymph differ from blood plasma?

5 How do lymphatic capillaries originate?

6 What is the cisterna chyli?

7 Where does lymph enter the blood vascular system?

8 In general, lymphatics resemble veins in structure. List three exceptions to this general rule of thumb.

9 What are the specialized lymphatics that originate in the villi of the small intestine called?

10 What is chyle? Where is it formed?

11 Briefly describe the anatomy of the lymphatic capillary wall.

12 Discuss the importance of valves in the lymphatic system.

13 How is lymph formed?

14 Discuss the "lymphatic pump."

15 List several important groups or clusters of lymph nodes.

16 Discuss how lymph nodes function in body defense and hemopoiesis.

17 If cancer cells from breast cancer were to enter the lymphatics of the breast, where do you think they might lodge and start new growths? Explain, using your knowledge of the anatomy of the lymphatic and circulatory systems.

18 Discuss the importance of lymphatic anastomoses in the spread of breast cancer.

19 Locate the thymus and describe its appearance and size at birth, at maturity and in old age.

20 Discuss the function of the thymus.

21 Describe the location and function of the spleen. What functions is it thought to perform?

Energy supply and waste excretion

chapter 17

The respiratory system

440

Functions and organs

The respiratory system functions as an air distributor and gas exchanger in order that oxygen may be supplied to and carbon dioxide be removed from the body's cells. Since most of our billions of cells lie too far distant from air to exchange gases directly with it, air must first exchange gases with blood, blood must circulate, and finally blood and cells must exchange gases. These events require the functioning of two systems, namely, the respiratory system and the circulatory system. All parts of the respiratory system—except its microscopic-sized sacs called alveoli—function as air distributors. Only the alveoli serve as gas exchangers.

The organs of the respiratory system are the nose, pharynx, larynx, trachea, bronchi, and lungs. Together they constitute the lifeline, the air supply line of the body. We shall first describe the structure and functions of these organs and then discuss the physiology of the respiratory system as a whole.

Organs

Nose
Structure

The nose consists of an internal and an external portion. The external portion, that is, the part that protrudes from the face, is considerably smaller than the internal portion, which lies over the roof of the mouth. The interior of the nose is hollow and is separated by a partition, the *septum* (Fig. 17-1), into a right and a left cavity. The palatine bones, which form both the floor of the nose and the roof of the mouth, separate the nasal cavities from the mouth cavity. Sometimes the palatine bones fail to unite completely, producing a condition known as *cleft palate*. When this abnormality exists, the mouth is only partially separated from the nasal cavity, and difficulties arise in swallowing.

Each nasal cavity is divided into three passageways (superior, middle, and inferior meati) by the projection of the turbinates (conchae) from the lateral walls of the internal portion of the nose (Figs. 17-2 and 17-3). The superior and middle turbinates are processes of the ethmoid bone, whereas the inferior turbinates are separate bones.

The external openings into the nasal cavities (nostrils) have the technical name of *anterior nares*. They open into an area just below the inferior meatus called the *vestibule*. The posterior nares (or choanae) are openings from an area of the internal nasal cavity above the superior meatus, called the *sphenoethmoidal recess*, into the nasopharynx.

If one were to "trace" the movement of air through the nose into the pharynx, it would pass, through several structures on the way. The se-

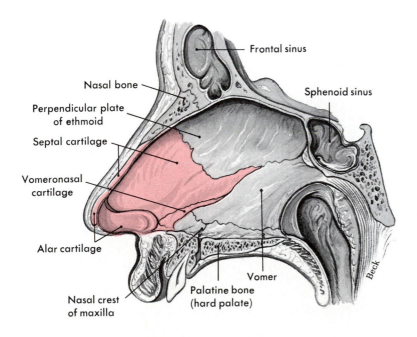

Labels: Frontal sinus · Nasal bone · Perpendicular plate of ethmoid · Septal cartilage · Vomeronasal cartilage · Alar cartilage · Nasal crest of maxilla · Palatine bone (hard palate) · Vomer · Sphenoid sinus · Beck

Fig. 17-1 The nasal septum consists of the perpendicular plate of the ethmoid bone, the vomer, and the septal and vomeronasal cartilages. The cartilages are shown in pink.

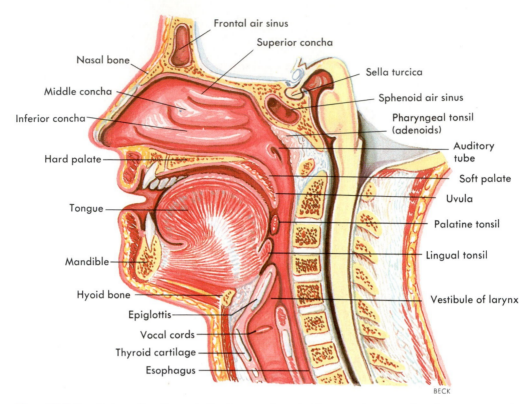

Fig. 17-2 Sagittal section through the face and neck. The nasal septum has been removed, exposing the lateral wall of the nasal cavity. Note the position of the conchae (turbinates).

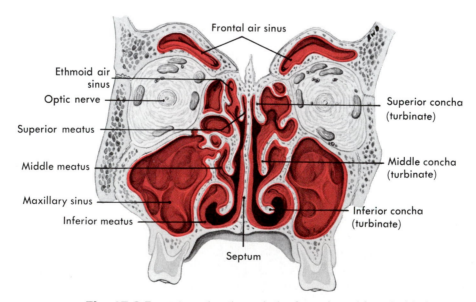

Fig. 17-3 Frontal section through the face viewed from behind.

quence is as follows:

1 Anterior nares (nostrils)
2 Vestibule
3 Inferior, middle, and superior meati, simultaneously
4 Posterior nares (choanae) and sphenoethmoidal recess, simultaneously

Ciliated mucous membrane lines the nose and the rest of the respiratory tract down as far as the smaller bronchioles.

Four pairs of sinuses drain into the nose. These paranasal sinuses are the frontal, maxillary, ethmoidal, and sphenoidal. They drain as follows:

1 Into the middle meatus (passageway below middle turbinate)—frontal, maxillary, and anterior ethmoidal sinuses
2 Into the superior meatus—posterior ethmoidal sinuses
3 Into the space above the superior turbinates (sphenoethmoidal recess)—sphenoidal sinuses

Functions

The nose serves as a passageway for air going to and from the lungs, filtering it of impurities and warming, moistening, and chemically examining it for substances that might prove irritating to the mucous lining of the respiratory tract. It serves as the organ of smell, since olfactory receptors are located in the nasal mucosa, and it aids in phonation.

Pharynx
Structure

Another name for the pharynx is the throat. It is a tubelike structure about 12.5 cm (5 inches) long that extends from the base of the skull to the esophagus and lies just anterior to the cervical vertebrae. It is made of muscle, is lined with mucous membrane, and has three anatomical divisions: the *nasopharynx* located behind the nose and extending from the posterior nares to the level of the soft palate; the *oropharynx*, lo-

cated behind the mouth from the soft palate above to the level of the hyoid bone below; and the *laryngopharynx*, which extends from the hyoid bone to its termination in the esophagus. Fig. 17-4 shows the divisions of the pharynx when viewed from behind (coronal section).

Seven openings are found in the pharynx (Fig. 17-2):

1 Right and left auditory (eustachian) tubes opening into the nasopharynx
2 Two posterior nares into the nasopharynx
3 The opening from the mouth, known as the *fauces*, into the oropharynx
4 The opening into the larynx from the laryngopharynx
5 The opening into the esophagus from the laryngopharynx

The *adenoids* or pharyngeal tonsils are located in the nasopharynx on its posterior wall opposite the posterior nares. Although the cavity of the nasopharynx differs from the oral and laryngeal divisions in that it does not collapse, it may become obstructed. If the adenoids become enlarged, they fill the space behind the posterior nares and make it difficult or impossible for air to travel from the nose into the throat. When this happens, the individual keeps his mouth open to breathe and is described as having an "adenoidy" appearance.

Two pairs of organs are found in the oropharynx: the faucial or *palatine tonsils*, located behind and below the pillars of the fauces, and the *lingual tonsils*, located at the base of the tongue. The palatine tonsils are the ones most commonly removed by a tonsillectomy. Only rarely are the lingual ones also removed.

Functions

The pharynx serves as a hallway for the respiratory and digestive tracts, since both air and food must pass through this structure before reaching the appropriate tubes. It also plays an important part in phonation. For example, only by the pharynx changing its shape can the different vowel sounds be formed.

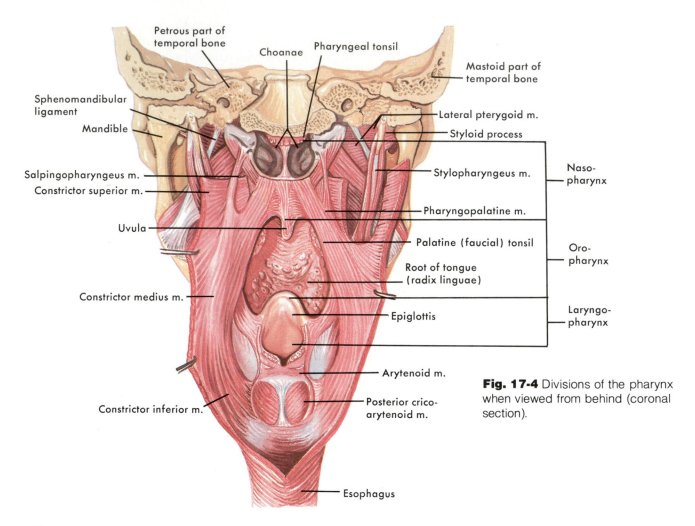

Fig. 17-4 Divisions of the pharynx when viewed from behind (coronal section).

Labels on figure:
Petrous part of temporal bone
Choanae
Pharyngeal tonsil
Mastoid part of temporal bone
Sphenomandibular ligament
Mandible
Lateral pterygoid m.
Styloid process
Salpingopharyngeus m.
Constrictor superior m.
Stylopharyngeus m.
Naso-pharynx
Pharyngopalatine m.
Uvula
Palatine (faucial) tonsil
Oro-pharynx
Root of tongue (radix linguae)
Constrictor medius m.
Epiglottis
Laryngo-pharynx
Arytenoid m.
Constrictor inferior m.
Posterior crico-arytenoid m.
Esophagus

Larynx

Location

The larynx or voice box lies between the root of the tongue and the upper end of the trachea just below and in front of the lowest part of the pharynx. It might be described as a vestibule opening into the trachea from the pharynx. It normally extends between the fourth, fifth, and sixth cervical vertebrae but is often somewhat higher in females and during childhood.

Structure

The larynx consists largely of cartilages and muscles (Fig. 17-5). It is lined by a mucous membrane that forms two pairs of folds that jut in-ward into its cavity. The upper pair is called the false vocal folds for the rather obvious reason that they play no part in vocalization. The lower pair serves as the true vocal cords. The slitlike space between the true vocal cords—the rima glottidis or glottis—is the narrowest part of the larynx. Edema of the mucosa covering the vocal cords is a potentially lethal condition. Even a moderate amount of swelling can obstruct the glottis so that air cannot get through, and asphyxiation results.

Cartilages of the larynx

Nine cartilages form the framework of the larynx. The three largest of these—the thyroid

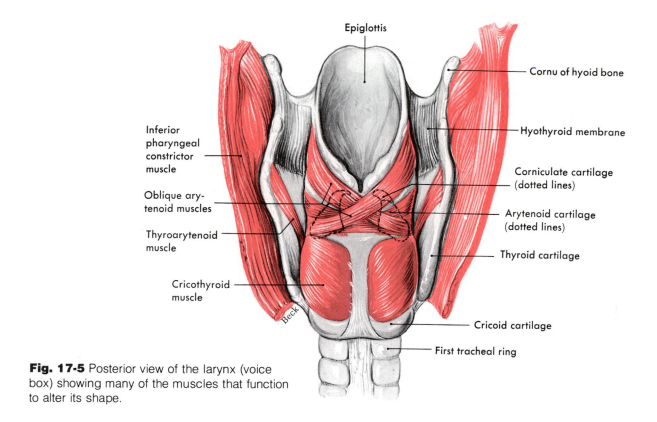

Epiglottis

Cornu of hyoid bone

Inferior pharyngeal constrictor muscle

Hyothyroid membrane

Oblique ary-tenoid muscles

Corniculate cartilage (dotted lines)

Thyroarytenoid muscle

Arytenoid cartilage (dotted lines)

Thyroid cartilage

Cricothyroid muscle

Cricoid cartilage

First tracheal ring

Beck

Fig. 17-5 Posterior view of the larynx (voice box) showing many of the muscles that function to alter its shape.

cartilage, and epiglottis, and the cricoid cartilage—are single structures. There are three pairs of smaller accessory cartilages, namely, the arytenoid, corniculate, and cuneiform cartilages. The three main cartilages are described here.

1 The thyroid cartilage (Adam's apple) is the largest cartilage of the larynx and is the one that gives the characteristic triangular shape to its anterior wall. It is usually larger in men than in women and has less of a fat pad lying over it—two reasons why a man's Adam's apple protrudes more than a woman's.

2 A small cartilage attached along one edge to the thyroid cartilage but free on its other borders is named the *epiglottis*.

3 The *cricoid* or signet ring cartilage, so called because its shape resembles a signet ring (turned so the signet forms part of the posterior wall of the larynx), is the most inferiorly placed of the nine cartilages.

Muscles of the larynx

Muscles of the larynx are often divided into intrinsic and extrinsic groups. Intrinsic muscles have both their origin and insertion on the larynx. They are important in controlling vocal cord length and tension and in regulating the shape of the laryngeal inlet. Extrinsic muscles insert on the larynx but have their origin on some other structure—such as the hyoid bone.

Therefore contraction of the extrinsic muscles actually moves or displaces the larynx as a whole. Muscles in both groups play important roles in respiration, vocalization, and swallowing. During swallowing, for example, contraction of the intrinsic aryepiglottic muscles (those that connect the arytenoid cartilages with the epiglottis) prevents "swallowing down the wrong throat" by squeezing the laryngeal inlet shut.

Two other pairs of intrinsic laryngeal muscles function to open and close the glottis. The posterior cricoarytenoid muscles (between cricoid and arytenoid cartilages) open the glottis by abducting the true vocal cords. The lateral cricoarytenoid muscles close the glottis by adducting

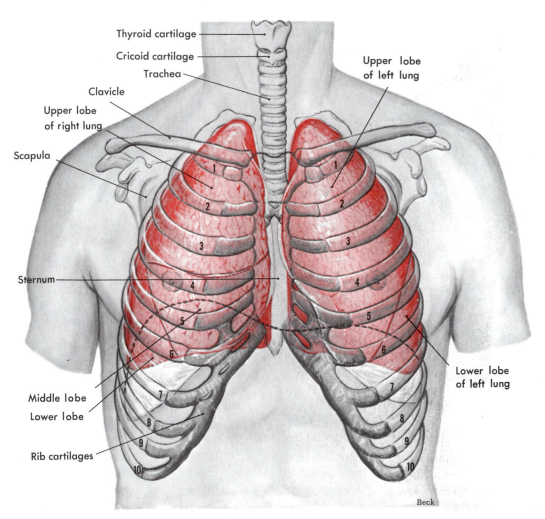

Fig. 17-6 Projection of the lungs and trachea in relation to the rib cage and clavicles. Dotted line indicates location of the dome-shaped diaphragm at the end of expiration and before inspiration. Note that apex of each lung projects above the clavicle. Ribs 11 and 12 are not visible in this view.

the cords. These events are crucial to both respiration and voice production. Certain other intrinsic muscles of the larynx function to influence the pitch of the voice by either lengthening and tensing or shortening and relaxing the vocal cords.

Functions

The larynx functions in respirations, since it constitutes part of the vital airway to the lungs. It protects the airway against the entrance of solids or liquids during swallowing. It also serves as the organ of voice production—hence its popular name, the voice box. Air being expired through the glottis, narrowed by partial adduction of the vocal cords, causes them to vibrate. Their vibration produces the voice. Several other structures besides the larynx contribute to the sound of the voice by acting as sounding boards or resonating chambers. Thus the size and shape of the nose, mouth, pharynx, and bony sinuses help to determine the quality of the voice.

Trachea
Structure

The trachea or windpipe is a tube about 11 cm (4½ inches) long that extends from the larynx in the neck to the bronchi in the thoracic cavity (Fig. 17-6). Its diameter measures about 2.5 cm (1 inch). Smooth muscle, in which are embedded C-shaped rings of cartilage at regular intervals, fashions the walls of the trachea. The cartilaginous rings are incomplete on the posterior surface. They give firmness to the wall, tending to prevent it from collapsing and shutting off the vital airway. Often a tube is placed in the trachea (endotracheal intubation) before patients leave the operating room, especially if they have been given a muscle relaxant. The purpose of the tube is to ensure an open airway. Another procedure done frequently in today's modern hospitals is a tracheostomy, that is, the cutting of an opening into the trachea. A surgeon may do this in order that a suction device can be used to remove secretions from the bronchial tree, or he may do it so that a machine such as the intermittent positive pressure breathing (IPPB) machine can be used to improve ventilation of the lung.

Function

The trachea performs a simple but vital function—it furnishes part of the open passageway through which air can reach the lungs from the outside. Obstruction of this airway for even a few minutes causes death from asphyxiation.

Bronchi
Structure

The trachea divides at its lower end into two *primary bronchi*, of which the right bronchus is slightly larger and more vertical than the left. This anatomical fact explains why aspirated foreign objects frequently lodge in the right bronchus. In structure the bronchi resemble the trachea. Their walls contain incomplete cartilaginous rings before the bronchi enter the lungs, but they become complete within the lungs. Ciliated mucosa lines the bronchi, as it does the trachea.

Each primary bronchus enters the lung on its respective side and immediately divides into smaller branches called *secondary bronchi*. The secondary bronchi continue to branch, forming small *bronchioles*. The trachea and the two primary bronchi and their many branches resemble an inverted tree trunk with its branches and are, therefore, spoken of as the bronchial tree. The bronchioles subdivide into smaller and smaller tubes, eventually terminating in microscopic branches that divide into *alveolar ducts*, which terminate in several alveolar sacs, the walls of which consist of numerous *alveoli* (Figs. 17-7 to 17-9). The structure of an alveolar duct with its branching alveolar sacs can be likened to a

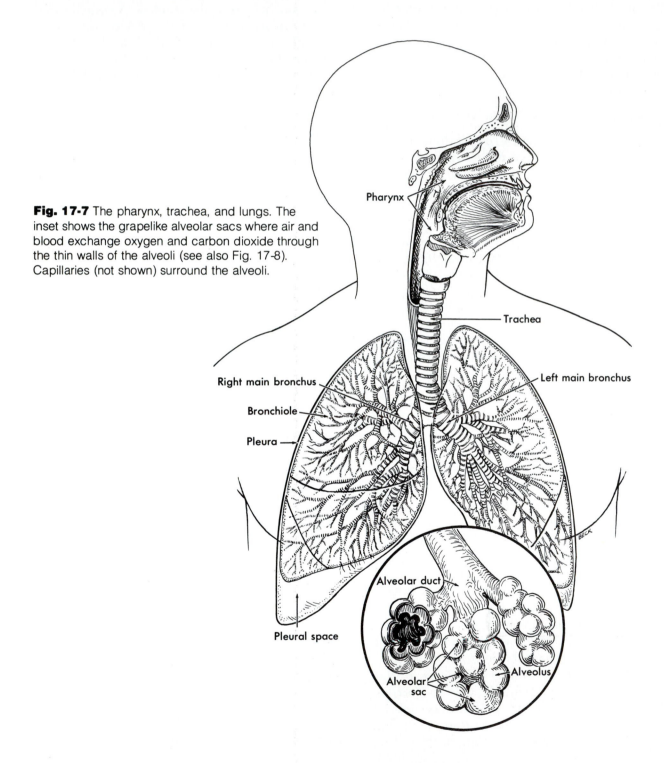

Fig. 17-7 The pharynx, trachea, and lungs. The inset shows the grapelike alveolar sacs where air and blood exchange oxygen and carbon dioxide through the thin walls of the alveoli (see also Fig. 17-8). Capillaries (not shown) surround the alveoli.

Pharynx

Trachea

Right main bronchus

Left main bronchus

Bronchiole

Pleura

Pleural space

Alveolar duct

Alveolar sac

Alveolus

Fig. 17-8 Metal cast of air spaces of the lungs of a dog. The inset shows a cast of clusters of alveoli at the terminations of tiny air tubes. The magnification of the inset is about 11 times the actual size. (From Carlson, A. J., and Johnson, V. E.: The machinery of the body, Chicago, The University of Chicago Press.)

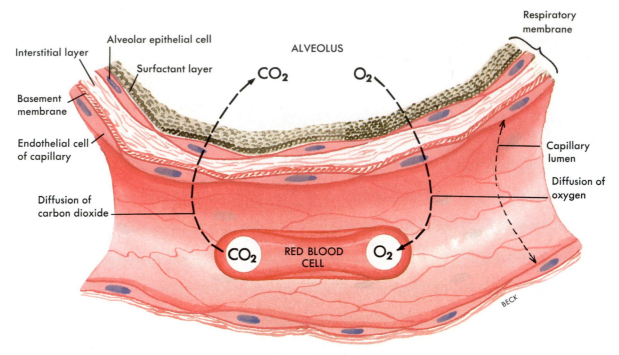

Fig. 17-9 Respiratory membrane greatly magnified to show epithelial and surfactant layers of alveolar membrane and basement membrane and endothelial layer of capillary membrane.

bunch of grapes—the stem represents the alveolar duct, each cluster of grapes represents an alveolar sac, and each grape represents an alveolus. Some 300 million alveoli are estimated to be present in our two lungs.

The structure of the secondary bronchi and bronchioles shows some modification of the primary bronchial structure. The cartilaginous rings become irregular and disappear entirely in the smaller bronchioles. By the time the branches of the bronchial tree have dwindled sufficiently to form the alveolar ducts and sacs and the alveoli, only the internal surface layer of cells remains. In other words, the walls of these microscopic structures consist of a single layer of simple, squamous epithelial tissue. As we shall see, this structural fact makes possible the performance of their functions.

Functions

The tubes composing the bronchial tree perform the same function as the trachea—that of distributing air to the lung's interior. The alveoli, enveloped as they are by networks of capillaries, accomplish the lung's main and vital function, that of gas exchange between air and blood. Someone has observed that "the lung passages all serve the alveoli" just as "the circulatory system serves the capillaries." Certain diseases may block the passage of air through the bronchioles or alveoli. For example, in pneumonia the alveoli become inflamed, and the accompanying wastes plug up these minute air spaces, making the affected part of the lung solid. Whether the victim survives depends largely on the extent of the solidification.

Lungs

Structure

The lungs are cone-shaped organs, large enough to fill the pleural portion of the thoracic cavity completely. (See Fig. 17-6.) They extend from the diaphragm to a point slightly above the clavicles and lie against the ribs both anteriorly and posteriorly. The medial surface of each lung is roughly concave to allow room for the mediastinal structures and for the heart, but concavity is greater on the left than on the right because of the position of the heart. The primary bronchi and pulmonary blood vessels (bound together by connective tissue to form what is known as the *root* of the lung) enter each lung through a slit on its medial surface called the *hilum*.

The broad inferior surface of the lung, which rests on the diaphragm, constitutes the *base*, whereas the pointed upper margin is the *apex*. Each apex projects above a clavicle.

The left lung is partially divided by fissures into two *lobes* (upper and lower) and the right lung into three lobes (superior, middle, and inferior). The interior of each lung consists of the almost innumerable tubes of dwindling diameters that make up the bronchial tree and serve as air distributors. The smallest tubes terminate in the smallest but functionally most important structures of the lung—the alveoli or "gas exchangers."

Visceral pleura covers the outer surfaces of the lungs and adheres to them much as the skin of an apple adheres to the apple.

Functions

The lungs perform two functions—air distribution and gas exchange. Air distribution to the alveoli is the function of the tubes of the bronchial tree. Gas exchange between air and blood is the joint function of the alveoli and the networks of blood capillaries that envelop them. These two structures—one part of the respiratory system and the other part of the circulatory system—together serve as highly efficient gas exchangers. Why? Because they provide an enormous surface area, the respiratory membrane, where the very thin-walled alveoli and equally thin-walled pulmonary capillaries come in contact (Fig. 17-9). This makes possible extremely rapid diffusion of gases between alveolar air and pulmonary capillary blood. Someone has estimated that if the lungs' 300 million or so alveoli could be opened up flat, they would form a surface about the size of a tennis court, that is, about 70 square meters,* or some 40 times the surface area of the entire body! No wonder such large amounts of oxygen can be so quickly loaded into the blood while large amounts of carbon dioxide are rapidly being unloaded from it.

Thorax (chest)

Structure

As described on p. 17, the thoracic cavity has three divisions, separated from each other by partitions of pleura. The parts of the cavity occupied by the lungs are the pleural divisions. The space between the lungs occupied mainly by the esophagus, trachea, large blood vessels, and heart is the mediastinum.

The parietal layer of the pleura lines the entire thoracic cavity. It adheres to the internal surface of the ribs and the superior surface of the diaphragm, and it partitions off the mediastinum. A separate pleural sac thus encases each lung. Since the outer surface of each lung is covered by the visceral layer of the pleura, the visceral pleura lies against the parietal pleura, separated only by a potential space (pleural space) that contains just enough pleural fluid for lubrication. Thus when the lungs inflate with air, the smooth, moist visceral pleura coheres to the smooth, moist parietal pleura. Friction is thereby avoided, and respirations are painless. In

*Comroe, J. H., Jr.: The lung, Sci. Am. **214**:57-66, Feb., 1966.

pleurisy, on the other hand, the pleura is inflamed and respirations become painful.

Functions

The thorax plays a major role in respirations. Because of the elliptical shape of the ribs and the angle of their attachment to the spine, the thorax becomes larger when the chest is raised and smaller when it is lowered. It is these changes in thorax size that bring about inspiration and expiration (discussed on p. 453). Lifting up the chest raises the ribs so that they no longer slant downward from the spine, and because of their elliptical shape, this enlarges both depth (from front to back) and width of the thorax. (If this does not sound convincing to you, examine a skeleton to see why it is so.)

Physiology

Definition of pulmonary ventilation

Pulmonary ventilation is a technical term for what most of us call breathing. One phase of it, inspiration, moves air into the lungs and the other phase, expiration, moves air out of lungs.

Mechanism of pulmonary ventilation

Air moves in and out of the lungs for the same basic reason that any fluid, that is, a liquid or a gas, moves from one place to another—briefly, because its pressure in one place is different from that in the other place. Or stated differently, the existence of a pressure gradient (a pressure difference) causes fluids to move. A fluid always moves down its pressure gradient. This means that a fluid moves from the area where its pressure is higher to the area where its pressure is lower. Under standard conditions, air in the atmosphere exerts a pressure of 760 mm Hg. Air in the alveoli at the end of one expiration and before the beginning of another inspiration also

exerts a pressure of 760 mm Hg. This fact explains why at that moment air is neither entering nor leaving the lungs. The mechanism that produces pulmonary ventilation is one that establishes a gas pressure gradient between the atmosphere and the alveolar air.

When atmospheric pressure is greater than pressure within the lung, air flows down this gas pressure gradient. Then air moves from the atmosphere into the lungs. In other words, inspiration occurs. When pressure in the lungs becomes greater than atmospheric pressure, air again moves down a gas pressure gradient. But now, this means that it moves in the opposite direction. This time, air moves out the lungs into the atmosphere. The pulmonary ventilation mechanism, therefore, must somehow establish these two gas pressure gradients—one in which intrapulmonic pressure (pressure within the lungs) is lower than atmospheric pressure to produce inspiration and one in which it is higher than atmospheric pressure to produce expiration.

These pressure gradients are established by changes in the size of the thoracic cavity, which in turn are produced by contraction and relaxation of respiratory muscles. An understanding of Boyle's law is important for understanding the pressure changes that occur in the lungs and thorax during the breathing cycle. It is a familiar principle stating that the volume of a gas varies inversely with pressure at constant temperature. Application: Expansion of the thorax (increase in volume) results in a decreased intrapleural (intrathoracic) pressure. This leads to a decreased intrapulmonic pressure that causes air to move from the outside into the lungs.

Inspiration

Contraction of the diaphragm alone, or of the diaphragm and the external intercostal muscles, produces quiet inspiration. As the diaphragm contracts, it descends, and this makes the thoracic cavity longer. Contraction of the external intercostal muscles pulls the anterior end of each

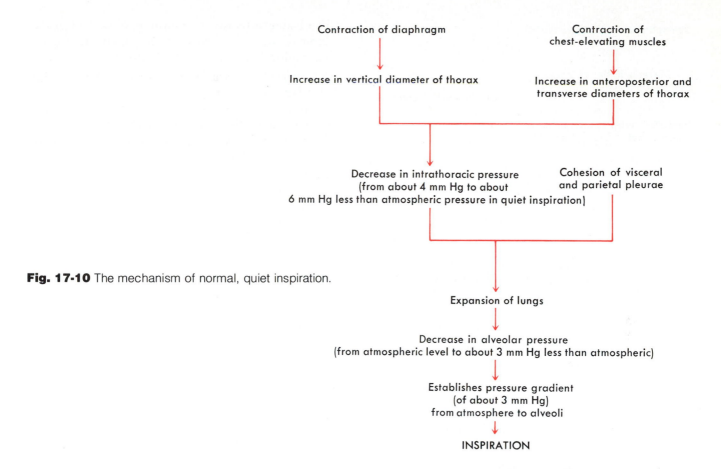

Contraction of diaphragm   Contraction of
chest-elevating muscles

Increase in vertical diameter of thorax   Increase in anteroposterior and
transverse diameters of thorax

Decrease in intrathoracic pressure
(from about 4 mm Hg to about
6 mm Hg less than atmospheric pressure in quiet inspiration)

Cohesion of visceral
and parietal pleurae

Expansion of lungs

Decrease in alveolar pressure
(from atmospheric level to about 3 mm Hg less than atmospheric)

Establishes pressure gradient
(of about 3 mm Hg)
from atmosphere to alveoli

INSPIRATION

Fig. 17-10 The mechanism of normal, quiet inspiration.

rib up and out. This also elevates the attached sternum and enlarges the thorax from front to back and from side to side. In addition, contraction of the sternocleidomastoid and serratus anterior muscles can aid in elevation of the sternum and rib cage during forceful inspiration. As the size of the thorax increases, the intrapleural (intrathoracic) and intrapulmonic pressures decrease (Boyle's law) and inspiration occurs. At the end of an expiration and before the beginning of the next inspiration, intrathoracic pressure is about 4 mm Hg less than atmospheric pressure (frequently written −4 mm Hg). During quiet inspiration, intrathoracic pressure decreases further to −6 mm Hg. As the thorax enlarges, it pulls the lungs along with it because of cohesion between the moist pleura covering

the lungs and the moist pleura lining the thorax. Thus the lungs expand, and the pressure in their tubes and alveoli necessarily decreases. Intrapulmonic (intra-alveolar) pressure decreases from an atmospheric level to a subatmospheric level—typically to about −3 mm Hg. The moment intrapulmonic pressure becomes less than atmospheric pressure, a pressure gradient exists between the atmosphere and the interior of the lungs. Air then necessarily moves into the lungs. For a diagram of the mechanism of inspiration just described, see Fig. 17-10.

Expiration

Quiet expiration is ordinarily a passive process that begins when those pressure changes or gradients that resulted in inspiration are re-

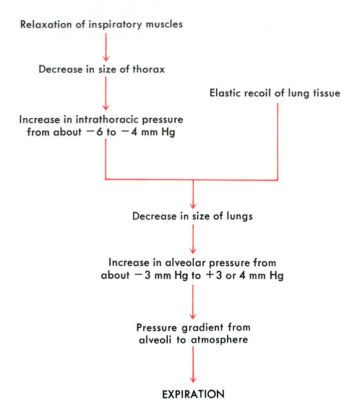

Relaxation of inspiratory muscles

↓

Decrease in size of thorax

↓

Elastic recoil of lung tissue

Increase in intrathoracic pressure
from about −6 to −4 mm Hg

↓

Decrease in size of lungs

↓

Increase in alveolar pressure from
about −3 mm Hg to +3 or 4 mm Hg

↓

Pressure gradient from
alveoli to atmosphere

↓

EXPIRATION

Fig. 17-11 The mechanism of normal, quiet expiration.

versed. The inspiratory muscles relax, causing a decrease in the size of the thorax and an increase in intrapleural (intrathoracic) pressure from about −6 mm Hg to a preinspiration level of −4 mm Hg. It is important to understand that this pressure between the parietal and visceral pleura is always negative, that is, less than atmospheric pressure. The negative intrapleural (intrathoracic) pressure is required to overcome the so-called "collapse tendency" of the lungs caused by surface tension of the fluid lining the alveoli and the stretch of elastic fibers that are constantly attempting to recoil.

As alveolar pressure increases from about −3 mm Hg to +3 or +4 mm Hg, a positive pressure gradient is established from alveoli to atmosphere and expiration occurs as air flows outward through the respiratory passageways. (Fig.

17-11 diagrams this mechanism of normal, quiet expiration.) In forced expiration, contraction of the abdominal and internal intercostal muscles can increase intra-alveolar pressure to over 100 mm Hg.

To apply some of the information just discussed about the respiratory mechanism, let us suppose that a surgeon makes an incision through the chest wall into the pleural space, as he would in doing one of the dramatic, modern open-chest operations. Air would then be present in the thoracic cavity, a condition known as *pneumothorax*. What change, if any, can you deduce would take place in respirations? Compare your deductions with those in the next paragraph.

Intrathoracic pressure would, of course, immediately increase from its normal subatmospheric level to the atmospheric level. More pressure than normal would therefore be exerted on the outer surface of the lung and would cause its collapse. It could even collapse the other lung. Why? Because the mediastinum is a mobile rather than a rigid partition between the two pleural sacs. This anatomical fact allows the increased pressure in the side of the chest that is open to push the heart and other mediastinal structures over toward the intact side, where they exert pressure on the other lung. Pneumothorax results in many respiratory and circulatory changes. They are of great importance in determining medical and nursing care but lie beyond the scope of this book.

Volumes of air exchanged in pulmonary ventilation

The volumes of air moved in and out of the lungs and remaining in them are matters of great importance. They must be normal in order that a normal exchange of oxygen and carbon dioxide can take place between alveolar air and pulmonary capillary blood.

An apparatus called a *spirometer* is used to measure the volume of air exchanged in breathing. The volume of air exhaled normally after a

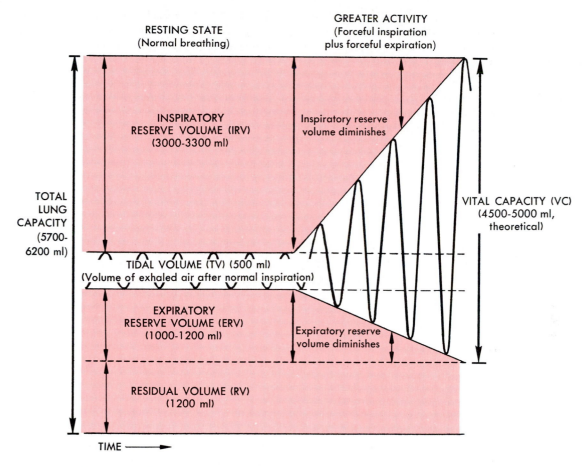

RESTING STATE
(Normal breathing)

GREATER ACTIVITY
(Forceful inspiration
plus forceful expiration)

INSPIRATORY
RESERVE VOLUME (IRV)
(3000-3300 ml)

Inspiratory reserve
volume diminishes

TOTAL
LUNG
CAPACITY
(5700-
6200 ml)

VITAL CAPACITY (VC)
(4500-5000 ml,
theoretical)

TIDAL VOLUME (TV) (500 ml)
(Volume of exhaled air after normal inspiration)

EXPIRATORY
RESERVE VOLUME (ERV)
(1000-1200 ml)

Expiratory reserve
volume diminishes

RESIDUAL VOLUME (RV)
(1200 ml)

TIME ⟶

Fig. 17-12 During normal, quiet respirations the atmosphere and lungs exchange about 500 ml of air *(TV)*. With a forcible inspiration, about 3,300 ml more air can be inhaled *(IRV)*. After a normal inspiration and normal expiration, approximately 1,000 ml more air can be forcibly expired *(ERV)*. Vital capacity is the amount of air that can be forcibly expired after a maximum inspiration and indicates, therefore, the largest amount of air that can enter and leave the lungs during respiration. Residual volume is the air that remains trapped in the alveoli.

normal inspiration is termed *tidal volume* (TV). As you can see in Fig. 17-12, the normal volume of tidal air for an adult is approximately 500 ml. After an individual has expired tidal air, he can force still more air out of his lungs. The largest additional volume of air that one can forcibly expire after expiring tidal air is called the *expiratory reserve volume* (ERV). An adult, as Fig. 17-12 shows, normally has an ERV of between 1,000

and 1,200 ml. *Inspiratory reserve volume* (IRV) is the amount of air that can be forcibly inspired over and above a normal inspiration. It is measured by having the individual exhale normally after a forced inspiration. The normal IRV is about 3.3 liters. No matter how forcefully an individual exhales, he cannot squeeze all the air out of his lungs. Some of it remains trapped in the alveoli. This amount of air that cannot be forcibly expired is known as *residual volume* (RV) and amounts to about 1.2 liters. Between breaths an exchange of oxygen and carbon dioxide occurs between the trapped residual air in the alveoli and the blood. This process helps to "level off" the amounts of oxygen and carbon dioxide in the blood during the breathing cycle. In pneumothorax the RV is eliminated when the lung collapses. Even after the RV is forced out the collapsed lung has a porous, spongy texture and will float in water because of trapped air called the *minimal volume*, which is about 40% of the RV.

Notice in Fig. 17-12 that vital capacity is the sum of IRV + TV + ERV. It represents the largest volume of air an individual can move in and out of his lungs. It is determined by measuring the largest possible expiration after the largest possible inspiration. How large a vital capacity one has depends on many factors—the size of one's thoracic cavity, one's posture, and various other factors. In general, a larger person has a larger vital capacity than a smaller one. An individual has a larger vital capacity when he is standing erect than when he is stooped over or lying down. The volume of blood in the lungs also affects the vital capacity. If the lungs contain more blood than normal, alveolar air space is encroached on and vital capacity accordingly decreases. This becomes a very important factor in congestive heart disease. Excess fluid in the pleural or abdominal cavities also decreases vital capacity. So too does the disease emphysema. In the latter condition, alveolar walls become stretched, that is, lose their elasticity, and are unable to recoil normally for expiration. This leads to an increased RV. In severe emphysema, the RV may increase so much that the chest occupies the inspiratory position even at rest. Excessive muscular effort is therefore necessary for inspiration, and because of the loss of elasticity of lung tissue, greater effort is required, too, for expiration.

In diagnosing lung disorders a physician may need to know the inspiratory capacity and the functional residual capacity of the patient's lungs. *Inspiratory capacity* (IC) is the maximum amount of air an individual can inspire after a normal expiration. From Fig. 17-12, you can deduce that IC = TV + IRV. With the volumes given in the figure, how many milliliters is the IC? *Functional residual capacity* is the amount of air left in the lungs at the end of a normal expiration. Therefore, as Fig. 17-12 indicates, FRC = ERV + RV. With the volumes given, the functional residual capacity is 2,200 to 2,400 ml. The total volume of air a lung can hold is called the *total lung capacity*. It is, as Fig. 17-12 indicates, the sum of all four lung volumes.

The term *alveolar ventilation* means the volume of inspired air that actually reaches, "ventilates," the alveoli. Only this volume of air takes part in the exchange of gases between air and blood. (Alveolar air exchanges some of its oxygen for some of blood's carbon dioxide.) With every breath we take, part of the entering air necessarily fills our air passageways—nose, pharynx, larynx, trachea, and bronchi. This portion of air does not descend into any alveoli, so it cannot take part in gas exchange. In this sense, it is "dead air." And appropriately, the larger air passageways it occupies are said to constitute the *anatomical dead space*. One rule of thumb estimates the volume of air in the anatomical dead space as the same number of milliliters as the individual's weight in pounds. Another generalization says that the anatomical dead space approximates 30% of the TV. TV minus dead space volume equals alveolar ventilation volume. Suppose you have a normal TV of 500 ml and that 30% of this, or 150 ml, fills the anatomical dead space. The amount of air reaching your alveoli—your alveolar ventilation volume—is then 350 ml per breath, or 70% of your TV. Emphysema and certain other abnormal conditions, in effect, increase the amount of dead space air. Consequently, alveolar ventilation decreases and this, in turn, decreases the amount of oxygen that can enter blood and the amount of carbon dioxide that can leave it. Inadequate air–blood gas exchange, therefore, is the inevitable result of inadequate alveolar ventilation. Stated differently, the alveoli must be adequately ventilated in order for an adequate gas exchange to take place in the lungs.

Types of breathing

eupnea—normal quiet breathing

hyperpnea—increased breathing, usually increased tidal volume with or without an increased rate of breathing

apnea—cessation of breathing at the end of a normal expiration

apneusis—cessation of breathing in the inspiratory position

Cheyne-Stokes respirations—gradually increasing tidal volume for several breaths, followed by several breaths with gradually decreasing tidal volume; cycle repeats itself

Biot's respirations—repeated sequences of deep gasps and apnea

Some principles about gases

Before discussing respirations further, we need to understand the following principles.

1 *Dalton's law* (or the law of partial pressures). The term *partial pressure* means the pressure exerted by any one gas in a mixture of gases or in a liquid. The partial pressure of a gas in a mixture of gases is directly related to the concentration of that gas in the mixture and to the total pressure of the mixture. Suppose we apply this principle to compute the partial pressure of oxygen in the atmosphere. The concentration of oxygen in the atmosphere is 20.96% and the total pressure of the atmosphere is 760 mm Hg under standard conditions. Therefore

Atmospheric P_{O_2} = 20.96% × 760 = 159.2 mm Hg

The symbol used to designate partial pressure is the capital letter P preceding the chemical symbol for the gas. Examples: alveolar air P_{O_2} is about 100 mm Hg; arterial blood P_{O_2} is also about 100 mm Hg; venous blood P_{O_2} is about 37 mm Hg. The word *tension* is often used as a

synonym for the term partial pressure—oxygen tension means the same thing as P_{O_2}.

2 The partial pressure of a gas in a liquid is directly determined by the amount of that gas dissolved in the liquid, which in turn is determined by the partial pressure of the gas in the environment of the liquid. Gas molecules diffuse into a liquid from its environment and dissolve in the liquid until the partial pressure of the gas in solution becomes equal to its partial pressure in the environment of the liquid. Alveolar air constitutes the environment of blood moving through pulmonary capillaries. Standing between the blood and the air are only the very thin alveolar and capillary membranes, and both of these are highly permeable to oxygen and carbon dioxide. By the time blood leaves the pulmonary capillaries as arterial blood, diffusion and approximate equilibration of oxygen and carbon dioxide across the membranes have occurred. Arterial blood P_{O_2} and P_{CO_2}, therefore, usually equal or very nearly equal alveolar P_{O_2} and P_{CO_2} (Table 17-1).

Exchange of gases in lungs (external respiration)

The exchange of gases in the lungs takes place between alveolar air and venous blood flowing through lung capillaries. Gases move in both directions through the alveolar-capillary membrane (Fig. 17-9). Oxygen enters blood from the

Table 17-1 Oxygen and carbon dioxide pressure gradients

	Atmosphere	Alveolar air	Arterial blood	Venous blood
P_{O_2}	160*	100	100	37
P_{CO_2}	0.3	40	40	46

*All figures indicate approximate mm Hg pressure under usual conditions.

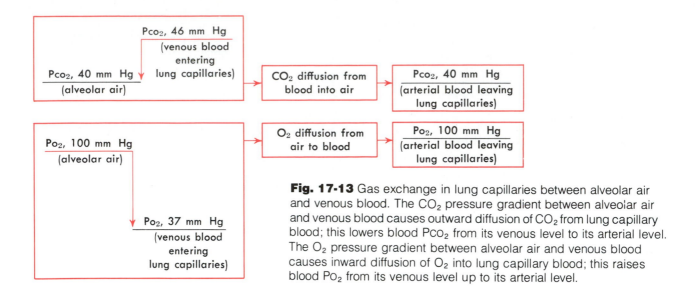

Fig. 17-13 Gas exchange in lung capillaries between alveolar air and venous blood. The CO_2 pressure gradient between alveolar air and venous blood causes outward diffusion of CO_2 from lung capillary blood; this lowers blood P_{CO_2} from its venous level to its arterial level. The O_2 pressure gradient between alveolar air and venous blood causes inward diffusion of O_2 into lung capillary blood; this raises blood P_{O_2} from its venous level up to its arterial level.

alveolar air because the P_{O_2} of alveolar air is greater than the P_{O_2} of venous blood. Another way of saying this is that oxygen diffuses "down" its pressure gradient. Simultaneously, carbon dioxide molecules exit from the blood by diffusing down the carbon dioxide pressure gradient out into the alveolar air. The P_{CO_2} of venous blood is much higher than the P_{CO_2} of alveolar air. This two-way exchange of gases between alveolar air and venous blood converts venous blood to arterial blood (examine Fig. 17-13).

The amount of oxygen that diffuses into blood each minute depends on several factors, notably these four:

1 The oxygen pressure gradient between alveolar air and venous blood (alveolar P_{O_2} − venous blood P_{O_2})

2 The total functional surface area of the alveolar-capillary membrane

3 The respiratory minute volume (respiratory rate per minute times volume of air inspired per respiration)

4 Alveolar ventilation (discussed on p. 456)

All four of these factors bear a direct relation to oxygen diffusion. Anything that decreases alveolar P_{O_2}, for instance, tends to decrease the alveolar-venous oxygen pressure gradient and therefore tends to decrease the amount of oxygen entering the blood. Application: Alveolar air P_{O_2} decreases as altitude increases, and therefore less oxygen enters the blood at high altitudes. At a certain high altitude, alveolar air P_{O_2} equals venous blood P_{O_2}. How would this affect oxygen diffusion into blood?

Anything that decreases the total functional surface area of the alveolar-capillary membrane also tends to decrease oxygen diffusion into the blood (by functional surface area is meant that which is freely permeable to oxygen). Application: In emphysema the total functional area decreases and is one of the factors responsible for poor blood oxygenation in the condition.

Anything that decreases the respiratory minute volume also tends to decrease blood oxygenation. Application: Morphine slows respirations and therefore decreases the respiratory minute volume (volume of air inspired per minute) and tends to lessen the amount of oxygen entering the blood. The main factors influencing blood oxygenation are shown in Fig. 17-14.

Several times we have stated the principle that structure determines functions. You may

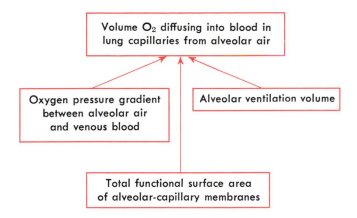

Fig. 17-14 Some major factors determining volume of oxygen entering lung capillary blood. An increase in any of the factors tends to increase oxygenation of blood. A decrease in any of them tends to decrease blood oxygenation. Alveolar ventilation volume means the volume of air that reaches the alveoli; it equals the volume of air inspired minus the volume of the anatomical dead space.

find it interesting to note the application of this principle to gas exchange in the lungs. Several structural facts facilitate oxygen diffusion from the alveolar air into the blood in lung capillaries:

1 The fact that the walls of the alveoli and of the capillaries together form a very thin barrier for the gases to cross (estimated at not more than 0.004 mm thick—see Fig. 17-9)

2 The fact that both alveolar and capillary surfaces are extremely large

3 The fact that the lung capillaries accommodate a large amount of blood at one time (about 90 ml) (The lung capillaries of a small individual—one who has a body surface area of 1.5 square meters—contain about 90 ml of blood at one time under resting conditions. Studies done in the 1960s reported that lung capillaries contain about 60 ml of blood per square meter of body surface.)

4 The fact that the blood is distributed through the capillaries in a layer so thin (equal only to the diameter of one red corpuscle) that each corpuscle comes close to alveolar air

How blood transports gases

Blood transports oxygen and carbon dioxide as solutes and as parts of molecules of certain chemical compounds. Immediately on entering the blood, both oxygen and carbon dioxide dissolve in the plasma. But, because fluids can hold only small amounts of gas in solution, most of the oxygen and carbon dioxide rapidly form a chemical union with some other blood constituent. In this way, comparatively large volumes of the gases can be transported. With a P_{O_2} of 100 mm Hg, only 0.3 ml of oxygen dissolves in 100 ml of arterial blood. Many times that amount, however, combines with the hemoglobin in 100 ml of blood to form oxyhemoglobin. Since each gram of hemoglobin can unite with 1.34 ml of oxygen, the exact amount of oxygen in blood depends mainly on the amount of hemoglobin present. Normally, 100 ml of blood contains about 15 gm of hemoglobin. If 100% of it combines with oxygen, then 100 ml of blood will contain 15×1.34, or 20.1 ml oxygen in the form of oxyhemoglobin.

Perhaps a more common way of expressing blood oxygen content is in terms of volume percent. Normal arterial blood, with P_{O_2} of 100 mm

Hg, contains about 20 vol% O_2 (meaning 20 ml of oxygen in each 100 ml of blood).

Blood that contains more hemoglobin can, of course, transport more oxygen and that which contains less hemoglobin can transport less oxygen. Hence hemoglobin deficiency anemia decreases oxygen transport and may produce marked cellular hypoxia (inadequate oxygen supply).

In order to combine with hemoglobin, oxygen must, of course, diffuse from plasma into the red blood cells where millions of hemoglobin molecules are located. Several factors influence the rate at which hemoglobin combines with oxygen in lung capillaries. For instance, as the following equation and Fig. 17-17 show, an increasing blood PO_2 and a decreasing PCO_2 both accelerate hemoglobin association with oxygen.

$$Hb + O_2 \xrightarrow[\text{(Decreasing PCO}_2)]{\text{(Increasing PO}_2)} HbO_2$$

Decreasing PO_2 and increasing PCO_2, on the other hand, accelerate oxygen dissociation from oxyhemoglobin, that is, the reverse of the preceding equation. Oxygen associates with hemoglobin rapidly—so rapidly, in fact, that about 97% of the blood's hemoglobin has united with oxygen by the time the blood leaves the lung capillaries to return to the heart. In other words, the average *oxygen saturation* of arterial blood is about 97%.

Carbon dioxide is carried in the blood in several ways, the most important of which are described briefly as follows.

1 A small amount of carbon dioxide dissolves in plasma and is transported as a solute (dissolved carbon dioxide produces the PCO_2 of blood).

2 More than half of the carbon dioxide is carried in the plasma as bicarbonate ions (Fig. 17-18).

3 Somewhat less than one third of blood carbon dioxide unites with the NH_2 group of hemoglobin and certain other proteins to form carbamino compounds. Most of these are formed and transported in the red blood cells, since hemoglobin is the main protein to combine with carbon dioxide. The compound formed has a tongue-twisting name—carbaminohemoglobin. Carbon dioxide association with hemoglobin is accelerated by an increased PCO_2 and a decreasing PO_2 and is slowed by the opposite conditions. How do the conditions that accelerate carbon dioxide association affect the rate of oxygen association with hemoglobin?

Exchange of gases in tissues (internal respiration)

The exchange of gases in tissues takes place between arterial blood flowing through tissue capillaries and cells (Fig. 17-18). It occurs because of the principle already noted that gases move down a gas pressure gradient. More specifically, in the tissue capillaries, oxygen diffuses out of arterial blood because the oxygen pressure gradient favors its outward diffusion (Fig. 17-15). Arterial blood PO_2 is about 100 mm Hg, interstitial fluid PO_2 is considerably lower, and intracellular fluid PO_2 is still lower. Although interstitial fluid and intracellular fluid PO_2 are not definitely established, they are thought to vary considerably—perhaps from around 60 mm Hg down to about 1 mm Hg. As

Table 17-2 Blood oxygen		
	Venous blood	**Arterial blood**
PO_2	37 mm Hg	100 mm Hg
Oxygen saturation	75%	97%
Oxygen content	15 ml O_2 per 100 ml blood	20 ml O_2 per 100 ml blood*

*Oxygen utilization by tissues = Difference between oxygen contents of arterial and venous blood (20-15) = 5 ml O_2 per 100 ml blood circulated per minute.

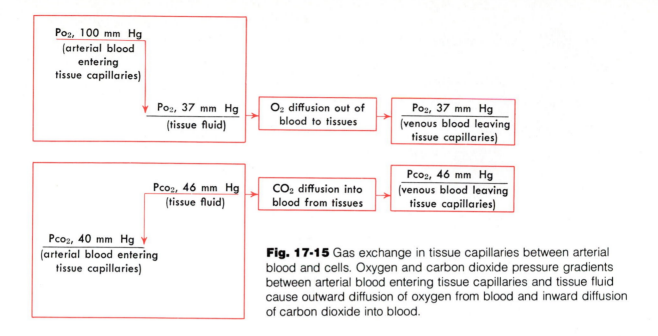

Fig. 17-15 Gas exchange in tissue capillaries between arterial blood and cells. Oxygen and carbon dioxide pressure gradients between arterial blood entering tissue capillaries and tissue fluid cause outward diffusion of oxygen from blood and inward diffusion of carbon dioxide into blood.

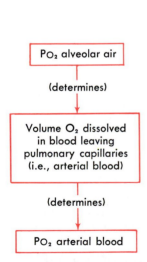

Fig. 17-16 Determinants of partial pressure of oxygen in arterial blood.

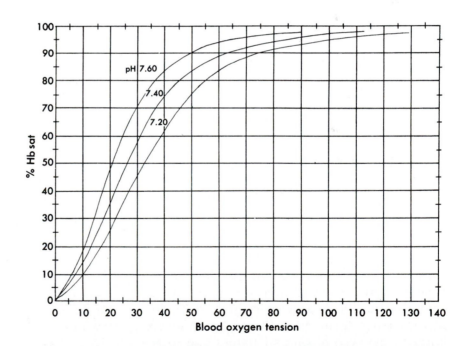

Fig. 17-17 Oxygen dissociation curve of blood at 37° C, showing variations at three pH levels. For a given oxygen tension the higher the blood pH, the more the hemoglobin holds onto its oxygen, maintaining a higher saturation. (From Egan, D. F.: Fundamentals of respiratory therapy, ed. 2, St. Louis, 1973, The C. V. Mosby Co.)

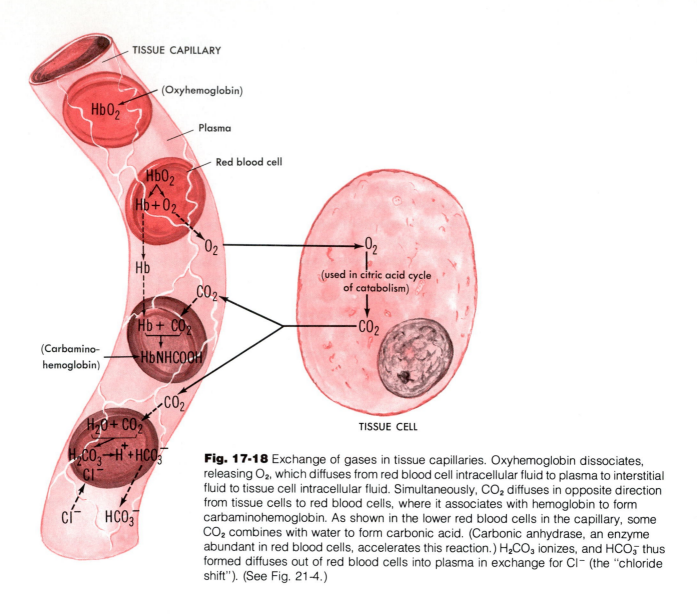

TISSUE CAPILLARY

(Oxyhemoglobin)

HbO_2

Plasma

Red blood cell

HbO_2

$Hb + O_2$

O_2

Hb

CO_2

Hb + CO_2

(Carbamino-hemoglobin)

HbNHCOOH

CO_2

$H_2O + CO_2$

$H_2CO_3 \rightarrow H^+ + HCO_3^-$

Cl^-

Cl^- HCO_3^-

O_2

(used in citric acid cycle of catabolism)

CO_2

TISSUE CELL

Fig. 17-18 Exchange of gases in tissue capillaries. Oxyhemoglobin dissociates, releasing O_2, which diffuses from red blood cell intracellular fluid to plasma to interstitial fluid to tissue cell intracellular fluid. Simultaneously, CO_2 diffuses in opposite direction from tissue cells to red blood cells, where it associates with hemoglobin to form carbaminohemoglobin. As shown in the lower red blood cells in the capillary, some CO_2 combines with water to form carbonic acid. (Carbonic anhydrase, an enzyme abundant in red blood cells, accelerates this reaction.) H_2CO_3 ionizes, and HCO_3^- thus formed diffuses out of red blood cells into plasma in exchange for Cl^- (the "chloride shift"). (See Fig. 21-4.)

activity increases in any structure, its cells necessarily utilize oxygen more rapidly. This decreases intracellular and interstitial P_{O_2}, which in turn tends to increase the oxygen pressure gradient between blood and tissues and to accelerate oxygen diffusion out of the tissue capillaries. In this way, the rate of oxygen utilization by cells automatically tends to regulate the rate of oxygen delivery to cells. As dissolved oxygen diffuses out of arterial blood, blood P_{O_2} decreases, and this accelerates oxyhemoglobin dissociation to release more oxygen into the plasma for diffusion out to cells, as indicated in the following equation.

$$Hb + O_2 \xrightleftharpoons[\text{(Increasing } P_{CO_2})]{\text{(Decreasing } P_{O_2})} HbO_2$$

Because of oxygen release to tissues from tissue capillary blood, P_{O_2}, oxygen saturation, and total oxygen content are less in venous blood than in arterial blood, as shown in Table 17-2.

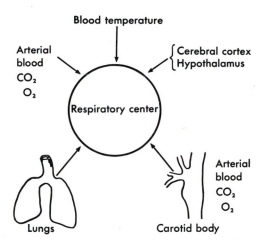

Fig. 17-19 Respiratory control mechanism. Scheme to show the main factors that influence the respiratory center and thereby control respirations. (See text for discussion.)

Carbon dioxide exchange between tissues and blood takes place in the opposite direction from oxygen exchange. Catabolism produces large amounts of carbon dioxide inside cells. Therefore intracellular and interstitial Pco_2 are higher than arterial blood Pco_2. This means that the carbon dioxide pressure gradient causes diffusion of carbon dioxide from the tissues into the blood flowing along through tissue capillaries (Fig. 17-15). Consequently, the Pco_2 of blood increases in tissue capillaries from its arterial level of about 40 mm Hg to its venous level of about 46 mm Hg. This increasing Pco_2 and decreasing Po_2 together produce two effects—they favor both oxygen dissociation from oxyhemoglobin and carbon dioxide association with hemoglobin to form carbaminohemoglobin (interpret the graphs in Fig. 17-17).

Regulation of respirations

The mechanism for the control of respirations has many parts (Fig. 17-19). A brief description of its main features follows.

1 The Pco_2, Po_2, and pH of *arterial blood* all influence respirations. The Pco_2 acts on chemoreceptors located in the medulla. Chemoreceptors

in this case are cells sensitive to changes in arterial blood's carbon dioxide and hydrogen ion concentrations. The normal range for arterial Pco_2 is about 38 to 40 mm Hg. When it increases even slightly above this, it has a stimulating effect mainly on central chemoreceptors (postulated to be present in the medulla). Large but tolerable increases in arterial Pco_2 stimulate peripheral chemoreceptors present in the carotid bodies and aorta. Stimulation of chemoreceptors by increased arterial Pco_2 results in faster breathing, with a greater volume of air moving in and out of the lungs per minute. Decreased arterial Pco_2 produces opposite effects—it inhibits central and peripheral chemoreceptors, which leads to inhibition of medullary respiratory centers and slower respirations. In fact, breathing stops entirely for a few moments (apnea) when arterial Pco_2 drops moderately—to about 35 mm Hg, for example.

A decrease in arterial blood pH (increase in acid), within certain limits, has a stimulating effect on chemoreceptors located in the carotid and aortic bodies.

The role of *arterial blood* Po_2 in controlling respirations is not entirely clear. Presumably, it has little influence as long as it stays above a certain level. But neurons of the respiratory centers, like all body cells, require adequate amounts of oxygen in order to function optimally. Consequently, if they become hypoxic, they become depressed and send fewer impulses to respiratory muscles. Respirations then decrease or fail entirely. This principle has important clinical significance. For example, the respiratory centers cannot respond to stimulation by an increasing blood CO_2 if, at the same time, blood Po_2 falls below a critical level—a fact that may become of life and death importance during anesthesia.

However, a decrease in arterial blood Po_2 below 70 mm Hg but not so low as the critical level stimulates chemoreceptors in the carotid and aortic bodies and causes reflex stimulation of the inspiratory center. This constitutes an emergency respiratory control mechanism. It does not help regulate respirations under usual conditions when arterial blood P_{O_2} remains considerably higher than 70 mm Hg—the level necessary to stimulate the chemoreceptors.

2 *Arterial blood pressure* helps control respirations through the respiratory pressoreflex mechanism. A sudden rise in arterial pressure, by acting on aortic and carotid baroreceptors, results in reflex slowing of respirations. A sudden drop in arterial pressure brings about a reflex increase in rate and depth of respirations. The pressoreflex mechanism is probably not of great importance in the control of respirations. It is, however, of major importance in the control of circulation.

3 The *Hering-Breuer reflexes* help control respirations, particularly their depth and rhythmicity. They are believed to regulate the normal depth of respirations (extent of lung expansion), and therefore the volume of tidal air, in the following way. Presumably, when the tidal volume of air has been inspired, the lungs are expanded enough to stimulate baroreceptors located within them. The baroreceptors then send inhibitory impulses to the inspiratory center, relaxation of inspiratory muscles occurs, and expiration follows—the Hering-Breuer expiratory reflex. Then, when the tidal volume of air has been expired, the lungs are sufficiently deflated to inhibit the lung baroreceptors and allow inspiration to start again—the Hering-Breuer inspiratory reflex.

4 The *pneumotaxic center* in the upper part of the pons is postulated to function mainly to maintain rhythmicity of respirations. Whenever the inspiratory center is stimulated, it sends impulses to the pneumotaxic center as well as to inspiratory muscles. The pneumotaxic center, after a moment's delay, stimulates the expiratory center, which then feeds back inhibitory impulses to the inspiratory center. Inspiration, therefore, ends, and expiration starts. Lung deflation soon initiates the Hering-Breuer inspiratory reflex, and inspiration starts again. In short, the pneumotaxic center and Hering-Breuer re-

flexes together constitute an automatic device for producing rhythmic respirations.

5 The *cerebral cortex* helps control respirations. Impulses to the respiratory center from the motor area of the cerebrum may either increase or decrease the rate and strength of respirations. In other words, an individual may voluntarily speed up or slow down his breathing rate. This voluntary control of respirations, however, has certain limitations. For example, one may will to stop breathing and do so for a few minutes. But holding the breath results in an increase in the carbon dioxide content of the blood, since it is not being removed by respirations. Carbon dioxide is a powerful respiratory stimulant. So when arterial blood P_{CO_2} increases to a certain level, it stimulates the inspiratory center both directly and reflexly to send motor impulses to the respiratory muscles, and breathing is resumed even though the individual may still will contrarily. This knowledge that the carbon dioxide content of the blood is a more powerful regulator of respirations than cerebral impulses is of practical value when dealing with a child who holds his breath to force the granting of his wishes. The best treatment is to ignore such behavior, knowing that respirations will start again as soon as the amount of carbon dioxide in arterial blood increases to a certain level.

6 Miscellaneous factors also influence respirations. Among these are blood temperature and sensory impulses from skin thermal receptors and from superficial or deep pain receptors.

a *Sudden painful stimulation* produces a reflex apnea, but continued painful stimuli cause faster and deeper respirations.

b *Sudden cold stimuli* applied to the skin cause reflex apnea.

c *Afferent impulses* initiated by stretching the anal sphincter produce reflex acceleration and deepening of respirations. Use has sometimes been made of this mechanism as an emergency measure to stimulate respirations during surgery.

d *Stimulation of the pharynx or larynx* by irritating chemicals or by touch causes a temporary apnea. This is the choking reflex, a valuable protective device. It operates, for example, to prevent aspiration of food or liquids during swallowing.

Control of respirations during exercise. Respirations increase abruptly at the beginning of exercise and decrease even more markedly as it ends. This much is known. The mechanism that accomplishes this, however, is not known. It is not identical to the one that produces moderate increases in breathing. A number of studies have shown that arterial blood P_{CO_2}, P_{O_2}, and pH do not change enough during exercise to produce the degree of hyperpnea (faster, deeper respirations) observed. Presumably, many chemical and nervous factors and temperature changes operate as a complex, but still unknown, mechanism for regulating respirations during exercise.

Outline summary

Functions and organs

Exchange of gases between blood and air; nose, pharynx, larynx, trachea, bronchi, lungs

Organs

Nose

A Structure
1 Portions—internal, in skull, above roof at mouth; external, protruding from face
2 Cavities
   a Divisions—right and left
   b Meati—superior, middle, and lower; named for turbinate located above each meatus
   c Opening—anterior nares that open into the vestibule from the exterior; posterior nares that permit air to flow from sphenoethmoidal recess into the nasopharynx
   d Turbinates (conchae)—superior and middle are processes of ethmoid bone; inferior turbinates are separate bones; divide internal nasal cavities into three passageways or meati
   e Floor—formed by palatine bones that also act as roof of mouth
3 Lining—ciliated mucous membrane
4 Sinuses draining into nose (or paranasal sinuses)—frontal, maxillary (or antrum of Highmore), sphenoidal, and ethmoidal
B Functions
1 Serves as passageway for incoming and outgoing air, filtering, warming, moistening, and chemically examining it
2 Organ of smell—olfactory receptors located in nasal mucosa
3 Aids in phonation

Pharynx

A Structure—made of muscle with mucous lining
1 Divisions—nasopharynx located behind the nose and extending from the posterior nares to the level of the soft palate; oropharynx, located behind the mouth from the soft palate above to the level of the hyoid bone below; laryngopharynx from the level of the hyoid bone above to its termination in the esophagus; see Fig. 17-4
2 Openings—four in nasopharynx: two auditory tubes and two posterior nares; one in oropharynx; fauces from mouth; and two in laryngopharynx; open into esophagus and larynx
3 Organs in pharynx—adenoids or pharyngeal

tonsils in nasopharynx; palatine and lingual tonsils in oropharynx

4 Cavity of nasopharynx differs from oral and laryngeal divisions in that it does not collapse

B Functions—serve both respiratory and digestive tracts as passageway for air, food, and liquids; aids in phonation

Larynx

A Location—at upper end of trachea, just below pharynx; normally extends between fourth, fifth, and sixth cervical vertebrae but often somewhat higher in females and during childhood

B Structure

1 Cartilages—nine pieces arranged in boxlike formation; thyroid largest, known as "Adam's apple"; epiglottis, "lid" cartilage; cricoid, "signet ring" cartilage

2 Vocal cords—false cords, folds of mucous lining; true cords, fibroelastic bands stretched across hollow interior of larynx; glottis, opening between true vocal cords

3 Lining—ciliated mucous membrane

4 Sexual differences—male larynx larger, covered with less fat, and therefore more prominent than female larynx

5 Muscles—intrinsic, have both origin and insertion on larynx; extrinsic, insert on larynx but have origin on some other structure

C Function—expired air causes true vocal cords to vibrate, producing voice; pitch determined by length and tension of cords

Trachea

A Structure

1 Walls—smooth muscle; contain C-shaped rings of cartilage at intervals, which keeps the tube open at all times; lining—ciliated mucous membrane

2 Extent—from larynx to bronchi; about 4½ inches long

B Function—furnishes open passageway for air going to and from lungs

Bronchi

A Structure—formed by division of trachea into two tubes; right bronchus slightly larger and more vertical than left; same structure as trachea; each primary bronchus branches as soon as enters lung into secondary bronchi, which branch into bronchioles, which branch into microscopic alveolar ducts, which terminate in cluster of blind sacs called alveoli; trachea and two primary bronchi and all their branches compose "bronchial tree"; alveolar walls composed of single layer of cells

B Function—bronchi and their many branching tubes furnish passageway for air going to and from lungs; alveoli provide large, thin-walled surface area where blood and air can exchange gases

Lungs

A Structure

1 Size, shape, and location—large enough to fill pleural division of thoracic cavity; cone shaped; extend from base, on diaphragm, to apex, located slightly above clavicle

2 Divisions—three lobes in right lung, two in left; root of lung consists of primary bronchus and pulmonary artery and veins, bound together by connective tissue; hilum is vertical slit on mesial surface of lung through which root structures enter lung; base is broad, inferior surface of lung; apex is pointed upper margin

3 Covering—visceral layer of pleura

B Function—furnish place where large amounts of air and blood can come in close enough contact for rapid exchange of gases to occur

Thorax (chest)

A Structure

1 Three divisions

a Right and left pleural portions—contain lungs

b Mediastinum—area between two lungs; contains heart, esophagus, trachea, great blood vessels, etc.

2 Lining

a Parietal layer of pleura lines entire chest cavity and covers superior surface of diaphragm

b Forms separate sac encasing each lung

c Parietal pleura separated from visceral pleura, covering lungs, only by potential space—pleural space—that contains few drops of pleural fluid

3 Shape of ribs and angle of their attachment to spine such that elevation of rib cage enlarges two dimensions of thorax, its width and depth from front

B Function

1 Increase in size of thorax leads to inspiration

2 Decrease in size of thorax leads to expiration

Physiology

A Definition pulmonary ventilation—breathing

B Mechanism of pulmonary ventilation

1 Contraction of diaphragm and chest elevating muscles enlarges thorax, thereby decreases intrathoracic pressure, which causes expansion of lungs, which decreases intrapulmonic pressure to subatmospheric level, which establishes gas pressure gradient, which causes air to move into lungs

2 Relaxation of inspiratory muscles produces opposite effects; see Fig. 17-11

C Volumes of air exchanged in pulmonary ventilation—directly related to gas pressure gradient between atmosphere and lung alveoli and inversely related to resistance opposing air flow

1 Tidal volume (TV)—average amount expired after normal inspiration; approximately 500 ml or 1 pint

2 Inspiratory reserve volume (IRV)—amount that can be forcibly inspired after normal inspiration; measured by having individual expire normally after forced inspiration

3 Expiratory reserve volume (ERV)—additional amount of air that can be forcibly expired after a normal inspiration and expiration

4 Residual volume (RV)—amount of air that cannot be forcibly expired from lungs

5 Minimal volume (MV)—trapped air that remains in the lungs after the residual volume is eliminated (as in pneumothorax)

6 Vital capacity (VC)—largest volume of air an individual can move in and out of his lungs; equals sum of inspiratory reserve volume, tidal volume, and expiratory reserve volume

7 Anatomical dead space—volume of air that fills nose, pharynx, larynx, trachea, bronchi, and smaller tubes but does not descend into alveoli, therefore takes no part in gas exchange; typically, about 30% of TV, or 150 ml, but varies under different conditions

8 Alveolar ventilation—volume of inspired air that actually reaches alveoli; computed by subtracting dead space air volume from tidal air volume

D Types of breathing

1 Eupnea—normal quiet breathing

2 Hyperpnea—abnormally increased breathing

3 Apnea—cessation of breathing at end of normal expiration

4 Apneusis—cessation of breathing in inspiratory position

5 Cheyne-Stokes respirations—gradually increasing tidal volume for several breaths, followed by several breaths with gradually decreasing tidal volume; cycle repeats itself

6 Biot's respirations—repeated sequence of deep gasps and apnea

B Some principles about gases

1 Dalton's law—partial pressure of gas in mixture of gases directly related to concentration of that gas in mixture and to total pressure of mixture

2 Partial pressure of gas in liquid directly related to amount of gas dissolved in liquid; becomes equal to partial pressure of that gas in environment of liquid

C Exchange of gases in lungs (between alveolar air and venous blood)

1 Where it occurs—in lung capillaries; across alveolar-capillary membrane

2 What exchange consists of—oxygen diffuses out of alveolar air into venous blood; carbon dioxide diffuses in opposite direction

3 Why it occurs—oxygen pressure gradient causes inward diffusion of oxygen; carbon dioxide pressure gradient causes outward diffusion of carbon dioxide

	Alveolar air	Venous blood
P_{O_2}	100 mm Hg	37 mm Hg
P_{CO_2}	40 mm Hg	46 mm Hg

4 Results of gas exchange

a P_{O_2} of blood increases to arterial blood level as blood moves through lung capillaries

b P_{CO_2} of blood decreases to arterial blood level as blood moves through lung capillaries

c Increasing P_{O_2} and decreasing P_{CO_2} accelerate both oxygen association with hemoglobin to form oxyhemoglobin and carbon dioxide dissociation from carbaminohemoglobin both accelerated by increasing P_{O_2} and decreasing P_{CO_2}

D How blood transports gases

1 Oxygen

a About 0.5 ml transported as *solute*, that is, dissolved in 100 ml blood

b About 19.5 ml O_2 per 100 ml blood transported as *oxyhemoglobin* in red blood cells

c About 20.0 ml = *Total O_2 content* per 100 ml blood (100% saturation of 15 gm hemoglobin)

2 Carbon dioxide
  a Small amount dissolves in plasma and transported as true *solute*
  b More than half of CO_2 transported as *bicarbonate ion* in plasma
  c Somewhat less than one third of CO_2 transported in red blood cells as *carbaminohemoglobin*
E Exchange of gases in tissues (between arterial blood and cells)
  1 Where it occurs—tissue capillaries
  2 What exchange consists of—oxygen diffuses out of arterial blood into interstitial fluid and on into cells, whereas carbon dioxide diffuses in opposite direction
  3 Why it occurs—O_2 pressure gradient causes diffusion of O_2 out of blood; CO_2 pressure gradient causes diffusion of CO_2 into blood

	Arterial blood	**Interstitial fluid**
P_{O_2}	100 mm Hg	60(?) mm Hg down to 1(?) mm Hg
P_{CO_2}	46 mm Hg	50(?) mm Hg

  4 Results of oxygen diffusion out of blood and carbon dioxide diffusion into blood
  a P_{O_2} blood decreases as blood moves through tissue capillaries; arterial P_{O_2}, 100 mm Hg, becomes venous P_{O_2}, 40 mm Hg (figures vary)
  b P_{CO_2} blood increases; arterial P_{CO_2}, 40 mm Hg, becomes venous P_{CO_2}, 46 mm Hg (figures vary)
  c Oxygen dissociation from hemoglobin and carbon dioxide association with hemoglobin to form carbaminohemoglobin both accelerated by decreasing P_{O_2} and increasing P_{CO_2}

F Regulation of respirations—see Fig. 17-19
  1 Respiratory centers—inspiratory and expiratory centers in medulla; pneumotaxic center in pons
  2 Control of respiratory centers
  a Carbon dioxide major regulator of respirations; increased blood carbon dioxide content, up to certain level, stimulates respiration and above this level depresses respirations; decreased blood carbon dioxide decreases respirations
  b Oxygen content of blood influences respiratory center—decreased blood O_2, down to a certain level, stimulates respirations and below this critical level depresses them; O_2 control of respirations nonoperative under usual conditions
  c Hering-Breuer mechanism helps control rhythmicity of respirations; increased alveolar pressure inhibits inspiration and starts expiration; decreased alveolar pressure stimulates inspiration and ends expiration
  d Pneumotaxic center acts with Hering-Breuer reflexes to produce rhythmic respirations
  e Cerebral cortex impulses to respiratory centers provide voluntary control, within limits, of rate and depth of respirations
  3 Control of respirations during exercise—not yet established

Review questions

1 What anatomical feature favors the spread of the common cold through the respiratory passages and into the middle ear and mastoid sinus?

2 How are the turbinates arranged in the nose? What are they?

3 What organs are found in the nasopharynx?

4 What tubes open into the nasopharynx?

5 Make a diagram showing the termination of a bronchiole in an alveolar duct with alveoli.

6 What kind of membrane lines the respiratory system?

7 What is the serous covering of the lung called? Where else, besides covering the lungs, is this same membrane found?

8 The pharynx is common to what two systems?

9 Are the lungs active or passive organs during breathing? Explain

10 What is the main inspiratory muscle?

11 How is inspiration accomplished? Expiration?

12 If an opening is made into the pleural cavity from the exterior, what happens? Why?

13 What is the pleural space? What does it contain?

14 What substance found in blood is the natural chemical stimulant for the respiratory center?

15 Normally, about what percentage of the tidal volume fills the anatomical dead space?

16 Normally, about what percentage of the tidal volume is useful air, that is, ventilates the alveoli?

17 What is the voice box? Of what is it composed? What is the Adam's apple?

18 What is the epiglottis? What is its function?

19 What are the true vocal cords? What are they? What name is given to the opening between the cords?

20 Name the three divisions of the thorax and their contents.

21 One gram of hemoglobin combines with how many milliliters of oxygen?

22 Suppose your blood has a hemoglobin content of 15 gm per 100 ml and an oxygen saturation of 97%. How many milliliters of oxygen would 100 ml of your arterial blood contain?

23 Compare the mechanisms that accelerate respiration with those that accelerate circulation during exercise.

24 Make a generalization about the effect of a moderate increase in the amount of blood CO_2 on circulation and respiration. What advantage can you see in this effect?

25 Make a generalization about the effect of a moderate decrease in blood O_2 on circulation and respiration.

26 Define the following terms briefly: alveolus, apnea, asphyxia, complemental air, cyanosis, dyspnea, minimal air, orthopnea, P_{CO_2}, P_{O_2}, pleurisy, residual air, respiration, spirometer, supplemental air, thorax, tidal air, vital capacity.

chapter 18

The digestive system

Functions and importance

Functions and importance

The organs of the digestive system together perform a vital function—that of preparing food for absorption and for use by the millions of body cells. Most food when eaten is in a form that cannot reach the cells (because it cannot pass through the intestinal mucosa into the bloodstream) nor could it be used by the cells even if it could reach them. It must, therefore, be modified as to both chemical composition and physical state. This process of altering the chemical and physical composition of food so that it can be absorbed and utilized by body cells is known as digestion and is the function of the digestive system. Part of the digestive system, the large intestine, serves also as an organ of elimination, ridding the body of the wastes resulting from the digestive process.

Organs

The main organs of the digestive system (Fig. 18-1) form a tube all the way through the ventral cavities of the body. It is open at both ends. This tube is usually referred to as the *alimentary canal* (or tract) or the *gastrointestinal* or GI tract. The following organs form the gastrointestinal tract: mouth, pharynx, esophagus, stomach, and intestines. Several accessory organs are located in the main digestive organs or open into them. They are the salivary glands, teeth, liver, gallbladder, pancreas, and vermiform appendix.

Walls of organs
Coats

The alimentary canal is essentially a tube whose walls are fashioned of four layers of tissues: a mucous lining, a submucous coat of connective tissue in which are embedded the main blood vessels of the tract, a muscular coat, and a fibroserous coat.

Modifications of coats

Although the same four tissue coats form the various organs of the alimentary tract, their structure varies in different organs. Some of these modifications are listed in Table 18-1 and should be referred to when each of those organs is studied in detail.

Mouth (buccal cavity)

The following structures form the buccal cavity: the cheeks (side walls), the tongue and its muscles (floor), and the hard and soft palates (roof). Of these, only the palates and the tongue will be included in this discussion.

The *hard palate* consists of portions of four bones: two maxillae and two palatines (Fig. 5-10, p. 94). The *soft palate*, which forms a partition between the mouth and nasopharynx (see Fig. 17-2, p. 442), is fashioned of muscle arranged in the shape of an arch. The opening in the arch leads from the mouth into the oropharynx and is named the *fauces*. Suspended from the midpoint of the posterior border of the arch is a small cone-shaped process, the *uvula*.

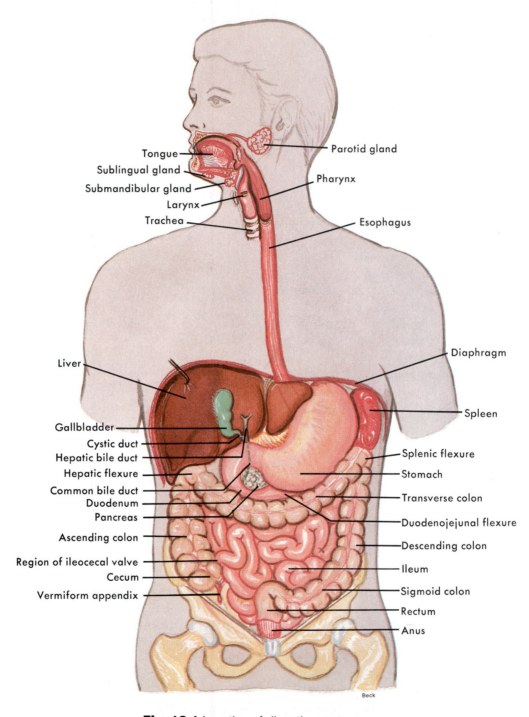

Tongue

Sublingual gland

Submandibular gland

Larynx

Trachea

Parotid gland

Pharynx

Esophagus

Liver

Gallbladder

Cystic duct

Hepatic bile duct

Hepatic flexure

Common bile duct

Duodenum

Pancreas

Ascending colon

Region of ileocecal valve

Cecum

Vermiform appendix

Diaphragm

Spleen

Splenic flexure

Stomach

Transverse colon

Duodenojejunal flexure

Descending colon

Ileum

Sigmoid colon

Rectum

Anus

Beck

Fig. 18-1 Location of digestive system organs.

Table 18-1 Modifications of coats of the digestive tract

Organ	Mucous coat	Muscle coat	Fibroserous coat
Esophagus		Two layers—inner one of circular fibers and outer one of longitudinal fibers; striated muscle in upper part and smooth in lower part of esophagus and in rest of tract	Outer coat fibrous; serous around part of esophagus in thoracic cavity
Stomach	Arranged in temporary longitudinal folds called *rugae;* allow for distention (Fig. 18-7) Contains gastric pits with microscopic gastric glands	Has three layers instead of usual two —circular, longitudinal, and oblique fibers; two sphincters—cardiac at entrance of stomach and pyloric at its exit, formed by circular fibers	Outer coat visceral peritoneum; hangs in double fold from lower edge of stomach over intestines, forming apronlike structure or "lace apron," *greater omentum* (Fig. 18-11); *lesser omentum* connects stomach to liver
Small intestine	Contains permanent circular folds, *plicae circulares* (Fig. 18-2) Microscopic fingerlike projections, *villi* (Fig. 18-8), with brush border (Fig. 18-9) Crypts of *Lieberkühn* Microscopic duodenal (Brunner's) mucous glands Clusters of lymph nodes, *Peyer's patches* Numerous single lymph nodes called solitary nodes	Two layers—inner one of circular fibers and outer one of longitudinal fibers	Outer coat, visceral peritoneum, continuous with *mesentery*
Large intestine	Solitary nodes Intestinal mucous glands *Anal columns* form in anal region	Outer longitudinal coat condensed to form three tapelike strips (taenia coli); small sacs (haustra) give rest of wall of large intestine puckered appearance (Fig. 18-10); internal anal sphincter formed by circular smooth fibers and external anal sphincter by striated fibers	Outer coat, visceral peritoneum, continuous with *mesocolon*

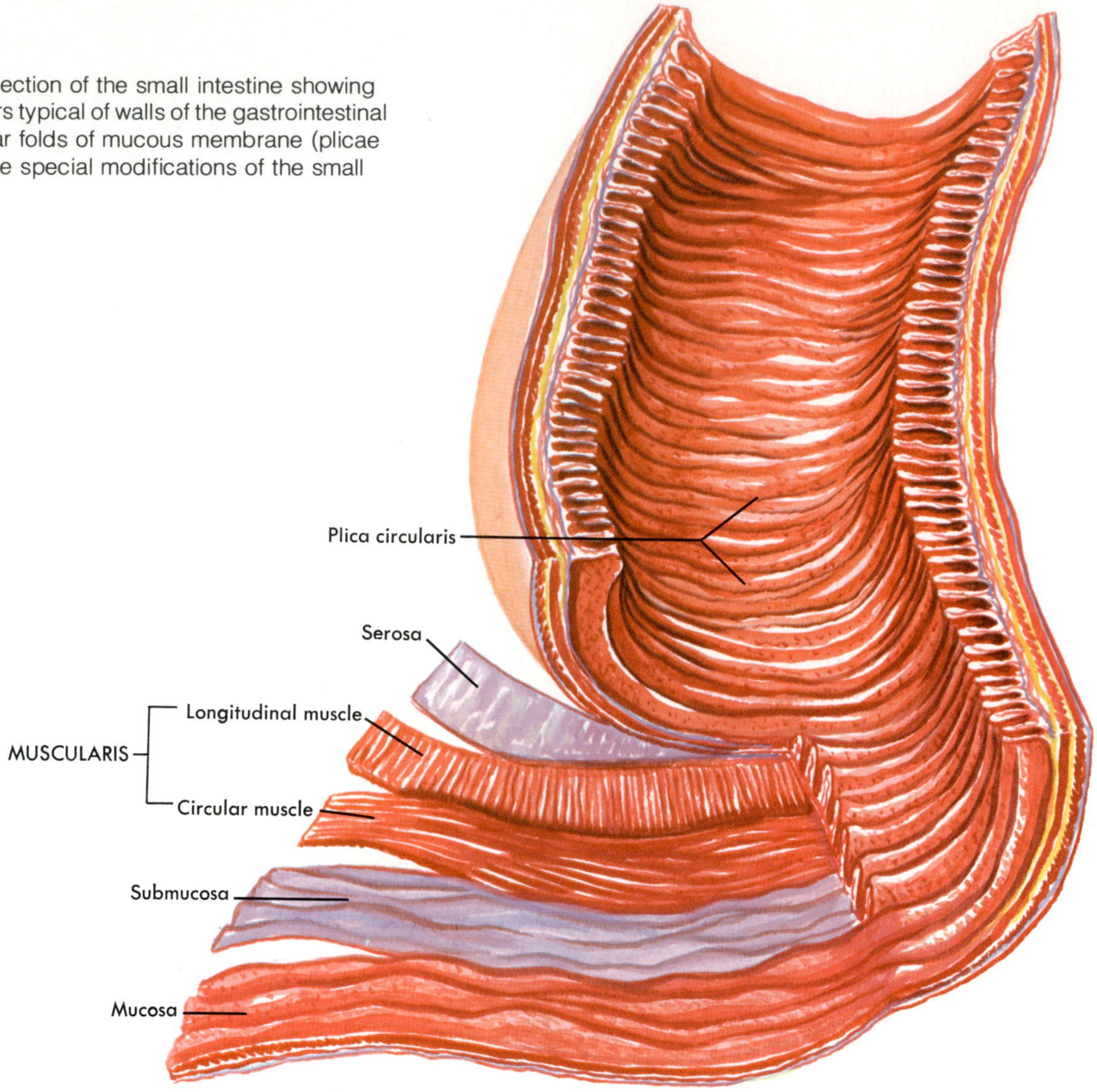

Fig. 18-2 Section of the small intestine showing the four layers typical of walls of the gastrointestinal tract. Circular folds of mucous membrane (plicae circularis) are special modifications of the small intestine.

Plica circularis

Serosa

MUSCULARIS — Longitudinal muscle

Circular muscle

Submucosa

Mucosa

Skeletal muscle covered with mucous membrane composes the *tongue.* Several muscles that originate on skull bones insert into the tongue. The rough elevations on the tongue's surface are called *papillae.* There are three types of them—filiform, fungiform, and circumvallate. You can readily distinguish them if you look at your own tongue. The filiform papillae are numerous threadlike structures distributed over the anterior two thirds of the tongue. Fungiform papillae, as the name indicates, are mushroom-shaped elevations most numerous near the edges of the tongue. Circumvallate papillae form an inverted V at the posterior part of the tongue. Taste buds are located on the sides of fungiform and circumvallate papillae.

The *frenulum* is a fold of mucous membrane in the midline of the undersurface of the tongue that helps to anchor the tongue to the floor of the mouth. If the frenum is too short for freedom of tongue movements, the individual is said to be tongue-tied, and his speech is faulty.

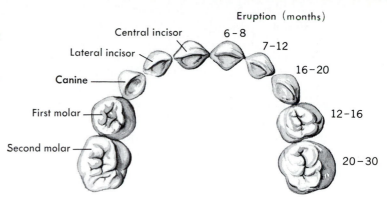

Fig. 18-3 The deciduous arch. Note that in the set of 20 temporary primary teeth there are no premolars (bicuspids) and there are only two pairs of molars in each jaw. Compare with permanent teeth (Fig. 18-4).

Eruption (months)

Central incisor — 6–8
Lateral incisor — 7–12
Canine — 16–20
First molar — 12–16
Second molar — 20–30

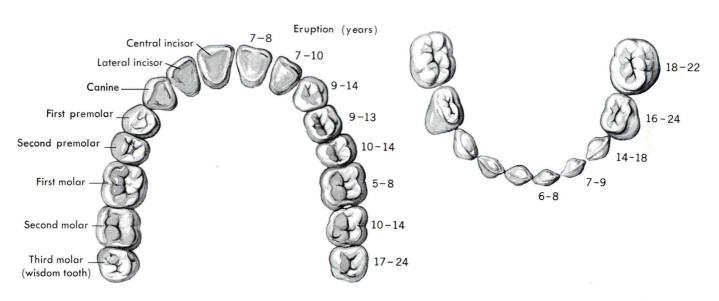

Eruption (years)

Central incisor — 7–8
Lateral incisor — 7–10
Canine — 9–14
First premolar — 9–13
Second premolar — 10–14
First molar — 5–8
Second molar — 10–14
Third molar (wisdom tooth) — 17–24

18–22
16–24
14–18
7–9
6–8

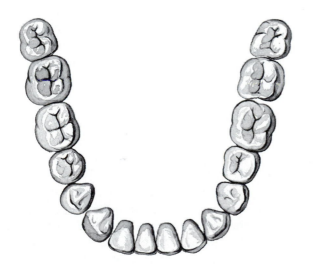

Fig. 18-4 The 32 permanent teeth. Generally the lower teeth erupt before the corresponding upper teeth, and all teeth usually erupt earlier in girls than in boys.

Salivary glands

Three pairs of glands secrete saliva into ducts that lead to the mouth (Fig. 18-1). The sublingual and submandibular glands are small glands under the tongue. The largest pair are the parotid glands located below and in front of the ears. The parotid (Stensen's) ducts empty into the mouth opposite the second molars.

Mumps is an acute infection of the parotid glands characterized by swelling of the glands. The act of opening the mouth causes pain because it squeezes that part of the gland that projects between the temporomandibular joint and the mastoid process.

Teeth

Twenty *deciduous teeth*, the so-called baby teeth (Fig. 18-3), appear early in life and are later replaced by 32 *permanent teeth* (Fig. 18-4). The names and numbers of teeth present in both sets are given in Table 18-2.

The first deciduous tooth erupts usually at the age of about 6 months. The rest follow at the rate of 1 or more a month until all 20 have appeared. There is, however, great individual variation in the age at which teeth erupt. Deciduous teeth are shed generally between the ages of 6 and 13 years. The third molars (wisdom teeth) are the last to appear, erupting usually sometime after 17 years of age.

Fig. 18-5 shows typical tooth structure.

Intact enamel resists bacterial attack, but once it is broken, the softer dentine decays. Periodontitis is an inflammation of the gums (gingivae) and periodontal membrane (Fig. 18-5).

Pharynx

Food passes from mouth to esophagus by way of the pharynx (p. 443).

Esophagus

The esophagus, a collapsible tube about 25 cm (10 inches) long, extends from the pharynx to the stomach, piercing the diaphragm in its descent from the thoracic to the abdominal cavity. It lies posterior to the trachea and heart.

Unlike the trachea, the esophagus is a collapsible tube, its muscle walls lacking the cartilaginous rings found in the trachea.

Stomach

Size, shape, and position

Just below the diaphragm, the digestive tube dilates into an elongated pouchlike structure, the stomach (Fig. 18-6), the size of which varies according to several factors, notably sex and the amount of distention. In general, the female

Table 18-2 Dentition		
	Number per jaw	
Name of tooth	**Deciduous set**	**Permanent set**
Central incisors	2	2
Lateral incisors	2	2
Cuspids (canines)	2	2
Premolars (bicuspids)	0	4
First molars (tricuspids)	2	2
Second molars	2	2
Third molars (wisdom teeth)	0	2
Total per jaw	10	16
Total per set	20	32

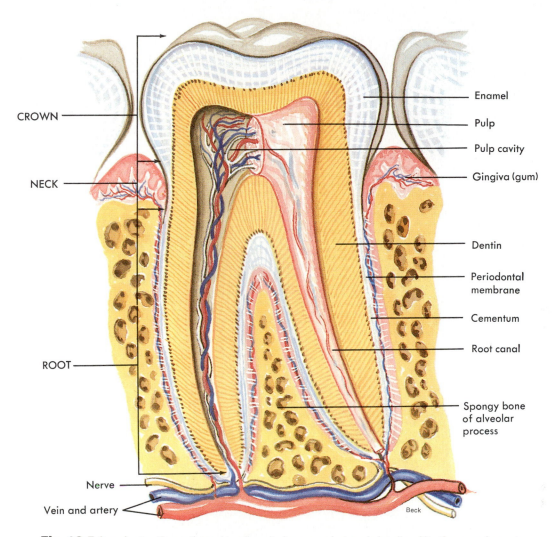

CROWN

NECK

ROOT

Nerve

Vein and artery

Enamel

Pulp

Pulp cavity

Gingiva (gum)

Dentin

Periodontal membrane

Cementum

Root canal

Spongy bone of alveolar process

Beck

Fig. 18-5 A molar tooth sectioned to show its bony socket and details of its three main parts: crown, neck, and root. Enamel (over the crown) and cementum (over the neck and root) surround the dentin layer. The pulp contains nerves and blood vessels.

stomach is usually more slender and smaller than the male stomach. For some time after a meal the stomach is enlarged because of distention of its walls, but as food leaves, the walls partially collapse, leaving the organ about the size of a large sausage.

The stomach lies in the upper part of the ab-

dominal cavity under the liver and diaphragm, with approximately five sixths of its mass to the left of the median line. In other words, it is described as lying in the epigastrium and left hypochondrium (Fig. 1-17, p. 19). Its position, however, alters frequently. For example, it is pushed downward with each inspiration and upward

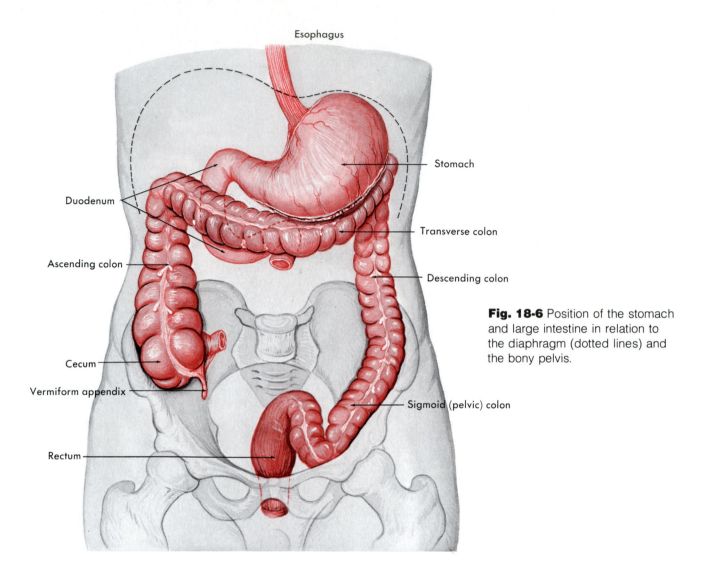

Esophagus

Duodenum

Ascending colon

Cecum

Vermiform appendix

Rectum

Stomach

Transverse colon

Descending colon

Sigmoid (pelvic) colon

Fig. 18-6 Position of the stomach and large intestine in relation to the diaphragm (dotted lines) and the bony pelvis.

with each expiration. When it is greatly distended from an unusually large meal, its size interferes with the descent of the diaphragm on inspiration, producing the familiar feeling of dyspnea that accompanies overeating. In this state the stomach also pushes upward against the heart and may give rise to the sensation that the heart is being crowded.

Divisions

The *fundus*, the *body*, and the *pylorus* are the three divisions of the stomach. The fundus is the enlarged portion to the left and above the open-

ing of the esophagus into the stomach. The body is the central part of the stomach, and the pylorus is its lower portion (Fig. 18-7).

Curves

The upper right border of the stomach presents what is known as the *lesser curvature* and the lower left border the *greater curvature*.

Sphincter muscles

Sphincter muscles guard both stomach openings. A sphincter muscle consists of circular fibers so arranged that there is an opening in

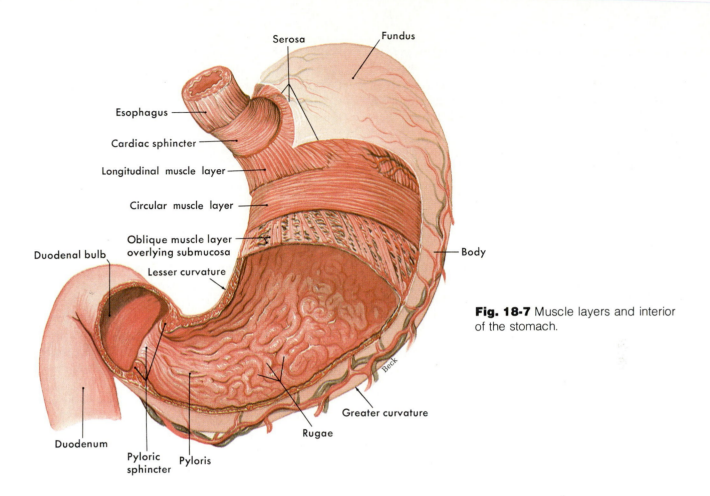

Serosa

Fundus

Esophagus

Cardiac sphincter

Longitudinal muscle layer

Circular muscle layer

Oblique muscle layer
overlying submucosa

Duodenal bulb

Lesser curvature

Body

Fig. 18-7 Muscle layers and interior of the stomach.

Duodenum

Pyloric sphincter

Pyloris

Rugae

Greater curvature

the center of them (like the hole in a doughnut) when they are relaxed and no opening when they are contracted.

The *cardiac sphincter* guards the opening of the esophagus into the stomach, and the *pyloric sphincter* guards the opening from the pyloric portion of the stomach into the first part of the small intestine (duodenum). This latter muscle is of clinical importance because *pylorospasm* is a fairly common condition in infants. The pyloric fibers do not relax normally to allow food to leave the stomach, and the infant vomits his food instead of digesting and absorbing it. The condition is relieved by the administration of a drug that relaxes smooth muscles. Another abnormality of the pyloric sphincter is *pyloric stenosis*, an obstructive narrowing of its opening.

Coats

See Table 18-1 and Fig. 18-7 for information on coats of the stomach.

Glands

Numerous microscopic tubular glands are embedded in pits in the gastric mucosa, particularly in the fundus and body of the stomach. The glands secrete most of the gastric juice, a fluid composed of mucus, enzymes, and hydrochloric acid. *Epithelial cells* that form the surface of the gastric mucosa (next to the lumen of the stomach) secrete mucus. *Chief cells* (zymogenic cells) secrete the enzymes of gastric juice. Parietal cells secrete hydrochloric acid and are also thought to produce the mysterious protein known as intrinsic factor. Atrophy of the gastric mucosa causes the condition known as *pernicious anemia*. As you might expect, there is a decrease in gastric enzymes and in hydrochloric acid. Achlorhydria is a characteristic finding in pernicious anemia. The absence of intrinsic factor is responsible for anemia caused by a breakdown in the erythropoietic mechanism. Intrinsic factor is necessary for vitamin B_{12} absorption, and vitamin B_{12} is essential for the normal development of red blood cells. Review the role of intrinsic factor in erythropoiesis on p. 354.

Functions

The stomach carries on the following functions.

1 It serves as a reservoir, storing food until it can be partially digested and moved farther along the gastrointestinal tract.

2 It secretes gastric juice to aid in the digestion of food.

3 Through contractions of its muscular coat, it churns the food, breaking it into small particles and mixing them well with the gastric juice. In due time, it moves the gastric contents on into the duodenum.

4 It secretes the intrinsic factor just mentioned.

5 It carries on a limited amount of absorption—of some water, alcohol, and certain drugs.

6 It produces the hormone *gastrin* in cells in the pyloric region.

Small intestine
Size and position

The small intestine is a tube measuring approximately 2.5 cm (1 inch) in diameter and 6 m (20 feet) in length. Its coiled loops fill most of the abdominal cavity.

Divisions

The small intestine consists of three divisions: the duodenum, the jejunum, and the ileum. The *duodenum** is the uppermost division and is the part to which the pyloric end of the stomach attaches. It is about 25 cm (10 inches) long and is shaped roughly like the letter C. The duodenum becomes *jejunum* at the point where the tube turns abruptly forward and downward. The jejunal portion continues for approximately the next 2.5 m (8 feet), where it becomes *ileum*, but without any clear line of demarcation between the two divisions. The ileum is about 3.5 m (12 feet) long.

Coats

Villi are important modifications of the mucosal layer of the small intestine. Millions of these projections, each about 1 mm in height, give the intestinal mucosa a velvety appearance. Each villus contains an arteriole, venule, and lymph vessel (lacteal) (Fig. 18-8). Epithelial cells on the surface of villi can be seen by microscopy (Fig. 18-9) to have a surface resembling a fine brush. This so-called *brush border* is formed by about 1,700 ultrafine *microvilli* per cell. Intestinal digestive enzymes, previously believed to be produced in crypts (of Lieberkühn) between villi, are now found to be produced in these brush border cells toward the top of villi. The presence of villi and microvilli increases the surface area of the small intestine by about 160 times, making this organ the main site of digestion and absorption. Mucus-secreting goblet cells (Fig. 18-9) are found in large numbers on villi and in crypts. See Table 18-1 and Figs. 18-2 and 18-8 for more information on the coats of the small intestine.

Functions

The small intestine carries on three main functions. A summary of them, in brief terms, follows on the next page.

---

*Derivation of the word "duodenum" may interest you. It comes from words that mean 12 fingerbreadths, a distance of about 11 inches, the approximate length of the duodenum.

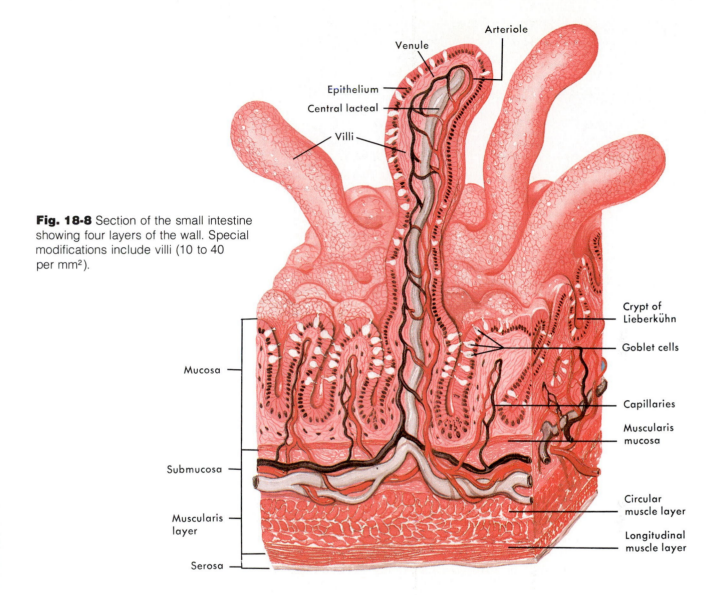

Fig. 18-8 Section of the small intestine showing four layers of the wall. Special modifications include villi (10 to 40 per mm²).

Labels on figure:
Venule
Arteriole
Epithelium
Central lacteal
Villi
Crypt of Lieberkühn
Goblet cells
Capillaries
Muscularis mucosa
Circular muscle layer
Longitudinal muscle layer
Mucosa
Submucosa
Muscularis layer
Serosa

1 It completes the digestion of foods. The intestinal juice contains mucus from the intestine, digestive enzymes from the pancreas, and bile received from the liver.
2 It absorbs the end products of digestion into blood and lymph.
3 It secretes hormones, for example, some that help control the secretion of pancreatic juice, bile, and intestinal juice (p. 499).

Large intestine (colon)

Size

The lower part of the alimentary canal bears the name *large intestine* because its diameter is noticeably larger than that of the small intestine. Its length, however, is much less, being about 1.5 to 1.8 m (5 or 6 feet). Its average diameter is approximately 6 cm (2½ inches), but this decreases toward the lower end of the tube.

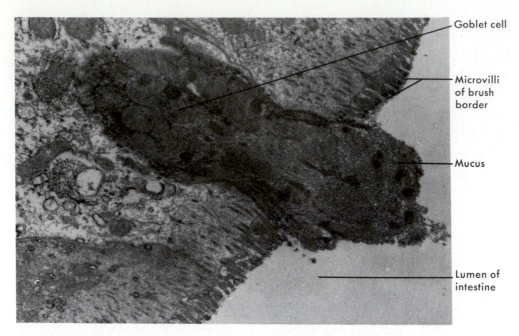

Goblet cell

Microvilli
of brush
border

Mucus

Lumen of
intestine

Fig. 18-9 Electron micrograph of a goblet cell in a villus of the jejunum.
The cell is discharging mucus. Note the microvilli of the brush border on the
adjacent epithelial cell. It is estimated that there are over 200 million
microvilli per mm² on intestinal mucosa. (×22,500.)
Courtesy Prof. Ivan Stotz, Department of Veterinary Science, South Dakota
State University, Brookings, S.D.

Divisions

The large intestine is divided into the cecum,
colon, and rectum.

Cecum. The first 5 to 8 cm (2 or 3 inches) of the
large intestine are named the cecum. It is located in the lower right quadrant of the abdomen (Fig. 18-6).

As its name suggests, the *vermiform appendix*
resembles a large worm in size and shape. The
blind tube averages 8 to 10 cm (3 or 4 inches) in
length and extends from the lower portion of the
cecum (Fig. 18-10). The structure of the appendix walls is similar to that of the rest of the intestine. Its mucous lining may become inflamed,
a condition well known as *appendicitis*.

Colon. The colon is divided into the following
portions: ascending, transverse, descending, and
sigmoid (Fig. 18-1).

1 The *ascending colon* lies in the vertical position, on the right side of the abdomen, extending
up to the lower border of the liver. The ileum
joins the large intestine at the junction of the
cecum and ascending colon, the place of attachment resembling the letter **T** in formation (Fig.
18-10). The ileocecal valve permits material to
pass from the ileum into the large intestine but
not in the reverse direction.

2 The *transverse colon* passes horizontally
across the abdomen, below the liver, stomach,
and spleen. Note that this part of the colon is
above the small intestine (Fig. 18-1). The transverse colon extends from the hepatic flexure to
the splenic flexure, the two points at which the
colon bends on itself to form 90 degree angles.

3 The *descending colon* lies in the vertical position, on the left side of the abdomen, extending

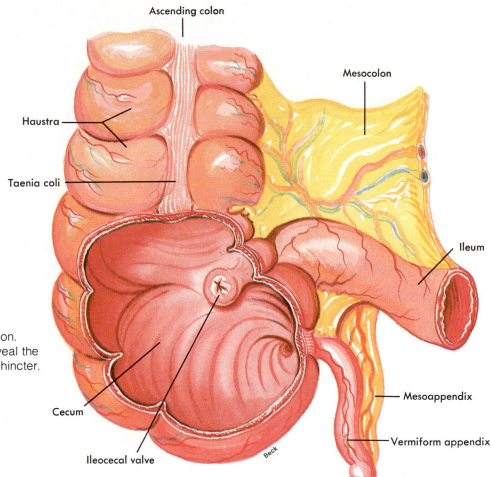

Ascending colon

Mesocolon

Haustra

Taenia coli

Ileum

Fig. 18-10 The vermiform appendix and ileocecal region. The cecum is opened to reveal the papillary form of ileocecal sphincter.

Mesoappendix

Vermiform appendix

Cecum

Ileocecal valve

Beck

from a point below the stomach and spleen to the level of the iliac crest.

4 The *sigmoid colon* is that portion of the large intestine that courses downward below the iliac crest. It describes an S-shaped curve. The lower part of the curve, which joins the rectum, bends toward the left, the anatomical reason for placing a patient on the left side when giving an enema. In this position, gravity aids the flow of the water from the rectum into the sigmoid flexure.

Rectum. The last 7 or 8 inches of the intestinal tube are called the rectum. The terminal inch of the rectum is called the *anal canal.* Its mucous lining is arranged in numerous vertical folds known as *anal columns,* each of which contains an artery and a vein. *Hemorrhoids* (or piles) are enlargements of the veins in the anal canal. The opening of the canal to the exterior is guarded by two sphincter muscles—an internal one of smooth muscle and an external one of striated muscle. The opening itself is called the *anus.* The general direction of the rectum is shown in Fig. 24-1, p. 608. Note that the anus is directed slightly anteriorly and is therefore at approximately right angles to the rectum. These anatomical relations should also be kept in mind during administration of an enema.

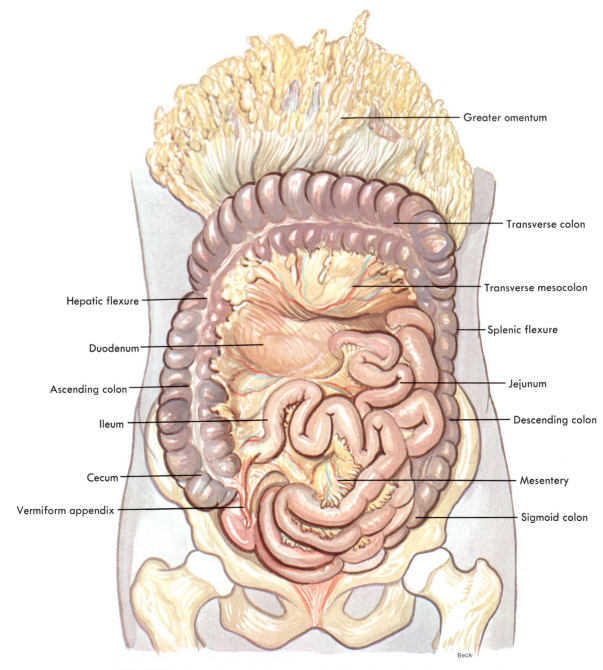

Fig. 18-11 Abdominal viscera from the front. The transverse colon and the greater omentum are elevated to reveal the flexures of the colon and the loops of the small intestine.

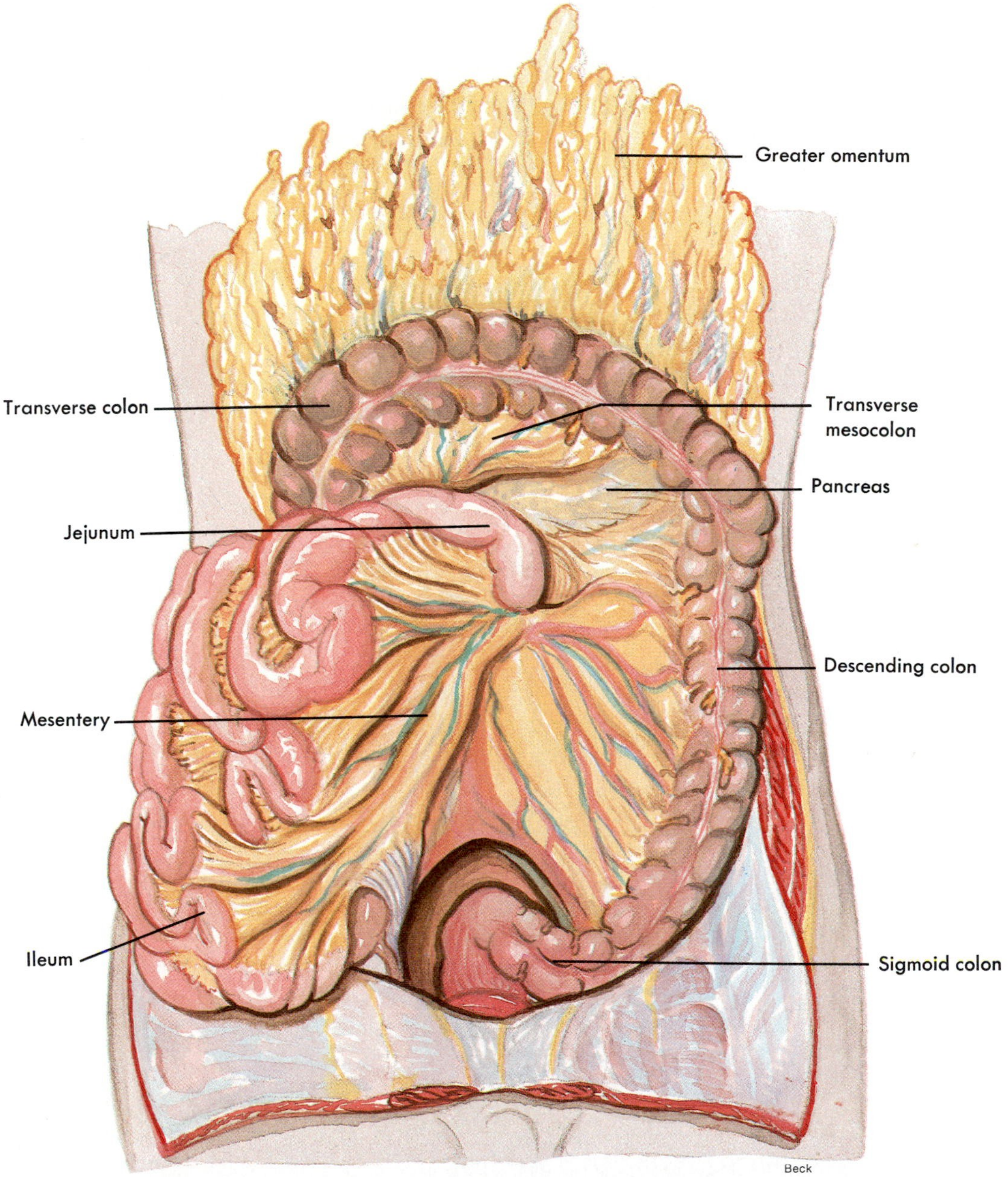

Greater omentum

Transverse colon

Transverse mesocolon

Pancreas

Jejunum

Descending colon

Mesentery

Ileum

Sigmoid colon

Beck

Fig. 18-12 The transverse colon and greater omentum are raised and the small intestine is pulled to the side to show the transverse mesocolon and mesentery.

Coats

See Table 18-1 for information on the coats of the large intestine.

Functions

The main functions of the large intestine are absorption of water, secretion of mucus, and elimination of the wastes of digestion.

Peritoneum

Now that you have completed the route through the digestive tube, consider for a moment, the membrane covering most of these organs and holding them loosely in place. The peritoneum is a large, continuous sheet of serous membrane. It lines the walls of the entire abdominal cavity (parietal layer) and also forms

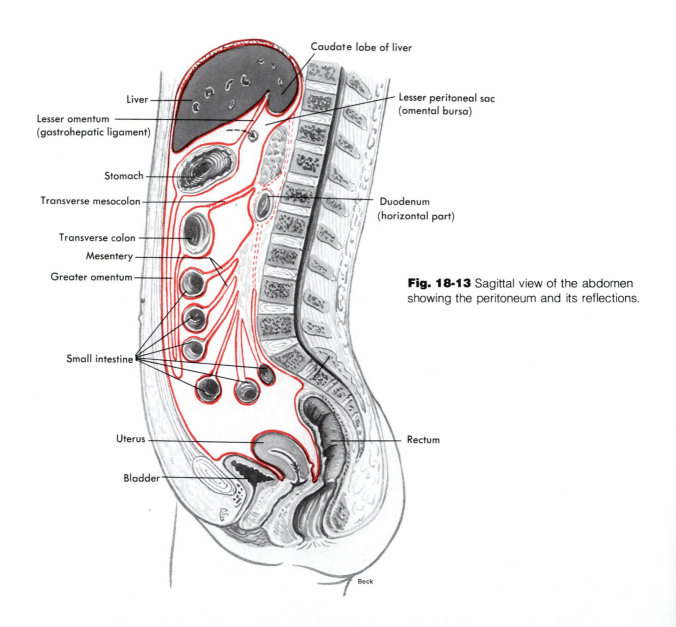

Caudate lobe of liver

Liver

Lesser omentum (gastrohepatic ligament)

Stomach

Transverse mesocolon

Transverse colon

Mesentery

Greater omentum

Small intestine

Uterus

Bladder

Lesser peritoneal sac (omental bursa)

Duodenum (horizontal part)

Rectum

Beck

Fig. 18-13 Sagittal view of the abdomen showing the peritoneum and its reflections.

the serous outer coat of the organs (visceral layer). In several places the peritoneum forms reflections, or extensions, that bind abdominal organs together (Figs. 18-11 to 18-13). The *mesentery* is a fan-shaped projection of the parietal peritoneum from the lumbar region of the posterior abdominal wall. The attached posterior border of this great fan is just 15 to 20 cm long, yet the loose outer edge enclosing the jejunum and ileum is 6 m long. The mesentery allows free movement of each coil of the intestine and helps prevent the long tube from strangulation. A similar but less extensive fold of peritoneum called the *transverse mesocolon* attaches the transverse colon to the posterior abdominal wall. The *greater omentum* is a continuation of the serosa of the greater curvature of the stomach and first part of the duodenum to the transverse colon. Spotty deposits of fat accumulate in the omentum and give it the appearance of a lacy apron hanging down loosely over the intestines. In cases of localized abdominal inflammation, such as appendicitis, the greater omentum envelops the inflamed area, walling it off from the rest of the abdomen. The *lesser omentum* attaches from the liver to the lesser curvature of the stomach and first part of the duodenum. The falciform ligament extends from the liver to the anterior abdominal wall. Examine the relations of peritoneal extensions in Fig. 18-13.

Liver

Location and size

The liver is the largest gland in the body. It weighs about 1.5 kg (3 to 4 pounds), lies immedi-

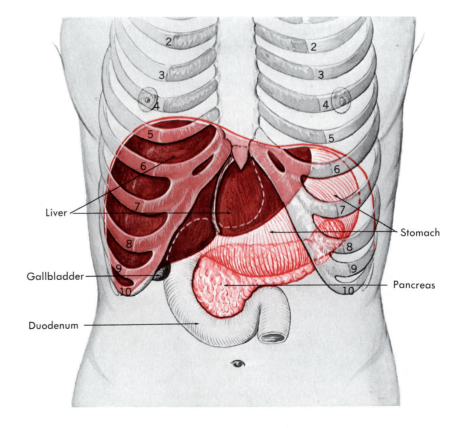

Fig. 18-14 The liver and pancreas in their normal positions relative to the rib cage, diaphragm, and stomach.

Liver

Gallbladder

Duodenum

Stomach

Pancreas

ately under the diaphragm, and occupies most of the right hypochondrium and part of the epigastrium (Fig. 18-14).

Lobes and lobules

The liver consists of two lobes separated by the falciform ligament. The left lobe forms about one sixth of the liver, whereas the right lobe comprises the remainder. The right lobe has three parts designated as the right lobe proper, the caudate lobe (a small four-sided area on the posterior surface), and the quadrate lobe (an approximately oblong section on the undersurface). Each lobe is divided into numerous lobules by small blood vessels and by fibrous strands that form a supporting framework (the capsule of Glisson) for them. The capsule of Glisson is an extension of the heavy connective tissue capsule that envelops the entire liver.

The *hepatic lobules* (Fig. 18-15), the anatomical units of the liver, are tiny hexagonal or pentagonal cylinders about 2 mm high and 1 mm in diameter. A small branch of the hepatic vein extends through the center of each lobule. Around this central (intralobular) vein, in plates or irregular walls radiating outward, are arranged the hepatic cells. Around the periphery of each lobule, several sets of three tiny tubes—branches of the hepatic artery, of the portal vein (interlobular veins), and of the hepatic duct (interlobular bile ducts)—are arranged. From these, irregular branches (sinusoids) of the interlobular veins extend between the radiating plates of hepatic cells to join the central vein. Minute bile canaliculi form networks around each cell.

Consider the function of the hepatic lobule while carefully examining Fig. 18-15. Blood enters a lobule from branches of the hepatic artery and portal vein. Arterial blood oxygenates the hepatic cells, whereas blood from the portal system passes through the liver for "inspection." Sinusoids in the lobule are lined with reticuloendothelial cells (mainly Kupffer cells). These phagocytic cells can remove toxic materials from the bloodstream. Ingested vitamins and other nutrients to be stored or metabolized by liver cells also enter the hepatic cell "bricks," forming radiating walls of the lobule. Blood continues along sinusoids to a vein at the center of the lobule. Such intralobular veins eventually lead to the two main hepatic veins. Bile formed by hepatic cells passes through canaliculi to the periphery of the lobule to join small bile ducts.

Ducts

The small bile ducts within the liver join to form two larger ducts that emerge from the undersurface of the organ as the right and left hepatic ducts. These immediately join to form one *hepatic duct*. The hepatic duct merges with the *cystic duct* from the gallbladder, forming the *common bile duct* (Fig. 18-16), which opens into the duodenum in a small raised area, called the major duodenal papilla. This papilla is located 7 to 10 cm below the pyloric opening from the stomach.

Functions

The liver is one of the most vital organs of the body. Here, in brief, are its main functions.

1 Liver cells detoxify a variety of substances.
2 Liver cells secrete about a pint of bile a day.
3 Liver cells carry on a number of important steps in the metabolism of all three kinds of foods—proteins, fats, and carbohydrates.
4 Liver cells store several substances—iron, for example, and vitamins A, B_{12}, and D.

Detoxification by liver cells. A number of poisonous substances enter the blood from the intestines. They circulate to the liver where, through a series of chemical reactions, they are changed to nontoxic compounds. Both ingested substances—alcohol, marijuana, and various other drugs, for example—and toxic substances formed in the intestines are detoxified in the liver.

Bile secretion by liver. The main components of bile are bile salts, bile pigments, and cholesterol. Bile salts (formed in the liver from cholesterol) are the most essential part of bile. They aid in

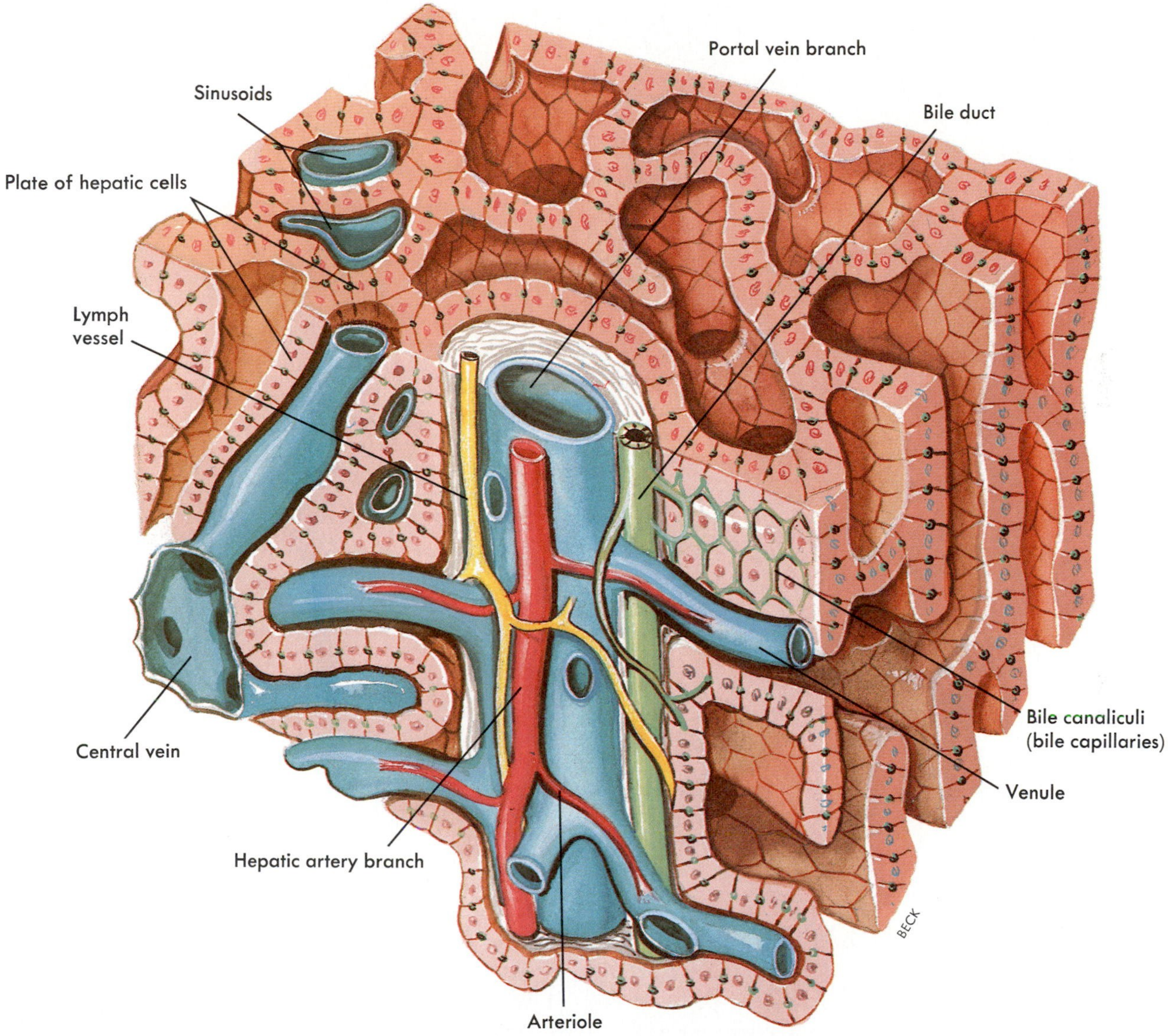

Fig. 18-15 Liver lobule. Blood from branches of the portal vein and hepatic artery passes through sinusoids between plates of hepatic cells. Sinusoidal blood empties into the central vein, which leads to the hepatic veins. Bile canaliculi empty bile into bile ducts.
Courtesy Dr. Hans Elias, Chicago.

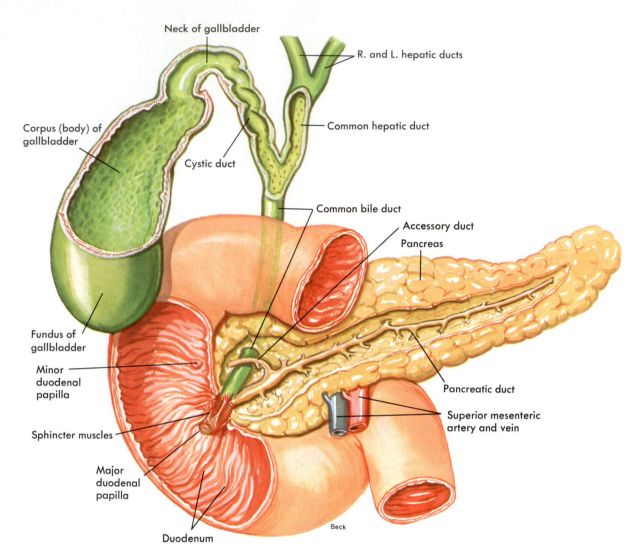

Neck of gallbladder

R. and L. hepatic ducts

Corpus (body) of gallbladder

Common hepatic duct

Cystic duct

Common bile duct

Accessory duct

Pancreas

Fundus of gallbladder

Minor duodenal papilla

Pancreatic duct

Sphincter muscles

Superior mesenteric artery and vein

Major duodenal papilla

Beck

Duodenum

Fig. 18-16 The gallbladder and its divisions: fundus, body, and neck. Obstruction of either the hepatic or the common bile duct by stone or spasm prevents bile from being ejected into the duodenum.

absorption of fats and then are themselves absorbed into the ileum. Eighty percent of bile salts are recycled in the liver to again become part of bile. Bile also serves as a pathway for elimination of certain breakdown products of red blood cells. The pigments bilirubin (red) and biliverdin (green), derived from hemoglobin, give bile its greenish color. Because it secretes bile into ducts, the liver qualifies as an exocrine gland.

Liver metabolism. Although all liver functions are important for healthy survival, some of its metabolic processes are crucial for survival itself. A fairly detailed description of the role of the liver in metabolism will be given in Chapter 19.

Gallbladder

Size, shape, and location

The gallbladder is a pear-shaped sac from 7 to 10 cm (3 to 4 inches) long and 3 cm broad at its widest point (Fig. 18-16). It can hold 30 to 50 ml of bile. It lies on the undersurface of the liver and is attached there by areolar tissue.

Structure

Serous, muscular, and mucous coats compose the wall of the gallbladder. The mucosal lining is arranged in rugae, similar in structure and function to those of the stomach.

Functions

The gallbladder stores the bile that enters it by way of the hepatic and cystic ducts. During this time the gallbladder concentrates bile 5- to 10-fold. Then later, when digestion is going on in the stomach and intestines, the gallbladder contracts, ejecting the concentrated bile into the duodenum.

Correlations

Inflammation of the lining of the gallbladder is called *cholecystitis. Cholecystectomy* is the surgical removal of the gallbladder, often necessitated by *cholelithiasis* or stones in this organ. *Jaundice,* a yellow discoloration of the skin and mucosa, results whenever obstruction of the hepatic or common bile duct occurs. Bile is thereby denied its normal exit from the body in the feces. Instead, it is absorbed into the blood. An excess of bile pigments in the blood gives it a yellow hue. Feces, deprived of their normal amount of bile pigments, become a grayish, so-called clay color.

Pancreas

Size, shape, and location

The pancreas is a grayish pink–colored gland about 12 to 15 cm (6 to 9 inches) long, weighing about 60 gm. It resembles a fish with its head

and neck in the C-shaped curve of the duodenum, its body extending horizontally behind the stomach, and its tail touching the spleen (Fig. 18-14). According to an old anatomical witticism, the "romance of the abdomen" is the pancreas lying "in the arms of the duodenum."

Structure

The pancreas is composed of two quite different types of glandular tissue, one exocrine and one endocrine. Most of the tissue is exocrine, with a compound acinar arrangement. The word acinar means that the cells are in a grapelike formation and that they release their secretions into a microscopic duct within each unit (Fig. 18-17, *B*). The word compound indicates that the ducts have branches. These tiny ducts unite to form larger ducts that eventually join the main pancreatic duct, which extends throughout the length of the gland from its tail to its head. It empties into the duodenum at the same point as the common bile duct, that is, at the major duodenal papilla. An accessory duct is frequently found extending from the head of the pancreas into the duodenum about 2 cm above the major papilla (Fig. 18-16).

Embedded between the exocrine units of the pancreas, like so many little islands, lie clusters of endocrine cells called *pancreatic islets* (or islands of Langerhans) (Fig. 18-17). Although there are about a million of these tiny islets, they constitute only about 2% of the total mass of the pancreas. Special staining techniques have revealed that two kinds of cells—alpha cells and beta cells—chiefly compose the islets. They are secreting cells, but their secretion passes into blood capillaries rather than into ducts. Thus the pancreas is a dual gland—an exocrine or duct gland because of the acinar units and an endocrine or ductless gland because of the islands of Langerhans.

Functions

1 The acinar units of the pancreas secrete the digestive enzymes found in pancreatic juice.

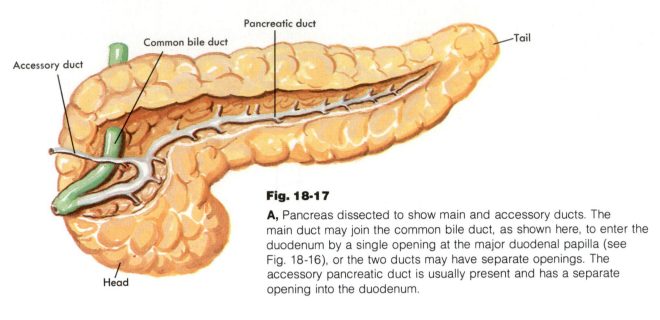

Accessory duct

Common bile duct

Pancreatic duct

Tail

Head

Fig. 18-17

A, Pancreas dissected to show main and accessory ducts. The main duct may join the common bile duct, as shown here, to enter the duodenum by a single opening at the major duodenal papilla (see Fig. 18-16), or the two ducts may have separate openings. The accessory pancreatic duct is usually present and has a separate opening into the duodenum.

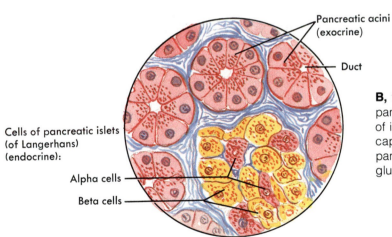

Pancreatic acini (exocrine)

Duct

Cells of pancreatic islets (of Langerhans) (endocrine):

Alpha cells

Beta cells

B, Exocrine glandular cells (around small pancreatic ducts) and endocrine glandular cells of islands of Langerhans (adjacent to blood capillaries). Exocrine pancreatic cells secrete pancreatic juice, alpha endocrine cells secrete glucagon, and beta cells secrete insulin.

Hence the pancreas plays an important part in digestion (Table 18-4, p. 495).

2 Beta cells of the pancreas secrete *insulin,* a hormone that exerts a major control over carbohydrate metabolism (p. 516 and Fig. 19-12).

3 Alpha cells secrete *glucagon.* It is interesting to note that glucagon, which is produced so close to insulin, has a directly opposite effect on carbohydrate metabolism (p. 516 and Fig. 19-13).

Digestion

Definition

Digestion is the sum of all the changes food undergoes in the alimentary canal.

Purpose

The purpose of digestion is the conversion of ingested food into a state in which it can be absorbed and used by the body. Digestion is necessary because foods, as eaten, are too complex in physical and chemical composition to pass through the intestinal mucosa into the blood or for cells to utilize them for energy and tissue building. In other words, digestion is the necessary preliminary to both absorption and metabolism of foods.

Kinds

Since both the physical and chemical composition of ingested food makes its absorption impossible, two kinds of digestive changes are necessary, *mechanical* and *chemical.*

Mechanical digestion. Mechanical digestion consists of all those movements of the alimentary tract that bring about the following:

1 Change the physical state of ingested food from comparatively large solid pieces into minute dissolved particles, thereby facilitating chemical digestion
2 Churn the intestinal contents in such a way that it becomes well mixed with the digestive juices and that all parts of it come in contact with the surface of the intestinal mucosa, thereby facilitating absorption
3 Propel the food forward along the alimentary tract, finally eliminating the digestive wastes from the body

A list of definitions of the different processes involved in mechanical digestion, together with the organs that accomplish them, are given in Table 18-3. Deglutition (swallowing), the movement of food out of the stomach, and defecation are considered in more detail here.

Deglutition. The process of swallowing may be considered as consisting of three main steps. The first step, which is voluntary, involves movement of food from the mouth to the pharynx. The second and third steps, both involuntary, consist of movement of food from the pharynx into the esophagus and finally into the stomach.

In the first step, food moistened with saliva is formed into a mass, or bolus, as the tongue presses it against the hard palate. Further tongue pressure moves the bolus back into the oropharynx. The first step in under control of the cerebral cortex.

In order to propel food from the pharynx into the esophagus, three openings must be blocked: mouth, nasopharynx, and larynx. Continued elevation of the tongue seals off the mouth. The soft palate is elevated and tensed, causing the nasopharynx to be closed off. Food is denied entrance into the larynx by muscle action that causes the epiglottis to block this opening. The mechanism involves raising of the larynx, a process easily noted by palpation of the thyroid cartilage during swallowing. As a result, the bolus slips over the back of the epiglottis to enter the laryngopharynx. A combination of gravity and contractions of pharynx and esophagus compress the bolus into and through the esophageal tube. These steps are involuntary and under control of the deglutition center in the medulla. The presence of a bolus stimulates sensory receptors in the mouth and pharynx, thus initiating reflex pharyngeal contractions. Consequently, anesthesia of sensory nerves from the mucosa of the mouth and pharynx by a drug such as procaine (Novocain) makes swallowing impossible.

Swallowing is a complex process requiring the coordination of many muscles and other structures in the head and neck. Not only does the process occur smoothly, but it must also take place rapidly, since respiration is inhibited during 1 to 3 seconds for each swallowing, while food clears the pharynx.

Emptying the stomach. The process of emptying the stomach takes about 2 to 6 hours after a meal, depending on the contents of the meal. Gastric juices mixed with food form a milky white material known as *chyme,* which is ejected about every 20 seconds into the duodenum. Since the volume of the stomach is large and that of the duodenum is small, gastric emptying must be regulated to prevent overburdening of the duodenum. Such control occurs by two mechanisms, one hormonal and one nervous.

Table 18-3 Processes in the mechanics of digestion

Organ	Mechanical process	Nature of process
Mouth (teeth and tongue)	Mastication	Chewing movements—reduce size of food particles and mix them with saliva
	Deglutition	Swallowing—movement of food from mouth to stomach
Pharynx	Deglutition	
Esophagus	Deglutition	
	Peristalsis	Wormlike movements that squeeze food downward in tract; constricted ring forms first in one section, the next, etc., causing waves of contraction to spread along entire canal
Stomach	Churning	Forward and backward movement of gastric contents, mixing food with gastric juices to form chyme
	Peristalsis	Waves start in body of stomach about three times per minute and sweep toward closed pyloric sphincter; at intervals, strong peristaltic waves press chyme past sphincter into duodenum
Small intestine	Segmentation (mixing contractions)	Forward and backward (nonprogressive) movement within segment of intestine; purpose, to mix food and digestive juices thoroughly and to bring all digested food in contact with intestinal mucosa to facilitate absorption; purpose of peristalsis, on the other hand, to propel intestinal contents along digestive tract
	Peristalsis	
Large intestine Colon	Segmentation	Churning movements within haustral sacs
	Peristalsis	
Descending colon	Mass peristalsis	Entire contents moved into sigmoid colon and rectum; occurs three or four times a day, usually after a meal
Rectum	Defecation	Emptying of rectum, so-called bowel movement

Fats present in the duodenum stimulate the intestinal mucosa to release a hormone called *enterogastrone* into the bloodstream. When it reaches the stomach wall via the circulation, enterogastrone has an inhibitory effect on gastric muscle, decreasing its peristalsis and thereby slowing down passage of food into the duodenum. Nervous control results from receptors in the duodenal mucosa that are sensitive both to the presence of acid and to distention. Sensory and motor fibers in the vagus nerve then cause a reflex inhibition of gastric peristalsis. This mechanism is known as the *enterogastric reflex.*

Defecation. Defecation is a reflex brought about by stimulation of receptors in the rectal mucosa. Normally the rectum is empty until mass peristalsis moves fecal matter out of the colon into the rectum. This distends the rectum and produces the desire to defecate. Also, it stimulates colonic peristalsis and initiates reflex relaxation of the internal sphincter of the anus. Voluntary straining efforts and relaxation of the external

Table 18-4 Chemical digestion

Digestive juices and enzymes	Food enzyme digests (or hydrolyzes)	Resulting product*
Saliva Amylase (ptyalin)	Starch (polysaccharide or complex sugar)	Dextrins (short polysaccharides) and maltose (a disaccharide or double sugar)
Gastric juice Protease (pepsin: pepsinogen + hydrochloric acid)	Proteins, including casein	Proteoses and peptones (partially digested proteins)
Lipase (of little importance)	Emulsified fats (butter, cream, etc.)	**Fatty acids and glycerol**
Bile (contains no enzymes)	Large fat droplets (unemulsified fats)	Small fat droplets or emulsified fats
Pancreatic juice Protease (trypsin and chymotrypsin)†	Proteins (either intact or partially digested)	Proteoses, peptides, and **amino acids**
Lipase	Bile-emulsified fats	**Fatty acids and glycerol**
Amylase	Starch	Maltose
Intestinal juice: in brush border of cells of villi Peptidases‡	Peptides	**Amino acids**
Sucrase	Sucrose (cane sugar)	**Glucose and fructose** (simple sugars or monosaccharides)
Lactase	Lactose (milk sugar)	**Glucose and galactose** (simple sugars)
Maltase	Maltose (malt sugar)	**Glucose** (dextrose)

*Substances in boldface type are end products of digestion or, in other words, completely digested foods ready for absorption.
†Trypsin secreted in inactive form (trypsinogen) and converted to active form by trypsin itself and intestinal enzyme, enterokinase. Chymotrypsin secreted in inactive form (chymotrypsinogen) and converted to active form by action of trypsin and chymotrypsinogen.
‡Secreted in inactive form (pepsinogen); converted to active form (pepsin) by action of pepsin itself (autocatalysis) in acid environment provided by hydrochloric acid.

anal sphincter may then follow as a result of the desire to defecate. Together these several responses bring about defecation. Note that this is a reflex partly under voluntary control. If one voluntarily inhibits it, the rectal receptors soon become depressed and the urge to defecate usually does not recur until hours later, when mass peristalsis again takes place.

Constipation occurs when the contents of the lower colon and rectum move at a rate that is slower than normal. Extra water is absorbed from the fecal mass, producing a hardened or constipated stool. *Diarrhea* occurs as a result of increased motility of the small intestine. Chyme moves through the small intestine too quickly, reducing the amount of absorption of water and electrolytes there. The large volume of material arriving in the large intestine exceeds the quite limited capacity of the colon for absorption, so a watery stool results. Prolonged diarrhea can be particularly serious even fatal, in infants, since they have a minimal reserve of water and electrolytes.

Chemical digestion. Chemical digestion consists of all the changes in chemical composition that foods undergo in their travel through the alimentary canal. These changes result from the hydrolysis of foods. *Hydrolysis* is a chemical process in which a compound unites with water and then splits into simpler compounds. Numerous enzymes* present in the various digestive juices catalyze the hydrolysis of foods (Table 18-4).

---

*Enzymes are usually defined simply as "organic catalysts," that is, they are organic compounds, and they accelerate chemical reactions without appearing in the final products of the reaction. Enzymes are vital substances. Without them, the chemical reactions necessary for life could not take place. So important are they that someone has even defined life as the "orderly functioning of hundreds of enzymes."

Chemical structure: Enzymes are proteins. Frequently their molecules also contain a nonprotein part called the *prosthetic group* of the enzyme molecule (if this group readily detaches from the rest of the molecule, it is spoken of as the *coenzyme*). Some prosthetic groups contain inorganic ions (Ca^{++}, Mg^{++}, Mn^{++}, etc.). Many of them contain vitamins. In fact, every vitamin of known function constitutes part of a prosthetic group of some enzyme. Nicotinic acid, thiamine, riboflavin, and other B complex vitamins, for example, function in this way.

Classification and naming: Two of the systems used for naming enzymes are as follows: suffix *-ase* is used either with the root name of the substance whose chemical reaction is catalyzed (the substrate chemical, that is) or with the word that describes the kind of chemical reaction catalyzed. Thus according to the first method, sucrase is an enzyme that catalyzes a chemical reaction in which sucrose takes part. According to the second method, sucrase might also be called hydrolase because it hydrolyzes sucrose. Enzymes investigated before these methods of nomenclature were adopted still are called by older names, such as ptyalin, pepsin, and trypsin.

Classified according to the kind of chemical reactions catalyzed, enzymes fall into several groups:

1 *Oxidation-reduction enzymes.* These are known as oxidases, hydrogenases, and dehydrogenases. Energy release for muscular contraction and all physiological work depends on these enzymes.
2 *Hydrolyzing enzymes* or hydrolases. Digestive enzymes belong to this group. These are generally named after the substrate acted on, for example, lipase, sucrase, and maltase.
3 *Phosphorylating enzymes.* These add or remove phosphate groups and are known as phosphorylases or phosphatases.
4 *Enzymes that add or remove carbon dioxide.* These are known as carboxylases or decarboxylases.
5 *Enzymes that rearrange atoms within a molecule.* These are known as mutases or isomerases.
6 *Hydrases.* These add water to a molecule without splitting it, as hydrolases do.

Enzymes are also classified as intracellular or extracellular, depending on whether they act within cells or outside of them in the surrounding medium. Most enzymes act intracellularly in the body, an important exception being the digestive enzymes.

Properties: In general, the properties of enzymes are the same as those of proteins, since enzymes are mainly proteins. For example, they form colloidal solutes in water and are precipitated or coagulated by various agents, such as high temperatures and salts of heavy metals. Hence these agents inactivate enzymes. Other important enzyme properties are as follows.

1 Most enzymes are *specific in their action*, that is, act only on a specific substrate. This is attributed to a "key-in-a-lock" kind of action, the configuration of the enzyme molecule fitting the configuration of some part of the substrate molecule.
2 Enzymes *function optimally at a specific pH* and become inactive if this deviates beyond narrow limits.
3 A *variety of physical and chemical agents inactivate or inhibit enzyme action*, for example, x-rays and ionizing radiation (this presumably accounts for some of the ill effects of excessive radiation), certain antibiotic drugs, or unfavorable pH.
4 *Most enzymes catalyze a chemical reaction in both directions*, the direction and rate of the reaction being governed by the law of mass action. An accumulation of a product slows the reaction and tends to reverse it. A practical application of this fact is the slowing of digestion when absorption is interfered with and the products of digestion accumulate.
5 *Enzymes are continually being destroyed in the body* and therefore have to be continually synthesized, even though they are not used up in the reactions they catalyze.
6 *Many enzymes are synthesized* as inactive proenzymes. Substances that convert proenzymes to active enzymes are often called kinases, for example, enterokinase changes inactive trypsinogen into active trypsin.

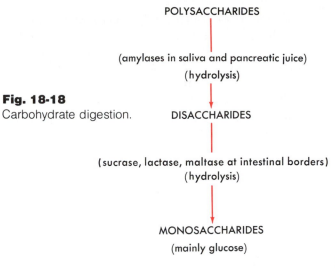

Fig. 18-18
Carbohydrate digestion.

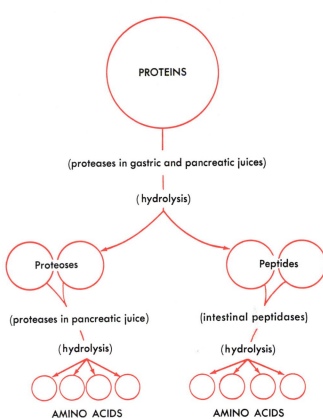

Fig. 18-19 Protein digestion. Gastric juice protease (pepsin) and pancreatic juice protease (trypsin and chymotrypsin) hydrolyze proteins to proteoses and peptides. Protein digestion is then completed by pancreatic proteases, which hydrolyze proteoses to amino acids, and intestinal peptidases, which hydrolyze peptides to amino acids.

Different enzymes require different hydrogen ion concentrations in their environment for optimal functioning. Ptyalin, the main enzyme in saliva, functions best in the neutral to slightly acid pH characteristic of saliva. It is gradually inactivated by the marked acidity of gastric juice. In contrast, pepsin, an enzyme in gastric juice, is inactive unless sufficient hydrochloric acid is present. Therefore in diseases characterized by gastric hypoacidity (pernicious anemia, for example), hydrochloric acid is given orally before meals.

Although we eat six kinds of chemical substances (carbohydrates, proteins, fats, vitamins, mineral salts, and water), only the first three named have to be chemically digested in order to be absorbed.

Carbohydrate digestion (Fig. 18-18). Carbohydrates are saccharide compounds. This means that their molecules contain one or more saccharide groups ($C_6H_{10}O_5$). Polysaccharides, notably starches and glycogen, contain many of these groups. Disaccharides (sucrose, lactose, and maltose) contain two of them, and monosaccharides (glucose, fructose, and galactose) contain only one. Polysaccharides are hydrolyzed to disaccharides by enzymes known as *amylases* found in saliva and pancreatic juice (salivary amylase is also called ptyalin). The enzymes that

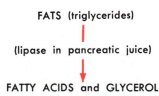

Fig. 18-20 Fat digestion (hydrolysis) by lipase, facilitated first by emulsion of fats by bile.

catalyze the final steps of carbohydrate digestion are *sucrase*, *lactase*, and *maltase*. These enzymes are located in the cell membrane of epithelial cells covering villi and therefore lining the intestinal lumen. The substrates (disaccharides) bind onto the enzymes at the surface of the brush border, giving the name *"contact digestion"* to the process. The resulting end products of digestion mainly glucose, are conveniently located at the site of absorption (and are not floating around somewhere in the lumen).

Protein digestion (Fig. 18-19). Protein compounds have very large molecules made up of entangled chains of amino acids, often hundreds in number. Enzymes called proteases catalyze the hydrolysis of proteins into intermediate compounds, for example, proteoses and peptides, and finally into amino acids. The main proteases are pepsin in gastric juice, trypsin in pancreatic juice, and peptidases in intestinal brush border.

Fat digestion (Fig. 18-20). Because fats are insoluble in water, they must be emulsified, that is, dispersed as very small droplets, before they can be digested. Bile emulsifies fats in the small intestine. This facilitates fat digestion by providing a greater contact area between fat molecules and pancreatic lipase, the main fat-digesting enzyme. For a summary of the actions of each digestive juice, see Table 18-4.

Residues of digestion. Certain components of food resist digestion and are eliminated from the intestines in the feces. These *residues of digestion* are cellulose from carbohydrates, undigested connective tissue and toxins from meat proteins, and undigested fats. In addition to these wastes, feces consist of bacteria, pigments, water, and mucus.

Control of digestive gland secretion

Digestive glands secrete when food is present in the alimentary tract or when it is seen, smelled, or imagined. Complicated reflex (nerve) and chemical (hormonal) mechanisms control the flow of digestive juices in such a way that they appear in proper amounts when and for as long as needed.

Saliva

As far as is known, only reflex mechanisms control the secretion of saliva. Chemical, mechanical, olfactory, and visual stimuli initiate afferent impulses to centers in the brain stem that send out efferent impulses to salivary glands, stimulating them. Chemical and mechanical stimuli come from the presence of food in the mouth. Olfactory and visual stimuli come, of course, from the smell and sight of food.

Gastric secretion

Stimulation of gastric juice secretion occurs in three phases controlled by reflex and chemical mechanisms. Because stimuli that activate these mechanisms arise in the head, stomach, and intestines, the three phases are known as the cephalic, gastric, and intestinal phases, respectively.

The *cephalic phase* is also spoken of as the psychic phase because psychic factors activate the mechanism. For example, the sight, smell, taste, or even thought of food that is pleasing to an individual stimulates various head receptors and thereby initiates stimulation of the gastric glands. Parasympathetic fibers in branches of the vagus nerve conduct the stimulating efferent impulses to the glands. The vagus also stimulates production of gastrin (discussed subsequently).

During the *gastric phase* of gastric juice secretion the following chemical control mechanism dominates. Products of protein digestion in foods that have reached the pyloric portion of the stomach stimulate its mucosa to release a hormone called *gastrin* into the blood in stomach capillaries. When it circulates to the gastric glands, gastrin greatly accelerates their secretion of gastric juice, which has a high pepsin and hydrochloric acid content (Table 18-5).

Table 18-5 Actions of some digestive hormones summarized

Hormone	Source	Action
Gastrin	Formed by gastric mucosa in presence of partially digested proteins	Stimulates secretion of gastric juice rich in pepsin and hydrochloric acid
Enterogastrone	Formed by intestinal mucosa in presence of fats	Inhibits gastric secretion and motility
Secretin	Formed by intestinal mucosa in presence of acid, partially digested proteins, and fats	Stimulates secretion of pancreatic juice low in enzymes and high in alkalinity (bicarbonate) Stimulates secretion of bile by liver
Cholecystokinin-pancreozymin (CCK-PZ)	Formed by intestinal mucosa in presence of fats, partially digested proteins, and acids	Stimulates ejection of bile from gallbladder and secretion of pancreatic juice high in enzymes

Hence this seems to be a device for ensuring that when food is in the stomach there will be enough enzymes there to digest it. Gastrin release is also stimulated by distention of the stomach (caused by the presence of food here) as well as by vagal nerve impulses to the pylorus.

The *intestinal phase* of gastric juice secretion is less clearly understood than the other two. A chemical control mechanism, however, is believed to operate. It is known, too, that the hormone *enterogastrone* (released by intestinal mucosa when fat is in the intestine) causes a lessening of both gastric secretion and motility.

Pancreatic secretion

Two hormones released by intestinal mucosa are known to stimulate pancreatic secretion. One of these, secretin,* evokes production of pancreatic fluid low in enzyme content, but high in bicarbonate. This alkaline fluid acts to neutralize the acid chyme entering the duodenum.

*Secretin has an interesting claim to fame. Not only was secretin the first hormone to be discovered, but its discovery gave rise to the broad concept of hormonal control of body activities.

As you might expect, the presence of acid in the duodenum serves as the most potent stimulator of secretin. (Additional control involving the same hormone is shown by the fact that fats in the duodenum also elicit secretin, which then influences the liver to increase its output of the fat emulsifier bile.)

The other intestinal hormone, known as cholecystokinin-pancreozymin (CCK-PZ) was originally thought to be two separate substances. It has now been identified as one chemical with two functions: It causes the pancreas to increase its exocrine secretions, high in enzyme content, and also stimulates contraction of the gallbladder so that bile can pass into the duodenum.

Secretion of bile

The hormones secretin and CCK-PZ, as just described, stimulate secretion of bile by the liver and ejection of bile from the gallbladder (Table 18-5).

Intestinal secretion

Relatively little is known about the regulation of intestinal secretions. Some evidence suggests that the intestinal mucosa, stimulated by hydrochloric acid and food products, releases into the blood a hormone, *enterocrinin*, which brings about increased production of digestive enzymes within intestinal mucosa. Presumably, neural mechanisms also help control the secretion of this digestive juice.

Absorption

Definition

Absorption is the passage of substances (notably digested foods, water, salts, and vitamins) through the intestinal mucosa into the blood or lymph.

How accomplished

Absorption of some substances may occur on the basis of the physical laws of diffusion, osmosis, and filtration alone. However, some substances depend on more complex mechanisms in order to be absorbed. As an example, imagine for a moment that you are a glucose molecule. You are waiting outside the brush border membrane of an intestinal mucosa cell. You have just completed the (exhausting) process of digestion, so you are considered an "end product of digestion." But will you really be permitted to enter? Do they need glucose inside that cell? (Maybe they are full.) Are you small enough to pass across the membrane and into that cell? (Actually you are a little large, so you'll need some tugboat assistance.) But are you suitable otherwise? Recall that the outer layer of that cell membrane facing you is lipid (Fig. 2-3, p. 32). But alas, only lipid-soluble (hydrophobic) molecules of your size could freely (passively) enter the lipid cell barrier. Well, remember your motto: "When faced with a barrier, call on a carrier." You must cloak yourself in a covering that is lipid soluble. You may thus conceal your hydrophilic nature within the confines of the en-

folding tugboat-carrier, which swiftly spirits you across the membrane, delivering you safely to the interior of the cell!

The above (real-life) fantasy of glucose absorption describes the active transport of this molecule utilizing a carrier. This energy-requiring process can move glucose into cells already concentrated in glucose, that is, against a concentration gradient. Fructose, another product of carbohydrate digestion, as well as amino acids, may be absorbed by facilitated diffusion (p. 43). By such a process, these molecules require carrier assistance but do not move against concentration gradients. Simple sugars and amino acids eventually pass through mucosal cells to reach blood capillaries in the villus.

Fatty acids (products of fat digestion) and cholesterol are transported with the aid of bile salts from the watery intestinal lumen to absorbing cells on villi. Bile salts form *micelles* that surround the lipid, making it temporarily water soluble. As micelles approach the brush border of absorbing cells, lipids are released to pass through the cell membrane (since its lipid bilayer is receptive to lipids) by simple diffusion. Once inside the cell, fatty acids are rapidly reunited with glycerol to form triglycerides (neutral fats). The final step in lipid transport by the intestine is the formation of *chylomicrons*, which are composed mainly of neutral fats and some cholesterol covered by a delicate protein envelope. This important envelope allows fats to be transported through lymph and into the bloodstream (Table 18-6).

Vitamins A, D, E, and K, known as the "fat-soluble vitamins," also depend on bile salts for their absorption. Many water-soluble vitamins, such as certain of the B group, are small enough to be absorbed by simple diffusion. Most drugs (sedatives, analgesics, antibiotics) appear to be absorbed by simple diffusion also.

Active transport mechanisms seem to be available for water and salt absorption. Certain salts (highly ionized ones) cannot be absorbed into intestinal cells. Magnesium sulfate (epsom salts),

Table 18-6 Food absorption

Form absorbed	Structures into which absorbed	Circulation
■ Protein—as amino acids Perhaps minute quantities of some whole proteins absorbed, for example, some antibodies	Blood in intestinal capillaries	Portal vein, liver, hepatic vein, inferior vena cava to heart, etc.
■ Carbohydrates—as simple sugars	Same as amino acids	Same as amino acids
■ Fats Glycerol	Lymph in intestinal lacteals	During absorption, that is, while in epithelial cells of intestinal mucosa, glycerol and fatty acids recombine to form microscopic particles of fats (chylomicrons); lymphatics carry them by way of thoracic duct to left subclavian vein, superior vena cava, heart, etc.; some fats transported by blood in form of phospholipids or cholesterol esters
Fatty acids combine with bile salts to form water-soluble substance	Lymph in intestinal lacteals	
Some finely emulsified undigested fats absorbed	Small fraction enters intestinal blood capillaries	

for example, is not absorbed from the intestine even though its molecules are smaller than glucose molecules. In fact, it is this nonabsorbability of magnesium sulfate that makes it an effective cathartic. (Since magnesium sulfate ions do not diffuse freely through the intestinal mucosa, do you think their presence in intestinal fluid would create an osmotic pressure gradient between the intestinal fluid and blood? If so, what effect would this have on water movement between these two fluids? Reread pp. 43-48 if you need help in answering these questions.)

Note that after absorption food does not pass directly into the general circulation. Instead it first travels by way of the portal system to the liver. During intestinal absorption, blood entering the liver via the portal vein contains greater concentrations of glucose, amino acids, and fats than does blood leaving the liver via the hepatic vein for the systemic circulation. Clearly the excess of these food substances over and above the normal blood levels has remained behind in the liver.

What the liver does with them is part of the story of metabolism, our next topic for discussion.

Outline summary

Functions and importance

A Prepare food for absorption and metabolism
B Absorption
C Elimination of wastes
D Vital importance

Organs

A Main organs—compose alimentary canal: mouth, pharynx, esophagus, stomach, and intestines
B Accessory organs—salivary glands, teeth, liver, gallbladder, pancreas, and vermiform appendix

Walls of organs

A Coats
  1 Mucous lining
  2 Submucous coat of connective tissue—main blood vessels here
  3 Muscular coat
  4 Fibroserous coat

B Modifications of coats
  1 Mucous lining
    a Rugae and microscopic gastric and hydrochloric acid glands in stomach
    b Circular folds, villi, intestinal glands, Peyer's patches, and solitary lymph nodes in small intestine
    c Solitary nodes and mucous glands in large intestine
  2 Muscle coat
    a Three layers (circular, longitudinal, oblique) in stomach instead of only two layers as in rest of tract
    b Three tapelike strips in outer, longitudinal layer pucker large intestine into small sacs called haustra
  3 Fibroserous coat
    a Peritoneum covers stomach and intestines
    b Greater omentum or lace apron—double fold of peritoneum that hangs from lower edge of stomach like an apron over intestines; should not be confused with mesentery, which also is double fold of peritoneum but fan shaped and attached at short side to posterior wall of abdominal cavity; small intestines anchored to posterior abdominal wall by means of mesentery

Mouth (buccal cavity)

A Formed by cheeks, hard and soft palates, tongue, and muscles
B Hard palate—formed by parts of two palatine bones and two maxillary bones
C Soft palate—formed of muscle in shape of arch; forms partition between mouth and nasopharynx; fauces is archway or opening from mouth to oropharynx; uvula is conical-shaped process suspended from midpoint of arch
D Tongue
  1 Many rough elevations on tongue's surface called papillae; contain taste buds
  2 Frenulum—fold of mucous membrane that helps anchor tongue to mouth floor

Salivary glands

A Parotid—below and in front of ear; duct opens on inside of cheek, opposite upper second molar tooth
B Submandibular—posterior part of mouth floor
C Sublingual—anterior part of mouth floor

Teeth

A Deciduous or baby teeth—10 per jaw or 20 in set
B Permanent—16 per jaw or 32 per set
C Structure of typical tooth—see Fig. 18-5

Pharynx

See p. 443.

Esophagus

A Position and extent
  1 Posterior to trachea and heart; pierces diaphragm
  2 Extends from pharynx to stomach, distance of approximately 10 inches
B Structure—collapsible, muscle tube

Stomach

A Size, shape, and position
  1 Size varies in different individuals; also according to whether distended or not
  2 Elongated pouch
  3 Lies in epigastric and left hypochondriac portions of abdominal cavity
B Divisions
  1 Fundus—portions above esophageal opening
  2 Body—central portion
  3 Pylorus—constricted, lower portion
C Curves
  1 Lesser—upper, right border
  2 Greater—lower, left border
D Sphincter muscles
  1 Cardiac—guarding opening of esophagus into stomach
  2 Pyloric—guarding opening of pylorus into duodenum
E Coats—see Table 18-1
F Glands
  1 Epithelial cells of gastric mucosa secrete mucus
  2 Parietal cells secrete hydrochloric acid and intrinsic factor
  3 Chief cells (zymogenic cells) secrete enzymes of gastric juice
G Functions
  1 Serves as food reservoir
  2 Secretes gastric juice
  3 Contractions break food into small particles, mix them well with gastric juice, and move contents on into duodenum
  4 Secretes the antianemic intrinsic factor
  5 Carries on a limited amount of absorption—some water, alcohol, and certain other drugs
  6 Secretes hormone gastrin

Small intestine

A Size and position
  1 Approximately 2.5 cm in diameter and 6 m in length
  2 Its coiled loops fill most of abdominal cavity
B Divisions
  1 Duodenum
  2 Jejunum
  3 Ileum
C Coats—see Table 18-1; note villi containing capillaries and lacteal and covered with epithelial cells having microvilli (brush border); greatly increase surface area for digestion and absorption
D Functions
  1 Completes digestion of foods
  2 Absorbs end products of digestion
  3 Secretes hormones that help control secretion of pancreatic juice, bile, and intestinal juice

Large intestine (colon)

A Size—approximately 6 cm in diameter and 1.5 to 1.8 m in length
B Divisions
  1 Cecum—vermiform appendix is blind-end tube off cecum; size and shape of large angleworm
  2 Colon
    a Ascending
    b Transverse
    c Descending
    d Sigmoid
  3 Rectum
C Coats—see Table 18-1
D Functions
  1 Absorption of water
  2 Elimination of digestive wastes

Peritoneum

A Extension of fibroserous coat over abdominal organs (visceral layer) and wall (parietal layer)
B Mesentery is fan shaped, attaches small intestine to posterior abdominal wall; transverse mesocolon transverse colon to posterior abdominal wall; greater omentum, or lace apron, hangs from stomach over intestines and attaches to transverse colon; lesser omentum connects stomach to liver; falciform ligament connects liver to anterior abdominal wall

Liver

A Location and size
  1 Occupies most of right hypochondrium and part of epigastrium
  2 Largest gland in body, weighs about 1.5 kg
B Lobes and lobules
  1 Right lobe, subdivided into three smaller lobes—right lobe proper, caudate, and quadrate
  2 Left lobe
  3 Lobes divided into lobules by blood vessels and fibrous partitions
  4 Lobules composed of plates of hepatic cells radiating from central vein; portal and hepatic artery blood flows through sinusoids to central vein; bile collects in tiny ducts
C Ducts
  1 Hepatic duct from liver
  2 Cystic duct from gallbladder
  3 Common bile duct formed by union of hepatic and cystic ducts and opens into duodenum at major duodenal papilla
D Functions
  1 Secretes bile
  2 Plays essential role in metabolism of carbohydrates, proteins, and fats, for example, carries on glycogenesis, glycogenolysis, and gluconeogenesis; also deamination and ketogenesis; synthesizes various blood proteins; detoxifies various substances; stores iron, vitamins A, B_{12}, and D

Gallbladder

A Size, shape, and location
  1 Approximately size and shape of small pear
  2 Lies on undersurface of liver
B Structure—sac of smooth muscle with mucous lining arranged in rugae
C Functions
  1 Concentrates and stores bile
  2 During digestion, ejects bile into duodenum

Pancreas

A Size, shape, and location
  1 Is 12 to 15 cm long, weighs 60 gm
  2 Shaped something like a fish with head, body, and tail
  3 Lies in C-shaped curve of duodenum
B Structure—similar to salivary glands
  1 Divided into lobes and lobules
  2 Pancreatic cells pour their secretion into duct that runs length of gland and empties into duo-

denum at major duodenal papilla; may be accessory duct

    3 Clusters of cells, not connected with any ducts, lie between pancreatic cells—called pancreatic islets or islands of Langerhans—composed of alpha and beta type cells

C Functions

    1 Acinar units secrete pancreatic juice

    2 Beta cells of islands of Langerhans secrete insulin

    3 Alpha cells of islands of Langerhans secrete glucagon

Digestion

A Definition—all changes food undergoes in alimentary canal

B Purpose—conversion of foods into chemical and physical forms that can be absorbed and metabolized

C Kinds

    1 Mechanical—movements that change physical state of foods, facilitate absorption, propel food forward in alimentary tract, and eliminate digestive wastes from tract (see Table 18-3 for description of processes involved in mechanical digestion)

      a Mastication (chewing)

      b Swallowing (deglutition)

        (1) Movement of food through mouth into pharynx—voluntary act

        (2) Movement of food through pharynx into esophagus—involuntary or reflex act initiated by stimulation of mucosa of back of mouth, pharynx, or laryngeal region; paralysis of receptors here, for example, by procaine (Novocain), makes swallowing impossible

        (3) Movement of food through esophagus into stomach; accomplished by esophageal peristalsis—reflex initiated by stimulation of esophageal mucosa

      c Peristalsis—wormlike movements that squeeze food downward in tract

      d Emptying of stomach—gastric peristalsis inhibited by two mechanisms: fats in chyme entering duodenum evoke intestinal hormone enterogastrone; acid and distention of duodenum stimulate enterogastric vagal reflex;

both mechanisms act to prevent too rapid emptying of stomach

      e Churning, segmentation, and mixing contractions

      f Mass peristalsis in colon

      g Defecation—reflex initiated by stimulation of rectal mucosa

    2 Chemical—series of hydrolytic processes dependent on specific enzymes (see Table 18-4 and Figs. 18-18 and 18-19 for description of chemical changes)

D Control of digestive gland secretion—see Table 18-5

    1 Saliva—secretion is reflex initiated by stimulation of taste buds, other receptors in mouth and esophagus, olfactory receptors, and visual receptors

    2 Gastric secretion—controlled reflexly by same stimuli that initiate salivary secretion; also controlled chemically by hormone, gastrin, released by pyloric mucosa in presence of partially digested proteins; enterogastrone (hormone just mentioned as slowing stomach emptying) also has inhibitory effect on gastric secretion

    3 Pancreatic secretion—controlled chemically by hormones secretin and CCK-PZ formed by intestinal mucosa especially when fats or acid enters duodenum

    4 Bile

      a Secretion—controlled chemically by same hormone (secretin) that regulates pancreatic secretion

      b Ejection into duodenum—contolled chemically by hormone CCK-PZ

    5 Intestinal secretion—control still obscure, although believed to be both reflex and chemical

Absorption

A Definition—passage of substances through intestinal mucosa into blood or lymph

B How accomplished—cell membrane consists of lipid bilayer, so lipids diffuse passively across and into interior of absorptive cells; water-soluble substances (hydrophilic) like sugars or amino acids use carriers to help them enter cells; some absorptive processes require energy (active transport); others do not (facilitated diffusion)

Review questions

1 Name and describe the coats that compose the walls of the esophagus, stomach, and intestines.
2 Differentiate between the peritoneum, the mesentery, and the omentum.
3 Give the names and number of deciduous teeth and of permanent teeth. Tell which is the first permanent tooth to erupt.
4 Discuss the functions of gastric juice.
5 What is chyme?
6 What juices digest proteins? Carbohydrates? Fats?
7 What functions does the liver perform? Which of these are vital functions?
8 What digestive functions does the pancreas perform?
9 What is the purpose of digestion?
10 Compare absorption of the three kinds of foods.
11 The doctor tells a patient that he has obstructive jaundice. Which duct or ducts might be obstructed to produce the symptom jaundice? Explain.
12 With obstructive jaundice, would digestion or absorption or both be affected? Explain.
13 What vitamin might the doctor prescribe for a patient with obstructive jaundice? Why?
14 Differentiate chief cells from parietal cells; villi from microvilli; exocrine from endocrine; cystic duct from hepatic duct; trypsin from pepsin; enterokinase from enterogastrone.
15 If you were a piece of indigestible cellulose, what would be the route you would take from mouth to anus? List the organs in the order you would pass through them.
16 Name each hormone that controls production of bile; ejection of bile; stimulation of gastric enzymes; inhibition of gastric emptying; secretion of alkaline fluid from pancreas; secretion of pancreatic enzymes.

Metabolism

Foods are first digested, then absorbed, and finally metabolized. Metabolism means the changes absorbed foods undergo within cells, or, more briefly, metabolism is the utilization of foods. This chapter begins by defining the major phases of metabolism and stating some important principles about it. It then goes on to give basic information about metabolism of the three kinds of foods. It also discusses metabolic rates, regulation of food intake, and homeostasis of body temperature.

Definitions

Metabolism consists of two major phases, catabolism and anabolism. *Catabolism* is a decomposition process in which relatively large food molecules are broken down to yield smaller molecules and energy. *Anabolism* is a synthesis process that uses energy to build relatively small molecules up into larger molecules, for example, enzymes, hormones, and antibodies.

Some important generalizations about metabolism

If you are studying physiology for the first time, you may find it difficult to fit together the many details of metabolism to form a clear picture of this complex process. We hope that the following generalizations will help you accomplish this challenging task.

1 Catabolism and anabolism both consist of many specific sequences of chemical reactions (sometimes called metabolic pathways) that go on continually and concurrently inside cells. Certain sequences of metabolic reactions occur in the utilization of all three kinds of foods, so are called common pathways.

2 Enzymes catalyze both catabolic and anabolic chemical reactions. Because metabolic enzymes are not only synthesized in cells but also function in them, they are called intracellular enzymes. (Digestive enzymes, in contrast, are extracellular enzymes because, although they are made in cells, they function outside of them.)

3 Energy changes accompany metabolic reactions, the reactions of both catabolism and anabolism. Whereas energy release accompanies

catabolism, energy use accompanies anabolism. (See Chapter 1, p. 7 for more about this.) Catabolism releases energy in two forms, heat and chemical energy. The relatively large amount of heat generated in catabolism is released in frequent small spurts, not in one large burst. (Imagine what would happen to living cells if a large amount of heat were released at one time. It would hard-boil them!) Heat is not utilizable for driving biological reactions, so is useless for anabolism or other kinds of cell work. In contrast, chemical energy released by catabolism is utilizable but cannot be used directly for biological reactions. It must first be transferred to high-energy bonds (~) of adenosine triphosphate (ATP) molecules. High-energy bonds bear this name because they store a higher amount of

Fig. 19-1 ATP/ADP system and its role in metabolism (explained on p. 508).

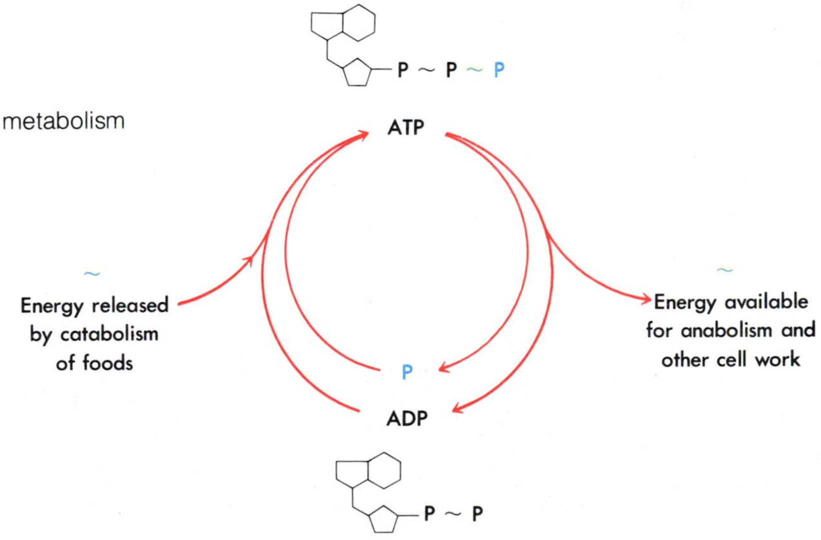

energy than ordinary chemical bonds. In addition, they are more labile (easily broken).

4 ATP is one of the most important compounds in the world. Why? Because it supplies energy directly to the energy-using reactions of all cells in all kinds of living organisms from one-celled plants to billion-celled humans. ATP functions as the universal biological currency. It pays the energy bills for all cells and is as important in the world of cells as money is in the world of contemporary society.

Look now at Fig. 19-1. The structural formula at the top of the diagram shows three phosphate groups attached to the rest of the ATP molecule, two of them by high-energy bonds. The breaking of the last one of these bonds yields a phosphate group (P), adenosine diphosphate (ADP), and energy, which, as the diagram indicates, is used for anabolism and other cell work. The diagram also shows that P and ADP then use energy released by catabolism to recombine and form ATP. This cycle is called the ATP/ADP system.

5 All cells are not created equal as far as metabolism is concerned. Different metabolic reactions take place in some cells than occur in others. In liver cells, for example, certain anabolic reactions synthesize blood proteins and various other compounds that are not made in any other kinds of cells. Metabolism also goes on at different rates in different kinds of cells. In general, the more active the cell, the higher its metabolic rate.

Carbohydrate metabolism

Carbohydrates serve as the primary source of energy in the human diet. Very little carbohydrate in absorbed food is stored or used for structural purposes. Instead, most is quickly catabolized for the release of energy. Carbohydrate foods, particularly glucose, serve as the "preferred energy fuel" of human cells. That is to say,

cells obtain their energy first from glucose—for as long as it continues to enter them—then next from fats, and last from proteins.

Carbohydrate metabolism starts with the movement of glucose through cell membranes. Immediately on reaching the interior of a cell, glucose reacts with ATP to form glucose-6-phosphate. This step, which activates the glucose for further reactions, is called phosphorylation of glucose and is catalyzed by the enzyme glucokinase. This particular reaction is reversible in only a few cells: in intestinal mucosa cells, liver cells (Fig. 19-10), and kidney tubule cells. In these cells a phosphatase enzyme is present that removes the phosphate and allows free glucose to leave the cell. But in most body cells, and this is especially important in brain and muscle cells, phosphatase is lacking. As a result, glucose, once inside the cell, is "trapped" as glucose-6-phosphate. This molecule may then be stored temporarily (notice that it is on the anabolic pathway to glycogen formation, Fig. 19-8), or it may be immediately catabolized to release energy (see glucose-6-phosphate in glycolysis, Fig. 19-3).

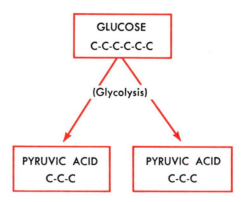

Fig. 19-2 Glycolysis. Glucose, a molecule containing 6 carbon atoms, is split by a series of chemical reactions into 2 molecules of pyruvic acid, each of which contains only 3 carbon atoms. Fig. 19-3 shows the series of intermediate reactions in brief form.

Glucose catabolism

Three successive processes or phases are involved in the total catabolism of glucose: glycolysis, citric acid cycle, and a third and final phase we shall call phase 3. Phase 3 consists of electron transport coupled to oxidative phosphorylation. The first step, glycolysis, does not use oxygen and so is called the anaerobic phase of glucose catabolism. The other steps do require oxygen and are collectively called the aerobic phase of catabolism, or cellular respiration.

Glycolysis. Stated very briefly, glycolysis is the process that breaks apart one molecule of glucose to form two molecules of pyruvic acid (Fig. 19-2). Actually, glycolysis consists of a series of about 10 reactions, most of them shown in Fig. 19-3. The specific chemical changes, however, are not the most important facts to remember about glycolysis. The most important facts to keep in mind are the following.

1 Glycolysis is a method of providing cells with energy when their oxygen supply is inadequate or even absent. During strenuous exercise, therefore, glycolysis becomes especially important. Respiration and circulation cannot then deliver as much oxygen to muscle cells as they need to use for the aerobic (oxygen-using) reactions of catabolism. In short, muscle cells must incur an "oxygen debt." Hyperventilation occurs immediately following exercise to pay this oxygen debt by providing enough oxygen to convert the stored lactic acid back to pyruvic acid, a compound that can be further catabolized in the citric acid cycle.

2 Glycolysis is relatively unimportant for providing cells with energy, since it releases only about 5% of the energy in glucose; 95% of the energy still remains locked in the chemical bonds of pyruvic acid. Glycolysis is important, however, as preparation for the final phases of catabolism. The relatively large molecules of glucose must first be converted into the smaller molecules of pyruvic acid in order to enter the citric acid cycle.

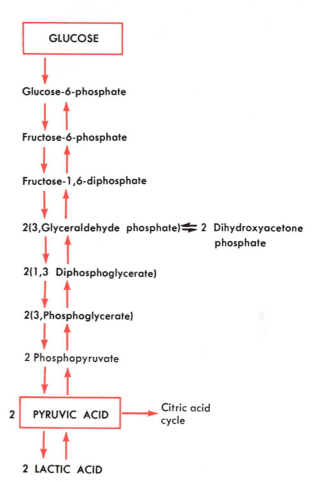

Fig. 19-3 Glycolysis showing intermediate products formed.

Citric acid cycle. The citric acid cycle is the second phase in the catabolism of glucose. Citric acid, the first chemical formed, is an acid with three carboxyl groups. This accounts for the names citric acid cycle and tricarboxylic acid (TCA) cycle for this sequence of reactions. An additional name, Krebs' cycle, commemorates the brilliant work of Sir Hans Krebs, which merited him the Nobel Prize for postulating this pathway (1937).

The citric acid cycle continues carbohydrate catabolism by converting pyruvic acid to carbon

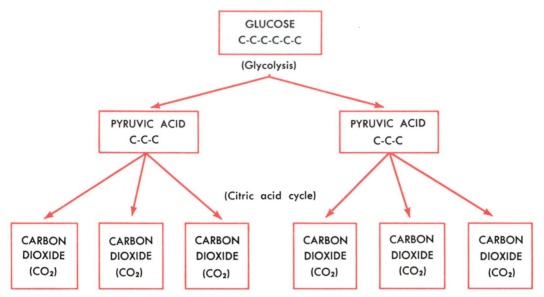

Fig. 19-4 Scheme to show that catabolism breaks larger molecules down into smaller ones. *Glycolysis* splits 1 molecule of glucose (6 carbon atoms) into 2 molecules of pyuvic acid (3 carbon atoms each). The citric acid cycle converts each pyruvic acid molecule to 3 carbon dioxide molecules.

dioxide (CO_2). Fig. 19-4 shows this in an abbreviated form. If you recall the chemical formula of glucose ($C_6H_{12}O_6$), you will note that the conversion of glucose to CO_2 requires that much hydrogen be removed from the carbohydrate. A look at a more detailed representation of the citric acid cycle in Fig. 19-5 will reveal five reactions in which hydrogen (H) atoms are removed. Although the cycle presents a sequence of complicated events, the essence of it is in the removal of H atoms that will be used in phase 3 for the all-important purpose of forming ATP. The CO_2 formed by the cycle has important functions in acid-base balance and in controlling respiration. But otherwise the CO_2 can be considered just a by-product of cell metabolism. ATP, resulting from the use of those removed H atoms, will be the real protagonist of this catabolic drama.

The importance of the cycle warrants brief further discussion of the steps involved. At the "entrance" to the citric acid cycle, pyruvic acid

is changed to acetylcoenzyme A (called acetyl-CoA). This is shown in reaction 1 in Fig. 19-5. The acetyl-CoA results from removal of CO_2 and H atoms and the addition of a molecule of coenzyme A (CoA). Coenzyme A activates the molecule, acetyl-CoA, launching it into the citric acid cycle. Step 2 shows acetyl-CoA combining with the 4-carbon oxaloacetic acid to form citric acid (6 carbons). A detailed study of steps 3 to 10 is not necessary here. The overall picture of the cycle should be noted, however. It shows the sequential rearrangement and "disassembly" of citric acid by the removal of H and CO_2 to ultimately regenerate 4-carbon oxaloacetic acid. The cycle may then begin again.

Electron transport system and oxidative phosphorylation (phase 3). The third and final phase in the catabolism of glucose climaxes the efforts of the first two phases, making it all worthwhile! Hydrogen atoms are removed during the citric acid cycle in an ionized form: as hydrogen ions (H^+) and electrons (e^-). The electrons are

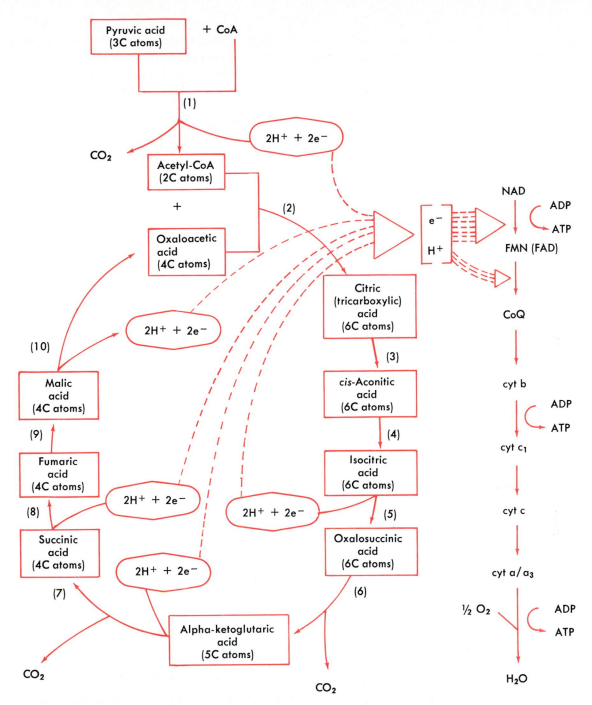

Fig. 19-5 The citric acid cycle and phase 3. The citric acid cycle releases hydrogen ions (H^+) and electrons (e^-). These pass along the electron transport system, releasing energy to form ATP. The citric acid cycle is shown to left of dividing line. Phase 3 (electron transport system and oxidative phosphorylation) is shown at right. Note that CO_2 is a byproduct of citric acid cycle. H_2O and ATP result from phase 3.

shuttled down a chain of carriers, called the electron transport system (ETS), much like buckets of water passed along a bucket brigade. In this case, however, the carriers (named nicotinamide adenine dinucleotide [NAD], flavin mononucleotide [FMN], or flavin adenine dinucleotide [FAD], coenzyme Q, and cytochromes) pick up some of the energy released by the electrons. At three points along the carrier chain the energy is trapped in ATP atoms (Fig. 19-5). Oxygen serves as the final acceptor of the electrons and hydrogen ions to form water. Although oxygen enters in the final scene, its role is vital in enticing electrons along the chain and therefore ensuring ATP formation. The lengthy name "oxidative phosphorylation" describes the formation of ATP at the three points along the electron transport chain. The "oxidative" indicates the role of oxygen in pulling electrons along the chain so that some of their energy can be drawn off. The energy is, of course, stored as the high energy phosphate bond ($\sim$P) added to adenosine diphosphate (ADP). The formation of ATP in this way is truly a "phosphorylation" process.

Summary. Summarizing the changes of glucose catabolism in equation form:

Glycolysis:

Glucose $\rightarrow$ 2 Pyruvic acid + 2 ATP + Heat

Citric acid cycle, and phase 3:

2 Pyruvic acid + 6 O_2 $\rightarrow$
6 CO_2 + 6 H_2O + 34 ATP + Heat

We can even summarize the long series of chemical reactions in glucose catabolism with one short equation:

$C_6H_{12}O_6$ + 6 O_2 $\rightarrow$ 6 CO_2 + 6 H_2O + 36 ATP + Heat
Glucose

Fig. 19-6 Catabolism releases energy stored in chemical bonds of glucose molecules. 180 gm or 1 mole of glucose contains 690 kilocalories (kcal) of stored energy. Catabolism releases about 310 kcal of this energy as heat and puts about 380 kcal back in storage in high-energy bonds of 36 moles of ATP. Thus the efficiency of glucose catabolism as a mechanism for supplying cells with usable energy is about 55% (380/690). Efficiency of 20% to 25% is typical for machines.

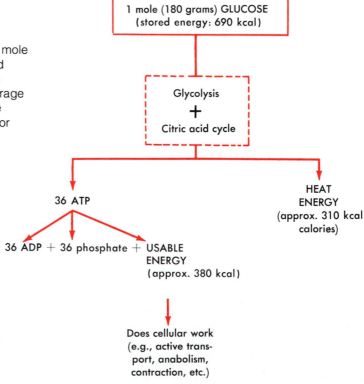

A summary of the energy changes within the entire process might be:

$$\text{Energy in glucose} \xrightarrow{\text{(glycolysis, citric acid cycle)}} \text{ATP} + \text{Heat}$$

The last equation emphasizes that energy originally present in the stable, relatively low energy chemical bonds of glucose was released in the form of hydrogen ions and electrons removed during glycolysis and the citric acid cycle. The stepwise transfer of the electrons along carriers allowed the capturing of energy from the electrons into "packages" usable by the cell, that is, in the unstable, high energy bonds of ATP. Fig. 19-6 shows that at least half of the energy in glucose is captured in ATP and available for cell work, with the remainder lost as heat. Few machines compare so favorably with the efficiency of the cell's energy-supplying mechanism.

The arrangement of catabolic "machinery" within the cell, as it is currently viewed, is shown in Fig. 19-7. Glycolytic enzymes present in the cytoplasm produce pyruvic acid, which diffuses into mitochondria. The enzymes of the citric acid cycle have been localized mostly in the soluble matrix. The hydrogen ions and electrons then pass to the "sphere"-studded cristae of the inner membrane, where the electron transport carriers and mechanism for phosphorylation are located. Since so many of the cell's energy-releasing enzymes are located within the mitochondria, these tiny structures are aptly described as the "power plants" of the cell.

This complex organization, a marvel in miniaturization, can boast also of its amazing internal control. Many researchers are concentrating on the regulation of catabolism, that is, how cells know how fast to catabolize. The control seems

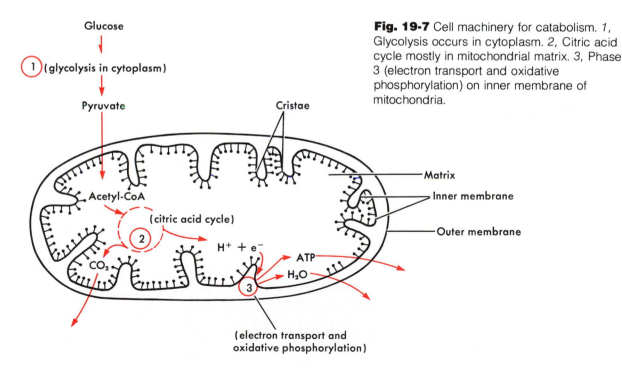

Fig. 19-7 Cell machinery for catabolism. *1*, Glycolysis occurs in cytoplasm. *2*, Citric acid cycle mostly in mitochondrial matrix. *3*, Phase 3 (electron transport and oxidative phosphorylation) on inner membrane of mitochondria.

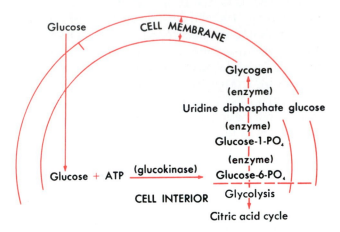

Fig. 19-8 Glycogenesis. Conversion of glucose-6-PO_4 into glycogen occurs when glycolytic and citric acid cycle pathways are saturated.

to lie in a negative feedback mechanism. In other words, a product accumulating in sufficient amounts signals the pathway to slow down. In the glycolytic and citric acid pathways, certain enzymes early in the pathways (shown in Figs. 19-3 and 19-5) act as "pacemakers." They are sensitive to and inhibited by the buildup of excess amounts of citric acid and ATP. Exactly how citric acid and ATP excesses notify the enzymes at these steps will be a matter for investigation for some time.

In looking back over the cells' accomplishments in catabolism, we are struck by a sense of wonder, well expressed by Wayne Becker*: "Respiratory metabolism can be regarded as a marvel of design and engineering. No transis-

*From Becker, W. M.: Energy and the living cell, Philadelphia, 1977, J. B. Lippincott Co., pp. 125-126.

Fig. 19-9 Homeostasis of blood glucose level. When blood glucose level starts to decrease toward lower normal, liver cells increase the rate at which they convert glycogen, amino acids, and fats to glucose (glycogenolysis and gluconeogenesis) and release it into blood. But when blood glucose level increases, liver cells increase the rate at which they remove glucose molecules from blood and convert them to glycogen for storage (glycogenesis). At still higher levels, glucose leaves blood for tissue cells to be anabolized into adipose tissue and at still higher levels is excreted in the urine.

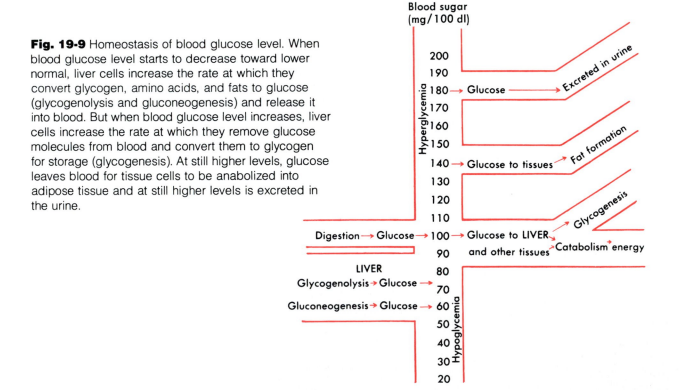

tors, no mechanical parts, no noise, no pollution—and all done in units of organization that require an electron microscope to visualize. Yet the process goes on routinely and continuously in almost every living cell with a degree of integration, efficiency, fidelity, and control that we can scarcely understand well enough to appreciate, let alone aspire to reproduce in our testtubes."

Glycogenesis

Imagine what happens in a cell if glucose catabolism is proceeding at maximum rate. What do you do if you see a traffic jam ahead, with no possible way to get through it? Probably take an alternate route, if available. Similarly, if glycolytic pathways are "saturated" because of high levels of glucose entering the cell, a "traffic jam" of glucose-6-phosphate will result. Unable to enter glycolysis, glucose-6-phosphate will begin an alternate route, that is, it will enter the anabolic pathway of glycogen formation. This process, called *glycogenesis* (Fig. 19-8), is a series of chemical reactions that may appear complex. But glycogen turns out to be just a necklace-like structure made up completely of stored glucose "beads."

The process of glycogenesis is part of a homeostatic mechanism that operates when the blood glucose level increases above the midpoint of its normal range (80 to 100 mg per 100 ml of blood), as indicated in Fig. 19-9. Example: Soon after a meal, while glucose is being absorbed rapidly, blood is quickly shunted to the liver via the portal system. Here a great many glucose molecules leave the blood for storage as glycogen. As a result of glycogenesis, blood glucose level decreases, ordinarily enough to reestablish its normal level. Muscle cells, like liver cells, have a high rate of glycogenesis. Certain cells, however, cannot carry out the process at all. Brain cells, for instance, are especially vulnerable, since they cannot form or store glycogen and are therefore dependent on a constant blood glucose supply.

Glycogenolysis

Glycogen molecules do not remain permanently in the cell but are eventually broken apart (hydrolyzed). This process of "splitting glycogen" is called glycogenolysis and is, in essence, a reversal of glycogenesis (Fig. 19-10). What are the products of glycogenolysis? That depends on the cell. Although all cells presumably have the enzymes to break glycogen to glucose-6-phosphate, only a few cell types (liver, kidney, intestinal mucosa) have phosphatase, which allows free glucose to form and possibly leave the cell. So the term glycogenolysis means different things in different cells. In muscles, glucose-6-phosphate is the product, which then undergoes glycolysis. But liver glycogenolysis (Fig. 19-10) results in free glucose that can leave the cell and increase the blood glucose level. Accordingly, liver glycogenolysis acts as a part of the homeostatic mechanism to maintain blood glucose level. Example: A few hours after a meal, when the blood glucose level decreases, liver glycogenolysis accelerates (Fig. 19-9). However, glycogenolysis alone can probably maintain homeostasis of blood glucose concentration for only a few hours, since the body can store only modest amounts of glycogen.

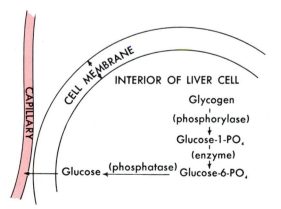

Fig. 19-10 Glycogenolysis in a liver cell.

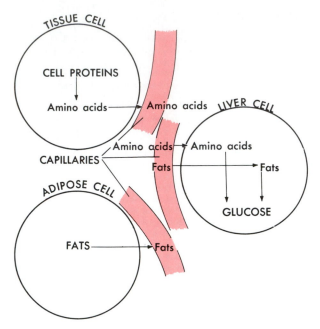

Fig. 19-11 Liver gluconeogenesis from mobilized tissue proteins and fats.

Gluconeogenesis

Gluconeogenesis means literally the formation of "new" glucose—"new" in the sense that it is made from proteins or the glycerol of fats, not from carbohydrates. The process occurs chiefly in the liver. It consists of many complex chemical reactions. The new glucose produced from fats or proteins by gluconeogenesis (Fig. 19-11) diffuses out of liver cells into the blood. Gluconeogenesis, therefore, can add glucose to the blood when needed, as can the process of liver glycogenolysis. Obviously, then, the liver is a most important organ for maintaining blood glucose homeostasis.

Control of glucose metabolism

The complex mechanism that normally maintains homeostasis of blood glucose concentration consists of hormonal and neural devices. At least five endocrine glands—islands of Langerhans, anterior pituitary gland, adrenal cortex, adrenal medulla, and thyroid gland—and at least

eight hormones secreted by those glands function as key parts of the glucose homeostatic mechanism.

Beta cells of the islands of Langerhans in the pancreas secrete the most famous sugar-regulating hormone of them all, *insulin*. Insulin decreases blood glucose level. Although it is not yet known exactly how insulin acts, several of its effects are well known. For instance, insulin is known to act in some way to accelerate glucose transport through cell membranes. It also increases the activity of the enzyme glucokinase. As shown in Fig. 19-8, glucokinase catalyzes glucose phosphorylation, the reaction that must occur before either glycogenesis or glucose catabolism can take place. By applying these facts, you can deduce the main methods by which insulin decreases blood glucose level: increased glycogenesis and increased catabolism of glucose. Fig. 19-12 shows these effects of insulin. By studying this figure, you can also deduce for yourself some of the prominent metabolic defects resulting from insulin deficiency such as occurs in diabetes mellitus. Slow glycogenesis with resulting low glycogen storage, decreased glucose catabolism, and increased blood glucose all result from insulin deficiency.

The islands of Langerhans secrete two sugar-regulating hormones—insulin from the beta cells and glucagon from the alpha cells (Fig. 18-17, *B*, p. 492). Whereas insulin tends to decrease the blood glucose level, glucagon tends to increase it. *Glucagon* increases the activity of the enzyme phosphorylase (Fig. 19-10), and this necessarily accelerates liver glycogenolysis and releases more of its product, glucose, into the blood. Fig. 19-13 shows the role of glucagon in raising blood glucose level.

Epinephrine is a hormone secreted in large amounts by the adrenal medulla in times of emotional or physical stress. Like glucagon, epinephrine increases phosphorylase activity (Fig. 19-10). This makes glycogenolysis go on at a faster rate. Epinephrine accelerates both liver and muscle glycogenolysis, whereas glucagon

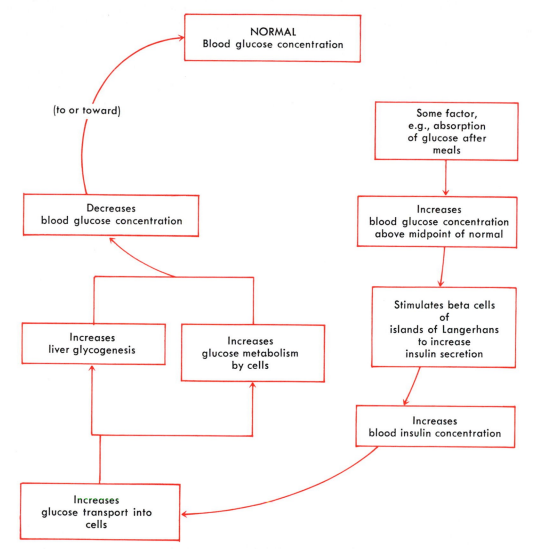

Fig. 19-12 Function of insulin. Homeostatic mechanism, which tends to prevent blood glucose concentration from increasing above the upper limit of normal. This insulin mechanism and the glucagon mechanism shown in Fig. 19-13 operate together to maintain homeostasis of blood glucose under usual circumstances in the normal body.

accelerates only liver glycogenolysis. Both hormones increase the blood glucose level. Epinephrine has the distinction of being the only hormone whose release into the systemic circulation (and therefore effects on metabolism) are directly under the control of the nervous system.

Adrenocorticotropic hormone (ACTH) and glucocorticoids are two more hormones that increase blood glucose concentration. ACTH stimulates the adrenal cortex to increase its secretion of glucocorticoids. Glucocorticoids accelerate gluconeogenesis. They do this by mobilizing pro-

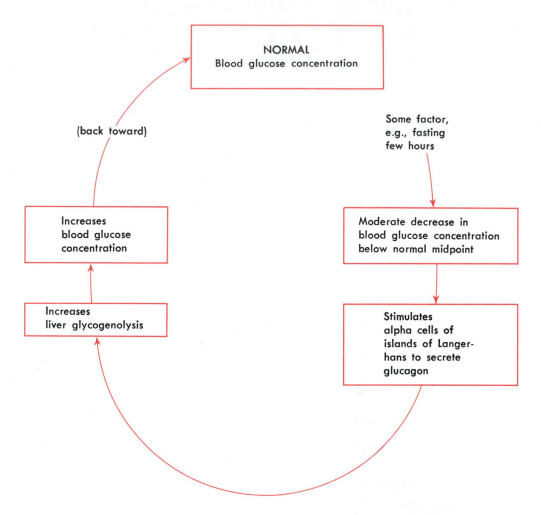

Fig. 19-13 Function of glucagon. Homeostatic mechanism that under usual conditions is chiefly responsible for preventing blood glucose level from falling below the lower limit of normal. This glucagon mechanism and the insulin mechanism shown in Fig. 19-12 work together to maintain homeostasis of blood glucose under usual circumstances in the normal body.

teins, that is, the breakdown or hydrolysis of tissue proteins to amino acids. More amino acids enter the circulation and are carried to the liver. Liver cells step up their production of "new" glucose from the mobilized amino acids. More glucose streams out of liver cells into the blood and adds to the blood glucose level.

The last four hormones we have discussed are able to increase glucose blood concentration by causing the formation of glucose, either from glycogen or from amino acids. Growth hormone, made by the anterior pituitary, also increases blood glucose level, but by a different mechanism. Growth hormone causes a shift from carbohydrate to fat catabolism. It does this by limiting the storage of fat in fat depots.

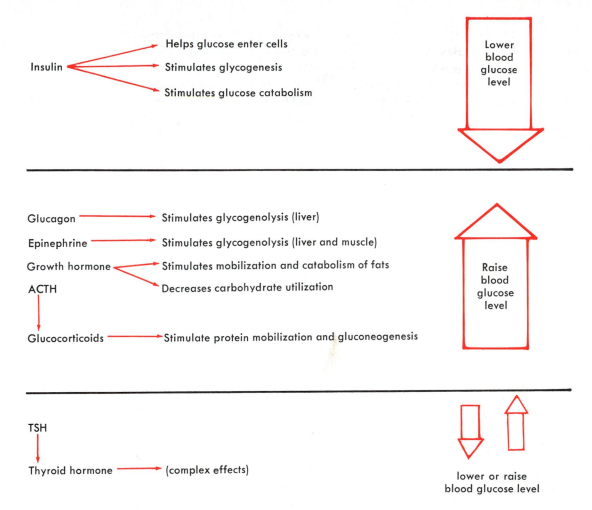

Fig. 19-14 Hormonal control of blood glucose level. Insulin lowers blood glucose level, so is hypoglycemic. Most hormones raise blood glucose level and are called hyperglycemic or anti-insulin hormones.

Instead, more fats are mobilized and catabolized. In this way, growth hormone "spares" carbohydrates from catabolism, and the level of glucose in the blood is increased.

A look at a summary of hormone control shown in Fig. 19-14 indicates that most hormones cause the blood level of glucose to rise. These hormones are called hyperglycemic. The one notable exception is insulin, which is hypoglycemic, or tends to decrease blood glucose. Thyroid-stimulating hormone (TSH) from the anterior pituitary gland and its target secretion, thyroid hormone, have complex effects on metabolism. Some of these raise, and some lower, the glucose level. One of the effects of thyroid hormone is to accelerate catabolism, and since glucose is the body's "preferred fuel," the result may be a decrease in blood glucose level.

Fat metabolism

Body cells both catabolize and anabolize fats. Fats constitute a more concentrated energy food than carbohydrates. Catabolism of 1 gm of fat yields 9 kilocalories (kcal) of heat; catabolism of 1 gm of carbohydrates yields only 4.1 kcal. *Fat catabolism*, like carbohydrate catabolism, consists of several processes, each of which, in turn, consists of a series of chemical reactions. The last steps are the same final reactions as in carbohydrate catabolism. But, because fat molecules cannot enter these final reactions, they must first be broken apart to form molecules than can. Specifically, fats must first be hydrolyzed into fatty acids and glycerol.

Glycerol can be converted into a compound that enters the glycolytic pathway. Fatty acids are broken down by a process called beta-oxidation into 2-carbon pieces, the familiar acetyl-CoA. These, of course, may be immediately catabolized in the citric acid cycle. Fig. 19-15 shows the steps in fat catabolism.

When fat catabolism occurs at an accelerated rate, as in diabetes mellitus (when glucose cannot enter cells) or fasting, excessive numbers of acetyl-CoA are formed. Liver cells then temporarily condense acetyl-CoA units together to form 4-carbon acetoacetic acids. Acetoacetic acid is classified as a ketone body and can be converted to two other ketone bodies, namely, acetone and beta-hydroxybutyric acid—hence the name ketogenesis for this process. Liver cells oxidize a small portion of the ketone bodies for their own energy needs, but most of them are transported by the blood to other tissue cells for the change back to acetyl-CoA and oxidation via the citric acid cycle (Fig. 19-15). Large amounts of ketone bodies may be present in the blood of a person with uncontrolled diabetes mellitus.

This condition is known as ketosis. Signs of it are acetone breath and ketonuria.

Fat anabolism (fat deposition or lipogenesis) consists of the synthesis of fats from fatty acids and glycerol or from compounds resulting from excess glucose or amino acids. So it is possible to "get fat" from foods other than fat. Fats are stored mainly in adipose tissue. These fat depots constitute the body's largest reserve energy source—too large, too often, unfortunately. Fat anabolism also includes synthesis of important complex compounds, for example, phospholipid, which is an important component of cell membranes (see Fig. 2-3, p. 32).

Only certain fatty acids (saturated ones) can be synthesized by the body. Others, the unsaturated fatty acids, must be provided by the diet and so are called *essential fatty acids*. Certain essential fatty acids serve as a source within the body for synthesis of an important group of lipids called prostaglandins. These hormonelike compounds, first discovered in the 1930s in fluid from the prostate gland, have in recent years gained increasing recognition for their occurrence in a variety of tissues, with a spectrum of biological activities (see p. 320 [Chapter 12] and p. 322). They are noted for their ability to lower blood pressure; they also play a role in fever production (p. 533). Since they can cause contraction of smooth muscle, their possible effects on uterine contraction (as in contraception and abortion) are being investigated. Prostaglandins apparently serve as intermediates between the "first messenger" (hormone) and "second messenger" (cyclic AMP), specifically affecting the enzyme adenylate cyclase (Fig. 12-2 and p. 321).

Control

Fat metabolism is controlled mainly by the following hormones: insulin, growth hormone, ACTH, and glucocorticoids. You probably recall from our discussion of these hormones in connection with carbohydrate metabolism that they regulate fat metabolism in such a way that the rate of fat catabolism is inversely related to the

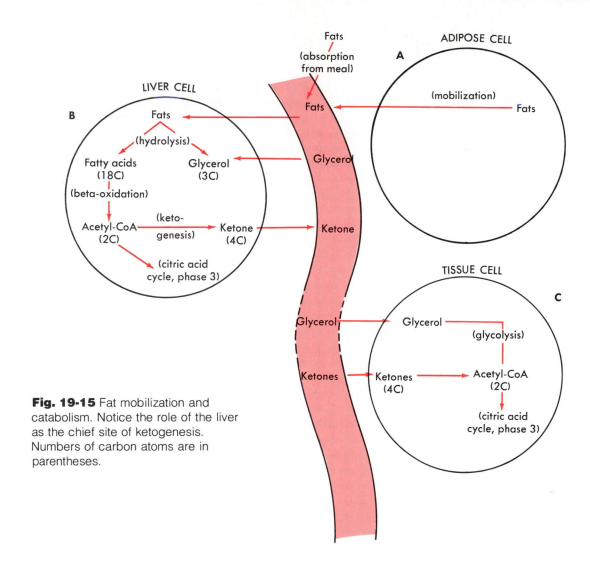

Fig. 19-15 Fat mobilization and catabolism. Notice the role of the liver as the chief site of ketogenesis. Numbers of carbon atoms are in parentheses.

rate of carbohydrate catabolism. If some condition such as diabetes mellitus causes carbohydrate catabolism to decrease below energy needs, increased secretion of growth hormone, ACTH, and glucocorticoids soon follows. These hormones, in turn, bring about an increase in fat catabolism. But when carbohydrate catabolism equals energy needs, fats are not mobilized out of storage and catabolized. Instead, they are spared and stored in adipose tissue. "Carbohydrates have a 'fat-sparing' effect," so says an old physiological maxim. Or, stating this truth more descriptively: "Carbohydrates have a 'fat-storing' effect."

Protein metabolism

In protein metabolism, anabolism is primary and catabolism is secondary. In carbohydrate and fat metabolism the opposite is true—catabolism is primary and anabolism is secondary. Proteins are primarily tissue-building foods.

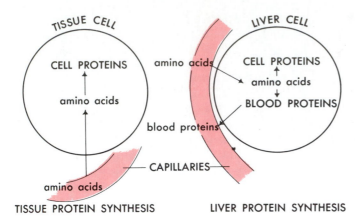

Fig. 19-16 Protein synthesis (anabolism). Growth hormone and testosterone tend to accelerate the processes shown and so are called anabolic hormones.

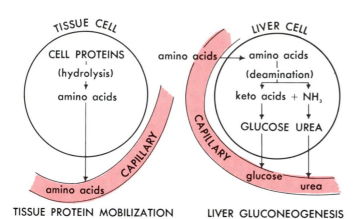

Fig. 19-17 Protein mobilization and catabolism. Glucocorticoids tend to accelerate these processes, so are classed as protein catabolic hormones. (Also see Fig. 19-18.)

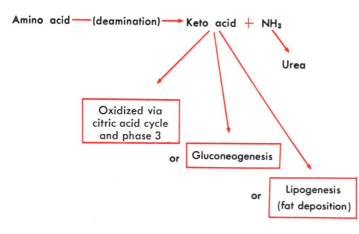

Fig. 19-18 Protein catabolism. First, as shown in Fig. 19-17, liver cells carry on deamination, a process that converts amino acids to keto acids and ammonia. Then keto acids may be changed to glucose by liver cells (gluconeogenesis), or liver and tissue cells may oxidize them (citric acid cycle) or convert them to fat (lipogenesis).

Carbohydrates and fats are primarily energy-supplying foods.

The cellular process called protein synthesis, or protein anabolism, produces many substances necessary for healthy survival. Essential among these are the proteins present in plasma—the so-called plasma proteins, 95% of which are synthesized in the liver. All plasma proteins contribute to the maintenance of water balance affecting osmotic pressure; they also increase the viscosity of blood, therefore helping to maintain normal blood pressure. The plasma proteins prothrombin and fibrinogen play crucial roles in blood coagulation. Gamma globulins, the only plasma proteins not synthesized in the liver, are antibodies essential in defense. Protein anabolism plays a major role in the growth, reproduction, and control of cells and of the body as a whole. Enzymes, for example, are proteins, as are many hormones. Protein anabolism is also the chief process of repair. It accomplishes the healing of wounds, the formation of scar tissue, and the replacement of cells destroyed by daily wear and tear. Red blood cell replacement alone, for instance, runs into the millions of cells per second. Protein anabolism is truly "big business" in the body.

A brief synopsis of the process of protein anabolism as now visualized appears on p. 599.

Protein catabolism, like the catabolism of fats, consists of three processes. The first takes place mainly in liver cells and the second and third are the citric acid cycle and phase 3, which occur in all cells. The first step in protein catabolism is known as *deamination*, a reaction in which an amino (NH_2) group is split off from an amino acid molecule to form a molecule of ammonia and one of keto acid (Fig. 19-18). Most of the ammonia is converted to urea and is excreted via the urine. The keto acid may be oxidized via the citric acid cycle and phase 3 or may be converted to glucose (gluconeogenesis) or to fat (lipogenesis). Both protein catabolism and anabolism go on continually. Only their rates differ from time to time. With a protein-deficient diet,

for example, protein catabolism exceeds protein anabolism. Various hormones, as we shall see, also influence the rates of protein catabolism and anabolism.

Usually a state of *protein balance* exists in the normal healthy adult body, that is, the rate of protein anabolism equals or balances the rate of protein catabolism. When the body is in protein balance, it is also in a state of *nitrogen balance.* For then the amount of nitrogen taken into the body (in protein foods) equals the amount of nitrogen in protein catabolic waste products excreted in the urine, feces, and sweat. Two kinds of protein or nitrogen imbalance exist. When protein catabolism exceeds protein anabolism, the amount of nitrogen in the urine exceeds the amount of nitrogen in the protein foods ingested. The individual is then said to be in a state of *negative nitrogen balance,* or in a state of "tissue wasting"—because more of his tissue proteins are being catabolized than are being replaced by protein synthesis. Protein-poor diets, starvation, and wasting illnesses, for example, produce a negative nitrogen balance.

A *positive nitrogen balance* (nitrogen intake in foods greater than nitrogen output in urine) indicates that protein anabolism is going on at a faster rate than protein catabolism. A state of positive nitrogen balance, therefore, characterizes any condition in which large amounts of tissue are being synthesized, such as during growth, pregnancy, and convalescence from an emaciating illness.

The main facts about protein metabolism are summarized in Figs. 19-16 to 19-18. Compare them with Figs. 19-4, 19-8, 19-10, 19-11, and 19-15. Note the important part played by the liver in the metabolism of all three kinds of foods.

Control

Protein metabolism, like that of carbohydrates and fats, is controlled largely by hormones rather than by the nervous system. Growth hormone and the male hormone testosterone both have a stimulating effect on protein

Table 19-1 Metabolism

Food	Anabolism	Catabolism
Carbohydrates	Temporary excess changed into glycogen by liver cells in presence of insulin; stored in liver and skeletal muscles until needed and then changed back to glucose (Figs. 19-10 and 19-11)	Oxidized, in presence of insulin, to yield energy (4.1 kcal per gram) and wastes (carbon dioxide and water)
	True excess beyond body's energy requirements converted into adipose tissue; stored in various fat depots of body (Fig. 19-19)	$C_6H_{12}O_6 + 6\ O_2 \rightarrow$ Energy $+ 6\ CO_2 + 6\ H_2O$
Fats	Built into adipose tissue; stored in fat depots of body	Fatty acids $\qquad$ Glycerol $\downarrow$ (beta-oxidation) $\qquad \downarrow$ (glycolysis) Acetyl-CoA $\rightleftarrows$ Ketones $\qquad$ Acetyl-CoA $\downarrow$ (tissues; citric acid cycle) Energy (9.3 kcal per gram) $+ CO_2 + H_2O$
Proteins	Temporary excess stored in liver and skeletal muscles	Deaminated by liver forming ammonia (which is converted to urea) and keto acids (which are either oxidized or changed to glucose or fat)
	Synthesized into tissue proteins, blood proteins, enzymes, hormones, etc.	

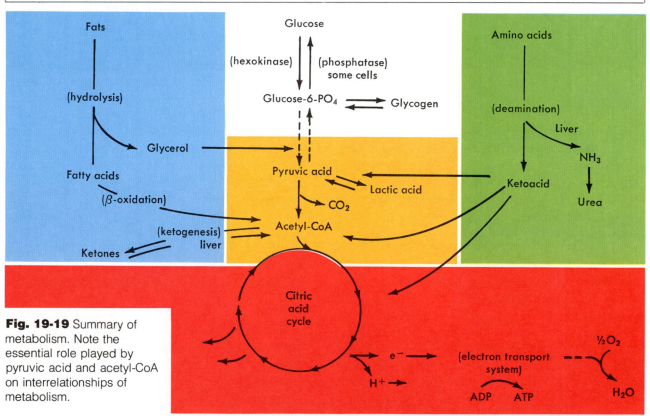

Fig. 19-19 Summary of metabolism. Note the essential role played by pyruvic acid and acetyl-CoA on interrelationships of metabolism.

synthesis or anabolism. For this reason, they are referred to as anabolic hormones. The protein catabolic hormones of greatest consequence are glucocorticoids. They are thought to act in some way, still unknown, to speed up tissue protein mobilization, that is, the hydrolysis of cell proteins to amino acids, their entry into the blood, and their subsequent catabolism (Fig. 19-17). ACTH functions indirectly as a protein catabolic hormone because of its stimulating effect on glucocorticoid secretion.

Thyroid hormone is necessary for and tends to promote protein anabolism and therefore, growth when plenty of carbohydrates and fats are available for energy production. On the other hand, under different conditions, for example, when the amount of thyroid hormone is excessive or when the energy foods are deficient, this hormone may then promote protein mobilization and catabolism.

▪ ▪ ▪

Some of the facts about metabolism set forth in the preceding paragraphs are summarized in Table 19-1 and Fig. 19-19. You may find it helpful to read once again the section on important generalizations about metabolism (pp. 506-508). As you do so, try to cite specific examples encountered in your study of metabolism.

Metabolic rates

Meaning

The term *metabolic rate* means the amount of energy released in the body in a given time by catabolism. It represents energy expended or used for accomplishing various kinds of work. In short, metabolic rate actually means catabolic rate or rate of energy release.

Ways of expressing

Metabolic rates are expressed in either of two ways: (1) in terms of the number of kilocalories* of heat energy expended per hour or per day or (2) as normal or as a definite percentage above or below normal.

Basal metabolic rate

The basal metabolic rate (BMR) is the body's rate of energy expenditure under "basal conditions," namely, when the individual:

1 Is awake but resting, that is, lying down and, as far as possible, not moving a muscle
2 Is in the postabsorptive state (12 to 18 hours after the last meal)
3 Is in a comfortably warm environment

Note that the BMR is not the minimum metabolic rate. It does not indicate the smallest

*One *kilocalorie* (a so-called "large" Calorie) is the amount of heat used to raise the temperature of 1 kg (liter) of water 1° Celsius. The 100 Calories (kcal) in a slice of bread would provide enough heat to raise a liter of water from 0° to 100° C.

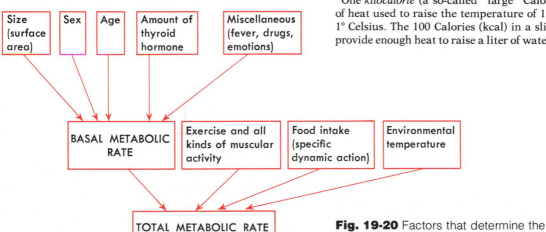

Fig. 19-20 Factors that determine the basal and total metabolic rates.

amount of energy that must be expended to sustain life. It does, however, indicate the smallest amount of energy expenditure that can sustain life and also maintain the waking state and a normal body temperature in a comfortably warm environment.

Factors influencing

The BMR is not identical for all individuals because of the influence of various factors (Fig. 19-20), some of which are described in the following paragraphs.

Size. In computing the BMR, size is usually indicated by the amount of the body's surface area. It is computed from the individual's height and weight. A large individual has the same BMR as a small one per square meter of body surface if other conditions are equal. However, because a large individual has more square meters of surface area, his BMR is greater than that of a small individual. For example, the BMR for a man in his twenties is about 40 kcal per square meter of body surface per hour (Table 19-2). A large man with a body surface area of 1.9 square meters would, therefore, have a BMR of 76 kcal per hour, whereas a smaller man with a surface area of perhaps 1.6 square meters would have a BMR of only 64 kcal per hour. The average surface area for American adults is 1.6 square meters for women and 1.8 square meters for men.

Sex. Men oxidize their food approximately 5% to 7% faster than women. Therefore their BMRs are about 5% to 7% higher for a given size and age. A man 5 feet, 6 inches tall, weighing 140 pounds, for example, has a 5% to 7% higher BMR than a woman of the same height, weight, and age.

Age. That the fires of youth burn more brightly than those of age is a physiological as well as a psychological fact. In general the younger the individual, the higher his BMR for a given size and sex (Table 19-2). Exception: The BMR is slightly lower at birth than a few years later. That is to say, the rate increases slightly during the first 3 to 6 years and then starts to decrease and continues to do so throughout life.

Thyroid hormone. Thyroid hormone stimulates basal metabolism. Without a normal amount of this hormone in the blood, a normal BMR cannot be maintained. When an excess of thyroid hormone is secreted, foods are catabolized faster, much as coal is burned faster when a furnace draft is open. Deficient thyroid secretion, on the other hand, slows the rate of metabolism.

Body temperature. Fever increases the BMR. According to DuBois, for every degree Celsius increase in body temperature, metabolism increases about 13%.* A decrease in body temperature (hypothermia) has the opposite effect. Metabolism decreases, and because it does, cells use less oxygen than they normally do. This knowledge has been applied clinically by the use of hypothermia in certain situations, for example, in open-heart surgery. Because circulation is reduced or interrupted during this procedure, oxygen supply necessarily decreases. Cells can tolerate this decreased oxygen supply reason-

Table 19-2 Basal metabolism (Aub-DuBois)		
Age (yr)	Kilocalories per hour per square meter body surface	
	Male	Female
10-12	51.5	50.0
12-14	50.0	46.5
14-16	46.0	43.0
16-18	43.0	40.0
18-20	41.0	38.0
20-30	39.5	37.0
30-40	39.5	36.5
40-50	38.5	36.0
50-60	37.5	35.0
60-70	36.5	34.0

*Mountcastle, V. B., editor: Medical physiology, ed. 13, St. Louis, 1974, The C. V. Mosby Co.

ably well, however, if their oxygen need has also decreased. Induced hypothermia decreases their rate of metabolism and thereby decreases their use of oxygen.

Drugs. Certain drugs, such as caffeine, amphetamine (Benzedrine), and dinitrophenol, increase the BMR.

Other factors. Other factors, such as *emotions*, *pregnancy*, and *lactation*, also influence basal metabolism. All of these factors increase the BMR.

How determined

BMR is determined in most hospitals, and in many doctors' offices and nutrition laboratories, by a method called *indirect calorimetry*. The rationale underlying this rapid, inexpensive method is the fact that the BMR (expressed as the number of kilocalories of heat produced per unit of time) can be calculated from the amount of oxygen consumed in a given time. The BMR can then be expressed as normal, or as a definite percent above or below normal, by dividing the actual kilocalorie rate by the known average kilocalorie rate for normal individuals of the same size, sex, and age. Statistical tables, based on research, list these normal BMRs. (If the BMR

were calculated to be 10% above normal, for example, it would be reported as +10.) Are you curious to know the average BMR for a person of your size, sex, and age? If so, take the following steps.

1 Start with your weight in kilograms and your height in centimeters. (Convert pounds to kilograms by dividing pounds by 2.2. Convert inches to approximate centimeters by multiplying inches by 2.5.) For example, 110 pounds = 50 kg; 5 feet, 3 inches = 158 cm.

2 Convert your weight and height to square meters, using Fig. 19-21. For example, weight 50 kg and height 158 cm = about 1.5 square meters surface area of body.

3 Find your age and sex in Table 19-2, and then multiply the number of kilocalories per square meter per hour given there by your square meters of surface area and then by 24. For example, average BMR per day for a 25-year-old female, weight 110 pounds and height 5 feet, 3 inches = 1,332 kcal (37 × 1.5 × 24).

A quick rule of thumb for estimating a young woman's BMR is to multiply her weight in pounds by 12.

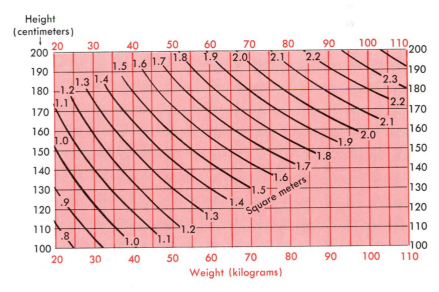

Fig. 19-21 Chart for determining surface area of man in square meters from weight in kilograms and height in centimeters according to the following formula: area (m²) = wt$^{0.425}$ × ht$^{0.725}$ × 7184. (Redrawn after DuBois, D., and DuBois, E. F.: Clinical calorimetry, Arch. Intern. Med. [Chicago] **17**:863-871, 1916).

Total metabolic rate

The *total metabolic rate* is the amount of energy used or expended by the body in a given time. It is expressed in kilocalories per hour or per day. Most of the many factors that together determine the total metabolic rate are shown in Fig. 19-20. Of these, the main direct determinants are the following:

1 The basal metabolic rate, that is, the energy used to do the work of maintaining life under the basal conditions previously described, plus

2 The energy used to do all kinds of skeletal muscle work—from the simplest activities such as feeding oneself or sitting up in bed to the most strenuous kind of physical labor or exercise.

3 The energy used for specific dynamic action (SDA) of foods. The metabolic rate increases for several hours after a meal, apparently because of the energy needed for metabolizing foods. Carbohydrates and fats have a SDA of about 5%. Proteins have a much higher SDA, about 30%. This means that for 100 kcal of protein, 30 kcal are used for such processes as deamination and oxidation of the protein, leaving just 70 kcal available for other cell work. For this reason, proteins are "worth" fewer calories and are popular as diet foods.

Energy balance and its relation to body weight

When we say that the body maintains a state of energy balance, we mean that its energy input equals its energy output. Energy input per day equals the total calories (kilocalories) in the food ingested per day. Energy output equals the total metabolic rate expressed in kilocalories. But you may be wondering what energy intake, output, and balance have to do with body weight. "Everything" would be a fairly good one-word answer. Or, to be somewhat more explicit, the following basic principles describe the relationships between these factors.

1 Body weight remains constant (except for possible variations in water content) when the body maintains energy balance—when the total calories in the food ingested equals the total metabolic rate, that is. Example: If you have a total metabolic rate of 2,000 kcal per day and if the food you eat per day yields 2,000 kcal, your body will be maintaining energy balance and your weight will stay constant.

2 Body weight increases when energy input exceeds energy output—when the total calories of food intake per day is greater than the total calories of the metabolic rate. A small amount of the excess energy input is used to synthesize glycogen for storage in the liver and muscles. But the rest of it is used for synthesizing fat and storing it in adipose tissue. If you were to eat 3,000 kcal each day for a week and if your total metabolic rate were 2,000 kcal per day, you would gain weight. How much you would gain you can discover by doing a little simple arithmetic:

Total energy input for week = 21,000 kcal
Total energy output for week = 14,000 kcal
Excess energy input for week = 7,000 kcal

Approximately 3,500 kcal are used to synthesize 1 pound of adipose tissue. Hence at the end of this one week of "overeating"—of eating 7,000 kcal over and above your total metabolic rate—you would have gained about 2 pounds.

3 Body weight decreases when energy input is less than energy output—when the total number of calories in the food eaten is less than the total metabolic rate. Suppose you were to eat only 1,000 kcal a day for a week and that you have a total metabolic rate of 2,000 kcal per day. By the end of the week your body would have used a total of 14,000 kcal of energy for maintaining life and doing its many kinds of work. All 14,000 kcal of this actual energy expenditure had to come from catabolism of foods, since this is the body's only source of energy. Catabolism of ingested food supplied 7,000 kcal and catabolism of stored food supplied the remaining 7,000 kcal.

That week your body would not have maintained energy balance, nor would it have maintained weight balance. It would have incurred an energy deficit paid out of the energy stored in approximately 2 pounds of body fat. In short, you would have lost about 2 pounds.

■ ■ ■

Anyone who wants to reduce should remember this cardinal principle: eat fewer calories than your total metabolic rate. To apply this principle, do one of two things or, better yet, do both: Decrease your caloric intake; increase your caloric output, that is, total metabolic rate, by increasing your physical activity. In fewer words, eat less, exercise more.

Foods are stored as glycogen, fats, and tissue proteins. As you will recall, cells catabolize them preferentially in this same order: carbohydrates, fats, and proteins. If there is no food intake, almost all of the glycogen is estimated to be used up in a matter of 1 or 2 days. Then, with no more carbohydrate to act as a fat sparer, fat is catabolized. How long it takes to deplete all of this reserve food depends, of course, on how much adipose tissue the individual has when he starts his starvation diet. Finally, with no more fat available as a protein sparer, tissue proteins are catabolized rapidly, and death soon ensues.

Mechanisms for regulating food intake

Mechanisms for regulating food intake are still not clearly established. That the hypothalamus plays a part in these mechanisms, however, seems certain. A number of data seem to indicate that a cluster of neurons in the lateral hypothalamus function as an *appetite center*—meaning that impulses from them bring about in-

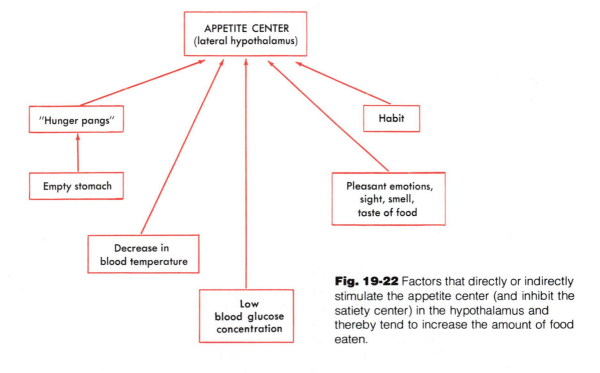

Fig. 19-22 Factors that directly or indirectly stimulate the appetite center (and inhibit the satiety center) in the hypothalamus and thereby tend to increase the amount of food eaten.

creased food intake. Other data suggest that a group of neurons in the ventral medial nucleus of the hypothalamus functions as a *satiety center*—meaning that impulses from these neurons inhibit food intake. What acts directly on these centers to simulate or depress them is still a matter of theory rather than fact. One theory (the "thermostat theory") holds that it is the temperature of the blood circulating to the hypothalamus that influences the centers. A moderate decrease in blood temperature stimulates the appetite center (and inhibits the satiety center). Result: The individual has an appetite. He wants to eat—and probably does. An increase in blood temperature produces the opposite effect, a depressed appetite (anorexia). One well-known instance of this is the loss of appetite in persons who have a fever.

Another theory (the "glucostat theory") says that it is the blood glucose concentration and rate of glucose utilization that influences the hypothalamic feeding centers. A low blood glucose concentration or low glucose utilization stimulates the appetite center, whereas a high blood glucose concentration inhibits it. Unquestionably, a great many factors operate together as a complex mechanism for regulating food intake. Some of these factors are indicated in Fig. 19-22.

Homeostasis of body temperature

Warm-blooded animals, such as man, maintain a remarkably constant temperature despite sizable variations in environmental temperatures.

Normally in most people, body temperature moves up and down very little in the course of a day. It hovers close to a midpoint of about 37° C, increasing perhaps to 37.6° C by late afternoon and decreasing to around 36.2° C by early morning. This homeostasis of body temperature is of the utmost importance. Why? Because healthy survival depends on biochemical reactions taking place at certain rates. And these rates, in turn, depend on normal enzyme functioning, which depends on body temperature staying within the narrow range of normal.

In order to maintain an even temperature, the body must, of course, balance the amount of heat it produces with the amount it loses. This means that if extra heat is produced in the body, this same amount of heat must then be lost from it. Obviously if this does not occur, if increased heat loss does not follow close on increased heat production, body temperature will climb steadily upward.

The heat-regulating centers are present at birth but do not function well for a short time after birth. Therefore newborns need to be kept somewhat warmer than adults. If the infant is born prematurely, the heat-regulation centers do not function for a longer time, perhaps several weeks.

Heat production

Heat is produced by one means—catabolism of foods. Because the muscles and glands (liver, especially) are the most active tissues, they carry on more catabolism and therefore produce more heat than any of the other tissues. So the chief determinant of how much heat the body produces is the amount of muscular work it does. During exercise and shivering, for example, catabolism and heat production increase greatly. But during sleep, when very little muscular work is being done, catabolism and heat production decrease. The metabolic rate drops during sleep to about 10% below BMR.

Heat loss

Heat is lost from the body by the physical processes of evaporation, radiation, conduction, and convection. Some 80% or more of this heat transfer occurs through the skin. The rest takes place through the mucous membranes of the respiratory, digestive, and urinary tracts.

Evaporation

Heat energy must be expended to evaporate any fluid. Evaporation of water, therefore, constitutes one method by which heat is lost from the body, especially from the skin. Evaporation is especially important at high environmental temperatures when it constitutes the only method by which heat can be lost from the skin. A humid atmosphere necessarily retards evaporation and therefore lessens the cooling effect derived from it—the explanation for the fact that the same degree of temperature seems hotter in humid climates than in dry ones. At moderate temperatures, evaporation accounts for about half as much heat loss as does radiation.

Radiation

Radiation is the transfer of heat from the surface of one object to that of another without actual contact between the two. Heat radiates from the body surface to nearby objects that are cooler than the skin and radiates to the skin from those that are warmer than the skin. This is, of course, the principle of heating and cooling systems. The amount of heat lost by radiation from the skin is made to vary as needed by dilation of surface blood vessels when more heat needs to be lost and by vasoconstriction when heat loss needs to be decreased. In cool environmental temperatures, radiation accounts for a greater percentage of heat loss from the skin than both conduction and evaporation combined. However, in hot environments no heat is lost by radiation but instead may be gained by radiation from warmer surfaces to the skin.

Conduction

Conduction means the transfer of heat to any substance actually in contact with the body—to clothing or jewelry, for example, or even to cold foods or liquids ingested. This process accounts for a relatively small amount of heat loss.

Convection

Convection is the transfer of heat away from a surface by movement of heated air or fluid particles. Usually, convection causes very little heat loss from the body's surface. But it can account for considerable heat loss—as you know from experience if you have ever stepped from your bath into even slightly moving air from an open window.

Thermostatic control of heat production and loss

The control mechanism that normally maintains homeostasis of body temperature consists of two parts:

1 A *heat-dissipating mechanism* that acts to increase heat loss when blood temperature increases above a certain point (Fig. 19-23). This mechanism, therefore, prevents body temperature from rising above normal under usual circumstances.

2 A *heat-gaining mechanism* that acts to accelerate catabolism and thereby to increase heat production when blood temperature decreases below a certain point. Under ordinary conditions, this mechanism prevents body temperature from falling below normal.

Heat-dissipating mechanism

In the anterior part of the hypothalamus, behind the sphenoid sinuses, lies a group of cells referred to collectively as the "human thermostat." These neurons are thermal receptors, that is, they are stimulated by a very slight increase in the temperature of the blood above the point at which the human thermostat is set—normally about 37° C. In a sense, one might say that these cells of the hypothalamus take the temperature of the blood circulating to them. Whenever it increases by as little as 0.01° above 37° C* (or some

*See Benzinger, T. H.: The human thermostat, Sci. Am. **204:**134-147, Jan., 1961.

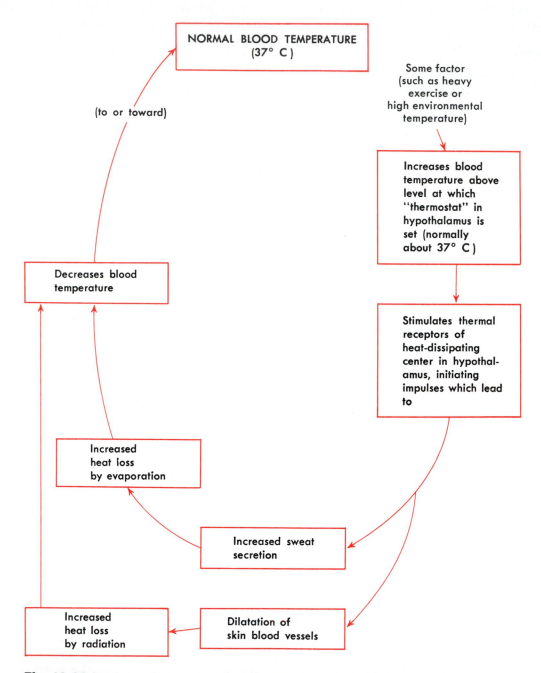

Fig. 19-23 Scheme to show how heat-dissipating mechanism operates to maintain normal body temperature. In principle, it cancels out any heat gain by bringing about an equal heat loss. Under usual circumstances, this mechanism succeeds in preventing body temperature from rising above the upper limit of normal.

other set point), these neurons send out impulses that eventually, via sympathetic nerves, reach sweat glands and blood vessels of the skin. They stimulate the body's 2 million or more sweat glands to increase their rate of secretion. Evaporation of the larger amount of sweat causes a greater heat loss from the skin. Dilation of surface blood vessels brings a larger quantity of blood close to the surface for more heat loss by radiation. Blood vessel supply to the skin is profuse and far exceeds the needs of the skin. Such an abundant blood supply serves primarily for regulation of body temperature.

Heat-gaining mechanism

In a cold environment the mechanism that tries to maintain homeostasis of body temperature includes two kinds of responses—those that decrease heat loss and those that increase heat production. Together, they almost always succeed in preventing a decrease in blood temperature below the lower limit of normal.

Decrease of heat loss results from reduced amounts of sweat secretion and from vasoconstriction of the skin blood vessels. Increased heat production occurs as a result of shivering and voluntary muscle contractions. Both kinds of muscle work accelerate catabolism and heat production.

Skin thermal receptors

In addition to the thermal receptors in the hypothalamus, there are many heat and cold receptors located in the skin. Impulses initiated in skin thermal receptors travel to the cerebral cortex sensory area. Here, they give rise to sensations of skin temperature and are relayed out over voluntary motor paths to produce skeletal muscle movements that affect skin temperature. For example, on a hot day you "feel hot" because of stimulation of your skin heat receptors. Often you make some sort of movements to "cool yourself off." You may start fanning yourself, or perhaps you turn on an air conditioner or go swimming. As a result, your skin temperature decreases to a more comfortable level. Thus the thermal receptors in the skin will have taken part in a conscious mechanism that helps regulate skin temperature. Convincing evidence supports the view that this is their only function and that impulses from them do not travel to the heat-regulating centers in the hypothalamus and therefore do not take part in the automatic regulation of internal body temperature.

Fever

Fever differs in certain respects from high body temperature brought about by either heavy exercise or exposure to high environmental conditions. In these situations the higher-than-normal temperature results both from an increase in heat production and a temporary inability of the heat-loss mechanisms to compensate for the heat gained. These mechanisms soon succeed in restoring normal temperature (Fig. 19-23). In fever the higher-than-normal temperature results from a resetting of the thermostat by the action on the hypothalamus of chemicals called pyrogens. The thermostat consists of neurons. Since they function to maintain a set temperature, the high temperature of fever lasts longer than the high temperature from strenuous exercise or a hot environment. Many compounds act as pyrogens, but the most common ones come from bacteria and viruses. All pyrogens accelerate the synthesis and release of prostaglandins. Also, their presence in the region of the hypothalamus appears to reset the thermostat. Aspirin, it is now widely accepted, reduces fever by preventing the synthesis of prostaglandins. Other antipyretics (fever-reducing drugs) act similarly. Heat-dissipating mechanisms that operate in an effort to maintain normal body temperature are shown in Fig. 19-23.

Outline summary

Definitions

A Catabolism—decomposition process in which relatively large food molecules break down to yield smaller molecules and energy

B Anabolism—synthesis process that uses energy to build relatively small molecules up into larger molecules, for example, enzymes, hormones, and antibodies

Some important generalizations about metabolism

A Many specific sequences of chemical reactions (metabolic pathways) make up both catabolism and anabolism; common pathways are those that occur in utilization of all three kinds of foods

B Enzymes catalyze both catabolic and anabolic chemical reactions

C Energy changes accompany metabolic reactions; catabolism releases energy, anabolism uses energy; nonutilizable heat released in small spurts in catabolism; chemical energy released in catabolism becomes utilizable for anabolism and other cell work after transferral to high energy bonds of ATP molecules

D ATP functions as universal biological currency; see Fig. 19-1

E All cells not created equal with regard to metabolism; different kinds of cells differ as to specific metabolic pathways and as to metabolic rates

Carbohydrate metabolism

A Carbohydrates—"preferred energy fuel"; cells catabolize glucose first, sparing fats and proteins; when their glucose supply becomes inadequate, cells next metabolize fats, lastly proteins

B Glucose transport through cell membranes and phosphorylation

  1 Insulin promotes this transport through cell membranes

  2 *Glucose phosphorylation*—conversion of glucose to glucose-6-phosphate, catalyzed by enzyme glucokinase; *insulin* increases activity of glucokinase, so promotes glucose phosphorylation

C Glucose catabolism

  1 Consists of complex series of chemical reactions that take place inside cells and yield energy, carbon dioxide, and water; about half of energy released from food molecules by catabolism is put back in storage in unstable high-energy bonds of ATP molecules and the rest is transformed to

heat; energy in high-energy bonds of ATP is released with explosive rapidity

  2 Purpose to continually provide cells with utilizable energy, that is, energy supplied instantaneously to energy-consuming mechanisms that do cellular work

  3 Glycolysis—series of anaerobic (nonoxygen-utilizing) chemical reactions that convert one glucose molecule to two pyruvic acid molecules and yield slightly less than 5% of ATP produced during glucose catabolism

  4 Citric acid cycle and phase 3—series of aerobic chemical reactions that utilize oxygen to oxidize two pyruvic acid molecules to six carbon dioxide molecules and six water molecules and yield about 95% of all the ATP formed during catabolism

D Glucose anabolism—synthesis of larger molecule compounds from glucose; an important kind of cellular work that uses some of energy made available by catabolism

E Glycogenesis—one kind of glucose anabolism; conversion of glucose to glycogen for storage; occurs in muscle cells, also in liver cells when blood glucose level exceeds 120 to 140 mg per deciliter

F Glycogenolysis

  1 In liver cells—when blood glucose level decreases below midpoint of normal, glycogenolysis accelerates to raise blood glucose level; enzyme, glucose phosphatase, present in liver cells catalyzes final step of glycogenolysis, changing glucose-6-phosphate to glucose; this enzyme lacking in most other cells; *glucagon* increases activity of phosphorylase, so accelerates liver glycogenolysis; *epinephrine* accelerates liver and muscle glycogenolysis

  2 In muscle cells—glycogen changed back to glucose-6-phosphate, preliminary to catabolism

G Gluconeogenesis—sequence of chemical reactions carried on in liver cells; converts protein or fat compounds into glucose; growth hormone, ACTH, and glucocorticoids have stimulating effect on rate of gluconeogenesis

H Control of glucose metabolism

  1 By hormones that cause a decrease in blood glucose level, that is, have hypoglycemic effect

    a Insulin (from beta cells of pancreatic islets)—tends to accelerate glucose utilization by cells because it enhances glucose transport through cell membranes and glucose phosphorylation

    b Thyrotropin (from anterior pituitary gland)—

stimulates thyroid gland to increase secretion of thyroid hormone, which accelerates catabolism, usually glucose catabolism, since glucose "preferred fuel"

2 By hormones that cause increase in blood glucose level, that is, have hyperglycemic effect

   a Glucagon (from alpha cells of pancreatic islets)—increases activity of enzyme phosphorylase, thereby accelerating liver glycogenolysis with release of glucose into blood.

   b Epinephrine (from adrenal medulla)—also increases phosphorylase activity, but in both liver and muscle cells

   c ACTH (from anterior pituitary gland)—stimulates adrenal cortex to increase secretion of glucocorticoids, which accelerate tissue protein mobilization and subsequent liver gluconeogenesis from mobilized proteins

   d Growth hormone (from anterior pituitary gland)—decreases fat deposition, increases fat mobilization and catabolism: hence tends to bring about shift to fat utilization from "preferred" glucose utilization

Fat metabolism

A Catabolism

   1 Hydrolysis of fats to fatty acids and glycerol, primarily in liver cells

   2 Glycerol oxidized same as carbohydrates

   3 Fatty acids converted by beta-oxidation to acetyl-CoA; when fat catabolism accelerated, acetyl-CoA condenses to form ketone bodies (ketogenesis); occurs mainly in liver; largest proportion of ketones enters blood from liver cells to be transported to tissues for oxidation to carbon dioxide and water via citric acid cycle and phase 3

B Anabolism—for tissue synthesis and for building various compounds: phospholipids important constituents of membranes; prostaglandins derived from essential fatty acids, have hormonelike effects on variety of tissues; fats deposited in connective tissue converts it to adipose tissue

C Fat mobilization—release of fats from adipose tissue cells, followed by their catabolism; occurs when blood contains less glucose than normal or when it contains less insulin than normal; if excessive, leads to ketosis

D Control—by following major factors

   1 Rate of glucose catabolism one of main regulators of fat metabolism; in general, normal or high rates of glucose catabolism accompanied by low rates of fat mobilization and catabolism and high rates of fat deposition; converse also true

   2 Insulin helps control fat metabolism by its effects on glucose metabolism; in general, normal amounts of insulin and blood glucose tend to decrease fat mobilization and catabolism and to increase fat deposition; insulin deficiency increases fat mobilization and catabolism; converse also true

   3 Growth hormone—decreases fat deposition and increases fat mobilization and utilization, that is, growth hormone tends to bring about shift from glucose to fat utilization

   4 Glucocorticoids help control fat metabolism; in general, when blood glucose level lower than normal and in various stress situations, more glucocorticoids secreted and accelerate fat mobilization and gluconeogenesis from them; when blood glucose level higher than normal but rate of glucose catabolism low (as in diabetes mellitus), glucocorticoids also increase fat mobilization, but it is followed by ketogenesis from them; when blood glucose level higher than normal, and provided its insulin content adequate, glucocorticoids accelerate fat deposition

Protein metabolism

A Anabolism of proteins of primary importance, their catabolism, secondary; amino acids used to synthesize all kinds of tissue, for example, for growth and repair; also used to synthesize many other substances such as enzymes, hormones, antibodies, blood proteins, and structural components of cellular membranes

B Catabolism

   1 Deamination of amino acid molecule to form ammonia and keto acid; mainly in liver cells

   2 Ammonia converted to urea (mainly in liver) and excreted via urine

   3 Keto acids may be converted to glucose in liver, or to fat, or oxidized in liver or tissue cells via tricarboxylic acid cycle with phase 3

C Control

   1 Growth hormone and testosterone both have stimulating effect on protein synthesis or anabolism

   2 ACTH and glucocorticoids—protein catabolic hormones; accelerate tissue protein mobilization, that is, hydrolysis of tissue proteins to amino acids and their release into blood; liver

converts amino acids to glucose (gluconeogenesis) or deaminates them

3 Thyroid hormone promotes protein anabolism when nutrition adequate and amount of hormone normal; therefore adequate amounts necessary for normal growth

Metabolic rates

A Meaning—amount of energy expended in given time

B Ways of expressing—in kilocalories or as "normal" or as a definite percentage above or below normal, for example, +10% or −10%

C Basal metabolic rate—amount of heat produced (energy expended) in waking state when body at complete rest, 12 to 18 hours after last meal, in comfortably warm environment
  1 Factors influencing
    a Size—greater surface area, higher BMR (surface area computed from height and weight)
    b Sex—approximately 5% higher in males
    c Age—higher in youth than in aged
    d Abnormal functioning of certain endocrines, particularly thyroid gland
    e Fever—each Celsius degree rise in temperature increases BMR approximately 13%
    f Certain drugs, for example, dinitrophenol, increase BMR
    g Other factors, for example, pregnancy and emotions, increase BMR
  2 How determined—indirect calorimetry based on fact that definite amount of heat is produced for each liter of oxygen consumed; average BMR can be computed using Table 19-2 and Fig. 19-21

D Total metabolic rate—amount of heat produced by body in average 24 hours; equal to basal rate plus number of kilocalories produced chiefly by muscular work and adjusting to cool temperatures; expressed in kilocalories per 24 hours

E Energy balance and its relationship to body weight
  1 Energy balance means that energy input (total kilocalories in food ingested) equals energy output, that is, total metabolic rate expressed in kilocalories
  2 In order for body weight to remain constant (except for variations in water content), energy balance must be maintained—total kilocalories ingested must equal total metabolic rate
  3 Body weight increases when energy input exceeds energy output—when total kilocalories ingested greater than total metabolic rate
  4 Body weight decreases when energy input less than energy output—when total kilocalories ingested less than total metabolic rate; no diet will reduce weight unless it contains fewer kilocalories than the total metabolic rate of the individual eating diet

Mechanisms for regulating food intake

A Thermostat theory holds that moderate decrease in blood temperature acts as stimulant to appetite center, so increases appetite; increase in blood temperature (fever) produces opposite effect

B Glucostat theory postulates that low blood glucose concentration or low rate of glucose utilization, for example, diabetes mellitus, acts as stimulant to appetite center, so increases appetite; high blood glucose level produces opposite effect

Homeostasis of body temperature

In order to maintain homeostasis of body temperature, heat production must equal heat loss

Heat production

By catabolism of foods in skeletal muscles and liver especially

Heat loss

A By physical processes of evaporation, radiation, conduction, and convection

B About 80% of heat loss occurs through skin; rest takes place through mucosa of respiratory, digestive, and urinary tracts

Thermostatic control of heat production and loss

A Heat-dissipating mechanism—see Fig. 19-23, p. 532

B Heat-gaining mechanism
  1 Details not established but mechanism activated by decrease in blood temperature
  2 Responses—skin blood vessel constriction, shivering, and voluntary muscle contractions

Skin thermal receptors

Stimulation of skin thermal receptors gives rise to sensations of heat or cold; also initiates voluntary movements to reduce these sensations, for example, fanning oneself to cool off or exercising to warm up

Fever

Caused by "resetting of thermostat" in hypothalamus to higher temperature; apparently caused by activity of prostaglandins, which increase because of presence of pyrogens in body

Review questions

1 What is metabolism?
2 What two processes make up the process of metabolism?
3 Contrast the function or purpose of digestion with the function or purpose of metabolism.
4 Describe carbohydrate catabolism briefly.
5 Explain why mitochondria are called "power plants" of the cell.
6 Tell what processes are involved in the maintenance of blood glucose level.
7 Describe protein catabolism briefly.
8 Describe fat catabolism briefly.
9 Compare proteins, carbohydrates, and fats as to functions they serve in the body.
10 Describe the main facts about hormonal control of metabolism.
11 Differentiate between digestive and metabolic wastes.
12 What does the term *metabolic rate* mean?
13 Differentiate between basal and total metabolic rates.
14 Discuss possible effects of advanced liver disease on digestion, absorption, and metabolism.

Situation: Mrs. A., who weighs 160 pounds, is 5 feet, 5 inches tall, is 50 years of age, has been given a reducing diet by her doctor. She complains to you that she "knows it won't work" because "it isn't what she eats that makes her fat." As proof, she says that both her husband and son "eat twice as much" as she and so does her younger sister.

15 How would you answer Mrs. A.?
16 Suppose that Mrs. A. has a total metabolic rate of 2,200 kcal per day and that her diet contains 1,700 kcal. How many pounds can she expect to lose per week?
17 Rapid bicycle riding for 15 minutes expends about 100 kcal. This alone, without decreasing your calorie intake at all, would cause you to lose a pound in how many days? In 50 weeks, how much could you lose this way?
18 Explain two mechanisms postulated to regulate the amount of food a person eats.
19 Explain briefly the mechanism for maintaining homeostasis of body temperature.
20 Explain a current concept about the cause of fever.

chapter 20

The urinary system

The urinary system consists of those organs that produce urine and eliminate it from the body. There are two kidneys, two ureters, one bladder, and one urethra (Fig. 20-1). The excretion of urine and its elimination from the body are vital functions, since together they constitute one of the most important mechanisms for maintaining homeostasis. As one individual has phrased it: "The composition of the blood (and internal environment) is determined not by what the mouth ingests but by what the kidney keeps."*

The substances excreted from the kidneys and other excretory organs are listed in Table 20-1.

Kidneys

Size, shape, and location

The kidneys resemble lima beans in shape. An average-sized kidney measures approximately 11.25 cm in length, 5 to 7.5 cm in width, and 2.5 cm in thickness. Usually the left kidney is slightly larger than the right. The kidneys lie behind the parietal peritoneum. They are located on either side of the vertebral column and extend from the level of the last thoracic vertebra to the third lumbar vertebra—in short, the kidneys are located just above the waistline. Usually the right kidney is a little lower than the left (Fig. 20-1), presumably because the liver

*From Smith, H. W.: Lectures on the kidney, Lawrence, Kan., 1943, University of Kansas, p. 3.

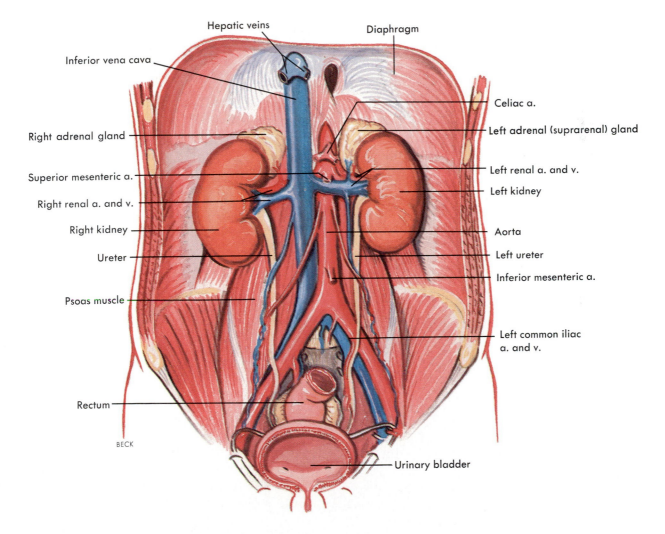

Hepatic veins

Diaphragm

Inferior vena cava

Celiac a.

Right adrenal gland

Left adrenal (suprarenal) gland

Superior mesenteric a.

Left renal a. and v.

Right renal a. and v.

Left kidney

Right kidney

Aorta

Ureter

Left ureter

Inferior mesenteric a.

Psoas muscle

Left common iliac
a. and v.

Rectum

BECK

Urinary bladder

Fig. 20-1 Location of urinary system organs.

takes up some of the space above the right kidney. A heavy cushion of fat normally encases each kidney and holds it up in position. Very thin individuals may suffer from ptosis (dropping) of one or both of these organs. Connective tissue (renal fasciae) anchors the kidneys to surrounding structures and helps to maintain their normal position.

External structure

The mesial surface of each kidney presents a concave notch called the *hilum*. Structures enter the kidneys through this notch just as they enter the lung through its hilum. A tough white fibrous capsule encases each kidney.

Table 20-1 Excretory organs of the body

Excretory organ	Substance excreted
Kidneys	Nitrogenous wastes (from protein catabolism)
	Toxins (for example, from bacteria)
	Water (from ingestion and from catabolism)
	Mineral salts
Skin (sweat glands)	Water
	Mineral salts
	Small amounts of nitrogenous wastes
Lungs	Carbon dioxide (from catabolism)
	Water
Intestine	Wastes from digestion (cellulose, connective tissue, etc.)
	Some metabolic wastes (for example, bile pigments; also salts of calcium and other heavy metals)

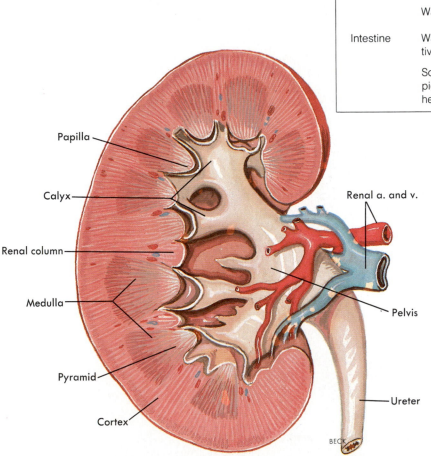

Fig. 20-2 Coronal section through right kidney.

Papilla

Calyx

Renal column

Medulla

Pyramid

Cortex

Renal a. and v.

Pelvis

Ureter

BECK

Internal structure

Fig. 20-2, a coronal section through the kidney, shows its internal structure. Identify the *cortex*, or outer portion, and the *medulla*, or inner portion. Observe the shape and location of the *pyramids;* a dozen or so of these triangular wedges make up the medulla. The *base* or wide margin of a pyramid faces out toward the cortex, and its narrow end or *papilla* juts into a division of the kidney's *pelvis* called a *calyx*. Identify these renal structures in Fig. 20-2. Another internal structural feature is that the cortex dips in between each two pyramids, forming the so-called *renal columns*. The pyramids have a striated appearance, as contrasted with the smooth texture of the cortical substance.

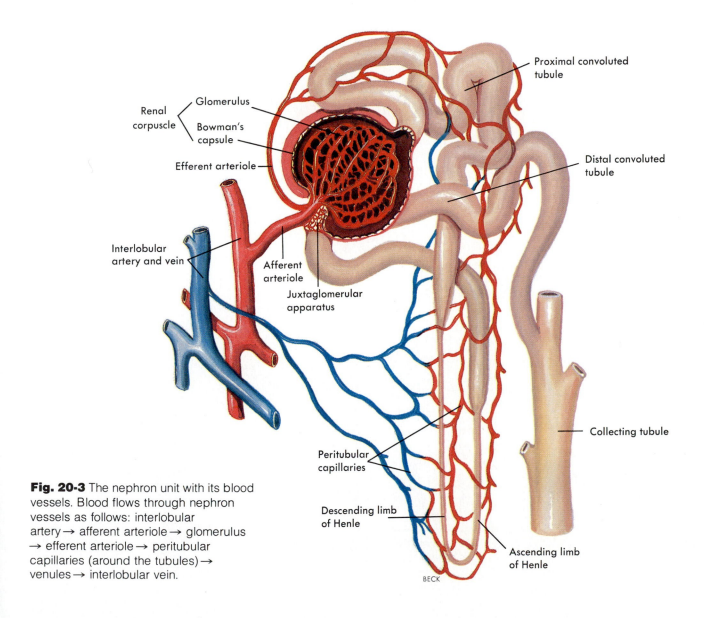

Fig. 20-3 The nephron unit with its blood vessels. Blood flows through nephron vessels as follows: interlobular artery → afferent arteriole → glomerulus → efferent arteriole → peritubular capillaries (around the tubules) → venules → interlobular vein.

Microscopic structure

Microscopic structures, named *nephrons* and numbering about 1¼ million per kidney, make up the bulk of kidney substance. The shape of a nephron is unique, unmistakable, and admirably suited to its functions. It resembles a tiny funnel with a very long stem, but a most unusual one in that it is not straight but highly convoluted

in parts. Look now at Fig. 20-3 to identify each of the following parts of a nephron.

1 *Bowman's capsule*—the cup-shaped mouth of the nephron formed by two layers of flat epithelial cells with a space between the layers. Invaginated in the Bowman's capsule is one of the body's most famous networks of capillaries and one of the most im-

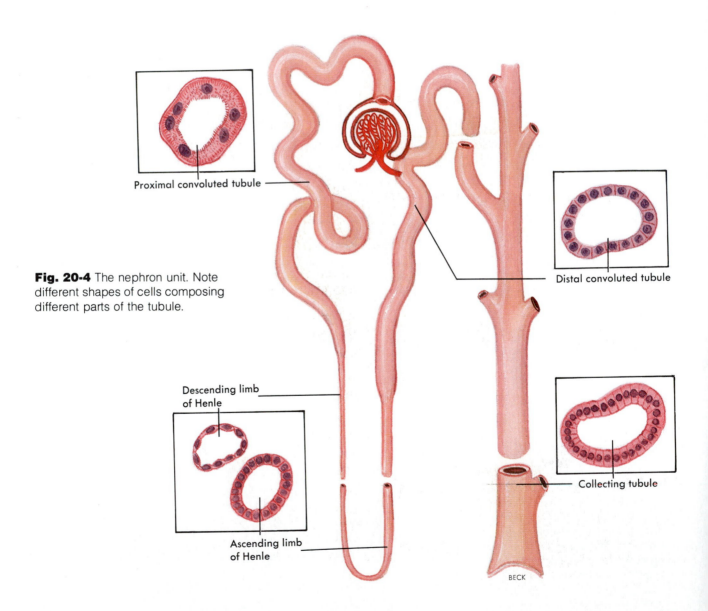

Proximal convoluted tubule

Distal convoluted tubule

Fig. 20-4 The nephron unit. Note different shapes of cells composing different parts of the tubule.

Descending limb of Henle

Ascending limb of Henle

Collecting tubule

BECK

portant ones for survival. Remember its name—*glomerulus*. A Bowman's capsule with its partially encased glomerulus is called a *renal* (or malpighian) *corpuscle*. Some renal corpuscles lie in the cortex of the kidney and others in its medulla.

2 *Proximal tubule*—the first part of the tubule, that is, the part nearest or proximal to Bowman's capsule. Note its winding, convoluted course. Its wall, as you can see in Fig. 20-4, consists of one layer of epithelial cells that have a brush border facing the lumen (interior) of the tubule. Thousands of microvilli form the brush border of a proximal tubule cell and greatly increase its luminal surface area—a fact related to its function, as we shall see.

3 *Loop of Henle*—the portion of the tubule just beyond the proximal tubule, consisting of a descending and an ascending limb.

a *Descending limb*—a straight extension of the tubule down into the medulla of the kidney. Observe in Fig. 20-4 how the flat, squamous epithelial cells that form the wall of the descending limb differ from those of the ascending limb and from proximal tubule cells. Notice, too, that the descending limb is a thin segment of tubule. It loops back upward in the medulla to become the ascending limb.

b *Ascending limb*—the thick segment of the loop of Henle.

4 *Distal tubule*—a convoluted portion of the tubule located distally to Bowman's capsule.

5 *Collecting tubule*—a straight tubule joined by the distal tubules of several nephrons.

Collecting tubules join larger tubules, and all the larger collecting tubules of one renal pyramid converge to form one tube that opens at a renal papilla into one of the small calyces. Bowman's capsules and both convoluted tubules lie in the cortex of the kidney, whereas the loops of Henle and collecting tubules lie in its medulla.

Blood vessels of kidneys

The kidneys are highly vascular organs. Every minute about 1,200 ml of blood flows through them. Stated another way, approximately one fifth of all the blood pumped by the heart per minute goes to the kidneys. From this fact, one might guess, and correctly so, that the kidneys process the blood in some way before returning it to the general circulation. A large branch of the abdominal aorta—the renal artery—brings blood into each kidney. Between the pyramids of the kidney's medulla the renal artery branches to form interlobar arteries that extend out toward the cortex, then arch over the bases of the pyramids to form the arcuate arteries. From the arcuate arteries, interlobular arteries penetrate the cortex. Branches of the interlobular arteries are the afferent arterioles shown in Fig. 20-3. As you can see in this illustration, afferent arterioles branch into the capillary networks called glomeruli. Recall that the usual direction of blood flow is arteries → arterioles → capillaries → venules → veins. Not so in the kidneys. Here, blood flows as follows: arteries → afferent arteriole → capillaries (of the glomeruli) → efferent arterioles → capillaries (peritubular capillaries) → venules → veins. Identify these blood vessels in Fig. 20-3. Also find the *juxtaglomerular apparatus*, a small structure but an important one because it helps regulate the volume of blood flowing into the glomerulus.

Functions

The function of the kidneys is to excrete urine, a life-preserving function because homeostasis depends on it. For example, the kidneys are the most important organs in the body for maintaining fluid and electrolyte and acid-base balance. They do this by varying the amounts of water and electrolytes leaving the blood in the urine so that they equal the amounts of these substances entering the blood from various avenues. Nitrogenous wastes from protein metabolism, notably urea, leave the blood by way of the kidneys.

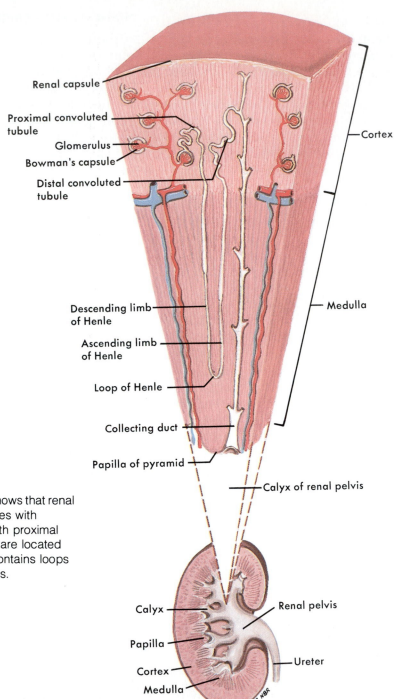

Renal capsule

Proximal convoluted
tubule

Glomerulus

Bowman's capsule

Distal convoluted
tubule

Cortex

Descending limb
of Henle

Ascending limb
of Henle

Loop of Henle

Collecting duct

Papilla of pyramid

Medulla

Calyx of renal pelvis

Fig. 20-5 Magnified wedge shows that renal
corpuscles (Bowman's capsules with
invaginated glomeruli) and both proximal
and distal convoluted tubules are located
in cortex of kidney. Medulla contains loops
of Henle and collecting tubules.

Calyx

Papilla

Cortex

Medulla

Renal pelvis

Ureter

AFTER NBR

Here are just a few of the blood constituents that cannot be held to their normal concentration ranges if the kidneys fail: sodium, potassium, chloride, and nitrogenous wastes from protein metabolism such as urea. In short, kidney failure means homeostasis failure and, if not relieved, inevitable death.

In addition to excreting urine, the kidneys are now known to influence blood pressure (see p. 555).

How kidneys form urine

The two main parts of a nephron, that is, Bowman's capsule with its invaginated glomerulus and the renal tubules, form urine by means of three processes: filtration, reabsorption, and

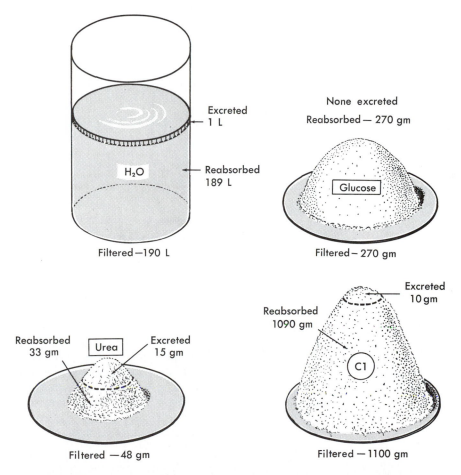

Fig. 20-6 Glomerular filtration and tubular reabsorption volumes. Note the enormous volume of water filtered out of glomerular blood per day—190 liters or many times the total volume of blood in the body. Only a very small proportion of this, however, is excreted in the urine. More than 99% of it (189 liters) is reabsorbed into tubular blood. One other substance shown is also filtered and reabsorbed in large amounts. Which one? Another is normally entirely reabsorbed. Which one? Somewhat more than half of the filtered amount of which substance is reabsorbed?

secretion. The following paragraphs discuss these processes. Fig. 20-6 indicates amounts of water and certain solutes filtered and reabsorbed.

Glomerular filtration. Filtration, the first step in urine formation, is a physical process that occurs in the kidneys' 2½ million or so renal corpuscles. As blood flows through glomerular capillaries, water and solutes filter out of the blood into Bowman's capsules. This filtration takes place through the glomerular-capsular membrane. It consists of the following layers: *endothelium* (a single layer of cells that forms the walls of the glomerular capillaries), *basement membrane* (a thin layer of highly permeable, nonliving material that completely surrounds the endothelium), and *epithelium* (a single layer of cells that forms the inner surface of Bowman's capsules, that is, the surface adjacent to the glomeruli). Filtration occurs more rapidly out of glomerular capillaries into Bowman's capsules than out of ordinary tissue

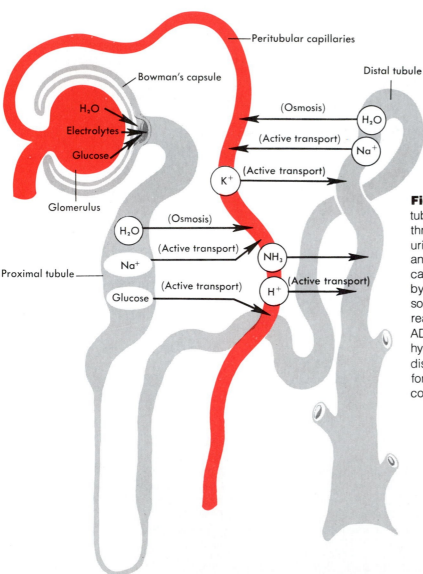

Fig. 20-7 Diagram showing glomerular filtration, tubular reabsorption, and tubular secretion—the three processes by which the kidneys excrete urine. From proximal convoluted tubules, sodium and glucose are reabsorbed into peritubular capillaries by active transport. Water reabsorption by osmosis follows. From distal convoluted tubules, sodium is reabsorbed by active transport. Osmotic reabsorption of water from them occurs when ADH is present. Secretion of ammonia and hydrogen occurs from peritubular capillaries into distal tubules by active transport. See Fig. 20-9 for reabsorption from loops of Henle by the countercurrent mechanism.

capillaries into tissue spaces. One reason for this is a structural difference between the endothelium of glomeruli and that of tissue capillaries. Glomerular endothelium has many more pores in it, so is more permeable than tissue capillary endothelium. Another reason for more rapid glomerular than tissue capillary filtration is that different types of vessels carry blood away from these two kinds of capillaries. An efferent arteriole carries blood out of a glomerulus, and it is smaller in diameter than the afferent arteriole that brings blood into it. This offers a relatively high resistance to blood flow out of the glomeruli, which, in turn, produces a relatively high hydrostatic pressure in them. In contrast, venules carry blood away from tissue capillaries, and they have a somewhat larger diameter than do the arterioles that carry blood into them.

Fluid filters out of glomeruli into Bowman's capsules for the same reason that it moves out of other capillaries into interstitial fluid or moves from any area to another—because a pressure

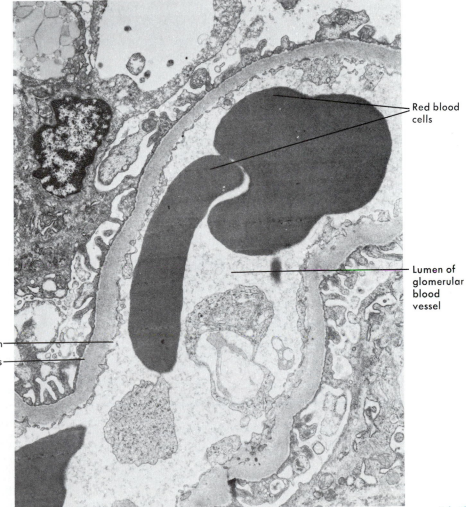

Fig. 20-8 Electron micrograph of glomerular wall (×13,600). The large dark bodies are red blood cells. Note the many small pores in the capillary endothelium. The wall of Bowman's capsule surrounds the glomerulus.

Courtesy Dr. James Goforth and Dr. William A. Anthony, Amarillo, Texas.

Red blood cells

Lumen of glomerular blood vessel

Pore in endothelium

Bowman's capsule

gradient exists between the two areas. Normally, glomerular hydrostatic pressure, blood colloidal osmotic pressure, and capsular hydrostatic pressure (that is, in Bowman's capsules) together determine the pressure gradient (commonly spoken of as the *effective filtration pressure* or EFP) between glomeruli and capsules. Of these three pressures the main driving force, the main determinant of the effective filtration pressure, is glomerular hydrostatic pressure. It tends to move fluid out of the glomeruli. In contrast, the *capsular* hydrostatic pressure and the blood colloidal osmotic pressure both exert force in the opposite direction. (If you do not recall why osmostic pressure is a "water-pulling" rather than a "water-pushing" force, reread p. 45.) Suppose that there is a glomerular hydrostatic pressure of 60 mm Hg and that it is opposed by a capsular hydrostatic pressure of 18 and a blood colloidal osmotic pressure of 32. There would then be a net or effective filtration pressure of 10 mm Hg (60 − 18 − 32). In other words, an effective filtration pressure of 10 mm Hg would cause fluid to filter out of the glomeruli into the Bowman's capsules. Since 1 mm Hg effective filtration pressure produces a glomerular filtration rate of 12.5 ml per minute from both kidneys, the glomerular filtration rate with 10 mm Hg effective filtration pressure is (10 × 12.5) or 125 ml per minute, a normal rate.

Table 20-2 Normal and abnormal glomerulocapsular pressures

	Hydrostatic pressure	Colloidal osmotic pressure
■ Normal		
Glomerular blood	60 mm Hg	32 mm Hg
Capsular filtrate	18 mm Hg	—
■ Abnormal		
Glomerular blood	50 mm Hg	32 mm Hg
Capsular filtrate	18 mm Hg	—

Another factor, *capsular* colloidal osmotic pressure, operates when disease has increased glomerular permeability enough to allow blood protein molecules to diffuse out of the blood into Bowman's capsule. Under these circumstances the capsular filtrate exerts an osmotic pressure in opposition to blood osmotic pressure. Capsular osmotic pressure tends to draw water out of the blood and so constitutes another force to be added to glomerular hydrostatic pressure in determining the effective filtration pressure.

By using the following formula and figures given in Table 20-2, you will be able to figure out for yourself how changes in the various pressures cause changes in the glomerular filtration rate.

$$
\begin{aligned}
\text{Effective} \\
\text{filtration} \\
\text{pressure}
\end{aligned}
=
\left(
\begin{aligned}
\text{Glomerular} \\
\text{hydrostatic} \\
\text{pressure}
\end{aligned}
+
\begin{aligned}
\text{Capsular} \\
\text{osmotic} \\
\text{pressure}
\end{aligned}
\right)
-
\left(
\begin{aligned}
\text{Glomerular} \\
\text{osmotic} \\
\text{pressure}
\end{aligned}
+
\begin{aligned}
\text{Capsular} \\
\text{hydrostatic} \\
\text{pressure}
\end{aligned}
\right)
$$

For example, glomerular hydrostatic pressure may decrease sharply under stress conditions such as that following severe hemorrhage. Table 20-2 gives pressures that might exist under these circumstances. Using these figures, what do you compute the effective glomerular filtration pressure to be? Do you think it is true, after doing this computation, that the effective filtration pressure falls to 0 and glomerular filtration ceases entirely when glomerular hydrostatic pressure falls below a certain critical level?

Kidney disease, as previously noted, sometimes causes a loss of blood proteins in the urine. Blood protein concentration then decreases and causes blood colloidal osmotic pressure to decrease. Suppose that blood colloidal osmotic pressure were to decrease to 20 mm Hg while the other normal figures given in Table 20-2 remained the same. Use these figures to compute the effective filtration pressure. If, based on this example, you were to formulate a principle, would you say that a decrease in blood protein concentration tends to increase or decrease the glomerular filtration rate? (Check your answer in the paragraph after the next one.)

Glomerular hydrostatic pressure is regulated by mechanisms that change the size of the afferent and efferent arterioles and is also influenced by changes in systemic blood pressure. For example, sympathetic impulses cause constriction of both afferent and efferent arterioles. But with intense sympathetic stimulation the afferent arteriole becomes much more constricted than the efferent. Consequently, glomerular hydrostatic pressure falls. Sometimes under severe stress conditions, it drops to a level too low to maintain filtration, and the kidneys "shut down" completely. In more technical language, renal suppression occurs, for example, see Table 20-2. Glomerular hydrostatic pressure and filtration are directly related to systemic blood pressure. By this, we mean that a decrease in blood pressure tends to produce a decrease in glomerular pressure and in the filtration rate. The converse is also true. When arterial pressure increases, however, a smaller increase in glomerular pressure follows because the afferent arterioles automatically constrict, thereby decreasing blood flow into the glomeruli and preventing a marked rise both in glomerular pressure and in the glomerular filtration rate. For instance, when the mean arterial pressure doubles, the glomerular filtration rate reportedly increases only 15% to 20%.

Glomerular filtration is inversely related to blood colloidal osmotic pressure and blood protein concentration. Specifically, a decrease in blood protein concentration causes a decrease in blood osmotic pressure, which in turn causes an increase in glomerular filtration. Normally the glomerular filtration rate averages about 125 ml per minute in men and somewhat less in women.

Reabsorption from proximal convoluted tubules. Reabsorption, the second step in urine formation, takes place by means of both passive and active transport mechanisms from all parts of the renal tubules. A major portion of water and electrolytes and normally all nutrients are, however, reabsorbed from proximal convoluted tubules.

Proximal tubules reabsorb sodium in this manner. First, sodium ions diffuse into the epithelial cells that form the walls of the proximal tubules. (They diffuse through the microvilli or brush border of these cells, that is, the border next to the lumen of the tubules, into their cytoplasm.) Almost immediately the sodium ions are actively transported through the basal and lateral borders of the epithelial cells and move on through the basement membrane and the endothelium of the peritubular capillaries into peritubular blood. As positive sodium ions move out of the tubular cells, their cytoplasm becomes momentarily electronegative to the fluid in the tubules and the cytoplasm's sodium concentration becomes less than that of the tubular fluid. The effect of these two changes is to create an electrochemical gradient that causes further diffusion of sodium ions into epithelial cells of the tubules, followed by further active transport of sodium out of them into peritubular blood. The addition of sodium ions to peritubular blood makes it momentarily electropositive to tubular fluid and causes an equal number of negative ions, notably chloride, to diffuse out of the tubular fluid into the peritubular blood.

Proximal tubules reabsorb water by osmosis. The movement of sodium out of tubular fluid into peritubular blood creates, momentarily, a higher osmotic pressure in peritubular blood than in tubular fluid. Therefore water, obeying the law of osmosis, moves rapidly into the peritubular blood to reestablish osmotic equilibrium between the two fluids. In short, sodium transport out of the proximal tubules causes water osmosis out of them. This is obligatory water reabsorption—obligatory because it is demanded by the law of osmosis.

Proximal tubules reabsorb nutrients, notably glucose and amino acids, into peritubular blood by active transport mechanisms. Normally, all the glucose that filters out of the glomerular blood is returned by the proximal tubules to

Fig. 20-9 Countercurrent mechanism for making possible hypotonic or hypertonic urine excretion. Basically, it is a device for producing and maintaining high osmolality in the interstitial fluid of the renal medulla. How does it do this? See text, p. 552. (Modified from Mountcastle, V. B., editor: Medical physiology, ed. 13, St. Louis, 1974, The C. V. Mosby Co.)

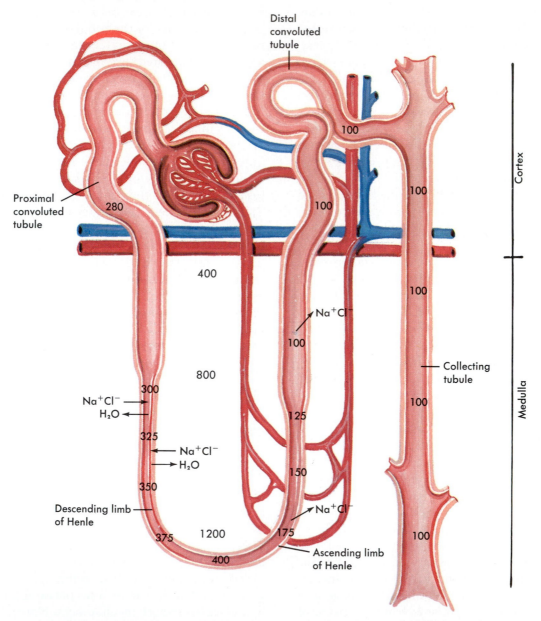

A, Hypotonic urine excretion, no ADH in blood. Figures on diagram indicate milliosmols of pressure (see footnote p. 552).

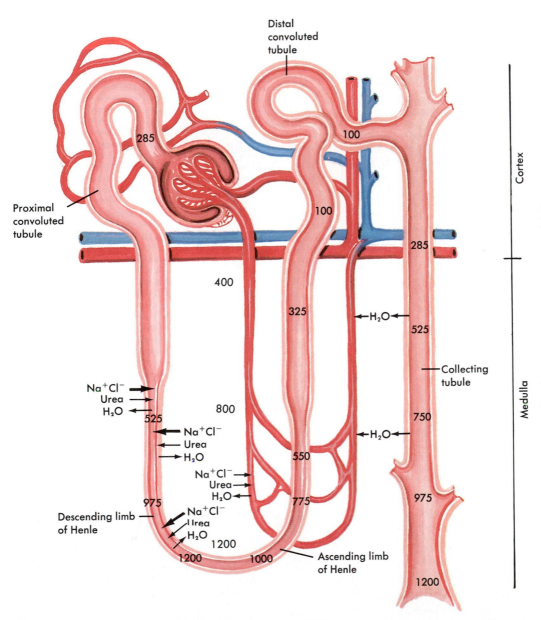

B, Hypertonic urine excreted (ADH present in blood). Figures on diagram indicate milliosmols of solute per liter of fluid.

the peritubular blood, so normally, no glucose at all is lost in the urine. If, however, blood glucose level exceeds a certain threshhold amount (often around 150 mg per 100 ml of blood), the glucose transport mechanism cannot reabsorb all of it, and the excess remains in the urine. In other words, this mechanism has a maximum capacity for moving glucose molecules back into the blood. Occasionally, this maximum transfer capacity is greatly reduced and glucose appears in the urine (glycosuria), even though the blood sugar level may be normal. This condition is known as renal diabetes or renal glycosuria. It is a congenital defect.

Reabsorption from the loop of Henle. The loops of Henle lie in the medulla of the kidney and are long, narrow, U-shaped tubes that function as a countercurrent mechanism. Basically, this countercurrent mechanism is a device for producing and maintaining high solute concentration (high osmolality) in the interstitial fluid in the kidney's medulla. It does this by actively transporting sodium chloride into the medullary interstitial fluid from the tubular fluid in the thick segment of the ascending limb of the loop of Henle. Water, however, does not follow sodium chloride into the interstitium for a very good reason—because the thick ascending limb is virtually impermeable to water. With sodium chloride but not water added to the interstitial fluid, its osmolality necessarily increases. And although some of the sodium chloride diffuses out of the interstitial fluid into the descending limb of the loop, it quickly returns to the interstitium. Here is the reason. Just as soon as the tubular fluid flows the minute distance around the bottom of the loop and into the ascending limb, sodium chloride is again transported out into the interstitial fluid. In essence, then, the countercurrent mechanism traps large amounts of sodium chloride in the medulla's interstitial fluid and thereby maintains high osmolality in it. The countercurrent mechanism also accomplishes something else. It lowers the osmolality of the tubular fluid as it moves up the ascending

limb and on into the distal tubules. Fig. 20-9 illustrates these changes. Is the countercurrent mechanism important? And if it is, why? The answer to the first question is an emphatic "Yes." The answer to the second question we shall postpone until after we have considered reabsorption from the distal and collecting tubules.

Reabsorption from distal and collecting tubules. Distal tubules, like proximal tubules, reabsorb sodium but in much smaller amounts. (Proximal tubules, under usual conditions, reabsorb about two thirds and distal tubules reabsorb about one tenth of the amount of sodium filtered from the glomeruli.) Distal tubules also reabsorb water. The amount of water reabsorbed from them and from the collecting tubules varies markedly and is regulated mainly by the amount of antidiuretic hormone (ADH) present in blood. If no ADH is present, both distal and collecting tubules are practically impermeable to water, so little or no water is reabsorbed from them. Consequently, osmolality of the fluid in these tubules does not change and the urine excreted in the absence of ADH has about the same osmolality* as the fluid leaving the ascending limb of the loop of Henle (100 milliosmols, according to Fig. 20-9). It is strongly hypotonic to blood, which has a normal osmolality of about 300 milliosmols.

When ADH is present in the blood, it acts on the kidneys in some way, not well understood, to make their distal and collecting tubules permeable to water. Therefore water osmoses out of the urine in these tubules into the hypertonic

---

*Osmolality is the osmotic pressure of a solution expressed as the number of osmols of pressure per kilogram of water. One osmol of pressure is produced when 1 gm molecular weight of a nonionizing substance is dissolved in 1 kg of water. An ionizing substance produces 1 osmol of pressure when 1 gm molecular weight of the substance, divided by the number of ions it forms per molecule, is dissolved in 1 kg of water. Example: Sodium chloride forms 2 ions per molecule and has a molecular weight of 58. Therefore 29 (58/2) grams of sodium chloride dissolved in 1 kg of water produces 1 osmol of pressure.

interstitial fluid in the renal medulla and eventually returns to the blood in the vasa recta (blood vessels shaped like and adjacent to the loops of Henle). This reabsorption of water from the distal and collecting tubules continues until the osmolality of the urine in them equilibrates with that of the medullary interstitial fluid (about 1,200 milliosmols). A highly concentrated urine, one hypertonic to blood, is therefore excreted in the presence of ADH. Also, a smaller volume of urine is excreted when ADH is present—a fact implied by its name antidiuretic hormone; antidiuretic means against the production of a large volume of urine.

Now we shall try to answer the question previously posed as to why the countercurrent mechanism is important. Briefly because by cre-

ating and maintaining a high degree of osmolality in the kidney's medullary interstitium, the countercurrent mechanism enables the kidney to excrete urine that is hypertonic to blood. By so doing, it conserves water for the body. Vertebrates whose kidneys have no loops of Henle cannot perform this function. It can be a life-saving one, for example, if blood volume decreases precipitously, as following hemorrhage.

Tubular secretion. In addition to reabsorption, tubule cells also secrete certain substances. Tubular secretion (or tubular excretion, as it is also called) means the movement of substances out of the blood into the tubular fluid. Tubular reabsorption, you recall, means the movement of substances in the opposite direction, that is, out of the tubular fluid into the blood. In both

Table 20-3 Functions of different parts of nephron in urine formation

Part of nephron	Function	Substance moved
Glomerulus	Filtration	Water
		All solutes except colloids such as blood proteins
Proximal tubule	Reabsorption by active transport	Na^+ and probably some other ions; nutrients—glucose and amino acids
	Reabsorption by diffusion (secondary to active transport)	Cl^-, HCO_3^-, and probably some other ions; also about 50% of urea
	Obligatory water reabsorption by osmosis	Water
Loop of Henle Descending limb	Reabsorption by diffusion	NaCl
Ascending limb	Reabsorption by active transport into interstitial fluid of renal medulla	NaCl (not followed by water reabsorption)
Distal and collecting tubules	Reabsorption by active transport	Na^+ and probably some other ions
	Facultative water reabsorption by osmosis (ADH controlled)	Water
	Secretion by diffusion	Ammonia
	Secretion by active transport	K^+, H^+, and some drugs

reabsorption and secretion, some substances are moved by active transport and some by passive mechanisms (diffusion and osmosis). Example: Sodium ions and glucose molecules are reabsorbed; potassium and hydrogen ions are secreted by active transport mechanisms. But water is reabsorbed by osmosis, and ammonia is secreted by diffusion—both passive processes. Tubule cells also secrete certain drugs, for example, penicillin and para-aminohippuric acid (PAH). Table 20-3 and Fig. 20-7 summarize the functions of the different parts of the nephron in forming urine.

Regulation of urine volume

Several factors regulate the volume of urine produced. Paramount among them is the presence or absence of ADH in the blood. As stated on p. 552, ADH makes the distal and collecting tubules permeable to water. This necessarily increases water reabsorption from these tubules, which, in turn, decreases urine volume. Another hormone that tends to decrease urine volume is aldosterone from the adrenal cortex. It increases distal tubule absorption of sodium and other ions, and this leads to increased water reabsorption by osmosis and thereby decreases urine volume.

Urine volume generally relates directly to extracellular fluid (ECF) volume. If ECF volume decreases, urine volume also decreases. Conversely, if ECF volume increases, urine volume increases. This presumably comes about indirectly by the change in ECF volume acting to stimulate or inhibit ADH secretion. The exact mechanism by which it accomplishes this is still unproved. It may operate through its effect on ECF sodium concentration. When ECF volume decreases, its sodium concentration tends to increase and this change triggers ADH secretion (Fig. 20-10), followed by a decrease in urine volume. On the other hand, when ECF volume increases, its sodium concentration tends to decrease and to inhibit ADH secretion, followed by an increase in urine volume. The common experience of increased output following rapid ingestion of a large amount of fluid attests to the fact than an increased ECF volume is soon followed by increased urinary output. Oliguria or even anuria in dehydrated patients gives evidence that decreased urinary output follows a decrease in ECF volume.

Urine volume relates directly to the total amount of solutes excreted in the urine—the more solutes, the more urine. Higher solute concentration in urine osmotically draws more water into renal tubules and thus tends to increase urine volume. Probably the best known example of this occurs in untreated diabetes mellitus. The symptom that often brings an unknown diabetic to a physician is the voiding of abnormally large amounts of urine. Excess glucose "spills over" into urine, increasing its solute concentration and leading to diuresis.

Urine volume is not normally altered by changes in the glomerular filtration rate. It remains remarkably constant despite changes in renal blood flow and in glomerular hydrostatic pressure. Under certain pathological conditions, however, the glomerular filtration rate may change enough to alter urine volume.

In summary, then, the volume of urine produced is regulated, not by changes in the amount of water filtered from the glomeruli, but by changes in the amount reabsorbed from the distal and collecting tubules. Chief among the factors that regulate water reabsorption from these tubules are the posterior pituitary hormone ADH and the adrenal cortex hormone aldosterone.

Tests used to evaluate kidney function

The test most commonly used by physicians to evaluate kidney function of patients is probably the measurement of serum creatinine. Normally, no significant changes occur in the amount of creatinine in blood serum because it is determined by a factor that does not readily change, namely, skeletal muscle mass. Elevation of the serum creatinine (above 1.5 mg per

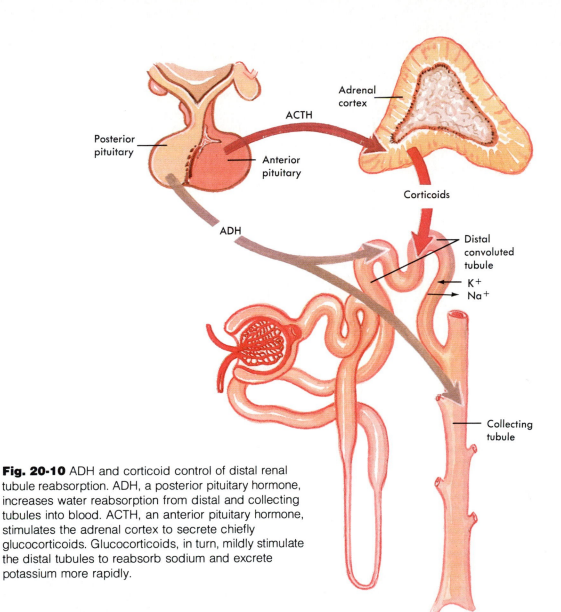

Fig. 20-10 ADH and corticoid control of distal renal tubule reabsorption. ADH, a posterior pituitary hormone, increases water reabsorption from distal and collecting tubules into blood. ACTH, an anterior pituitary hormone, stimulates the adrenal cortex to secrete chiefly glucocorticoids. Glucocorticoids, in turn, mildly stimulate the distal tubules to reabsorb sodium and excrete potassium more rapidly.

100 ml), therefore, is considered a reliable indication of depressed renal function.

Influence of kidneys on blood pressure

Clinical observation and animal experiments have established the fact that destruction of a large proportion of total kidney tissue usually results in the development of hypertension. This happens frequently, for example, in patients who have severe renal arteriosclerosis or glomerulonephritis. Many experiments have been performed and various theories devised to explain the mechanism responsible for "renal hypertension." Ischemic kidneys are known to produce a proteolytic enzyme, *renin*, which hydrolyzes one of the blood proteins to produce *angiotensin* (Fig. 12-14). Angiotensin tends to increase blood pressure by constricting arterioles.

Ureters

Location and structure

The two ureters are tubes from 10 to 12 inches long. At their widest point, they measure less than ½ inch in diameter. They lie behind the parietal peritoneum and extend from the kidneys to the posterior surface of the bladder. As the upper end of each ureter enters the kidney, it enlarges into a funnel-shaped basin named the *renal pelvis*. The pelvis expands into several branches called *calyces*. Each calyx contains a renal papilla. As urine is secreted, it drops out of the collecting tubules, whose opening are in the papillae, into the calyces, then into the pelvis, and down the ureters into the bladder.

The walls of the ureters are composed of three coats: a lining coat of mucous membrane, a middle coat of two layers of smooth muscle, and an outer fibrous coat.

The ureters enter the bladder through an oblique tunnel that functions as a valve, preventing backflow of urine into the ureters during bladder contraction.

Function

The ureters, together with their expanded upper portions, the pelves and calyces, collect the urine as it forms and drain it into the bladder. Peristaltic waves (about one to five per minute) force the urine down the ureters and into the bladder.

Correlations

Stones known as *renal calculi* sometimes develop within the kidney. Urine may wash them into the ureter, where they cause extreme pain if they are large enough to distend its walls.

Bladder

Structure and location

The urinary bladder is a collapsible bag located directly behind the symphysis pubis. It lies below the parietal peritoneum, which covers only its superior surface. Smooth muscle fibers fashion the wall of the bladder. Some extend lengthwise of it, some obliquely, and some more or less circularly. Oblique, intersecting fibers at the sides of the bladder constitute the detrusor muscle. Mucous membrane, arranged in rugae, forms the bladder lining. Because of the rugae and the elasticity of its walls, the bladder is capable of considerable distention, although its capacity varies greatly with individuals. There are three openings in the floor of the bladder—two from the ureters and one into the urethra. The ureter openings lie at the posterior corners of the triangular-shaped floor (the trigone) and the urethral opening at the anterior and lower corner (Fig. 20-11).

Functions

The bladder performs two functions.
1 It serves as a reservoir for urine before it leaves the body.
2 Aided by the urethra, it expels urine from the body.

The mechanism for voiding urine starts with voluntary relaxation of the external sphincter muscle of the bladder. In rapid succession, reflex contraction of linear smooth muscle fibers along the urethra and then of the detrusor muscle follow. The combination of these events squeezes urine out of the bladder. Parasympathetic fibers transmit the impulses that cause contractions of the bladder and relaxation of the internal sphincter. Voluntary contraction of the external sphincter to prevent or terminate micturition is learned. Voluntary control of micturition is possible only if the nerves supplying the bladder and urethra, the projection tracts of the cord and brain, and the motor area of the cere-

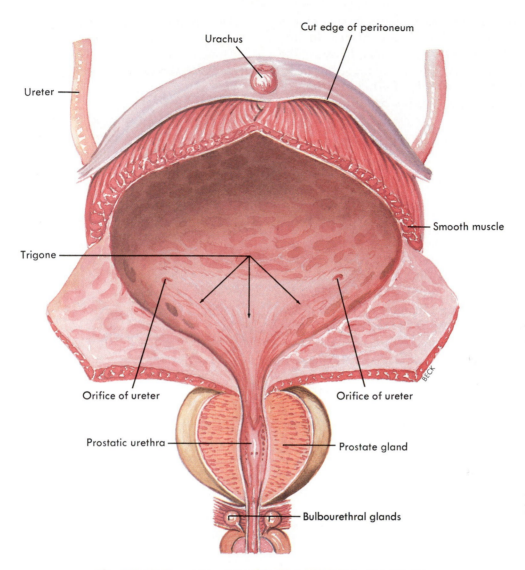

Ureter

Urachus

Cut edge of peritoneum

Smooth muscle

Trigone

Orifice of ureter

Orifice of ureter

Prostatic urethra

Prostate gland

Bulbourethral glands

BECK

Fig. 20-11 The male urinary bladder cut to show the interior.

brum are all intact. Injury to any of these parts of the nervous system, by a cerebral hemorrhage or cord injury, for example, results in involuntary emptying of the bladder at intervals. Involuntary micturition is called *incontinence*. In the average bladder, 250 ml of urine will cause a moderately distended sensation and therefore the desire to void.

Occasionally an individual is unable to void even though the bladder contains an excessive amount of urine. This condition is known as *retention*. It often follows pelvic operations and childbirth. Catheterization (introduction of a rubber tube through the urethra into the bladder to remove urine) is used to relieve the discomfort accompanying retention. A more serious complication, which is also characterized by the inability to void, is called *suppression*. In this condition the patient cannot void because the kidneys are not secreting any urine, and therefore the bladder is empty. Catheterization, of course, gives no relief for this condition.

Urethra

Structure and location

The urethra is a small tube leading from the floor of the bladder to the exterior. In the female, it lies directly behind the symphysis pubis and anterior to the vagina. It extends up, in, and back for a distance of about 1 to 1½ inches. The male urethra follows a tortuous course for a distance of approximately 8 inches. Immediately below the bladder, it passes through the center of the prostate gland, then between two sheets of white fibrous tissue connecting the pubic bones, and then through the penis, the external male reproductive organ. These three parts of the urethra are known, respectively, as the prostatic portion, the membranous portion, and the cavernous portion.

The opening of the urethra to the exterior is named the *urinary meatus*.

Mucous membrane lines the urethra as well as the rest of the urinary tract.

Functions

Since it is the terminal portion of the urinary tract, the urethra serves as the passageway for eliminating urine from the body. In addition, the male urethra is the terminal portion of the reproductive tract and serves as the passageway for eliminating the reproductive fluid (semen) from the body. The female urethra serves only the urinary tract.

Urine

Physical characteristics

The physical characteristics of normal urine are listed in Table 20-4.

Chemical composition

Urine is approximately 95% water, in which are dissolved several kinds of substances, the most important of which are discussed here.

Table 20-4 Physical characteristics of normal urine

Amount (24 hours)	Three pints (1,500 ml) but varies greatly according to fluid intake, amount of perspiration, and several other factors
Clearness	Transparent or clear; on standing, becomes cloudy
Color	Amber or straw colored; varies according to amount voided—less voided, darker the color, usually; diet also may change color, for example, reddish color from beets
Odor	"Characteristic"; on standing, develops ammonia odor from formation of ammonium carbonate
Specific gravity	1.015 to 1.020; highest in morning specimen
Reaction	Acid but may become alkaline if diet is largely vegetables; high-protein diet increases acidity; stale urine has alkaline reaction from decomposition of urea forming ammonium carbonate; normal range for urine pH 4.8 to 7.5; average, about 6; rarely becomes more acid than 4.5 or more alkaline than 8

1 *Nitrogenous wastes* from protein metabolism—such as urea (most abundant solute in urine), uric acid, ammonia, and creatinine

2 *Electrolytes*—mainly the following ions: sodium, potassium, ammonium, chloride, bicarbonate, phosphate, and sulfate; amounts and kinds of minerals vary with diet and other factors

3 *Toxins*—during disease, bacterial poisons leave the body in the urine—an important reason for "forcing fluids" on patients suffering with infectious diseases, so as to dilute the toxins that might damage the kidney cells if they were eliminated in a concentrated form

4 *Pigments*

5 *Hormones*

6 Various *abnormal constituents* sometimes found in urine—such as glucose, albumin, blood, casts, or calculi

Definitions

glycosuria or *glucosuria*—sugar (glucose) in the urine

hematuria—blood in the urine

pyuria—pus in the urine

casts—substances, such as mucus, that harden and form molds inside the uriniferous tubules and then are washed out into the urine; microscopic in size

dysuria—painful urination

polyuria—unusually large amounts of urine

oliguria—scanty urine

anuria—absence of urine

Outline summary

Kidneys

A Size, shape, and location
  1 11.25 cm × 5 to 7.5 cm × 2.5 cm
  2 Shaped like lima beans
  3 Lie against posterior abdominal wall, behind peritoneum, at level of last thoracic and first lumbar vertebrae; right kidney slightly lower than left

B External structure
  1 Hilum, concave notch on mesial surface
  2 Enveloping capsule of white fibrous tissue

C Internal structure
  1 Outer layer called cortex
  2 Inner portion called medulla
  3 Renal pyramids triangular wedges of medullary substance, apices of which called papillae
  4 Renal columns inward extensions of cortex between pyramids

D Microscopic structure
  1 Nephron—microscopic functional unit of kidneys; consists of renal corpuscle, and tubules
  2 Renal corpuscle—consists of glomerulus, a network of capillaries, and Bowman's capsule, in which glomerulus is invaginated
  3 Renal tubules—consist of proximal convoluted tubule, loop of Henle, distal convoluted tubule, and collecting tubule

E Functions
  1 Excrete urine, by which various toxins and metabolic wastes excreted and composition and volume of blood regulated
  2 Influence blood pressure

F How kidneys form urine (see Table 20-3 and Figs. 20-6, and 20-7)
  1 Filtration of substances from blood in glomeruli into Bowman's capsules
  2 Reabsorption of most of water and part of solutes from tubular filtrate into blood
  3 Secretion of K^+, H^+, NH_3, and some other substances into tubular filtrate from blood

G Regulation of urine volume by factors that change
  1 Amount of water reabsorbed from tubular filtrate into blood—most important determinant of amount of urine formed; posterior pituitary ADH increases water reabsorption from distal and collecting tubules; corticoids (especially aldosterone) increase sodium reabsorption and therefore also increase water reabsorption
  2 Rate of filtration from glomeruli—quite constant at about 125 ml per minute except in certain kidney diseases

Ureters

A Location and structure
  1 Lie retroperitoneally

2 Extend from kidneys to posterior part of bladder floor

3 Ureter expands as it enters kidney, becoming renal pelvis, which is subdivided into calyces, each of which contains renal papilla

4 Walls of smooth muscle with mucous lining and fibrous outer coat

B Function—collect urine and drain it into bladder

Bladder

A Structure and location
   1 Collapsible bag of smooth muscle lined with mucosa
   2 Lies behind symphysis pubis, below parietal peritoneum
   3 Three openings—one into urethra and two into ureters

B Functions
   1 Reservoir for urine
   2 Expels urine from body by way of urethra, called micturition, urination, or voiding; retention, inability to expel urine from bladder; suppression, failure of kidneys to form urine

Urethra

A Structure and location
   1 Musculomembranous tube lined with mucosa
   2 Lies behind symphysis, in front of vagina in female
   3 Extends through prostate gland, fibrous sheet, and penis in male
   4 Opening to exterior called urinary meatus

B Functions
   1 Female—passageway for expulsion of urine from body
   2 Male—passageway for expulsion of urine and of reproductive fluid (semen)

Urine

A Physical characteristics—see Table 20-4

B Chemical composition—consists of approximately 95% water in which are dissolved
   1 Wastes from protein metabolism (urea, uric acid creatinine, etc.)
   2 Mineral salts (sodium chloride main one but various others according to diet)
   3 Toxins—for example, from bacteria
   4 Pigments
   5 Sex hormones
   6 Abnormal constituents—for example, glucose, albumin, blood, casts, and calculi

C Definitions
   1 Glycosuria—sugar in urine
   2 Hematuria—blood in urine
   3 Pyuria—pus in urine
   4 Casts—microscopic bits of substances that harden and form molds inside tubules
   5 Dysuria—painful urination
   6 Polyuria—excessive amounts of urine
   7 Oliguria—scanty urine
   8 Anuria—absence of urine

Review questions

1 What four organs are excretory organs?
2 Which organs eliminate wastes of protein metabolism? Of digestion? Of carbohydrate and fat metabolism?
3 Name, locate, and give main function(s) of each organ of the urinary system.
4 How far and in which direction must a catheter be inserted to reach the bladder in the female? In the male?
5 Describe the microscopic structure of the kidney.
6 Describe the mechanism of urine formation, relating each step to part of the nephron that performs it.

Situation: An artificial kidney consists of a device in which blood flows directly from a patient's body through cellophane tubing immersed in a dialyzing fluid that contains prescribed amounts of various electrolytes and other substances. The two columns below show the composition of one patient's blood and of the dialyzing fluid used for his treatment with the artificial kidney.

	Blood plasma (in coiled tube) mEq/L	Dialyzing fluid (around coiled tube) mEq/L
Na^+	142	126
K^+	5	5
Ca^{++}	5	0
Mg^{++}	3	0
Cl^-	103	110
HCO_3^-	27	25
$HPO_4^=$	2	0
$SO_4^=$	1	0

	mg/100 ml	mg/100 ml
Glucose	100	1,750
Urea	26	0
Uric acid	4	0
Creatinine	1	0

7 What body structures do you think the cellophane tube substitutes for?

8 Which, if any, of the foregoing substances diffuse out of the blood into the dialyzing fluid? Give your reasons.

9 Which, if any, of these substances diffuse into the blood from the dialyzing fluid? Give your reasons.

10 Net diffusion of which, if any, of these substances does not pass through the cellophane membrane in either direction? Give your reasons.

11 What reasons can you see for having the dialyzing fluid contain the concentration of each substances given? What does it accomplish?

Situation: Results of a blood urea clearance test indicate that the glomerular filtration rate of a patient who has had a severe hemorrhage is less than 50% of normal.

12 Explain the mechanism responsible for the drop in the glomerular filtration rate.

13 Do you consider this a homeostatic mechanism? Does it serve a useful purpose? If so, what purpose?

14 What is the normal glomerular filtration rate?

15 What would you expect to be true of the volume of urine this patient would excrete—normal, polyuria, oliguria?

16 Define the following terms briefly: Bowman's capsule, calculi, calyces, casts, cystitis, dysuria, glomerulus, glycosuria, hematuria, incontinence, nephritis, oliguria, polyuria, ptosis, pyelitis, renal capsule, renal cortex, renal hilum, renal medulla, renal papilla, renal pelvis, renal pyramids, retention suppression.

Fluid, electrolyte, and acid-base balance

chapter 21

Fluid and electrolyte balance

Some general principles about fluid balance

Avenues by which water enters and leaves body

Mechanisms that maintain homeostasis of total fluid volume
Regulation of urine volume
Factors that alter fluid loss under abnormal conditions
Regulation of fluid intake

Mechanisms that maintain homeostasis of fluid and electrolyte distribution
Comparison of plasma, interstitial fluid, and
  intracellular fluid
Importance, distribution, and measurement of
  electrolytes in body fluids
Control of water and electrolyte movement between
  plasma and interstitial fluid
Control of water and electrolyte movement through
  cell membranes between interstitial and intracellular
  fluids

The term *fluid balance* means several things. It, of course, means the same thing as homeostasis of fluids. To say that the body is in a state of fluid balance is to say that the total amount of water in the body is normal and that it remains relatively constant. But fluid balance also means something more. It also means relative constancy of the distribution of water in the body's three fluid compartments. The volumes of water inside the cells, in the interstitial spaces, and in the blood vessels all remain relatively constant when a condition of fluid balance exists. Fluid imbalance, then, means that both the total volume of water in the body and the amount in one or more of its fluid compartments have increased or decreased beyond normal limits.

The bonds between molecules of certain organic substances such as glucose are such that they do not permit the compound to break up or dissociate in solution. Such compounds are called *nonelectrolytes*. Chemical compounds such as sodium chloride (NaCl) that do break up or dissociate in solution into separate particles (Na^+ and Cl^-) are known as *electrolytes*. The dissociated particles of an electrolyte are called *ions* and carry an electrical charge (discussed in Chapter 1).

Many electrolytes and their dissociated ions are of critical importance in fluid balance. Fluid balance and electrolyte balance are so interdependent that if one deviates from normal, so does the other. A discussion of one, therefore, necessitates a discussion of the other.

Modern medicine attaches great importance to fluid and electrolyte balance. Today a large proportion of hospital patients receive some kind of fluid and electrolyte therapy. To help you understand the rationale underlying such treatment, this chapter is included. We shall consider successively some general principles about fluid balance, the avenues by which water enters and leaves the body, the mechanisms that maintain homeostasis of total volume, and the mechanisms that maintain homeostasis of fluid and electrolyte distribution.

Some general principles about fluid balance

1 The cardinal principle about fluid balance is this: fluid balance can be maintained only if intake equals output. Obviously, if more water or less leaves the body than enters it, imbalance will result. Total fluid volume will increase or decrease but cannot remain constant under those conditions.

2 Devices for varying output so that it equals intake constitute the most crucial mechanism for maintaining fluid balance, but mechanisms for adjusting intake to output also operate. Fig. 21-1 illustrates the aldosterone mechanism for decreasing fluid output (urine volume) to compensate for decreased intake. Fig. 21-2 diagrams a postulated mechanism for adjusting intake to compensate for excess output.

3 Mechanisms for controlling water movement between the fluid compartments of the body constitute the most rapid-acting fluid balance devices. They serve first of all to maintain normal blood volume at the expense of interstitial fluid volume.

Avenues by which water enters and leaves body

Water enters the body, as everyone knows, from the digestive tract—in the liquids one drinks and in the foods one eats. But, in addition, and less universally known, water enters the body, that is, is added to its total fluid volume, from its billions of cells. Each cell produces water by catabolizing foods, and this water enters the bloodstream. Water normally leaves the body by four exits: kidneys (urine), lungs (water in expired air), skin (by diffusion and by sweat), and intestines (feces). In accord with the cardinal principle of fluid balance, the total volume of water entering the body normally equals the total volume leaving. In short, fluid intake normally equals fluid output. Fig. 21-3 illustrates the portals of water entry and exit, and Table 21-1 gives their normal volumes. These, however, can vary considerably and still be considered normal.

Mechanisms that maintain homeostasis of total fluid volume

Under normal conditions, homeostasis of the total volume of water in the body is maintained or restored primarily by devices that adjust output (urine volume) to intake and secondarily by mechanisms that adjust fluid intake.

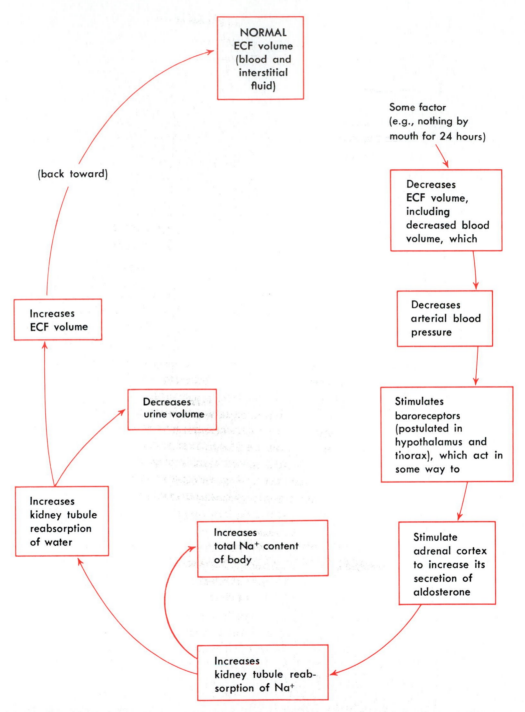

Fig. 21-1 Aldosterone mechanism that tends to restore normal extracellular fluid (ECF) volume when it decreases below normal. Excess aldosterone, however, leads to excess extracellular fluid volume, that is, excess blood volume (hypervolemia) and excess interstitial fluid volume (edema), and also to an excess of the total Na⁺ content of the body.

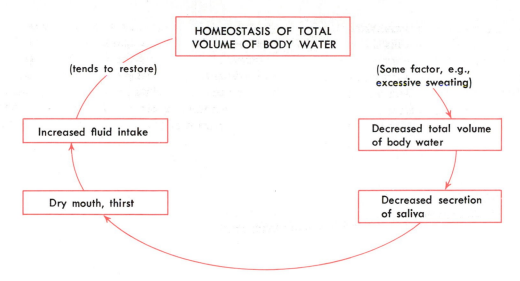

Fig. 21-2 The basic principle of a postulated homeostatic mechanism for adjusting intake to compensate for excess output.

Regulation of urine volume

Two factors together determine urine volume: the glomerular filtration rate and the rate of water reabsorption by the renal tubules. The glomerular filtration rate, except under abnormal conditions, remains fairly constant—hence it does not normally cause urine volume to fluctuate. The rate of tubular reabsorption of water, on the other hand, fluctuates considerably. The rate of tubular reabsorption, therefore, rather than the glomerular filtration rate, normally adjusts urine volume to fluid intake. The amount of ADH and of aldosterone secreted regulates the amount of water reabsorbed by the kidney tubules (discussed on p. 554; see also Fig. 21-1). In other words, urine volume is regulated chiefly by hormones secreted by the posterior lobe of the pituitary gland (ADH) and by the adrenal cortex (aldosterone).

Although changes in the volume of fluid loss via the skin, the lungs, and the intestines also affect the fluid intake-output ratio, these volumes are not automatically adjusted to intake volume, as is the volume of urine.

Factors that alter fluid loss under abnormal conditions

Both the rate of respiration and the volume of sweat secreted may greatly alter fluid output under certain abnormal conditions. For example, a patient who hyperventilates for an extended time loses an excessive amount of water via the expired air. If, as frequently happens, he also takes in less water by mouth than normal, his fluid output then exceeds his intake and he develops a fluid imbalance, namely, dehydration (that is, a decrease in total body water). Other abnormal conditions such as vomiting, diarrhea, or intestinal drainage also cause fluid and electrolyte output to exceed intake and so produce fluid and electrolyte imbalances.

Regulation of fluid intake

Physiologists disagree about the details of the mechanism for controlling intake so that it increases when output increases and decreases when output decreases. In general, it operates in this way: When dehydration starts to develop, salivary secretion decreases, producing a "dry

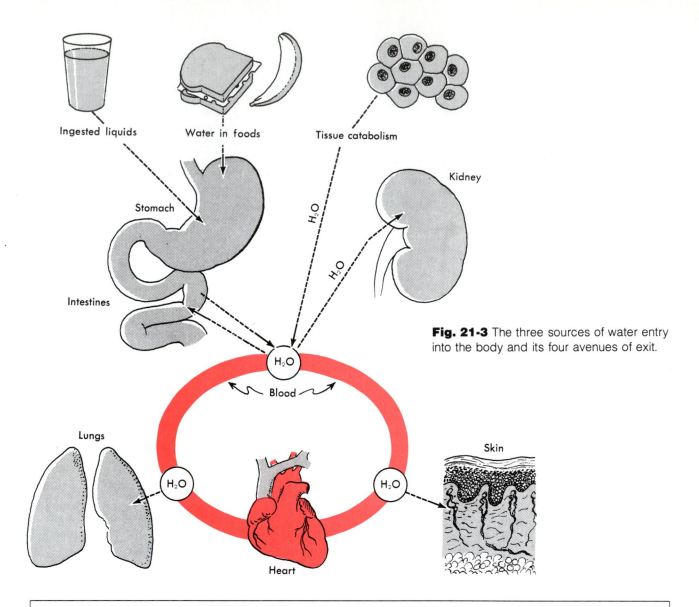

Fig. 21-3 The three sources of water entry into the body and its four avenues of exit.

Table 21-1 Typical normal values for each portal of water entry and exit (with wide variations)

Intake		Output	
Ingested liquids	1,500 ml	Kidneys (urine)	1,400 ml
Water in foods	700 ml	Lungs (water in expired air)	350 ml
Water formed by catabolism	200 ml	Skin	
		By diffusion	350 ml
		By sweat	100 ml
		Intestines (in feces)	200 ml
Totals	2,400 ml		2,400 ml

mouth feeling" and the sensation of thirst. The individual then drinks water, thereby increasing his fluid intake to offset his increased output, and this tend to restore fluid balance (Fig. 21-2). If, however, an individual takes nothing by mouth for several days, fluid balance cannot be maintained despite every effort of homeostatic mechanisms to compensate for the zero intake. Obviously, under this condition, the only way balance could be maintained would be for fluid output to also decrease to zero. But this cannot occur. Some output is obligatory. Why? Because as long as respirations continue, some water leaves the body by way of the expired air. Also, as long as life continues, an irreducible minimum of water diffuses through the skin.

Mechanisms that maintain homeostasis of fluid and electrolyte distribution
Comparison of plasma, interstitial fluid, and intracellular fluid

Structurally speaking, body fluids occupy three fluid compartments: blood vessels, tissue spaces, and cells. Thus we speak of the plasma, interstitial fluid, and intracellular fluid. Functionally, however, these three fluids may be considered as only two fluids: that which lies outside the cells and that which lies within them. Functionally, then, we speak of the extracellular fluid (ECF), meaning both the plasma and interstitial fluid, and the intracellular fluid (ICF), meaning the water inside all the cells. Extracellular fluid, as you already know, constitutes the internal environment of the body. It therefore serves the dual vital functions of providing a relatively constant environment for cells and öf transporting substances to and from them. Intracellular fluid, on the other hand, because it is a solvent, functions to facilitate intracellular chemical reactions that maintain life. Compared as to volume, intracellular fluid is the largest, plasma the smallest, and interstitial fluid in be-

tween. Fig. 21-4 gives typical normal fluid volumes in a young man. Note that intracellular fluid constitutes 40% of body weight, interstitial fluid 16%, and blood plasma 4%. Or, expressed differently, for every kilogram of its weight, the body contains about 400 ml of intracellular fluid, 160 ml of interstitial fluid, and 40 ml of plasma. Normal fluid volumes, however, vary considerably—mainly according to the fat content of the body. Fat people have a lower water content per kilogram of body weight than slender ones. Women have a relatively lower water content than men—but chiefly because a

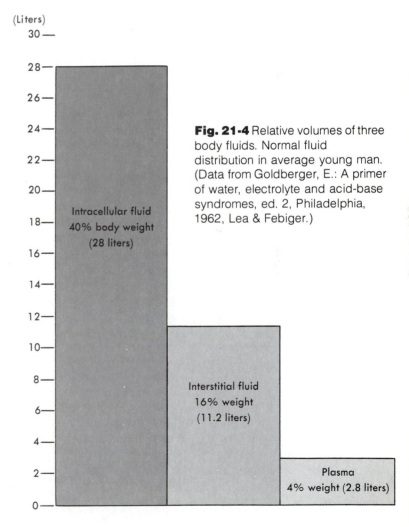

Fig. 21-4 Relative volumes of three body fluids. Normal fluid distribution in average young man. (Data from Goldberger, E.: A primer of water, electrolyte and acid-base syndromes, ed. 2, Philadelphia, 1962, Lea & Febiger.)

Table 21-2 Electrolyte composition of blood plasma*

Cations	Anions
142 mEq Na⁺	102 mEq Cl⁻
4 mEq K⁺	26 HCO₃⁻
5 mEq Ca⁺⁺	17 protein⁻
2 mEq Mg⁺⁺	6 other
	2 HPO₄⁼
153 mEq/L plasma	153 mEq/L plasma

*Data from Ruch, T. C., and Patton, H. D.: Physiology and biophysics, ed. 19, Philadelphia, 1965, W. B. Saunders Co.

woman's body contains a higher percentage of fat. Total fluid volume and fluid distribution also vary with age. As a person grows older, the amount of water in his body decreases. Fluid makes up a smaller percent of his body weight. Extracellular fluid volume is proportionately larger in infants and children than in adults.

Compared chemically, plasma and interstitial fluid (the two extracellular fluids) are almost identical. Intracellular fluid, on the other hand, shows striking differences from either of the two extracellular fluids. Let us examine first the chemical structure of plasma and interstitial fluid as shown in Fig. 21-5 and Table 21-2.

Importance, distribution, and measurement of electrolytes in body fluids

We have defined an electrolyte as a compound that will break up or dissociate into charged particles called ions when placed in solution. Sodium chloride, when dissolved in water, provides a positively charged sodium ion (Na⁺) and a negatively charged chloride ion (Cl⁻).

If two electrodes charged with a weak current are placed in an electrolyte solution, the ions will move or migrate in opposite directions according to their charge. Positive ions such as Na⁺ will be attracted to the negative electrode

or cathode and are called *cations*. Negative ions such as Cl⁻ will migrate to the positive electrode or anode and are called *anions*. A variety of anions and cations serve critical nutrient or regulatory roles in the body. Important cations include sodium (Na⁺), calcium (Ca⁺⁺), potassium (K⁺), and magnesium (Mg⁺⁺). Important anions include chloride (Cl⁻), bicarbonate (HCO₃⁻), phosphate (HPO₄⁼), and many proteins.

The importance of electrolytes in controlling the movement of water between the body fluid compartments will be discussed in this chapter. Their role in maintaining acid-base balance will be examined in Chapter 22.

Perhaps the first difference between the two extracellular fluids to catch your eye (in Fig. 21-5) is that blood contains a slightly larger total of electrolytes (ions) than do interstitial fluids. If you compare the two fluids, ion for ion, you will discover the most important difference between blood plasma and interstitial fluid. Look at the anions (negative ions) in these two extracellular fluids. Note that blood contains an appreciable amount of protein anions. Interstitial fluid, in contrast, contains hardly any protein anions. This is the only functionally important difference between blood and interstitial fluid. It exists because the normal capillary membrane is practically impermeable to proteins. Hence almost all protein anions remain behind in the blood instead of filtering out into the interstitial fluid. Because proteins remain in the blood, certain other differences also exist between blood and interstitial fluid—notably, blood contains more sodium ions and fewer chloride ions than does interstitial fluid.*

*According to the Donnan equilibrium principle, when nondiffusible anions (negative ions) are present on one side of a membrane, there are on that side of the membrane fewer diffusible anions and more diffusible cations (positive ions) than on the other side. Applying this principle to the blood and interstitial fluid: because blood contains nondiffusible protein anions, it contains fewer chloride ions (diffusible anions) and more sodium ions (diffusible cations) than does interstitial fluid.

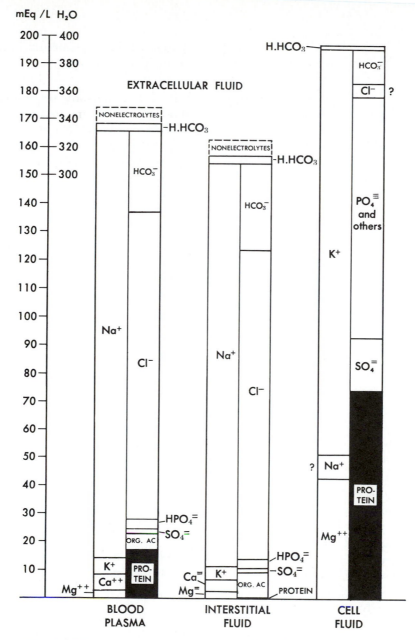

Fig. 21-5 Chief chemical constituents of three fluid compartments. The column of figures on the left (200, 190, 180, etc.) indicates amounts of cations or of anions, whereas the figures on the right (400, 380, 360, etc.) indicate the sum of cations and anions. Note that chloride and sodium values in cell fluid are questioned. It is probable that at least muscle intracellular fluid contains some sodium but no chloride. (Slightly modified from Mountcastle, V. B., editor: Medical physiology, St. Louis, The C. V. Mosby Co.; after Gamble, J. L.: Physiological information gained from studies on the life raft ration, Harvey Lect. **42:**247-273, 1946-1947.)

Extracellular fluids and intracellular fluid are more unlike than alike chemically. Chemical difference predominates between the extracellular and intracellular fluids. Chemical similarity predominates between the two extracellular fluids. Study Fig. 21-5 and make some generalizations about the main chemical differences between the extracellular and intracellular fluids. For example: What is the most abundant cation in the extracellular fluids? In the intracellular fluid? What is the most abundant anion in the extracellular fluids? In the intracellular fluid? What about the relative concentrations of protein anions in extracellular fluids and intracellular fluid?

The only reason we have called attention to the chemical structure of the three body fluids is that here, as elsewhere, structure determines function. In this instance the chemical structure of the three fluids helps control water and electrolyte movement between them. Or, phrased differently, the chemical structure of body fluids, if normal, functions to maintain homeostasis of fluid distribution and, if abnormal, results in fluid imbalance. Hypervolemia (excess blood volume) is a case in point. Edema, too, frequently stems from changes in the chemical structure of body fluids. But before discussing mechanisms that control water and electrolyte movement between blood, interstitial fluid, and intracellular fluid, we shall digress briefly to explain the units used for measuring electrolytes.

Once the important electrolytes and their constituent ions in the body fluid compartments had been established, physiologists needed to measure changes in their levels in order to understand the mechanisms of fluid balance. To have meaning, measurement units used to report electrolyte levels must be related to actual physiological activity. In the past, only the weight of an electrolyte in a given amount of solution was measured. The number of milligrams per 100 ml of solution (mg%) was one of the most frequently used units of measurement. Simply reporting the weight of an important electrolyte

such as sodium or calcium in milligrams per 100 ml of blood (mg%) gives no direct information about its chemical combining power or physiological activity in body fluids. The importance of valence and electrovalent or ionic bonding in chemical reactions was discussed in Chapter 1 (see Fig. 1-3). The reactivity or combining power of an electrolyte depends not just on the number of molecular particles present but also on the total number of ionic charges (valence). Univalent ions such as sodium (Na^+) carry only a single charge but the divalent calcium ion (Ca^{++}) carries two units of electrical charge.

The need for a unit of measurement more related to activity has resulted in increasing use of a more meaningful measurement yardstick—the *milliequivalent*. Milliequivalents measure the number of ionic charges or electrovalent bonds in a solution and therefore serve as an accurate measure of the chemical (physiological) combining power or reactivity of a particular electrolyte solution. The number of milliequivalents of an ion in a liter of solution (mEq/L) can be calculated from its weight in 100 ml (mg%) using a convenient conversion formula.*

We are ready now to try to answer the following question: How does the chemical structure of body fluids control water movement between them and thereby control fluid distribution in the body?

Control of water and electrolyte movement between plasma and interstitial fluid

Over 60 years ago, Starling advanced a hypothesis about the nature of the mechanism that

---

*Conversion of milligrams per 100 ml (mg%) to milliequivalents per liter (mEq/L):

$$mEq/L = \frac{mg/100 \ ml \times 10 \times Valence}{Atomic \ weight}$$

Example: Convert 15.6 mg% K^+ to mEq/L

Atomic weight of K^+ = 39
Valence of K^+ = 1

$$mEq/L = \frac{15.6 \times 10 \times 1}{39} = \frac{156}{39} = 4$$

Therefore 15.6 mg/100 ml K^+ = 4 mEq/L

controls water movement between plasma and interstitial fluid, that is, across the capillary membrane. This hypothesis has since become one of the major premises of physiology and is often spoken of as Starling's "law of the capillaries." According to this law, the control mechanism for water exchange between plasma and interstitial fluid consists of four pressures: blood hydrostatic and colloid osmotic pressures* on one side of the capillary membrane and interstitial fluid hydrostatic and colloid† osmotic pressures on the other side.

According to the physical laws governing filtration and osmosis, blood hydrostatic pressure (HP) tends to force fluid out of capillaries into interstitial fluid (IF), but blood colloid osmotic pressure (OP) tends to draw it back into them. Interstitial fluid hydrostatic pressure, in contrast, tends to force fluid out of the interstitial fluid into the capillaries, and interstitial fluid colloid osmotic pressure tends to draw it back out of capillaries. In short, two of these pressures constitute vectors in one direction and two in the opposite direction. Does this remind you of another mechanism studied earlier? (To check your answer, see p. 548.)

The difference between the two sets of opposing forces obviously represents the net or effective filtration pressure—in other words, the effective force tending to produce the net fluid movement between blood and interstitial fluid. In general terms, therefore, we may state Starling's law of the capillaries this way: The rate and direction of fluid exchange between capillaries and interstitial fluid is determined by the hydrostatic and colloid osmotic pressures of the

Table 21-3 Pressures at arterial end of tissue capillaries

Arterial end of capillary	Hydrostatic pressure	Colloid osmotic pressure
Blood	35 mm Hg	25 mm Hg
Interstitial fluid	2 mm Hg	0 mm Hg

Table 21-4 Pressures at venous end of tissue capillaries

Venous end of capillary	Hydrostatic pressure	Colloid osmotic pressure
Blood	15 mm Hg	25 mm Hg
Interstitial fluid	1 mm Hg	3 mm Hg*

*A small amount of blood proteins passes through the capillary membrane and tends to concentrate around the venous ends of capillaries—hence this pressure.

two fluids. Or, we may state it more specifically as a formula:

$$(BHP + IFOP) - (IFHP + BOP) = EFP*$$

Note that the factors enclosed in the first set of parentheses tend to move fluid out of capillaries and that those in the second set oppose this movement—they tend to move fluid into the capillaries.

To illustrate operation of Starling's law, let us consider how it controls water exchange at the arterial ends of tissue capillaries. Table 21-3

---

*Osmotic pressure from concentrations of protein in blood and interstitial fluid. Since the capillary membrane is permeable to other plasma solutes, they quickly diffuse through the membrane, so cause no osmotic pressure to develop against it. Only the proteins, to which the capillary membrane is practically impermeable, cause an actual osmotic pressure against the capillary membrane.

†A small amount of blood protein passes through the capillary membrane and tends to concentrate around the venous end of capillaries, hence this pressure.

---

*BHP, Blood hydrostatic pressure; IFOP, interstitial fluid osmotic pressure; IFHP, interstitial fluid hydrostatic pressure; BOP, blood osmotic pressure; EFP, effective filtration pressure between blood and interstitial fluid.

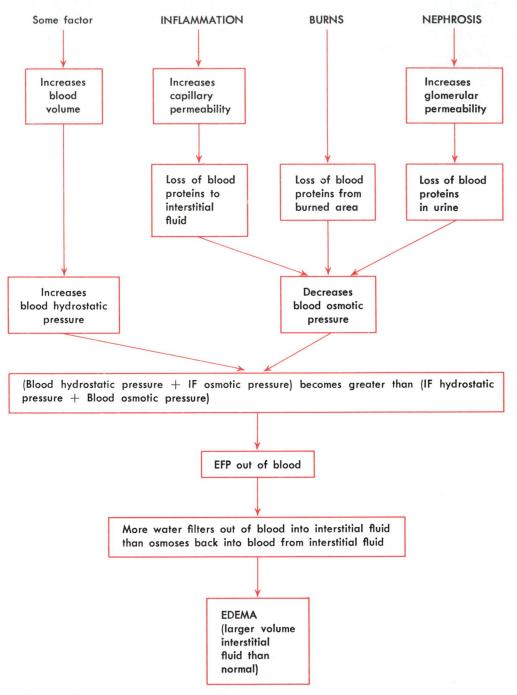

Fig. 21-6 Mechanisms of edema formation in some common conditions. *IF,* Interstitial fluid; *EFP,* effective or net filtration pressure. (Also see Fig. 21-7.)

gives typical normal pressures. Using these figures in Starling's law of the capillaries:

$(35 + 0) - (2 + 25) = 8$ mm Hg net pressure (EFP), causing water to filter out of blood at arterial ends of capillaries into interstitial fluid

The same law operates at the venous end of capillaries (Table 21-4). Again, apply Starling's law of the capillaries. What is the net effective pressure at the venous ends of capillaries? In which direction does it cause water to move? Assuming that the figures given in Table 21-4 are normal, do you agree that theoretically "the same amount of water returns to the blood at the venous ends of the capillaries as left it from the arterial ends"?

On the basis of our discussion thus far, we can formulate some principles about the transfer of water between blood and interstitial fluid.

1 No net transfer of water occurs between blood and interstitial fluid as long as the effective filtration pressure (EFP) equals 0, that is, when

(Blood HP + IFOP) = (IFHP + Blood OP)

2 A net transfer of water, a "fluid shift," occurs between blood and interstitial fluid whenever the EFP does not equal 0, that is, when

(Blood HP + IFOP)
does not equal
(IFHP + Blood OP)

3 Since (Blood HP + IFOP) is a force that tends to move water out of capillary blood, fluid shifts out of blood into interstitial fluid whenever

(Blood HP + IFOP)
is greater than
(IFHP + Blood OP)

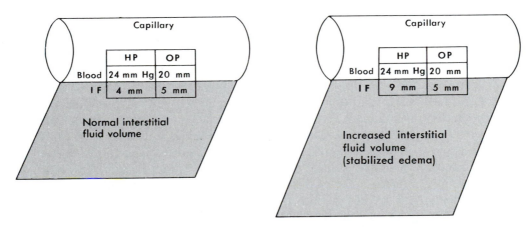

Fig. 21-7 The mechanism of edema formation initiated by a decrease in blood protein concentration and, therefore, in blood osmotic pressure. Left diagram, blood osmotic pressure has just decreased to 20 from the normal 25 mm Hg. This increases the effective filtration pressure (EFP) to 5 mm Hg from a normal of 0 (see Starling's formula, p. 573). The EFP of 5 mm Hg causes fluid to shift out of blood into interstitial fluid (IF) until the EFP again equals 0—in this case, when the interstitial fluid volume has increased enough to raise interstitial fluid hydrostatic pressure to 9 mm Hg, as shown in the diagram on the right. At this point a new equilibrium is established, and equal amounts of water once more are exchanged between the blood and interstitial fluid. Thus the increased interstitial fluid volume, that is, the edema, becomes stabilized.

4 Since (IFHP + Blood OP) is a force that tends to move water out of interstitial fluid into capillary blood, fluid shifts out of interstitial fluid into blood whenever

(IFHP + Blood OP)
is greater than
(Blood HP + IFOP)

Or, stated the other way around, whenever

(Blood HP + IFOP)
is less than
(IFHP + Blood OP)

To apply these principles, answer questions 8 to 11 on p. 581 and examine Figs. 21-6 and 21-7.

Control of water and electrolyte movement through cell membranes between interstitial and intracellular fluids

The mechanism that regulates water movement through cell membranes is similar to the one that regulates water movement through capillary membranes. In other words, interstitial fluid and intracellular fluid hydrostatic and osmotic pressures regulate water transfer between these two fluids. But because the osmotic pressures of interstitial and intracellular fluids vary more than do their hydrostatic pressures, their osmotic pressures serve as the chief regulators of water transfer across cell membranes. Their osmotic pressures, in turn, are directly related to the electrolyte concentration gradients—notably sodium and potassium—maintained across cell membranes. As Fig. 21-5 shows, most of the body sodium is outside the cells. A concentration of 138 to 143 mEq/L makes sodium the chief electrolyte by far in interstitial fluid. The intracellular fluid's main electrolyte is potassium salt. Therefore a change in the sodium or the potassium concentrations of either of these fluids causes the exchange of fluid between them to become unbalanced. For example, a decrease in interstitial fluid sodium concentration immediately decreases interstitial fluid osmotic pressure, making it hypotonic to intracellular fluid osmotic pressure. In other words a decrease in

interstitial fluid sodium concentration establishes an osmotic pressure gradient between interstitial and intracellular fluids. This causes net osmosis to occur out of interstitial fluid into cells. In short, interstitial fluid and intracellular fluid electrolyte concentrations are the main determinants of their osmotic pressures; their osmotic pressures regulate the amount and direction of water transfer between the two fluids, and this regulates their volumes. Hence fluid balance depends on electrolyte balance. Conversely, electrolyte balance depends on fluid balance. An imbalance in one produces an imbalance in the other (Fig. 21-8).

Normal sodium concentration in interstitial fluid and potassium concentration in intracellular fluid depend on many factors but especially on the amount of ADH and aldosterone secreted. As shown in Fig. 21-9, ADH regulates extracellular fluid electrolyte concentration and osmotic pressure by regulating the amount of water reabsorbed into blood by renal tubules. Aldosterone, on the other hand, regulates extracellular fluid volume by regulating the amount of sodium reabsorbed into blood by renal tubules (Fig. 21-1).

If for any reason conservation of body sodium is required, the normal kidney is capable of excreting an essentially sodium-free urine and is therefore considered the chief regulator of sodium levels in body fluids. Sodium lost in sweat can become appreciable with elevated environmental temperatures or fever. However, the thirst that results may lead to replacement of water but not the lost sodium and, as a result of the increased fluid intake, the remaining sodium pool may be diluted even more. Sweat loss of sodium is not, therefore, considered a normal means of regulation.

In addition to the well-regulated movement of sodium into and out of the body and between the three primary fluid compartments, there is a continuous movement or circulation of this important electrolyte between a number of internal secretions. Over 8 *liters* of various internal

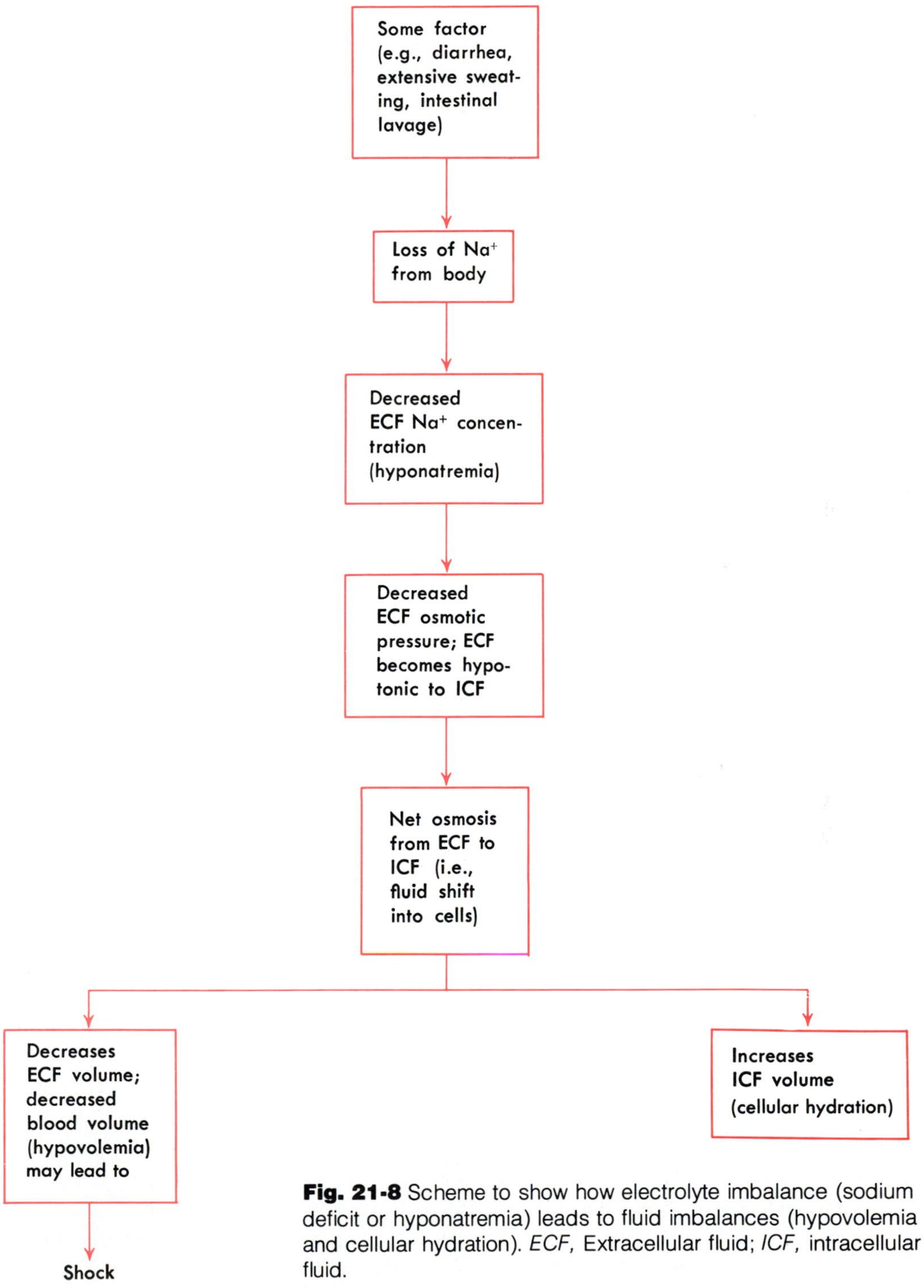

Fig. 21-8 Scheme to show how electrolyte imbalance (sodium deficit or hyponatremia) leads to fluid imbalances (hypovolemia and cellular hydration). *ECF*, Extracellular fluid; *ICF*, intracellular fluid.

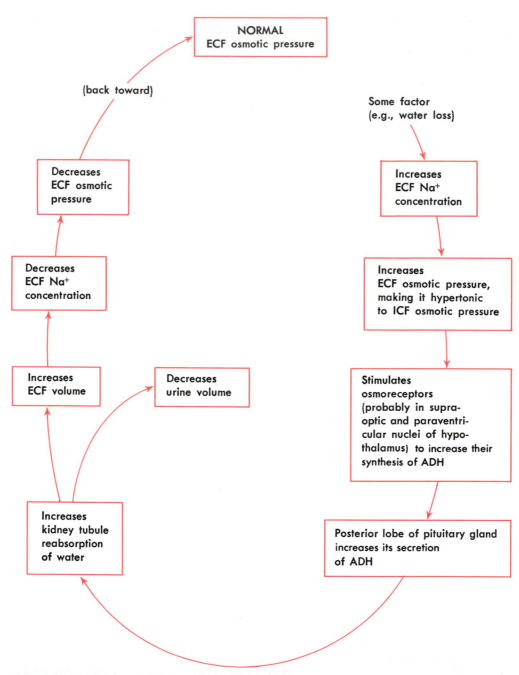

Fig. 21-9 Antidiuretic hormone (ADH) mechanism that helps maintain homeostasis of extracellular fluid (ECF) osmotic pressure by regulating its volume and thereby its electrolyte concentration, that is, mainly ECF Na⁺ concentration.

secretions such as saliva, gastric and intestinal secretions, bile, and pancreatic fluid are produced every day. The total daily secretion of sodium into these alimentary tract fluids alone will average between 1,200 and 1,400 mEq. A 70 kg (154 lb) adult has a total body sodium pool of only 2,800 to 3,000 mEq. Precise regulatory and conservation mechanisms are required for survival.

Chloride is the most important extracellular anion and is almost always linked to sodium. Generally ingested together, they provide in large part for the isotonicity of extracellular fluid. Chloride ions are generally excreted in the urine as a potassium salt, and therefore chloride deficiency *(hypochloremia)* is often found in cases of potassium loss.

Total body potassium content in the average-sized adult is approximately 4,000 mEq. Because the vast majority of body potassium is intracellular, serum determinations, which normally fall between 4.0 to 5.0 mEq/L, may not be the best index to reflect imbalances. The body may lose one third to one half of its intracellular potassium reserves before the loss is reflected in lowered serum potassium levels.

Potassium deficit or *hypopotassemia* occurs whenever there is cell breakdown, as in starvation, trauma, or dehydration. As individual cells disintegrate, potassium enters the extracellular fluid and is rapidly excreted because it is not reabsorbed efficiently by the kidney.

Outline summary

Introduction

A Meaning of fluid balance
  1 Same as homeostasis of fluids, that is, total volume of water in body normal and remains relatively constant
  2 Volume of blood plasma, interstitial fluid, and intracellular fluid all remain relatively constant, that is, homeostasis of distribution of water as well as of total volume
B Fluid balance and electrolyte balance interdependent—see Figs. 21-1 and 21-9
C Chemical compounds such as NaCl that break up or dissociate in solution (Na^+ and Cl^-) are known as electrolytes
D Dissociation particles of an electrolyte are called ions and carry an electrical charge

Some general principles about fluid balance

A Cardinal principle—intake must equal output
B Fluid and electrolyte balance maintained primarily by mechanisms that adjust output to intake; secondarily by mechanisms that adjust intake to output
C Fluid balance also maintained by mechanisms that control movement of water between fluid compartments

Avenues by which water enters and leaves body

A Water enters body through digestive tract in
  1 Liquids
  2 Foods
B Water formed in body by metabolism of foods
C Water leaves body via kidneys, lungs, skin, and intestines

Mechanisms that maintain homeostasis of total fluid volume

A Regulation of urine volume—under normal conditions, by factors that control reabsorption of water by distal and collecting tubules
  1 Extracellular fluid electrolyte concentration (crystalloid osmotic pressure) controls ADH secretion, which controls tubule water reabsorption
  2 Extracellular fluid volume controls aldosterone secretion, which controls tubule sodium ion reabsorption and therefore water reabsorption
B Factors that alter fluid loss under abnormal conditions—hyperventilation, hypoventilation, vomiting, diarrhea, circulatory failure, etc.
C Regulation of fluid intake—see Fig. 21-2
  1 Mechanism by which intake adjusted to output not completely known

2 One controlling factor seems to be degree of moistness of mucosa of mouth—if output exceeds intake, mouth feels dry, sensation of thirst occurs, and individual ingests liquids

Mechanisms that maintain homeostasis of fluid distribution

A Comparison of plasma, interstitial fluid, and intracellular fluid
   1 Plasma and interstitial fluid constitute extracellular fluid (ECF), internal environment of body or, in other words, environment of cells
   2 Intracellular fluid (ICF) volume largest, plasma volume smallest; ICF about 40% of body weight, IF about 16%, and plasma about 4%, or ICF volume about 10 times that of plasma and IF volume about 4 times plasma volume
   3 Chemically, plasma and IF almost identical except that plasma contains slightly more electrolytes and considerably more proteins than IF; also blood contains somewhat more sodium and fewer chloride ions
   4 Chemically, ECF and ICF strikingly different; sodium main cation of ECF, potassium main cation of ICF; chloride main anion of ECF; phosphate main anion of ICF; protein concentration much higher in ICF than in IF
B Importance, distribution, and measurement of electrolytes in body fluids
   1 Positive ions such as sodium (Na^+) are attracted to negative electrode or cathode—called cations
   2 Negative ions such as chloride (Cl^-) migrate to positive electrode or anode—called anions
   3 Important cations: sodium (Na^+), calcium Ca^{++}), potassium (K^+), and magnesium (Mg^{++})
   4 Important anions: chloride (Cl^-), bicarbonate (HCO_3^-), phosphate ($HPO_4^=$), and many proteins
   5 Electrolytes serve important roles in fluid balance and in acid-base balance

6 Physiological reactivity or combining power of an electrolyte depends not just on the number of molecular particles present but also on total number of ionic charges (valence)
7 Milliequivalents (mEq) measure the number of ionic charges or electrovalent bonds in a solution and therefore serve as an accurate measure of the chemical (physiological) combining power or reactivity of a particular electrolyte
8 Conversion formula for mg% to mEq/L:

$$mEq/L = \frac{mg/100\ ml \times 10 \times Valence}{Atomic\ weight}$$

C Control of water movement between plasma and IF
   1 By four pressures—blood hydrostatic and colloid osmotic pressures and IF hydrostatic and colloid osmotic pressures
   2 Effect of these pressures on water movement between plasma and IF expressed in Starling's "law of the capillaries"; only when (Blood hydrostatic pressure + IF colloid osmotic pressure) − (Blood colloid osmotic pressure + IF hydrostatic pressure) = 0, do equal amounts of water filter out of blood into IF and osmose back into blood from IF; in other words, water balance exists between these two fluids under these conditions
D Control of water and electrolyte movement through cell membranes between IF and ICF—primarily by relative crystalloid osmotic pressures of ECF and ICF, which depend mainly on sodium concentration of ECF and potassium concentration of ICF, which in turn depend on intake and output of sodium and potassium and on sodium-potassium transport mechanisms

Review questions

1 Explain in your own words the meaning of the term *fluid balance.*

2 How is total volume of body fluid kept relatively constant, that is, what other factors must be controlled in order to keep the total volume of water in the body relatively constant?

3 What, if any, are functionally important differences between the chemical composition of plasma and interstitial fluid?

4 What, if any, are functionally important differences between the chemical composition of extracellular and intracellular fluids?

5 Are plasma and interstitial fluid more accurately described as "similar" or "different" as to chemical composition? Volume?

6 Support your answer to question 5 with some specific facts.

7 Are interstitial fluid and intracellular fluid more accurately described as "similar" or "different" as to chemical composition? Volume?

8 Explain Starling's law of the capillaries in your own words. Be as brief and clear as you can. This law describes the mechanism for controlling what?

9 Suppose that in one individual the normal average or mean pressures (in mm Hg) are capillary blood hydrostatic pressure 24 and osmotic pressure 25; interstitial fluid hydrostatic pressure 4 and osmotic pressure 5. Following a hemorrhage, this patient's blood hydrostatic pressure falls to 18. (a) Assuming that the other pressures momentarily stay the same, what is the EFP now? (b) The new EFP causes a fluid shift in which direction? (c) When will the fluid shift stop and an even exchange of water again go on between blood and interstitial fluids?

10 Formulate a principle by filling in the blanks in the following sentences: Anything that decreases blood hydrostatic pressure in the capillaries tends to cause a fluid shift from _____ into _____. Conversely, anything that increases blood hydrostatic pressure in the capillaries tends to cause a fluid shift out of _____ into _____.

11 Formulate a principle by completing the blanks in the following sentence: A decrease in blood protein concentration tends to decrease blood _____ pressure and therefore tends to cause a fluid shift from _____ to _____.

Situation: A patient has had marked diarrhea for several days.

12 Explain the homeostatic mechanisms that would tend to compensate for this excessive fluid loss.

13 Do you think they could succeed in maintaining fluid balance or would fluid therapy probably be necessary?

14 What, if any, abnormality of fluid distribution do you think would occur in this patient without fluid therapy? Explain your reasoning.

15 Why does dehydration necessarily develop if an individual takes nothing by mouth for several days and receives no fluid therapy? Why cannot homeostatic mechanisms prevent this?

16 Define the following terms: cation, anion, milliequivalent, electrolyte, hypochloremia, hypopotassemia.

17 What is considered as the chief regulator of sodium levels in body fluids?

18 Is loss of sodium through sweat a normal means for regulation of this electrolyte?

19 Why does excessive loss of potassium in the urine often cause serious depletion of chloride from the extracellular fluid?

20 What important cation is lost from the intracellular fluid compartment as a result of cell breakdown as in starvation or trauma?

chapter 22

Acid-base balance

Acid-base balance is one of the most important of the body homeostatic mechanisms. The term refers to regulation of hydrogen ion concentration in the body fluids. Even slight deviations from the normal pH range will result in pronounced, potentially fatal changes in metabolic activity. Precise regulation of pH at the cellular level is necessary for survival.

Mechanisms that control pH of body fluids

Meaning of term pH

The term pH is a symbol used to mean the hydrogen ion concentration of a solution. Actually, pH stands for the negative logarithm of the hydrogen ion concentration.* pH indicates the degree of acidity or alkalinity of a solution. As the concentration of hydrogen ions increases, the pH goes down and the solution becomes more acid; a decrease in hydrogen ion concentration makes the solution more alkaline and the pH goes up. A pH of 7 indicates neutrality (equal amounts of H^+ and OH^-), a pH of less than 7 indicates acidity (more H^+ than OH^-), and one greater than 7 indicates alkalinity (more OH^- than H^+). The

---

*A pH of 7, for example, means that a solution contains 10^{-7} gm of hydrogen ions per liter. Or, translating this logarithm into a number, a pH of 7 means that a solution contains 0.0000001 (that is, 1/10,000,000) gm of hydrogen ions per liter. A solution of pH 6 contains 0.000001 (1/1,000,000) gm of hydrogen ions per liter and one of pH 8 contains 0.00000001 (1/100,000,000) gm of hydrogen ions per liter. Note that a solution with pH 7 contains 10 times as many hydrogen ions as a solution with pH 8 and that pH decreases as hydrogen ion concentration increases.

overall pH range is often expressed numerically on a logarithmic scale of 1 to 14. Keep in mind that a change of 1 pH unit on this type of scale represents a 10-fold difference in actual concentration of hydrogen ions.

The slight increase in acidity of venous blood (pH 7.35) compared to arterial blood (pH 7.45) results primarily from carbon dioxide entering venous blood as a waste product of cellular metabolism. The lungs remove the equivalent of over 30 *liters* of 1-normal carbonic acid each day from the venous blood by elimination of carbon dioxide, and yet 1 liter of venous blood contains only about 1/100,000,000 gm more hydrogen ions than does 1 liter of arterial blood. What incredible constancy! The pH homeostatic mechanism does indeed control effectively—astonishingly so.

Types of pH control mechanisms

Since various acids and bases continually enter the blood from absorbed foods and from the catabolism of foods, some kind of mechanism for neutralizing or eliminating these substances is necessary if blood pH is to remain constant. Actually, three different devices operate together to maintain constancy of pH. Collectively, these devices—buffers, respirations, and kidney excretion of acids and bases—might be said to constitute the pH homeostatic mechanism.

Effectiveness of pH control mechanisms; range of pH

The most eloquent evidence of the effectiveness of the pH control mechanism is the extremely narrow range of blood pH, normally 7.35 to 7.45.

Buffer mechanism for controlling pH of body fluids
Buffers defined

In terms of action a buffer is a substance that prevents marked changes in the pH of a solution when an acid or a base is added to it. Let us suppose that a small amount of the strong acid, hydrochloric acid, is added to a solution that contains a buffer (blood, for example) and that its pH decreases from 7.41 to 7.27. But if the same amount of hydrochloric acid were added to pure water containing no buffers, its pH would decrease much more markedly, from 7 to perhaps 3.4. In both instances, pH decreased on addition of the acid, but much less so with buffers present than without them.

In terms of chemical composition, buffers consist of two kinds of substances and are, therefore, often referred to as "buffer pairs."

Buffer pairs present in body fluids

Most of the body fluid buffer pairs consist of a weak acid and a salt of that acid. The main buffer pairs present in body fluids are as follows:

Bicarbonate pairs $\dfrac{NaHCO_3}{H_2CO_3}$, $\dfrac{KHCO_3}{H_2CO_3}$, etc.

Plasma protein pair $\dfrac{Na \cdot Proteinate}{Proteins \text{ (weak acids)}}$

Hemoglobin pairs $\dfrac{K \cdot Hb}{Hb}$ and $\dfrac{K \cdot HbO_2}{HbO_2}$
(Hb and HbO_2 are weak acids)

Phosphate buffer pair $\dfrac{Na_2HPO_4 \text{ (basic phosphate)}}{NaH_2PO_4 \text{ (acid phosphate)}}$

Action of buffers to prevent marked changes in pH of body fluids

Buffers react with a relatively strong acid (or base) to replace it by a relatively weak acid (or base). That is to say, an acid that highly dissociates to yield many hydrogen ions is replaced by one that dissociates less highly to yield fewer hydrogen ions. Thus by the buffer reaction, instead of the strong acid remaining in the solution and contributing many hydrogen ions to drastically lower the pH of the solution, a weaker acid takes its place, contributes fewer additional hydrogen ions to the solution, and thereby lowers its pH only slightly. Therefore because blood contains buffer pairs, its pH fluctuates much less widely than it would without them. In other words, blood buffers constitute one of the devices for preventing marked changes in blood pH.

Let us consider, as a specific example of buffer action, how the sodium bicarbonate ($NaHCO_3$)–carbonic acid (H_2CO_3) system works in the presence of a strong acid or base.

Addition of a strong acid, such as hydrochloric acid (HCl), to the sodium bicarbonate–carbonic acid buffer system would initiate the reaction shown in Fig. 22-1. Note how this reaction between HCl and the base bicarbonate ($NaHCO_3$) applies the principle of buffering. As a result of the buffering action of $NaHCO_3$, the weak acid, $H \cdot HCO_3$, replaces the very strong acid, HCl, and therefore the hydrogen ion concentration of the blood increases much less than it would have if HCl were not buffered.

If, on the other hand, a strong base such as sodium hydroxide (NaOH) is added to the same buffer system, the reaction shown in Fig. 22-2 would take place. The hydrogen ion of $H \cdot HCO_3$, the weak acid of the buffer pair, combines with the hydroxyl ion (OH^-) of the strong base NaOH to form water. Note what this accomplishes. It decreases the number of hydroxyl ions added to the solution, and this, in turn, prevents the drastic rise in pH that would occur in the absence of buffering.

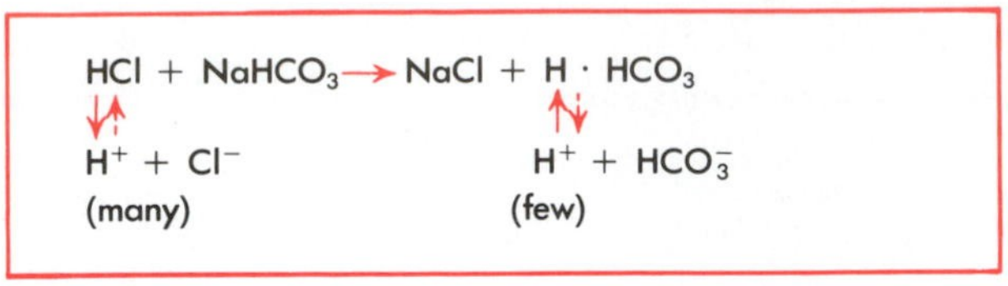

$$HCl + NaHCO_3 \longrightarrow NaCl + H \cdot HCO_3$$
$$H^+ + Cl^- \qquad\qquad H^+ + HCO_3^-$$
$$\text{(many)} \qquad\qquad \text{(few)}$$

Fig. 22-1 Buffering of acid HCl by sodium bicarbonate.

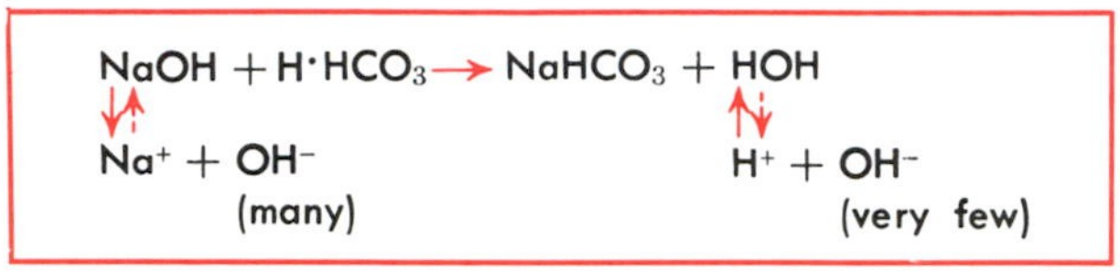

$$NaOH + H \cdot HCO_3 \longrightarrow NaHCO_3 + HOH$$
$$Na^+ + OH^- \qquad\qquad H^+ + OH^-$$
$$\text{(many)} \qquad\qquad\qquad \text{(very few)}$$

Fig. 22-2 Buffering of base NaOH by carbonic acid.

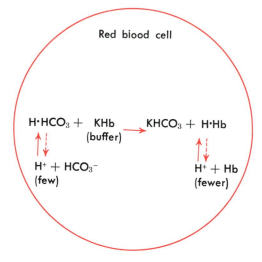

Fig. 22-3 Buffering of the volatile carbonic acid ($H \cdot HCO_3$) inside red blood cell by potassium salt of hemoglobin. Note that each molecule of carbonic acid is replaced by a molecule of the acid hemoglobin. Since hemoglobin is a weaker acid than carbonic acid, fewer of these hemoglobin molecules dissociate to form hydrogen ions. Hence fewer hydrogen ions are added to red blood cell intracellular fluid than would have been added by unbuffered carbonic acid. Also, since some of the carbonic acid in the red blood cell has come from plasma, fewer hydrogen ions remain in blood than would have if there were no buffering of carbonic acid.

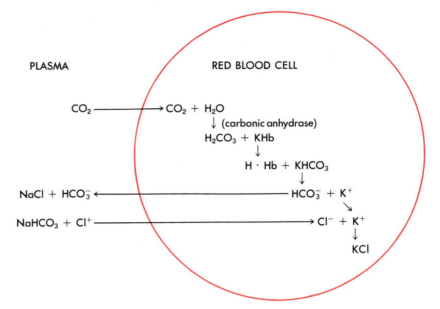

Fig. 22-4 The chloride shift.

PLASMA

RED BLOOD CELL

$CO_2 \longrightarrow CO_2 + H_2O$
$\downarrow$ (carbonic anhydrase)
$H_2CO_3 + KHb$
$\downarrow$
$H \cdot Hb + KHCO_3$
$\downarrow$
$NaCl + HCO_3^- \longleftarrow \qquad\qquad HCO_3^- + K^+$
$\searrow$
$NaHCO_3 + Cl^+ \longrightarrow \qquad\qquad Cl^- + K^+$
$\downarrow$
KCl

The principles of buffer action illustrated by the reaction of HCl and NaOH with the sodium bicarbonate buffer pair can be applied equally to the plasma protein, hemoglobin, and phosphate buffer systems.

Carbon dioxide and other acid waste products are continuously being formed as a result of cellular metabolism. The formation of carbonic acid from carbon dioxide and water requires the enzyme carbonic anhydrase, which is found in the red blood cells. Therefore carbonic acid is buffered primarily by the potassium salt of hemoglobin inside the red blood cell, as shown in Fig. 22-3.

It is interesting to note in Fig. 22-4 that the $KHCO_3$, formed by the buffering of carbonic acid, dissociates in the red blood cell and the bicarbonate ion returns to the blood plasma. Here it is exchanged for a chloride ion. The process of exchanging a bicarbonate ion formed in the red blood cell with a chloride ion from the plasma is called the *chloride shift*. This process makes it possible for carbon dioxide to be buffered in the red blood cell and then carried as bicarbonate in the plasma. Fig. 22-4 summarizes the reactions of the chloride shift.

Nonvolatile or fixed acids, such as hydrochloric acid, lactic acid, and ketone bodies, are buffered mainly by sodium bicarbonate (Fig. 22-5).

Normal blood pH and acid-base balance depend on a base bicarbonate to carbonic acid buffer pair ratio of 20 to 1 in the extracellular fluid. Actually, in a state of acid-base balance a liter of plasma contains 27 mEq of $NaHCO_3$ as base bicarbonate (BB)—ordinary baking soda—and 1.3 mEq of carbonic acid (CA):

$$\frac{27 \text{ mEq } NaHCO_3}{1.3 \text{ mEq } H_2CO_3} = \frac{(BB)}{(CA)} = \frac{20}{1} = \text{pH } 7.4$$

The *ratio* of base to acid is critical. If the ratio is maintained, acid-base balance (pH) will remain near normal despite changes in the absolute amounts of either component of the buffer pair. For example, a BB/CA ratio of 40/2 or 10/0.5 would result in a compensated state of acid-base balance. However, an increase in the ratio causes an increase in pH (uncompensated alkalosis) and a decrease in the ratio a decrease in pH (uncompensated acidosis). The ability of the body to regulate the amount of either component of the bicarbonate buffer pair, in order to

$$H \cdot lactate + NaHCO_3 \longrightarrow Na \cdot lactate + H \cdot HCO_3$$

$$H^+ + lactate^-$$
(few)

$$H^+ + HCO_3^-$$
(fewer)

Fig. 22-5 Lactic acid (H · lactate) and other nonvolatile or "fixed" acids are buffered by sodium bicarbonate, the most abundant base bicarbonate in the blood. Carbonic acid (H · HCO$_3$ or H$_2$CO$_3$, a weaker acid than lactic acid) replaces lactic acid. Result: fewer hydrogen ions are added to blood than would be if lactic acid were not buffered. Ketone bodies, for example, acetoacetic acid, from fat catabolism are also buffered by base bicarbonate.

maintain the correct ratio for acid-base balance, makes this system one of the most important and unpredictable for controlling pH of body fluids.

The relationship between the hydrogen ion concentration of body fluids and the ratio of base bicarbonate to carbonic acid has been expressed as a mathematical formula called the *Henderson-Hasselbalch equation.** This equation is useful in clinical medicine to predict the blood pH changes that will occur if the sodium bicarbonate–carbonic acid buffer system ratio is altered by drugs or disease.

Evaluation of role of buffers in pH control

Buffering alone cannot maintain homeostasis of pH. As we have seen, hydrogen ions are added continually to capillary blood despite buffering. If even a few more hydrogen ions were added every time blood circulated and no way were provided for eliminating them, blood hydrogen ion concentration would necessarily increase and thereby decrease blood pH. "Acid blood," in other words, would soon develop. Respiratory

*Henderson-Hasselbalch equation for the bicarbonate buffer system is as follows:

$$pH = 6.1 + \frac{\log (HCO_3^-)}{(CO_2)} \text{ (molar concentrations of bicarbon-}$$
ate ion and of dissolved carbon dioxide)

and urinary devices must, therefore, function concurrently with buffers in order to remove from the blood and from the body the hydrogen ions continually being added to blood. Only then can the body maintain constancy of pH.

Respiratory mechanism of pH control
Explanation of mechanism

Respirations play a vital part in controlling pH. With every expiration, carbon dioxide and water leave the body in the expired air. The carbon dioxide has come from the venous blood —has diffused out of it as it moves through the lung capillaries. Less carbon dioxide remains, therefore, in the arterial blood leaving the lung capillaries. The lower P_{CO_2} in arterial blood reduces the amount of carbonic acid and the number of hydrogen ions that can be formed in red blood cells by the following reactions:

$$CO_2 + H_2O \xrightarrow{\text{(carbonic anhydrase)}} H_2CO_3$$
$$H_2CO_3 \longrightarrow H^+ + HCO_3^-$$

Arterial blood, therefore, has a lower hydrogen ion concentration and a higher pH than venous blood. A typical average pH for venous blood is 7.36, and 7.41 is a typical average pH for arterial blood.

Adjustment of respirations to pH of arterial blood

Obviously, in order for respirations to serve as a mechanism of pH control, there must be some mechanism for increasing or decreasing respirations as needed to maintain or restore normal pH. Suppose that blood pH has decreased, that is, hydrogen ion concentration has increased. Respirations then need to increase in order to eliminate more carbon dioxide from the body and thereby leave less carbonic acid and fewer hydrogen ions in the blood.

One mechanism for adjusting respirations to arterial blood carbon dioxide content or pH op-

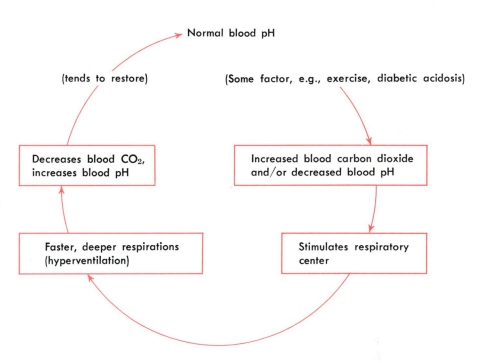

Fig. 22-6 Respiratory mechanism of pH control. A rise in arterial blood CO_2 content or a drop in its pH (below about 7.38) stimulates respiratory center neurons. Hyperventilation results. Less CO_2 and therefore less carbonic acid and fewer hydrogen ions remain in the blood so that blood pH increases, often reaching the normal level.

erates in this way. Neurons of the respiratory center are sensitive to changes in arterial blood carbon dioxide content and to changes in its pH. If the amount of carbon dioxide in arterial blood increases beyond a certain level, or if arterial blood pH decreases below about 7.38, the respiratory center is stimulated and respirations accordingly increase in rate and depth. This, in turn, eliminates more carbon dioxide, reduces carbonic acid and hydrogen ions, and increases pH back toward the normal level (Fig. 22-6). The carotid chemoreflexes (p. 463) are also devices by which respirations adjust to blood pH and, in turn, adjust pH.

Some principles relating respirations and pH of body fluids

1 A decrease in blood pH below normal, that is, acidosis, tends to stimulate increased respirations (hyperventilation), which tends to increase pH back toward normal. In other words, acidosis causes hyperventilation, which in turn acts as a compensating mechanism for the acidosis.

2 Prolonged hyperventilation may increase blood pH enough to produce alkalosis.

3 An increase in blood pH above normal (or alkalosis) causes hypoventilation, which serves as a compensating mechanism for the alkalosis by decreasing blood pH back toward normal.

4 Prolonged hypoventilation may decrease blood pH enough to produce acidosis.

Urinary mechanism of pH control
General principles about mechanism

Because the kidneys can excrete varying amounts of acid and base, they, like the lungs, play a vital role in pH control. Kidney tubules, by excreting many or few hydrogen ions, in exchange for reabsorbing many or few sodium ions, control urine pH and thereby help control blood pH. If, for example, blood pH decreases below normal, kidney tubules secrete more hydrogen ions from blood to urine and, in exchange for each hydrogen ion, reabsorb a sodium ion from the urine back into the blood. This, of course, decreases urine pH. But simultaneously —and of far more importance—it increases blood pH back toward normal. This urinary mechanism of pH control is a device for excreting varying amounts of hydrogen ions from the body to match the amounts entering the blood. It constitutes a much more effective device for adjusting hydrogen output to hydrogen input than does the body's only other mechanism for expelling hydrogen ions, namely, the respiratory mechanism previously described. But abnormalities of any one of the three pH control mechanisms soon throw the body into a state of acid-base imbalance. Only when all three parts of this complex mechanism—buffering, respirations, and urine secretion—function adequately can acid-base balance be maintained.

Let us turn our attention now to the mechanisms that adjust urine pH to counteract changes in blood pH.

Mechanisms that control urine pH

A decrease in blood pH accelerates the renal tubule ion-exchange mechanisms that both acidify urine and conserve blood's base and

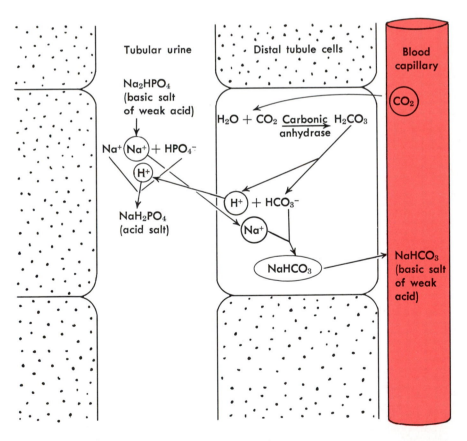

Fig. 22-7 Acidification of urine and conservation of base by distal renal tubule excretion of H ions (discussed on p. 587).

thereby tend to increase blood pH back to normal. The following paragraphs describe these mechanisms.

1 Distal and collecting tubules secrete hydrogen ions into the urine in exchange for basic ions, which they reabsorb. Refer to Fig. 22-7 as you read the rest of this paragraph. Note that carbon dioxide diffuses from tubule capillaries into distal tubule cells, where the enzyme carbonic anhydrase accelerates the combining of carbon dioxide with water to form carbonic acid. The latter dissociates into hydrogen ions and bicarbonate ions. The hydrogen ions then diffuse into the tubular urine, where they displace basic ions (most often sodium) from a basic salt of a weak acid and thereby change the basic salt to an acid salt or to a weak acid that is eliminated in the urine. While this is happening, the displaced sodium or other basic ion diffuses into a tubule cell. Here, it combines with the bicarbonate ion left over from the carbonic acid dissociation to form sodium bicarbonate. The sodium bicarbonate then diffuses—is reabsorbed, that is—into the blood. Consider the various results of this mechanism. Sodium bicarbonate (or other base bicarbonate) is conserved for the

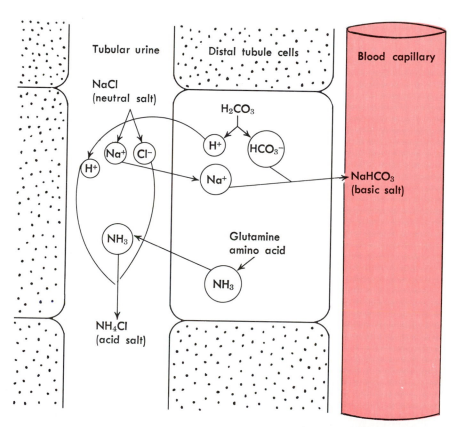

Fig. 22-8 Acidification of urine by tubule excretion of ammonia (NH_3). An amino acid (glutamine) leaves blood, enters a tubule cell, and is deaminized to form ammonia, which is excreted into urine. In exchange the tubule cell reabsorbs a basic salt (mainly $NaHCO_3$) into blood from urine.

body. Instead of all the basic salts that filter out of glomerular blood leaving the body in the urine, considerable amounts are recovered into peritubular capillary blood. In addition, extra hydrogen ions are added to the urine and thereby eliminated from the body. Both the reabsorption of base bicarbonate into blood and the excretion of hydrogen ions into urine tend to increase the ratio of the bicarbonate buffer pair $B \cdot HCO_3/H \cdot HCO_3$ (BB/CA) present in blood. This automatically increases blood pH. In short, kidney tubule base bicarbonate reabsorption and hydrogen-ion excretion both tend to alkalinize blood by acidifying urine.

Renal tubules can excrete either hydrogen or potassium in exchange for the sodium they reabsorb. Therefore, in general, the more hydrogen ions they excrete, the fewer the potassium ions they can excrete. For example, in acidosis, tubule excretion of hydrogen ions increases markedly and potassium ion excretion decreases—an important fact because it may lead to *hyperkalemia* (excessive blood potassium), a dangerous condition because it can cause heart block and death.

2 Distal and collecting tubule cells excrete ammonia into the tubular urine. As Fig. 22-8 shows, the ammonia combines with hydrogen to

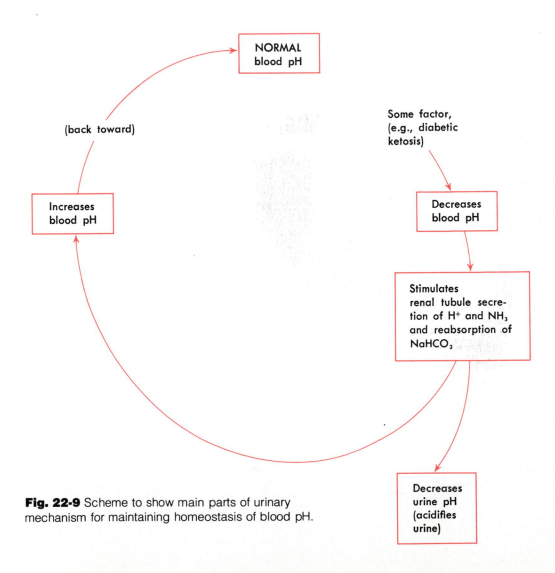

Fig. 22-9 Scheme to show main parts of urinary mechanism for maintaining homeostasis of blood pH.

form an ammonium ion. The ammonium ion displaces sodium or some other basic ion from a salt of a fixed (nonvolatile) acid to form an ammonium salt. The basic ion then diffuses back into a tubule cell and combines with bicarbonate ion to form a basic salt, which in turn diffuses into tubular blood. Thus, like the renal tubules' excretion of hydrogen ions, their excretion of ammonia and its combining with hydrogen to form ammonium ions also tends to increase the blood bicarbonate buffer pair ratio and, therefore, tends to increase blood pH. Quantitatively, however, ammonium ion excretion is more important than hydrogen ion excretion.

Renal tubule excretion of hydrogen and ammonia is controlled at least in part by the blood pH level. As indicated in Fig. 22-9, a decrease in blood pH accelerates tubule excretion of both hydrogen and ammonia. An increase in blood pH produces the opposite effects.

Acid-base imbalances

All of the buffer pairs present in body fluids play an important role in acid-base balance. However, only in the bicarbonate system can the body regulate quickly and with precision the levels of both chemical components in the buffer pair. Carbonic acid levels can be regulated by the respiratory system and bicarbonate ion by the kidneys. Recall that a 20 to 1 ratio of base bicarbonate to carbonic acid (BB/CA) will, according to the Henderson-Hasselbalch equation, maintain acid-base balance and normal blood pH. Therefore, from a clinical standpoint, disturbances in acid-base balance can be considered dependent on the relative quantities of carbonic acid and base bicarbonate present in the extracellular fluid. Two types of disturbances, metabolic and respiratory, can alter the proper ratio of these components. Metabolic disturbances affect the bicarbonate and respiratory disturbances the carbonic acid element of the buffer pair.

Metabolic disturbances

Metabolic acidosis (bicarbonate deficit)

During the course of certain diseases, such as diabetic ketosis, or during starvation, abnormally large amounts of acids enter the blood. The ratio of BB/CA is altered as the base bicarbonate component of the buffer pair reacts with the acids. The result may be a new ratio near 10 to 1. The decreasing ratio will lower the blood pH, and the respiratory center will be stimulated. The resulting hyperventilation will result in a "blow-off" of carbon dioxide, with a decrease in carbonic acid. This compensatory action of the respiratory system, coupled with excretion of H^+ and NH_3 in exchange for reabsorbed Na^+ by the kidneys, may be sufficient to adjust the *ratio* of BB/CA, and therefore blood pH, to normal. (The compensated BB/CA ratio may approach 10/0.5.) If, despite these compensating homeostatic devices, the ratio and pH cannot be corrected, uncompensated metabolic acidosis develops.

Increased blood hydrogen ion concentration, that is, decreased blood pH, as we have noted, stimulates the respiratory center. For this reason, hyperventilation is an outstanding clinical sign of acidosis. Increases in hydrogen ion concentration above a certain level depress the central nervous system and, therefore, produce such symptoms as disorientation and coma.

Metabolic alkalosis (bicarbonate excess)

Patients suffering from chronic stomach problems such as hyperacidity sometimes ingest large quantities of alkali, often plain baking soda or sodium bicarbonate, for extended periods of time. Such improper use of antacids or excessive vomiting can produce metabolic alkalosis. Initially the condition results in an increase in the BB/CA ratio to perhaps 40 to 1. Compensatory mechanisms are aimed at both increasing carbonic acid and decreasing the bicarbonate load. With breathing suppressed and the kidneys excreting bicarbonate ions, a compensated ratio of 30 to 1.5 might result. Such a ratio would restore acid-base balance and blood pH to normal. In uncompensated metabolic alkalosis the ratio, and therefore the pH, remain increased.

Respiratory disturbances

Respiratory acidosis (carbonic acid excess)

Clinical conditions such as pneumonia or emphysema tend to cause retention of carbon dioxide in the blood. Also, drug abuse or overdose, such as barbiturate poisoning, will suppress breathing and result in respiratory acidosis. The carbonic acid component of the bicarbonate buffer pair increases above normal in respira-

tory acidosis. Body compensation, if successful, increases the bicarbonate fraction so that a new BB/CA ratio (perhaps 40/2) will return blood pH to normal or near normal levels.

Respiratory alkalosis (carbonic acid deficit)

Hyperventilation caused by fever or mental disease (hysteria) can result in excessive loss of carbonic acid and lead to respiratory alkalosis with a bicarbonate buffer pair ratio of 20 to 0.5. Compensatory mechanisms may adjust the ratio to 10/0.5 and return blood pH to near normal.

Outline summary

Mechanisms that control pH of body fluids

A Meaning of term pH—negative logarithm of H ion concentration of solution
B Types of pH control mechanisms
   1 Buffers
   2 Respirations
   3 Kidney excretion of acids and bases
C Effectiveness of pH control mechanisms; range of pH—extremely effective, normally maintain pH within very narrow range of 7.35 to 7.45

Buffer mechanism for controlling pH of body fluids

A Buffers defined
   1 Substances that prevent marked change in pH of solution when acid or base added to it
   2 Consist of weak acid (or its acid salt) and basic salt of that acid
B Buffer pairs present in body fluids—mainly carbonic acid, proteins, hemoglobin, acid phosphate, and sodium and potassium salts of these weak acids

C Action of buffers to prevent marked changes in pH of body fluids
   1 Nonvolatile acids, such as hydrochloric acid, lactic acid, and ketone bodies, buffered mainly by sodium bicarbonate
   2 Volatile acids, chiefly carbonic acid, buffered mainly by potassium salts of hemoglobin and oxyhemoglobin
   3 The chloride shift makes it possible for carbonic acid to be buffered in the red blood cell and then carried as bicarbonate in the plasma
   4 Bases buffered mainly by carbonic acid

$$\left(\text{Ratio } \frac{\text{B} \cdot \text{HCO}_3}{\text{H}_2\text{CO}_3} = \frac{20}{1}\right.$$

   when homeostasis of pH at 7.4 exists)
   5 The Henderson-Hasselbalch equation is a mathematical formula that explains the relationship between hydrogen ion concentration of body fluids and the ratio of base bicarbonate to carbonic acid
D Evaluation of role of buffers in pH control—cannot maintain normal pH without adequate functioning of respiratory and urinary pH control mechanisms

Respiratory mechanism of pH control

A Explanation of mechanism

1 Amount of blood carbon dioxide directly related to amount of carbonic acid and, therefore, to concentration of H ions

2 With increased respirations, less carbon dioxide remains in blood, hence less carbonic acid and fewer H ions; with decreased respirations, more carbon dioxide remains in blood, hence more carbonic acid and more H ions

B Adjustment of respirations to pH of arterial blood —see Fig. 22-6

C Some principles relating respirations and pH of body fluids

1 Acidosis → hyperventilation
↓
increases elimination of CO_2
↓
decreases blood CO_2
↓
decreases blood H_2CO_3
↓
decreases blood H ions, that is, increases blood pH
↓
tends to correct acidosis, that is, to restore normal pH

2 Prolonged hyperventilation, by decreasing blood H ions excessively, may produce alkalosis

3 Alkalosis causes hypoventilation, which tends to correct alkalosis by increasing blood CO_2 and, therefore, blood H_2CO_3 and H ions

4 Prolonged hypoventilation, by eliminating too little CO_2, causes increase in blood H_2CO_3 and, consequently, in blood H ions, thereby may produce acidosis

Urinary mechanism of pH control

A General principles about mechanism—plays vital role in acid-base balance because kidneys can eliminate more H ions from body while reabsorbing more base when pH tends toward acid side and eliminate fewer H ions while reabsorbing less base when pH tends toward alkaline side

B Mechanisms that control urine pH

1 Secretion of H ions into urine—when blood CO_2, H_2CO_3, and H ions increase above normal, distal tubules secrete more H ions into urine to displace basic ion (mainly sodium) from a urine salt and then reabsorb sodium into blood in exchange for the H ions excreted

2 Secretion of NH_3—when blood hydrogen ion concentration increases, distal tubules secrete more NH_3, which combines with H ion of urine to form NH_4 ion, which displaces basic ion (mainly sodium) from a salt; basic ion then reabsorbed back into blood in exchange for ammonium ion excreted

Acid-base imbalances

Metabolic disturbances affect the bicarbonate and respiratory disturbances the carbonic acid element of the bicarbonate buffer pair

Metabolic disturbances

A Metabolic acidosis (bicarbonate deficit)—results in decreased alkaline reserve (mainly $NaHCO_3$), but ratio of BB to CA maintained at normal 20 to 1 by proportionately decreasing blood carbonic acid by hyperventilation coupled with excretion of H^+ and NH_3 in exchange for reabsorbed Na^+ by the kidneys

B Metabolic alkalosis (bicarbonate excess)—initially the condition results in an increase in the BB to CA ratio; compensatory mechanisms are aimed at both increasing carbonic acid and decreasing the bicarbonate load

Respiratory disturbances

A Respiratory acidosis (carbonic acid excess)—the carbonic acid component of the bicarbonate buffer pair increases above normal in respiratory acidosis; compensatory mechanisms are aimed at increasing the bicarbonate fraction in order to return blood pH to normal or near normal levels

B Respiratory alkalosis (carbonic acid deficit)—characterized by excessive loss of carbonic acid caused by hyperventilation

Review questions

1 Explain, in your own words, what the term pH means.

2 What is the normal range for pH of body fluids?

3 Explain what a buffer is in terms of its chemical composition and in terms of its function. Cite specific equations to illustrate your explanation.

4 What is the numerical value of the ratio of B HCO_3/H_2CO_3 when blood pH is 7.4 and the body is in a state of acid-base balance?

5 Is the ratio B HCO_3/H_2CO_3 necessarily abnormal when an acid-base disturbance is present? Give reasons to support your answer.

6 Is blood pH always abnormal when an acid-base disturbance is present? Give reasons for your answer.

7 Explain how the chloride shift makes it possible for carbon dioxide to be buffered in the red blood cell and then carried as bicarbonate in the plasma.

8 Explain how the "respiratory mechanism of pH control" operates.

9 Why is hyperventilation a characteristic clinical sign in acidosis?

Situation: A patient has suffered a severe head injury. Respirations are markedly depressed.

10 Which, if either, do you think is a potential danger —acidosis or alkalosis? State your reasons.

11 Describe the homeostatic mechanisms that would operate to try to maintain acid-base balance in this patient.

Situation: A mother brings her baby to the hospital and reports that he has seemed very sick for the past 24 hours and that he has not eaten during that time and has passed no urine.

12 Do you think acidosis or alkalosis may be present in this baby?

13 Why?

chapter 23

Reproduction of cells

Deoxyribonucleic acid (DNA)
Recombinant DNA

Mitosis—interphase, prophase, metaphase,
anaphase, and telophase

Meiosis

Spermatogenesis

Oogenesis

Cell reproduction is one of the most funda-
mental of all living functions. On it depend the creation of all individual organisms
and the continued existence of all complex, multicellular organisms. The survival, therefore, of
each one of us and of our human species depends
on cell reproduction. In addition, cell reproduction is one of the most fascinating stories that
physiology has to tell. Advances in our understanding of this most basic but enormously complex phenomenon have been described by many
as a scientific revolution with implications more
far-reaching and important to the human species than the birth of the atomic age. We shall
start our version of it by telling about DNA—
"the most golden of all molecules," Watson
called it.*

Deoxyribonucleic acid (DNA)

The deoxyribonucleic acid molecule is a giant
among molecules. Both its size and the complexity of its shape exceed those of most molecules. The importance of its function—in a word,
heredity—surpasses that of any other molecule
in the world. A little more than 25 years ago,
an American, James D. Watson, and two British
scientists, Francis H. C. Crick and Maurice H. F.
Wilkins, solved the puzzle of DNA's molecular
structure. Watson and Crick announced their
discovery with an understated, one-page report
that appeared in the April 25, 1953, issue of the

*Watson, J. D.: The double helix, New York, 1969, The New
American Library, p. 21.

British journal *Nature*. Nine years later the coveted Nobel Prize in Medicine was awarded all three scientists for their brilliant and significant work—hailed as the greatest biological discovery of our times. Watson tells how they accomplished this in *The Double Helix*, now a classic, a book you will probably find hard to put down once you start it.

To try to visualize the shape of the DNA molecule, picture to yourself an extremely long, narrow ladder made of a pliable material. Now see it twisting round and round on its axis and taking on the shape of a steep spiral staircase millions of turns long. This is the shape of the DNA molecule—a double helix (Greek word for spiral).

As to its structure, the DNA molecule is a polymer. This means that it is a large molecule made up of many smaller molecules joined together in sequence. DNA is a polymer of millions of pairs of nucleotides. A nucleotide is a compound formed by combining phosphoric acid with a sugar and a nitrogenous base. In the DNA molecule there are four different kinds of nucleotides. Each nucleotide consists of a phosphate group that attaches to the sugar deoxyribose that attaches to one of four bases. Nucleotides differ, therefore, in their nitrogenous base component—containing either adenine or guanine (purine bases) or cytosine or thymine (pyrimidine bases). (Deoxyribose is a sugar that is not sweet and one whose molecules contain only five carbon atoms.) Notice what the name deoxyribonucleic acid tells you—that this compound contains deoxyribose, that it occurs in nuclei, and that it is an acid.

Fig. 23-1 reveals additional and highly significant facts about DNA's molecular structure. First, observe which compounds form the sides of the DNA spiral staircase—a long line of phosphate and deoxyribose units joined alternately one after the other. Look next at the stair steps. Notice two facts about them: that two bases join (loosely bound by a hydrogen bond) to form each step, and that only two combinations of bases occur. The same two bases invariably pair off with each other in a DNA molecule—like teenagers "going steady." Adenine always goes with thymine (or vice versa, thymine with adenine), and guanine always goes with cytosine (or vice versa). This fact about DNA's molecular structure is called *obligatory base-pairing*. Pay particular attention to it, for it is the key to understanding how a DNA molecule is able to duplicate itself. DNA duplication, or replication as it is usually called, is one of the most important of all biological phenomena because it is an essential and crucial part of the mechanism of heredity.

Another fact about DNA's molecular structure that has great functional importance is the sequence of its base pairs. Although the base pairs in all DNA molecules are the same, the sequence of these base pairs is not the same in all DNA molecules. For instance, the sequence of the base pairs composing the seventh, eighth, and ninth steps of one DNA molecule might be cytosine-guanine, adenine-thymine, and thymine-adenine. Such a sequence of three bases forms a code word or "triplet" called a *codon*. In another DNA molecule the coding sequence of the base pairs making up these same steps might be en-

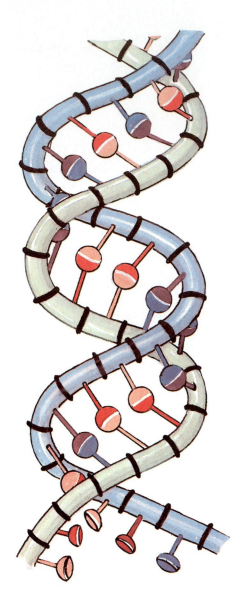

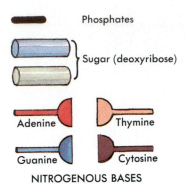

Phosphates

Sugar (deoxyribose)

Adenine — Thymine

Guanine — Cytosine

NITROGENOUS BASES

Fig. 23-1 Watson-Crick structure of DNA molecule. Note that each side of the DNA molecule consists of alternating sugar and phosphate groups. Each sugar group is united to the sugar group opposite it by a pair of nitrogenous bases, adenine-thymine or thymine-adenine and cytosine-guanine or guanine-cytosine. Differences in the sequences of base pairs establish the identity of the many different kinds of DNA.

tirely different, perhaps thymine-adenine, guanine-cytosine, and cytosine-guanine. Perhaps these seem to be minor details. But nothing could be further from the truth, since it is the sequence of the base pairs in the nucleotides composing DNA molecules that identifies each gene. Hence it is the sequence of base pairs that determines all hereditary traits.

A human *gene* is a segment of a DNA molecule. One gene consists of a chain of about 1,000 pairs of nucleotides joined one after the other in a precise sequence. One gene controls the production within the cell of one polypeptide chain. Two or more polypeptides make up enzymes and structural proteins. Each enzyme catalyzes one chemical reaction. Therefore, as Fig. 23-2 indicates, the millions of genes constituting a cell's DNA determine the cell's structure and its functions.

Genes control protein synthesis (anabolism). A brief synopsis of this process as now visualized

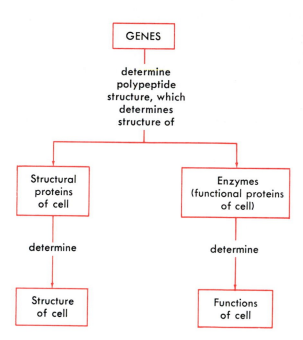

Fig. 23-2 Brief scheme to show how genes determine hereditary characteristics.

follows. First, a strand of RNA (ribonucleic acid) forms along a segment of one strand of a DNA molecule. RNA differs from DNA in certain respects. Its molecules are smaller than those of DNA, and RNA contains ribose instead of deoxyribose. Also, one of the four bases in RNA is uracil instead of thymine. As a strand of RNA is forming along a strand of DNA, uracil attaches to adenine and guanine attaches to cytosine. The process is known as complementary pairing. Thus a single-strand molecule of messenger RNA (abbreviated mRNA) is formed. The name "messenger RNA" describes its function. As soon as it is formed, it separates from the DNA strand, diffuses out of the nucleus, and carries a "message" to a ribosome in the cell's cytoplasm, directing its synthesis of a specific protein. Ribosomes contain another type of RNA called appropriately "ribosomal RNA" (abbreviated rRNA). Next, a molecule of a third type of RNA

—transfer RNA, or tRNA—present in cytoplasm attaches itself to a molecule of a specific amino acid and transfers it to a ribosome. Here the amino acid fits into the position indicated by the mRNA. More tRNA molecules, one after the other in rapid sequence, bring more amino acids to the ribosome and fit them into their proper positions. Result? A chain of amino acids joined to each other in a definite sequence—a protein, in other words—is formed. In short, protein anabolism has occurred. It is one of the major kinds of cellular work. One human cell is estimated to synthesize perhaps 2,000 different enzymes! In addition to this staggering work load, it produces many different protein compounds that help form its own structures and many cells also synthesize special proteins. Liver cells are a notable example; they synthesize prothrombin, fibrinogen, albumins, and globulins.

Recombinant DNA

Each time fertilization occurs, a new and unique package of genetic material (DNA), and therefore a new individual, comes into being. Sexual reproduction involves this recombination or mingling of DNA from sperm and egg. The term "recombinant DNA," however, means something entirely different. It refers to the joining together of hereditary material, often from different species, into new, biologically functional combinations.

Recombination or gene-splicing techniques permit the joining of DNA from different organisms—inside bacteria! Scientists split small circular rings of bacterial DNA called *plasmids* by using specialized "restriction enzymes" as chemical cutters. These same enzymes are used to cut segments of DNA from some other animal source (perhaps a mouse), and the foreign DNA fragment is then annealed into the opened plasmid ring. The recombinant plasmid is then reinserted into a bacterial "host." The result is a type of hybrid that never before existed in nature. Each time the host organism divides, so does the recombinant DNA. Such rapidly reproducing synthetic bacteria have already made human insulin and somatostatin in the laboratory and give promise for future uses.

In 1971 Dr. Paul Berg of Stanford University discussed his plans to insert a DNA fragment that caused cancer in mammals into *E. coli* bacteria. He terminated the experiment at the request of a group of scientists who believed that there was at least the potential for some kind of biological catastrophe in such a study—the creation of a new mutant organism with malignant powers that might escape from the laboratory and cause widespread disease.

Further advances in recombinant DNA technology aroused both public and scientific appre-

hension, and in 1974 Berg and a group of other molecular biologists called for a moratorium on certain types of recombinant DNA experiments and brought to public attention the potential dangers of genetic engineering in bacteria.* Not since the congressional investigations of atomic energy and nuclear radiation dangers had science sparked such heated controversy. As a result of discussions that followed the moratorium, an elaborate set of guidelines has now

*Berg, P., Baltimore, D., Boyer, H. W., Cohen, S. N., Davis, R. W., Hogness, D. S., Nathans, D., Roblin, R. O., Watson, J. D., Weissman, S., and Zinder, N. D.: Potential biohazards of recombinant DNA molecules, Science **185:**303, 1974.

Table 23-1 Mitosis

Interphase	Prophase	Metaphase	Anaphase	Telophase
1 Period when cell prepares for division and also grows in size 2 Chromosomes elongate and become too thin to be visible as such, but chromatin granules become visible 3 DNA of each chromosome replicates itself, forming two chromatids attached only at centromere	1 Chromosomes shorten and thicken (from coiling of DNA molecules that compose them); each chromosome consists of two *chromatids* attached at *centromere* 2 Centrioles move to opposite poles of cell; spindle fibers appear and, under control of centrioles, begin to orient between opposing poles	1 Chromosomes align across equator of spindle fibers; each pair of chromatids attached to spindle fiber at its centromere 2 Nucleoli and nuclear membrane disappear	1 Each centromere divides into two, thereby detaching two chromatids that compose each chromosome from each other 2 Divided centromeres start moving to opposite poles, each pulling its chromatid (now called a chromosome) along with it; there are now twice as many chromosomes as there were before mitosis started	1 Changes occurring during telophase essentially reverse of those taking place during prophase; new chromosomes start elongating (DNA molecules start uncoiling) 2 Nucleoli and two new nuclear membranes appear, enclosing each new set of chromosomes 3 Spindle fibers disappear 4 *Cytokinesis,* or dividing of cytoplasm, usually occurs during telophase; starts as pinching in along equator of old cell and ends with division of old cell into two new cells 5 Centrioles replicate

been adopted to ensure safety for experiments in this area.

The ultimate goal of recombinant DNA research is to increase our understanding of mammalian gene regulation. Genes initiate and control the enormously complex mechanism that permits development of a fertilized ovum into a fully differentiated individual.

Mitosis

Human cells, other than sex cells, reproduce by the process of mitosis. In this process a cell divides in order to multiply. One cell divides to form two cells. Mitosis consists of a sequence of events plainly visible in suitably stained cells viewed with the light microscope. The events of mitosis occur in five phases: interphase, prophase, metaphase, anaphase, and telophase.

DNA replication occurs during the interphase of mitosis (cell reproduction). The tightly coiled DNA molecules uncoil except in small segments. Since these remaining tight little coils are denser than the thin elongated sections, they absorb more stain and appear as chromatin granules under the microscope. The thin uncoiled sections, in contrast, are invisible because they absorb so little stain. As the DNA molecule uncoils, its two strands come apart. Then, along

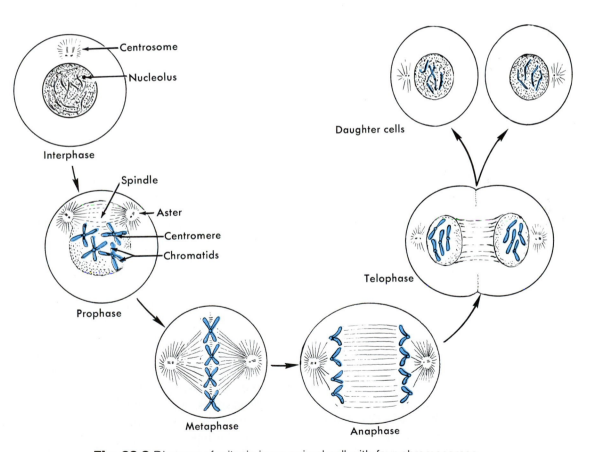

Fig. 23-3 Diagram of mitosis in an animal cell with four chromosomes.

each of the two separated strands of nucleotides, a complementary strand forms. Intracellular fluid contains many molecules of deoxyribose and nitrogenous bases and many phosphate ions. By the mechanism of obligatory base-pairing, these substances become attached at their right places along each DNA strand. Interpreted, this means that new thymine, that is, from the intracellular fluid, attaches to the "old" adenine in the original DNA strand. Conversely, new adenine attaches to old thymine. Also, new guanine joins old cytosine and, vice versa, new cytosine joins old guanine. By the end of the interphase, each of the two DNA strands of the original DNA molecules has a complete new complementary strand attached to it. Each half DNA molecule, in other words, has duplicated itself to create a whole new DNA molecule. Thus two new chromosomes now replace each original chromosome. However, at this stage (end of interphase, beginning of prophase), they are called chromatids instead of chromosomes. The two chromatids formed from each original chromosome contain duplicate copies of DNA and, therefore, the same genes as the chromosome from which they were formed. Chromatids are present as attached pairs. Centromere is the name of their point of attachment. By the time the parent cell divides to form two daughter cells, chromatids have separated and become chromosomes and one set of them has become part of the nucleus of one daughter cell and the other set has become part of the nucleus of the other daughter cell. Hence each of the two new cells formed by mitosis contains a complete set of genes identical to those in its parent cell. Both daughter cells have the potentialities to become like their parent cell. Thus the function of mitosis is to enable cells to reproduce their own kind. By means of the mechanism of mitosis, a new generation of cells inherits both the structural and the functional characteristics of the preceding generation (Fig. 23-2). In short, mitosis is the mechanism of heredity for cells. Table 23-1 (p. 600) summarizes the events that char-acterize the five phases of mitosis. As you read the events, try to correlate them with the drawings in Fig. 23-3 (p. 601).

Meiosis

Meiosis is the type of cell division that occurs only in primitive sex cells during the process of their becoming mature sex cells. The scientific name of the primitive male sex cells that reproduce by meiosis is *primary spermatocytes*. *Primary oocytes* is the name of the primitive female sex cells. The nucleus of each primary spermatocyte and oocyte contains 23 pairs of chromosomes. This total of 46 chromosomes per cell is known as the *diploid* number of chromosomes for human cells. The scientific name for either a male or female mature sex cell is gamete. Male gametes are named spermatozoa but are usually called sperm. Female gametes are named ova (singular, ovum).

Meiosis consists of two cell divisions that take place one after the other in quick succession. They are referred to as meiotic division I and meiotic division II, and in both, an interphase, prophase, metaphase, anaphase, and telophase occur. In the interphase that precedes prophase I (of meiotic division I) the same events occur as take place in the interphase of mitosis. Specifically, the DNA of each chromosome replicates itself and thereby changes each chromosome into two chromatids, attached only at the centromere. Prophase I of meiosis consists of several stages. You can find the names of these stages in Fig. 23-4. For simplicity's sake, only two pairs of chromosomes—one member of each pair black and one green—are shown in this figure. Look closely at the diplotene stage and you will notice that a segment of one of the chromatids of the green chromosome and a segment of one of the chromatids of the black chromosome have exchanged places. A chromatid segment of each chromosome has crossed over and become part of the other chromosome. This is a highly signifi-

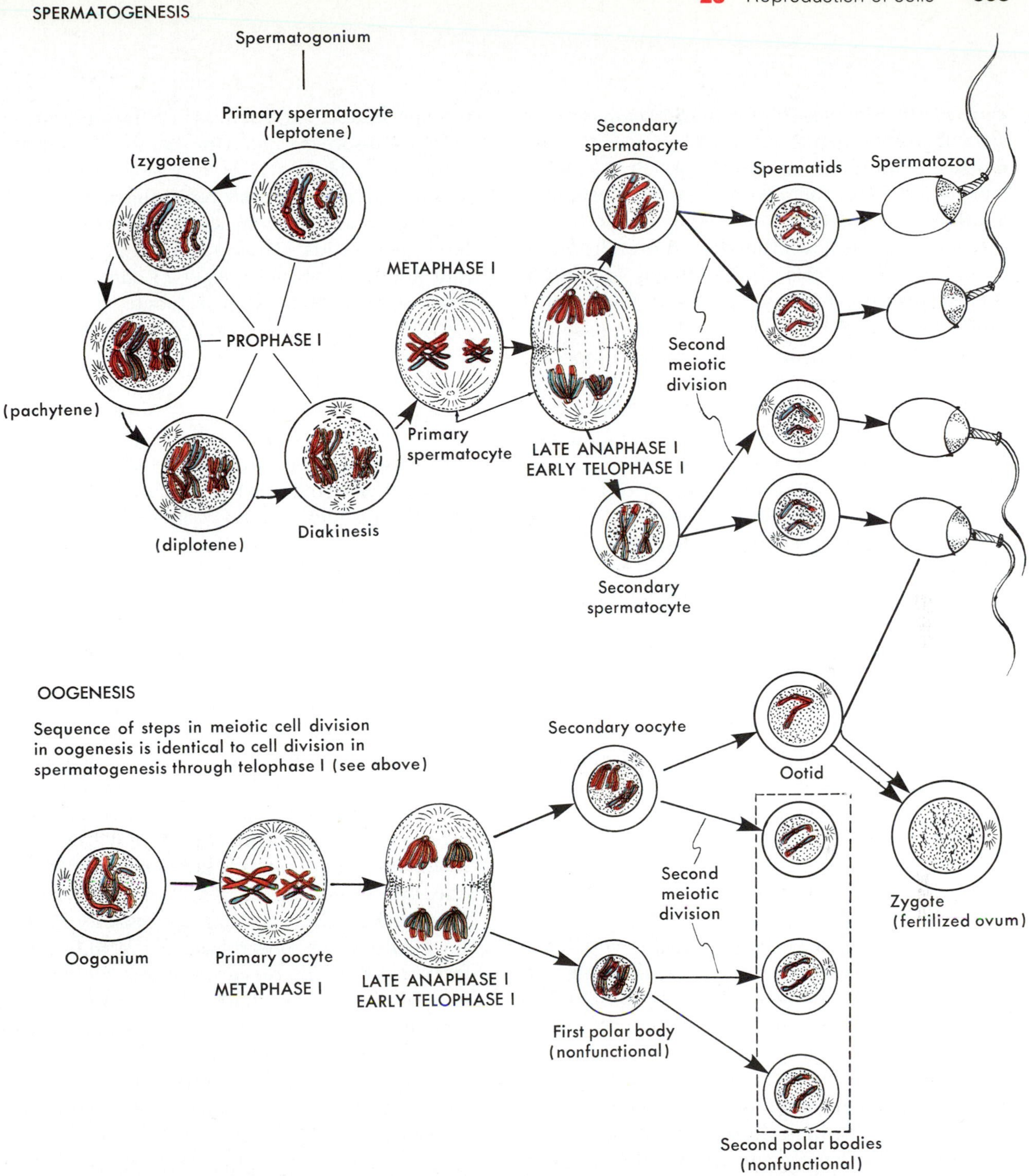

SPERMATOGENESIS

OOGENESIS

Sequence of steps in meiotic cell division in oogenesis is identical to cell division in spermatogenesis through telophase I (see above)

Fig. 23-4 Diagram representing meiotic sequence preceding maturation of sex cells (gametes) in male and female animals. Left to right: Major stages of meiosis in spermatogenesis (top), resulting in formation of four spermatids, which become mature spermatozoa. Oogenesis (bottom) results in formation of one ootid capable of being fertilized and three nonfunctional polar bodies.

cant event. Since each chromatid segment consists of specific genes, the crossing-over of chromatids reshuffles the genes, so to speak—it transfers some of them from one chromosome to another.

Metaphase I follows the last stage of prophase I, and, as in mitosis, the chromosomes align themselves along the equator of the spindle fibers as Fig. 23-4 shows. But in anaphase the two chromatids that make up each chromosome do not separate from each other as they do in mitosis to form two new chromosomes out of each original one. Therefore in anaphase I of Fig. 23-4, only two chromosomes move to each pole of the parent cell. When the parent cell divides to form two cells, each daughter cell contains two chromosomes or half as many as the parent cell had. The daughter cells formed by meiotic division I contain a *haploid* number of chromosomes, which means half as many as the diploid number present in the parent cell.

The right side of Fig. 23-4 shows meiotic division II; it is essentially the same as mitosis. Notice that it reproduces each of the two cells formed by meiotic division I and so forms four cells, each with the haploid number of chromosomes.

Spermatogenesis

Spermatogenesis is the process by which the primitive sex cells (spermatogonia) present in the seminiferous tubules of a newborn baby boy become transformed into mature sperm. Spermatogenesis starts at about the time of puberty and usually continues throughout a man's life. Fig. 23-4, *A*, diagrams some of the major steps of spermatogenesis. Meiotic division I reproduces one primary spermatocyte to form two second-ary spermatocytes, each with a haploid number of chromosomes (23). One of the 23 chromosomes is either an X or a Y sex chromosome. Meiotic division II then reproduces each of the two secondary spermatocytes to form a total of four spermatids. Thus spermatogenesis forms four sperm—each with only 23 chromosomes—from one primary spermatocyte that has 23 pairs or 46 chromosomes.

Oogenesis

Oogenesis is the process by which primitive female sex cells (oogonia) become mature ova. During the fetal period, oogonia reproduce in the ovaries by mitosis to form primary oocytes—about a half million of them by the time a baby girl is born. Most of the primary oocytes develop to prophase I before birth. There they stay until puberty. Then the other steps of oogenesis shown in Fig. 23-4, *B*, take place about once a month for the next 30 or 40 years, that is, until a woman reaches her menopause. Note that oogenesis forms only one mature ovum (egg) from each primitive oocyte. How many sperm are formed from each primary spermatocyte by spermatogenesis? An ovum, like a sperm, contains only 23 chromosomes, but unlike sperm, half of which have an X and half of which a Y sex chromosome, all ova have an X chromosome. This distinction between sperm and ovum makes a difference—the difference between whether the new life formed by their union will be male or female. If a sperm with an X chromosome fertilizes an ovum, the baby will be a girl. If a Y-bearing sperm fertilizes an ovum, it will be a boy. Chapter 25 will give more information about this.

Outline summary

Deoxyribonucleic acid (DNA)

A Structure—see Fig. 23-1
B Relation to genes
  1 One gene is a segment of a DNA molecule, consisting of a chain of about 1,000 pairs of nucleotides joined in a precise sequence
  2 DNA of one human chromosome made up of about 175,000 genes according to one cytologist's estimate
C Replication—DNA of each chromosome duplicates itself during interphase of mitosis, thereby forming two chromatids out of each chromosome
D Function—DNA replication makes possible cellular heredity

Recombinant DNA

A Definition—joining together of hereditary material, often from different species, into new, biologically functional combinations
B Techniques
  1 Bacterial DNA—plasmid rings—are split
  2 "Restriction enzymes" used to open plasmid rings and to cut segments of DNA from other sources
  3 Foreign DNA then annealed into open plasmid ring
  4 Recombinant plasmid then reinserted into a bacterial "host"
  5 Each time synthetic host organism divides, so does recombinant DNA

Mitosis

A Definition—process by which somatic cells reproduce
B Phases—interphase, prophase, metaphase, anaphase, and telophase; see Table 23-1 for summary of events during each phase
C Function—transmits to daughter cells the same (diploid) number of chromosomes composed of same genes as present in parent cell and thereby makes possible inheritance of parent cell's traits

Meiosis

A Definition—process by which primitive sex cells reproduce in the process of becoming mature sex cells (gametes) with haploid number of chromosomes
B Stages
  1 Meiotic division I—consists of interphase, prophase I, metaphase I, anaphase I, and telophase I; primary sex cell reproduces to form two secondary sex cells containing haploid number of chromosomes, that is, 23
  2 Meiotic division II—consists of same phases as meiotic division I—4 sperm or 1 ovum result from this division; see Fig. 23-4
C Function—formation of gametes with haploid number of chromosomes

Spermatogenesis

Process by which spermatogonia present at birth in seminiferous tubules undergo meiosis to become mature sperm; spermatogenesis begins at puberty and usually continues throughout a man's life

Oogenesis

Process by which oogonia become mature ova; oogonia in ovaries of fetus reproduce by mitosis to form primary oocytes at prophase I stage; beginning at puberty, and recurring about once a month for the next 30 or 40 years, remaining steps of oogenesis take place as shown in Fig. 23-4, *B;* ovum contains 23 chromosomes, one of which is an X chromosome; sperm also contain 23 chromosomes but one of them is either an X or a Y sex chromosome

Review questions

1 Describe the size and shape of a DNA molecule.
2 Where are DNA molecules located?
3 Explain the steps of DNA replication.
4 When does DNA replication occur?
5 Watson called DNA "the most golden of all molecules," primarily because of its function. What is it?
6 Define the term "recombinant DNA."
7 What is a "plasmid"?
8 What are "restriction enzymes" and how are they used in recombinant DNA research?
9 Discuss the possible applications of recombinant DNA techniques in human medicine.
10 Define mitosis and name its phases.
11 Define meiosis and name its stages.
12 When does the reduction of chromosomes from the diploid to haploid number take place? Meiosis is a part of what larger male process? Of what larger female process?
13 Define spermatogenesis. When and where does it occur?
14 Define oogenesis. When and where does it occur?

chapter 24

The male reproductive system

Meaning and function

Reproductive system functions differ notably from the functions of any other system of the body. The proper functioning of the reproductive system and of its enormously complex control mechanisms ensures survival, not of the individual, but of the species. In the male the system consists of those organs whose functions are to produce, transfer, and ultimately introduce mature sperm into the female reproductive tract where fertilization can occur. Also, the testes secrete androgens or male sex hormones, notably testosterone.

Male reproductive organs

Organs of the male reproductive system may be classified as either *essential organs* for the production of gametes (sex cells) or *accessory organs* that play some type of supportive role in the reproductive process.

In both sexes the essential organs of reproduction that produce the gametes or sex cells (sperm or ova) are called *gonads*. The gonads of the male are the testes.

The accessory organs of reproduction in the male include a number of genital ducts, glands, and supporting structures.

Genital ducts serve to convey sperm to the outside of the body. The ducts are a pair of epididymides (singular: epididymis), a pair of vas deferens (ductus deferens), a pair of ejaculatory ducts, and the urethra.

Accessory glands in the reproductive system produce secretions that serve to nourish, transport, and mature sperm. The glands are a pair of seminal vesicles, one prostate, and a pair of bulbourethral (Cowper's) glands.

The supporting structures are the scrotum, the penis, and a pair of spermatic cords.

Testes

Structure and location

The testes are small ovoid glands that are somewhat flattened from side to side, measure about 4 or 5 cm in length, and weigh 10 to 15 gm each. The left testis is generally located about 1 cm lower in the scrotal sac than the right. Both are suspended by attachment to scrotal tissue and the spermatic cords (Fig. 24-7, p. 615). A dense white fibrous capsule called the *tunica albuginea* encases each testis and then enters the gland, sending out partitions that radiate through its interior, dividing it into 200 or more cone-shaped lobules.

Each lobule of the testis contains one to three tiny, coiled seminiferous tubules and numerous interstitial cells (of Leydig). If unraveled, each seminiferous tubule measures about 75 cm in length. The tubules from each lobule come together to form a plexus called the *rete testis*. A

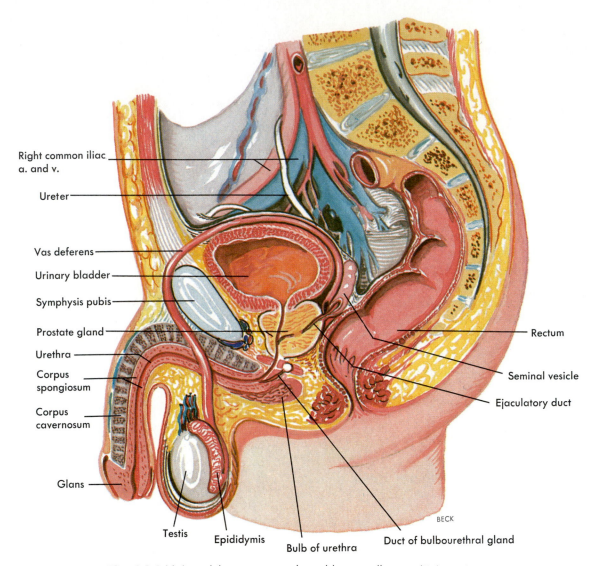

Right common iliac
a. and v.

Ureter

Vas deferens

Urinary bladder

Symphysis pubis

Prostate gland

Urethra

Corpus
spongiosum

Corpus
cavernosum

Glans

Testis

Epididymis

Bulb of urethra

Duct of bulbourethral gland

Rectum

Seminal vesicle

Ejaculatory duct

BECK

Fig. 24-1 Male pelvic organs as viewed in a median sagittal section.

series of ducts called *efferent ductules* drain the rete testis and then pierce the tunica albuginea and enter the head of the epididymis (Fig. 24-2).

Functions

The testes perform two primary functions: spermatogenesis and secretion of hormones.

1 Spermatogenesis, the production of spermatozoa (sperm), the male gametes or reproductive cells. The seminiferous tubules produce the sperm (Fig. 24-2).

2 Secretion of hormones, chiefly testosterone (androgen or masculinizing hormone) by interstitial cells (Leydig cells). Testosterone serves the following general functions:

a It promotes "maleness," that is, the development and maintenance of male secondary sex characteristics, of male accessory organs such

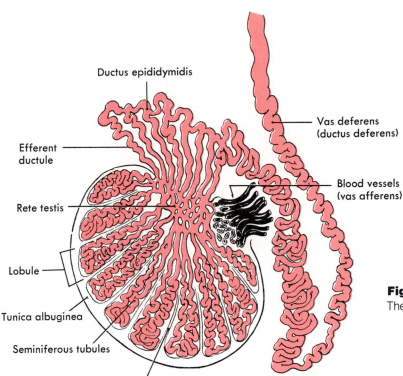

Ductus epididymidis

Efferent
ductule

Rete testis

Lobule

Tunica albuginea

Seminiferous tubules

Septum

Vas deferens
(ductus deferens)

Blood vessels
(vas afferens)

Fig. 24-2 Tubules of the testis and epididymis. The ducts and tubules are exaggerated in size.

as the prostate, seminal vesicles, etc., and of adult male sexual behavior.

b It helps regulate metabolism and is sometimes referred to as "the anabolic hormone" because of its marked stimulating effect on protein anabolism. By stimulating protein anabolism, testosterone promotes growth of skeletal muscles (responsible for greater muscular development and strength of male) and growth of bone. However, testosterone also promotes closure of the epiphyses. Early sexual maturation, therefore, generally leads to early epiphyseal closure and shortness. The converse also holds true: late sexual maturation, delayed epiphyseal closure, and tallness tend to go together.

c It plays a part in fluid and electrolyte metabolism. Testosterone has a mild stimulating effect on kidney tubule reabsorption of sodium and therefore water; it also promotes kidney tubule excretion of potassium.

d It inhibits anterior pituitary secretion of gonadotropins, namely, FSH and ICSH (interstitial cell–stimulating hormone; called LH or luteinizing hormone in the female).

The anterior pituitary gland controls the testes by means of its gonadotropic hormones—specifically, FSH and ICSH just mentioned. FSH stimulates the seminiferous tubules to produce sperm more rapidly. ICSH stimulates interstitial cells to increase their secretion of testosterone. Soon the blood concentration of testosterone reaches a high level that inhibits anterior pituitary secretion of FSH and ICSH. Thus a negative feedback mechanism operates between the anterior pituitary gland and the testes. A high blood concentration of gonadotropins stimulates testosterone secretion. But a high blood concentration of testosterone inhibits (has a negative effect on) gonadotropin secretion (Fig. 24-3).

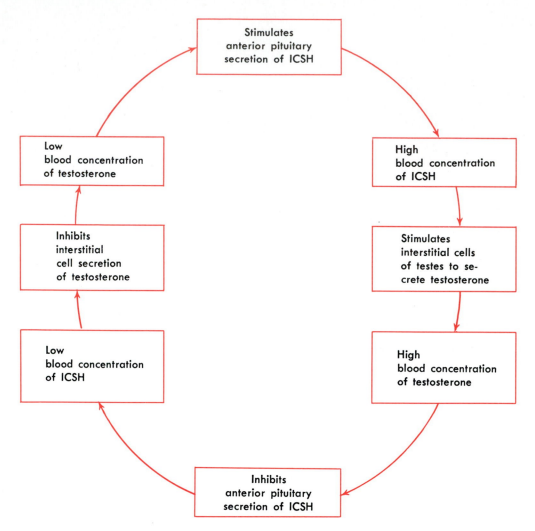

Fig. 24-3 The negative feedback mechanism that controls anterior pituitary gland secretion of ICSH and interstitial cell secretion of testosterone.

Structure of spermatozoa

Sperm found in the seminiferous tubules appear fully formed. We know, however, that they undergo a process of "ripening" or maturation as they pass through the genital ducts prior to ejaculation. Although anatomically complete and highly motile when ejaculated, sperm must still undergo a complex process called *capacitation* before they are capable of actually fertilizing an ovum. Normally, capacitation occurs in sperm only after they have been introduced into the vagina of the female.

Fig. 24-4 shows the characteristic parts of a spermatozoon: head, neck, middle piece, and elongated, lashlike tail.

The head of a spermatozoon is, in essence, a highly compact package of genetic chromatin material covered by a specialized *acrosomal cap* designed to aid in penetration of the ovum. The cylindrical middle piece is characterized by a

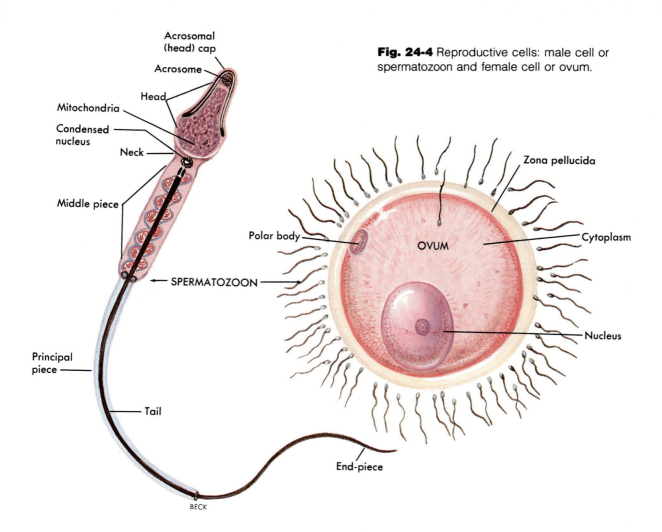

Fig. 24-4 Reproductive cells: male cell or spermatozoon and female cell or ovum.

helical arrangement of mitochondria. The tail is divided into a principal piece and short end-piece, both typical in appearance of all flagella capable of motility.

Ducts of testes

Epididymis
Structure and location

Each epididymis consists of a single tightly coiled tube enclosed in a fibrous casing. The tube has a very small diameter (just barely mac-

roscopic) but measures approximately 20 feet in length. It lies along the top and side of the testis. The epididymis is divided into a blunt superior *head* (which is connected to the testis by the efferent ductules), a central *body*, and a tapered inferior portion that is continuous with the vas deferens called the *tail*. (See Figs. 24-2, p. 609, and 24-7, p. 615.)

Functions

The epididymis serves the following functions:
1 It serves as one of the ducts through which sperm pass in their journey from the testis to the exterior.

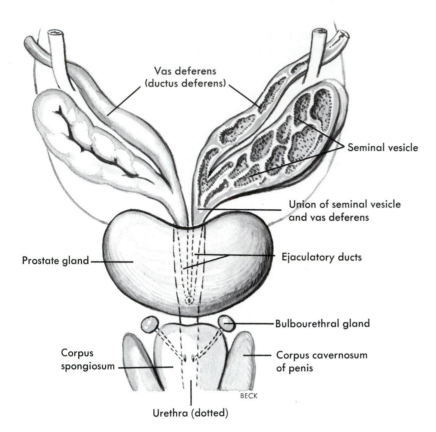

Vas deferens
(ductus deferens)

Seminal vesicle

Union of seminal vesicle
and vas deferens

Prostate gland

Ejaculatory ducts

Bulbourethral gland

Corpus
spongiosum

Corpus cavernosum
of penis

BECK

Urethra (dotted)

Fig. 24-5 Formation of the ejaculatory ducts by union of the seminal vesicles with the vas deferens just before entrance into the prostate gland. The ejaculatory ducts open into the prostatic portion of the urethra.

2 It stores a small quantity of sperm prior to ejaculation.

3 It secretes a small part of the seminal fluid (semen).

Vas deferens (ductus deferens)

Structure and location

The vas deferens, like the epididymis, is a tube. In fact, the duct of the vas is an extension of the tail of the epididymis. It ascends from the scrotum and passes through the inguinal canal as part of the spermatic cord—enclosed by fibrous connective tissue with blood vessels, nerves, and lymphatics—into the abdominal cavity. Here it extends over the top and down the posterior surface of the bladder where an enlarged and tortuous portion called the *ampulla* joins the duct from the seminal vesicle to form the ejaculatory duct (Figs. 24-5 and 24-6).

Function

The vas deferens serves as one of the ducts for the testis, connecting the epididymis with the ejaculatory duct.

Correlation

Severing of the vas deferens—that is, a vasectomy, usually done through incision in the scrotum—makes a man sterile. Why? Because it interrupts the route to the exterior from the epididymis. To leave the body, sperm must journey in succession through the epididymis, vas deferens, ejaculatory duct, and urethra.

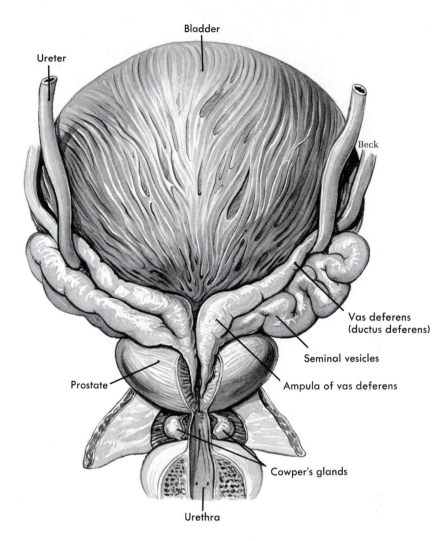

Fig. 24-6 The urinary bladder and male pelvic reproductive organs viewed from behind.

Labels on figure:
Ureter
Bladder
Beck
Vas deferens (ductus deferens)
Seminal vesicles
Ampula of vas deferens
Prostate
Cowper's glands
Urethra

Ejaculatory duct

The two ejaculatory ducts are short tubes that pass through the prostate gland to terminate in the urethra. As Fig. 24-5 shows, they are formed by the union of the vas deferens with the ducts from the seminal vesicles.

Urethra

For a discussion of the urethra, see p. 558.

Accessory reproductive glands

Seminal vesicles
Structure and location

The seminal vesicles are convoluted pouches that lie along the lower part of the posterior surface of the bladder, directly in front of the rectum (Fig. 24-1, p. 608).

Function

The seminal vesicles secrete a viscous liquid component of the semen rich in fructose. As a simple sugar, fructose serves as an energy source for sperm motility after ejaculation. The thick yellowish secretion also contains prostaglandins, substances postulated to influence cyclic AMP formation (p. 322). Normal secretory activity of the seminal vesicles depends on adequate levels of testosterone.

Prostate gland
Structure and location

The prostate is a compound tubuloalveolar gland that lies just below the bladder and is shaped like a doughnut. The fact that the urethra passes through the small hole in the center of the prostate is a matter of considerable clinical significance. Many older men suffer from enlargement of this gland. As it enlarges, it squeezes the urethra, frequently closing it so completely that urination becomes impossible. Urinary retention results. Surgical removal of the gland (prostatectomy) is resorted to as a cure for this condition when other less radical methods of treatment fail.

Function

The prostate secretes a thin alkaline substance that constitutes the largest part of the seminal fluid. Its alkalinity helps protect the sperm from acid present in the male urethra and female vagina and thereby increases sperm motility. (Acid depresses or, if strong enough, kills sperm. Sperm motility is greatest in neutral or slightly alkaline media.)

Bulbourethral glands
Structure and location

The two bulbourethral or Cowper's glands resemble peas in both size and shape. You can see the location of these compound tubuloalveolar glands below the prostate glands in Fig. 24-6. A duct approximately 1 inch long connects them with the penile portion of the urethra.

Function

Like the prostate, the bulbourethral glands secrete an alkaline fluid that is important for counteracting the acid present in both the male urethra and the female vagina.

Supporting structures

Scrotum (external)

The scrotum is a skin-covered pouch suspended from the perineal region. Internally, it is divided into two sacs by a septum, each sac containing a testis, epididymis, and lower part of a spermatic cord.

Penis (external)
Structure

Three cylindrical masses of erectile or cavernous tissue, enclosed in separate fibrous coverings and held together by a covering of skin, compose the penis. The two larger and uppermost of these cylinders are named the *corpora cavernosa penis*, whereas the smaller, lower one, which contains the urethra, is called the *corpus cavernosum urethrae* (corpus spongiosum).

The distal part of the corpus cavernosum urethrae overlaps the terminal end of the two corpora cavernosa penis to form a slightly bulging structure, the *glans penis*, over which the skin is folded doubly to form a more or less loose-fitting, retractable casing known as the *prepuce* or foreskin. If the foreskin fits too tightly about the glans, a circumcision is usually performed to prevent irritation. The opening of the urethra at the tip of the glans is called the *external urinary meatus*.

Fig. 24-7 Lateral view of the *left* spermatic cord and testis.

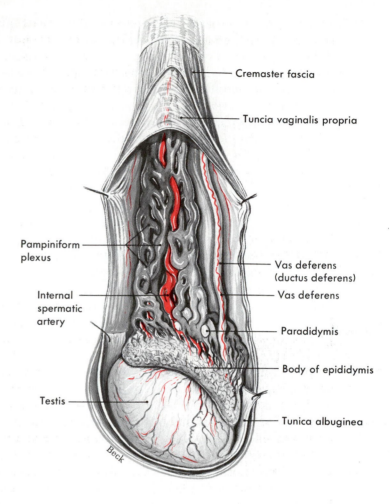

Cremaster fascia

Tuncia vaginalis propria

Pampiniform plexus

Internal spermatic artery

Testis

Vas deferens (ductus deferens)

Vas deferens

Paradidymis

Body of epididymis

Tunica albuginea

Beck

Functions

The penis contains the urethra, the terminal duct for both urinary and reproductive tracts. It is the copulatory organ by means of which spermatozoa are introduced into the female vagina. The scrotum and penis together constitute the *external genitals* of the male.

Spermatic cords (internal)

The *spermatic cords* are cylindrical casings of white fibrous tissue located in the inguinal canals between the scrotum and the abdominal cavity. They enclose the vas deferens, blood vessels, lymphatics, and nerves (Fig. 24-7).

Composition and course of seminal fluid

The following structures secrete the substances that, together, make up the seminal fluid or semen:

1 Testes and epididymides—their secretions, according to one estimate, constitute less than 5% of the seminal fluid volume.

2 Seminal vesicles—their secretions are reported to contribute about 30% of the seminal fluid volume.

3 Prostate gland—its secretions constitute the bulk of the seminal fluid volume, reportedly about 60%.

4 Bulbourethral glands—their secretions are said to constitute less than 5% of the seminal fluid volume.

Besides contributing slightly to the fluid part of semen, the testes also add hundreds of millions of sperm. In traversing the distance from their place of origin to the exterior, the sperm must pass from the testis through the epididymis, vas deferens, ejaculatory duct, and urethra. Note that sperm originate in the testes, glands located outside the body (that is, not within a body cavity), travel inside, and finally are expelled outside.

Early in fetal life the testes are located in the abdominal cavity but normally descend through the inguinal canals into the scrotum about 2 months before birth. Occasionally a baby is born with undescended testes, a condition called *cryptorchidism*, which is readily observed by palpation of the scrotum. Because the higher temperature inside the abdominal cavity inhibits spermatogenesis, measures must be taken to bring the testes down into the scrotum in order to prevent permanent sterility.

Male fertility

Male fertility relates to many factors—most of all to the number of sperm ejaculated but also to their size, shape, and motility. Fertile sperm have a uniform size and shape. They are highly motile. Although only one sperm fertilizes an ovum, millions of sperm seem to be necessary for fertilization to occur. According to one estimate, when the sperm count falls below about 50 million per milliliter of semen, sterility results.

One hypothesis suggested to explain this puzzling fact is this: Semen that contains an adequate number of sperm also contains enough hyaluronidase to liquefy the intercellular substance between the cells that encase each ovum. Without this, a single sperm cannot penetrate the layers of cells (corona radiata) around the ovum and hence cannot fertilize it.

Male functions in reproduction

Recall that all body functions but one have for their ultimate goal survival of the individual. Only the function of reproduction serves a different, a longer range, and, no doubt in nature's scheme, a more important purpose—survival of the human species. Male functions in reproduction consist of the production of male sex cells (spermatogenesis, discussed in Chapter 23) and introduction of these cells into the female body (coitus, copulation, or sexual intercourse). In order for coitus to take place, erection of the penis must first occur, and in order for sperm to enter the female body, both the sex cells and secretions from the accessory glands must be introduced into the urethra (emission) and semen ejaculated from the penis.

Erection is a parasympathetic reflex initiated mainly by certain tactile, visual, and mental stimuli. It consists of dilation of the arteries and arterioles of the penis, which in turn floods and distends spaces in its erectile tissue and compresses its veins. Therefore more blood enters the penis through the dilated arteries than leaves it through the constricted veins. Hence, it becomes larger and rigid, or in other words, erection occurs.

Emission is the reflex movement of sex cells, spermatozoa, and secretions from the genital ducts and accessory glands into the prostatic urethra. Once emission has occurred, ejaculation will follow.

Ejaculation of semen is also a reflex response. It is the usual outcome of the same stimuli that initiate erection. Ejaculation and various other responses—notably accelerated heart rate, increased blood pressure, hyperventilation, dilated skin blood vessels, and intense sexual excitement—characterize the climax of coitus known as an *orgasm*.

Outline summary

Meaning and function

Consists of organs that, together, produce new individual

Male reproductive organs

A Essential—testes (male gonads)
B Accessory
  1 Glands
    a Seminal vesicles (paired)
    b Prostate gland
    c Bulbourethral (Cowper's) glands (paired)
  2 Ducts of testes
    a Epididymis (paired)
    b Vas deferens (ductus deferens) (paired)
    c Ejaculatory ducts (paired)
    d Urethra
  3 Supporting structures
    a External—scrotum and penis
    b Internal—spermatic cords (paired)

Testes

A Structure and location
  1 Several lobules composed of seminiferous tubules and interstitial cells (of Leydig), separated by septa, encased in fibrous capsule called the tunica albuginea
  2 Seminiferous tubules in testis open into a plexus called rete testis, which is drained by a series of efferent ductules that emerge from top of organ and enter head of epididymis
  3 Located in scrotum, one testis in each of two scrotal compartments
B Functions
  1 Spermatogenesis—formation of mature male gametes (spermatozoa) by seminiferous tubules
  2 Secretion of hormone (testosterone) by interstitial cells
C Structure of spermatozoa—consists of a head covered by acrosome, neck, middle piece, and tail divided into a principal piece and a short end-piece

Ducts of testes

A Epididymis
  1 Structure and location
    a Single tightly coiled tube enclosed in fibrous casing
    b Lies along top and side of each testis
    c Anatomic divisions include head, body, and tail

  2 Functions
    a Duct for seminal fluid
    b Also secretes part of seminal fluid
    c Sperm become capable of motility while they are stored in epididymis
B Vas deferens (seminal duct; ductus deferens)
  1 Structure and location
    a Tube, extension of epididymis
    b Extends through inguinal canal, into abdominal cavity, over top and down posterior surface of bladder
    c Enlarged terminal portion called ampulla—joins duct of seminal vesicle
  2 Function
    a One of excretory ducts for seminal fluid
    b Connects epididymis with ejaculatory duct
C Ejaculatory duct
  1 Formed by union of vas deferens with duct from seminal vesicle
  2 Passes through prostate gland, terminating in urethra
D Urethra—See p. 558.

Accessory reproductive glands

A Seminal vesicles
  1 Structure and location—convoluted pouches on posterior surface of bladder
  2 Function—secrete prostaglandins and viscous nutrient-rich part of seminal fluid
B Prostate gland
  1 Structure and location
    a Doughnut-shaped
    b Encircles urethra just below bladder
  2 Function—adds alkaline secretion to seminal fluid
C Bulbourethral glands
  1 Structure and location
    a Small, pea-shaped structures with 1-inch long ducts leading into urethra
    b Lie below prostate gland
  2 Function—secrete alkaline fluid that is part of semen

Supporting structures

A External
  1 Scrotum
    a Skin-covered pouch suspended from perineal region
    b Divided into two compartments
    c Contains testis, epididymis, and first part of vas deferens

2 Penis—composed of three cylindrical masses of erectile tissue, one of which contains urethra

B Internal

  1 Spermatic cords

    a Fibrous cylinders located in inguinal canals

    b Enclose seminal ducts, blood vessels, lymphatics, and nerves

Composition and course of seminal fluid

A Consists of secretions from testes, epididymides, seminal vesicles, prostate, and bulbourethral glands

B Each drop contains millions of sperm

C Passes from testes through epididymis, vas deferens, ejaculatory duct, and urethra

Male functions in reproduction

A Spermatogenesis

B Coitus (copulation or sexual intercourse)

  1 Erection—a parasympathetic reflex initiated by various kinds of stimuli and consisting of dilation of arteries and arterioles of the penis, causing its enlargement and erection

  2 Emission—reflex movement of sperm and secretions of genital ducts and accessory organs into prostatic urethra

  3 Ejaculation of semen—one of several responses that characterize orgasm, the climax of coitus

Review questions

1 How does the function of reproduction differ from all other body functions?

2 Name the accessory glands of the male reproductive system.

3 List the genital ducts in the male.

4 List the supporting structures of the male reproductive system.

5 What is the tunica albuginea? How does it aid in dividing the testis into lobules?

6 What is the relationship between the rete testis, seminiferous tubules, and efferent ductules?

7 What are the two primary functions of the testes?

8 What are the general functions of testosterone?

9 Discuss the structure of a mature spermatozoon.

10 List three functions of the epididymis.

11 What is meant by the term "capacitation"?

12 List the anatomical divisions of the epididymis.

13 Discuss the formation of the ejaculatory ducts.

14 What is the relationship of the prostate gland to the urethra?

15 What and where are the bulbourethral (Cowper's) glands?

16 What is the spermatic cord? From what to what does it extend, and what does it contain?

17 Discuss the type of secretion typical of the prostate gland and seminal vesicles.

18 Name the three cylindrical masses of erectile or cavernous tissue in the penis.

19 What and where is the glans penis? The prepuce or foreskin?

20 Define the following terms: cryptorchidism, external urinary meatus, coitus, erection, emission, ejaculation.

21 Of what is the seminal fluid composed? Trace its course from its formation in the gonads to the exterior.

chapter 25

The female reproductive system

Female reproductive organs

Examine Fig. 25-1 to identify the following organs of the female reproductive system:

1 *Essential sex organs*—the two ovaries (female gonads)
2 *Accessory sex organs*—two uterine tubes (fallopian tubes or oviducts), one uterus, one vagina, one vulva (pudendum or external genitalia), and two breasts or mammary glands

Uterus

Structure

Size, shape, and divisions. The uterus is pear shaped in its virginal state and measures approximately 7.5 cm (3 inches) in length, 5 cm (2 inches) in width at its widest part, and 3 cm (1 inch) in thickness. Note in Fig. 25-2 that the uterus has two main parts: an upper portion, the *body*, and a lower, narrow section, the *cervix*. Did you notice that the body of the uterus rounds into a bulging prominence above the level at which the uterine tubes enter? This bulging upper component is the *fundus*.

Wall. Three coats compose the walls of the uterus: the inner endometrium, a middle myometrium, and an outer incomplete layer of parietal peritoneum.

1 The lining of mucous membrane, called the *endometrium*, is composed of three layers of tissues: a compact surface layer of partially ciliated simple columnar epithelium called the *stratum compactum*, a spongy middle or intermediate layer of loose connective tissue—the *stratum spongiosum*, and a dense inner layer termed the *stratum basale* that attaches the endometrium to the underlying myometrium. During menstruation and following delivery of a baby, the compact and spongy layers slough off. The endometrium varies in thickness from 0.5 mm just after the menstrual flow to about 5 mm near the end of the endometrial cycle.

2 A thick, middle coat (the *myometrium*) consists of three layers of smooth muscle fibers that extend in all directions, longitudinally, transversely, and obliquely, and give the uterus great strength. The bundles of smooth muscle fibers interlace with elastic and connective tissue components and generally blend into the endometrial lining with no sharp line of demarcation between the two layers. The myometrium is thickest in the fundus and thinnest in the cervix —a good example of the principle of structural adaptation to function. In order to expel a fetus, that is, move it down and out of the uterus, the fundus must contract more forcibly than the lower part of the uterine wall and the cervix must be stretched or dilated.

3 An external coat of serous membrane, the *parietal peritoneum*, is incomplete, since it covers none of the cervix and only part of the body (all

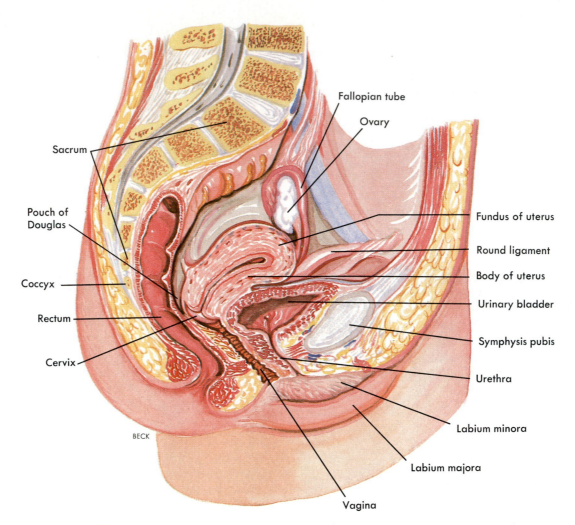

Sacrum

Pouch of
Douglas

Coccyx

Rectum

Cervix

BECK

Fallopian tube

Ovary

Fundus of uterus

Round ligament

Body of uterus

Urinary bladder

Symphysis pubis

Urethra

Labium minora

Labium majora

Vagina

Fig. 25-1 Female pelvic organs as viewed in a median sagittal section.

except the lower one fourth of its anterior surface). The fact that the entire uterus is not covered with peritoneum has clinical value because it makes it possible to perform operations on this organ without the risk of infection that attends cutting into the peritoneum.

Cavities. The cavities of the uterus are small because of the thickness of its walls (see Fig. 25-2). The body cavity is flat and triangular. Its apex is directed downward and constitutes the *internal os,* which opens into the *cervical canal.* The cervical canal is constricted on its lower

end also, forming the *external os,* which opens into the vagina. The uterine tubes open into the body cavity at its upper, outer angles.

Blood supply. The uterus receives a generous supply of blood from uterine arteries, branches of the internal iliac arteries. In addition, blood from the ovarian and vaginal arteries reaches the uterus by anastomosis with the uterine vessels. Tortuous arterial vessels enter the layers of the uterine wall as arterioles and then break up into capillaries between the endometrial glands.

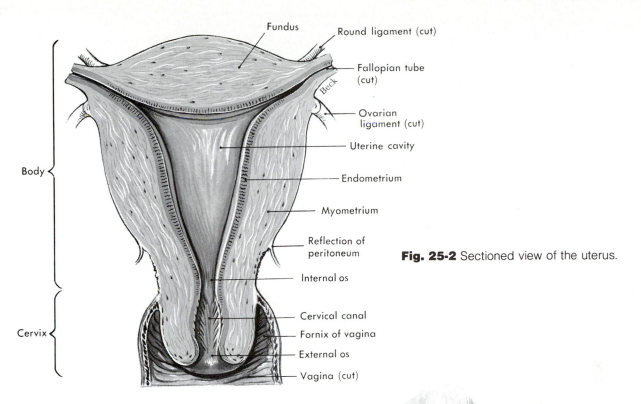

Fundus

Round ligament (cut)

Fallopian tube (cut)

Ovarian ligament (cut)

Uterine cavity

Endometrium

Myometrium

Reflection of peritoneum

Internal os

Cervical canal

Fornix of vagina

External os

Vagina (cut)

Body

Cervix

Beck

Fig. 25-2 Sectioned view of the uterus.

Fig. 25-3 Female pelvic contents as seen from above and in front.

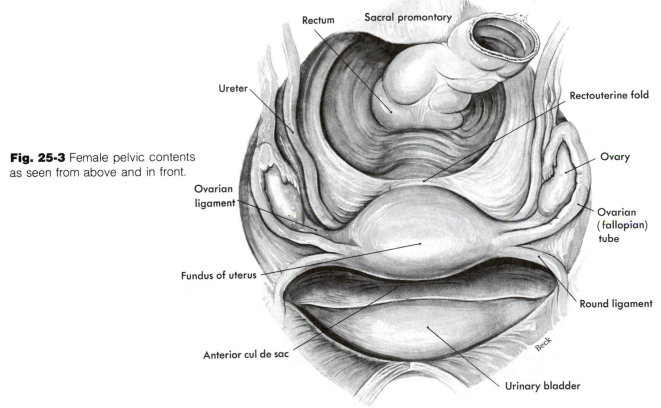

Rectum

Sacral promontory

Ureter

Rectouterine fold

Ovary

Ovarian (fallopian) tube

Round ligament

Ovarian ligament

Fundus of uterus

Anterior cul de sac

Urinary bladder

Beck

Uterine, ovarian, and vaginal veins return venous blood from the uterus to the internal iliac veins.

Location

Fig. 25-3 shows the location of the uterus in the pelvic cavity between the urinary bladder in front and the rectum behind. Age, pregnancy, and distention of related pelvic viscera such as the bladder will alter the position of the uterus.

Between birth and puberty the uterus descends gradually from the lower abdomen into the true pelvis. At menopause the uterus begins a process of involution that results in a decrease in size and a position deep in the pelvis.

Position

1 *Normally* the uterus is flexed between the body and cervix, with the body lying over the superior surface of the bladder, pointing forward and slightly upward (Fig. 25-4). The cervix points downward and backward from the point of flexion, joining the vagina at approximately a right angle. Several ligaments hold the uterus in place but allow its body considerable movement, a characteristic that often leads to malpositions of the organ. Fibers from several muscles that form the pelvic floor (see Fig. 7-36 on p. 179) converge to form a node called the *perineal body*, which also serves an important role in support of the uterus.

2 The uterus may lie in any one of several *abnormal positions*. A common one is retroversion or backward tilting of the entire organ.

3 *Eight ligaments* (three pairs, two single ones) anchor the uterus in the pelvic cavity: broad (paired), uterosacral (paired), posterior (single), anterior (single), and round (paired). Six of these so-called ligaments are actually extensions of the parietal peritoneum in different directions. The round ligaments are fibromuscular cords.

a The *two broad ligaments* are double folds of parietal peritoneum that form a kind of partition across the pelvic cavity. The uterus is suspended between these two folds.

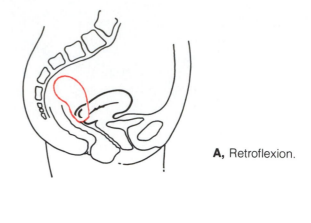

A, Retroflexion.

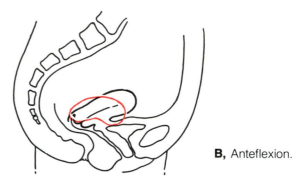

B, Anteflexion.

Fig. 25-4 Normal and abnormal positions of the uterus. Red lines show the abnormal positions.

b The *two uterosacral ligaments* are foldlike extensions of the peritoneum from the posterior surface of the uterus to the sacrum, one on each side of the rectum.

c The *posterior ligament* is a fold of peritoneum extending from the posterior surface of the uterus to the rectum. This ligament forms a deep pouch known as the *cul-de-sac of Douglas* (or rectouterine pouch) between the uterus and rectum. Since this is the lowest point in the pelvic cavity, pus collects here in pelvic inflammations. To secure drainage, an incision may be made at the top of the posterior wall of the vagina (posterior colpotomy).

d The *anterior ligament* is the fold of peritoneum formed by the extension of the peritoneum on the anterior surface of the uterus to the posterior surface of the bladder. This fold also forms a cul-de-sac but one that is less deep than the posterior pouch.

e The *two round ligaments* (Fig. 25-3) are fibromuscular cords extending from the upper, outer angles of the uterus through the inguinal canals and terminating in the labia majora.

Functions

The uterus or womb plays a role in the accomplishment of three functions vital for survival of the human species but not for individual survival: menstruation, pregnancy, and labor.

1 *Menstruation* is a sloughing away of the compact and spongy layers of the endometrium, attended by bleeding from the torn vessels.

2 In *pregnancy* the embryo implants itself in the endometrium and there lives as a parasite throughout the fetal period.

3 *Labor* consists of powerful, rhythmic contractions of the muscular uterine wall that result in expulsion of the fetus or birth.

Uterine tubes—fallopian tubes or oviducts

Location

The uterine tubes are about 10 cm (4 inches) long and are attached to the uterus at its upper outer angles (see Figs. 25-1 and 25-3). They lie in the upper free margin of the broad ligaments and extend upward and outward toward the sides of the pelvis and then curve downward and backward.

Structure

Wall. The same three coats (mucous, smooth muscle, and serous) of the uterus compose the tubes. The mucosa of the tubes, however, is fully ciliated and directly continuous with the peritoneum—a fact of great clinical significance because the tubal mucosa is continuous with that of the uterus and vagina and, therefore, often

becomes infected by gonococci or other organisms introduced into the vagina. Inflammation of the tubes (salpingitis) may readily spread to become inflammation of the peritoneum (peritonitis), a serious condition. In the male, there is no such direct route by which microorganisms can reach the peritoneum from the exterior.

Divisions. Each uterine tube consists of three divisions:

1 A medial third that extends from the upper outer angle of the uterus called the *isthmus*.

2 An intermediate dilated portion called the *ampulla* that follows a tortuous path over the ovary.

3 A funnel-shaped terminal component called the *infundibulum* that opens directly into the peritoneal cavity. The open outer margin of the infundibulum resembles a fringe in its irregular outline. The fringelike projections are known as *fimbriae* (Fig. 25-5).

Function

The tubes serve as ducts for the ovaries, although they are not attached to them. Ovaries are the organs that produce ova (the female gametes). Fertilization, the union of a spermatozoon with an ovum, normally occurs in the tubes.

Ovaries—female gonads

Location and size

The female gonads or ovaries are homologous (homology—similarity in origin) with the testes in the male. They are nodular glands that after puberty present a puckered uneven surface, resemble large almonds in size and shape, and are located one on either side of the uterus, below and behind the uterine tubes. Each ovary weighs about 3 gm and is attached to the posterior surface of the broad ligament by the mesovarian ligament. The ovarian ligament anchors it to the uterus. The distal portion of the tube curves about the ovary in such a way that the fimbriae

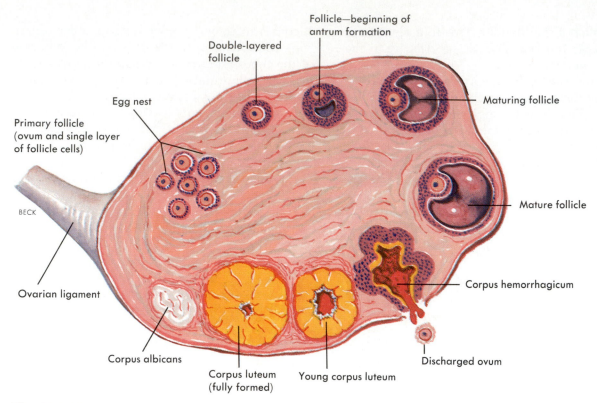

Fig. 25-5 Mammalian ovary showing successive stages of ovarian follicle and ovum development. Begin with the first stage (egg nest) and follow around clockwise to the final stage (corpus albicans).

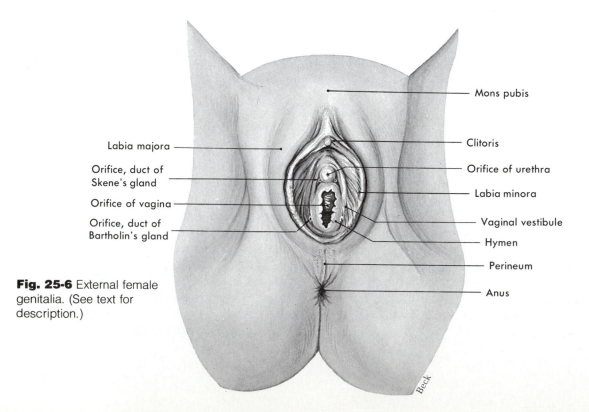

Fig. 25-6 External female genitalia. (See text for description.)

cup over the ovary but do not actually attach to it. Here, then, is a gland whose duct is detached from it, a fact that makes possible pregnancy in the pelvic cavity instead of in the uterus, as is normal.

Microscopic structure

The surface of the ovary is covered by a layer of small epithelial cells called the *germinal epithelium*. The term "germinal" is confusing because the epithelial cells of this layer do not give rise to ova. Deep to the surface layer of epithelial cells thousands of microscopic structures known as *ovarian follicles* are embedded in a connective tissue matrix. The follicles contain the female sex cells or ova and after puberty are present in varying stages of development. The primordial follicles consist of an oocyte surrounded by follicular cells. Refer to Fig. 25-5 and trace the development of a female sex cell from its most primitive state through ovulation.

Functions

The ovaries perform two functions: ovulation and hormone secretion. Ova develop and mature in the ovaries and are discharged from them into the pelvic cavity. The ovaries also secrete the female hormones—estrogens (chiefly estradiol and estrone) and progesterone. More details about their secretion and their functions appear on pp. 633-636.

Vagina

Location

The vagina is a tubular organ situated between the rectum, which lies posterior to it, and the urethra and bladder, which lie anterior to it. It extends upward and backward from its external orifice in the vestibule between the labia minora of the vulva (see p. 628) to the cervix (see Figs. 25-1, 25-2, and 25-6).

Structure

The vagina is a collapsible tube about 7 or 8 cm long that is capable of great distention. It is composed mainly of smooth muscle and is lined with mucous membrane arranged in rugae. It should be noted that the anterior wall of the vagina is shorter than the posterior wall because of the way the cervix protrudes into the uppermost portion of the tube (see Fig. 25-1). In the virginal state, a fold of mucous membrane, the *hymen* ("maidenhead"), forms a border around the external opening of the vagina, partially closing the orifice. Occasionally, this structure completely covers the vaginal outlet, a condition referred to as *imperforate hymen*. Perforation has to be done before the menstrual flow can escape.

Functions

The vagina constitutes an essential part of the reproductive tract because of the following:
1 It is the organ that receives the seminal fluid from the male.
2 It serves as the lower part of the birth canal.
3 It acts as the excretory duct for uterine secretions and the menstrual flow.

Vulva

Fig. 25-6 shows the structures that, together, constitute the female external genitals (reproductive organs) or vulva: mons veneris, labia majora, labia minora, clitoris, urinary meatus (urethral orifice), vaginal orifice, and Bartholin's or the greater vestibular glands.

1 The *mons pubis* is a skin-covered pad of fat over the symphysis pubis. Coarse hairs appear on this structure at puberty and persist throughout life.

2 The *labia majora* or "large lips" are covered with pigmented skin and hair on the outer surface and are smooth and free from hair on the

inner surface. They are composed mainly of fat and numerous glands.

3 The *labia minora* or "small lips" are located within the labia majora and are covered with modified skin. These two lips come together anteriorly in the midline. The area between the labia minora is the *vestibule*.

4 The *clitoris* is a small organ composed of erectile tissue, located just behind the junction of the labia minora, and homologous to the corpora cavernosa and glans of the penis. The *prepuce* or foreskin covers the clitoris, as it does the glans penis in the male.

5 The *urinary meatus* (urethral orifice) is the small opening of the urethra, situated between the clitoris and the vaginal orifice.

6 The *vaginal orifice* is an opening that, in the virginal state, is usually only slightly larger than the urinary meatus because of the constricting border formed by the hymen. In the nonvirginal state the vaginal orifice is noticeably larger than the urinary meatus. It is located posterior to the meatus.

7 *Bartholin's* or the *greater vestibular glands* are two bean-shaped glands, one on either side of the vaginal orifice. Each gland opens by means of a single, long duct into the space between the hymen and the labium minus. These glands are of clinical importance because they are frequently infected (bartholinitis or Bartholin's abscess), particularly by gonococci. They are homologous to the bulbourethral glands in the male, and they secrete a lubricating fluid. Opening into the vestibule near the urinary meatus by way of two small ducts is a group of tiny mucous glands, the *lesser vestibular* or *Skene's glands*. These have clinical interest because gonococci that lodge there are difficult to eradicate.

Perineum

The perineum is the skin-covered muscular region between the vaginal orifice and the anus. This area has great clinical importance because

of the danger of its being torn during childbirth. Such tears are often deep, have irregular edges, and extend all the way through the perineum, the muscular perineal body, and even through the anal sphincter, resulting in involuntary seepage from the rectum until the laceration is repaired. In addition, injuries to the perineal body can result in partial uterine or vaginal prolapse if this important support structure is weakened. To avoid these possibilities a surgical incision known as an *episiotomy* is usually made in the perineum, particularly at the birth of a first baby.

Breasts

Location and size

The breasts lie over the pectoral muscles and are attached to them by a layer of connective tissue (fascia). Estrogens and progesterone, two ovarian hormones, control their development during puberty. Estrogens stimulate growth of the ducts of the mammary glands, whereas progesterone stimulates development of the alveoli, the actual secreting cells. Breast size is determined more by the amount of fat around the glandular tissue than by the amount of glandular tissue itself. Hence the size of the breast does not relate to its functional ability.

Structure

Each breast consists of several lobes separated by septa of connective tissue. Each lobe consists of several lobules, which, in turn, are composed of connective tissue in which are embedded the secreting cells (alveoli) of the gland, arranged in grapelike clusters around minute ducts. The ducts from the various lobules unite, forming a single lactiferous duct for each lobe, or between 15 and 20 in each breast. These main ducts converge toward the nipple, like the spokes of a wheel. They enlarge slightly before reaching the nipple into ampullae or small

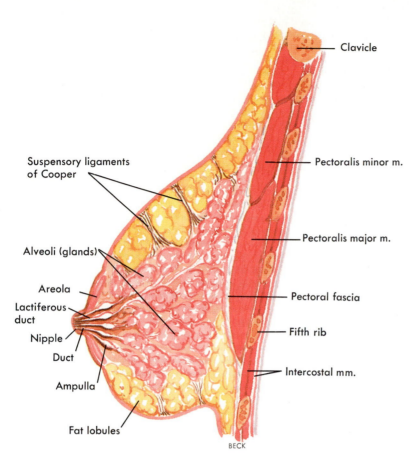

Suspensory ligaments of Cooper

Alveoli (glands)

Areola

Lactiferous duct

Nipple

Duct

Ampulla

Fat lobules

Clavicle

Pectoralis minor m.

Pectoralis major m.

Pectoral fascia

Fifth rib

Intercostal mm.

BECK

Fig. 25-7 Lateral view of the breast (sagittal section). The gland is fixed to the overlying skin and the pectoral muscles by the suspensory ligaments of Cooper. Each lobule of secretory tissue is drained by a lactigenous duct that opens through the nipple.

"reservoirs" (Fig. 25-7). Each of these main ducts terminates in a tiny opening on the surface of the nipple. Adipose tissue is deposited around the surface of the gland, just under the skin, and between the lobes. The nipples are bordered by a circular pigmented area, the *areola*. It contains numerous sebaceous glands that appear as small nodules under the skin. In Caucasians (other than those with very dark complexions) the areola and nipple change color from delicate pink to brown early in pregnancy —a fact of value in diagnosing a first pregnancy. The color decreases after lactation has ceased but never entirely returns to the virginal hue. In darker-skinned women, no noticeable color change in the areola or nipple heralds the first pregnancy.

A knowledge of the lymphatic drainage of the breast is critically important in clinical medicine because cancerous cells from malignant breast tumors often spread to other areas of the body through the lymphatics. Details of the lymphatic drainage of the breast are presented in Chapter 16. (See also Fig. 16-6.)

Function

The function of the mammary glands is lactation, that is, the secretion of milk for the nourishment of newborn infants.

Mechanism controlling lactation

Very briefly, lactation is controlled as follows:
1 The ovarian hormones, estrogens and progesterone, act on the breasts to make them struc-

turally ready to secrete milk. Estrogens promote development of the ducts of the breasts. Progesterone acts on the estrogen-primed breasts to promote completion of the development of the ducts and development of the alveoli; the secreting cells of the breasts. A high blood concentration of estrogens, for example, during pregnancy, inhibits anterior pituitary secretion of lactogenic hormone.

2 Shedding of the placenta following delivery of the baby cuts off a major source of estrogens. The resulting rapid drop in the blood concentration of estrogens stimulates anterior pituitary secretion of lactogenic hormone. Also, the suckling movements of a nursing baby act in some way to stimulate both anterior pitui-

tary secretion of lactogenic hormone and posterior pituitary secretion of oxytocin (Fig. 25-8).

3 Lactogenic hormone stimulates lactation, that is, stimulates alveoli of the mammary glands to secrete milk. Milk secretion starts about the third or fourth day after delivery of a baby, supplanting a thin, yellowish secretion called *colostrum*. With repeated stimulation by the suckling infant, plus various favorable mental and physical conditions, lactation may continue almost indefinitely.

4 Oxytocin stimulates the alveoli of the breasts to eject milk into the ducts, thereby making it accessible for the infant to remove by suckling.

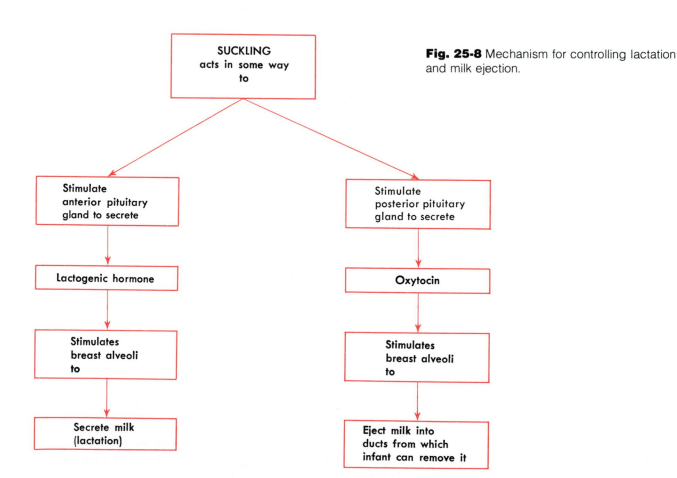

Fig. 25-8 Mechanism for controlling lactation and milk ejection.

Female sexual cycles

Recurring cycles

Many changes recur periodically in the female during the years between the onset of the menses (menarche) and their cessation (menopause or climacteric). Most obvious, of course, is menstruation—the outward sign of changes in the endometrium. Most women also note periodic changes in the breasts. But these are only two of many changes that occur over and over again at fairly uniform intervals during the approximately 30 years of female reproductive maturity. Rhythmic changes also take place in the ovaries, in the amount and consistency of the cervical mucus produced during each cycle, in the myometrium, in the vagina, in gonadotropin secretion, in body temperature, and even in mood or "emotional tone." Many of these recurring events that a woman may recognize on almost a monthly schedule during her reproductive years are called "fertility signs" and are manifestations of those body changes required for successful reproductive function.

First we shall describe the major cyclical changes, and then we shall discuss the mechanisms that produce them.

Ovarian cycles

Once each month, on about the first day of menstruation, several primitive graafian follicles and their enclosed ova begin to grow and develop. The follicular cells proliferate and start to secrete estrogens (and minute amounts of progesterone). Usually, only one follicle matures and migrates to the surface of the ovary. The surface of the follicle degenerates, causing expulsion of the mature ovum into the pelvic cavity (ovulation).

When does ovulation occur? This is a question of great practical importance and one that in the past was given many answers. Today it is known that ovulation usually occurs 14 days before the next menstrual period begins. (Only in a 28-day

menstrual cycle is this also 14 days after the beginning of the preceding menstrual cycle, as explained subsequently.) A few women experience pain within a few hours after ovulation. This is referred to as *mittelschmerz*—German for middle pain. It has been ascribed to irritation of the peritoneum by hemorrhage from the ruptured follicle.

Shortly before ovulation the ovum undergoes meiosis, the process by which its number of chromosomes is reduced by half. Immediately after ovulation, cells of the ruptured follicle enlarge and, because of the appearance of lipoid substances in them, become transformed into a golden-colored body, the *corpus luteum*. The corpus luteum grows for 7 or 8 days. During this time, it secretes progesterone in increasing amounts. Then, provided fertilization of the ovum has not taken place, the size of the corpus luteum and the amount of its secretions gradually diminish. In time, the last components of each nonfunctional corpus luteum are reduced to a white scar called the *corpus albicans*, which moves into the central portion of the ovary and eventually disappears (see Fig. 25-5).

Endometrial or menstrual cycle

During menstruation, necrotic bits of the compact and spongy layers of the endometrium slough off, leaving denuded bleeding areas. Following menstruation, the cells of these layers proliferate, causing the endometrium to reach a thickness of 2 or 3 ml by the time of ovulation. During this period, endometrial glands and arterioles grow longer and more coiled—two factors that also contribute to the thickening of the endometrium. (See Fig. 25-12.) After ovulation, the endometrium grows still thicker (reaching a maximum of about 4 to 6 ml). Most of this increase, however, is believed to be caused by swelling produced by fluid retention rather than by further proliferation of endometrial cells. The increasingly tortuous endometrial glands start to secrete during the time between ovulation and the next menses. Then, the day before menstruation starts again, the tightly coiled arterioles constrict, producing endometrial ischemia. This leads to necrosis, sloughing, and, once again, menstrual bleeding.

The menstrual cycle is customarily divided into phases, named for major events occurring in them: menses, postmenstrual or preovulatory phase, ovulation, and postovulatory or premenstrual phase.

1 The *menses* or *menstrual period* occurs on cycle days 1 to 5. There is some individual variation, however.

2 The *postmenstrual phase* occurs between the end of the menses and ovulation. Therefore it is the preovulatory as well as the postmenstrual phase. In a 28-day cycle, it usually includes cycle days 6 to 13 or 14. But the length of this phase varies more than do the others. It lasts longer in long cycles and ends sooner in short ones. This phase is also called the *estrogenic* or *follicular phase* because of the high blood estrogen level resulting from secretion by the developing follicle. *Proliferative phase* is still another name for it because proliferation of endometrial cells occurs at this time.

3 *Ovulation*, that is, rupture of the mature follicle with expulsion of its ovum into the pelvic cavity, occurs frequently on cycle day 15 in a 28-day cycle. However, it occurs on different days in different length cycles, depending on the length of the preovulatory phase. For example, in a 32-day cycle the preovulatory phase would probably last until cycle day 18 and ovulation would then occur on cycle day 19 instead of 15. In short, because the vast majority of women show some month-to-month variation in the length of their cycles, the day of ovulation in a current or future cycle can not be predicted with accuracy based on the length of previous cycles. Simply knowing the length of previous cycles can not ensure with any degree of accuracy how many days the preovulatory phase will last in the next or some future cycle. This physiological fact probably accounts for most of the

unreliability of the calendar rhythm method of contraception and its replacement by other more sophisticated "natural" methods of family planning that are not based on a knowledge of previous cycle lengths to predict the day of ovulation. Instead, such natural methods base their judgments about fertility at any point in a woman's cycle on other changes. Examples: measurement of basal body temperature and recognition of cyclic changes in the amount and consistency of cervical mucus, both of which occur in response to changes in circulating hormones that control ovulation.

4 The *postovulatory* (or *premenstrual*) *phase* occurs between ovulation and the onset of the menses. This phase is also called the *luteal phase* because the corpus luteum secretes only during this time and *progesterone phase* because it secretes mainly this hormone. The length of the premenstrual phase is pretty constant, lasting usually 14 days—or cycle days 15 to 28 in a 28-day cycle. Differences in length of the total menstrual cycle, therefore, exist mainly because of differences in duration of the preovulatory rather than of the premenstrual phase.

Myometrial cycle

The myometrium contracts mildly but with increasing frequency during the 2 weeks preceding ovulation. Contractions decrease or disappear between ovulation and the next menses, thereby lessening the probability of expulsion of an implanted ovum.

Gonadotropic cycles

The adenohypophysis (anterior pituitary gland) secretes two hormones called gonadotropins that influence female reproductive cycles. Their names are follicle-stimulating hormone (FSH) and luteinizing hormone (LH). The amount of each gonadotropin secreted varies with a rhythmic regularity that can be related, as we shall see, to the rhythmic ovarian and uterine changes just described.

Fig. 25-9 Follicle-stimulating hormone's principal effects on ovaries.

Control

Physiologists agree that hormones play the major role in producing the cyclic changes characteristic in the female during the years of reproductive maturity. The recent development of a method called radioimmunoassay has made it possible to measure blood levels of gonadotropins. By correlating these with the monthly ovarian and uterine changes, investigators have worked out the main features of the control mechanisms.

A brief description follows of the mechanisms that produce cyclical changes in the ovaries and uterus and in the amounts of gonadotropins secreted.

Control of cyclical changes in ovaries

Cyclical changes in the ovaries result from cyclical changes in the amounts of gonadotropins secreted by the anterior pituitary gland.

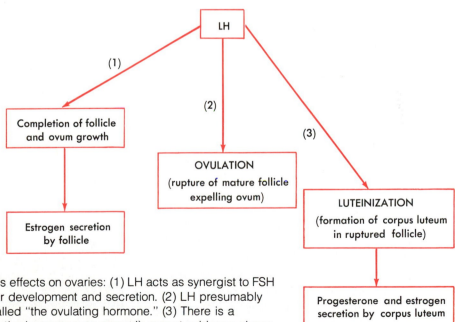

Fig. 25-10 Luteinizing hormone's effects on ovaries: (1) LH acts as synergist to FSH to enhance its effects on follicular development and secretion. (2) LH presumably triggers ovulation—hence it is called "the ovulating hormone." (3) There is a luteinizing effect of LH (for which the hormone was named); recent evidence shows that FSH is also necessary for luteinization.

An increasing FSH blood level has two effects: it stimulates one or more primitive graafian follicles and ova to start growing, and it stimulates the follicles to secrete estrogens. (Developing follicles also secrete very small amounts of progesterone.) Because of the influence of FSH on follicle secretion, the level of estrogens in blood increases gradually for a few days during the postmenstrual phase. Then suddenly, on about the twelfth cycle day, it leaps upward to a maximum peak. Scarcely 12 hours after this "estrogen surge," an "LH surge" occurs and presumably triggers ovulation a day or two later.* The control of cyclical ovarian changes by the gonadotropins FSH and LH is summarized in Figs. 25-9 and 25-10. As Fig. 25-10 shows, LH brings about the following changes:

1 Completion of growth of the follicle and ovum with increasing secretion of estrogens before ovulation. LH and FSH act as synergists to produce these effects.
2 Rupturing of the mature follicle with expulsion of its ripe ovum (process known as *ovulation*). Because of this function, LH is sometimes called the ovulating hormone.
3 Formation of a golden body, the corpus luteum, in the ruptured follicle (process called *luteinization*). The name luteinizing hormone refers, obviously, to this LH function —a function to which, according to recent evidence, FSH also contributes. The corpus luteum functions as a temporary endocrine gland. It secretes only during the luteal (postovulatory or premenstrual) phase of the menstrual cycle. Its hormones are progestins (the important one of which is progesterone) and also estrogens. The blood level of progesterone rises rapidly after the "LH surge" described before. It remains at a high level for about a week, then it decreases to a very low level approximately 3 days before menstruation starts again.

*Mountcastle, V. B., editor: Medical physiology, ed. 13, St. Louis, 1974, The C. V. Mosby Co., p. 1752 and Fig. 72-5.

This low blood level of progesterone persists during both the menstrual and postmenstrual phases. Its sources? Not from the corpus luteum, which secretes only during the luteal phase, but from the developing follicles and the adrenal cortex. Blood's estrogen content increases during the luteal phase but to a lower level than develops before ovulation.

If pregnancy does not occur, the corpus luteum regresses in about 14 days and is replaced by the corpus albicans. Fig. 25-5 shows the cyclical changes in the ovarian follicles.

Control of cyclical changes in uterus

Cyclical changes in the uterus are brought about by changing blood concentrations of estrogens and progesterone. As blood estrogens increase during the postmenstrual phase of the menstrual cycle, they produce the following main changes in the uterus.

1 Proliferation of endometrial cells, producing a thickening of the endometrium
2 Growth of endometrial glands and of its spiral arteries
3 Increase in the water content of the endometrium
4 Increased myometrial contractions

Increasing blood progesterone concentration during the premenstrual phase of the menstrual cycle produces progestational changes in the uterus, that is, changes favorable for pregnancy, specifically:

1 Secretion by endometrial glands, thereby preparing the endometrium for implantation of a fertilized ovum
2 Increase in the water content of the endometrium
3 Decreased myometrial contractions

Control of cyclical changes in amounts of gonadotropins secreted

Both negative and positive feedback mechanisms help control anterior pituitary secretion of the gonadotropins FSH and LH. These mechanisms involve the ovaries' secretion of both estrogens and progesterone and the hypothalamus' secretion of releasing hormones (until recently, called releasing factors). Fig. 25-11 describes a negative feedback mechanism that controls gonadotropin secretion. Examine it carefully. Note particularly the effects of a high blood concentration of estrogens on anterior pituitary gland secretion and the effect of a low blood concentration of FSH on the development of a graafian follicle and ovum. Establishment of these two facts led eventually to the development of "the pill" for preventing pregnancy. Contraceptive pills contain synthetic estrogen-like or progesterone-like compounds or both. By building up a high blood concentration of these substances, they prevent the development of a follicle and its ovum that month. With no mature ovum to be expelled, ovulation does not occur, and therefore pregnancy cannot occur.

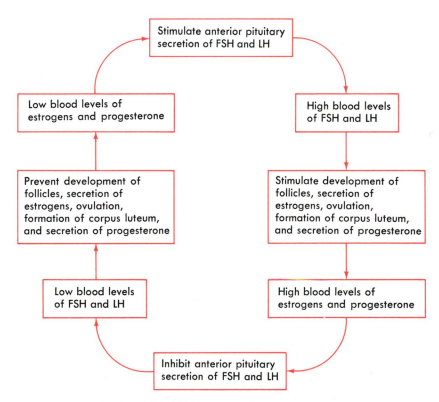

Fig. 25-11 A negative feedback mechanism that controls anterior pituitary secretion of follicle-stimulating hormone (FSH) and ovarian secretion of estrogens. A high blood level of FSH stimulates estrogen secretion, whereas the resulting high estrogen level inhibits FSH secretion. How does this compare with interstitial cell–stimulating hormone (ICSH)–testosterone feedback mechanism? (See Fig. 24-3 if you want to check your answer.) According to the above diagram, what effect does a high blood concentration of estrogens have on anterior pituitary secretion of FSH? of LH?

The next menses, however, does take place—because the progesterone and estrogen dosage is stopped in time to allow their blood level to decrease as they normally do near the end of the cycle to bring on menstruation. Actually the effects of contraceptive pills—estrogens and progesterone—are much more complex than our explanation indicates. They have widespread effects on the body quite independent of their action on the reproductive and endocrine systems and still are not completely understood. Side effects that may limit or prohibit use of "the pill" often result because of the nonreproductive system–related activity of these complex hormones.

Several observations and animal experiments strongly suggest that high blood levels of estrogens and progesterone may also act indirectly to inhibit pituitary secretion of FSH and LH. These ovarian hormones appear to inhibit certain neurons of the hypothalamus (part of the central nervous system) from secreting FSH-releasing and LH-releasing hormones into the pituitary portal vessels. Without the stimulating effects of these releasing hormones the pituitary's secretion of FSH and LH decreases.

A positive feedback mechanism has also been postulated to control LH secretion. The rapid and marked increase in blood's estrogen content that occurs late in the follicular phase of the menstrual cycle is thought to stimulate the hypothalamus to secrete LH-releasing hormone into the pituitary portal vessels. As its name suggests, this hormone stimulates the release of LH by the anterior pituitary, which in turn accounts for the "LH surge" that triggers ovulation. The fact that a part of the brain—the hypothalamus—secretes FSH- and LH-releasing hormones has interesting implications. This may be the crucial part of the pathway by which changes in a woman's environment or in her emotional state can alter her menstrual cycle. That this occurs is a matter of common observation. For example, intense fear of either becoming or not becoming pregnant often delays menstruation.

Functions served by cycles

The major function seems to be to prepare the endometrium each month for a pregnancy. If it does not occur, the thick vascular lining, no longer needed, is shed.

If fertilization of the ovum (pregnancy) occurs, the menstrual cycle is modified as follows:

1 The corpus luteum does not disappear but persists and continues to secrete progesterone and estrogens for 6 months or more of pregnancy. If it is removed by any means during the early months of pregnancy, spontaneous abortion results.

2 The fertilized ovum, which immediately starts developing into an embryo, travels down the tube and implants itself in the endometrium, so carefully prepared for this event.

Menarche and menopause

The menstrual flow first occurs (menarche) at puberty, at about the age of 13 years, although there is wide individual variation according to race, nutrition, health, heredity, etc. Normally, it recurs about every 28 days for about 40 years, except during pregnancy, and then ceases (menopause or climacteric). The average age at which menstruation ceases is reported to have increased markedly—from about age 40 a few decades ago to between ages 45 and 50 more recently.

Embryology

Meaning and scope

Embryology is the science of the development of the individual before birth. It is a story of miracles, describing the means by which a new human life is started and the steps by which a

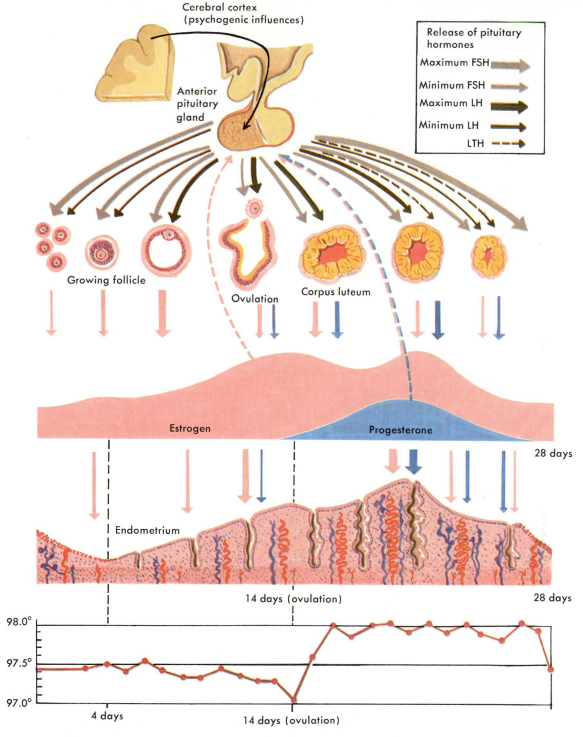

Fig. 25-12 Diagram illustrating the interrelationships among the cerebral, hypothalamic, pituitary, ovarian, and uterine functions throughout a usual 28-day menstrual cycle. The variations in basal body temperature are also illustrated.

single microscopic cell is transformed into a complex human being. In a work of this kind, it seems feasible to include only a few of the main points in the development of the new individual, since the facts amassed by the science of embryology are so many and so intricate that to tell them requires volumes, not paragraphs.

Value of knowledge of embryology

The value to the medical profession of knowing the steps by which a new individual is evolved is evidenced by rapid advances in antenatal (from the Latin *ante*, before, and *natus*, birth) diagnosis and treatment. Pediatric medicine acquired a new dimension when Dr. Albert W. Liley of New Zealand, a world-renowned fetologist, developed techniques by which Rh babies could be given transfusions prior to birth. Now, new techniques including fetal electrocardiography and ultrasound permit pediatricians to diagnose and treat illness in the fetus much like any other patient. A knowledge of embryology also provides an explanation of various congenital deformities. Harelip, to mention only one example, is among the most common malformations; it is a condition that results from imperfect fusion of the frontal and maxillary processes during embryonic development.

Steps in development of new individual
Preliminary processes

Production of a new human being starts with the union of a spermatozoon and an ovum to form a single cell. Two preliminary steps, however, are necessary before such a union can take place: maturation of the sex cells (meiosis) and ovulation and insemination.

Maturation of sex cells (meiosis*)—reduction in number of chromosomes to half original number. Only when maturation has occurred are the ovum and spermatozoon mature and ready to unite with each other. The necessity for chromosome reduction as a preliminary to union of the sex cells is explained by the fact that the cells of each species of living organisms contain a specific number of chromosomes. Human cells, for example, contain 23 pairs or a total of 46 chromosomes. If the male and female cells united without first halving their respective chromosomes, the resulting cell would contain twice as many chromosomes as is specific for human beings. Mature ova and sperm, therefore, contain only 23 chromosomes or half as many as other human cells. Of these, one is the sex chromosome and may be either one of two types, known as X or Y. All ova contain an X chromosome. Sperm, on the other hand, have either an X or a Y chromosome. A female child results from the union of an X chromosome–bearing sperm with an ovum and a male child from the union of a Y chromosome–bearing sperm with an ovum.

Ovulation and insemination. The second preliminary step necessary for conception of a new individual consists in bringing the sperm and ovum into proximity with each other so that the union of the two can take place. Two processes are involved in the accomplishment of this step.

1 Ovulation or expulsion of the mature ovum from the graafian follicle into the pelvic cavity, from which it enters one of the uterine tubes.

2 Insemination or expulsion of the seminal fluid from the male urethra into the female vagina. Several million sperm enter the female reproductive tract with each ejaculation of semen. By lashing movements of their flagella-like tails, assisted somewhat by muscular contractions of surrounding structures, the sperm make their way into the external os of the cervix,

*For details of the mechanism of meiosis, see pp. 602-604.

through the cervical canal and uterine cavity, and into the tubes.

Research has shown that X sperm swim more slowly than do Y sperm. An interesting hypothesis stems from this fact. If insemination occurs on the day of ovulation, Y sperm, being faster than X sperm, should reach the ovum first. Therefore a Y sperm would be more apt to fertilize the ovum. And since Y sperm produce males, there should be a greater probability of having a baby boy when insemination occurs on the day of ovulation. Statistical evidence supports this view.

Developmental processes

Fertilization or union of male and female gametes to produce one-celled individual called zygote. The sperm "swim" up the tube toward the ovum. Although numerous sperm surround the ovum, only one penetrates it. As soon as the head and neck of one spermatozoon enter the ovum (the tail drops off), the remaining sperm seem to be repulsed. The sperm head then forms itself into a nucleus that approaches and eventually fuses with the nucleus of the ovum, producing, at that

moment, a new single-celled individual or *zygote*. Half of the 46 chromosomes in the zygote nucleus have come from the sperm and half from the ovum. Since chromosomes are composed of *genes*, the new being inherits half of its genes and the characteristics they determine from its father and half from its mother. Genetically, the zygote is complete. Time and nourishment are all that is needed for expression of those characteristics such as sex, body build, and skin color that were determined at fertilization.

Normally, fertilization occurs in the outer third of the uterine tube. Occasionally, however, it takes place in the pelvic cavity, as evidenced by pregnancies that start to develop in the pelvic cavity instead of in the uterus.

Inasmuch as the ovum lives only a short time (probably less than 48 hours) after leaving the graafian follicle, fertilization can occur only around the time of ovulation (p. 632). Sperm may live a few days after entering the female tract.

Cleavage and implantation. Cleavage consists of repeated mitotic divisions, first of the zygote to form two cells, then of those two cells to form

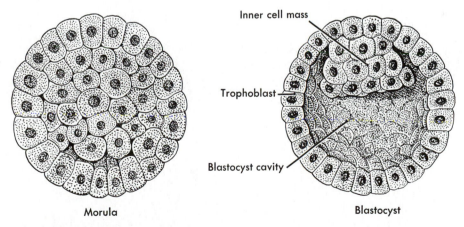

Morula

Inner cell mass

Trophoblast

Blastocyst cavity

Blastocyst

Fig. 25-13 Stages in the development of the human embryo. The morula consists of an almost solid spherical mass of cells. The embryo reaches this stage about 3 days after fertilization. The blastocyst (hollowing) stage develops later, after implantation in the uterine lining.

four cells,* etc., resulting in about 3 days' time in the formation of a solid, spherical mass of cells known as a *morula* (Fig. 25-13). A day or so later the embryo reaches the uterus, where it starts to implant itself in the endometrium. Occasionally, implantation occurs in the tube or pelvic cavity instead of in the uterus. The condition is known as an *ectopic pregnancy*. Fig. 25-14 shows the appearance of the uterus and uterine tubes from fertilization to implantation.

As the cells of the morula continue to divide, a hollow ball of cells or *blastocyst*, consisting

*On July 25, 1978, the world's first test-tube baby (a girl) was born in Oldham, England. Nine months earlier a mature ovum had been removed from the mother's body by laparoscopy ("belly button surgery") and fertilized in a laboratory dish by her husband's sperm. After 2½ days' growth in a controlled environment, physicians returned the fertilized ovum (now at the eight-cell stage) to the mother's uterus.

of an outer layer of cells and an inner cell mass, is formed. Implantation in the uterine lining is now complete. About 10 days have elapsed since fertilization. The cells that compose the outer wall of the blastocyst are known as trophoblasts. They eventually become part of the placenta, the structure that anchors the fetus to the uterus. The placenta is derived in part from both the developing embryo and from maternal tissues. Note in Fig. 25-15 the close relationship of maternal and fetal blood vessels. Although in proximity, maternal and fetal blood does not mix. Exchange of nutrients occurs across an important fetal membrane called the *chorion*.

Differentiation. As the cells composing the inner mass of the blastocyst continue to divide, they arrange themselves into a structure shaped like a figure eight, containing two cavities separated by a double-layered plate of cells known as the *embryonic disk*. Young human embryos have

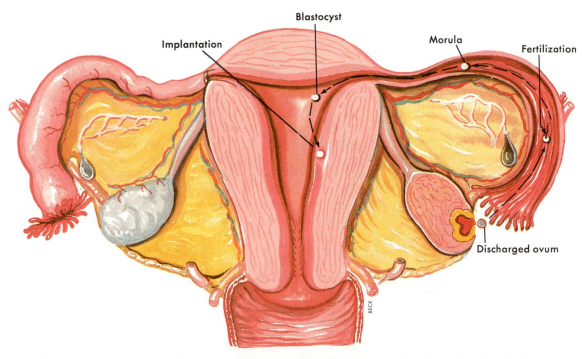

Fig. 25-14 Appearance of uterus and uterine tubes—fertilization to implantation. Fertilization occurs in the outer third of the uterine tube. Development reaches the blastocyst stage after the embryo has entered the uterus.

been examined at this stage, or about 2 weeks old dated from the time of fertilization. The cells that form the cavity above the embryonic disk eventually become a fluid-filled, shock-absorbing sac (the amnion) in which the fetus floats. The cells of the lower cavity form the yolk sac, a small vesicle attached to the belly of the embryo until about the middle of the second month, when it breaks away. Only the double layer of cells that compose the embryonic disk is destined to form the new individual. The upper layer of cells is called the *ectoderm* and the lower layer the *entoderm*. Refer to Fig. 25-16 and follow the zygote from fertilization to implanta-

Fig. 25-15 Placenta showing fetal and maternal blood vessels.

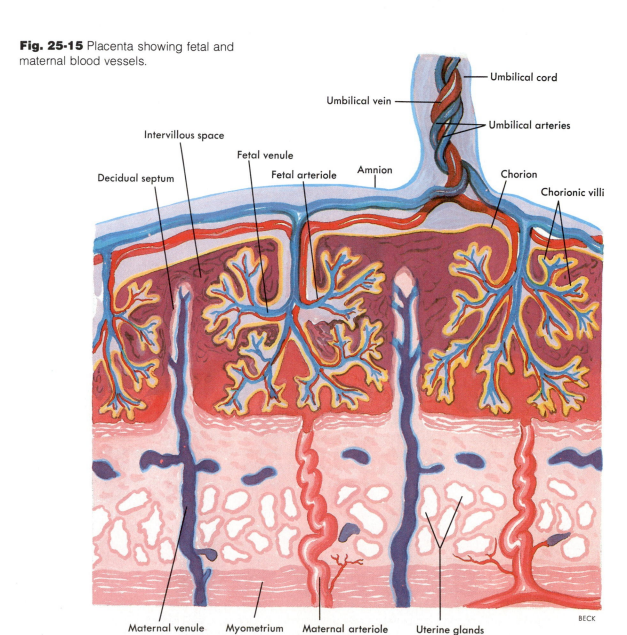

BECK

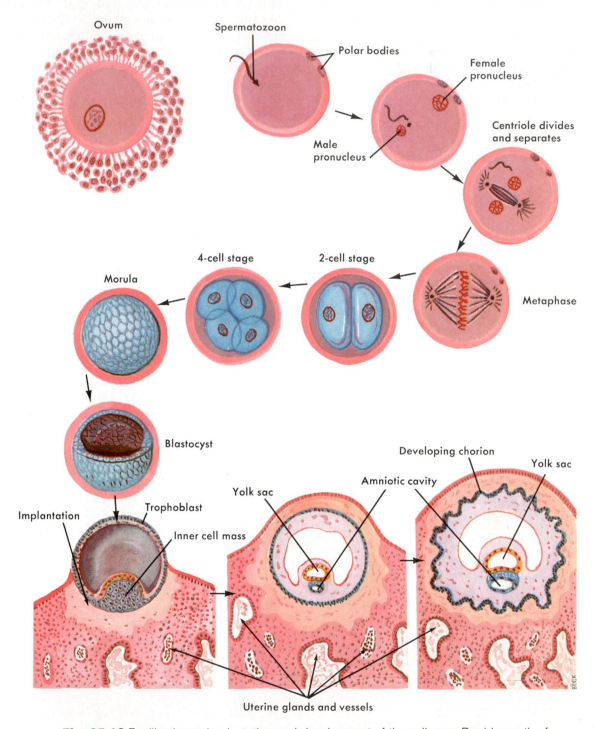

Fig. 25-16 Fertilization to implantation and development of the yolk sac. Rapid growth of uterine glands and vessels covers the developing blastocyst at the time of implantation.

tion and development of the yolk sac. A third layer of cells, known as the *mesoderm*, develops between the ectoderm and entoderm. Up to this time, all the cells have appeared alike, but now they are differentiated into three distinct types, ectodermal, mesodermal, and entodermal, known as the *primary germ layers*, each of which will give rise to definite structures. For example, the ectoderm cells will form the skin and its appendages and the nervous system; the mesoderm, the muscles, bones, and various other connective tissues; the entoderm, the epithelium of the digestive and respiratory tracts, etc.

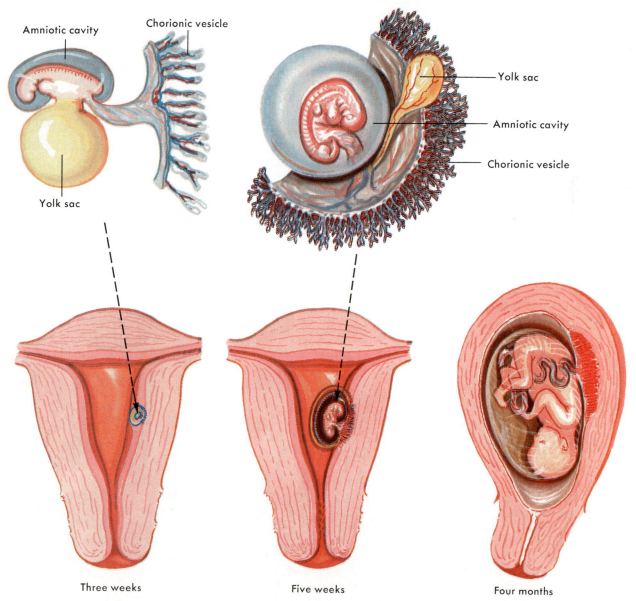

Three weeks   Five weeks   Four months

Fig. 25-17 Development of the embryo—yolk sac to 4 months of gestation.

Histogenesis and organogenesis. The story of how the primary germ layers develop into many different kinds of tissues (histogenesis) and how those tissues arrange themselves into organs (organogenesis) is long and complicated. Its telling belongs to the science of embryology. But for the beginning student of anatomy it seems sufficient to appreciate that life begins when two sex cells unite to form a single cell and that the new human body evolves by a series of processes consisting of cell multiplication, cell growth, cell differentiation, and cell rearrangements, all of which take place in definite, orderly sequence. Fig. 25-17 shows development of the embryo from the yolk sac stage to 4 months of gestation.

By the third week of development the third germ layer—the mesoderm—appears between the two initial layers of the embryonic disk. The developing heart is beating by the fifth week, and the tiny baby, only about 8 mm in length, is showing rapid development of every major organ system. By the end of the second month, a recognizable human figure has been formed and at the end of another month the sex is clearly distinguishable. Development of structure and function go hand in hand, and from 4 months of gestation, when every organ system is in place and functioning, until term (about 280 days), development is mainly a matter of growth.

Outline summary

Female reproductive organs

A Essential sex organs
  1 Ovaries (gonads)
B Accessory sex organs
  1 Accessory glands
    a Bartholin's (greater vestibular)
    b Skene's (lesser vestibular)
    c Mammary (breasts)
  2 Ducts
    a Uterine (fallopian) tubes
    b Uterus
    c Vagina
  3 External genitalia (vulva or pudendum)

Uterus

A Structure
  1 Size, shape, and divisions
    a In virginal state, 7.5 cm (3 inches) × 5 cm (2 inches) × 2.5 cm (1 inch)
    b Pear-shaped
    c Consists of body and cervix; fundus bulging upper surface of body
  2 Wall—lining of mucosa called endometrium; thick middle coat of muscle called myometrium; partial external coat of peritoneum
    a Endometrium itself composed of three layers
      (1) Outer stratum compactum
      (2) Intermediate stratum spongiosum
      (3) Dense inner stratum basale

  3 Cavities—body cavity small and triangular in shape with three openings—two from tubes and one (internal os) into cervical canal; external os opening of cervical canal into vagina
  4 Blood supply
    a Arterial—generous from uterine arteries (branches of internal iliac) and by anastomosis from ovarian and vaginal arteries
    b Venous—to internal iliac veins by way of uterine, ovarian, and vaginal veins
B Location—in pelvic cavity between bladder and rectum
C Position
  1 Flexed between body and cervix, with body lying over bladder pointing forward and slightly upward
  2 Cervix joins vagina at right angles
  3 Capable of considerable mobility; therefore often in abnormal positions, such as retroverted
  4 Eight ligaments anchor it—two broad, two uterosacral, one posterior, one anterior, and two round
  5 Important support structure for uterus—perineal body—muscular node in perineum
D Functions
  1 Menstruation
  2 Pregnancy
  3 Labor and expulsion of fetus

Uterine tubes—fallopian tubes or oviducts

A Location—about 10 cm (4 inches) long and attached to uterus at upper, outer angles; lie in upper free margin of broad ligaments

B Structure
1 Same three coats as uterus
2 Distal ends open with fimbriated margins
3 Mucosa and peritoneum in direct contact here

C Divisions
1 Isthmus—medial third attached to uterus
2 Ampulla—intermediate dilated portion
3 Infundibulum—funnel-shaped terminal component

D Function
1 Serve as ducts for ovaries, providing passageway by which ova can reach uterus
2 Fertilization occurs here normally

Ovaries—female gonads

A Location and size
1 Size and shape of large almonds
2 Lie behind and below uterine tubes
3 Anchored to uterus and broad ligament

B Microscopic structure
1 Consists of several thousand ovarian follicles embedded in connective tissue base
2 Follicles in all stages of development; usually each month one matures, ruptures, and expels its ovum into abdominal cavity

C Functions
1 Oogenesis—formation of mature female gametes (ova)
2 Secretion of hormones—estrogens and progesterone

Vagina

A Location—between rectum and urethra

B Structure
1 Collapsible, musculomembranous tube, capable of great distention
2 External outlet protected by fold of mucous membrane, hymen

C Functions
1 Receive seminal fluid
2 Is lower part of birth canal
3 Is excretory duct for uterine secretions and menstrual flow

Vulva

A Consists of numerous structures that, together, constitute external genitalia

B Main structures
1 Mons pubis—skin-covered pad of fat over symphysis pubis
2 Labia majora—hairy, skin-covered lips
3 Labia minora—small lips covered with modified skin
4 Clitoris—small mound of erectile tissue just below junction of two labia minora
5 Urinary meatus—just below clitoris; opens into urethra
6 Vaginal orifice—below urethra
7 Bartholin's or greater vestibular glands
  a Comparable to bulbourethral glands of male
  b Open by means of long duct in space between hymen and labia minora
  c Ducts from lesser vestibular of Skene's glands open near urinary meatus

Perineum

A Region between vaginal orifice and anus

B Location of perineal body—important muscular support structure for uterus and vagina

C Surgical incision—episiotomy—often made in perineum at childbirth to decrease chances of possible laceration

Breasts

A Location and size
1 Just under skin, over pectoral muscles
2 Size depends on deposits of adipose tissue

B Structure
1 Divided into lobes and lobules; latter composed of glands
2 Single excretory duct per lobe opens in nipple
3 Circular, pigmented area called areola borders nipple

C Function—secrete milk for infant

D Mechanism controlling lactation—see Fig. 25-8

Female sexual cycles
Recurring cycles

A Ovarian cycles
1 Each month, follicle and ovum develop
2 Follicle secretes estrogens
3 Ovum matures and follicle ruptures (ovulation)
4 Corpus leteum forms and secretes progesterone and estrogens

5 If no pregnancy, corpus luteum gradually degenerates and is replaced by fibrous tissue; remains if there is pregnancy

B Endometrial or menstrual cycle

 1 Surface of endometrium sloughs off during menses, with bleeding from denuded area

 2 Regeneration and proliferation of lining occurs

 3 Endometrial glands and arterioles become longer and more tortuous

 4 Glands secrete viscous mucus and cycle repeats

C Myometrial cycle—contractility increases before ovulation and subsides following

D Gonadotropic cycles

 1 FSH—a high rate of secretion for about 10 days after menses

 2 LH secretion increasing from ninth cycle day to day of ovulation (Fig. 25-10)

Control

A Cyclical changes in ovaries controlled as follows

 1 Increasing blood level of FSH during postmenstrual phase stimulates one or more primitive graafian follicles to start growing and to start secreting estrogens (Fig. 25-9); "estrogen surge" day or two before ovulation, lowest blood level of estrogen on day of ovulation

 2 Increasing blood concentration of LH (from cycle day 9 to day of ovulation) causes completion of growth of follicle and ovum, stimulates follicle to secrete estrogens, causes ovulation and luteinization, which in turn leads to progesterone secretion (Fig. 25-10)

B Cyclical changes in uterus controlled as follows

 1 Increasing blood concentration of estrogens during preovulatory phase causes proliferation of endometrial cells with resultant thickening of endometrium, growth of endometrial glands, increased water content of endometrium, and increased contractions by myometrium

 2 Increasing blood concentration of progesterone after ovulation, that is, during premenstrual phase, causes secretion by endometrial glands and preparation of endometrium for implantation, increased water content of endometrium, and decreased contractions by myometrium

C Cyclical changes in FSH and LH secretion—controlled by both negative and positive feedback mechanisms that include releasing hormones from hypothalamus

Functions served by cycles

A Preparation of endometrium for pregnancy during preovulatory and premenstrual periods

B Shedding of progestational endometrium if no pregnancy occurs

Menarche and menopause

A Menarche—onset of menses; often between 11 and 14 years of age

B Menopause (climacteric)—cessation of menses; usually between 45 and 50 years of age

Embryology
Meaning and scope

A Science of development of individual before birth

B Long and complex study

Value of knowledge of embryology

A Provides information for advances in antenatal diagnosis and treatment

B Interprets many abnormal formations

Steps in development of new individual

A Preliminary processes

 1 Maturation of ovum and sperm or reduction of their chromosomes to half original number, that is, half of 46

 2 Ovulation and insemination

  a Ovulation once every 28 days, usually about 14 days before beginning of next menstrual period

  b Insemination, indefinite occurrence; sperm introduced into vagina, swim up to meet ovum in tube

B Developmental processes (Fig. 25-14)

 1 Fertilization or union of ovum and sperm

  a Normally occurs in tube

  b Only one sperm penetrates ovum

  c One-celled new individual called zygote formed by union

 2 Cleavage or segmentation (Figs. 25-13 and 25-16)

  a Multiplication of cells from zygote by repeated mitosis

  b Morula or solid ball of cells formed; becomes hollow with a cluster of cells attached at one point on inner surface of sphere; called blastocyst at this stage; now implanted in endometrium

 3 Development of placenta from both fetal and maternal tissues provides a structure for transfer of nutrients, waste products, etc. (Fig. 25-15)

4 Differentiation of cells into three primary germ layers—ectoderm, mesoderm, and entoderm

5 Histogenesis or formation of various tissues from primary germ layers; organogenesis or formation of various organs by rearrangements of tissues, such as fusions, shiftings, foldings, etc. (Figs. 25-17 and 25-18)

Review questions

1 Name the female sex glands. Name all the internal female reproductive organs.

2 What is the perineum? Of what clinical importance is it in the female?

3 Name the three openings to the exterior from the female pelvis. How many openings to the exterior are there in the male?

4 What is a graafian follicle? What does it contain?

5 How many ova mature in a month, usually?

6 When, in the menstrual cycle, is ovulation thought to occur?

7 Discuss the mechanism thought to control menstruation.

8 Name the periods in the menstrual cycle with the approximate length in days of each and the main events.

9 What two organs are necessary for menstruation to occur?

10 Name the divisions of the uterus.

11 What and where are the fallopian tubes? Approximately how long are they? With what are they lined? Their lining is continuous on their distal ends with what? On their proximal ends? Why is an infection of the lower part of the female reproductive tract likely to develop into a very serious condition?

12 What is the cul-de-sac of Douglas?

13 Describe briefly the hormonal control of breast development and lactation.

14 Explain how contraceptive pills act to prevent pregnancy.

15 On which day of the menstrual cycle would ovulation most probably occur in a 26-day cycle? In a 32-day cycle?

16 Where does fertilization normally take place?

17 Which of the following patients will no longer menstruate? Why? (1) One who has had a bilateral salpingectomy (removal of the tubes)? (2) One who has had a panhysterectomy (removal of the entire uterus)? (3) One who has had a bilateral oophorectomy (removal of both ovaries)? (4) One who has had a cervical hysterectomy (body and fundus removed)? (5) One who has had a unilateral oophorectomy?

18 Which of the patients described in question 16 could no longer become pregnant? Explain.

19 Following menopause, which of the following hormones—estrogens, FSH, progesterone—would you expect to have a high blood concentration? Which a low blood concentration? Explain your reasoning.

20 Define or make an identifying statement about each of the following: adolescence, climacteric, colostrum, corpus luteum, ectopic pregnancy, embryology, endometrium, fertilization, fimbriae, gamete, genes, gonad, hysterectomy, maturation, meiosis, menarche, mitosis, morula, oophorectomy, ovum, ovulation, puberty, salpingitis, semen, spermatozoon, vulva, zygote.

Defense and adaptation

The immune system

Enemies of many kinds and great numbers assault the body during a lifetime. Most threatening are hordes of microorganisms. We live our lives out in a virtual sea of bacteria and viruses. So ever-present and potentially lethal are these small but formidable foes that no newborn baby could live through infancy, much less survive to adulthood or old age, without effective defenses against them. This chapter presents information about the system that provides such defenses—the immune system. An enormous amount of research aimed at learning more about this system is going on during this decade of the seventies.

Definitions

Because the immune system has a vocabulary of its own, we shall begin by defining some of its terms.

antigens—foreign macromolecules, that is, very large molecules not normally present in the body, which, when introduced into the body, induce it to make certain responses. Most antigens are foreign proteins. Some, however, are foreign polysaccharides and some are nucleic acids. Many antigens that enter the body are macromolecules located in the surface membranes of microorganisms.

antibodies—proteins of the class called immunoglobulins. Unlike antigens, antibodies are native molecules, that is, they are normally present in the body.

antigenic determinants—variously shaped, small patches on the surface of an antigen molecule; a less cumbersome name is *epi-*

topes. In a protein molecule, for instance, an epitope consists of a sequence of only about 10 amino acids that are part of a much longer, folded chain of amino acids. The sequence of the amino acids in an epitope determines its shape. Since the sequence differs in different kinds of antigens, each kind of antigen has specific and uniquely shaped epitopes.

combining sites—small concave regions on the surface of an antibody molecule. Like epitopes, combining sites have specific and unique shapes. An antibody's combining sites are so shaped than an antigen's epitope that has a complementary shape can fit into the combining site and thereby bind the antigen to the antibody to form an *antigen-antibody complex*. Because combining sites receive and bind antigens, they are also called *antigen receptors* and *antigen-binding sites*.

clone—family of cells, all of which have descended from one cell.

Major components of the immune system

Names

A vast army of cells and molecules make up the immune system. Lymphocytes are the main cells, and antibodies are the chief molecules. Reportedly, the number of lymphocytes in the body is about 1 trillion (10^{12})—a gigantic number and yet a paltry one compared to the 100 million trillion (10^{20}) antibodies estimated![*] But even these enormous numbers of cells and molecules cannot by themselves do the vital work of defending the body against invaders. Still other cells (chiefly leukocytes and macrophages) and other molecules (notably complement) must also play important roles.

Functions

Lymphocytes and antibodies together search out, recognize, and destroy three kinds of enemies—microorganisms that have invaded the body tissues, cells of tissues or organs that have been transplanted into the body, and the body's own cells that have become transformed and able to develop into cancer.

Lymphocytes

Formation and types

Lymphocytes derive from hemopoietic stem cells, that is, primitive ancestor cells not only of lymphocytes but also of other kinds of white and red blood cells. Hemopoietic stem cells originate early in the life of a human embryo in the yolk sac. Later, but before birth, they migrate through the fetal liver and spleen and take up residence in the bone marrow. Those stem cells destined to become lymphocytes follow two developmental paths and differentiate into two major classes of lymphocytes—B-lymphocytes

[*]Jerne, N. K.: The immune system, Sci. Am. **229**:53, July, 1973.

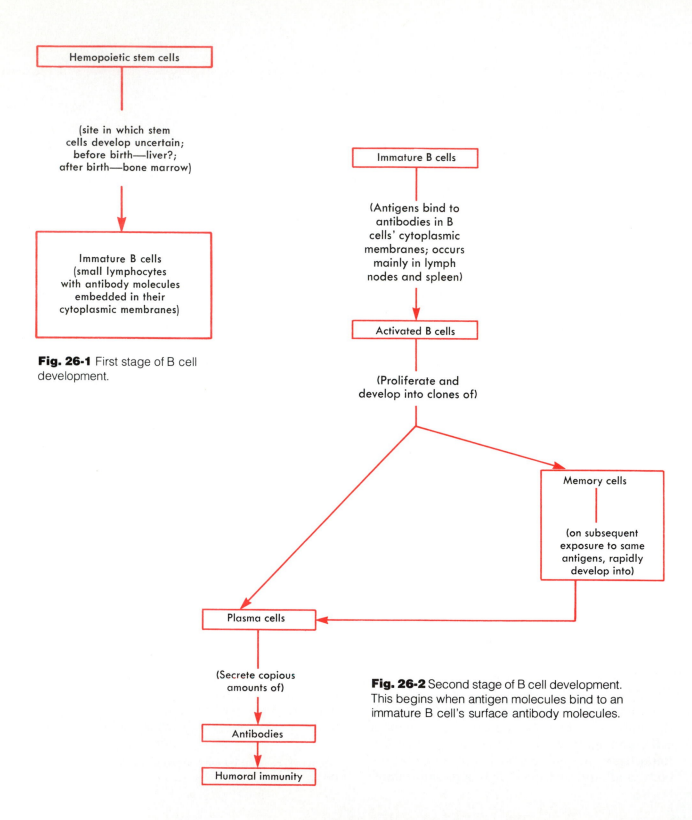

Fig. 26-1 First stage of B cell development.

Fig. 26-2 Second stage of B cell development. This begins when antigen molecules bind to an immature B cell's surface antibody molecules.

and T-lymphocytes, or simply, B cells and T cells.

Development, activation, and functions of B cells

The development of B cells occurs in two stages. In chickens the first stage of B cell development occurs in the bursa of Fabricus—hence the name B cells. Since humans do not have a bursa of Fabricus, another organ must serve as the site for the first stage of B cell development. Some evidence suggests that the fetal liver serves this purpose. At any rate, by the time a human infant is a few months old, its B cells have completed the first stage of their development (Fig. 26-1). They are then called immature B cells, and they constitute the small lymphocytes of the blood. Immature B cells synthesize antibody molecules and insert about 100,000 of them into their cytoplasmic membranes, but do not secrete them—at least in any significant amount. The combining sites of the antibody molecules located in the cytoplasmic membranes of immature B cells are able to serve eventually as receptors for a specific antigen. A single immature B cell, by undergoing repeated mitosis, forms a clone, or family of many identical B cells. Because all the cells in one clone have descended from the same B cell, they synthesize and display the same antibodies on their surface membranes.

The second stage of B cell development (Fig. 26-2) usually takes place in the lymph nodes and spleen, but only under certain conditions. It must be initiated by an encounter between an immature B cell and its specific antigen, that is, one whose epitopes fit the combining sites of the B cell's surface antibodies. The antigen selects these antibodies to bind to, and this usually activates the B cell. Specifically, the binding of antigen to antibodies on a B cell's surface usually stimulates it both to divide repeatedly and to differentiate, forming two kinds of clones—one made up of plasma cells and the other of memory cells. *Plasma cells* synthesize and secrete copious amounts of antibody, reportedly 2,000 antibody molecules per second for every second of the few days that plasma cells live.* All the cells in one clone of plasma cells necessarily secrete identical antibodies because they have all descended from the same B cell. *Memory cells* do not secrete antibodies, but on a later exposure to the same antigen that led to their formation, memory cells very quickly develop into plasma cells that secrete the appropriate antibody. In effect, memory cells seem to remember an earlier encounter with this antigen. Briefly then, the ultimate function of B cells is to serve as ancestors of the antibody-secreting plasma cells and of memory cells, the partly differentiated cells that can quickly complete their differentiation into antibody-secreting plasma cells.

Development, activation, and functions of T cells

T cells, by definition, are lymphocytes that have made a detour through the thymus gland before migrating to the lymph nodes and spleen (Fig. 26-3). During their residence in the thymus, stem cells develop into *thymocytes*, cells that proliferate as rapidly as any in the body. Thymocytes undergo mitosis three times a day, and as a result their numbers increase enormously in a relatively short time. They stream out of the thymus into the blood and find their way to a new home in areas of the lymph nodes and spleen called thymus-dependent zones. From this time on, they are known as T cells. Each T cell, like each B cell, displays antigen receptors on its surface membrane, but these receptors are sites on as yet unidentified molecules—presumably, not immunoglobulins but perhaps compounds similar to them. B cell receptors, in contrast, are known to be the combining sites of immunoglobulin molecules. When an antigen encounters a T cell whose surface receptors fit the antigen's epitopes, the antigen selects that T cell by bind-

*Jerne, N. K.: The immune system, Sci. Am. **229**:56, July, 1973.

ing to its receptors. This activates or sensitizes the T cell, causing it to initiate a series of reactions. The first reaction is that the sensitized T cell divides repeatedly to form a clone of identical sensitized T cells. All of these sensitized cells

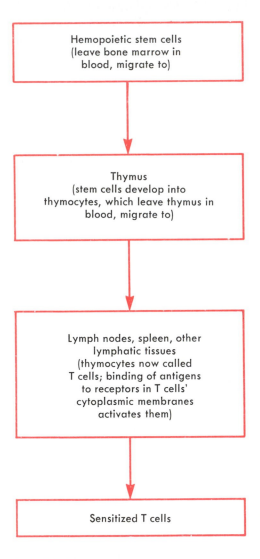

Fig. 26-3 Development of T cells. To become T cells, stem cells must pass through the thymus, a fact suggested by the designation "T cells." Compare with Fig. 26-1. Sensitized T cells release various chemicals collectively known as lymphokines. These act in several ways to destroy foreign cells.

release chemical compounds into the surrounding tissues that are collectively known as *lymphokines*. Names of some individual lymphokines are these: chemotactic factor, migration inhibition factor, macrophage activating factor, and lymphotoxin. *Chemotactic factor* attracts macrophages, causing hundreds of them to migrate into the vicinity of the antigen-bound, sensitized T cell. *Migration inhibition factor* halts macrophage migration. *Macrophage activating factor* prods the assembled macrophages to destroy antigens by phagocytosing them at a rapid rate. *Lymphotoxin* is a powerful poison that quickly kills any cell it attacks. It is this lethal substance that has earned sensitized T cells the nickname of "killer cells."

T cells perform another important function: they act as B cell helpers. Their influence on B cells seems to be essential for B cell functioning.

To summarize very briefly—the first function of T cells, like that of B cells, is to search out, recognize, and bind to appropriate antigens. The ultimate function of T cells is to kill cells that have foreign antigens on their surface membranes, be they microorganisms, cells of transplanted tissues or organs, or tumor cells.

Distribution of lymphocytes

The densest populations of lymphocytes occur in the structures where they are formed and multiply, namely, in the bone marrow, thymus gland, lymph nodes, and spleen. From these structures, a continuous stream of lymphocytes pours into the blood. They are distributed to tissues all over the body by leaving the blood through the thin walls of the capillaries and entering the interstitial fluid. After wandering about through the tissue spaces, they eventually find their way into lymphatic capillaries. Lymph flow transports the lymphocytes through a succession of lymph nodes and lymphatic vessels and empties them by way of the thoracic and right lymphatic ducts into the subclavian veins located behind the clavicles. Thus returned to the blood, the lymphocytes embark on still another long journey—through blood, tissue spaces, lymph, and back to blood. The survival value of the continued recirculation of lymphocytes and of their widespread distribution throughout body tissues seems apparent. It provides these major cells of the immune system ample opportunity to perform their functions of searching out, recognizing, and destroying foreign invaders.

Antibodies (immunoglobulins)
Structure

Antibodies are proteins of the family called immunoglobulins. Like all proteins, they are very large molecules and are composed of long chains of amino acids (polypeptides). Each immunoglobulin molecule consists of four polypeptide chains—two heavy chains and two light chains. Each polypeptide chain is intricately folded to form globular regions that are joined together in such a way that the immunoglobulin molecule as a whole is Y-shaped. Look now at Fig. 26-4. The two shorter black lines in the diagram represent the light chains, and the two longer lines represent the heavy chains. Each light chain consists of 214 amino acids. Heavy chains consist of 446 amino acids each, so are about twice as long and weigh about twice as much as light chains.

The colored ovals that lie over the black lines in Fig. 26-4 represent different regions in the polypeptide chains of the antibody molecule. The more vividly colored ovals represent *variable regions*, that is, regions in which the sequence of amino acids varies in different antibody molecules. Note the relative positions of the variable regions of the light and heavy chains; they lie directly opposite each other. Because amino acid sequence determines conformation or shape, and because different sequences of amino acids occur in the variable regions of different antibodies, the shapes of the sites between the variable regions also differ. These are the anti-

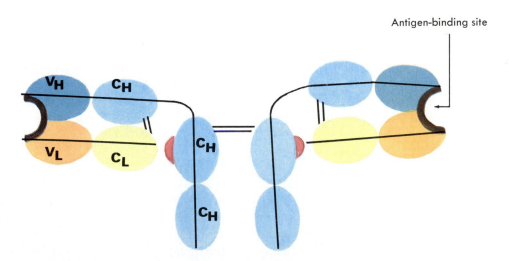

Antigen-binding site

Fig. 26-4 Structure of antibody (IgG) molecule. Each of the two heavy polypeptide chains in the molecule consist of one variable region (V_H) and three constant regions (C_H). The light chains consist of one variable region (V_L) and one constant region (C_L). The red areas represent complement-binding sites.

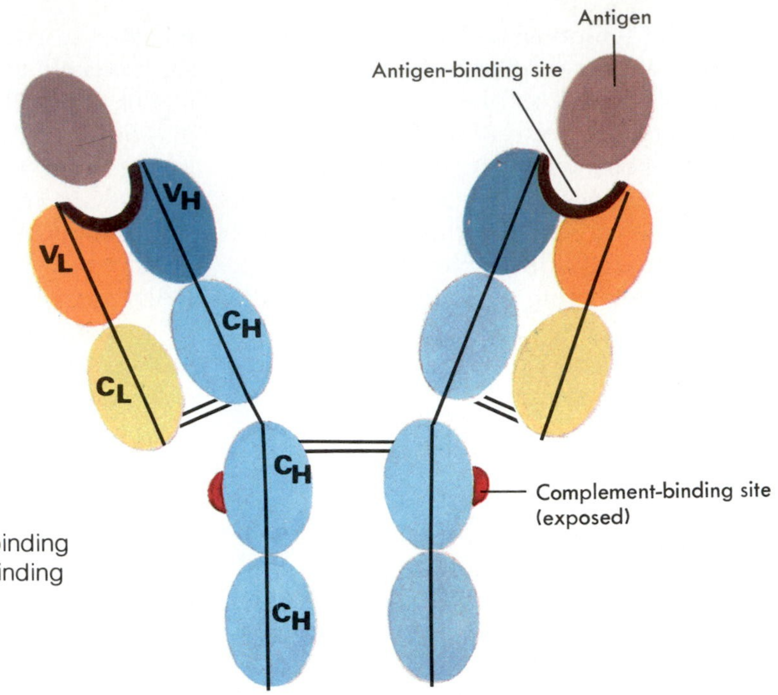

Fig. 26-5 Exposure of complement-binding site by binding of antigen to antigen-binding site of antibody. (See p. 658.)

body's combining sites (defined on p. 651), and there are two such sites on each antibody molecule. Locate these in Fig. 26-4. Each one of us is thought to have millions of different kinds of antibody molecules in our bodies. And each one of these, almost unbelievably, has its own uniquely shaped combining sites. It is this structural feature that enables antibodies to recognize and combine with specific antigens, both of which are crucial first steps in the body's defense against invading microbes and other foreign cells.

In addition to its variable region, each light chain in an antibody molecule also has a constant region. The constant region consists of 106 amino acids whose sequence is identical in all antibody molecules. Each heavy chain of an antibody molecule consists of three constant regions in addition to its one variable region. Identify the constant and variable regions of the light and heavy chains in Figs. 26-4 and 26-5. Be sure to observe the location of two *complement binding sites* on the antibody molecule.

In summary, an immunoglobulin or antibody molecule consists of two heavy and two light polypeptide chains. Each light chain consists of one variable region and one constant region. Each heavy chain consists of one variable region and three constant regions. Disulfide bonds join the two heavy chains to each other; they also bind each heavy chain to its adjacent light chain. An antibody has two antigen-binding sites—one at the tip of each pair of variable regions—and two complement binding sites located as shown in Figs. 26-4 and 26-5.

Diversity

Every normal baby is born with an enormous number of different clones of B cells populating its bone marrow, lymph nodes, and spleen. All the cells of each clone are committed to synthesizing a specific antibody with a sequence of amino acids in its variable regions that is different from the sequence synthesized by any other of the innumerable clones of B cells. How does this astounding diversity originate? A complete answer to this question has not yet been found. However, the short, generally agreed upon answer is that antibody diversity results from mutations in the B cell genes that code for the sequence of amino acids in the variable regions of antibodies. Lymphocytes that contain these mutated genes do not synthesize antibodies identical to those made by their parent cells. Instead, they synthesize antibodies with different amino acid sequences in their variable regions. Lymphocytes are known to have one of the most rapid mitosis rates of any human cells. Also, a rapid mitosis rate is known to be accompanied by a high mutation rate. These two facts together are believed to account for the apparently unlimited diversity of antibodies. In the words of one noted immunologist,* "If at each of 50 positions of both (heavy and light) chains there were an independent choice between just two amino acids, there would be 2^{100} (or 10^{30}) potentially different molecules!" In short, the body's millions of diverse antibodies originate from mutations that lead to amino acid replacements in various positions of their variable regions.

Classes

There are five classes of immunoglobulins identified by letter names as immunoglobulins M, G, A, E, and D. *IgM* (abbreviation for immunoglobulin M) is the antibody that immature B cells synthesize and insert into their cytoplasmic membranes. It is also the predominant class of antibody produced after initial contact with an antigen. The most abundant circulating antibody, the one that normally makes up about 75% of all the antibodies in the blood, is *IgG*. It is the predominant antibody of the secondary antibody response, that is, following subsequent contacts with a given antigen. *IgA* is the major class of antibody present in the mucous membrane lining the intestines and the bronchi, in

*Jerne, N. K.: The immune system, Sci. Am. **229:**52, July, 1973.

saliva, and in tears. *IgE*, although minor in amount, can produce major harmful effects, those associated with allergies. *IgD* is present in the blood in very small amounts, and its function is as yet unknown.

Functions

Antibody molecules—some 100 million trillion of them—play a major role in defending the body against invaders. To do this, they must first recognize them as invaders, as foreign substances rather than native components of the body. Recognition occurs when an antigen's epitopes (small patches on its surface) fit into and bind to the antibody molecule's antigen-binding sites. The binding of the antigen to antibody forms an antigen-antibody complex that may produce one or more of several effects. For example, it transforms antigens that are toxins (chemicals poisonous to cells) into harmless substances. It agglutinates antigens that are molecules on the surfaces of microorganisms. In other words, it makes them stick together in clumps, and this, in turn, makes it possible for macrophages and other phagocytes to dispose of them more rapidly by ingesting and digesting large numbers of them at one time. The binding of antigen to antibodies frequently produces still another effect—it alters the shape of the antibody molecule, not very much, but enough to expose the molecule's previously hidden complement-binding sites. This seems a trivial enough change, but not so. It initiates an astonishing series of reactions (described on p. 658) that culminate in the destruction of microorganisms and other foreign cells.

Self-tolerance

The term self-tolerance means that normally one's own antibodies do not combine with epitopes present on one's own cells and macromolecules. In effect, antibodies seem to distinguish between macromolecules native to the body and those foreign to it. Sir Macfarlane Burnet described this phenomenon as the differentiation between "self" and "not-self" and called it the central process in immunology. To explain it, he postulated the following events. During the early stages of embryonic life, ancestor cells of lymphocytes mutate at a high rate, producing lymphocytes capable (because of certain of their genes) of producing all possible antibody patterns—including many that are complementary to patterns present on the body's own cells and macromolecules. Such lymphocytes, he postulated, are killed off before birth by contacting and binding to macromolecules having their complementary patterns. By the time of birth, therefore, no lymphocytes remain that can produce antibodies capable of combining with the body's own components. In short, self-tolerance has developed.

Clonal selection theory

The clonal selection theory, proposed in 1959 by Sir Macfarlane Burnet, has two basic tenets. First, it holds that the body contains an enormous number of diverse clones of cells, each committed by certain of its genes to synthesize a different antibody. Second, the clonal selection theory postulates that when an antigen enters the body, it selects the clone whose cells are committed to synthesizing its specific antibody and stimulates these cells to proliferate and to thereby produce more antibody. We now know that the clones selected by antigens consist of lymphocytes. We also know how antigens select lymphocytes—by the shape of antigen receptors on the lymphocyte's cytoplasmic membrane. An antigen recognizes receptors that fits its epitopes and combines with them. By thus selecting the precise clone committed to making its specific antibody, each antigen provokes its own destruction.

Complement

Complement, a component of blood serum, consists of a set of several proteins. They are in-

active enzymes that are activated in a definite sequence to catalyze an intricately linked series of reactions. The binding of an antigen located on the surface of a microorganism (or of a tumor or transplant cell) to an antibody alters the shape of the antibody molecule in such a way as to expose its complement-binding sites (Fig. 26-5). Complement protein 1 immediately binds to these sites, is thereby activated, and touches off the catalytic activity of the next complement protein in the series. This is followed in rapid sequence by the activity of the next protein, then the next, and the next, and so on until the entire series of enzymes has functioned. The end result of this rapid-fire activity is something that in all probability not even the most far-ranging imagination would conceive. Molecules formed by these reactions have assembled themselves on the enemy cell's surface in such a way as to form a doughnut-shaped structure—complete with a hole in the middle!* We might even say that complement has drilled a hole through the foreign cell's surface membrane. This kills the cell because ions, followed by water, move into the cell and cause it to swell and burst. The technical name for this is *cytolysis*. Briefly, then, complement functions to produce cytolysis.

Properdin

Like complement, properdin consists of a set of proteins present in blood serum as inactive enzymes. When activated, they assemble on the surface of a foreign cell, where they initiate the complement attack sequence just described. Properdin enzymes, however, activate protein 3, whereas antibody bound to a cell surface antigen activates complement protein 1. Properdin provides an alternate method of activating complement, one in addition to the usual antibody method of activating it.

*Mayer, M. M.: The complement system, Sci. Am. p. 60, Nov., 1973.

Interferon

Interferon is another protein compound that plays a part in producing immunity. It is synthesized by body cells under certain conditions, notably after viruses have invaded them. Released from the cells that produce it, interferon acts on other body cells to defend them against viruses. Recent evidence suggests that interferon retards the growth of cancer cells. Considerable research is currently under way to learn more about how this compound acts.

Applications

Applications of knowledge about the immune system are numerous and varied. They include the following: identification of several conditions involving abnormalities of the immune system, explanation of transplant rejection, development of methods of inducing immunosuppression, and postulated explanations of cancer.

Conditions involving abnormalities of the immune system

Agammaglobulinemia is the name of one of the diseases that has been identified as an abnormality of the immune system. The blood of an individual with this disease contains no gamma globulins and, therefore, no antibodies, since antibodies are gamma globulins. Without antibodies a crucial defense against microorganisms is missing, and victims of agammaglobulinemia are highly susceptible to fatal infections.

Autoimmune diseases, by definition, are those in which the immune system responds to the body's own cells as it does to foreign cells—specifically, it produces antibodies that attack its own cells. Normally the immune system differentiates between "self" and "not-self," to use Burnet's words. It recognizes which cells are its own and which are foreign and does not attack its own cells.

Question: How do the cells of one body differ from all other cells? Answer: All cells of one body possess unique "surface markers," that is, certain antigens present in their cytoplasmic membranes that are different from antigens present in the membranes of any other cells. These molecules, unique in every individual, are glycoprotein compounds called *histocompatibility antigens*. Knowledge of histocompatibility antigens is applied in the laboratory procedure used to match tissues and organs for transplantation surgery.

Question: How does the immune system recognize differences in cell surface antigens so that normally it tolerates its own cells but destroys foreign cells? Answer: Briefly, lymphocytes recognize foreign cells by their histocompatibility antigens. Epitopes on these surface antigens fit combining sites on antibodies present in the lymphocyte's surface membranes. Normally, blood contains no lymphocytes with combining sites that fit antigens on the body's own cells. (Reread the section on self-tolerance, p. 657.)

Autoimmune disease indicates failure in the system's recognition function. It mistakenly identifies its own cells' surface antigens as foreign instead of native.

Transplant rejection

The immune system normally rejects, that is, kills, cells of transplanted tissues and organs. Surface molecules on lymphocytes recognize as foreign the histocompatibility antigens on transplanted cells, combine with them, and thereby initiate reactions that end in the destruction of these foreign cells. To combat this bugaboo of transplant surgery, methods have been devised to suppress the immune system. For example, certain drugs, for example, azathioprine (Imuran), and radiation are used as immunosuppressants.

Cancer

Among the currently held concepts about cancer are the following:

1 Cell mutations occur frequently in the normal body, and many of the mutated cells formed are cancer cells. Cancer cells, according to a simple definition, are cells capable of proliferating wildly to form a tumor—unless they are killed before this can happen.

2 Abnormal antigens, called tumor-specific antigens, are present in the surface membranes of cancer cells in addition to the histocompatibility antigens present in all body cells.

3 Lymphocytes, in their repeated wanderings through body tissues, are almost sure to come in contact with newly formed cancer cells. Normally, they recognize these cells as foreign to the body by the tumor-specific antigens that so label them. They then quickly initiate reactions that kill the cancer cells.

4 It sometimes happens that newly formed cancer cells escape destruction by cells of the immune system. Various ingenious theories have been proposed to explain how this might happen but so far, none has been proved.

Outline summary

Definitions

A Antigens—foreign macromolecules, usually proteins, which, when introduced into the body, induce it to make certain responses; many antigens are macromolecules located in surface membranes of microorganisms

B Antibodies—proteins of the class called immunoglobulins; antibodies are natives of body, are present at birth

C Antigenic determinants, or epitopes—small regions on surfaces of antigen molecules that have various and specific shapes

D Combining sites (also called antigen-binding sites and antigen receptors)—regions on antibody molecule; shape of each combining site is complementary to the shape of a specific antigen's epitopes

E Clone—family of cells descended from one cell and therefore identical to it

Major components of immune system

A Names
1 Lymphocytes, about 1 trillion of them, are the main cells of the immune system; macrophages and leukocytes also perform functions that contribute to immunity

2 Antibodies, an estimated 100 million trillion of them, are the main molecules of the immune system; complement, properdin, and interferon are others

B Functions—lymphocytes and antibodies search out, recognize, and destroy foreign cells, notably microorganisms, tumor cells, and transplanted tissue cells

Lymphocytes

A Formation and types—lymphocytes are formed from hemopoietic stem cells present in the embryo in the yolk sac and, after birth, in the bone marrow; stem cells follow one developmental path to become B-lymphocytes (B cells) or another path to become T-lymphocytes (T cells)

B Development, activation, and function of B cells
1 Development and activation
  a First stage—organ in which occurs not yet positively identified, but some evidence points to fetal liver; first stage of B cell development produces immature B cells and is completed by time infant a few months old; immature B cells are present after birth in bone marrow and lymphatic tissues (notably, lymph nodes

and spleen) and in blood, where they constitute some of the small lymphocytes (T cells make up rest of small lymphocytes); immature B cells synthesize and insert antibody molecules into their cytoplasmic membranes but do not secrete antibody molecules; immature B cells form clones of identical immature B cells, thereby multiplying their numbers

   b Second stage—initiated by immature B cell contacting its specific antigen; antigen selects and binds to B cell's surface antibodies, thereby activating B cell to enter second stage of its development; characterized by repeated cell divisions and differentiation of immature B cells to form clone of plasma cells and clone of memory cells

  2 Functions

   a Initial function—immature B cells search out, recognize, and bind to antigens that fit B cell's surface antibodies

   b Ultimate function—formation of fully differentiated plasma cells and partially differentiated memory cells from activated immature B cells; plasma cells secrete antibodies; memory cells quickly develop into plasma cells on subsequent exposure to appropriate antigen

C Development, activation, and functions of T cells

  1 Development and activation—stem cells migrate from bone marrow to thymus, where they develop into rapidly proliferating thymocytes; thymocytes migrate to thymus-dependent zones of lymph nodes and spleen; from this time on they are called T-lymphocytes or T cells; T cells activated by binding to their specific antigens, then are called sensitized T cells

  2 Functions

   a Initial—search out, recognize, and bind to antigens that fit T cell's surface receptors

   b Ultimate—kill foreign cells by releasing lymphokines that kill cells either directly or indirectly by stimulating macrophages to phagocytose them

D Distribution of lymphocytes—bone marrow, thymus, lymph nodes, spleen, other lymphatic tissues, blood, lymph, tissue spaces

Antibodies (immunoglobulins)

A Structure—immunoglobulin molecules composed of two heavy and two light polypeptide chains held together by disulfide bonds between the heavy chains and between each heavy chain and its ad-

jacent light chain; light chain consists of one variable and one constant region; heavy chain consists of one variable and three constant regions; two antigen-binding sites (combining sites) and two complement-binding sites present on each antibody molecule

B Diversity—normally, individual is born with enormous number of different clones of B cells, each clone committed to synthesizing antibodies different from those secreted by other clones; mutations in certain genes produces cells committed to synthesizing different sequences of amino acids in variable regions of antibodies

C Classes of immunoglobulines

  1 IgM—present in surface membranes of immature B cells; also, is the predominant antibody produced after initial contact of B cell with its specific antigen

  2 IgG—the most abundant circulating antibody (makes up about 75% of them); also is the predominant antibody produced after subsequent exposure to same antigen

  3 IgA—chief antibody in lining mucosa of intestines and bronchi, in saliva, and in tears

  4 IgE—antibody responsible for allergic effects

  5 IgD—only small amounts in blood; function unknown

D Functions of antibodies

  1 Recognition and binding to antigens to form antigen-antibody complex

  2 Antigen-antibody complex produces variety of effects, for example, agglutination of antigens and alteration of antibody molecule's shape so as to expose its complement-binding site, which leads to series of reactions that end in destruction of antigen

E Self-tolerance—means that one's own antibodies do not normally combine with one's own body's components; self-tolerance is acquired during embryonic period presumably because all immature lymphocytes that bind to macromolecules having complementary patterns to the lymphocytes' surface antibodies are killed by so binding

F Clonal selection theory

  1 Enormous number of diverse clones of lymphocytes present at birth, each clone committed to synthesizing a different antibody

  2 An antigen, on entering the body, selects the clone of lymphocytes committed to synthesizing its specific antibody and stimulates these cells to proliferate and produce more antibody

Complement

Normal constituent of blood; consists of several inactive enzymes, first of which is activated by binding of antibody to antigen located on surface of a cell (usually a microorganism, tumor cell, or transplant cell); shape of antibody molecule altered by antigen binding to it, thus exposing complement-binding site and making it possible for complement protein 1 to bind to it; other complement proteins activated in sequence and assemble on cell surface to form dough-nut-shaped structure; result, cell killed by process of cytolysis

Properdin

Set of inactive enzymes normally present in blood; activated, they assemble on a cell surface where they activate complement protein 3; result—complement action continued and cell killed by cytolysis

Interferon

Protein released from body cells following viral invasion of them; interferon acts in some way on other body cells to defend them against viral invasion

Applications
Conditions involving abnormalities of the immune system

A Agammaglobulinemia—disease characterized by absence of gamma globulins in blood
B Autoimmune diseases—plasma cells produce antibodies that attack body's own cells; apparently lymphocytes mistakenly identify them as foreign cells

Transplant rejection

Immune system cells, that is, lymphocytes, recognize as foreign the histocompatibility antigens on the surface of transplanted cells and initiate reactions that end in the destruction of transplanted cells

Cancer

A Cell mutations are common in normal body and produce cancer cells
B Surface membranes of cancer cells contain abnormal antigens called tumor-specific antigens
C Tumor-specific antigens label cancer cells as foreign, and normally lymphocytes recognize them as foreign and initiate reactions that kill them before they have time to proliferate to form tumor
D Newly formed cancer cells sometimes escape destruction by immune system; exactly how they escape is still uncertain

Review questions

1 Define the following terms: antigens, antibodies, antigenic determinants, combining sites, clone.
2 The word epitope is a synonym for which of the terms defined in question 1?
3 What two terms are synonyms for combining sites?
4 What kind of white blood cells constitute the main cells of the immune system?
5 When an atigen-antibody complex is formed, what region on the antigen molecule fits into what region on the antibody molecule?
6 Immune cells (lymphocytes) and immune molecules (antibodies) together perform what three general functions?
7 Lymphocytes, like all blood cells, originate before birth in the embryonic yolk sac. What structure forms the stem cells after birth and from that time one?
8 Explain the derivation of the name B-lymphocytes.
9 Explain the derivation of the name T-lymphocytes.
10 B cells are identified by the presence in their cytoplasmic membranes of what kind of molecules?
11 What activates an immature B cell to undergo the second stage of its development?
12 The second stage of B cell development ends in the formation of what two kinds of cells?
13 What cells synthesize and secrete antibodies?
14 Explain the function of memory cells.
15 How are T cells activated or sensitized?

16 Both B cells and T cells perform the same preliminary functions of searching out, recognizing, and binding to specific antigens. Describe several functions performed by T cells after they have been activated or sensitized.
17 What are killer cells? Why is this nickname appropriate?
18 Antibodies belong to what class of compounds? Diagram and describe the structure of an antibody molecule.
19 What is a synonym for immunoglobulins?
20 An enormous diversity of antibodies exists in our bodies. Explain briefly how this diversity originated.
21 Antibodies do not themselves destroy microorganisms and other foreign cells. They do, however, contribute to their destruction. How?
22 What does the term self-tolerance mean? Explain Burnet's theory to explain self-tolerance.
23 Explain the two basic tenets of Burnet's clonal selection theory.
24 What is complement and how does it function?
25 What is properdin and how does it function?
26 What is interferon and how does it function?
27 What does the term autoimmune disease mean?
28 What are histocompatibility antigens? What laboratory procedure applies knowledge about these antigens?
29 Explain briefly the phenomenon of transplant rejection.
30 Describe briefly some currently held concepts about the development of cancer.

Stress

Selye's concept of stress

In 1935 Hans Selye of McGill University in Montreal made an accidental discovery that launched him on a lifelong career and led him to conceive the idea of stress. This chapter will tell the story briefly of how Selye developed his stress concept and will describe the mechanism of stress that he postulated. It will then present some current ideas about stress.

Development of concept

Selye made his accidental discovery in 1935 when he was trying to learn whether there might be another sex hormone besides those already known. He had injected rats with various extracts made from ovaries and placenta, expecting to find that different changes had occurred in animals injected with different hormonal preparations. But to his surprise and puzzlement, he found the same three changes in all of the animals. The cortex of their adrenal glands were enlarged but their lymphatic organs—thymus glands, spleens, and lymph nodes—were atrophied, and bleeding ulcers of the stomach and duodenum had developed in every animal. Then he tried injecting many other substances, for example, extracts from pituitary glands, from kidneys, and from spleens, and even a poison, formaldehyde. Every time he found the same three changes: enlarged adrenals, shrunken lymphatic organs, and bleeding gastrointestinal ulcers. They seemed to be a syndrome, he thought. A syndrome, according to the classical definition, is a set of signs and symptoms that occur together and that are characteristic of one particular disease. The three changes Selye had observed occurred together but they seemed to be characteristic not of any one particular kind of injury but of any and all kinds of harmful stimuli. More experiments using a wide variety of chemicals and injurious agents confirmed for him that the three changes truly were a syndrome of injury. His first publication on the subject was a short paper entitled "A syndrome produced by diverse nocuous agents"; it appeared in the July, 1936, issue of the British journal, *Nature*. Years later, in 1956, he published his monumental technical treatise, *The Stress of Life*.

Definitions

Stress, according to Selye's usage of the word, is a state or condition of the body produced by "diverse nocuous agents" and manifested by a syndrome of changes. He named the agents that produce stress *stressors* and coined a name—*general adaptation syndrome*—for the syndrome or group of changes which make the presence of stress in the body known.

Stressors

A stressor is any agent or stimulus that produces stress. Precise classification of stimuli as stressors or non-stressors is not possible. We can, however, make the following few generalizations about the character of stressors.

1 Stressors are extreme stimuli—too much or too little of almost anything. In contrast, almost anything in moderation or mild stimuli are non-stressors. Thus, coolness, warmth, and soft music are non-stressors, whereas extreme cold, extreme heat, and extremely loud music almost always act as stressors. Not only extreme excesses, but also extreme deficiencies may act as stressors. One example of this kind of stressor is an extreme lack of social contact stimuli. Solitary confinement in a prison, space travel, social isolation due to blindness or deafness or, in some cases old age, have all been identified as stressors. But the opposite extreme, an excess of social contact stimuli, for example, due to overcrowding also acts as a stressor.

2 Stressors very often are injurious or unpleasant or painful stimuli—but not always. "A painful blow and a passionate kiss," Selye wrote, "can be equally stressful." *

3 Anything that an individual perceives as a threat, whether real or imagined, arouses fear or anxiety. These emotions act as stressors. So, too, does the emotion of grief.

4 Stressors differ in different individuals and in one individual at different times. A stimulus that is a stressor for you may not be a stressor for someone else. A stimulus that is a stressor for you today may not be a stressor for you tomorrow and might not have been a stressor for you yesterday. Many factors—including one's physical and mental health, heredity, past experiences, coping habits, and even one's diet—determine which stimuli are stressors for each individual.

*Selye, H.: The stress syndrome, Am. J. Nurs. **65**:97-99, Mar., 1965.

The general adaptation syndrome
Manifestations

Stress, like health or any other state or condition, is an intangible phenomenon. It cannot itself be seen, heard, tasted, smelled, felt, or measured directly. How, then, can we know that stress exists? It can be inferred to exist when certain visible, tangible, measurable responses occur. Selye, for example, inferred that the animals he experimented on were in a state of stress when he found the syndrome of the three changes before noted—hypertrophied adrenals, atrophied lymphatic organs, and bleeding gastrointestinal ulcers. Because this syndrome indicated the presence of stress and consisted of three changes, he called them the "stress triad." Eventually he found that many other changes also took place as a result of stress. The entire group of changes or responses he named the *general adaptation syndrome (GAS)*. For each word in this name he gave his reasons. By the word "general" he wanted to suggest that the syndrome was "produced only by agents which have a general effect upon large portions of the body."* The word "adaptation" was meant to imply that the syndrome of changes made it possible for the body to adapt, to cope successfully with stress. Selye thought that these responses seemed to protect the animals from serious damage by extreme stimuli and to promote their healthy survival. He looked upon the general adaptation syndrome as a crucial part of the body's complex defense mechanism.

Stages

The changes that make up the general adaptation syndrome do not all take place simultaneously, but over a period of time in three stages. Selye named these stages the alarm reaction, the stage of resistance or adaptation, and the stage of exhaustion. A different syndrome of changes, he noted, characterized each stage. Among the

*Selye, H.: The stress of life, New York, 1956, McGraw-Hill Book Co., p. 32.

responses characteristic of the *alarm reaction*, for example, were the stress triad already described—hypertrophied adrenal cortex, atrophied lymphatic organs (thymus, spleen, lymph nodes), and bleeding gastric and duodenal ulcers. In addition, the adrenal cortex increased its secretion of glucocorticoids, the number of lymphocytes decreased markedly, and so, too, did the number of eosinophils. Also, the sympathetic nervous system and the adrenal medulla greatly increased their activity. Each of these changes, in turn, produced other widespread changes. Fig. 27-1 indicates some of the changes stemming from adrenal cortical hypertrophy. Fig. 27-2 shows responses produced by increased sympathetic activity and increased secretion by the adrenal medulla of its hormone, epinephrine (adrenaline).

Quite different responses characterize the *stage of resistance* of the general adaptation syndrome. For instance, the adrenal cortex and medulla both return to their normal rates of secreting hormones. The changes that had taken place in the alarm stage as a result of increased corticoid secretion disappear during the stage of resistance. All of us go through the first and second stages of the stress syndrome many times in our lifetimes. Stressors of one kind or another act on most of us every day. They may upset or alarm us but we soon resist them successfully. In short, we adapt, we cope.

The *stage of exhaustion* develops only when stress is extremely severe or when it continues over long periods of time. Otherwise, that is, when stress is mild and of short duration, it ends with a successful stage of resistance and adaptation to the stressor. If stress continues to the stage of exhaustion, corticoid secretion increases above the normal rate and may even surpass the high level of secretion during the alarm reaction. But despite the increased amounts of corticoids, adaptation does not increase but decreases during the stage of exhaustion. The body can no longer cope successfully with the stressor and death may ensue. For a brief summary of

Table 27-1 The three stages of the general adaptation syndrome

Alarm	Resistance stage	Exhaustion stage
Increased secretion of glucocorticoids and resultant changes (Fig. 27-1)	Glucocorticoid secretion returns to normal	Increased glucocorticoid secretion sometimes to higher levels than during alarm reaction
Increased activity of sympathetic nervous system	Sympathetic activity returns to normal	
Increased norepinephrine secretion by adrenal medulla	Norepinephrine secretion returns to normal	
"Fight or flight" syndrome of changes (Fig. 27-2)	"Fight or flight" syndrome disappears	
Low resistance to stressors	High resistance (adaptation) to stressor	Loss of resistance to stressor; may lead to death
Stress triad (hypertrophied adrenals, atrophied thymus and lymph nodes, bleeding ulcers in stomach and duodenum)		

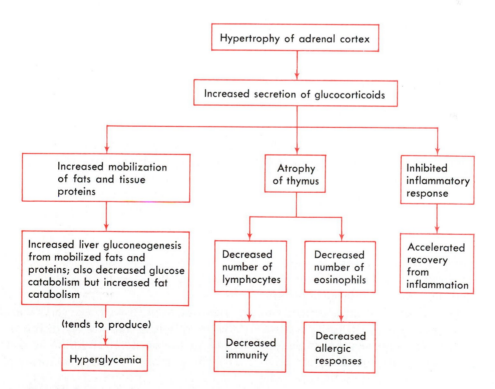

Fig. 27-1 Alarm reaction responses resulting from hypertrophy of adrenal cortex.

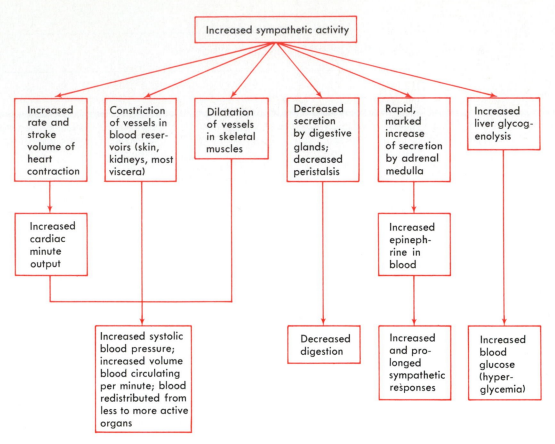

Fig. 27-2 Alarm reaction responses resulting from increased sympathetic activity. Note that these are the responses commonly referred to as the "fight-or-flight" reaction.

the changes characteristic of each of the three stages of the general adaptation syndrome, see Table 27-1.

Mechanism of stress

Stressors produce a state of stress. A state of stress, in turn, inaugurates a series of responses that Selye called the general adaptation syndrome. More simply, a state of stress turns on the stress response mechanism. It activates the organs that produce the responses that make up the general adaptation syndrome. But just how stress—a state of the body—does this is not clear. Selye could only guess about it and his terms were vague. For instance, he postulated

that by some unknown "alarm signals," stress "acted through the floor of the brain"* (presumably the hypothalamus) to stimulate the sympathetic nervous system and the pituitary gland. In Fig. 27-3, you can see Selye's hypothesis in diagram form.

Stress and disease

Stress, as we have observed several times, produces different results in different individuals and in the same individual at different times. In one person, a certain amount of stress may

*Selye, H.: The stress of life, New York, 1956, McGraw-Hill Book Co., p. 114.

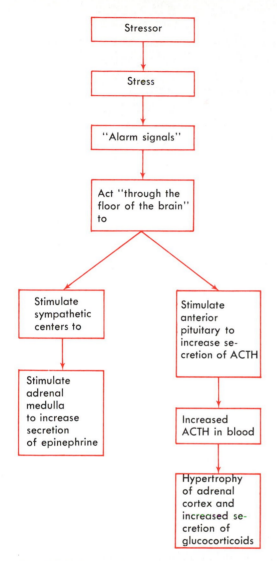

Fig. 27-3 Selye's hypothesis about activation of the stress mechanism.

many stressors that assail it. He held that the state of stress alerts physiological mechanisms to meet the challenge imposed by stressors. But a challenge issued does not necessarily mean a challenge successfully met. Selye proposed that sometimes the body's adaptive mechanisms fail to meet the challenge issued by stressors and that, when they fail, disease results—diseases of adaptation, he called them. He and his co-workers succeeded in inducing a variety of pathological changes by exposing sensitized animals to intense stimuli of various kinds. Their experimental animals developed a number of different diseases, among them hypertension, arthritis, arteriosclerosis, nephrosclerosis, and gastrointestinal ulcers. As a result of these experiments, many researchers have attempted to establish stress as a cause of disease in man. Their findings, however, have not been widely accepted as convincing evidence that stress, induced in a normal individual, can produce disease.

Around the middle of this century, one of the problems studied was the relationship of blood glucocorticoid concentration to disease. If stress is adaptive, the investigators reasoned, and helps the body combat the effects of many kinds of stressors (for example, infection, injury, burns, and the like), then possibly various diseases might be treated by adding to the body's natural output of glucocorticoids. In 1949, Hench and his colleagues* at the Mayo Clinic cautiously suggested that it might be helpful to give glucocorticoids to patients afflicted with various illnesses. Subsequently, they acted on their own suggestion and reported finding that a hormone of the adrenal cortex improved some of the clinical symptoms of rheumatoid arthritis. Since that time, cortisone has been administered to thousands of patients with widely different ailments. So often has it been used and so numer-

induce responses that maintain or even enhance his health. But in another person that same amount of stress appears to "make him sick." Whether stress is "good" or "bad" for you seems to depend more upon your own body's responses to it than upon the severity of the stressors inducing it. You may recall that Selye emphasized the adaptive nature of stress responses. He coined the term *general adaptation syndrome* because he believed that stress responses usually enable the body to adapt successfully to the

*Hench, P. S., Kendall, E. C., Slocumb, C. H., and Polley, H. F.: The effect of a hormone of the adrenal cortex (17-hydroxy-11-dehydrocorticosterone; compound E) and of pituitary adrenocorticotropic hormone on rheumatoid arthritis; preliminary report, Proc. Staff Meet. Mayo Clin. **24:**181-197, Apr., 1949.

ous have been the articles written about cortisone that today it would be almost impossible to find an adult who has never heard of this hormone.

Some current concepts about stress

Definitions

Some of today's physiologists use the terms stressors and stress as Selye did—that is, they define stressors as stimuli that produce stress, a state or condition of the body. In contrast, many physiologists now use the word stress to mean something much more specific than a state of the body. *Stress*, according to their operational definition is any stimulus that directly or indirectly stimulates neurons of the hypothalamus to release corticotropin-releasing hormone (CRH). CRH acts as a trigger that initiates many diverse changes in the body. Together, these changes constitute a syndrome now commonly called the *stress syndrome* or simply the *stress response*.

The stress syndrome

Look now at Fig. 27-4. It summarizes some major ideas currently held about the syndrome of stress responses. Starting at the top of the diagram, note that the initiator of the stress syndrome is stress—any factor that stimulates the hypothalamus to release CRH. Most often, stress consists of injurious or extreme stimuli. These may act directly on the hypothalamus to stimulate it. Instead, or in addition, they may act indirectly on the hypothalamus. An example of stress that stimulates the hypothalamus directly is hypoglycemia. A lower than normal concentration of glucose in the blood circulating to the hypothalamus stimulates it to release CRH. Indirect stimulation of the hypothalamus occurs

in this way: Stress stimulates the limbic lobe—the so-called "emotional brain"—and other parts of the cerebral cortex and these regions then send stimulating impulses to the hypothalamus causing it to release CRH. In addition to releasing CRH, note in Fig. 27-4 that the stress-stimulated hypothalamus sends stimulating impulses to sympathetic centers and to the posterior pituitary gland.

CRH stimulates the anterior pituitary gland to secrete increased amounts of ACTH. ACTH stimulates the adrenal cortex to secrete greatly increased amounts of cortisol and more moderately increased amounts of aldosterone. These two hormones induce various stress responses. Some important ones worth remembering are listed in Fig. 27-4 under cortisol effects and aldosterone effects.

Stimulation of sympathetic centers by impulses from the stress-stimulated hypothalamus leads to many stress responses, those known collectively as the fight-or-flight reaction. They include such important changes as an increase in the rate and strength of the heart beat, a rise in the blood pressure, and hyperglycemia. Other sympathetic stress responses are pallor and coolness of the skin, sweaty palms, and dry mouth.

Water-retention and an increase in blood volume are common stress responses. They stem, as you can see in Fig. 27-4, from both increased ADH and increased aldosterone secretion.

Indicators of stress

Whether or not an individual's body is responding to stress stimuli can be determined by certain measurements and observations. Examples are the following: an increase in the rate and force of the heart beat, a rise in systolic blood pressure, an increase in blood and urine concentration of epinephrine and norepinephrine, sweating of the palms of the hands, and dilatation of the pupils. The heart rate has been shown to increase in response to such varied

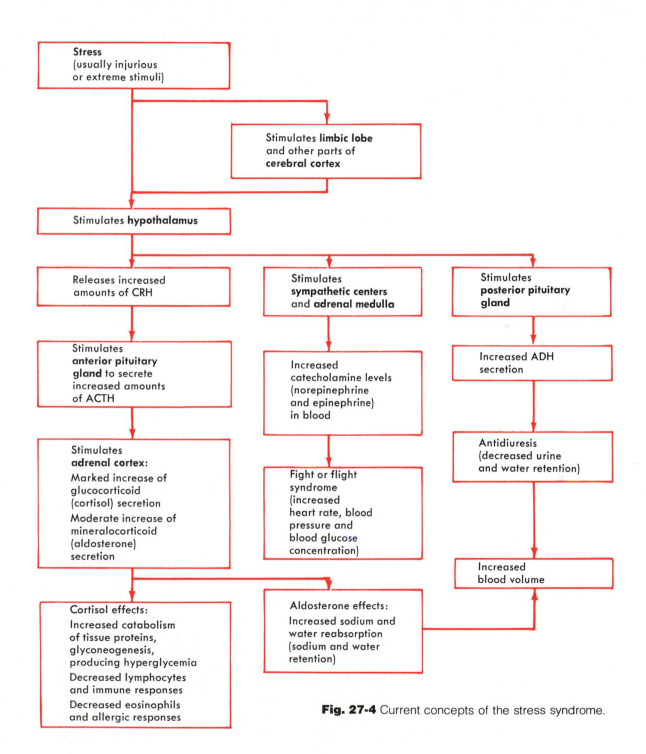

Fig. 27-4 Current concepts of the stress syndrome.

stress stimuli as anaesthesia, annoying sounds, and the entrance of an attractive nurse into a patient's room. Even anticipation by patients in a coronary care unit of their transferal to a less closely supervised convalescent unit has been identified as a stress stimulus that causes the heart rate to speed up. A decrease in the number of eosinophils and lymphocytes in the blood indicates that the individual is responding to stress stimuli. Soldiers stressed by prolonged marching, for example, have been found to have fewer circulating eosinophils than normal. This same stress indicator has been observed in college oarsmen when they were anticipating performing in an exhibition and in heart patients when they were anticipating transfer out of a coronary care unit to a convalescent unit.

The amount of urinary adrenocorticoids is often used as a measure of stress. It has been found to increase in depressed persons who feel hopeless and doomed, in test pilots, and in college students taking examinations or attending exciting movies. In contrast, urinary corticoids were found to drop markedly in persons watching unexciting nature study films.

The level of adrenocorticoids in the blood plasma of disturbed patients having acute psychotic episodes has been found to be 70% higher than that in normal individuals or in calm patients. Another study showed that the plasma corticoid levels of chronically depressed patients were significantly lower than those of acutely anxious patients. Smoking and exposure to nicotine have also been shown to be stressors that caused a marked rise in plasma adrenocorticoids—by as much as 77% in both human beings and experimental animals.

Corticoids and resistance to stress

Selye thought that the increase in corticoids that occurred in his stressed animals enabled them to adapt to and to resist stress. Today many physiologists doubt this. No one questions that adrenocortical hormones increase during stress. That fact has been clearly established. But what many do question is how essential this increase is for resisting stress. No one has proved by an unequivocal experiment that a higher than normal blood level of corticoids increases an animal's ability to adapt to stress. Some clinical evidence, however, seems to indicate that it does increase man's coping ability. For instance, patients who have been taking cortisol for some time are known to require increased doses of this hormone in order to successfully resist stresses such as surgery or severe injury.

Psychological stress

Stress as defined by Selye is physiological stress, that is, a state of the body. Psychological stress, in contrast, might be defined as a state of the mind. It is caused by psychological stressors and manifested by a syndrome. By definition, a *psychological stressor* is anything that an individual perceives as a threat—a threat either to his survival or to his self-image. Moreover, the threat does not need to be real—it needs only to be real to the individual. He must see it as a threat although in truth it may not be. Psychological stressors produce a syndrome of subjective and objective responses. Dominant among the subjective reactions is a feeling of anxiety. Other emotional reactions such as anger, hate, depression, fear, and guilt are also common subjective responses to psychological stressors. Some characteristic objective responses are restlessness, fidgeting, criticizing, quarreling, lying, and crying. Another objective indicator of psychological stress, discovered only a few years ago, is that the concentration of lactate in the blood increases. For an interesting account of this, read "The biochemistry of anxiety" by F. N. Pitts, Jr., in the February, 1969, issue of *Scientific American.*

Does psychological stress relate to physiological stress? The answer is clearly "yes." Physiological stress usually is accompanied by some degree of psychological stress. And, conversely, in most people, psychological stress produces some physiological stress responses. Ancient peoples intuitively recognized this fact. For example, it is said that in ancient times when the Chinese suspected a person of lying, they made him chew rice powder and then spit it out. If the powder came out dry, not moistened by saliva, they judged the suspect guilty. They seemed to know that lying makes a person "nervous"—and that nervousness makes a person's mouth dry. We moderns also know these facts but we describe them with more technical language. Lying, we might say, induces psychological stress and psychological stress acts in some way to cause the physiological stress response of decreased salivation.

Within the last few years, a new scientific discipline called psychophysiology has come into being. Psychophysiologists, using accepted research methods and sophisticated instruments—including polygraphs designed especially for this type of research—have investigated a wide variety of physiological responses made by individuals being subjected to psychological stressors. Their findings amply confirm the principle that psychological stressors often produce physiological stress responses. They have found, too, that identical psychological stressors do not necessarily induce identical physiological responses in different individuals. In the words of one experimenter, "for one person the cardiovascular system may quite regularly mirror emotion most sensitively, whereas another person may be primarily a pulmonary reactor."[*] Another of their interesting discoveries is that some organ systems become less responsive after they have been stimulated a number of times.

Summarizing, here are some principles to remember about psychological stress.

1 Physiological stress almost always is accompanied by some degree of psychological stress.
2 In most people, psychological stress leads to some physiological stress responses. Many of these are measurable autonomic responses, for example, accelerated heart rate, and increased systolic blood pressure.
3 Identical psychological stressors do not always induce identical physiological responses in different individuals.
4 In any one individual, certain autonomic responses are better indicators of psychological stress than others.

[*]Smith, B. M.: The polygraph, Sci. Am. **216**:25-31, Jan., 1967.

Outline summary

Selye's concept of stress
Development of concept

Through many experiments in which Selye exposed animals to wide variety of noxious agents and found that they all responded with the same syndrome of changes

Definitions

A Stress—a state or condition of the body produced by "diverse nocuous agents" and manifested by a syndrome of changes
B Stressors—the agents that produce stress
C General adaptation syndrome—the group of changes that manifest the presence of stress

Stressors

A Stressors are extreme stimuli—too much or too little of almost anything
B Stressors are very often injurious or painful stimuli
C Anything an individual perceives as a threat is a stressor for that individual
D Stressors are different for different individuals and for one individual at different times

The general adaptation syndrome

A Manifestations—stress triad (hypertrophied adrenals, atrophied thymus and lymph nodes, and bleeding ulcers) and many other changes
B Stages—three successive phases, namely, alarm reaction, stage of adaptation or resistance, and stage of exhaustion, each characterized by different syndrome of changes (Table 27-1)

Mechanism of stress

A Consists of group (syndrome) of responses to internal condition of stress; stress responses nonspecific in that same syndrome of responses occurs regardless of kind of extreme change that produced stress
B Stimulus that produces stress and thereby activates stress mechanism is nonspecific in that it can be any kind of extreme change in environment
C Stress responses, Selye thought, were adaptive; tend to enable body to adapt to and survive extreme change; Selye referred to this syndrome of stress responses as general adaptation syndrome
D Numerous factors influence stress responses—individual's physical and mental condition, age, sex, socioeconomic status, heredity, previous experi-

ence with similar stressor, and even religious affiliation
E Stress most often results successfully—i.e., results in adaptation, healthy survival, and increased resistance; sometimes, however, stress produces exhaustion and death
F See Figs. 27-3, then 27-1, and 27-2 for a summary of Selye's stress mechanism

Stress and disease

A Selye held that stress could result in disease instead of adaptation
B To date, research evidence that stress, induced in normal individual, can produce disease has not gained wide acceptance

Some current concepts about stress
Definitions

A Stress—any stimulus that directly or indirectly stimulates hypothalamus to release CRH
B Stress syndrome—also called the stress response, many diverse changes initiated by stress

The stress syndrome

See Fig. 27-4

Indicators of stress

A Changes due to increased sympathetic activity; e.g., faster, stronger heartbeat, higher blood pressure, sweaty palms, dilated pupils
B Changes due to increased corticoids—eosinopenia, lymphocytopenia, increased adrenocorticoids in blood and urine

Corticoids and resistance to stress

Still controversial; not proved that increased corticoids increase animal's or man's ability to resist stress but is proved that corticoid levels in blood increase in stress

Psychological stress

A Psychological stressors—anything that an individual perceives as a threat to either his survival or self-image
B Psychological stress—a mental state characterized by a syndrome of subjective and objective responses; dominant subjective response is anxiety; some characteristic objective responses are restlessness, quarrelsomeness, lying, crying
C Relation to physiological stress—see summary of principles on p. 673

Review questions

1 Describe the experimental results that led Selye to conceive his idea of stress.
2 Define the terms stress, stressor, and general adaptation syndrome as Selye used them.
3 Make a few generalizatons about the kinds of stimuli that constitute stressors.
4 According to Selye, what three stages make up the general adaptation syndrome? What changes characterize each stage?
5 What changes constituted Selye's "stress triad"?
6 According to Selye, an increase in what three hormones brought about the changes that he named the general adaptation syndrome?
7 What function, according to Selye, does the general adaptation syndrome serve?
8 What relation, if any, did Selye suggest that exists between stress and disease?
9 Stress, according to Selye, is a state or condition of the body. According to a current operational definition, what is stress?
10 What part of the brain, according to current ideas, plays the key role in initiating the stress syndrome of responses?
11 What parts of the nervous system other than that named in question 10 are involved in inducing the stress syndrome?
12 Briefly, what role, if any do each of the following hormones play when the body is subjected to stress: ACTH, ADH, aldosterone, cortisol, CRH, epinephrine, and norepinephrine?
13 Describe some changes that you might observe that would indicate that an individual was being subjected to stress.
14 What has been proved about corticoids in relation to stress?
15 What is the controversial issue about corticoids in relation to stress?
16 Define psychological stressor, psychological stress.
17 Cite several examples of psychological stressors.
18 Are psychological stressors the same for all individuals?
19 Give some examples of subjective indicators of psychological stress.
20 Give some examples of objective responses that are part of the syndrome of psychological stress.
21 State three or four principles about the relationship between psychological and physiological stress.

Supplementary readings

1 Organization of the body

Anthony, C. P.: Basic concepts in anatomy and physiology: a programmed presentation, ed. 3, St. Louis, 1974, The C. V. Mosby Co.

Cannon, W. B.: The wisdom of the body, ed. 2, New York, 1963, W. W. Norton & Co., Inc.

Laughlin, A.: Roe's principles of chemistry, ed. 12, St. Louis, 1976, The C. V. Mosby Co.

2 Cells

Anthony, C. P.: Basic concepts in anatomy and physiology: a programmed presentation, ed. 3, St. Louis, 1974, The C. V. Mosby Co.

Capaldi, R. A.: A dynamic model of cell membranes, Sci. Am. **230:**27-33, Mar., 1974.

Engelman, D. M., and Moore, P. B.: Neutron-scattering studies of the ribosome, Sci. Am. **230:**44-53, Oct., 1976.

Everhart, R. E., and Hayes, T. L.: The scanning electron microscope, Sci. Am. **226:**55-69, Jan., 1972.

Hayflick, L.: Human cells and aging, Sci. Am. **218:**32-37, Mar., 1968.

Howland, J. L.: Cell physiology, New York, 1973, Macmillan Publishing Co., Inc.

Luria, S. E.: Colicins and the energetics of cell membranes, Sci. Am. **233:**30-37, Dec., 1975.

Miller, O. L., Jr.: The visualization of genes in action, Sci. Am. **228:**34-41, Mar., 1973.

Satir, B.: The final steps in secretion, Sci. Am. **233:**29-37, Oct., 1975.

3 Tissues
4 Membranes and glands

Di Fiore, M. S. H.: An atlas of human histology, ed. 4, Philadelphia, 1974, Lea & Febiger.

Roberts, S. L.: Skin assessment for color and temperature, Am. J. Nurs. **75:**610-613, Apr., 1975.

5 The skeletal system—bones and cartilage

Hogan, L., and Beland, I.: Cervical spine syndrome, Am. J. Nurs. **76:**1104-1107, July, 1976.

6 The skeletal system—articulations

Sonstegard, D. A., Matthews, L. S., and Kaufer, H.: The surgical replacement of the human knee joint, Sci. Am. **238:**44-51, Jan., 1978.

7 Skeletal muscles

Carlson, R. D., and Wilkie, D. R.: Muscle physiology, Biological Science Series, Englewood Cliffs, N.J., 1974, Prentice-Hall, Inc.

Cohen, C.: The protein switch of muscle contraction, Sci. Am. **233:**36-45, Nov., 1975.

Hoyle, G.: How is muscle turned on and off? Sci. Am. **222:**84-93, Apr., 1970.

Komi, P. V.: Biomechanics V, vol. 1 and 2, Baltimore, 1976, University Park Press.

Lester, H. A.: The response to acetylcholine, Sci. Am. **236:**106-118, Feb., 1977.

Margaria, R.: The sources of muscular energy, Sci. Am. **226:**83-91, Mar., 1972.

Morehouse, L. E., and Miller, A. T., Jr.: Physiology of exercise, ed. 7, St. Louis, 1976, The C. V. Mosby Co.

Murray, J. M., and Weber, A.: The cooperative action of muscle proteins, Sci. Am. **230:**59-71, Feb., 1974.

Yagi, N., et al.: Return of myosin heads to thick filaments after muscle contraction. Science **197:**4304, Aug., 1977.

8 Nervous system cells

Eccles, J.: The synapse, Sci. Am. **212:**56-66, Jan., 1965.

Guyton, A. C.: Textbook of medical physiology, ed. 5, Philadelphia, 1976, W. B. Saunders Co., pp. 112-128.

Hyden, H.: Satellite cells in the nervous system, Sci. Am. **205:**62-70, Dec., 1961.

Willis, W. D., Jr., and Grossman, R. G.: Medical neurobiology, St. Louis, 1973, The C. V. Mosby Co., pp. 1-36.

9 The somatic nervous system

Kimura, D.: The asymmetry of the human brain, Sci. Am. **228:**70-77, Mar., 1973.

Lester, H. A.: The response to acetylcholine, Sci. Am. **236:**107-118, Feb., 1977.

Llinas, R. R.: The cortex of the cerebellum, Sci. Am. **232:**56-71, Jan., 1975.

Nathanson, J. A., and Greengard, P.: "Second messengers" in the brain, Sci. Am. **237:**108-119, Aug., 1977.

Ornstein, R. E.: The psychology of consciousness, San Francisco, 1972, W. H. Freeman & Co. Publishers.

Pappenheimer, J. R.: The sleep factor, Sci. Am. **235:**24-29, Aug., 1976.

Rosenzweig, M. R., Bennett, E. L., and Diamond, M. C.: Brain changes in response to experience, Sci. Am. **226:**22-29, Feb., 1972.

Snyder, S. H.: Opiate receptors and internal opiates, Sci. Am. **237:**44-56, Mar., 1977.

Wallace, R. K., and Benson, H.: The physiology of meditation, Sci. Am. **226:**85-90, Feb., 1972.

10   The autonomic nervous system

Benson, H.: The relaxation response, New York, 1975, William Morrow & Co., Inc.

Dicara, L. V.: Learning in the autonomic nervous system, Sci. Am. **222:**30-39, Jan., 1970.

Mountcastle, V. B., editor: Medical physiology, ed. 13, St. Louis, 1974, The C. V. Mosby Co., pp. 783-834.

Sterman, L. T.: Clinical biofeedback, Am. J. Nurs. **75:**2006-2012, Nov., 1975.

Trotter, R. J.: Transcendental meditation, Sci. News **104:**376-378, Dec., 1973.

Wallace, R. K., and Benson, H.: The physiology of meditation, Sci. Am. **226:**85-90, Feb., 1972.

11   Sense organs

Fernsebner, W.: Early diagnosis of acute angle-closure glaucoma, Am. J. Nurs. **75:**1154-1156, July, 1975.

Land, E. H.: The retinex theory of color vision, Sci. Am. **237:**108-128, Dec., 1977.

Pettigrew, J. D.: The neurophysiology of binocular vision, Sci. Am. **227:**84-95, Aug., 1972.

van Heyningen, R.: What happens to the human lens in cataract, Sci. Am. **233:**70-81, Dec., 1975.

Young, R. W.: Visual cells. Sci. Am. **223:**81-91, Oct., 1970.

12   The endocrine system

Binkley, S., Riebman, J. B., and Reilly, K. B.: Time keeping by the pineal gland, Science **197:**4309, Sept., 1977.

Gardener, L. I.: Deprivation dwarfism, Sci. Am. **227:**76-82, July, 1972.

Guillemin, R., and Burgus, R.: The hormones of the hypothalamus, Sci. Am. **227:**24-32, Nov., 1972.

Langs, D. A., Erman, M., and DeTitta, G. T.: Conformations of prostaglandin F_2 and recognition of prostaglandins by their receptors, Science **197:**4307, Sept., 1977.

McEwen, B. S.: Interactions between hormones and nervous tissue, Sci. Am. **235:**48-67, July, 1976.

McKusick, V. A., and Rimoin, D. L.: General Tom Thumb and other midgets, Sci. Am., **217:**102-106, July, 1967.

Mountcastle, V. B., editor: Medical physiology, ed. 13, St. Louis, 1974, The C. V. Mosby Co., pp. 1601-1802.

Pastan, I.: Cyclic AMP, Sci. Am. **227:**97-105, Aug., 1972.

Peart, W. S.: Renin-angiotensin system, N. Engl. J. Med. **292:**302, 1975.

Pike, J. E.: Prostaglandins, Sci. Am. **225:**84-91, Nov., 1971.

Ramwell, P. W., and Shaw, J. E., editor: Prostaglandins, Ann. N.Y. Acad. Sci. **180,** 1971.

Sterling, K., et al.: Thyroid hormone action: the mitochondrial pathway, Science **197:**4307, Sept., 1977.

13   Blood

Adolph, E. F.: The heart's pacemakers, Sci. Am. **213:**32-37, March, 1967.

Anthony, C. P.: Basic concepts in anatomy and physiology: a programmed presentation, ed. 3, St. Louis, 1974, The C. V. Mosby Co., pp. 108-144.

Bennett, B., et al.: The normal coagulation mechanism, Med. Clin. North Am. **59:**95, 1972.

Capra, J. D., and Edmundson, A. B.: The antibody combining site, Sci. Am. **236:**50-59, Jan., 1977.

Clarke, C. A.: The prevention of "Rhesus" babies, Sci. Am. **219:**46-52, Nov., 1968.

Conover, M. H., and Zalis, E. G.: Understanding electrocardiography: physiological and interpretive concepts, ed. 2, St. Louis, 1976, The C. V. Mosby Co.

Cooper, M. D., and Lawton, A. R., III: The development of the immune system, Sci. Am. **231:**58-72, Nov., 1974.

Effler, D. B.: Surgery for coronary disease, Sci. Am. **219:**36-43, Oct., 1968.

Hamilton, W. J., editor: Textbook of human anatomy, ed. 2, St. Louis, 1976, The C. V. Mosby Co., pp. 9-16.

Ingram, M., and Preston, K., Jr.: Automatic analysis of blood cells, Sci. Am. **223:**72-82, Nov., 1970.

Lerner, R. A., and Dixon, F. J.: The human lymphocyte in an experimental animal, Sci. Am. **228:**82-91, June, 1973.

Mountcastle, V. B., editor: Medical physiology, ed. 13, St. Louis, 1974, The C. V. Mosby Co., pp. 1027-1037.

Porter, R. R.: The structure of antibodies, Sci. Am. **217:**81-87, Oct., 1967.

Wood, E. J.: The venous system, Sci. Am. **218:**86-96, Jan., 1968.

14   Anatomy of the cardiovascular system

Benditt, E. P.: The origin of atherosclerosis, Sci. Am. **236:**74-85, Feb., 1977.

Coodley, E. L.: Anatomy of circulation, Consultant, May, 1972.

Folkman, J.: The vascularization of tumors, Sci. Am. **234:**58-73, May, 1976.

Hamilton, W. J., editor: Textbook of human anatomy, ed. 2, St. Louis, 1976, The C. V. Mosby Co., pp. 201-279.

Hurst, J. W., editor-in-chief: The heart: arteries and veins, ed. 3, New York, 1974, McGraw-Hill Book Co.

Warwick, R., and Williams, P. L., editors: Gray's anatomy ed. 35 (British), Philadelphia, 1973, W. B. Saunders, Co., pp. 588-712.

Wood, E. J.: The venous system, Sci. Am. **218:**86-96, Jan., 1968.

15   Physiology of the cardiovascular system

Adolph, E. F.: The heart's pacemakers, Sci. Am. **213:**32-37, Mar., 1967.

Berne, R. M. and Levy, M. N.: Cardiovascular physiology, ed. 3, St. Louis, 1977, The C. V. Mosby Co.

Conover, M. H., and Zalis, E. G.: Understanding electrocardiography: physiological and interpretive concepts, ed. 2, St. Louis, 1976, The C. V. Mosby Co.

Guyton, A. C., Jones, C. E., and Coleman, T. C.: Circulatory physiology: cardiac output and its regulation, Philadelphia, 1973, W. B. Saunders Co.

Mountcastle, V. B., editor: Medical physiology, ed. 13, St. Louis, 1974, The C. V. Mosby Co., pp. 839-1026.

Noble, D.: The initiation of the heartbeat, New York, 1975, Oxford University Press, Inc.

Smith, O.: Reflex and central mechanisms involved in the control of the heart and circulation, Annu. Rev. Physiol. **36:**93, 1974.

16   The lymphatic system

Hamilton, W. J., editor: Textbook of human anatomy, ed. 2, St. Louis, 1976, The C. V. Mosby Co., pp. 279-295.

Kinmonth, J. B.: Some general aspects of the investigation and surgery of the lymphatic system, J. Cardiovasc. Surg. (Torino) **5:**680-2, 1964.

Mayerson, H.: The lymphatic circulation, Sci. Am. **208:**80-93, June, 1963.

Rusznyak, I., Foldi, M., and Sazbo, G.: Lymphatics and lymph circulation, Elmsford, N.Y., 1960, Pergamon Press, Inc.

Warwick, R., and Williams, P. L., editors: Gray's anatomy, ed. 35 (British), Philadelphia, 1973, W. B. Saunders Co., pp. 42-52.

17   The respiratory system

Avery, M. E., Wang, N., and Taeusch, H. W.: The lung of the newborn infant, Sci. Am. **228:**74-85, Apr., 1973.

Cherniack, R.: Respiration in health and disease, Philadelphia, 1972, W. B. Saunders Co.

Comroe, J. H., Jr.: The lung, Sci. Am. **214:**57-66, Feb., 1966.

Comroe, J. H., Jr.: Physiology of respiration, Chicago, 1965, Year Book Medical Publishers, Inc.

Clements, J. A.: Surface tension in the lungs, Sci. Am. **207:**120-134, Dec., 1962.

Egan, D. F.: Fundamentals of respiratory therapy, ed. 3, St. Louis, 1977, The C. V. Mosby Co.

Mountcastle, V. B., editor: Medical physiology, ed. 13, St. Louis, 1974, The C. V. Mosby Co., pp. 1361-1597.

Slonim, N. B., and Hamilton, L. H.: Respiratory physiology, ed. 3, St. Louis, 1976, The C. V. Mosby Co.

18   The digestive system

Davenport, H. W.: Why the stomach does not digest itself, Sci. Am. **225:**87-93, Jan., 1972.

Dudrick, S. J., and Rhoades, J. E.: Total intravenous feeding, Sci. Am. **226:**73-79, May, 1972.

Kretchmer, N.: Lactose and lactase, Sci. Am. **227:**70-78, Oct., 1972.

Programmed instruction: potassium imbalance, Am. J. Nurs. **67:**343-366, Feb., 1967.

19   Metabolism

Brady, R. O.: Hereditary fat-metabolism diseases, Sci. Am. **229:**88-97, Aug., 1973.

Morehouse, L. E., and Miller, A. T., Jr.: Physiology of exercise, ed. 7, St. Louis 1976, The C. V. Mosby Co.

Young, V. R., and Schrimshaw, N. S.: The physiology of starvation, Sci. Am. **225:**14-21, Oct., 1971.

20   The urinary system

Anthony, C. P.: Basic concepts in anatomy and physiology: a programmed presentation, ed. 3, St. Louis, 1974, The C. V. Mosby Co.

Melber, S., Leonard, M., and Primack, W.: Hemodialysis at camp, Am. J. Nurs. **76:**938-940, June, 1976.

Mountcastle, V. B., editor: Medical physiology, ed. 13, St. Louis, 1974, The C. V. Mosby Co., pp. 1065-1102.

Schumann, B.: The renal donor, Am. J. Nurs. **74:**105-110, Jan., 1974.

Wolf, Z. R.: What patients awaiting kidney transplant want to know, Am. J. Nurs. **76:**92-94, Jan., 1976.

21 Fluid and electrolyte balance

Anthony, C. P.: Basic concepts in anatomy and physiology: a programmed presentation, ed. 3, St. Louis, 1974, The C. V. Mosby Co.

Deetjen, P., Boylan, J. W., and Kramer, K.: Physiology of the kidney and of water balance, New York, 1975, Springer-Verlag New York, Inc.

Gamble, J. L.: Chemical anatomy, physiology, and pathology of extracellular fluid, ed. 6, Cambridge, Mass., 1958, Harvard University Press.

Guyton, A. C., Taylor, A. E., and Granger, H. J.: Circulatory physiology II. Dynamics and control of the body fluids, Philadelphia, 1975, W. B. Saunders Co.

Mountcastle, V. B., editor: Medical physiology, ed. 13, St. Louis, 1974, The C. V. Mosby Co., pp. 1049-1064.

Shore, L., and Claybough, J. R.: Regulation of body fluids, Annu. Rev. Physiol. **34:**235, 1972.

Voda, A. M.: Body water dynamics, Am. J. Nurs. **70:**2594-2601, Dec., 1970.

22 Acid-base balance

Anthony, C. P.: Basic concepts in anatomy and physiology: a programmed presentation, ed. 3, St. Louis, 1974, The C. V. Mosby Co.

Burke, S. R.: The composition and function of body fluids, ed. 2, St. Louis, 1976, The C. V. Mosby Co.

Davenport, H. W.: The ABC of acid-base chemistry, ed. 6, Chicago, 1973, University of Chicago Press.

Pitts, R. F.: Renal regulation of acid-base balance, 1974, Year Book Medical Publishers, Inc.

Vander, A. J.: Renal physiology, New York, 1975, McGraw-Hill Book Co., pp. 96-163.

Weldy, N. J.: Body fluids and electrolytes: a programmed presentation, ed. 2, St. Louis, 1976, The C. V. Mosby Co.

23 Reproduction of cells

Chinn, P. L.: Child health maintenance: Concepts in family centered care, St. Louis, 1974, The C. V. Mosby Co., pp. 63-67.

Cohen, S. N.: The manipulation of genes, Sci. Am. **233:**24-33, July, 1975.

Cohen, S. N.: Recominant DNA: fact and fiction, Science **195:**654, 1977.

Jukes, T. H.: How many anticodons? Science **198:**4314, Oct., 1977.

Kornberg, A.: The synthesis of DNA, Sci. Am. **214:**64-78, Oct., 1968.

Levine, L.: Biology of the gene, ed. 2, St. Louis, 1973, The C. V. Mosby Co., pp. 46-57.

Mazia, D.: The cell cycle, Sci. Am. **231:**55-63, Jan., 1974.

Stern, C.: Principles of human genetics, ed. 3, San Francisco, 1973, W. H. Freeman & Co. Publishers.

24 The male reproductive system

Duckett-Racey, J. D., editor: The biology of the male gamete, New York, 1975, Academic Press, Inc.

Epel, D.: The program of fertilization, Sci. Am. **237:**128-139, Nov., 1977.

Mountcastle, V. B., editor: Medical physiology, ed. 13, St. Louis, 1974, The C. V. Mosby Co., pp. 1763-1770.

25 The female reproductive system

Beer, A. E., and Billingham, R. E.: The embryo as a transplant, Sci. Am. **230:**36-51, Apr., 1974.

Edwards, R. G., and Fowler, R. E.: Human embryos in the laboratory, Sci. Am. **222:**45-54, Dec., 1970.

Epel, D.: The program of fertilization, Sci. Am. **237:**128-139, Nov., 1977.

Goldberg, V. J., and Ramwell, P. W.: Role of prostaglandins in reproduction, Physiol. Rev. **55:**325, 1975.

Mountcastle, V. B., editor: Medical physiology, ed. 13, St. Louis, 1974, The C. V. Mosby Co., pp. 1741-1762.

26 The immune system

Burke, D. C.: The status of interferon, Sci. Am. **236:**42-50, Apr., 1977.

Burnet, M.: The mechanism of immunity, Sci. Am. **204:**58-67, Jan., 1961.

Capra, J. D., and Edmundson, A. B.: The antibody combining site, Sci. Am. **236:**50-59, Jan., 1977.

Cooper, M. D., and Lawton, A. R., III: The development of the immune system, Sci. Am. **231:**59-72, Nov., 1974.

Cunningham, B. A.: The structure and function of histocompatibility antigens, Sci. Am. **237:**96-107, Oct., 1977.

Dharan, M.: Immunoglobulin abnormalities, Am. J. Nurs. **76:**1626-1628, Oct., 1976.

Donley, D. L.: Nursing the patient who is immunosuppressed, Am. J. Nurs. **76:**1619-1625, Oct., 1976.

Edelman, G. M.: The structure and function of antibodies, Sci. Am. **223:**34-42, Aug., 1970.

Glasser, R. J.: The body is the hero, New York, 1976, Random House, Inc.

Jerne, N. K.: The immune system, Sci. Am. **229:**52-60, Nov., 1973.

Mayer, M. M.: The complement system, Sci. Am. **229:**54-66, Nov., 1973.

Nysather, J. O., Katz, A. E., and Lenth, J. L.: The immune system, its development and functions, Am. J. Nurs. **76:**1614-1616, Oct., 1976.

Old, L. J.: Cancer immunology, Sci. Am. **236:**62-79, May, 1977.

Raff, M. C.: Cell-surface immunology, Sci. Am. **234:**30-39, May, 1976.

27 Stress

Levine, S.: Stress and behavior, Sci. Am. **224:**26-31, Jan., 1971.

Marcinek, M. B.: Stress in the surgical patient, Am. J. Nurs. **77:**1809-1811, Nov., 1977.

Mountcastle, V. B., editor: Medical physiology, ed. 13, St. Louis, 1974, The C. V. Mosby Co., pp. 832-834, 1696-1698, 1700-1703.

Pitts, F. N., Jr.: The biochemistry of anxiety, Sci. Am. **220:** 69-75, Feb., 1969.

Selye, H.: The stress of life, New York, 1956, McGraw-Hill Book Co.

Smith, B. M.: The polygraph, Sci. Am. **216:**25-31, Jan., 1967.

Stephenson, C. A.: Stress in critically ill patients, Am. J. Nurs. **77:**1806-1809, Nov., 1977.

Trotter, R. J.: The biological depths of loneliness, Sci. News **103:**140-142, Mar., 1973.

Weiss, J. M.: The psychological factors in stress and disease, Sci. Am. **226:**104-113, June, 1972.

Additional references
Biochemistry

Leininger, A. L.: Biochemistry, New York, 1975, Worth Publishers, Inc.

Orten, J. M., and Neuhaus, O. W.: Human biochemistry, ed. 9, St. Louis, 1975, The C. V. Mosby Co.

Stryer, L.: Biochemistry, 1975, San Francisco, W. H. Freeman & Co. Publishers.

Developmental anatomy

Arey, L. B.: Developmental anatomy—a textbook and laboratory manual of embryology, ed. 7, Philadelphia, 1965, W. B. Saunders Co.

Gross anatomy

Goss, C. M., editor: Gray's anatomy of the human body, ed. 29, Philadelphia, 1973, Lea & Febiger.

Sobotta, J., and Figge, F. H. J.: Atlas of human anatomy, ed. 8, New York, 1963, Hafner Publishing Co., Inc. (3 vols.)

Spalteholz, W.: Atlas of human anatomy, ed. 16 (revised and re-edited by R. Spanner), New York, 1967, F. A. Davis Co.

Microscopic anatomy

Di Fiore, M. S. H.: An atlas of human histology, ed. 4, Philadelphia, 1974, Lea & Febiger.

Physiology

Cannon, W. B.: The wisdom of the body, rev. ed., New York, 1963, W. W. Norton & Co. Inc.

Guyton, A. D.: Textbook of medical physiology, ed. 5, Philadelphia, 1976, W. B. Saunders Co.

Mountcastle, V. B., editor: Medical physiology, ed. 13, St. Louis, 1974, The C. V. Mosby Co.

Periodicals

American Journal of Nursing
American Journal of Physiology
Annual Review of Physiology
Journal of the American Medical Association
Science News
Scientific American

Abbreviations, prefixes, and suffixes

Abbreviations

Å Angstrom unit
ACh acetylcholine
AChE acetylcholinesterase
ACTH adrenocorticotropic hormone
ADH antidiuretic hormone
ADP adenosine diphosphate
ASC altered state of consciousness
ATP adenosine triphosphate
BMR basal metabolic rate
BNA Basle Nomina Anatomica (see Glossary)
C Celsius, centigrade
CA catecholamines
Cal large calorie
cm centimeter
CNS central nervous system
COMT catechol-*O*-methyl transferase
CP creatine phosphate
CRF corticotropin-releasing factor
CVA cardiovascular accident
DA dopamine
DNA deoxyribonucleic acid
DPN diphosphopyridine nucleotide
ECF extracellular fluid
EFP effective filtration pressure
EPSP excitatory postsynaptic potential
ER endoplasmic reticulum
ERV expiratory reserve volume
F Fahrenheit
FRC functional respiratory capacity
FSH follicle-stimulating hormone
GABA gamma aminobutyrate
GH growth hormone
Hb hemoglobin
HbO$_2$ oxyhemoglobin
HP hydrostatic pressure
5-HT 5-hydroxytryptamine (serotonin)
IC inspiratory capacity

ICF intracellular fluid
ICSH interstitial cell–stimulating hormone
IF interstitial or intercellular fluid
Ig immunoglobulin
IPSP inhibitory postsynaptic potential
IRV inspiratory reserve volume
kcal kilocalorie
kg kilogram
LH luteinizing hormone
LSD lysergic acid diethylamide
MAO monoamine oxidase
mEq milliequivalents
mg milligram
μ micron
ml milliliter
mm millimeter
mm Hg pres millimeter mercury pressure
MSH melanocyte-stimulating hormone
mv millivolt
MW molecular weight
NE norepinephrine
OP osmotic pressure
PAH para-aminohippuric acid
PBI protein-bound iodine
Pco$_2$ partial pressure of carbon dioxide
PG prostaglandin
pH hydrogen ion concentration; negative logarithm of hydrogen ion concentration
Po$_2$ partial pressure of oxygen
PNS peripheral nervous system
RBC, rbc red blood cells
REM rapid eye movements
RNA ribonucleic acid
RV residual volume
SD systolic discharge
SDA specific dynamic action
STH somatotropic hormone
SV stroke volume

SWS slow-wave sleep
TH thyrotropic hormone
TMR total metabolic rate
TSH thyroid-stimulating hormone
TV tidal volume
VC vital capacity
WBC, wbc white blood cells

Prefixes

a- without
ab- away from
ad- to, toward
adeno- glandular
amphi- on both sides
an- without
ante- before, forward
anti- against
bi- two, double, twice
circum- around, about
contra- opposite, against
de- away from, from
dys- difficult
ecto- outside
endo- in, within
ento- inside, within
epi- on
eu- well
ex- from out of, from
extra- outside, beyond, in addition
hemi- half
hyper- over, excessive, above

hypo- under, deficient
infra- underneath, below
inter- between, among
intra- within, on the side
iso- equal, like
para- beside, to side of
peri- round about, beyond
post- after, behind
pre- before, in front of
pro- before, in front of
retro- backward, back
semi- half
sub- under, beneath
super- above, over
supra- above, on upper side
syn- with, together
trans- across, beyond

Suffixes

-algia pain, painful
-blast young cell
-cele swelling
-cide killer
-cyte cell
-ectomy cutting out
-emia blood
-itis inflammation
-kinin motion (action)
-ology knowledge of
-oma tumor

Glossary

abdomen body area between the diaphragm and pelvis.

abduct to move away from the midline; opposite of adduct.

absorption passage of a substance through a membrane, for example, skin or mucosa, into blood.

acapnia marked decrease in blood carbon dioxide content.

acetabulum socket in the hip bone (os coxae or innominate bone) into which the head of the femur fits.

acetone bodies ketone bodies, acids formed during the first part of fat catabolism, namely, acetoacetic acid, betahydroxybutyric acid, and acetone.

Achilles tendon tendon inserted on calcaneus; so called because of the Greek myth that Achilles' mother held him by the heels when she dipped him in the river Styx, thereby making him invulnerable except in this area.

acid substance that ionizes in water to release hydrogen ions.

acidosis condition in which there is an excessive proportion of acid in the blood.

acromion bony projection of the scapula; forms point of the shoulder.

actin one of the contractile proteins in muscle.

adduct to move toward the midline; opposite of abduct.

adenine nitrogenous base; component of RNA and DNA.

adenohypophysis anterior pituitary gland.

adenoids glandlike; adenoids or pharyngeal tonsils are paired lymphoid structures in the nasopharynx.

adolescence period between puberty and adulthood.

adrenergic fibers axons whose terminals release norepinephrine and epinephrine.

adventitia, externa outer coat of a tube-shaped structure such as blood vessels.

aerobic requiring free oxygen; opposite of anaerobic.

afferent neuron transmitting impulses to the central nervous system.

albuminuria albumin in the urine.

aldosterone a hormone secreted by the adrenal cortex.

alkali reserve bicarbonate salts present in body fluids; mainly sodium bicarbonate.

alkalosis condition in which there is an excessive proportion of alkali in the blood; opposite of acidosis.

alveolus literally a small cavity; alveoli of lungs are microscopic saclike dilations of terminal bronchioles.

ameboid movement movement characteristic of amebae, that is, by projections of protoplasm (pseudopodia) toward which the rest of the cell's protoplasm flows.

amenorrhea absence of the menses.

amino acid organic compound having an NH_3 and a COOH group in its molecule; has both acid and basic properties; amino acids are the structural units from which proteins are built.

amphiarthrosis slightly movable joint.

ampulla saclike dilation of a tube or duct.

anabolism synthesis by cells of complex compounds, for example, protoplasm, hormones, from simpler compounds (amino acids, simple sugars, fats, minerals); opposite of catabolism, the other phase of metabolism.

anaerobic not requiring free oxygen; opposite of aerobic.

anaphase stages of mitosis; duplicate chromosomes move to poles of dividing cell.

anastomosis connection between vessels, for example, the circle of Willis is an anastomosis of certain cerebral arteries.

androgen male sex hormone.

anemia deficient number of red blood cells or deficient hemoglobin.

anesthesia loss of sensation.

aneurysm blood-filled saclike dilation of the wall of an artery.

angina any disease characterized by spasmodic suffocative attacks, for example, angina pectoris, paroxysmal thoracic pain with feeling of suffocation.

Angstrom unit 0.1 mμ or $^1/_{10}$ millionth of a meter or about $^1/_{250}$ millionth of an inch.

anion negatively charged particle.

anorexia loss of appetite.

anoxemia deficient blood oxygen content.

anoxia deficient oxygen supply to tissues.

antagonistic muscles those having opposite action, for example, muscles that flex the upper arm are antagonistic to muscles that extend it.

anterior front or ventral; opposite of posterior or dorsal.

antibody, immune body substance produced by the body that destroys or inactivates a specific substance (antigen) that has entered the body, for example, diphtheria antitoxin is the antibody against diphtheria toxin.

antigen substance that, when introduced into the body, causes formation of antibodies against it.

antrum cavity, for example, the antrum of Highmore, the space in each maxillary bone, or the maxillary sinus.

anus distal end or outlet of the rectum.

apex pointed end of a conical structure.

aphasia loss of a language faculty such as the ability to use words or to understand them.

apnea temporary cessation of breathing.

aponeurosis flat sheet of white fibrous tissue that serves as a muscle attachment.

aqueduct tube for conduction of liquid, for example, the cerebral aqueduct conducts cerebrospinal fluid from the third to the fourth ventricle.

arachnoid delicate, weblike middle membrane of the meninges.

areola small space; the pigmented ring around the nipple.

arteriole small branch of an artery.

arteriosclerosis hardening of the arteries.

artery vessel carrying blood away from the heart.

arthrosis joint or articulation.

articular referring to a joint.

articulation joint.

arytenoid ladle-shaped; two small cartilages of the larynx.

ascites accumulation of serous fluid in the abdominal cavity.

asphyxia loss of consciousness from deficient oxygen supply.

aspirate to remove by suction.

asthenia bodily weakness.

astrocytes star-shaped neuroglia, connective tissue cells in brain and cord.

ataxia loss of power of muscle coordination.

atherosclerosis arteriosclerosis or hardening of artery walls characterized by lipid deposits in tunica intima.

atrium chamber or cavity, for example, atrium of each side of the heart.

atrophy wasting away of tissue; decrease in size of a part.

auricle part of the ear attached to the side of the head; earlike appendage of each atrium of heart.

autonomic self-governing, independent.

axilla armpit.

axon nerve cell process that transmits impulses away from the cell body.

B

baroreceptor receptor stimulated by change in pressure.

Bartholin seventeenth century Danish anatomist.

basophil white blood cell that stains readily with basic dyes.

biceps two headed.

bilirubin red pigment in the bile.

biliverdin green pigment in the bile.

blastocyst hollow cell mass; stage in development of ovum following morula.

BNA (Basle Nomina Anatomica) anatomic terminology accepted at Basle by the Anatomical Society in 1895.

Bowman nineteenth century English physician.

brachial pertaining to the arm.

bronchiectasis dilation of the bronchi.

bronchiole small branch of a bronchus.

bronchus one of the two branches of the trachea.

buccal pertaining to the cheek.

buffer compound that combines with an acid or with a base to form a weaker acid or base, thereby lessening the change in hydrogen ion concentration that would occur without the buffer.

bursa fluid-containing sac or pouch lined with synovial membrane.

buttock prominence over the gluteal muscles.

c

calcitonin a hormone secreted by the parathyroid glands.

calculus stone formed in various parts of the body; may consist of different substances.

calorie heat unit; a large Calorie is the amount of heat needed to raise the temperature of 1 kg of water 1° C.

calyx cup-shaped division of the renal pelvis.

canaliculus little canal.

capillary microscopic blood vessel; capillaries connect arterioles with venules; also, microscopic lymphatic vessels.

carbhemoglobin, carbaminohemoglobin compound formed by union of carbon dioxide with hemoglobin.

carbohydrate organic compounds containing carbon, hydrogen, and oxygen in certain specific proportions, for example, sugars, starches, cellulose.

carboxyhemoglobin compound formed by union of carbon monoxide with hemoglobin.

carcinoma cancer, a malignant tumor.

caries decay of teeth or of bone.

carotid from Greek word meaning to plunge into deep sleep; carotid arteries of the neck so called because pressure on them may produce unconsciousness.

carpal pertaining to the wrist.

casein protein in milk.

cast mold, for example, formed in renal tubules.

castration removal of testes or ovaries.

catabolism breakdown of food compounds or of protoplasm into simpler compounds; opposite of anabolism, the other phase of metabolism.

catalyst substance that alters the speed of a chemical reaction.

cataract opacity of the lens of the eye.

catecholamines norepinephrine and epinephrine.

cation positively charged particle.

caudal pertaining to the tail of an animal; opposite of cephalic.

cecum blind pouch; the pouch at the proximal end of the large intestine.

celiac pertaining to the abdomen.

cellulose polysaccharide, the main plant carbohydrate.

centimeter $1/100$ of a meter, about $2/5$ of an inch.

centriole organelle necessary for spindle formation during mitosis.

centromere structure that separates chromatids during mitosis.

centrosphere, centrosome spherical area or body near center of cell.

cephalic pertaining to the head; opposite of caudal.

cerumen earwax.

cervix neck; any necklike structure.

chemoreceptor distal end of sensory dendrites especially adapted for chemical stimulation.

chiasm crossing; specifically, a crossing of the optic nerves.

cholecystectomy removal of the gallbladder.

cholesterol organic alcohol present in bile, blood, and various tissues.

cholinergic fibers axons whose terminals release acetylcholine.

cholinesterase enzyme; catalyzes breakdown of acetylcholine.

chorion outermost fetal membrane derived from the blastocyst; contributes to placenta formation.

choroid, chorioid skinlike.

chromatids newly formed chromosomes.

chromatin deep-staining substance in the nucleus of cells; divides into chromosomes during mitosis.

chromosomes deep-staining, rod-shaped bodies in cell nucleus; composed of genes.

chyle milky fluid; the fat-containing lymph in the lymphatics of the intestine.

chyme partially digested food mixture leaving the stomach.

cilia hairlike projections of protoplasm.

circadian daily.

cistron gene; segment of DNA molecule that transmits an inheritable trait.

cleavage mitotic cell division that begins after fertilization.

clone family of cells descended from one cell and therefore identical to it.

cochlea snail shell or structure of similar shape.

coenzyme nonprotein substance that activates an enzyme.

collagen principal organic constituent of connective tissue.

colloid solute particles with diameters of 1 to 100 mμ.

colostrum first milk secreted after childbirth.

combining sites, antigen-binding sites, antigen receptors regions on antibody molecule; shape of each combining site is complementary to shape of a specific antigen's epitopes.

commissure bundle of nerve fibers passing from one side to the other of the brain or cord.

complement several inactive enzymes normally present in blood; act to kill foreign cells by cytolysis.

concha shell-shaped structure, for example, bony projections into the nasal cavity.

condyle rounded projection at the end of a bone.

congenital present at birth.

contralateral on the opposite side.

coracoid like a raven's beak in form.

corium true skin or derma.

coronal of or like a crown.

coronary encircling; in the form of a crown.

corpus body.

corpuscles very small body or particle.

cortex outer part of an internal organ, for example, of the cerebrum and of the kidneys.

cortisol, hydrocortisone, compound F glucocorticoid secreted by adrenal cortex.

costal pertaining to the ribs.

covalent bond chemical bond formed by two atoms sharing a pair of electrons.

crenation, plasmolysis shriveling of a cell because of water withdrawal.

cretinism dwarfism caused by hypofunction of the thyroid gland.

cribriform sievelike.

cricoid ring shaped; a cartilage of this shape in the larynx.

cruciate cross shaped.

crystalloid solute particle less than 1 mμ in diameter.

cubital pertaining to the forearm.

cutaneous pertaining to the skin.

cyanosis bluish appearance of the skin from deficient oxygenation of blood.

cytokinesis dividing of cytoplasm to form two cells.

cytology study of cells.

cytoplasm the protoplasm of a cell exclusive of the nucleus.

D

deamination chemical reaction by which the amino group NH_3 is split from an amino acid.

deciduous temporary; shedding at a certain stage of growth, for example, deciduous teeth.

decomposition chemical reaction involving breaking of chemical bonds; forms two or more smaller molecule compounds from larger molecule compounds.

decussation crossing over like an X.

defecation elimination of waste matter from the intestines.

deferens carrying away.

deglutition swallowing.

deltoid triangular, for example, deltoid muscle.

dendrite, dendron branching or treelike; a nerve cell process that transmits impulses toward the cell body.

dens tooth.

dentate having toothlike projections.

dentine main part of a tooth, under the enamel.

dentition teething; also, number, shape, and arrangement of the teeth.

dermatome area of skin supplied by sensory fibers of a single dorsal root.

dermis, corium true skin.

dextrose glucose, a monosaccharide, the principal blood sugar.

dialysis separation; the separation of crystalloids from colloids by the faster diffusion of the former through a membrane.

diapedesis passage of blood cells through intact blood vessel walls.

diaphragm membrane or partition that separates one thing from another; the muscular partition between the thorax and abdomen; the midriff.

diaphysis shaft of a long bone.

diarthrosis freely movable joint.

diastole relaxation of the heart interposed between its contractions; opposite of systole.

diencephalon "tween" brain; parts of the brain between the cerebral hemispheres and the mesencephalon or midbrain.

diffusion spreading, for example, scattering of solute particles.

digestion conversion of food into assimilable compounds.

diplopia double vision; seeing one object as two.

disaccharide sugar formed by the union of two monosaccharides; contains 12 carbon atoms.

distal toward the end of a structure; opposite of proximal.

diuresis increased urine production.

diverticulum outpocketing from a tubular organ such as the intestine.

dorsal, posterior pertaining to the back; opposite of ventral.

Douglas Scottish anatomist of the late seventeenth and early eighteenth centuries.

dropsy accumulation of serous fluid in a body cavity, in tissues; edema.

duct canal or passage.

dura mater literally strong or hard mother; outermost layer of the meninges.

dyspnea difficult or labored breathing.

dystrophy faulty nutrition.

E

ectopic displaced; not in the normal place, for example, extrauterine pregnancy.

edema excessive fluid in tissues; dropsy.

effector responding organ, for example, voluntary and involuntary muscle, the heart, and glands.

efferent carrying from, as neurons that transmit impulses from the central nervous system to the periphery; opposite of afferent.

electrocardiogram graphic record of heart's action potentials.

electroencephalogram graphic record of brain's action potentials.

electrolyte substance that ionizes in solution, rendering the solution capable of conducting an electric current.

electron minute, negatively charged particle.

electrovalent (ionic) bond chemical bond formed by transfer of electron.

elimination expulsion of wastes from the body.

embolism obstruction of a blood vessel by foreign matter carried in the bloodstream.

embryo animal in early stages of intrauterine development; the human fetus the first 3 months after conception.

emesis vomiting.

emphysema dilation of pulmonary alveoli.

empyema pus in a cavity, for example, in the chest cavity.

encephalon brain.

endocrine secreting into the blood or tissue fluid rather than into a duct; opposite of exocrine.

endomysium connective tissue between individual muscle fibers.

endoplasm cytoplasm located toward center of cell, as distinguished from ectoplasm located nearer periphery of cell.

endoplasmic reticulum network of tubules and vesicles in cytoplasm.

energy power or ability to do work; *kinetic* or *active*, caused by moving particles, for example, mechanical energy, heat; *potential* or *stored*, caused by attraction between particles, for example, chemical energy.

enteron intestine.

enzyme catalytic agent formed in living cells.

eosinophil, acidophil white blood cell readily stained by eosin.

epidermis "false" skin; outermost layer of the skin.

epimysium connective tissue sheath that envelops a skeletal muscle.

epinephrine secretion of the adrenal medulla.

epiphyses ends of a long bone.

epitope uniquely shaped region of surface of antigen.

erythrocyte red blood cell.

erythropoiesis formation of red blood cells.

ethmoid sievelike.

eupnea normal respiration.

Eustachio Italian anatomist of the sixteenth century.

exocrine secreting into a duct; opposite of endocrine.

exophthalmos abnormal protrusion of the eyes.

extrinsic coming from the outside; opposite of intrinsic.

extrinsic factor vitamin B_{12}.

F

facilitation decrease in a neuron's resting potential to a point above its threshold of stimulation.

facilitated diffusion carrier-mediated diffusion.

Fallopius sixteenth century Italian anatomist.

fascia sheet of connective tissue.

fasciculus little bundle.

fetus unborn young, especially in the later stages; in human beings, from third month of intrauterine period until birth.

fiber threadlike structure.

fibrillation ineffective, uncoordinated contraction of cardiac muscle fibers.

fibrin insoluble protein in clotted blood.

fibrinogen soluble blood protein that is converted to insoluble fibrin during clotting.

fibroblasts connective tissue cells that synthesize interstitial fibers and gels.

fibrocytes old fibroblasts.

filtration passage of water and solutes through a membrane from hydrostatic pressure gradient.

fimbria fringe.

fissure groove.

flaccid soft, limp.

follicle small sac or gland.

fontanel "soft spots" of the infant's head; unossified areas in the infant's skull.

foramen small opening.

fossa cavity or hollow.

fovea small pit or depression.

fundus base of a hollow organ, for example, the part farthest from its outlet.

G

gamete sex cell; sperm or ovum.

ganglion cluster of nerve cell bodies outside the central nervous system.

gasserian named for Gasser, a sixteenth century Austrian surgeon.

gastric pertaining to the stomach.

gene part of the chromosome that transmits a given hereditary trait.

genitalia reproductive organs.

gestation pregnancy.

gland secreting structure.

glomerulus compact cluster, for example, of capillaries in the kidneys.

glossal of the tongue.

glucagon hormone secreted by alpha cells of the islands of Langerhans.

glucocorticoids hormones that influence food metabolism; secreted by adrenal cortex.

glucokinase enzyme; catalyzes conversion of glucose to glucose-6-phosphate.

gluconeogenesis formation of glucose from protein or fat compounds.

glucose monosaccharide or simple sugar; the principal blood sugar.

gluteal of or near the buttocks.

glycerin, glycerol product of fat digestion.

glycogen "animal starch"; main polysaccharide stored in animal cells.

glycogenesis formation of glycogen from glucose or from other monosaccharides, fructose or galactose.

glycogenolysis hydrolysis of glycogen to glucose-6-phosphate or to glucose.

glyconeogenesis the formation of glycogen from protein or fat compounds.

gonad sex gland in which reproductive cells are formed.

graafian named for Graaf, a seventeenth century Dutch anatomist.

gradient a slope or difference between two levels, for example, concentration gradient—a difference between the concentrations of two substances.

gustatory pertaining to taste.

gyrus convoluted ridge.

H

haploid half the normal chromosome number.

haversian named for Havers, English anatomist of the late seventeenth century.

helix spiral; coil.

hematocrit volume (percent) of red blood cells in a sample of whole blood.

hemiplegia paralysis of one side of the body.

hemoglobin iron-containing protein in red blood cell.

hemolysis destruction of red blood cells with escape of hemoglobin from them into surrounding medium.

hemopoiesis blood cell formation.

hemorrhage bleeding.

hepar liver.

heparin substance obtained from the liver that inhibits blood clotting.

heredity transmission of characteristics from a parent to a child.

hernia, "rupture" protrusion of a loop of an organ through an abnormal opening.

hexose 6 carbon atom sugar.

hilus, hilum depression where vessels enter an organ.

His German anatomist of the late nineteenth century.

histology science of minute structure of tissues.

homeostasis relative uniformity of the normal body's internal environment.

hormone substance secreted by an endocrine gland.

hyaline glasslike.

hydrocortisone, cortisol, compound F hormone secreted by the adrenal cortex.

hydrolysis literally "split by water"; chemical reaction in which a compound reacts with water.

hymen Greek for skin; mucous membrane that may partially or entirely occlude the vaginal outlet.

hyoid shaped like the letter U; bone of this shape at the base of the tongue.

hypercapnia abnormally high blood CO_2 concentration.

hyperemia increased blood in a part.

hyperglycemia higher than normal blood glucose concentration.

hyperkalemia higher than normal blood potassium concentration.

hypernatremia higher than normal blood sodium concentration.

hyperopia farsightedness.

hyperplasia increase in the size of a part from an increase in the number of its cells.

hyperpnea abnormally rapid breathing; panting.

hypertension abnormally high blood pressure.

hyperthermia fever; body temperature above 37° C.

hypertonic solution solution having a higher osmotic pressure than blood.

hypertrophy increased size of a part from an increase in the size of its cells.

hypervolemia larger volume of blood than normal.

hypocalcemia lower than normal blood calcium concentration.

hypokalemia lower than normal blood potassium concentration.

hyponatremia lower than normal blood sodium concentration.

hypophysis Greek for undergrowth; hence the pituitary gland, which grows out from the undersurface of the brain.

hypothalamus part of the diencephalon; gray matter in the floor and walls of the third ventricle.

hypothermia subnormal body temperature; below 37° C.

hypotonic solution solution having a lower osmotic pressure than blood.

hypovolemia subnormal volume of blood.

hypoxia oxygen deficiency.

I

implantation embedding of developing blastocyst in uterine lining (endometrium).

inclusions any foreign or heterogenous substance contained in a cell or in any tissue or organ that was not introduced as a result of trauma.

incus anvil; the middle ear bone that is shaped like an anvil.

inferior lower; opposite of superior.

inguinal of the groin.

inhalation inspiration or breathing in; opposite of exhalation or expiration.

inhibition an increase in a neuron's resting potential above its usual level.

innominate not named, anonymous, for example, ossa coxae (hip bones) formerly known as innominate bones.

insulin hormone secreted by beta cells of the islands of Langerhans in the pancreas.

intercellular between cells; interstitial.

interneurons, internuncial or **intercalated neurons** conduct impulses from sensory to motor neurons; lie entirely within the central nervous system.

interferon protein released from body cells following viral invasion of them; acts in some way to defend other body cells against viral invasion.

interphase period between two cell divisions.

interstitial of or forming small spaces between things; intercellular.

intima innermost.

intrinsic not dependent on externals; located within something; opposite of extrinsic.

involuntary not willed; opposite of voluntary.

involution return of an organ to its normal size after enlargement; also retrograde or degenerative change.

ion electrically charged atom or group of atoms.

ipsilateral on the same side; opposite of contralateral.

irritability excitability; ability to react to a stimulus.

ischemia local anemia; temporary lack of blood supply to an area.

isometric muscle contraction that produces no change in length of muscle.

isotonic of the same tension or pressure.

isotopes atoms of same element with different atomic weights because their nuclei contain different numbers of neutrons.

K

keratin protein compound present in the human body, chiefly in hair and nails.

ketones acids (acetoacetic, beta-hydroxybutyric, and acetone) produced during fat catabolism.

ketosis excess amount of ketone bodies in the blood.

kilogram 1,000 gm, approximately 2.2 pounds.

kinesthesia "muscle sense," that is, sense of position and movement of body parts.

L

labia lips.

lacrimal pertaining to tears.

lactation secretion of milk.

lactose milk sugar, a disaccharide.

lacuna space of cavity, for example, lacunae in bone contain bone cells.

lamella thin layer, as of bone.

lateral of or toward the side; opposite of medial.

lemniscus, medial a flat band of sensory fibers extending up from the medulla, through the pons and midbrain to the thalamus.

leukocyte white blood cell.

ligament bond or band connecting two objects; in anatomy, a band of white fibrous tissue connecting bones.

limbic lobe or **system** cerebral cortex on the medial surface of the brain that forms a border around the corpus callosum; older name rhinencephalon.

lipid fats and fatlike compounds.

loin part of the back between the ribs and hip bones.

lumbar of or near the loins.

lumen passageway or space within a tubular structure.

luteum golden yellow.

lymph watery fluid in the lymphatic vessels.

lymphocyte one type of white blood cells.

lysis the destruction of cells and other antigens by a specific lysin antibody.

lysosomes membranous organelles containing various enzymes that can dissolve most cellular compounds; hence called "digestive bags" or "suicide bags" of cells.

M

malleolus small hammer; projections at the distal ends of the tibia and fibula.

malleus hammer; the tiny middle ear bone that is shaped like a hammer.

Malpighii seventeenth century Italian anatomist.

maltose disaccharide or "double" sugar.

mamillary like a nipple.

mammary pertaining to the breast.

manometer instrument used for measuring the pressure of fluids.

manubrium handle; upper part of the sternum.

mastication chewing.

matrix ground substance in which cells are embedded.

meatus passageway.

medial of or toward the middle; opposite of lateral.

mediastinum middle section of the thorax, that is, between the two lungs.

medulla Latin for marrow; hence the inner portion of an organ in contrast to the outer portion or cortex.

meiosis nuclear division in which the chromosomes are reduced to half their original number before the cell divides in two.

melatonin hormone secreted by pineal gland.

membrane thin layer or sheet.

menstruation monthly discharge of blood from the uterus.

mesencephalon midbrain.

mesentery fold of peritoneum that attaches the intestine to the posterior abdominal wall.

mesial situated in the middle; median.

metabolism complex process by which food is utilized by a living organism.

metabolite any substance produced by metabolism.

metacarpus "after" the wrist; hence the part of the hand between the wrist and fingers.

metatarsus "after" the instep; hence the part of the foot between the tarsal bones and toes.

meter about 39.5 inches.

microglia one type of connective tissue cell found in the brain and cord.

micron 0.001 ml; about $1/25{,}000$ of an inch.

micturition urination, voiding.

milliequivalent a unit of chemical equivalence computed by the following formula:

$$\frac{mg}{Atomic\ weight} \times Valence = mEq$$

For example: $\dfrac{3{,}266\ mg\ Na^+}{23\ (At\ wt\ Na)} \times 1 = 142\ mEq\ Na^+$

millimeter 0.001 meter; about $1/25$ of an inch.

mineralocorticoids hormones that influence mineral salt metabolism; secreted by adrenal cortex.

mitochondria threadlike structures.

mitosis indirect cell division involving complex changes in the nucleus.

mitral shaped like a miter.

molar concentration number of grams solute per liter of solution divided by the solute's molecular weight.

mole molecular weight of a compound in grams; also called gram molecular weight.

monosaccharide simple sugar.

Monro eighteenth century English surgeon.

morphology study of shape and structure of living organisms.

morula early stage in development of embryo; a solid, spherical mass of cells.

motoneurons, motor or **efferent neurons** transmit nerve impulses away from the brain or spinal cord.

myelin lipoid substance found in the myelin sheath around some nerve fibers.

myocardium muscle of the heart.

myopia nearsightedness.

myosin contractile protein present in muscle cells.

N

nares nostrils.

neurilemma nerve sheath.

neurohypophysis posterior pituitary gland.

neuron nerve cell, including its processes.

neurovesicles microscopic sacs in axon terminals; contain transmitter substance.

neutrophil white blood cell that stains readily with neutral dyes.

nuchal pertaining to the nape of the neck.

nucleotide component of DNA and RNA; a nucleotide is composed of a sugar (deoxyribose or ribose), a nitrogenous base (adenine, thymine, cytosine, or guanine), and a phosphate group.

nucleus spherical structure within a cell; a group of neuron cell bodies in the brain or cord.

O

occiput back of the head.

olecranon elbow.

olfactory pertaining to the sense of small.

oligodendroglia a type of connective tissue cell found in the brain and cord.

ophthalmic pertaining to the eyes.

organelle cell organ; one of the specialized parts of a single-celled organism (protozoon), serving for the performance of some individual function.

os Latin for mouth and for bone.

osmoreceptor receptor stimulated by change in osmotic pressure.

osmosis movement of a fluid through a semipermeable membrane.

ossicle little bone.

oxidation loss of hydrogen or electrons from a compound or element.

oxyhemoglobin a compound formed by the union of oxygen with hemoglobin.

P

palate roof of the mouth.

palpebrae eyelids.

papilla small nipple-shaped elevation.

paralysis loss of the power of motion or sensation, especially voluntary motion.

parenchyma the distinguishing, functional cells of an organ.

parietal of the walls of an organ or cavity.

parotid located near the ear.

parturition act of giving birth to an infant.

patella small, shallow pan; the kneecap.

Pavlov Russian physiologist of the late nineteenth and early twentieth centuries.

pectineal pertaining to the pubic bone.

pectoral pertaining to the chest or breast.

pelvis basic or funnel-shaped structure.

pentose 5 carbon sugar.

perimysium connective tissue between bundles of muscle fibers.

peripheral pertaining to an outside surface.

peroneus, peroneal of or near the fibula.

petrous rocklike.

Peyer Swiss anatomist of the late seventeenth and early eighteenth centuries.

pH hydrogen ion concentration; the negative logarithm of hydrogen ion concentration.

phagocytosis process by which a segment of cell membrane forms a small pocket around a bit of solid matter outside the cell, breaks off from rest of the membrane, and moves into the cell; briefly, ingestion and digestion of particles by a cell.

phalanges finger or toe bones.

phrenic pertaining to the diaphragm.

pia mater gentle mother; the vascular innermost covering (meninges) of the brain and cord.

pilomotor mover of a hair.

pineal shaped like a pine cone.

pinocytosis process by which a segment of cell membrane forms a small pocket around a bit of fluid outside the cell, breaks off from rest of membrane, and moves into the cell.

piriformis pear shaped.

pisiform pea shaped.

pituicyte cell type of posterior pituitary gland.

plantar pertaining to the sole of the foot.

plasma liquid part of the blood.

plasmolysis shrinking of a cell from water loss by osmosis.

plexus network.

plica fold.

polymorphonuclear having many-shaped nuclei.

polypeptide compound formed of many amino acids linked together by peptide bonds (between carboxyl group of one amino acid and amino group of another).

polyribosomes group of ribosomes working together to synthesize proteins.

polysaccharide carbohydrate containing a large number of saccharide groups ($C_6H_{10}O_5$), for example, starch, glycogen.

pons bridge.

popliteal behind the knee.

posterior following after; hence located behind; opposite of anterior.

potential or **potential difference** difference in electrical charges, for example, on outer and inner surfaces of cell membrane.

potential, action difference in electrical charges on inner and outer surfaces of the cell membrane during impulse conduction; outer surface negative to inner.

potential, resting difference in electrical charges on inner and outer surfaces of the cell membrane when it is not conducting impulses; outer surface positive to inner.

Poupart seventeenth century French anatomist.

presbyopia "oldsightedness"; farsightedness of old age.

pressoreceptors receptors stimulated by a change in pressure; baroreceptors.

pronate to turn palm downward.

properdin group of enzymes normally present in blood; function to activate complement.

prophase first stage of mitosis; chromosomes become visible.

proprioreceptors receptors located in the muscles, tendons, and joints.

prostaglandins (PG) potent, widespread biocompounds (fatty acids); may regulate hormone activity at cellular level.

protein compound composed of amino acids joined by peptide bonds; polypeptide.

proton positively charged particle in atomic nucleus.

protoplasm living substance.

proximal next or nearest; located nearest the center of the body or the point of attachment of a structure.

psoas pertaining to the loin, the part of the back between the ribs and hip bones.

psychosomatic pertaining to the influence of the mind (notably the emotions) on body functions.

pterygoid wing-shaped.

puberty age at which the reproductive organs become functional.

purine bases adenine and guanine.

pyrimidine bases cytosine and thymine.

R

racemose like a cluster of grapes.

radioactivity emission from atomic nucleus of particles or rays known as alpha, beta, and gamma rays.

ramus branch.

Ranvier French pathologist of the late nineteenth and early twentieth centuries.

receptor peripheral ending of a sensory neuron.

reflex involuntary action.

refraction bending of a ray of light as it passes from a medium of one density to one of a different density.

refractory resisting stimulation.

renal pertaining to the kidney.

reticular netlike.

reticulum network.

rhinencephalon *see* **limbic lobe**.

ribosomes organelles in cytoplasm of cells; synthesize proteins so nicknamed "protein factories."

rugae wrinkles or folds.

S

sagittal like an arrow; longitudinal.

salpinx tube; oviduct.

sarcolemma cytoplasmic membrane of muscle cell.

sarcomere contractile unit of myofibril between two C lines.

sarcoplasm cytoplasm of muscle cell.

sartorius tailor; hence the thigh muscle used to sit cross-legged like a tailor.

sciatic pertaining to the ischium.

sclera from Greek for hard.

scrotum bag.

sebum Latin for tallow; secretion of sebaceous glands.

sella turcica Turkish saddle; saddle-shaped depression in the sphenoid bone.

semen Latin for seed; male reproductive fluid.

semilunar half-moon shaped.

senescence old age.

serratus saw toothed.

serum any watery animal fluid; clear, yellowish liquid that separates from a clot of blood.

sesamoid shaped like a sesame seed.

sigmoid S shaped.

sinus cavity.

soleus pertaining to a sole; a muscle in the leg shaped like the sole of a shoe.

solute material dissolved or suspended in a solution.

solvent dissolving medium or fluid in a solution.

somatic of the body framework or walls, as distinguished from the viscera or internal organs.

sphenoid wedge shaped.

sphincter ring-shaped muscle.

splanchnic visceral.

squamous scalelike.

stapes stirrup; tiny stirrup-shaped bone in the middle ear.

Starling English physiologist of the late nineteenth and early twentieth centuries.

stereognosis awareness of the shape of an object by means of touch.

stimulus agent that causes a change in the activity of a structure.

stratum layer.

stress, physiological according to Selye, a condition in the body produced by all kinds of injurious factors that he calls "stressors" and manifested by a syndrome.

stressor any injurious factor that produces biological stress, for example, emotional trauma, infections, or severe exercise.

striated marked with parallel lines.

stroma Greek for mattress or bed; hence the framework or matrix of a structure.

sudoriferous secreting sweat.

sulcus furrow or groove.

superior higher; opposite of inferior.

supinate to turn the palm of the hand upward; opposite of pronate.

Sylvius seventeenth century anatomist.

symphysis Greek for a growing together.

synapse joining; point of contact between adjacent neurons.

synovia literally "with egg"; secretion of the synovial membrane resembles egg white.

synthesis chemical reaction in which new chemical bonds are formed to put together two or more substances to form a more complex whole.

systole contraction of the heart muscle.

T

T cells lymphocytes that undergo first stage of their development in thymus gland before migrating to lymph nodes and spleen.

talus ankle; one of the bones of the ankle.

tamponade (cardiac) accumulation of fluid in the pericardial sac.

target organ (cell) organ (cell) acted on by a particular hormone and responding to it.

tarsus instep.

tegmentum part of the brain stem that covers the posterior surface of the cerebral peduncles and the pons.

tendon band or cord of fibrous connective tissue that attaches a muscle to a bone or other structure.

tetanus sustained muscle contraction.

thorax chest.

thrombosis formation of a clot in a blood vessel.

thymosin hormone of the thymus gland.

tibia Latin for shin bone.

tonus continued, partial contraction of muscle.

tract bundle of axons located within the central nervous system.

trauma injury.

treppe progressive increase in strength of muscle contraction with repeated maximal stimuli.

triglycerides compounds formerly called neutral fats.

trochlear pertaining to a pulley.

trophic having to do with nutrition.

tropic hormones hormones that control secretions of other endocrine glands.

unica covering.

turbinate shaped like a cone or like a scroll or spiral.

twitch contraction quick jerky contraction in response to a single stimulus.

tympanum drum.

U

umbilicus navel.

utricle little sac.

uvula Latin for a little grape; projection hanging from the soft palate.

V

vagina sheath.

vagus Latin for wandering.

valence number of unpaired electrons in atom's outer shell.

valve structure that permits flow of a fluid in one direction only.

vas vessel or duct.

vastus wide, of great size.

Vater German anatomist of the late seventeenth and early eighteenth centuries.

vein vessel carrying blood to the heart.

ventral of or near the belly; in man, front or anterior; opposite or dorsal or posterior.

ventricle small cavity.

vermiform worm shaped.

villus hairlike projection.

viscera internal organs.

volar pertaining to the palm of the hand or the sole of the foot; palmar; plantar.

vomer ploughshare.

W

Willis seventeenth century English anatomist.

X

xiphoid sword shaped.

Y

yoga meditation; an altered state of consciousness.

Z

zygoma yoke.

zygote fertilized ovum.

Index

L